# HM-181 and HM-126F

## A Compliance Guide for DOT's New Hazmat Transportation Regulations

# HM-181 and HM-126F

## A Compliance Guide for DOT's New Hazmat Transportation Regulations

*Brian Karnofsky*

*John E. Maroney*

*Michael F. Randall*

**Environmental Resource Center**

JOHN WILEY & SONS, INC.

New York • Chichester • Weinheim • Brisbane • Singapore • Toronto

**A NOTE TO THE READER:**
**This book has been electronically reproduced from digital information stored at John Wiley & Sons, Inc. We are pleased that the use of this new technology will enable us to keep works of enduring scholarly value in print as long as there is a reasonable demand for them. The content of this book is identical to previous printings.**

This book is printed on acid-free paper. ♾



Published simultaneously in Canada



**Library of Congress Cataloging-in-Publication Data:**

HM-181 and HM-126F : a compliance guide for DOT's new hazmat transportation regulations / Environmental Resource Center.
p. cm.
Includes index.
ISBN 0-471-28844-6
1. Hazardous substances—Transportation—Law and legislation—United States. I. Environmental Resource Center.
KF3945.H6 1994
344.73'04622—dc20
[347.3044622] 93-33290

# Contents

# Preface

This book was developed in 1992 and 1993 in response to the United States Department of Transportation's issuance of a large new body of regulations governing the transportation of hazardous materials. These new regulations are generally referred to by the designation of the legal docket that was maintained for their adoption: HM–181.

The Department of Transportation (DOT) has authority from Congress, under the Hazardous Materials Transportation Act, to "issue regulations for the safe transportation of hazardous materials in intrastate, interstate, and foreign commerce ... [that] shall govern any aspect of hazardous materials transportation safety which the Secretary deems necessary or appropriate." Congress stated that the purpose for this authority is "to improve the regulatory and enforcement authority of the Secretary of transportation to protect the Nation adequately against the risks to life and property which are inherent in the transportation of hazardous materials in commerce."

In response to this mandate, and the development of complex and varied shipping networks by highway, rail, air, and water for many existing and new kinds of goods that could present public hazards in their transportation, DOT built up a body of rules called the Hazardous Materials Regulations. These regulations apply to the transportation of hazardous materials in commerce, and govern the safety aspects of transportation. They include requirements for classification of materials, packaging, hazard communication, transportation and handling procedures, and reporting of incidents involving hazardous materials.

However, as these rules developed over the years, they became difficult to use because of their length and complexity, and were relatively inflexible and outdated where non-bulk packaging technology was concerned. They were also considered deficient in terms of safety in the classification of some categories of hazardous materials, and were generally not in alignment with international regulations based on U.N. recommendations. To address these problems, DOT proposed in 1982 a package of regulatory changes based on its own initiative and the U.N. Committee of Experts' recommendations on the Transportation of Dangerous Goods. The completed package of regulations was published on 21 December 1990. Some urgent rules went into effect immediately. Other provisions of the new rules are to be phased in over a transition period that extends up to five years for some provisions.

The material in this book is intended to give the meaning of these new regulations in plain English for persons who send out, transport, or receive hazardous materials. It covers the background of the new rules, their relation to existing rules, provisions for making the transition to the new regulatory system, and procedures for determining how to describe, package, mark, label, and handle the materials in connection with shipping.

# Chapter 1
# Introduction to the New Requirements

## 1.1 Background: The Hazardous Materials Transportation Act and HM-181

The U.S. Department of Transportation (DOT) is one of the federal agencies that Congress has directed to control certain aspects of the management of hazardous materials. Congress gave DOT the authority to "issue regulations for the safe transportation of hazardous materials in intrastate, interstate, and foreign commerce ... [that] shall govern any aspect of hazardous materials transportation safety which the Secretary [of Transportation] deems necessary or appropriate." This authority was granted in the Hazardous Materials Transportation Act (HMTA), first adopted in 1974, and amended through November 1990 (the Hazardous Materials Transportation Uniform Safety Act of 1990 [HMTUSA], P.L. 101–615). The Congressional policy declared in HMTA is "to improve the regulatory and enforcement authority of the Secretary of transportation to protect the Nation adequately against the risks to life and property which are inherent in the transportation of hazardous materials in commerce." Within DOT, the hazardous materials regulations are the special responsibility of the Research and Special Projects Administration (RSPA).

The DOT hazardous materials regulations (HMR) apply to the transportation of hazardous materials in commerce, and govern the safety aspects of transportation. They include requirements for—

- classification of materials
- packaging:
  - manufacture
  - continuing qualification
  - maintenance
- hazard communication:
  - package marking
  - labeling
  - placarding
  - shipping documentation
- transportation and handling
- incident reporting

As the HMR developed over the years, they became difficult to use because of their length and complexity, and relatively inflexible and outdated where non-bulk packaging technology was concerned. They were also deficient in terms of safety in the classification of some categories of hazardous materials, and were generally not in alignment with international regulations based on U.N. recommendations.

To address the problems that both DOT and the regulated community saw with the existing HMR, RSPA proposed in 1982 a package of regulatory changes based on its own initiative and the U.N. Committee of Experts' Recommendations on the Transportation of Dangerous Goods. The initial proposal was assigned a Docket Number of **HM-181**, and this name has been generally applied to all the related changes in this area, even though many of them were proposed in other dockets. These HM-181 regulations, which are the subject of this manual, went through an unusually long period of public comment and revision before they were published as final regulations on 21 December 1990 (55 Fed. Reg. 52402). Although the final rules were effective 1 October 1991, DOT took the unusual step of allowing persons regulated to begin complying with the new regulations on 1 January 1991. It was important to get certain rules affecting international shipments into place as soon as possible, because a transitional provision of the international regulations, allowing non-complying U.S. shipments, expired on 31 December 1990. Other provisions of the HM-181 rules are to be phased in over a transition period that extends up to five years for some provisions (further discussed below).

### 1.2 What The Standard Covers

The HMR are contained in Title 49 of the Code of Federal Regulations (CFR), Subtitle B, Chapter 1, Subchapter C, Parts 171 through 180. Each numbered part deals with a single subject, and is divided into sections numbered with the part number and a decimal. For example, the definitions and abbreviations used in the HMR are listed in section *171.8* in Part 171—General information, regulations, and definitions. This section would be cited in abbreviated form as *49 CFR 171.8*, or, when it is clear that we are dealing with Title 49, *§171.8*.

The most important parts of the HMR for our purposes are the following:

**Part 171**: Includes definitions, reporting requirements, reference materials, and procedural requirements (including provisions permitting use of other regulations, like the international rules governing air and sea shipments).

**Part 172**: Includes the Hazardous Materials Table (§172.101), and hazard communication requirements—shipping paper descriptions, package marking and labeling, placarding of vehicles and bulk packagings, and emergency response communication.

**Part 173**: Contains the hazard class definitions for classifying materials, lists DOT packagings authorized for specific materials, and references the appropriate sections of Part 178 when DOT specification packagings are required.

Parts 174 through 177 deal with transportation by specific modes:

**Part 174**: Rail car.

**Part 175**: Aircraft.

**Part 176**: Vessel (ship, barge, boat, etc.).

**Part 177**: Motor vehicle.

**Part 178**: Addresses primarily container manufacturers (but affects handlers, shippers, and transporters who must decide on fitness of packagings). It contains specifications for a wide variety of approved packagings. Previously, these were detailed descriptions of required construction materials, dimensions, fasteners, etc. With the introduction of HM–181 provisions, the emphasis will be on performance-oriented packaging (POP)—packagings that can pass a series of tests of their ability to perform in a safe and satisfactory manner under conditions normally incident to transportation.

**Part 179**: Specifications for tank cars.

**Part 180**: Requirements for continuing qualification and maintenance of packagings—how they are to be maintained, repaired, and tested to assure they continue to meet the standards that Part 178 sets for new packagings.

The changes made by HM–181 affect all of these parts of the HMR. We will be dealing with these changes in greater detail in subsequent chapters, but the major features of the changes are the following:

- The older two Hazardous Materials Tables are combined into one.
- The Hazard Classes for materials are aligned to the U.N. numerical system.
- Materials are assigned to Packing Groups that determine the level of regulatory requirements, based on their hazards and compatibility.
- Specifications for containers are changed from detailed descriptions of their composition and construction to performance-oriented standards based on their ability to contain and protect materials in realistic transportation conditions. This Performance-Oriented Packaging (POP) is considered by DOT to be the central feature of HM–181.
- Standards for reuse of drums in shipment are based on a minimum wall thickness.
- New limits are set for the filling of bulk packages.
- Container manufacturers must notify their customers of any failures of packages to meet the stated performance specifications (this is not, however, a recall of these containers).

### 1.3 Who Is Covered By The Standard

All personnel involved in the packaging and labeling of hazardous materials (including hazardous wastes), preparation of shipping documents (e.g., bill of lading or hazardous waste manifest), and/or loading and unloading of vehicles must be familiar with and trained in the DOT regulations affecting their activities. Any employee who offers a hazardous material for transportation, and each carrier by air, highway, rail, or water who transports a hazardous material, is responsible for complying with these regulations.

## 1.4 Other Related Laws And Regulations Applicable To Hazardous Materials Handlers

Hazardous materials, as defined in the HMR (see Chapter III), are only one part of the universe of regulated types of matter classed as hazardous. There are three principal federal agencies that deal with these:

| | Agency | Hazardous ... | Federal Law | Regulations |
|---|---|---|---|---|
| 1 | Dept. of Labor, Occupational Safety and Health Administration (OSHA) | Chemical | Occupational Safety and Health Act (OSHA) | Hazard Communication Standard (HAZCOMM), 29 CFR 1910.1200 |
| 2 | Environmental Protection Agency (EPA) | Waste | Resource Conservation and Recovery Act (RCRA) | Hazardous Waste Management Regulations, 40 CFR 261–271 |
| | | Substance | Comprehensive Environmental Response, Compensation, and Liability Act (CERCLA, Superfund) | Designation, Reportable Quantities, and Notification, 40 CFR 302 |
| 3 | DOT, RSPA | Material | HMTA | HMR, 49 CFR 171–180 |

The OSHA Hazard Communication Standard applies to **hazardous chemicals**: any chemicals for which a Material Safety Data Sheet (MSDS) is required because of physical hazards or health hazards. The standard requires warning labels, MSDSs, training and information, and hazard communication programs in the workplace to assure that employees and contractors know the hazards associated with chemicals used there and how to protect themselves against them.

The EPA Hazardous Waste Management Regulations under RCRA deal strictly with **hazardous waste** as defined in the regulations. Each facility that generates any type of waste must determine if its wastes are hazardous. If they are, the facility must follow specific standards for accumulation of the wastes on site and for shipping them via an approved transporter to a designated treatment. storage, or disposal facility (TSDF). The wastes covered may be explicitly listed by EPA, hazardous by one of four characteristics, mixtures of hazardous and non-hazardous wastes, or residues from treatment of hazardous waste. Because all RCRA hazardous wastes are defined by the HMR as DOT hazardous materials, they must be handled, prepared for shipping, and transported in accordance with the HMR. The EPA (or state) **hazardous waste manifest** that is prepared for every hazardous waste shipment must describe the shipment by **proper shipping name(s)** from the Hazardous Materials Table (49 CFR 172.101).

**Hazardous substances** are designated by EPA under authority of the Comprehensive Environmental Response, Compensation, and Liability Act of 1980 (CERCLA), usually referred to as Superfund. The list of hazardous substances is intended to bring together elements, compounds, mixtures, solutions, etc., designated by EPA as hazardous to health or the environment under the Clean Water Act, Resource Conservation and Recovery Act, imminently hazardous chemical substance or mixture provisions of the Toxic Substances Control Act, hazardous air pollutant provisions of the Clean Air Act, and Superfund. These substances do not include petroleum, natural gas, or synthetic gas usable for fuel (some petroleum fractions or derivatives may be individually listed as exceptions). Spills or releases of any of these substances in excess of a **reportable quantity** (RQ), set for each substance by EPA, must be reported. Even if a hazardous substance does not appear elsewhere in DOT listings of hazardous materials, it must be treated as a hazardous material for all shipping purposes if any one package contains a quantity of the substance equal to or greater than the RQ.

Figure 1.1 illustrates the relationship of these categories as a set of overlapping areas (a Venn diagram), where the labeled circles and ellipses represent hazardous materials (HM), hazardous chemicals (HC), hazardous substances (HS), and hazardous wastes (HW).

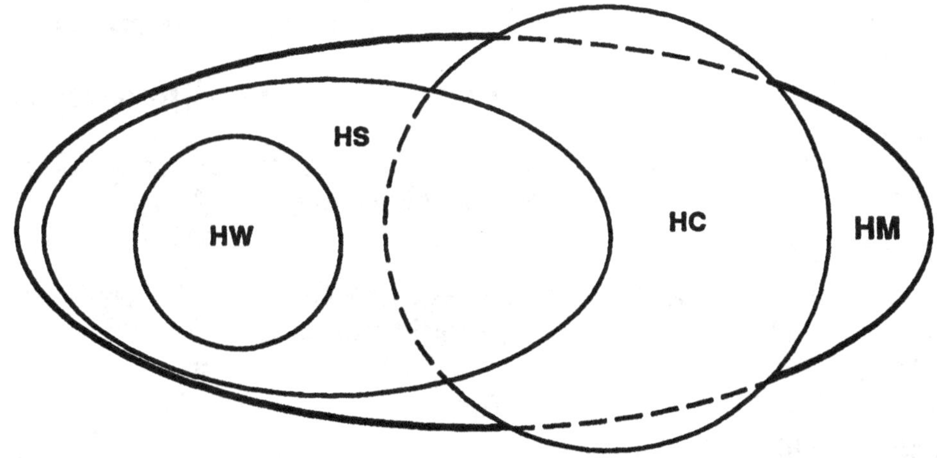


Figure 1.1 Relationships of Regulated Hazardous Classifications

### 1.5 Compliance Deadlines

The rule published in the *Federal Register* on 21 December 1990, effective 1 October 1991, resulted in a comprehensive revision of the HMR. At the same time, DOT published a new Section 171.14, with the purpose of providing an orderly transition to the new requirements, so as to minimize any burdens associated with them. During a transition period as provided in §171.14, persons may elect to comply with either the applicable old requirements of the HMR in effect on 30 September 1991, or the new HM–181 requirements. Since the rules were completely republished in their new form in the 1991 edition of 49 CFR, RSPA strongly recommended that persons who did not elect to comply with the new regulations immediately should retain the 1990 edition for reference to the then-existing regulations.

The transition dates are (there are detailed regulatory requirements associated with most of these dates; they are described further below and in the Appendices):

**1 October 1991**

- New explosives hazard classification procedures in Subpart C of Part 173, and (for new explosives only) hazard communication requirements (i.e., shipping papers, emergency response information, package markings, and labeling) in Part 172, except for vehicle placarding.
- Classification of materials poisonous by inhalation. For such materials, the words "Poison—Inhalation Hazard" or "Inhalation Hazard," as appropriate, must be entered on shipping papers in association with the basic description.

**1 October 1992**

- Hazard communication requirements of Part 172 (including placarding requirements of Subpart F) for all materials poisonous by inhalation.

**1 October 1993**

- Classification and hazard communication requirements in Part 172, other than Subpart F (placarding), and Part 173, that were not previously in effect.
- Packaging requirements for all materials meeting the criteria for poisonous by inhalation.
- Requirements to segregate materials by mode of transportation in §§174.81 and 177.848.
- All other requirements of the 21 December 1990 rule for which a longer transition period is not provided elsewhere in §171.14.

**1 January 1994**

- For infectious substances (etiologic agents), the hazard classification procedures in §173.134, packaging requirements in §173.196, and hazard communication requirements (i.e., shipping papers, emergency response information, package markings, and labeling) in Part 172.

**1 October 1994**

- Placarding requirements in Subpart F of Part 172 that were not previously in effect (except for domestic highway shipments—see 1 October 2001, below, and Chapter VII, sect. 7.4.4).
- Package manufacturing and marking requirements under the provisions of Subpart B of Part 173, and Parts 178 and 179. (DOT specification packagings removed from Part 178 by the 21 December 1990, rule may no longer be manufactured.)

**1 October 1996**

- Requirements in parts 172 and 173 for maintenance and use of packagings that were not previously in effect are effective. (DOT specification packagings removed from part 178 by the 21 December 1990 final rule and packaging authorizations removed from Part 173 by the 21 December 1990 final rule may no longer be used in place of new packaging requirements.)

**1 October 2001**

- All placarding requirements in Subpart F of Part 172 not previously in effect, including domestic highway shipments. Until this date, placards that conform to specifications in effect on 30 September 1991 or placards specified in the 21 December 1990 final rule may be used, for highway transportation only, in place of the placards specified in subpart F. Rules for choosing the correct placards during this transition are discussed in Chapter VII, sect. 7.4.4.
- All marking, labeling, and packaging requirements of HM–181, including packages filled prior to 1 October 1991. Until 1 October 2001, notwithstanding the marking and labeling provisions of Subparts D and E of Part 172, and the packaging provisions of Part 173 and Subpart B of Part 172, a package filled prior to 1 October 1991 may be offered for transportation and transported if it—
  - Conforms to the old requirements of the HMR in effect on 30 September 1991
  - Is filled with hazardous materials prior to 1 October 1991
  - Is marked "Inhalation Hazard," if appropriate, in accordance with §172.313 or Special Provision 13, as assigned in the §172.101 Table
  - Is not emptied and refilled on or after 1 October 1991.

### Intermixing old and new requirements

During the transition periods provided, RSPA recommends that hazard communication requirements be consistent where practicable, i.e., marking, labeling, placarding, and shipping paper descriptions should conform to either the old requirements of the HMR in effect on 30 September 1991, or new requirements added or revised by the 21 December 1990 rule, without intermixing of communication elements. However, some intermixing is permitted, during the applicable transition periods, for packaging, hazard communication, and handling provisions, subject to conditions in §171.14(c)(3).

## 1.6 Registration with DOT

Effective 15 September 1992, the Department of Transportation (DOT) requires certain persons who offer for shipment or transport hazardous materials to register and pay a fee. The new regulations, 49 CFR Part 107, Subpart G—Registration of persons who offer or transport hazardous materials (§§107.601–.620), apply to anyone who offers for transportation—

- Any highway route-controlled quantity of Class 7 (radioactive) materials
- More than 25 kg (55 lb) of a Division 1.1, 1.2, or 1.3 (explosive) material in a motor vehicle, rail car, or freight container
- More than 1 L (1.06 qt) per package of a material extremely toxic by inhalation (Division 2.3, Hazard Zone A; or Division 6.1, Packing Group I, Hazard Zone A)
- A hazardous material in a bulk packaging, container, or tank having a capacity equal to or greater than 13,248 L (3,500 gal) for liquids or gases, or more than 13.24 $m^3$ (468 $ft^3$) for solids

- A shipment other than in a bulk packaging of 2,268 kg (5,000 lb) gross weight or more of one class of hazardous materials for which placarding of a vehicle, rail car, or freight container is required for that class. In this rule, "shipment" means (and is limited to) the hazardous material being offloaded or loaded at **one loading facility**.

After 1 October 1993, the materials subject to these rules will include—

- Oils in bulk packagings, including animal and vegetable oils (but not edible oils), residual fuel oil, and lubricating oil
- Elevated temperature materials, such as asphalt or molten metals
- Designated marine pollutants

DOT estimates that the addition of these three types of materials will add about 1,000 new registrants to the 26,273 shippers and carriers registered as of April 1993.

Those required to register do not include federal, state, or state political subdivision agencies, their employees, hazmat employees, or (until 1 July 1994) persons domiciled outside of the U.S. offering materials for transportation solely from outside the U.S.

It is important to note that U.S. companies under the majority stock ownership of another company, or wholly-owned or -controlled subsidiaries of another company are required to register, even when the parent company is also subject to registration requirements. Also, anyone who acts as an intermediary in hazardous materials transportation, like freight forwarders, is subject to registration if they act as an offeror of hazardous materials.

If your facility falls into any of the regulated categories, then you must have filed a registration statement by **31 August 1992** (or before performing any action that requires registration), and pay a $300 fee. You must renew registration by **30 June** of each year, also with a $300 fee. If your facility is subject to these rules and has not registered, you will not be allowed to ship hazardous materials after **15 September 1992**. Registration years after 1992 begin on 1 July and end on 30 June.

For special cases (short-notice or temporary applicability status), you may get a temporary registration number, valid for 45 days, by calling (800)942–6990 or (617)494–2545 and providing the following information:

Company name
Principal place of business
Credit card (MasterCard or VISA) number for payment

There is an additional $50 expedited handling fee for a temporary registration number. You must formally register before the expiration date of the temporary number.

The registration form is DOT Form F 5800.2. A copy of the registration form updated by DOT in March 1993 appears below in Figure 1.2. You can also obtain a copy, with instructions for its completion, from the following address:

Hazardous Materials Registration Program, DHM–60
U.S. Department of Transportation
Washington, DC 20590–0001
Telephone: (617)494–2545 or (202)366–4109

**Note**: In a *Federal Register* notice on 29 April 1993, DOT stated that any person who has engaged in an activity requiring registration since 16 September 1992, but has not registered for the 1992–93 registration year, may be subject to civil penalties for each day a covered function is performed. DOT also stated that the legal obligation to register for a year in which any activity requiring registration was conducted does not end with the registration year—the shipper or transporter must file a registration form for the past activity. A late registration of this type may also involve paying interest, late-payment penalties, and administrative charges, as well as a civil penalty.

### Registration Records

Every person who is required to register must maintain the following records at their principal place of business for three years from the date a Certificate of Registration is issued:

- A copy of the registration statement filed (DOT Form F 5800.2)
- The Certificate of Registration issued by RSPA

In addition, motor carriers subject to registration requirements must carry on board **each vehicle** that is transporting a hazardous material requiring registration one of the following:

- A copy of the carrier's current Certificate of Registration
- Another document bearing the registration number identified as the "U.S. DOT Hazmat Reg. No."

Exempted from Paperwork Reduction Act by 49 App. U.S.C. 1805(c)

**U.S. DEPARTMENT OF TRANSPORTATION**
**HAZARDOUS MATERIALS REGISTRATION STATEMENT**
**REGISTRATION YEAR 19____ - 19____**

(Please Type or Print all Responses)

Initial Registration ____ Renewal of Registration ____ Amendment to Registration ____ Expedited Follow-up ____

Current Registration # _ _ _ _ _ _ _ _ _ _ _ _ _ _ _ _ _

1. **Registrant** ______________________________________
(Company Name)
[Place pre-printed label here if provided and if name and address are correct. Otherwise, provide correct information.]

2. **Mailing Address of Principal Place of Business**
Street or P.O. Box ______________________________ City ____________________
County ____________________ State __ Zip Code _ _ _ _ _ - _ _ _ _ Country ____________________

3. **Carrier's US DOT ID Number, ICC Number, or Reporting Railroad Alphabetic Code (if applicable)**
US DOT ID # ____________________ ICC # ____________________ Railroad Alphabetic Code ____________

4. **Mode(s) Used to Transport Hazardous Materials:** Highway ___ Rail ___ Water ___ Air ___

5. **Industrial Classification:** Check the primary industry in which the registrant operates. **Mark Only One.**

**Transportation**
- ☐ Carrier
- ☐ Warehousing
- ☐ Freight Forwarding, Agent Services
- ☐ Transportation Repair & Service Facilities

**Manufacturing**
- ☐ Petroleum Refining & Related Industries
- ☐ Apparel & Other Textile Products
- ☐ Paper, Wood & Allied Products
- ☐ Printing & Publishing
- ☐ Chemicals & Allied Products
- ☐ Food & Kindred Products
- ☐ Rubber & Miscellaneous Plastic Products
- ☐ Electric & Electronic Equipment
- ☐ Explosives
- ☐ Other Manufacturing Industries

**Other**
- ☐ Agriculture & Support Services
- ☐ Mining - Metal & Non-Metal
- ☐ Oil & Gas Extraction
- ☐ Gasoline, Fuel Oil, Propane Sales & Delivery
- ☐ Construction - All Types
- ☐ Wholesale or Retail Trade
- ☐ Non-Transportation Repair Facilities
- ☐ Hazardous Waste Services - Transportation, Disposal, Treatment
- ☐ Other ____________________

6. **Annual Registration Fee.** The combined registration and processing fee is $300.00. (Complete only when submitting initial or renewal registration.)

Total Amount Enclosed: ____________________

Make check or money order in U.S. funds, drawn on a U.S. bank, and payable to "U.S. Department of Transportation," and identified as payment for the "Hazmat Registration Fee."

**Method of Payment**

Check ____ Money Order ____ Credit Card: VISA ____ MasterCard ____

**Credit Card Users Please Provide the Following Information:**

Card Number: ☐☐☐☐☐☐☐☐☐☐☐☐☐☐☐☐ Expiration Date: ☐☐ ☐☐
MO YR

Name as it appears on the card ______________________________

Authorized Signature ______________________________

Cardholder acknowledges ordering goods or services in the amount of the total shown hereon and agrees to perform the obligations set forth in the Cardholder's agreement with the issuer. Credit card statement will list this payment as "US DOT Hazmat Regis."

NOTE: If completing an Expedited Registration, do not resubmit credit card information here.

Form DOT F 5800.2 (REVISED 03/93) THIS FORM MAY BE REPRODU[CED]

Figure 1.2 DOT Registration Form

# Chapter 2
# DOT Requirements for Training of Hazmat Employees

In a notice published at 57 *Fed. Reg.* 20953 (15 May 1992), creating a new Subpart H—Training (49 CFR 172.700–.704), DOT established new requirements for training of hazmat employees. This is the HM–126F Training for Safe Transportation of Hazardous Materials program. The rule was corrected on 27 May 1992 to incorporate requirements for testing, and amended at 58 *Fed. Reg.* 5850 (22 January 1993) to delay the deadline for training until 1 October 1993. Additional training requirements for the individual modes of transportation are prescribed in parts 174, 175, 176, and 177 of the HMR. The training requirements came into effect 1 July 1992.

The purpose of the training rules is to ensure that each hazmat employer trains its hazmat employees regarding safe loading, unloading, handling, storing, and transporting of hazardous materials and emergency preparedness for responding to accidents or incidents involving the transportation of hazardous materials.

## 2.1 Definitions

*Training* as defined in Subpart H means a **systematic program** that ensures a hazmat employee has **familiarity** with the general provisions of the HMR, is able to **recognize and identify** hazardous materials, has knowledge of specific requirements of the HMR **applicable to functions** performed by the employee, and has knowledge of **emergency response** information, self-protection measures and accident prevention methods and procedures.

Hazmat employees are defined in 49 CFR 172.8:

*Hazmat employee* means a person who is employed by a **hazmat employer** [defined in the next paragraph] and who in the course of employment directly affects hazardous materials transportation safety. This term includes an owner-operator of a motor vehicle which transports hazardous materials in commerce. This term includes an individual, including a self-employed individual, employed by a hazmat employer who, during the course of employment:

- Loads, unloads, or handles hazardous materials;
- Tests, reconditions, repairs, modifies, marks, or otherwise represents containers, drums, or packagings as qualified for use in the transportation of hazardous materials;
- Prepares hazardous materials for transportation;
- Is responsible for the safety of transporting hazardous materials; or
- Operates a vehicle used to transport hazardous materials.

*Hazmat employer* means a person (including a business or organization) that uses one or more of its employees in connection with:

- transporting hazardous materials in commerce;
- causing hazardous materials to be transported or shipped in commerce; or

- does any of the following with containers, drums, or packagings to qualify them for use in the transportation of hazardous materials:
  - represent
  - mark
  - certify
  - sell
  - offer
  - recondition
  - test
  - repair
  - modify

This term includes an owner-operator of a motor vehicle that transports hazardous materials in commerce.

This term also includes any department, agency, or instrumentality of the United States, a state, a political subdivision of a state, or an Indian tribe engaged in an activity described in the first sentence of this definition.

Based on these definitions, examples of the kinds of employees requiring hazmat training would include those who—

- determine if a material is a hazardous material
- design, produce and/or sell packagings for hazardous materials
- determine proper packaging for hazardous materials
- put hazardous materials in packagings
- mark and label hazardous material packages
- fill out shipping papers for hazardous materials
- load or unload hazardous materials
- move hazardous material packagings in a warehouse during the course of transportation
- operate vehicles transporting hazardous materials

## 2.2 General Training Requirements

A hazmat employer must ensure that each of its hazmat employees is trained in accordance with the Training for Safe Transportation of Hazardous Materials program.

A hazmat employee who performs any function regulated by the HMR may not perform that function unless he or she has received training as described in Subpart H of Part 172. Each hazmat employer has a duty to comply with the applicable requirements of the HMR and to thoroughly instruct each hazmat employee in how to comply with the regulations.

The required training may be provided by the hazmat employer or other public or private sources.

Every hazmat employee must receive both initial and recurrent training that includes—

- General awareness/familiarization training
- Function-specific training
- Safety training
- Testing

Each hazmat employer is responsible for compliance with the requirements of the DOT regulations for shippers, transporters, and manufacturers, regardless of whether the training required by the DOT hazmat training regulations has been completed. This means that compliance with the substantive rules cannot be put off until training is completed, and that providing training does not excuse violations of the rules.

**General awareness/familiarization training** is designed to provide familiarity with the DOT regulations for shippers, transporters, and manufacturers, and to enable the employee to recognize and identify hazardous materials consistent with the hazard communication standards of the DOT regulations.

**Function-specific training** relates to requirements of the DOT regulations for shippers, transporters, and manufacturers that are specifically applicable to the functions the employee performs.

If a hazmat employer conducts training authorized by DOT regulations to comply with requirements of the International Civil Aviation Organization (ICAO) Technical Instructions, or the International Maritime Dangerous Goods (IMDG) Code, and is actually subject to those requirements, that training may be used as appropriate for function-specific training. This is true only if, and to the extent that, the ICAO or IMDG training requirements can serve as substitutes for the DOT hazmat training requirements.

**Safety training** must cover—

- Emergency response information required by 49 CFR Part 172, Subpart G—Emergency Response Information
- Measures to protect the individual employee from the hazards associated with hazardous materials to which he or she may be exposed in the work place, including specific measures the hazmat employer has implemented to protect employees from exposure
- Methods and procedures for avoiding accidents, such as the proper procedures for handling packages containing hazardous materials

If an employer conducts training to comply with the hazard communication program required by the Occupational Safety and Health Administration (OSHA) (29 CFR 1910.120) or the Environmental Protection Agency (EPA) (40 CFR 311.1), to the extent that training addresses the elements just described for DOT safety training, it may be used to satisfy the DOT safety training requirements in order to avoid unnecessary duplication of training.

There is a limitation on the coverage of requirements for safety training. A hazmat employee who repairs, modifies, reconditions, or tests packagings as qualified for use in the transportation of hazardous materials, and who does not perform any other function subject to the requirements of the DOT regulations for shippers, transporters, and manufacturers, is not subject to the safety training requirement.

**Testing:** Each employer must test each of its hazmat employees by appropriate means on the training subjects covered by the general awareness, function-specific, and safety training.

## 2.3 Initial And Recurrent Training, Recordkeeping

**Initial training** is required for each hazmat employee subject to the following conditions:

- If employed on or before 2 July 1993, training must be completed before **1 October 1993**.
- If employed after 2 July 1993, training must be completed **within 90 days** after employment.
- Any hazmat employee who changes hazardous materials job functions must complete training in the new job function(s) within 90 days after the change. However, the employee may perform new hazardous materials job functions prior to the completion of training, provided the employee performs those functions under the supervision of a properly trained and knowledgeable hazmat employee.

**Recurrent training**: Hazmat employees must receive all the DOT training required for their positions at least once every **two years**.

Any relevant training received from a previous employer or other source may be used to satisfy the DOT requirements for initial or recurrent hazmat training, provided that a current record of training is obtained from the hazmat employee's previous employer.

**Training records**: Each hazmat employer must create and retain a record of current training for each hazmat employee. The record must include at least the **last two years**. It must be retained for as long as that employee is employed by that employer as a hazmat employee and for **90 days** after termination of employment as a hazmat employee. Although these are the minimum regulatory requirements, company policy and liability considerations may determine longer periods of coverage and retention.

The training records must include the following information:

- Hazmat employee's name
- Most recent training completion date of the hazmat employee's training
- Description, copy, or location of training materials used to meet the requirements for general/ familiarization, function-related, and safety training
- Name and address of the person providing the training
- Certification that the hazmat employee has been trained and tested as required by the DOT hazmat training regulations

## 2.4 Specialized Training For Drivers

In addition to other training requirements of the HMR, the regulations at 49 CFR 177.816 require that no carrier may transport, or cause to be transported, a hazardous material unless each hazmat employee who will operate a motor vehicle has been trained in the applicable requirements of 49 CFR parts 383, 387, 390—399, and the procedures necessary for the safe operation of that motor vehicle.

**Subject matter**: Driver training must include the following subjects:

- Pre-trip safety inspection
- Use of vehicle controls and equipment, including operation of emergency equipment

- Vehicle operation, including—
  - turning, backing, braking, parking, and handling
  - vehicle characteristics including those that affect vehicle stability, such as effects of braking and curves
  - effects of speed on vehicle control
  - dangers associated with maneuvering through curves
  - dangers associated with weather or road conditions that a driver may experience (e.g., blizzards, mountainous terrain, high winds)
  - dangers associated with high center of gravity
- Procedures for navigating tunnels, bridges, and railroad crossings
- Regulatory requirements connected with—
  - attendance of vehicles
  - parking
  - smoking
  - routing
  - incident reporting
- Loading and unloading of materials, including—
  - Compatibility and segregation of cargo in a mixed load
  - Package handling methods
  - Load securement.

Specialized requirements for cargo tanks and portable tanks. In addition to the basic driver training requirement outlined above, each person who operates a cargo tank or a vehicle with a portable tank with a capacity of 1,000 gallons or more must receive training applicable to the requirements of the HMR, and have the appropriate state-issued commercial driver's license required by 49 CFR 383. Specialized training for tanker drivers must include the following subjects:

- Operation of emergency control features of the cargo tank or portable tank
- Special vehicle handling characteristics, including—
  - high center of gravity
  - fluid load subject to surge
  - effects of fluid-load surge on braking
  - characteristic differences in stability among baffled, unbaffled, and multi-compartmented tanks
  - effects of partial loads on vehicle stability
- Loading and unloading procedures
- The properties and hazards of the material transported
- Retest and inspection requirements for cargo tanks.

Special requirements for radioactive materials shipments. No person may transport a package of highway route controlled quantity of Class 7 (radioactive) materials, as defined in §173.403(l) of the HMR, on a public highway unless the driver is trained as required in the Training for Safe

Transportation of Hazardous Materials program and §177.816 (driver training). In addition, a copy of the record of training required by §172.704 must be in the immediate possession of the driver.

The hazmat driver training described above may be satisfied by compliance with the current requirements for a commercial driver's license (**CDL**) with a tank vehicle or hazardous materials **endorsement**, as applicable.

Requirements for initial and recurring training and for recordkeeping for drivers are now identical with the §172.704 requirements described above for general hazmat employee training.

# Chapter 3
# What Is A Hazardous Material?

As pointed out in Chapter I, there are hazardous **chemicals** under OSHA, hazardous **wastes** under RCRA, hazardous **substances** under CERCLA, and hazardous **materials** under HMTA. These categories overlap to some extent (recall that all RCRA hazardous wastes and CERCLA hazardous substances are HMTA hazardous materials). When we deal with hazardous materials in preparing them for shipping or actually transporting them, we need to know which substances are in fact defined as hazardous materials, and how to describe, mark, label, package, and handle them in order to comply with the HMR.

These questions are the major topic of 49 CFR Part 172—Hazardous Materials Table, Special Provisions, Hazardous Materials Communications, Emergency Response Information, and Training Requirements. As its long title suggests, this part of the HMR covers a wide variety of topics. It takes up more than 160 pages in the format in which it is usually printed (see Appendix A). We dealt with Subpart H—Training, in Chapter II. In this chapter, we will deal with the concepts of DOT shipping **descriptions**, including proper **shipping names**, **hazard classes** or **divisions**, DOT **identification numbers**, and **packing groups**. These ideas will be basic to the use of the materials in later chapters on how to select and manage containers, label them, prepare shipping papers, load and unload vehicles, and follow safe work practices appropriate to the kinds of materials being handled.

## 3.1 Definitions

The two fundamental definitions we need are found in §171.8—Definitions and abbreviations (you will also find them quoted in full in the Glossary of this manual):

*Hazardous material*: a substance or material (including hazardous substances, hazardous wastes, marine pollutants, and elevated temperature materials), which has been determined by the Secretary of Transportation to be capable of posing an unreasonable risk to health, safety, and property when transported in commerce, and which has been so designated.

*Hazardous substance* for the purposes of the DOT regulations for shippers, transporters, and manufacturers: a material, including its mixtures and solutions, that—

- Is listed in Appendix A to 49 CFR 172.101 (List of hazardous substances and reportable quantities)
- Is in a quantity, in one package, which equals or exceeds the reportable quantity (RQ) listed in Appendix A to 49 CFR 172.101, and
- When in a mixture or solution—
  - For radionuclides, conforms to paragraph 6 of Appendix A to 49 CFR 172.101 (comparing level of radioactivity with RQ levels).
  - For other than radionuclides, is in a concentration by weight which equals or exceeds a specified concentration corresponding to the RQ of the material (there is a table of concentrations vs. RQ in §171.8)

This definition of a hazardous substance does not apply to petroleum products that are lubricants or fuels (see 40 CFR 300.5 in EPA's rules).

Another category of regulated materials was added in November 1992, effective 1 January 1993:

*Marine pollutant*: a hazardous material that when in a solution or mixture of one or more marine pollutants, is packaged in a concentration equal to—

- Ten percent or more by weight of the solution or mixture for materials listed in the appendix; or
- One percent or more by weight of the solution or mixture for materials that are identified as **severe marine pollutants** in the appendix.

We use the listed names of these substances to construct a basic description of the material to be shipped.

*Proper shipping name*: the name of the hazardous material shown in Roman print (not italics) in 49 CFR 172.101 (the Hazardous Materials Table). The words printed in italics in the Table are comments, explanations, or connecting words (*and*, *or*, etc.). As we shall see later, it is often necessary to exercise considerable thought to determine a proper shipping name for a material that is not listed in the Hazardous Materials Table under the name we ordinarily use for it.

*Hazard class*: the category of hazard assigned to a hazardous material according to the defining criteria of 49 CFR Part 173 and the Hazardous Materials Table. A material may **meet** the defining criteria for more than one hazard class, but it is **assigned** to only one hazard class.

There are 9 hazard classes under the HM–181 rules. Their numerical designations are the required method of referring to them on shipping papers (except for ORM–D). The class and division names given in § 173.2 (see the table of defining sections in Part 173 below) are used in the regulations for explanatory purposes.

A hazard class may be broken down into **divisions** (not all are).

*Division*: a subdivision of a hazard class. See the following two pages for the 9 hazard classes and their divisions.

*Identification number*: A distinctive number assigned to each proper shipping name in the Hazardous Materials Table. Those preceded by the letters "UN" are associated with proper shipping names considered appropriate for international transportation as well as domestic transportation. Those preceded by the letters "NA" are associated with proper shipping names not recognized for international transportation, except to and from Canada.

*Packing group*: a grouping according to the degree of danger presented by hazardous materials. Packing Group I indicates great danger; Packing Group II, medium danger; Packing Group III, minor danger. In the Hazardous Materials Table, Column 5 specifies one or more packing groups assigned to a material corresponding to the proper shipping name and hazard class for that material. Classes 2 and 7 materials and ORM-D materials do not have packing groups.

| Class | Division | Characterization |
|---|---|---|
| 1 | | **Explosive materials** |
| | 1.1 | Explosives having a mass explosion hazard |
| | 1.2 | Explosives having a projection hazard, but not a mass explosion hazard |
| | 1.3 | Explosives having a fire hazard, and either a minor blast hazard or a minor projection hazard, but not a mass explosion hazard |
| | 1.4 | Explosive devices (containing no more than 25 g [9 oz.] of detonating material) presenting a minor explosion hazard |
| | 1.5 | Very insensitive explosives |
| | 1.6 | Extremely insensitive articles that do not have a mass explosive hazard |
| 2 | | **Gases** |
| | 2.1 | Flammable gases |
| | 2.2 | Non-flammable, non-poisonous compressed gases (including compressed gas, liquefied gas, pressurized cryogenic gas and compressed gas in solution) |
| | 2.3 | Poisonous gases (by inhalation) |
| 3 | | **Flammable liquids** (*Combustible* liquids are specially defined. They have a higher flash point temperature than flammable liquids. Although "combustible liquid" is not a numerical division designation within Class 3, it is used to classify materials for some purposes—strictly for domestic shipments—in the HMR.) |
| 4 | | **Flammable solids** |
| | 4.1 | Wetted explosives, self-reactive materials, and readily combustible solids |
| | 4.2 | Spontaneously combustible materials |
| | 4.3 | Dangerous when wet materials |
| 5 | | **Oxidizers** |
| | 5.1 | Oxidizing materials |
| | 5.2 | Organic peroxides |
| 6 | | **Poisonous materials** |
| | 6.1 | Poisonous liquids or solids |
| | 6.2 | Infectious substances (etiologic agents) |
| 7 | | **Radioactive materials** |
| 8 | | **Corrosive materials** |
| 9 | | **Miscellaneous hazardous materials** (includes hazardous substances and wastes not otherwise classified) |
| ORM-D | | Materials such as consumer commodities that are subject to the HMR, but present a limited hazard during transportation because of their form, quantity, and packaging. There are specific exemptions for ORM–D materials in the Hazardous Materials Table. |

Table 3.1 The Nine Hazard Classes

The hazard classes and divisions are defined in the following sections of 49 CFR Part 173:

| Class No. | Division No. (if any) | Name of class or division | 49 CFR reference for definitions |
|---|---|---|---|
| None | ...... | Forbidden materials | §173.21 |
| None | ...... | Forbidden explosives | §173.54 |
| 1 | 1.1 | Explosives (with a mass explosion hazard) | §173.50 |
| 1 | 1.2 | Explosives (with a projection hazard) | §173.50 |
| 1 | 1.3 | Explosives (with predominately a fire hazard). | §173.50 |
| 1 | 1.4 | Explosives (with no significant blast hazard). | §173.50 |
| 1 | 1.5 | Very insensitive explosives; blasting agents | §173.50 |
| 1 | 1.6 | Extremely insensitive detonating substances | §173.50 |
| 2 | 2.1 | Flammable gas | §173.115 |
| 2 | 2.2 | Non-flammable compressed gas | §173.115 |
| 2 | 2.3 | Poisonous gas | §173.115 |
| 3 | ...... | Flammable (and combustible) liquid | §173.120 |
| 4 | 4.1 | Flammable solid | §173.124 |
| 4 | 4.2 | Spontaneously combustible material | §173.124 |
| 4 | 4.3 | Dangerous when wet material | §173.124 |
| 5 | 5.1 | Oxidizer | §173.127 |
| 5 | 5.2 | Organic peroxide | §173.128 |
| 6 | 6.1 | Poisonous materials | §173.132 |
| 6 | 6.2 | Infectious substance (Etiologic agent) | §173.134 |
| 7 | ...... | Radioactive material | §173.403 |
| 8 | ...... | Corrosive material | §173.136 |
| 9 | ...... | Miscellaneous hazardous material | §173.140 |
| None | ...... | Other regulated material: ORM-D | §173.144 |

**Brief definitions of the listed hazard classes and divisions:**

*Forbidden materials*: Any of the following materials (further described in §173.21):

- Materials that are designated "Forbidden" in Column 3 of the §172.101 Table
- Forbidden explosives as defined in §173.51
- Electrical devices which are likely to create sparks or generate a dangerous quantity of heat, unless packaged in a manner which precludes such an occurrence
- For carriage by aircraft, any package which has a magnetic field of more than 0.00525 gauss measured at 4.5 m (15 feet) from any surface of the package

- A material in the same packaging, freight container, or overpack with another material, the mixing of which is likely to cause a dangerous evolution of heat, or flammable or poisonous gases or vapors, or to produce corrosive materials
- A package containing a material which is likely to decompose with a self-accelerated decomposition temperature (SADT) of 50°C (122°F) or less with an evolution of a dangerous quantity of heat or gas when decomposing or polymerizing, unless the material is stabilized or inhibited in a manner to preclude such evolution. The SADT may be determined by any of the test methods described in Part II of the UN Recommendations on the Transport of Dangerous Goods, Tests and Criteria, Second Edition (1990).
  - A package meeting these criteria may be required to be shipped under controlled temperature conditions. The control temperature and emergency temperature for a package can be found in the table in §173.21(f) based upon the SADT of the material. The **control temperature** is the temperature above which a package of the material may not be offered for transportation or transported. The **emergency temperature** is the temperature at which, due to imminent danger, emergency measures must be initiated.
  - For self-reactive materials listed in the §173.224(b) Table, control and emergency temperatures, where required are shown in Columns 4a and 4b, respectively. For organic peroxides listed in The Organic Peroxides Table in §173.225 control and emergency temperatures, where required, are shown in Columns 7a and 7b, respectively.
  - Refrigeration may be used as a means of stabilization only when approved by the Associate Administrator for Hazardous Materials Safety. For status of approvals previously issued by the Bureau of Explosives, see §171.19.
- Packages which give off a flammable gas or vapor, released from a material not otherwise subject to the HMR, likely to create a flammable mixture with air in a transport vehicle
- Packages containing materials (other than those classed as explosive) which will detonate in a fire.
  - For purposes of this definition, "detonate" means an explosion in which the shock wave travels through the material at a speed greater than the speed of sound.
- A package containing a cigarette lighter, or other similar device, equipped with an ignition element and containing fuel; except that a cigarette lighter or similar device subject to this provision may be shipped if the design of the device and its inner packaging has been examined by the Bureau of Explosives and specifically approved by the Associate Administrator for Hazardous Materials Safety. For the status of approvals previously issued by the Bureau of Explosives, see §171.19.

- An organic peroxide of the "ketone peroxide" category which contains more than 9 percent available oxygen as calculated using the equation in §173.128(b)(4)(iii). The category ketone peroxide includes, but is not limited to—
  - Acetyl acetone peroxide
  - Cyclohexanone peroxide(s)
  - Diacetone alcohol peroxides
  - Methylcyclohexanone peroxide(s)
  - Methyl ethyl ketone peroxide(s)
  - Methyl isobutyl ketone peroxide(s)

*Forbidden explosives*: Any of the following materials (further described in §173.54):

- An explosive that has not been approved in accordance with §173.56
- An explosive mixture or device containing a chlorate and—
  - An ammonium salt, including a substituted ammonium or quaternary ammonium salt; or
  - An acidic substance, including a salt of a weak base and a strong acid
- A leaking or damaged package of explosives
- Propellants that are unstable, condemned or deteriorated
- Nitroglycerin, diethylene glycol dinitrate, or any other liquid explosives not specifically authorized in the HMR
- A loaded firearm (except as provided in 14 CFR 108.11)
- Fireworks that combine an explosive and a detonator
- Fireworks containing yellow or white phosphorus
- A toy torpedo, the maximum outside dimension of which exceeds 23 mm (0.906 inch), or a toy torpedo containing a mixture of potassium chlorate, black antimony (antimony sulfide), and sulfur, if the weight of the explosive material in the device exceeds 0.26 g (0.01 ounce)
- Explosives specifically forbidden in the §172.101 Table
- Explosives not meeting the acceptance criteria specified in §173.57.

*Explosives*: For the purposes of the HMR, any substance or article, including a device, that is designed to function by explosion (i.e., an extremely rapid release of gas and heat) or that, by chemical reaction within itself, is able to function in a similar manner even if not designed to function by explosion, unless the substance or article is otherwise classed under the provisions of the HMR.

- Division 1.1 explosives are those that have a mass explosion hazard. A mass explosion is one that affects almost the entire load instantaneously.
- Division 1.2 consists of explosives that have a projection hazard but not a mass explosion hazard.
- Division 1.3 consists of explosives that have a fire hazard and either a minor blast hazard or a minor projection hazard or both, but not a mass explosion hazard.

- Division 1.4 consists of explosives that present a minor explosion hazard. The explosive effects are largely confined to the package and no projection of fragments of appreciable size or range is to be expected. An external fire must not cause virtually instantaneous explosion of almost the entire contents of the package.
- Division 1.5 consists of very insensitive explosives. This division is comprised of substances that have a mass explosion hazard but are so insensitive that there is very little probability of initiation or of transition from burning to detonation under normal conditions of transport. (The probability of transition from burning to detonation is greater when large quantities are transported in a vessel.)
- Division 1.6 consists of extremely insensitive articles which do not have a mass explosive hazard. This division is comprised of articles which contain only extremely insensitive detonating substances and which demonstrate a negligible probability of accidental initiation or propagation. (The risk from articles of Division 1.6 is limited to the explosion of a single article.)

*Gases* (definitions of some technical terms, and measurement methods, can be found in § 173.115):

- *Flammable gas* (Division 2.1)—any material that is a gas at 20°C (68°F) or less and 101.3 kPa (14.7 psi) of pressure and that—
  - Is ignitable at 101.3 kPa (14.7 psi) when in a mixture of 13 percent or less by volume with air, or
  - Has a flammable range at 101.3 kPa (14.7 psi) with air of at least 12 percent regardless of the lower limit.
- *Non-flammable, non-poisonous compressed gas* (Division 2.2)—including compressed gas, liquefied gas, pressurized cryogenic gas and compressed gas in solution—means any material (or mixture) that—
  - Exerts in the packaging an absolute pressure of 280 kPa (41 psia) at 20 °C (68°F), and
  - Does not meet the definition of Division 2.1 or 2.3.
- *Gas poisonous by inhalation* (Division 2.3)—a material that is a gas at 20 °C (68°F) or less and a pressure of 101.3 kPa (14.7 psi) and that—
  - Is known to be so toxic to humans as to pose a hazard to health during transportation, or
  - In the absence of adequate data on human toxicity, is presumed to be toxic to humans because when tested on laboratory animals it has an $LC_{50}$ value of not more than 5,000 ml/m$^3$ (see § 173.116(a) for assignment of Hazard Zones A, B, C or D). $LC_{50}$ values for mixtures may be determined using the formula in § 173.133(b)(1)(i).
- *Non-liquefied compressed gas*—a gas, other than in solution, that in a packaging under the charged pressure is entirely gaseous at a temperature of 20 °C (68°F).
- *Liquefied compressed gas*—a gas that in a packaging under the charged pressure is partially liquid at a temperature of 20°C (68°F).

- *Compressed gas in solution*—non-liquefied compressed gas that is dissolved in a solvent.
- *Cryogenic liquid*—a refrigerated liquefied gas having a boiling point colder than –90 °C (–130 °F) at 101.3 kPa (14.7 psi) absolute. A material meeting this definition is subject to requirements of the HMR without regard to whether it meets the definition of a non-flammable, non-poisonous compressed gas in § 173.115(b).
- *Refrigerant gas or Dispersant gas*—all non-poisonous refrigerant gases, dispersant gases (fluorocarbons) listed in §§ 172.101, 173.304(a)(2), 173.314(c), 173.315(a)(1) and 173.315(h), and mixtures thereof, or any other compressed gas having a vapor pressure not exceeding 1792 kPa (260 psi) at 54 °C (130 °F), and restricted for use as a refrigerant, dispersant or blowing agent.

*Flammable and combustible liquids* (Class 3) (definitions of some technical terms, and measurement methods, can be found in § 173.120):

- *Flammable liquid*—a liquid having a flash point of not more than 60.5 °C (141 °F), or any material in a liquid phase with a flash point at or above 37.8 °C (100 °F) that is intentionally heated and offered for transportation or transported at or above its flash point in a bulk packaging, except for—
  - Any liquid meeting one of the definitions specified in § 173.115 (gases)
  - Any mixture having one or more components with a flash point of 60.5 °C (141 °F) or higher, that makes up at least 99 percent of the total volume of the mixture, if the mixture is not offered for transportation or transported at or above its flash point
  - For the purposes of the HMR, a distilled spirit of 140 proof or lower is considered to have a flash point of no lower than 23 °C (73 °F).
- *Combustible liquid*—any liquid that does not meet the definition of any other hazard class specified in the HMR, except Class 9, and has a flash point above 60.5 °C (141 °F) and below 93 °C (200 °F). This classification is for U.S. domestic use only. There is no corresponding international class.
  - A **flammable** liquid with a flash point at or above 38 °C (100 °F) that does not meet the definition of any other hazard class, except Class 9, may be **reclassed** as a combustible liquid. This provision does not apply to transportation by vessel or aircraft, except where other means of transportation is impracticable.

*Flammable Solid* (Division 4.1)—any of the following three types of materials: **wetted explosives, self-reactive materials**, or **readily combustible solids.**

- *Wetted explosives*—materials that—
  - When dry are Explosives of Class 1 other than those of compatibility group A, that are wetted with sufficient water, alcohol, or plasticizer to suppress explosive properties; and
  - Are specifically authorized by name either in the § 172.101 Table or have been assigned a shipping name and hazard class by the Associate Administrator for Hazardous Materials Safety under the provisions of—

- An exemption issued under subchapter B of the HMR, or
- An approval issued under § 173.56(i).

- *Self-reactive materials*—materials that are liable to undergo, at normal or elevated temperatures, a strongly exothermal decomposition caused by excessively high transport temperatures or by contamination.
- *Readily combustible solids*—materials that—
  - Are solids which may cause a fire through friction, such as matches
  - Show a burning rate faster than 2.2 mm (0.087 inches) per second when tested in accordance with paragraph 2.3 of appendix E to Part 273
  - Any metal powders that can be ignited and react over the whole length of a sample in 10 minutes or less, when tested in accordance with paragraph 2.c.(2) of appendix E to Part 273.

*Spontaneously Combustible Material* (Division 4.2)—a **pyrophoric material** or a **self-heating material.**

- *Pyrophoric material*—a liquid or solid that, even in small quantities and without an external ignition source, can ignite within 5 minutes after coming in contact with air when tested according to paragraph 3.a.(1) or 3.a.(2), as appropriate, of appendix E to Part 273.
- *Self-heating material*—a material that, when in contact with air and without an energy supply, is liable to self-heat. A material of this type that exhibits spontaneous ignition, or if the temperature of the sample exceeds 200°C (392°F) during the 24-hour test period when tested in accordance with paragraph 3.b.(1) of appendix E to Part 273, is classed as a Division 4.2 material.

*Dangerous when wet material* (Division 4.3)—a material that, by contact with water, is liable to become spontaneously flammable or to give off flammable or toxic gas at a rate greater than 1 liter per kilogram of the material, per hour, when tested in accordance with paragraph 4 of appendix E to Part 273.

*Oxidizer* (Division 5.1)—a material that may, generally by yielding oxygen, cause or enhance the combustion of other materials.

- A **solid** material is classed as a Division 5.1 material if, when tested in accordance with Appendix F to Part 273, in either concentration tested, the mean burning time of the test mixture, is equal to or less than that of the average of the three tests with ammonium persulfate mixture.
- A **liquid** is classed as a Division 5.1 material by analogy to existing entries in the § 172.101 Table.

*Organic peroxide* (Division 5.2)—any organic compound containing oxygen (O) in the bivalent **-O-O-** structure and which may be considered a derivative of hydrogen peroxide, where one or more of the hydrogen atoms have been replaced by organic radicals, unless any of the following apply:

- The material meets the definition of an explosive as prescribed in Subpart C of Part 173, in which case it must be classed as an explosive
- The material is forbidden from being offered for transportation according to § 172.101 or § 173.21

- The Associate Administrator for Hazardous Materials Safety has determined that the material does not present a hazard which is associated with a Division 5.2 material
- The material meets one of the following conditions:
  - For materials containing no more than 1.0 percent hydrogen peroxide, the available oxygen, as calculated using the equation in §173.128(a)(4)(ii), is not more than 1.0 percent
  - For materials containing more than 1.0 percent but not more than 7.0 percent hydrogen peroxide, the available oxygen, content ($O_a$) is not more than 0.5 percent, when determined using the equation in §173.128(a)(4)(ii).

  Division 5.2 organic peroxides are assigned to a generic system which consists of seven types (A–G). An organic peroxide identified by technical name in the Organic Peroxides Table in §173.225 is assigned to a generic type in accordance with that Table. Organic peroxides not identified in the Organic Peroxides Table are assigned to generic types under the procedures of §173.128(c).

*Poisonous material* (Division 6.1)—a material, other than a gas, that is known to be so toxic to humans as to afford a hazard to health during transportation, or which, in the absence of adequate data on human toxicity—

- Is presumed to be toxic to humans because it falls within any one of the following categories when tested on laboratory animals:
  - Oral Toxicity: a liquid with an $LD_{50}$ for acute oral toxicity of not more than 500 mg/kg or a solid with an $LD_{50}$ for acute oral toxicity of not more than 200 mg/kg.
  - Dermal Toxicity: a material with an $LD_{50}$ for acute dermal toxicity of not more than 1000 mg/kg.
  - Inhalation Toxicity:
    - A dust or mist with an $LC_{50}$ for acute toxicity on inhalation of not more than 10 mg/L, or
    - A material with a saturated vapor concentration in air at 20°C (68°F) of more than one-fifth of the $LC_{50}$ for acute toxicity on inhalation of vapors and with an $LC_{50}$ for acute toxicity on inhalation of vapors of not more than 5,000 ml/$m^3$

  $LD_{50}$ and $LC_{50}$ are lethal dose-50 and lethal concentration-50, respectively. These are the doses (or concentrations for airborne poisons) that will cause death to 50 percent of a sample population of experimental animals under specified conditions (see §173.132(b)).
- Is an irritating material, with properties similar to tear gas, which causes extreme irritation, especially in confined spaces.
- These toxicity categories do not apply if the Associate Administrator for Hazardous Materials Safety determines that the physical characteristics of the material or its probable hazards to humans as shown by

documented experience indicate that the material will not cause serious sickness or death.

*Infectious substance* (Division 6.2)—a viable microorganism, or its toxin, which causes or may cause disease in humans or animals, and includes those agents listed in 42 CFR 72.3 of the regulations of the Department of Health and Human Services or any other agent that causes or may cause severe, disabling or fatal disease. The terms "infectious substance" and "etiologic agent" are synonymous.

*Radioactive material*—any material having a specific activity greater than 0.002 microcuries per gram (μCi/g). ("Specific activity" of a radionuclide means the activity of the radionuclide per unit mass of that nuclide. The specific activity of a material in which the radionuclide is essentially uniformly distributed is the activity per unit mass of the material.) (See §173.403 for the long list of technical definitions associated with radioactive materials.)

*Corrosive material* (Class 8)—a liquid or solid that causes visible destruction or irreversible alterations in human skin tissue at the site of contact, or a liquid that has a severe corrosion rate on steel or aluminum, in accordance with the criteria in §173.36.

*Miscellaneous hazardous material* (class 9)—a material that—

- Has an anesthetic, noxious or other similar property which could cause extreme annoyance or discomfort to a flight crew member so as to prevent the correct performance of assigned duties
- Meets the definition in §171.8 for a hazardous substance or a hazardous waste.

*ORM–D material*—a material such as a consumer commodity, which, although otherwise subject to the HRM, presents a limited hazard during transportation due to its form, quantity and packaging. It must be a material for which exceptions are provided in the §172.101 Table. Each ORM–D material and category of ORM–D material is listed in the §172.101 Table.

## 3.2 The Hazardous Materials Table—General

The definition of hazardous material cited above said that it was a substance or material designated by the Secretary of Transportation. The Hazardous Materials Table in 49 CFR 172.101 is the place in the regulations where DOT officially designates the materials listed as hazardous materials for the purpose of transportation of those materials. For each listed material, the Table identifies the hazard class, or specifies that the material is forbidden in transportation; and gives the proper shipping name, or directs the user to the preferred proper shipping name. In addition, the Table specifies or references requirements in other parts of the HMR connected with labeling, packaging, quantity limits aboard aircraft, and stowage of hazardous materials aboard vessels. Before the introduction of HM–181, the Table was often referred to as the "Commodity List." Current usage tends to stick to calling it the Hazardous Materials Table or 172.101 Table.

One of the greatest potential sources of confusion during the transition period to the full implementation of the HM–181 rules is use of the Hazardous

Materials Table. It is central to the selection of a proper shipping name for materials (used on DOT shipping papers and/or hazardous waste manifests), selection of appropriate containers and their management, labeling of containers, procedures required for specific modes of transport, and safe work practices. Recall that, except for new explosives and some aspects of materials poisonous by inhalation, classification and hazard communication provisions of HM–181 do not come into full effect until 1 October 1994, and that the complete requirements for packagings will not be effective until 1 October 1996. In the interim, you have the option of complying with the new regulations or with those published in the 1990 edition of 49 CFR. Within limits, you can even mix practices complying with the old and new regulations.

### 3.3 The HM–181 Hazardous Materials Table

Both the old and new Hazardous Materials Tables are similarly organized. Each **row** of the table deals with a different hazardous material description and proper shipping name (sometimes only as a cross-reference to another row). Each **column** contains information about the material that is needed to make decisions about it to comply with the HMR.

There are ten columns in the HM–181 (new) Hazardous Materials Table:

**Column 1**: *Symbols,* may contain one or more of five symbols ('+', 'A', 'D', 'I', and 'W') When they appear (Column 1 is often blank), they mean—

+ fixes the proper shipping name, hazard class and packing group for that row entry without regard to whether the material meets the definition of that class or packing group or meets any other hazard class definition. However, an appropriate alternate proper shipping name and hazard class may be authorized by DOT. (See, e.g., Boron Tribromide in the Table reprinted in Appendix A).

**A** restricts the application of requirements of the HMR to materials offered or intended for transportation by aircraft, <u>unless</u> the material is a hazardous **substance** or a hazardous **waste.*** These exceptions are very important: you must check the tables in Appendix A to §172.101 to determine whether a substance is listed there, and is being shipped in an amount equal to or greater than its reportable quantity (RQ). If you are shipping the substance as a waste, then you must classify, package, and handle it according to the Table entries even if it is not to be shipped by air.

**D** identifies proper shipping names which are appropriate for describing materials for domestic transportation but may be inappropriate for international transportation under the provisions of international regulations. An important example is "Combustible liquid, n.o.s. [not otherwise specified]."

**I** identifies proper shipping names which are appropriate for describing materials in international transportation. An alternate proper shipping name may be selected when <u>only</u> domestic transportation is involved.

**W** restricts the application of requirements of the HMR to materials offered or intended for transportation by vessel (boat, ship, barge, etc.), unless the material is a hazardous **substance** or a hazardous **waste**. As with the symbol 'A', you must be very aware of when these exceptions apply.

**Column 2**: *Hazardous materials descriptions and proper shipping names,* lists the hazardous materials descriptions and proper shipping names of materials designated as hazardous materials. Modification of a proper shipping name may otherwise be required or authorized by §172.101.

- Proper shipping names are limited to those shown in Roman type (not italics).
- Proper shipping names may be used in the singular or plural and in either capital or lower case letters. Words may be alternatively spelled in the same manner as they appear in the ICAO Technical Instructions or the IMDG Code. For example "aluminum" may be spelled "aluminium" and "sulfur" may be spelled "sulphur." However, the word "inflammable" may not be used in place of the word "flammable."
- Punctuation marks and words in italics are <u>not</u> part of the proper shipping name, but <u>may</u> be used in addition to the proper shipping name. The word *"or"* in italics indicates that terms in the sequence may be used as the proper shipping name, as appropriate. (See, e.g., n-Pentanes *or* Isopentane.)
- The abbreviation "n.o.i." (not otherwise indicated) or "n.o.i.b.n." (not otherwise indicated by name) may be used interchangeably with "n.o.s." (not otherwise specified).
- Except for hazardous wastes (where the word "Waste" must come first), when qualifying words are used as part of the proper shipping name, their sequence in the package markings and shipping paper description is optional. However, the entry in the Table reflects the preferred sequence.
- When one entry references another entry by use of the word *"see,"* if <u>both</u> names are in Roman type, either name may be used as the proper shipping name (e.g., Ethyl alcohol, *see* Ethanol).

*In dealing with both the proper shipping name and the symbols 'A' and 'W,' the following definitions are important:

Hazardous **substances**: Appendix A to §172.101 lists materials that are listed or designated as hazardous substances under §101(14) of the Comprehensive Environmental Response, Compensation, and Liability Act (CERCLA; Superfund). Proper shipping names for hazardous substances are determined as follows:

- If the hazardous substance appears in the Table by technical name, then the technical name is the proper shipping name.
- If the hazardous substance does not appear in the Table and is not a forbidden material, then an appropriate generic or "n.o.s." shipping name must be selected corresponding to the hazard class (and packing group, if any) of the material as determined by the defining criteria of the HMR. For example, a hazardous substance that is listed in Appendix A but not in the Table and which meets the definition of a flammable liquid might be described as "Flammable liquid, n.o.s." or other appropriate shipping name corresponding to the flammable liquid hazard class.

Hazardous **wastes**: If the word "waste" is not included in the hazardous material description in Column 2 of the Table, the proper shipping name for a hazardous waste (as defined in §171.8), must include the word "Waste" preceding the proper shipping name of the material. For example: Waste acetone.

As you can see from the discussion above, the proper shipping name is the key to the table. It is difficult even to introduce the topic without getting into some of the technical particulars that we will treat further in Chapter VI. The ideas just introduced should help you find your way through the table until we take it up in more detail below.

**Column 3**: *Hazard class or Division*, contains a designation of the numerical hazard class or division corresponding to each proper shipping name, or the word "Forbidden."

- A material for which the entry in this column is "Forbidden" may not be offered for transportation or transported. This prohibition does not apply if the material is diluted, stabilized or incorporated in a device, and it is then classed in accordance with the definitions of hazardous materials contained in 49 CFR Part 173.
- You will find a basic description of each hazard class in the Definitions section (3.1) of this chapter. The section reference for class definitions appear in §173.2 of the HMR.
- Earlier, we mentioned the special case of combustible liquids. The HMR (§172.101(d)(4)) provide that when there is a reference to Class 3 (flammable liquids) in Column 3, the reference to Class 3 is modified to read "Combustible liquid" when that material is reclassified in accordance with §173.150(f) (dealing with exceptions for Class 3), or has a flash point above 60.5°C (141°F) but below 93°C (200°F).

**Column 4**: *Identification number*, lists the identification number assigned to each proper shipping name.

- Numbers preceded by the letters "UN" are associated with proper shipping names considered appropriate for international transportation as well as domestic transportation.
- Numbers preceded by the letters "NA" are associated with proper shipping names not recognized for international transportation, except to and from Canada.
- Identification numbers in the "NA9000" series are associated with proper shipping names not appropriately covered by international hazardous materials (dangerous goods) transportation standards, or not appropriately addressed by international transportation standards for emergency response information purposes, except for transportation between the United States and Canada.

**Column 5**: *Packing group*, specifies one or more packing groups assigned to a material corresponding to the proper shipping name and hazard class for that material.

- Classes 2 and 7 materials and ORM–D materials do not have packing groups.
- If more than one packing group is indicated for an entry, the packing group for the hazardous material is determined using the criteria for assignment of packing groups specified in Subpart D of Part 173, outlined below in Chapter IV.

- If a material is a hazardous **waste** or a hazardous **substance**, and has its proper shipping name preceded in Column 1 of the Table by the symbol "**A**" or "**W**", its packing group is changed to read "III" whenever the material is offered for transportation or transported by a mode in which its transportation is not otherwise regulated elsewhere in the HMR.

**Column 6**: *Labels*, specifies the hazard warning label(s) required for a package filled with a material belonging to the associated hazard class and having the listed proper shipping name, unless the package is excepted from labeling by special provisions in Subpart D of Part 172, or Part 173.

- The first label shown for each entry indicates the **primary hazard** of the material. Additional labels indicate **subsidiary** hazards.
- Provisions in §172.402 may require some other label than that specified in Column 6 to be affixed to the package in addition to the one specified.

**Column 7**: *Special provisions*, specifies codes for special provisions applicable to hazardous materials. When Column 7 contains a special provision code for a hazardous material, the meaning and requirements of that special provision are given in §172.102 of the HMR (immediately following the Table and its Appendices). In general—

- Codes consisting only of numbers (for example, "11") are multi-modal in application, and may apply to bulk and non-bulk packagings.
- Codes with the letter "**A**" apply only to transportation by aircraft.
- Codes with the letter "**B**" apply only to bulk packagings, but not to intermodal (IM) portable tanks.
- Codes with the letter "**H**" apply only to transportation by highway.
- Codes with the letter "**N**" apply only to non-bulk packaging requirements.
- Codes with the letter "**R**" apply only to transportation by rail.
- Codes with the letter "**T**" apply only to transportation in IM portable tanks.
- Codes with the letter "**W**" apply only to transportation by water.

**Column 8**: *Packaging authorizations*, consists of three subcolumns indicating the sections in Part 173 where requirements are spelled out for—

- Exceptions (Col. 8a)
- Non-bulk packaging requirements (Col. 8B)
- Bulk packaging requirements (Col. 8C)

Each of these columns is completed in a manner which indicates that "§173." precedes the designated numerical entry. For example, the entry "202" in Column 8B associated with the proper shipping name "Gasoline" indicates that non-bulk packagings for this material must conform to §173.202 of the HMR. When packaging requirements are specified, they are in addition to the standard requirements for all packagings prescribed in §173.24 and any other applicable requirements in Subparts A and B of Part 173.

**Column 9**: *Quantity limitations*, consists of two subcolumns specifying the maximum quantities that may be offered for transportation in one package by—

- passenger-carrying aircraft or passenger-carrying rail car (Column 9A)
- cargo aircraft only (Column 9B)

These limitations are subject to the following conditions:

- The quantity limitation is "net" except where otherwise specified (e.g., for "Consumer commodity" which specifies "30 kg gross").
- When **articles** or **devices** are specifically listed by name, the net quantity limitation applies to the entire article or device (less packaging and packaging materials), rather than only to its hazardous components.
- A package offered or intended for transportation by aircraft and that is filled with a material forbidden on passenger-carrying aircraft, but permitted on cargo aircraft only, or which exceeds the maximum net quantity authorized on passenger-carrying aircraft, must be labeled with the **CARGO AIRCRAFT ONLY** label specified in § 172.448.

**Column 10**: *Vessel stowage requirements*, also consists of two subcolumns specifying stowage locations and requirements on board cargo and passenger vessels.

- Column 10A (Vessel stowage) specifies the authorized stowage locations on board cargo and passenger vessels. These codes indicate stowage categories as follows:

  A—material may be stowed "on deck" or "under deck" on a cargo vessel and a passenger vessel.

  B—material may be stowed "on deck" or "under deck" on a cargo vessel, but must be stowed "on deck" on a passenger vessel.

  C—material must be stowed "on deck" on a cargo vessel and on a passenger vessel.

  D—material must be stowed "on deck" on a cargo vessel, but is prohibited on a passenger vessel.

  E—material may be stowed "on deck" or "under deck" on a cargo vessel, but is prohibited on a passenger vessel.

  - For bulk transportation by vessel, see 46 CFR Parts 30 to 40, 70, 98, 148, 151, 153 and 154.
- Column 10B (Other provisions) specifies codes for stowage requirements for specific hazardous materials.
  - In addition to the stowage requirements in Column 10A, further requirements for specific hazardous ,materials may be indicated by code in Column 10B For example, for Aniline the code "A" in 10A indicates the material may be stowed "on deck" or "under deck." However, there is a code "40" in 10B. From the Table of Provisions at § 176.84(b), you can determine that this means that if the material is stowed "under deck" it must be stowed "clear of living quarters."

### 3.4 The "Old" (1990) Hazardous Materials Table

The DOT Hazardous Material Table in the 1990 edition of 49 CFR is divided into **eight** columns (numbered 1–7; 3A is an addition). These columns present information about the materials in the rows generally similar to, but differently arranged from the HM–181 Table:

**Column 1**: *Applicability*, may contain one or more of three symbols ('+', 'A', and 'W') When they appear, they mean—

- \+ The proper shipping name and the hazard class of the material must be shown on the shipping papers and the package whether or not the material or its mixtures or solutions meet the definition of the class.
- **A** The material is subject to DOT regulations only when transported by air, unless the material is a hazardous substance or a hazardous waste.
- **W** The material is subject to DOT regulations only when transported by water, unless the material is hazardous substance or a hazardous waste.
- If neither "A" nor "W" appears, the substance is regulated for all modes of transportation.

**Column 2**: *Hazardous materials descriptions and proper shipping names*, lists the hazardous materials descriptions and proper shipping names of materials designated as hazardous materials, as it does in the HM–181 Table. Use of international spellings and "n.o.i." or "n.o.i.b.n." is not included.

**Column 3**: *Hazard Class*, gives the hazard class as words, not numerical codes. In some instances Column 3 will contain the word "Forbidden," which means the material may not be offered or accepted for transportation unless it is diluted, stabilized or incorporated in a device, and it is then classed in accordance with the definitions of hazardous materials contained in 49 CFR Part 173.

The older hazard classes are not broken down into divisions. There are 22 categories designated by words. Although there is no exact one-to-one correspondence between the HM–181 classes and divisions and the older classes, they can be put in rough correlation as follows:

| HM–181 Classes and Divisions | 1990 Hazard Classes |
|---|---|
| Class 1 (Explosive materials) | Explosives |
| 1.1 | EXPLOSIVE A |
| 1.2 | EXPLOSIVE A or B |
| 1.3 | EXPLOSIVE B |
| 1.4 | EXPLOSIVE C |
| 1.5 | BLASTING AGENT |
| 1.6 | |
| Class 2 (Gases) | Gases |
| 2.1 | FLAMMABLE GAS |
| 2.2 | NONFLAMMABLE GAS |
| 2.3 | POISON GAS |
| Class 3 (Flammable liquids) | Flammable and combustible liquids |
| 3 | FLAMMABLE LIQUID |
| | COMBUSTIBLE LIQUID |
| Class 4 (Flammable solids) | Flammable solids |
| 4.1 | FLAMMABLE SOLID |
| 4.2 | FLAMMABLE SOLID or FLAMMABLE LIQUID (spontaneously combustible or pyrophoric) |

| HM-181 Classes and Divisions | 1990 Hazard Classes |
|---|---|
| 4.3 | FLAMMABLE SOLID, Dangerous When Wet |
| Class 5 (Oxidizers) | Oxidizing Substances |
| 5.1 | OXIDIZER |
| 5.2 | ORGANIC PEROXIDE |
| Class 6 (Poisonous materials) | Poisons |
| 6.1 | POISON B |
| 6.2 | ETIOLOGIC AGENT (Infectious Substance) |
| Class 7 (Radioactive materials) | RADIOACTIVE MATERIAL |
| Class 8 (Corrosive materials) | CORROSIVE MATERIAL |
| Class 9 (Miscellaneous hazardous materials) | OTHER REGULATED MATERIAL (ORM–A through –E) |

This leaves two classes, POISON A and IRRITATING MATERIAL, without close new equivalents.

**Column 3A**: *Identification number*, shows the DOT identification number assigned to the hazardous material. These numbers prefixed by "UN" are recognized internationally and in the United States by emergency personnel. Those numbers prefixed by "NA" are only recognized in North America.

**Column 4**: *Label(s) required*, shows the labels that must be applied to the outside of the package of hazardous waste. There are additional labeling requirements not specified in Column 4 that must be complied with.

**Column 5**: *Packaging*, is divided into two subcolumns, showing which section of the regulations contains packaging requirements for the hazardous waste you are shipping. Column 5(a) will either indicate that there are no exceptions to the packaging requirements in Column 5(b) using the word "none," or will indicate which section of the regulations contains exceptions. Exceptions to specific packaging requirements are typically for very small or "limited quantities" of hazardous waste and materials. Column 5(b) lists the specific section of 49 CFR that describes the packaging requirements for the hazardous waste or material. From 1 October 1993 until 1 October 1996, you will need to refer to this column of the old HMT if you wish to choose a pre-POP packaging for a material for which such a packaging is still permitted.

**Column 6**: *Maximum net quantity in one package*, is also divided into two subcolumns, providing the restrictions on the transportation of hazardous material by air and rail. Column 6(a) shows the maximum amount of material that may be shipped in one package by passenger aircraft or passenger railcar. Column 6(b) shows the maximum amount of material that may be shipped in one package by cargo-only aircraft.

**Column 7**: *Water shipments*, is divided into three subcolumns, providing requirements for water shipments for cargo-only vessels 7(a) and passenger vessels 7(b). Column 7(c) contains additional requirements for water shipments such as "Keep dry" or "Keep cool."

## 3.5 Comparison of the Hazardous Materials Tables

To review the fairly complex content of this chapter, let us briefly compare the format and contents of the 1990 and HM–181 Hazardous Materials Tables. Remember that our focus here has been on the Table because it is the means by which DOT officially designates hazardous materials and directs shippers and transporters to the regulatory requirements for each designated material.

| Column | 1990 HM Table | HM–181 HM Table |
|---|---|---|
| 1 | + fixed proper shipping name | + fixed proper shipping name |
| | A applicable for air only | A applicable for air only |
| | ...... | D for domestic use |
| | ...... | I for international use |
| | W applicable to water shipment only | W applicable to water shipment only |
| 2 | Hazardous material description and proper shipping name | Hazardous material description and proper shipping name |
| 3 | Hazard Class | Hazard class or division |
| 3A | Identification number | ...... |
| 4 | Label(s) required if not excepted | Identification number |
| 5 | Packaging | Packing Group |
| 5(a) | Exceptions | ...... |
| 5(b) | Specific requirements | ...... |
| 6 | Maximum net quantity in one package | Label(s) required if not excepted |
| 6(a) | Passenger-carrying aircraft or railcar | ...... |
| 6(b) | Cargo-only aircraft | ...... |
| 7 | Water shipments | Special provisions |
| 7(a) | Cargo vessel stowage limitations | ...... |
| 7(b) | Passenger vessel stowage limitations | ...... |
| 7(c) | Other requirements | ...... |
| 8 | ...... | Packaging authorizations |
| 8A | ...... | Exceptions |
| 8B | ...... | Non-bulk packaging |
| 8C | ...... | Bulk packaging |
| 9 | ...... | Quantity limitations |
| 9A | ...... | Passenger aircraft or railcar |
| 9B | ...... | Cargo aircraft only |
| 10 | ...... | Vessel stowage requirements |
| 10A | ...... | Vessel stowages (locations) |
| 10B | ...... | Other stowage provisions |

# Chapter 4
# Container Selection and Management

Only containers which meet DOT standards may be used for the transportation of hazardous materials or hazardous waste. In order to avoid repackaging hazardous waste prior to shipment and to avoid compatibility problems between the waste and the container, it is wise to use DOT-specification containers for hazardous waste accumulation as well. DOT has recognized that not all materials are compatible with every type of container. Acids will destroy metal drums and some organic solvents will dissolve the materials in certain plastic containers. In addition, containers must be designed and constructed to hold hazardous materials and wastes during the stresses and rigors that occur during container handling and transportation. No one container has been designed which is appropriate for the transportation of all hazardous materials. Consequently, DOT has specified which containers are best suited for particular materials and wastes.

US DOT hazardous material regulations were amended on December 21, 1990 to align the regulations with United Nations recommendations and International Civil Aviation Organization Technical Instructions in the areas of classification, packaging, and hazard communication during the transport of hazardous materials. International performance-oriented packaging standards will replace the current DOT-specification system used for container selection and must be used for non-bulk, domestic shipments by October 1, 1996. New packaging requirements for materials that are poisonous by inhalation become mandatory on October 1, 1993. Under the new system, packaging requirements for a hazardous material are based on the material's packing group and vapor pressure, and the chemical compatibility between the packaging and the hazardous material.

The new regulations change the requirements on the selection, use, and reuse of plastic and metal drums as well as other containers. Use of the new hazardous materials table and associated classification, labeling, marking, and packaging requirements is mandatory on October 1, 1991 for explosives and most infectious substances. By October 1, 1992, hazard communication requirements are effective for all materials which are poisonous by inhalation. January 1, 1994 is the date after which most other provisions of this final rule become effective (certain performance-oriented packaging requirements become effective in 1994 and 1996 as noted above).

## 4.1 Container Selection

To select the proper container for transportation off-site, you must use the DOT Hazardous Materials Table found in 49 CFR 172.101. There are two versions of the table, either of which may be used depending on date of the shipment and the type of containers used. The newer (HM-181) DOT Hazardous Materials Table is reprinted in Appendix A of this handbook. These tables identify the regulatory requirements for hazardous material containers. A hazardous material is defined by DOT as a substance or material in a quantity and form which may pose an unreasonable risk to health, safety and property when transported in commerce. All hazardous wastes are DOT hazardous materials when transported in commerce.

The following procedure may be used to select the proper container for waste accumulation.

1. Select a proper shipping name for the hazardous material. For example, if you are shipping acetone, look up the word "Acetone" in column 2. If the hazardous material is a mixture containing more than one chemical compound or element, use the most descriptive proper shipping name (A procedure for determining the proper shipping names for mixtures is presented in Chapter 6).

2a. If using the HM-181 table, determine the packing group associated with the proper shipping name. The packing group is listed in column 5 of the HM-181 hazardous materials table. The more hazardous the material being shipped, the more stringent the packing requirements. There are three packing groups (PG): PG I = great danger; PG II = medium danger; PG III = minor danger.

2. In column 5 of the pre-HM-181 table, find the packaging specifications corresponding to the proper shipping name you have selected. Column 5(a) refers to packaging specifications for very small or limited quantities of hazardous materials. Column 5(b) is used to determine which regulations specify the proper non-bulk containers, such as drums, that can be used for transportation of hazardous materials. Or, in column 8 of the HM-181 table, find the packaging standards for the proper shipping name you selected. Column 8A contains specifications for exceptions, similar to column 5(a) of the pre-HM-181 table. Column 8B is used to determine which regulations specify the proper non-bulk containers that can be used for transportation of hazardous materials. Column 8C is used to determine the requirements for bulk packages.

3. Identify the specific packaging requirements and packaging codes for the material in the referenced section of 49 CFR 173.

4. Obtain containers that are marked with one of the specification codes identified in the referenced section of 49 CFR.

If a container is a DOT specification package, somewhere on the container (e.g., top or bottom) will be a mark with the DOT specification. This mark indicates the container manufacturer has followed DOT requirements when manufacturing and testing the container. An example mark and a complete explanation of the container manufacturer's marking system is presented in Chapter 5.

### 4.1.1 Assignment of Packing Groups under HM-181

When more than one packing group is identified on the hazardous materials table, the shipper must determine the appropriate packing group based on the physical and health hazards of the material being shipped. The following discussion explains how to determine the packing group for all hazard classes except for Class 1 explosives and Class 7 radioactive materials, for which the reader is referred to the DOT regulations in 49 CFR 173.

**Class 2 Gases**

In Class 2, only Division 2.3 gases have multiple packing groups. The packing group is based on the inhalation toxicity of the gas, which is separated into three hazard zones. The hazard zone and packing group may be determined using the table below:

| Hazard zone | Inhalation toxicity* |
|---|---|
| A............ | $LC_{50} \le 200$ ppm |
| B............ | 200 ppm $< LC_{50} \le$ 1000 ppm |
| C............ | 1000 ppm $< LC_{50} \le$ 3000 ppm |
| D............ | 3000 ppm $< LC_{50} \le$ 5000 ppm |

*$LC_{50}$ is lethal concentration-50. This is the concentration that will cause death to 50 percent of a sample population of experimental animals under specified conditions (see definitions for poisonous substances in Chapter III).

**Class 3 Flammable Liquids**

For Class 3, packing group is determined by the criteria on the following table:

| Packing Group | Flash Point (closed cup) | Initial Boiling Point |
|---|---|---|
| I | ........................................ | ≤ 35°C (95°F) |
| II | < 23°C (73°F) | > 35°C (95°F) |
| III | ≥ 23°C, ≤ 60.5°C (141°F) | > 35°C (95°F) |

**Class 4 Flammable Solids**

Packing groups are determined as follows when the hazardous materials table assigns more than one packing group:

Division 4.1

- For materials other than metal powders, PG II if the burning rate (when tested in accordance with Appendix E of 49 CFR 173) is greater than 2.2 mm/s and flame passes the wetted zone. PG III if the burning rate is greater than 2.2 mm/s and the wetted zone stops the flame.
- For metal powders, PG II if the zone of reaction spreads over the whole length of the sample in ≤ 5 minutes; PG III if the zone of reaction spreads over the whole length of the sample in more than 5 but less than 10 minutes.

Division 4.2

- Self heating materials are assigned to PG II if the material gives a positive test result when tested with a 2.5 cm cube size sample; PG III if the material gives a positive test result when tested with a 10 cm cube size sample but a negative test result with a 2.5 cm cube size sample.

Division 4.3

- PG I is assigned if spontaneous ignition occurs, or material demonstrates a tendency of spontaneous ignition, or the rate of evolution of flammable gases is ≥ 10 l/kg of material over any one minute.

- PG II is assigned the if rate of evolution of flammable gases is ≥ 20 l/kg of material per hour.
- PG III is assigned if the rate of evolution of flammable gases is greater than 1 l/kg of material per hour.

### Class 6 Poisonous Materials

Packing groups for Division 6.1 materials that are toxic in any way **except** by inhalation of vapors are determined by using the following table when the hazardous materials table assigns more than one packing group:

| Packing Group | Oral $LD_{50}$ mg/kg | Dermal $LD_{50}$ mg/kg | Inhalation $LD_{50}$ mg/kg (dusts and mists) |
|---|---|---|---|
| I | ≤5 | ≤40 | ≤0.5 |
| II | >5, ≤ 50 | >40, ≤200 | >0.5, ≤2 |
| III | solids: >50, ≤200<br>liquids: >50, ≤500 | >200, ≤1000 | >2, ≤10 |

The packing group and hazard zone assignments for liquids based on inhalation of vapors are based on the criteria listed in the following table and in Fig. 4.1.. These overlap with the Division 2.3 gases hazard zones as shown by the extensions at the right side of the graph (see also 49 CFR 173.133(2)(ii)).

| Packing Group | Vapor concentration* and toxicity |
|---|---|
| I (Hazard Zone A) | $V \geq 500 \times LC_{50}$ and $LC_{50} \leq 200$ ml/m$^3$ |
| I (Hazard Zone B) | $V \geq 10 \times LC_{50}$ ; $LC_{50} \leq 1000$ ml/m$^3$; and the criteria for PG I, HZ A are not met. |
| II | $V \geq LC_{50}$ ; $LC_{50} \leq 3000$ ml/m$^3$; and the criteria for PG I are not met. |
| III | $V \geq .2 \times LC_{50}$ ; $LC_{50} \leq 5000$ ml/m$^3$; and the criteria for PG I and II are not met. |

*V is the saturated vapor concentration in air of the material in ml/m3 at 20°C and standard atmospheric pressure

### Class 8

Packing groups for Class 8 are determined as using the following criteria when the hazardous materials table assigns more than one packing group:

- PG I for substances that cause visible destruction or irreversible alterations of the skin at the site of contact when tested on the intact skin of an animal for a period of not more than 3 minutes.
- PG II for substances, other than PG I, that cause visible destruction or irreversible alterations of the skin at the site of contact when tested on the intact skin of an animal for a period of not more than 60 minutes.
- PG III for substances other than PG I or II that cause visible destruction or irreversible alterations of the skin at the site of contact when tested on the intact skin of an animal for a period of not more than 4 hours or which have a corrosion rate on steel or aluminum exceeding 6.25 mm (0.246"). a year at a temperature of 55°C.

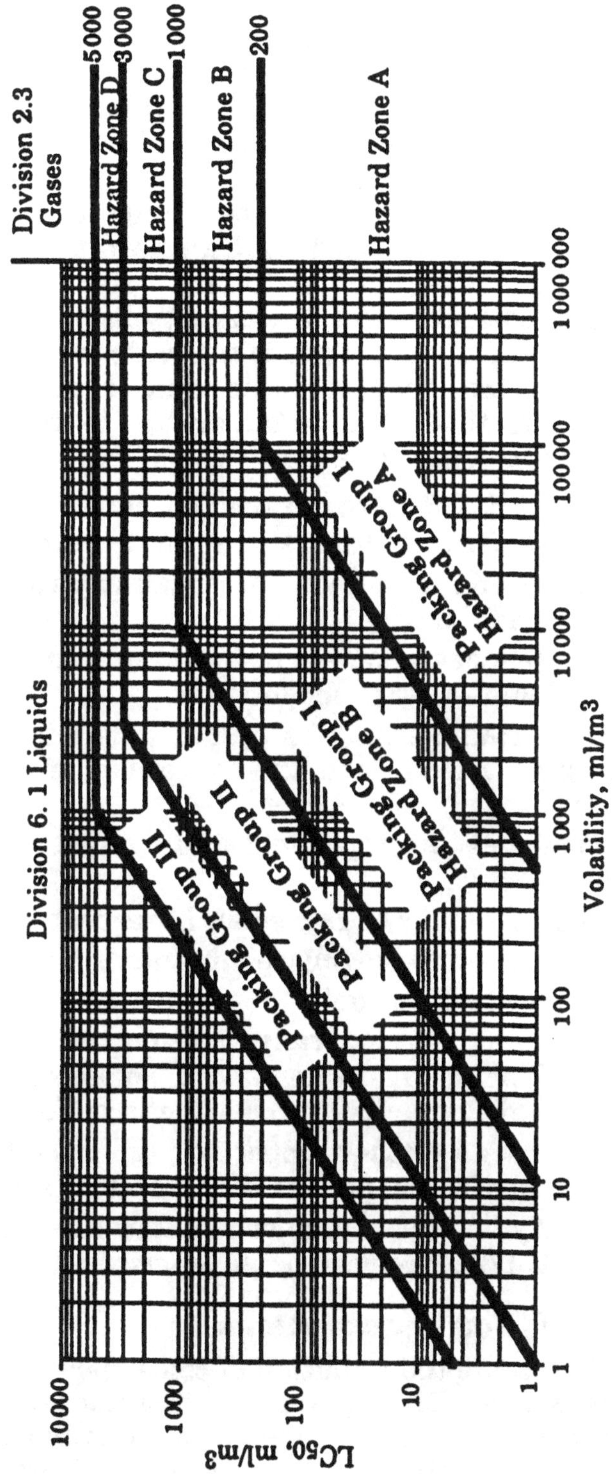


**Fig. 4.1 Inhalation Toxicity: Packing Group and Hazard Zone Borderlines**

### 4.1.1 Container Reuse

Under the pre-HM-181 system, a container may be marked "NRC" meaning non-reusable container or "STC" meaning single trip container. The implications are the same for containers marked with either acronym. These containers may be reused for the shipment of hazardous materials only if the following conditions are met: 1) the material must be packaged and offered for transportation in accordance with DOT regulations, 2) transportation is by highway only, 3) the container may not be offered for transportation less than 24 hours after it is finally closed for transportation, 4) the container must be inspected for leakage immediately before being offered for transportation, 5) each container must be loaded by the shipper and unloaded by the consignee, unless the motor carrier is a private or contract carrier, and 6) unless the container has been reconditioned or altered and retested according to prescribed methods, the container may be reused only once for the transportation of hazardous waste.

Under the HM-181 40 CFR 173.28 requirements, non-bulk packages (except those made of paper, plastic film or textile) may be re-used provided:

- There is no evidence of a reduction in container integrity (unless the container is then reconditioned)
- Packages subject to the leakproofness test are retested in accordance with 49 CFR 178.604 and marked with the letter "L"
- Metal and plastic containers may be reused only when marked with a minimum thickness for reuse and the containers meet this minimum thickness
- Plastic inner receptacles of composite packages must have a minimum thickness of 1 mm
- Used non-bulk containers may be reused for hazardous waste provided:
  - the container meets the packaging requirements for the waste shipped
  - transportation is by highway only
  - the container is not offered for transportation less than 24 hours after it is finally closed for transportation and it is inspected and found to be free from leaks immediately prior to being offered for shipment
  - the container is loaded by the shipper and unloaded by the consignee, unless the motor carrier is a private or contract carrier
  - The container may be reused only once under this exception unless it meets the 49 CFR 173.28 requirements described above.

### 4.1.2 Container Condition Requirements

Both EPA and DOT regulate the condition of containers used to accumulate and ship hazardous waste. During waste accumulation, 40 CFR 265.171 requires that hazardous waste in containers at both accumulation points and at satellite accumulation points must be in good condition. Good condition means that drums should have no severe rusting, no sharp-edged creases or dents, no bulging heads caused by overpressuring a container, and no severe structural defects. If a container is not in good condition or begins to leak, the hazardous waste must be immediately transferred to another container or must be overpacked in a salvage drum. Containers with pools of hazardous waste on the top are not in good condition and must be decontaminated or overpacked prior to transportation off-site.

All waste placed in containers must be compatible with the container. By following the container selection process outlined above, you will ensure that your hazardous waste is compatible with the container you have selected.

### 4.2 Container Management Procedures

A container holding hazardous materials or waste must always be kept closed except when it is necessary to add or remove waste from the container. Open head drums are considered to be closed when the lid is placed on the drum, the lid is secured to the drum with a retaining ring or other DOT-specification closure device, and the ring is bolted closed.

Containers must not be stored or handled in a manner which may cause them to rupture or leak. The following precautions are *recommended* to prevent container ruptures and leaks:

- Do not overfill container. For example, only fill a 55-gallon drum to 50 gallons. Liquids expand in containers as the temperature increases. A steel drum painted a dark color can easily rise to temperatures above 100°F and the pressure created by the expansion of the liquid causes bulging heads and damages the integrity of the container. Bulging containers also create a safety hazard for personnel expected to add waste to or handle the containers.
- Protect containers from freezing. Many materials go through a freeze/thaw cycle during changing weather conditions. This freeze/thaw cycle causes metal stress and can result in leaking containers.
- Ground containers bearing flammable and combustible materials to prevent a spark generated by static electricity to ignite flammable vapors which may be present. Use a bonding wire and a ground wire when transferring flammable liquids into containers to prevent sparks caused by the buildup of static electricity during pouring operations.
- Handle drums and other containers with equipment designed for the task. Drum grappler attachments may be purchased for tow motors to securely grab and move containers. Secure containers to pallets before moving pallets. Use drum carts designed for the types of containers used by your facility to reduce the likelihood of dropping a container during handling. Never balance drums on the forks of a forklift or tow motor.

### 4.3 Management of Empty Containers

It is very important to responsibly manage all *empty* containers of chemicals to comply with EPA regulations and to prevent contamination of the environment from residues left in *empty* containers.

#### 4.3.1 DOT Requirements for Empty Packages

DOT regulates the shipment of containers which previously held hazardous materials. 49 CFR 173.29 states that a container having a capacity of 110 gallons or less that previously contained a hazardous material may not be offered for transportation unless offered in the same manner as required when it previously contained the hazardous material. This requirement does not apply to a container that has been cleaned and purged of all residues to remove any potential hazard; contains only residue of an ORM-D material; or that has been filled with a material that is not hazardous. As discussed below, containers which contain less than one inch of residue may be empty per EPA regulations, but *are* still regulated by DOT during transportation.

Containers with hazardous material residues must be properly marked and labeled in accordance with DOT requirements for the material which was previously in the container. (DOT marking and labeling requirements are discussed in more detail in Chapter V). Shipment of these packages by common carrier must be accompanied by shipping papers, such as bills of lading, which contain a proper shipping description for the hazardous material. Shipping papers are not required if the empty packages are collected and transported by a contract or private carrier for reconditioning or reuse. Placards are required on vehicles transporting containers, if the residues in the containers meet the definitions of the hazard classes listed in Table 1 of the Placard Key in Chapter VII. Each empty container with residues of a hazardous material must be free from leaks, and each opening must be closed.

An empty container (cleaned and purged with no potential hazard) may not be marked and labeled as a hazardous material unless it contains some of the hazardous material that previously required display of the label or marking. However, marks and labels may be left on cleaned, empty containers if they are loaded on-site and the containers are placed into a closed vehicle so the containers are not visible during transportation.

Used drums may be handled in several different ways. They may be reconditioned and reused as shipping containers, they may be processed for steel scrap recycling, they may be crushed and buried at a permitted TSD facility, and they may, under certain circumstances, be sent to a solid waste disposal facility. Reconditioning or recycling are usually preferred methods for handling empties due to the cost involved in disposal of empties at a TSD facility, and the long term environmental liability associated with land disposal at a solid waste facility.

### 4.3.2 EPA Requirements for Empty Hazardous Waste Containers

Under EPA regulations, and most state environmental rules, certain *empty* containers (and the residues therein) are not subject to hazardous waste regulations. A container or an inner liner removed from a container that held hazardous waste is empty and is not regulated as a hazardous waste if:

- All waste has been removed that can be removed by using common practices such as pouring, pumping, and aspirating, and
- No more than one inch of residue remains on the bottom of the container or inner liner, or
- No more than 3 percent by weight of the total capacity of the container remains in the container or inner liner if the container is less than or equal to 110 gallon in size, or no more than 0.3 percent remains if the container is larger than 110 gallons in size.

If the container held an acute hazardous waste, to be considered *empty*, one of the following must occur:

- The container or inner liner must be triple rinsed using a solvent capable of removing the commercial chemical product or manufacturing chemical intermediate,
- The container or inner liner has been cleaned by another method that has been shown in the scientific literature, or by tests conducted by the generator, to achieve equivalent removal, or
- The case of the container, the inner liner that prevented contact of the commercial chemical product or manufacturing chemical intermediate with the container, has been removed.

A container that has held a hazardous waste that is a compressed gas is empty when the pressure in the container reaches atmospheric pressure.

If containers that held hazardous waste have not been emptied in accordance with this definition, the containers must be managed as hazardous waste.

# Chapter 5
# Container Labeling and Marking

It is the shipper's responsibility to determine what container labels and marks are required before offering the container for transportation. DOT requirements for container labeling and marking, including changes coming under HM-181 are discussed below.

## 5.1 Container Labeling

Any person or employee who offers a package for transportation that contains a hazardous material or hazardous waste must affix the appropriate DOT warning label (the diamond-shaped color-coded label). The following steps should be followed to determine the proper DOT label(s) for hazardous material and hazardous waste shipments.

1. Locate the proper shipping name in column 2 of the Hazardous Materials Table in 49 CFR 172.101.

2. Locate the hazard class (or classes) in column 3 of the Table.

3. Locate the hazard label required for the material in column 4 of the pre-HM-181 Table (or column 6 of the HM-181 Table.

Hazard warning labels must be placed next to the proper shipping name marked on the container.

Some hazardous material descriptions have more than one hazard class listed in the Hazardous Materials Table. If the container being shipped meets more than one hazard class, multiple labels are required. Multiple labels should be located next to one another (do not place labels over any other information).

### 5.1.1 Multiple Labeling Requirements (Pre-HM-181)

The following DOT rules apply for containers bearing more than one hazard class or for those bearing special hazards:

- When compatible hazardous materials of different hazard classes are in the same package, overpack, or container, the label for each class must be shown on the container.
- A material classed as an explosive A, poison A, or radioactive material that also meets the definition of another hazard class, must be labeled for each class.
- A poison B material that also meets the definition of a flammable liquid, flammable solid, corrosive material, or oxidizing material must be labeled "POISON" in addition to the appropriate label for the other class.
- A flammable solid that also meets the definition of a water reactive material must be labeled "FLAMMABLE SOLID" and "DANGEROUS WHEN WET."

- A flammable solid that also meets the definition of a corrosive material must be labeled "FLAMMABLE SOLID" and "CORROSIVE."

- An oxidizer that meets the definition of a corrosive material must be labeled "OXIDIZER" and "CORROSIVE."

- A flammable liquid that also meets the definition of a corrosive must be labeled "FLAMMABLE LIQUID" and "CORROSIVE."

### 5.1.2 Additional Labeling Requirements (Pre-HM-181)

In addition to the hazard labels specified by the DOT Hazardous Material Table, other labels are required under the circumstances described below:

- Empty radioactive materials packagings which have been emptied as far as practical must have previously applied labels removed and the "Empty" label applied.

- Wastes and materials which are poison inhalation hazards as described in the marking section of this chapter, must be labeled "POISON" in addition to any other labels which are required.

- "Cargo Aircraft Only" labels must be applied when shipping hazardous waste via air in quantities not permitted on a passenger carrying aircraft.

- A bung label must be applied to metal barrel or drums containing a flammable liquid having a vapor pressure between 16 and 40 pounds psia at 100°F. The following warning is required on a bung label:

> CAUTION: Unscrew This Bung SLOWLY. Do not unscrew entirely until all interior pressure has escaped through the loosened threads. REMOVE BUNG IN OPEN AIR. Keep all open flame lights and fires away. Enclosed Electric Lights are safe.

### 5.1.3 Multiple Labeling Requirements (HM-181)

In addition to the label specified in Column 6 of the Hazardous Materials Table, each package of Class 1 material that also meets the definition for Division 6.1 packing groups I or II shall be labeled POISON. Class 7 radioactive materials also have additional labeling requirements (see 49 CFR 173.403).

When column 6 of the hazardous materials table indicates that a subsidiary label is required, this label must be affixed to the container next to the primary hazard label. The subsidiary label may not display the hazard class number or division number. If a hazardous material does not have a subsidiary labeling requirement designated in column 6 of the hazardous material and it is not a Class 1 or Class 2 material, it must be labeled with a subsidiary label in accordance with the following table:

| Subsidiary hazard level (packing group) | Subsidiary Hazard (Class or Division) | | | | | | |
|---|---|---|---|---|---|---|---|
| | 3 | 4.1 | 4.2 | 4.3 | 5.1 | 6.1 | 8 |
| I | X | *** | *** | X | X | X | X |
| II | X | X | X | X | X | X | X |
| III | * | N | X | X | N | N | ** |

X Required for all modes
* Required for transport by vessel only
** Required for transport by aircraft and vessel only
*** Impossible as subsidiary hazard
N None required

### 5.1.4 Exceptions from Labeling Requirements (Pre-HM-181)

Labels are not always needed on every package containing a hazardous waste or hazardous material. For example a freight container having a volume of 640 cubic feet or more should not be labeled, but should be placarded. The same is true for portable tanks that are properly placarded. Combustible liquids in packages do not require labels. Other Regulated Materials or ORMs are not labeled if they do not contain any other hazardous material. These materials are identified by the ORM mark.

### 5.1.5 Exceptions from Labeling Requirements (HM-181)

Labels are not required for the following:

- Cylinders containing Division 2.1 or Division 2.2 gas that is not poisonous, carried by a private or contract motor carrier, not overpacked, and durably and legibly marked in accordance with Compressed Gas Association Pamphlet C-7, Appendix A.
- Military explosives, ammunition and certain other materials shipped by or on behalf of the DOD (see 49 CFR 172.400a for details)
- An overpack or unit load device in or on which labels representative of each hazardous material in the overpack or unit load device is visible.
- A package of low specific activity radioactivity when transported in accordance with the packaging requirements in 49 CFR 173.425(b).
- Small or limited quantities with listed exemptions in 49 CFR 173.

### 5.1.6 Changes under HM-181

Effective **1 October 1993**, a number of significant DOT labeling changes go into effect:

1. The new Hazardous Materials Table must be used to determine which labels are required on a container. You will note that the labeling requirements are in column 6 of the new table.

2. The new DOT labels must be used. A number of significant changes have been made to the labels, such as:
   - the addition of the numerical hazard class number in the lower corner of the square-on-point labels
   - there are revisions to, removals and additions of some labels:
     - Explosive A, Explosive B, Blasting Agent, and Irritant labels are removed
     - Explosive 1.1, 1.2, 1.3, 1.4, 1.5, 1.6 labels, Explosive subsidiary risk label,
       KEEP AWAY FROM FOOD and INFECTIOUS SUBSTANCE labels, and the CLASS 9 labels are added
     - the ETIOLOGIC AGENT label is removed

## 5.2 Container Marking by Hazardous Material Shippers

DOT marking requirements are concerned with the obligation to mark the exterior of a package or container with specific information. For shipments in **non-bulk packages** only, the following information must be marked by the shipper on each container:

- *Proper Shipping Name*—as it appears on the Hazardous Materials Table (49 CFR 172.101). The pre-HM-181table may be used until **1 October 1993**, after which the new HM-181 DOT table must be used.
- *Technical Name*—This must be marked on the container whenever the material is described with an "n.o.s." (not otherwise specified) entry.
- *Identification Number (UN or NA)*—this is to immediately follow the proper shipping name.
- *Names and Addresses of Consignee* or *Consignor*
- *ORM*—the Other Regulated Material designation is required on packages containing materials classed as Other Regulated Materials on the Hazardous Materials Table (such as ORM-D, indicating a consumer commodity).
- *This End Up / This Side Up*—required for packages containing inside packaging of liquid hazardous materials. These packages must be packed with the closures upward and marked with the words "This End Up" or "This Side Up." Arrows may be used in addition to indicate the proper orientation of the package, but this does not eliminate the requirement for marking the package with the words "This End Up" or "This Side Up."
- *Inhalation Hazard*—Materials that pose a health hazard through inhalation must be marked with the words "Inhalation Hazard." This is an HM-181 requirement that is in effect today.
- *Reportable Quantity*—if the material being shipped is a Hazardous Substance (listed in Appendix A to 49 CFR 172.101) with a reportable quantity in the individual package, the letters "RQ" must be marked in association with the proper shipping name (including the technical name or waste code, as applicable).
- *EPA Hazardous Waste Code*—If the material being shipped is a hazardous waste, the EPA hazardous waste code must be marked on the container.

For example, a container of ignitable hazardous waste (used paint thinner) must be marked "D001."

- *Radioactive materials*—Gross weight, Type A or Type B packaging, and USA (if package is destined for export shipment).

### 5.2.1 Changes under HM-181

In addition to the package marking requirements specified above, the following requirements go into effect **1 October 1993**:

1. *Exemption Numbers*—packagings authorized under an exemption must be plainly marked "DOT-E" followed by the exemption number.
2. All ORM marks are eliminated, with the exception of ORM-D (consumer commodities). Each package containing an ORM-D material must be marked on at least one side or end with the ORM-D designation immediately following or below the proper shipping name.
3. Combination packages that have inner packages containing liquid hazardous materials which must be packed with the closure upward must be marked on two opposite sides with arrows pointing in the correct upward direction.
4. The word POISON must be marked on each non-bulk outer package used as a single or composite package for Division 6.1 materials.
5. Class 1 explosives must be marked with the exception number for each substance, article, or device in the package.

### 5.2.2 Marine Pollutant Markings

For **vessel transportation** of each **non-bulk packaging** that contains a listed marine pollutant, the following requirements go into effect on **1 January 1993**:

- If the **proper shipping name** for a material that is a marine pollutant does not identify by name the component that makes the material a marine pollutant, the name of that component must be marked on the package **in parentheses** in association with the marked proper shipping name. Where two or more components that make a material a marine pollutant are present, the names of **at least two** of the components most predominantly contributing to the marine pollutant designation must appear in parentheses in association with the marked proper shipping name.
- The **MARINE POLLUTANT mark** must be placed in association with the hazard warning labels required, or, in the absence of any labels, in association with the marked proper shipping name. There are exceptions for certain **combination packagings**:
  - A combination packaging containing a **severe** marine pollutant in inner packagings each of which contains—
    - 0.5 liters (17 ounces) or less net capacity for liquids
    - 500 grams (17.6 ounces) or less net capacity for solids

- A combination packaging containing a marine pollutant, other than a severe marine pollutant, in inner packagings each of which contains—
  - 5 liters (1 gallon) or less net capacity for liquids
  - 5 kilograms (11 pounds) or less net capacity for solids

### 5.2.3 Bulk Packages

The following are required to be displayed when transporting hazardous materials in portable tanks, cargo tanks, and tank cars with a capacity greater than or equal to 450 liters or 400 kilograms:

1. **Hazardous material identification numbers**, displayed in one of several ways:
   - on an orange panel measuring 160 mm by 400 mm (6.3″ x 15.7″)
   - across the center of the primary hazard placard
   - for hazard classes for which placards are not required, the identification number must be displayed either on an orange panel or on a plain white square-on-point panel with the same outside dimensions as a placard.

   Identification numbers must be placed on each side of a package that has a capacity of < 3,785 liters (1,000 gal); and on each side and each end of a package that has a capacity of ≥ 3,785 liters, and any tube trailer motor vehicle.
2. **Portable tanks** with a capacity ≥ 450 liters or 400 kilograms must also (in addition to item 1) be marked with the proper shipping name on two opposite sides, and the owner or lessee's name.
3. **Cargo tanks** with a capacity ≥ 450 liters or 400 kilograms transporting Class 2 materials must also (in addition to item 1) be marked on each side with the proper shipping name and an appropriate common name.
4. **Tank cars and multi-unit tank cars** with a capacity ≥ 450 liters or 400 kilograms must also (in addition to item 1) be marked on each side with the proper shipping name and an appropriate common name.
5. **MARINE POLLUTANT mark** must be placed on each end and each side of a bulk packaging (capacity ≥ 450 liters or 400 kilograms) that contains a marine pollutant, and be visible from the direction the packaging faces. This mark may be displayed in black lettering on a white square-on-point configuration having the same outside dimensions as a placard (see following page). Except for transportation by vessel, a bulk packaging or freight container that bears a specified **label** or **placard** does not require a MARINE POLLUTANT mark.

The symbol, letters and border must be black and the background white, or the symbol, letters, border and background must be of contrasting color to the surface to which the mark is being affixed. For **non-bulk** packagings of marine pollutants, each side of the mark must be at least 100 mm (3.9″), except in the case of packagings which, because of their size, can only bear smaller marks. For **bulk** packagings, each side of the mark must be at least 250 mm (9.8″).

**MARINE POLLUTANT Mark:**

## 5.3 Manufacturer's Mark

Containers that are manufactured to meet the HM-181 performance oriented packaging requirements must be marked with the following information, as shown on the diagram below.

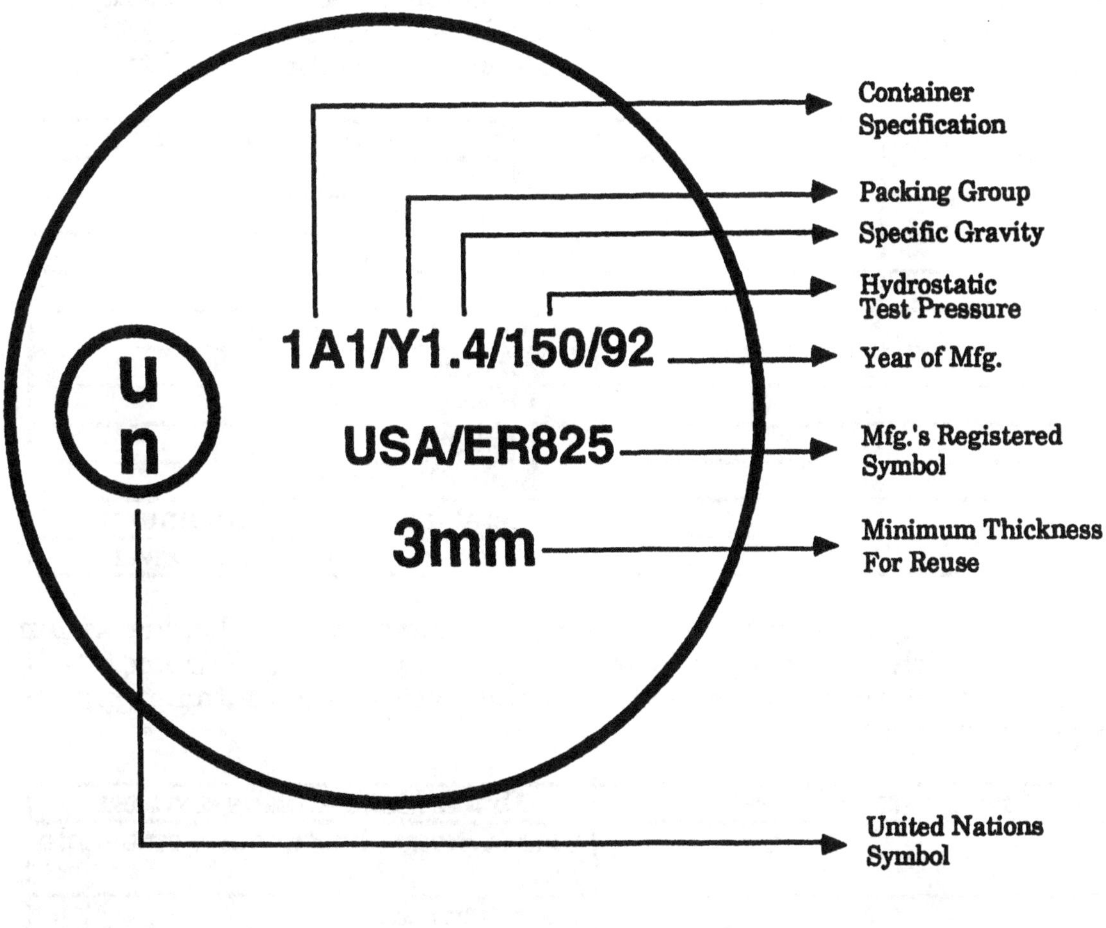

Container manufacturer's marks include the following information:

- The United Nations symbol
- The container specification code, such as **1A1**, where the first digit designates the type of container, the second item is a letter that identifies the materials of construction, and for drums, the third designates closed (1) or removable (2) head drum. Number designations for non-bulk containers are shown on the table below:

| 1st Container Code | Type of Container |
|---|---|
| 1 | Drum |
| 2 | Wooden Barrel |
| 3 | Jerrican |
| 4 | Box |
| 5 | Bag |
| 6 | Composite Packaging |
| 7 | Pressure Receptacle |

The second item designates the materials of construction, as shown below. Composite containers will be marked with two codes, with the first code identifying the material of construction of the inner layer and the second identifying the material of construction for the outer layer. For example, a non-composite, closed head steel drum may have the specification 1A1, and an open head steel drum with a plastic liner would have the code 1HA2.

| 2nd Container Code | Materials of Construction |
|---|---|
| A | Steel |
| B | Aluminum |
| C | Natural Wood |
| D | Plywood |
| F | Reconstituted Wood |
| G | Fiberboard |
| H | Plastic |
| L | Textile |
| M | Multi-Wall Paper |
| N | Metal (not steel or aluminum) |
| P | Porcelain, Glass, or Stoneware |

A third code may be a numeral indicating the category of packaging within the type to which the packaging belongs. For example, for steel drums ("1A"), the codes are as shown below (note that this varies from container type to type; many have no third code):

| 3rd Container Code | Drum Head Configuration |
|---|---|
| 1 | Closed head drum, non-removable head |
| 2 | Open head drum |

The container specification code may be followed by a letter designating the Packing Group performance standard, designated by the letters X, Y, or Z, as shown on the table below:

| Performance Standard | For Packages Meeting Packing Groups |
|---|---|
| X | I, II, and III |
| Y | II and III |
| Z | III |

For containers intended to hold liquids, the Packing Group performance standard is followed by a designation of the maximum rated specific gravity. The specific gravity may be omitted if it is less than 1.2. For containers intended to hold solids (or inner packages), the performance standard is followed by the maximum gross mass (in kg).

For packages intended to hold liquids, the specific gravity is followed by the hydrostatic test pressure, to the nearest 10 kPa. For packages intended to hold solids, the maximum gross mass is followed by the letter "S."

The last two digits of the year of manufacture are marked on all drums, except 1H (plastic drums) and 3H (jerricans) which must also be marked with the month of manufacture. This is followed by the abbreviation of the country where the container was manufactured and for which standards it meets, such as "USA," followed by the symbol of the container manufacturer (or the manufacturer's name and address).

Combination packages that are tested in accordance with a selective test procedure are marked with the letters "SP" to indicate that the container qualifies for this test exemption. The container is also marked with the minimum thickness (in millimeters) for reuse.

Reconditioned containers should be marked with additional markings near those described above, including the country (or abbreviation) where the container was reused, the reconditioner's symbol (or name and address), the month and last two digits of the year the container was reconditions, the letter "R," and the letter "L" if the container successfully passed a leakproofness test.

# Chapter 6
# DOT Shipping Papers

## 6.1 Introduction to Shipping Papers

It is the shipper's responsibility to properly complete the appropriate shipping documents (even though the transporter may do this for you, you are still responsible if the papers are not filled out properly). Shipping papers are completed for all shipments of hazardous materials. When they are hazardous wastes, the shipping paper is the Uniform Hazardous Waste Manifest. It is your responsibility to verify that the information on the shipping paper accurately and properly describes the materials being shipped, in accordance with DOT and EPA regulations.

### 6.1.1 Content of Hazardous Materials Shipping Papers

Typically, the hazardous material shipping paper is a bill of lading—a document issued by the shipper to a carrier that describes the hazardous material to be shipped, acknowledges its receipt, and states the terms of the contract for its carriage. Keep in mind, however, that shipping papers must be prepared for all shipments of hazardous materials over public highways, by air, rail, or water. If a commercial bill of lading is not required—for example, for shipments on a company's own truck from one facility to another within the same company—there must still be some type of shipping paper to describe the shipment. The paper may be a company memorandum or internal form, as long as it contains the information required by the HMR, and is carried on the vehicle (or other mode of transport) transporting the hazardous materials shipment.

At a minimum, the shipping papers must include the following information:

- *Name and address of the consignor and consignee*
- *Proper shipping name of the hazardous material.* If the material is a hazardous substance and the name of the hazardous substance is not identified in the proper shipping name, the constituents making it a hazardous substance must be entered in parentheses after the proper shipping name
- *Hazard Class* (if the proper shipping name includes the hazard class, it doesn't need to be repeated)
- *UN/NA identification number*
- *Total quantity of the shipment by weight or volume*
- *Emergency response telephone number.* If your organization has registered with CHEMTREC (Chemical Transportation Emergency Center), it is authorized to use CHEMTREC's 24-hour Emergency Notification Number - (800) 424-9300

  DOT regulations also require emergency response information to be included on or attached to the shipping papers. Photocopies of the appropriate guide from the DOT Emergency Response Guidebook or Material Safety Data Sheets may be attached to the shipping papers to satisfy this requirement, or write the DOT Guide number on the papers and **always** keep a DOT Guidebook readily accessible in the cab of the vehicle. If the DOT Guidebook is used, the driver must verify that the corresponding page of the Guidebook is clearly referenced **on** the shipping papers for each hazardous material.

- *Shipper's certification*. This is certification that the shipment is prepared in accordance with the applicable hazardous materials regulations. The specific wording required is "This is to certify that the above-named materials are properly classified, described, packaged, marked and labeled, and are in proper condition for transportation according to the applicable regulations of the Department of Transportation." This shipper's certification statement is not required when the hazardous material is offered for highway transport if it is in a cargo tank supplied by the carrier, or when the material is transported by the shipper operating as a private carrier, unless the material will be transferred from one carrier to another or reshipped.

NOTE: When a material that is not subject to these requirements (i.e. not a hazardous material) is described on the same shipping paper as a hazardous material, the hazardous material description must be:

- entered first, or
- entered in a color which clearly contrasts from the description of non-hazardous materials; or
- must be identified by an "X" placed before the proper shipping name in a column titled "HM."

Figure 6.1, at the end of this chapter, gives an example of a multi-purpose shipping paper that might be used by a company making frequent shipments of the same types of hazardous materials, plus other items that may or may not be hazardous. Note that this format **cannot** be used in place of a hazardous waste manifest.

### 6.1.2 Changes to Shipping Documents under HM-181

Effective **1 October 1993**, the first major change that affects shipping documents (e.g., the hazardous material bill of lading) is that the appropriate Packing Group must be described on the shipping document. Under HM-181, there have been significant changes to the manufacturing specifications for DOT packages. DOT is moving to performance-oriented packaging (POP) standards, and under this new system, packaging requirements for a material are based on:

- Packing Group—
  - Packing Group I—great danger
  - Packing Group II—medium danger
  - Packing Group III—minor danger
- its vapor pressure
- chemical compatibility between the packaging and the hazardous material.

In short, after 1 October 1993, all hazardous material shipping documents must indicate the Packing Group. The packing group is indicated in column 5 of the new DOT Hazardous Materials Table in 49 CFR 172.101.

Under HM–181 requirements, the proper shipping name, hazard class, identification number, packing group, and total quantity of material being shipped must also be in a prescribed sequence. The first entry must be the **proper shipping name.** The second entry is the **hazard class,** followed by the **identification number,** followed by the **Packing Group,** followed by the **total quantity** (this may also be placed before the proper shipping name).

## 6.2 Choosing a Proper Shipping Description for Hazardous Materials and Hazardous Wastes

The key to properly completing DOT shipping papers is to determine the best and most descriptive DOT proper shipping **description,** which will lead you to the entries in the Hazardous Materials Table that set out the regulatory requirements for your shipment. The proper shipping description consists of the proper **shipping name, hazard class** or **division** (as a number), DOT **identification number,** and **packing group** (I–III), in that order. For example:

Acetone, 3, UN 1090, PG II.

[Compare with the pre-HM–181: Acetone, Flammable liquid, UN 1090.]

Except for empty packagings, cylinders of compressed gases, and bulk packagings, you must also indicate the total quantity, including the unit of measurement. For example:

55 gal. *or* 208 L

Depending on the format of the shipping papers, the quantity may be placed before or after the description.

In order to determine the best shipping description, you must be familiar with the DOT Hazardous Materials Table and the appendices (Hazardous Substances and Marine Pollutants) to the Hazardous Materials Table in 49 CFR 172.101. (The HM–181 version of the Hazardous Materials Table is provided in Appendix A of this handbook.) The process will generally be more complex for hazardous wastes, which often consist of mixtures of varying composition. You should also know detailed information about your hazardous materials or wastes including:

- Process which generated the materials or wastes
- Chemical constituents present
- Concentration or percent of each chemical constituent present
- Physical state (e.g., solid, liquid, gas)
- Specific gravity or density
- Hazard characteristics (e.g., oxidizer, flammable solid, flammable gas, etc.)
- Flashpoint
- Intended use of the material
- pH

- Whether material is reactive or contains cyanides or sulfides
- If a waste, concentration of any Toxicity Characteristic constituents which are present
- Shipment method (e.g., bulk or container)
- Amount of material per container

You may obtain this information by reviewing the process generating the material to determine the actual or likely chemical constituents, by reviewing MSDSs and documents which provide chemical hazard information, and, for wastes, by analyzing them for hazard characteristics and chemical constituents. Once you have obtained this information, you can use the following key to determine the best US DOT description for your hazardous materials and wastes.

**Key To Determining the US DOT Description for Mixtures**

1. Does the material contain more than one chemical compound or element?

   **Yes:** Go to question 2.

   **No:** The material is not a mixture. Use table 172.101 to find the proper shipping name which most accurately describes the material. Go to question 13.

2. Is there a name on the Hazardous Materials Table that describes the mixture? See, for example, chlorate and magnesium chloride mixture, carbon dioxide-oxygen mixture, boron trifluoride-acetic acid complex, arsenical mixture, etc.

   **Yes:** Use the proper shipping name listed in table 172.101, column 2. Add the word "Waste" before the US DOT hazardous material description if the material is a hazardous waste. Go to question 9.

   **No:** Go to question 3.

3. Is there a name listed in the table that defines the material by a former use? See, for example, compounds, cleaning, liquid; coating solution; paint related material; etc.

   **Yes:** Use the proper shipping name listed in table 172.101, column 2. Add the word "Waste" before the US DOT hazardous material description if the material is a hazardous waste. Go to question 9.

   **No:** Go to question 4.

4. Is the material a mixture of one material listed in table 172.101, column 2 and one or more other materials which <u>are not</u> listed in 172.101, column 2 and are not DOT hazardous materials?

   **Yes:** Go to question 8.

   **No:** Go to question 5.

5. Is the material a mixture of at least two materials which <u>are</u> listed by name or hazard class in 172.101?

   **Yes:** Go to question 6.

   **No:** Go to question 7.

6. Does the material meet the description and the definition of more than one hazard class? For example, is the mixture both a Flammable Liquid (Class 3) and a Corrosive Material (Class 8)?

   **Yes:** Select a proper shipping name from Table 6.3 that includes both hazard classes in the proper priority as established by referring to the precedence tables (Tables 6.1 and 6.2). If both hazard classes are not represented by one shipping name, use the hazard precedence list in Table 6.2 to select a hazard class for the material. Precede this proper shipping name by the word "Waste" if the material is a waste. Add the highest priority hazard class/division number to the shipping name you selected. Check the 172.101 table for proper wording and punctuation. Go to question 10.

   **No:** Use the hazard class for the mixture as the proper shipping name (preceded by the word "Waste" if it is a waste) and followed by the letters "n.o.s." Check the Hazardous Materials Table for proper wording of the shipping name. Go to question 10.

   **Note**: If there is a plus (+) in column 1 of the hazardous materials table for one or more of the components of the material, then the hazard class and proper shipping name of this material cannot be modified.

7. Is the material a mixture of materials which <u>are not</u> listed by name or hazard class in 172.101, column 2 of the hazardous materials table?

   **Yes:** Use your knowledge of the material's characteristics to determine an appropriate hazard class. (DOT hazard classes are defined in Chapter III). Use the hazard class you select as the proper shipping name (preceded by the word "Waste" if it is a waste) and followed by the letters "n.o.s." Check the Hazardous Material Table for proper wording and punctuation. Go to question 10.

   If the hazardous material or waste does not meet the definition of any other hazard class other than Class 9 (Miscellaneous hazardous materials) use only one of the following shipping descriptions and go to Question 10:

   | | |
   |---|---|
   | For RCRA Hazardous Waste: | • Hazardous waste, liquid, n.o.s., 9, NA 3082, PG III<br>• Hazardous waste, solid, n.o.s., 9, NA 3077, PG III<br>**or** |
   | For Other Hazardous Materials: | • Environmentally hazardous substances, liquid, n.o.s., 9, UN 3082, PG III<br>• Environmentally hazardous substances, solid, n.o.s., 9, UN 3077, PG III |

   **No:** Error, return to question 1.

8. Is the mixture a homogeneous liquid consisting of two or more chemical compounds or elements that will not undergo any segregation under conditions normal to transportation?

   **Yes:** The material is a solution, according to DOT definitions. Refer to the Hazardous Materials Table and select the name listed for the hazardous component of the mixture. If the mixture is a waste, add the word "Waste" prior to the shipping name. Add the qualifying word "solution" after the name (for example, Waste Acetone Solution). Go to question 9.

**No:** The material is a mixture, according to DOT definitions. The qualifying word "mixture" should be inserted after any shipping name you select (for example, Waste Sodium Hydroxide Mixture). Go to question 9.

9. Does the material have just one hazard class and meet the definition of the hazard class associated with the proper shipping name you selected? (DOT hazard classes are defined in Chapter III).

   **Yes:** Use the shipping name you selected followed by the hazard class listed next to that shipping name in column 3 of table 172.101. Add the DOT identification number and Packing Group designation after the hazard class. Go to question 10.

   **No:** Discard the shipping name you previously selected. If the material has more than one hazard class, select a proper shipping name from Table 6.3 which includes both hazard classes (in precedence order), preceded by the word "Waste" if it is a waste. Add the hazard class from table 172.101 to the shipping name you selected. If both hazard classes are not represented by one shipping name, use the hazard precedence list in Tables 6.1 and 6.2 to select a hazard class for the material. Use the hazard class determined from the precedence list for the mixture as the proper shipping name (preceded by the word "Waste" if it is a waste) and followed by the letters "n.o.s." Check the Hazardous Material Table for proper wording and punctuation. Go to question 10.

   **Note:** If there is a plus (+) in column 1 of the hazardous materials table for one or more of the components of the material, then the hazard class of this material cannot be modified. For example, if column 3 of the hazardous materials table calls the material a Class 8 material, you may not change the hazard class to 3, or any other hazard class even though a different hazard class may more adequately describe the waste mixture.

10. Are the constituents of the material and their respective concentrations (percentages) known?

    **Yes:** Go to question 12.

    **No:** Is the material a waste?

    - **Yes:** Go to question 11.
    - **No:** Go to question 12.

11. Is the material a waste with a D, F, or K hazardous waste code?

    **Yes:** Look up the waste stream listing and RQ in Appendix A to 172.101. Go to question 13.

    **No:** Go to question 15.

12. Is any component of the mixture in Appendix A to 172.101?

    **Yes:** Go to question 13.

    **No:** Go to question 11.

13. Is the material a hazardous substance (refer to the definition of hazardous substance on page 65)?

**Yes:** Add the letters "RQ" either before or after the proper shipping description. Use the Hazardous Materials Table and, after the hazard class, add the DOT ID Number and packing group that corresponds to the proper shipping name unless it was entered previously.

After the US DOT description, add in parentheses the name or, for a waste with unknown constituents, the EPA hazardous waste number of the constituent(s) that caused the material to be a hazardous substance. Go to question 15.

**No:** Go to question 14.

14. Have you selected a proper shipping name listed in Table 6.3?

**Yes:** After the US DOT description, add in parentheses "contains _____ [the technical name of the hazardous material most predominantly contributing to the hazard]." If the material is a mixture of two or more hazardous materials, the technical names of at least two components most predominantly contributing to the hazards of the mixture or solution must be entered. Add the Hazard Class and DOT ID Number that correspond to the proper shipping name unless they were entered previously. Determine from the table, or from the criteria in 49 CFR Part 173, Subpart D, what packing group the material is in and enter the packing group number. For example, "Flammable liquid, corrosive, n.o.s., 3, UN 2924, PG II (contains Methanol, Potassium hydroxide)." Go to question 15.

**No:** Go to question 15.

15. Is the material a Marine Pollutant (refer to the definition on page 65)?

**Yes:** Add the words "Marine Pollutant" to the shipping description. If there are two or more listed marine pollutants, add the technical names of at least the two listed constituents that most predominantly contribute to the designation of marine pollutant. Go to question 16.

**No:** Go to question 16.

16. Does your material fall in any of the following categories?

- Class 7 Radioactive Material—Add the words "RADIOACTIVE MATERIAL" unless it is already in the shipping description. Add the physical and chemical form of the material. Enter the name of the radionuclide. Add the activity of each package. Enter the words "HIGHWAY ROUTE CONTROLLED QUANTITY" if applicable. Enter the category of label required (category I and II only). Enter the Transport Index if applicable.
- Division 4.3 Dangerous when wet materials—Add the words "DANGEROUS WHEN WET" to the shipping description.
- Division 6.1 Poisonous materials (primary or subsidiary hazard, Packing Groups I and II)—Add the word "POISON" to the shipping

- Inhalation Hazards (division 6.1 PG I or division 2.3)—Add the words "POISON—INHALATION HAZARD" and the appropriate hazard zone (ZONE A, B, C, or D) immediately following the shipping description.
- Elevated Temperature materials—Enter the word "HOT" immediately before the proper shipping name
- Empty Packagings—Containers with residue in them must have the words "RESIDUE: Last contained _____." and RQ entered in the description when appropriate.

**Yes:** Enter the appropriate wording described above. Check notes below.

**No:** Check notes below.

**Notes:** If the name of a compound that causes a mixture to meet the DOT definition of "poison" is not included in the proper shipping name for the mixture, the technical name of the compound must be incorporated into the description of the material. In addition, if a liquid or solid material meets the definition of a poison and the fact that it is a poison is not disclosed in the shipping name or hazard class entry, the word "Poison" must be entered in association with the shipping description.

The words "Poison Inhalation Hazard" or "Inhalation Hazard" and the appropriate hazard zone must be added to the basic shipping description for materials which are poisonous by inhalation. "Poison inhalation hazard" is defined below, and Table 6.3 contains a list of proper shipping names from HM-181 for materials designated as poison inhalation hazards. Hazard zones are defined in Chapter 4.

On completing the preceding key questions and notes, your shipping description should be complete.

## Definitions

Elevated Temperature Material
: A material which, when offered for transportation or transported in a bulk packaging:

(1) Is in a liquid phase and at a temperature at or above 100 °C (212 °F);

(2) Is in a liquid phase with a flash point at or above 37.8 °C (100 °F) that is intentionally heated and offered for transportation or transported at or above its flash point; **or**

(3) Is in a solid phase and at a temperature at or above 240 °C (464 °F).

Hazardous Material
: A material that is listed on the DOT Hazardous Materials Table, the appendices to the table, or a material which meets any of the hazard classifications identified in 49 CFR 173.

| | |
|---|---|
| Hazardous Substance | A material, including its mixtures and solutions that meets all of the following three criteria:<br><br>1) It (or any of its ingredients) is listed in the DOT reportable quantities table (Appendix A to 172.101),<br>2) It is in a quantity, in one package, which equals or exceeds the reportable quantity (RQ) listed in the table, **and**<br>3) When in a mixture or solution, is in a concentration by weight which equals or exceeds the concentration corresponding to the RQ of the material, as shown by the following table: |

| RQ (lbs) | per cent | ppm |
|---|---|---|
| 5,000 | 10 | 100,000 |
| 1,000 | 2 | 20,000 |
| 100 | 0.2 | 2,000 |
| 10 | 0.02 | 200 |
| 1 | 0.002 | 20 |

| | |
|---|---|
| Marine Pollutant | A hazardous material that when in a solution or mixture of one or more marine pollutants, is packaged in a concentration equal to or exceeding one of the following:<br>(1) 10% by weight of the solution or mixture<br>(2) 1% by weight of the solution or mixture for materials that are identified as severe marine pollutants |
| Mixture | A material composed of more than one chemical compound or element. |
| Solution | Any homogenous liquid mixture of two or more chemical compounds or elements that will not undergo any segregation under conditions normal to transportation. |
| Technical Names | A recognized chemical name currently used in scientific and technical handbooks, journals, and texts. Generic descriptions are authorized for use as technical names provided they readily identify the general chemical group. Examples of acceptable generic descriptions are organic phosphate compound, petroleum aliphatic hydrocarbons, and tertiary amines. Except for names which appear in Subchapter B of Part 172, trade names may not be used as technical names. |

| | |
|---|---|
| Poison Inhalation Hazards | "Division 2.3 Poisonous Gas," which means a material which is a gas at 20°C or less and a pressure of 101.3 kPa and which is known to be so toxic to humans as to pose a health hazard during transportation, or in the absence of adequate data on human toxicity, is presumed to be toxic to humans because when tested on laboratory animals it has an $LC_{50}$ value not more than 5,000 ppm.<br>Division 6.1, Poisonous Material," which means a material, other than a gas, which is known to be so toxic to humans as to afford a hazard to health during transportation, or which, in the absence of adequate data on human toxicity, is presumed to be toxic to humans because when tested on laboratory animals: (1) it is dust or mist with an $LC_{50}$ for acute toxicity on inhalation of not more than 10 mg/liter or, (2) it is a material with a saturated vapor concentration at 20°C (68°F) of more than one-fifth of the $LC_{50}$ for acute toxicity on inhalation of vapors of not more than 5,000 $ml/m^3$, or (3) it is an irritating material, with properties similar to tear gas, which causes extreme irritation, especially in confined spaces. |

A hazardous material having more than one hazard must be classed according to the following order of hazards*:

(1) Class 7 (radioactive materials, other than limited quantities)
(2) Division 2.3 (poisonous gases)
(3) Division 2.1 (flammable gases)
(4) Division 2.2 (nonflammable gases)
(5) Division 6.1 (poisonous liquids), Packing Group I, poisonous-by-inhalation only
(6) Division 4.2 (pyrophoric materials as defined in §173.124(b)(1))
(7) Division 4.1 (self-reactive materials as defined in §173.124(a)(2))
(8) One of the following classes and divisions†:
  Class 3 (flammable liquids)
  Class 8 (corrosive materials)
  Division 4.1 (flammable solids)
  Division 4.2 (spontaneously combustible materials)
  Division 4.3 (dangerous when wet materials)
  Division 5.1 (oxidizers)
  Division 6.1 (poisonous liquids or solids other than Packing Group I, poisonous-by-inhalation)

†If the material falls in more than one of these categories, its hazard class and packing group must be determined by using the detailed Precedence of Hazard Table (Table 6.2).

(9) Combustible liquids
(10) Class 9 (miscellaneous hazardous materials)

Table 6.1 Hazard Precedence List (49 CFR 173.2)

*The following materials, because of their unique properties, are not subject to being classed according to this precedence list:

- A Class 1 (explosive) material meeting any other hazard class or division must be assigned a division in Class 1. Class 1 materials are required to be classed and approved under § 173.56.
- A Division 5.2 (organic peroxide) material meeting any other hazard class or division must be assigned to Division 5.2.
- A Division 6.2 (infectious substance) material meeting any other hazard class or division must be assigned to Division 6.2.
- A Division 4.1 (wetted explosive) material. These are specifically listed in the § 172.101 Table, or are approved by the Associate Administrator for Hazardous Materials Safety.
- A limited quantity of a Class 7 (radioactive) material that meets the definition of more than one hazard class or division shall be classed as specified in § 173.421–2

## PRECEDENCE OF HAZARD TABLE
(Hazard Class & Packing Group)

| | 4.2 | 4.3 | 5.1 I[1] | 5.1 II[1] | 5.1 III[1] | 6.1, I dermal | 6.1, I oral | 6.1 II | 6.1 III | 8, I liquid | 8, I solid | 8, II liquid | 8, II solid | 8, III liquid | 8, III solid |
|---|---|---|---|---|---|---|---|---|---|---|---|---|---|---|---|
| 3 I | | | | | | 3 | 3 | 3 | 3 | 3 | (3) | 3 | (3) | 3 | (3) |
| 3 II | | | | | | 3 | 3 | 3 | 3 | 8 | (3) | 3 | (3) | 3 | (3) |
| 3 III | | | | | | 6.1 | 6.1 | 6.1 | 3[4] | 8 | (3) | 8 | (3) | 3 | (3) |
| 4.1 II[2] | 4.2 | 4.3 | 5.1 | 4.1 | 4.1 | 6.1 | 6.1 | 4.1 | 4.1 | (3) | 8 | (3) | 4.1 | (3) | 4.1 |
| 4.1 III[2] | 4.2 | 4.3 | 5.1 | 4.1 | 4.1 | 6.1 | 6.1 | 6.1 | 4.1 | (3) | 8 | (3) | 8 | (3) | 4.1 |
| 4.2 II | | 4.3 | 5.1 | 4.2 | 4.2 | 6.1 | 6.1 | 4.2 | 4.2 | (3) | 8 | (3) | 4.2 | (3) | 4.2 |
| 4.2 III | | 4.3 | 5.1 | 5.1 | 4.2 | 6.1 | 6.1 | 6.1 | 4.2 | (3) | 8 | (3) | 8 | (3) | 4.2 |
| 4.3 I | | | 5.1 | 4.3 | 4.3 | 6.1 | 4.3 | 4.3 | 4.3 | 4.3 | 4.3 | 4.3 | 4.3 | 4.3 | 4.3 |
| 4.3 II | | | 5.1 | 4.3 | 4.3 | 6.1 | 4.3 | 4.3 | 4.3 | 8 | 8 | 8 | 4.3 | 4.3 | 4.3 |
| 4.3 III | | | 5.1 | 5.1 | 4.3 | 6.1 | 6.1 | 6.1 | 4.3 | 8 | 8 | 8 | 8 | 4.3 | 4.3 |
| 5.1 I[1] | | | | | | 5.1 | 5.1 | 5.1 | 5.1 | 5.1 | 5.1 | 5.1 | 5.1 | 5.1 | 5.1 |
| 5.1 II[1] | | | | | | 6.1 | 5.1 | 5.1 | 5.1 | 8 | 8 | 8 | 5.1 | 5.1 | 5.1 |
| 5.1 III[1] | | | | | | 6.1 | 6.1 | 6.1 | 5.1 | 8 | 8 | 8 | 8 | 5.1 | 5.1 |
| 6.1 I, Dermal | | | | | | | | | | 8 | 6.1 | 6.1 | 6.1 | 6.1 | 6.1 |
| 6.1 I, Oral | | | | | | | | | | 8 | 6.1 | 6.1 | 6.1 | 6.1 | 6.1 |
| 6.1 II, Inhalation | | | | | | | | | | 8 | 6.1 | 6.1 | 6.1 | 6.1 | 6.1 |
| 6.1 II, Dermal | | | | | | | | | | 8 | 6.1 | 8 | 6.1 | 6.1 | 6.1 |
| 6.1 II, Oral | | | | | | | | | | 8 | 8 | 8 | 6.1 | 6.1 | 6.1 |
| 6.1 III | | | | | | | | | | 8 | 8 | 8 | 8 | 8 | 8 |

[1] There are at present no established criteria for determining Packing Groups for liquids in Division 5.1. For the time being, the degree of hazard is to be assessed by analogy with listed substances, allocating the substances to Packing Group I, great; II, medium; or III, minor danger.

[2] Substances of Division 4.1 other than self-reactive substances.

[3] Denotes an impossible combination.

[4] For pesticides only, where a material has the hazards of Class 3, Packing Group III, and Division 6.1, Packing Group III, the primary hazard is Division 6.1, Packing Group III.

Table 6.2 HM-181 Precedence of Hazard Table for Classes 3 and 8, and for Divisions 4.1, 4.2, 4.3, 5.1, and 6.1 (49 CFR 173.2a)

The Precedence of Hazard Table in 49 CFR 173.2a (our Table 6.2) is rather complicated. The key to understanding it is to recognize that it applies <u>only</u> when you must decide what hazard class to assign to a material that is a mixture of materials listed as number 8 in priority order in the Hazard Precedence List from § 173.2 (our Table 6.1). These are Classes 3 and 8, and Divisions 4.1, 4.2, 4.3, 5.1, and 6.1.

The row and column names along the left side and top of the table refer to up to three of the following characteristics of the materials: their **hazard class** or **division, packing group, route of administration** (the path through which the material presents a hazard to health: *dermal*, *oral*, or *inhalation*), and/or physical **state** (*solid* or *liquid*). At the intersection of a row for one class, packing group, route, and/or state with another, you will find the hazard class that you must assign to their mixture. The blank cells are either impossible combinations or the intersection of two identical material classifications. Where impossible combinations occur in cells surrounded by valid classifications, DOT has used a footnote ([3]) to avoid confusion.

To use the table, find the hazard class (and division, if applicable), plus other characteristics, of one of the components of your mixture in the row or column headings (note that the headings have been set up to duplicate each other as little as possible—for most materials you will find the classification along the top or left side only). Next, find the classification of the other component in the column or row headings. The cell where the chosen row and column intersect contains the class or division number that you should assign to the mixture. Be sure to take into account any footnotes before automatically applying the table categories (e.g., footnote 4 completely turns around the classification of a Class 3, PG III/Div. 6.1, PG III mixture if it is a pesticide).

Alcohols, toxic, n.o.s.
Articles, explosive, n.o.s.
Caustic alkali liquids, n.o.s.
Combustible liquid, n.o.s.
Compounds, cleaning liquid, *corrosive*
Compounds, cleaning liquid, *flammable*
Compounds, tree *or* weed killing, liquid, *corrosive*
Compounds, tree *or* weed killing, liquid, *flammable*
Compounds, tree *or* weed killing, liquid, *poisonous*
Components, explosive train, n.o.s.
Compressed *or* Liquefied gases, n.o.s.
Compressed *or* Liquefied gases, flammable, n.o.s.
Compressed *or* Liquefied gases, flammable, toxic, n.o.s.
Compressed *or* Liquefied gases, toxic, n.o.s.
Contrivances, water-activated
Corrosive liquids, flammable, n.o.s.
Corrosive liquids, n.o.s.
Corrosive liquids, oxidizing, n.o.s.
Corrosive liquids, poisonous, n.o.s.
Corrosive liquids, which in contact with water emit flammable gases, n.o.s.
Corrosive solids, flammable, n.o.s.
Corrosive solids, n.o.s.
Corrosive solids, oxidizing, n.o.s.
Corrosive solids, poisonous, n.o.s.
Corrosive solids, self heating, n.o.s.
Corrosive solids, which in contact with water emit flammable gases, n.o.s.
Disinfectants, corrosive, liquid, n.o.s.
Disinfectants, liquid, n.o.s.
Disinfectants, solids, n.o.s.
Dispersant gas, n.o.s.
Dyes, liquid, *or* Dye intermediates, liquid, n.o.s.
Dyes, solid, *or* Dye intermediates, solid, n.o.s.
Environmentally hazardous substances, liquid, n.o.s.
Environmentally hazardous substances, solid, n.o.s.
Flammable liquids, corrosive, n.o.s.
Flammable liquids, n.o.s.
Flammable liquids, poisonous, n.o.s.
Flammable solids, corrosive, n.o.s.
Flammable solids, n.o.s.
Flammable solids, poisonous, n.o.s.
Halogenated irritating liquids, n.o.s.
Hazardous waste, liquid, n.o.s.
Hazardous waste, solid, n.o.s.
Infectious substances, affecting animals
Infectious substances, affecting humans

Table 6.3 Proper Shipping Names Requiring Technical Names as Part of the Shipping Description

Insecticide gases, *flammable*, n.o.s.
Insecticide gases, n.o.s.
Insecticide gases, toxic, n.o.s.
Medicines, n.o.s.
Metal powders, flammable, n.o.s.
Organic peroxide type B, liquid
Organic peroxide type B, liquid, temperature controlled
Organic peroxide type B, solid
Organic peroxide type B, solid, temperature controlled
Organic peroxide type C, liquid
Organic peroxide type C, liquid, temperature controlled
Organic peroxide type C, solid
Organic peroxide type C, solid, temperature controlled
Organic peroxide type D, liquid
Organic peroxide type D, liquid, temperature controlled
Organic peroxide type D, solid
Organic peroxide type D, solid, temperature controlled
Organic peroxide type E, liquid
Organic peroxide type E, liquid, temp. controlled
Organic peroxide type E, solid
Organic peroxide type E, solid, temperature controlled
Organic peroxide type F, liquid
Organic peroxide type F, liquid, temperature controlled
Organic peroxide type F, solid
Organic peroxide type F, solid, temperature controlled
Other regulated substances, liquid, n.o.s.
Other regulated substances, solid, n.o.s.
Oxidizing substances, liquid, corrosive, n.o.s.
Oxidizing substances, liquid, n.o.s.
Oxidizing substances, liquid, poisonous, n.o.s.
Oxidizing substances, solid, corrosive, n.o.s.
Oxidizing substances, solid, n.o.s.
Oxidizing substances, solid, poisonous, n.o.s.
Pesticides, liquid, flammable, toxic, n.o.s.
Pesticides, liquid, toxic, flammable, n.o.s.
Pesticides, liquid, toxic, n.o.s.
Pesticides, solid, toxic, n.o.s.
Poisonous liquids, corrosive, n.o.s.
Poisonous liquids, flammable, n.o.s.
Poisonous liquids, n.o.s.
Poisonous liquids, oxidizing, n.o.s.
Poisonous liquids, which in contact with water emit flammable gases, n.o.s.
Poisonous solids, corrosive, n.o.s.
Poisonous solids, flammable, n.o.s.
Poisonous solids, n.o.s.

Table 6.3 Proper Shipping Names Requiring Technical Names as Part of the Shipping Description (Continued)

Poisonous solids, oxidizing, n.o.s.
Poisonous solids, self-heating, n.o.s.
Poisonous solids, which in contact with water emit flammable gases, n.o.s.
Pyrophoric liquids, n.o.s.
Pyrophoric solids, n.o.s.
Pyrophoric metals, n.o.s. *or* Pyrophoric alloys, n.o.s.
Refrigerant gas, n.o.s.
Refrigerant gases, n.o.s.
Rodenticides, n.o.s.
Samples, explosive (*other than initiating explosives*)
Self-heating substances, solid, corrosive, n.o.s.
Self-heating substances, solid, n.o.s.
Self-heating substances, solid, poisonous, n.o.s.
Self-reactive substances, Sample, n.o.s.
Self-reactive substances,
Trial quantities, n.o.s.
Substances , explosive, n.o.s.
Substances , explosive, very insensitive (substances, EVI) n.o.s.
Substances , which in contact with water emit flammable gases, liquid, corrosive, n.o.s.
Substances , which in contact with water emit flammable gases, liquid, n.o.s.
Substances, which in contact with water emit flammable gases, liquid, poisonous, n.o.s.
Substances, which in contact with water emit flammable gases, solid, corrosive, n.o.s.
Substances, which in contact with water emit flammable gases, solid, n.o.s.
Substances, which in contact with water emit flammable gases, solid, oxidizing, n.o.s.
Substances, which in contact with water emit flammable gases, solid, poisonous, n.o.s.
Substances, which in contact with water emit flammable gases, solid, self-heating, n.o.s.
Substances, which in contact with water emit flammable gases, solid, flammable n.o.s.
Tear gas substances, n.o.s. liquid *or* solid

Table 6.3 Proper Shipping Names Requiring Technical Names as Part of the Shipping Description (Continued)

Acetone cyanohydrin, stabilized
Acrolein, inhibited
Aerosols, poison, n.o.s.
Allyl alcohol
Allylamine
Allychloroformate
Allyl isothiocyanate, stabilized
Ammonia, anhydrous, liquified
Arsenic trichloride
Arsine
Boron tribromide
Boron trichloride
Boron trifluoride
Bromide chloride
Bromine
Bromine pentafluoride
Bromine solutions
Bromine trifluoride
Bromoacetone
n-Butylchloroformate
sec-Butylchloroformate
n-Butylisocyanate
tert-Butylisocyanate
Carbon dioxide and ethylene oxide mixtures
Carbon monoxide
Carbonyl fluoride
Carbonyl sulfide
Chlorine
Chlorine pentafluoride
Chlorine trifluoride
Chloroacetaldehyde
Chloroacetone, stabilized
Chloroacetonitrile
Chloroactyl chloride
Chloroformates, n.o.s.
Chloropicrin
Chloropicrin and methyl bromide mixtures
Chloropicrin and methyl chloride mixtures
Chloropicrin mixture, n.o.s.
Chloroplvailoyl chloride
Chlorosulfonic acid
Coal gas
Compressed gases, flammable, toxic, n.o.s.
Compressed gases, toxic, n.o.s.
Crotonaldehyde, stabilized
Cyanogen bromide

Table 6.4 HM-181 Proper Shipping Names for Materials which are Poison Inhalation Hazards

Cyanogen chloride, inhibited
Cyanogen, liquified
Cyclohexyl isocyanate
Diborane
Diborane mixtures
Dichlorodifluoromethane and ethylene oxide mixture
Dichlorosilane
3,5-Dichloro-2,4,6trifluoropyridine
Diketene, inhibited
Dimethyl hydrazine, symmetrical
Dimethyl hydrazine, unsymmetrical
Dimethyl sulfate
Dimethyl thiophospharyl chloride
Dinitrogen tetroxide, liquified (Nitrogen dioxide)
Ethylchloroformate
Ethyl chlorothioformate
Ethyl dichloroarsine
Ethylene chlorohydrin
Ethylene dibromide
Ethyleneimine, inhibited
Ethylene oxide
Ethyl isocyanate
Ethyl phosphonothioic dichloride, anhydrous
Ethyl phosponous dichloride, anhydrous
Ethyl phosphorodichloridate
Fluorine, compressed
Gas identification set
Germane
Hexachlorocyclopentadiene
Hexaethyl tetraphosphate and compressed gas mixtures
Hexafluoroacetone
Hydrocyanic acid, aqueous solutions
Hydrogen bromide, anhydrous
Hydrogen chloride, anhydrous
Hydrogen chloride, refrigerated liquid
Hydrogen cyanide, anhydrous, stabilized
Hydrogen selenide, anhydrous
Hydrogen sulfide, liquified
Insecticide gases, toxic, n.o.s.
Iron pentacarbonyl
Isobutyl chloroformate
Isobutyl isocyanate
Isocyanates, n.o.s.
Isophoronediisocyanate
Isopropyl chloroformate
Isopropyl isocyanate

Table 6.4 HM-181 Proper Shipping Names for Materials which are Poison Inhalation Hazards (Continued)

Methoxymethyl isocyanate
Mesitylene*
Methylamine, anhydrous
Methyl bromide
Methyl bromide and ethylene dibromide mixtures
Methyl chloroformate
Methylchloromethyl ether
Methylchlorosilane
Methyldichloroarsine
Methylhydrazine
Methyl iodide
Methyl isocyanate
Methyl isothiocyanate
Methyl mercaptan
Methyl orthosilicate
Methyl phosphonic dichloride
Monochloroacetic acid, liquid*
Nickel carbonyl
Nitric acid, red fuming
Nitric oxide
Nitric oxide and nitrogen dioxide mixtures
Nitrogen dioxide, liquified
Nitrogen trifluoride
Nitrogen trioxide
Organic phosphate compound; mixed with
compressed gas
Oxygen difluoride
Parathion and compressed gas mixture
Pentaborane
Perchloromethylmercaptan
Perchloryl fluoride
Phenylcarbylamine chloride
Phenyl isocyanate
Phenyl mercaptan
Phosgene
Phosphine
Phosphorus oxychloride
Phosphorous pentafluoride
Phosphorus trichloride
Poisonous liquids, corrosive, n.o.s. *inhalation hazard*
Poisonous liquids, flammable, n.o.s. *inhalation hazard*
Poisonous liquids, n.o.s. *inhalation hazard*
Poisonous liquids, oxidizing, n.o.s. *inhalation hazard*
n-Propyl chloroformate
n-Propyl isocyanate
Selenium hexafluoride

Table 6.4 HM-181 Proper Shipping Names for Materials which are Poison Inhalation Hazards (Continued)

Silicon tetrafluoride
Stibene
Sulfur chlorides
Sulfur dioxide, liquified
Sulfuric acid, fuming
Sulfur tetrafluoride
Sulfur trioxide, inhibited
Sulfur trioxide, uninhibited
Sulfuryl fluoride
Tellurium hexafluoride
Tetraethyl dithiopyrophosphate and gases in solution (and gas mixtures)
Tetraethyl pyrophosphate and compressed gas mixtures
Tetramethoxysilane*
Tetranitromethane
Thio-4-pentanal
Thionyl chloride
Thiophosgene
Titanium tetrachloride
Trichloroacetyl chloride
Trimethoxysilane
Trimethylacetyl chloride
Tungsten hexafluoride

Table 6.4 HM-181 Proper Shipping Names for Materials which are Poison Inhalation Hazards (Continued)

* Denotes chemicals which are not listed in the DOT Hazardous Materials Table.

**STRAIGHT BILL OF LADING**
**ORIGINAL—NOT NEGOTIABLE**

PAGE ___ OF ___
Shipper's No. ______
Carrier's No. ______

| | | |
|---|---|---|
| OUR TRUCK ☐<br>THEIR TRUCK ☐<br>COMMON CARRIER ☐ | CARRIER: SCAC DATE<br>Route: | Vehicle Number |

| TO: Consignee Street Destination ZIP | FROM: STUPENDOUS INDUSTRIES, INC. 666 INDUSTRIAL PARKWAY COMPLEX, NC 27999 |
|---|---|

| NUMBER OF PACKAGES | PACKAGE TYPE (Line out those not applicable) | HM (Enter X if the material is a Hazardous Material) | Description of Articles IF HAZARDOUS MATERIALS—PROPER SHIPPING NAME | Hazard Class or Division | Identification Number (UN or NA) | Packing Group | Total Quantity | Unit (Wt/Vol) | Shipping weight (subject to correction) |
|---|---|---|---|---|---|---|---|---|---|
| | DRUM PAIL BOX TT | X | Paint | 3 | UN1263 | II | | | |
| | DRUM PAIL BOX TT | X | Paint related material | 3 | UN1263 | III | | | |
| | DRUM PAIL BOX TT | X | Nitrocellulose with alcohol, ___% alcohol by mass and ___% nitrogen by dry mass | 4.1 | UN2556 | II | | | |
| | DRUM PAIL BOX TT | | Aqueous Paint Freezable Liquid | | | | | | |
| | DRUM PAIL BOX TT | | Aqueous Filler Freezable Liquid | | | | | | |
| | DRUM PAIL BOX TT | | | | | | | | |
| | | | | | | | | | |

| Remit C.O.D. to:<br>Address:<br>City: State ZIP: | C.O.D. Amt.:<br>$ | C.O.D. Fee:<br>☐ Prepaid<br>☐ Collect $ | FREIGHT CHARGES<br>☐ Prepaid<br>☐ Collect |
|---|---|---|---|

NOTE—Where the rate is dependent on value, shippers are required to state specifically in writing the agreed or declared value of the property. The agreed or declared value of the property is hereby specifically stated by the shipper to be not exceeding $ ________ Per ________.

Subject to Section 7 of Conditions of applicable bill of lading, if this shipment is to be delivered to the consignee without recourse on the consignor, the consignor shall sign the following statement:
The carrier shall not make delivery of this shipment without payment of freight and all other lawful charges.
(Signature of Consignor) ________

RECEIVED, subject to the classifications and lawfully filed tariffs in effect on the date of issue of this Bill of Lading, the property described above in apparent good order, except as noted (contents and conditions of contents of packages unknown), marked, consigned, and destined as indicated above, which said carrier (the word *carrier* being understood throughout this contract as meaning any person or corporation in possession of the property under the contract) agrees to carry to its usual place of delivery at said destination, if on its route, otherwise to deliver to another carrier on the route to said destination. It is mutually agreed as to each carrier of all, or any of, said property over all or any portion of said route to destination and as to each party at any time interested in all or any said property, that every service to be performed hereunder shall be subject to all the bill of lading terms and conditions in the governing classification on the date of shipment.
Shipper hereby certifies that he is familiar with all the bill of lading terms and conditions in the governing classification and the said terms and conditions are hereby agreed to by the shipper and accepted for himself and his assigns.

Where the applicable tariff provisions specify a limitation of the carrier's liability (NMFC Item 172), if there is no release of value declaration by the shipper, and the shipper does not declare a value or release the carrier's liability, that liability shall be limited to the extent provided by NMFC Item 172. California intrastate shipments must comply with NMFC Item 173.

THIS IS TO CERTIFY THAT THE ABOVE-NAMED MATERIALS ARE PROPERLY CLASSIFIED, DESCRIBED, PACKAGED, MARKED AND LABELED, AND ARE IN PROPER CONDITION FOR TRANSPORTATION ACCORDING TO THE APPLICABLE REGULATIONS OF THE DEPARTMENT OF TRANSPORTATION.

SIGNED: ________

PLACARDS REQUIRED:
*(Line out those not applicable)*
FLAMMABLE LIQUID
FLAMMABLE SOLID
________
________

PLACARDS SUPPLIED
☐ YES
☐ NO—FURNISHED BY CARRIER

DRIVER'S SIGNATURE: ________

SHIPPER: ________
PER: ________
DATE: ________

CARRIER: ________
PER: ________
DATE: ________

**EMERGENCY CONTACT:** FOR CHEMICAL EMERGENCY SPILL, LEAK, FIRE, EXPLOSION, OR ACCIDENT, CALL—
TELEPHONE NUMBER: (___) ___-______

Monitored at all times the Hazardous Material is in transportation including storage incidental to transportation (49 CFR 172.604).

ENVIRONMENTAL RESOURCE CENTER

SBL–HM (4/93

**Figure 6.1 Example of a Customized Bill of Lading/Shipping Paper Form**

# Chapter 7
# Vehicle Loading, Unloading, and Placarding

## 7.1 General Loading Requirements

The following are required for all shipments of hazardous materials:

- If tanks, barrels, drums, cylinders, or other packagings are not permanently attached to a motor vehicle, and they contain any flammable liquids (Class 3), compressed gases (Class 2), corrosive materials (Class 8), poisonous materials (Division 6.1), or radioactive materials (Class 7), they must be secured against movement within the vehicle during transportation. Brace these containers by using load bars or other devices that will prevent movement. Be careful to avoid damaging valves and fitting on containers during loading and stowage.
- Never load hazardous wastes or hazardous materials into pole trailers.
- Never smoke while loading explosive, flammable and oxidizing materials. Post signs or take other precaution to keep fire away and prevent anyone in the vicinity from smoking.
- Make sure the hand brake of the vehicle is set before loading or unloading all hazardous wastes and materials. The engine must be stopped whenever explosive or flammable materials are loaded unless the engine is needed to operate a pump on a cargo tank.
- Use tools designed for the containers that are loaded into the vehicle. Never use tools or equipment that could damage the containers or the closures on the containers. For example, a fork lift used to lift and move an open-head 55-gallon by the ring on the drum could damage the ring and result in a spill of hazardous waste during transportation.
- Cargo heaters are always prohibited in vehicles hauling explosives. Cargo heaters used in hauling flammable liquids must be catalytic and must meet the design and operational requirements specified in 49 CFR 177.
- Cargo tanks must be attended by a qualified person at all times during loading. "Attended" means throughout the process the person is awake, has an unobstructed view of the tank, and is within 25 feet of the tank or the delivery hose to the tank. A person is "qualified" if he is aware of the nature of the hazardous waste that is loaded, has been instructed on the procedures to follow in an emergency, and is authorized to move the tank and is able to do so.
- Follow bonding and grounding procedures when loading flammable liquids into cargo tanks through open filling holes. Bonding and grounding is not required by DOT when a cargo tank is loaded through a vapor tight top or bottom connection so that there is no release of vapor at a point where a spark could occur.
- The load on any motor vehicle transporting flammable solids (Class 4) or oxidizing materials (Class 5) must be contained entirely within the body of the motor vehicle and covered by the body, by tarpaulins, or other suitable means; and if the motor vehicle has a tailboard or tailgate, it must be closed and secured in place during transportation of the Class 4 or 5

materials. However, these precautions do not need to be applied to "pick-up and delivery" motor vehicles when they are used in no other transportation than in and around cities, towns, or villages. Shipments in water-tight bulk containers do not need to be covered by a tarpaulin or other means.

- Never load poisons into a motor vehicle that also contains foodstuffs, feed or any edible material meant for human or animal consumption. Hazardous waste should never be shipped with food and animal feed or in tank trucks used for shipping food stuffs. Hazardous waste should not be transported in the cab of the vehicle.

## 7.2 Stowing of Incompatible Materials

DOT prohibits the loading of certain combinations of hazardous materials on the same motor vehicle in order to prevent fires, explosions and spills caused by incompatible chemical reactions. When shipping hazardous materials by highway, the material segregation requirements in the following table must be followed:

To read the chart you must know the hazard class for the material being shipped. If you are shipping a hazardous material with a hazard class that is listed, follow the row on the chart associated with that hazard class and specific material. The symbols in the table have the following meanings:

- An "X" in the row indicates that materials may not be loaded, transported, or stored together.
- An "O" indicates that the materials may not be transported together unless separated by a distance of 1.2 meters (4 ft) in all directions, and packages maintained at a minimum height of 10 cm (4 in) off the floor of the vehicle, or separated in such a way that, in the event of leakage from packages under normal transportation conditions, the hazardous materials could not commingle.
- A "*" indicates the segregation of Class 1 materials is governed by the Compatibility Table for Class 1 materials in 49 CFR 177.848.
- The note "A" in the second column of the Table means that, notwithstanding the requirements of the letter "X", ammonium nitrate fertilizer may be loaded or stored with Division 1.1 (Class A explosive) materials.

In addition to the requirements shown on the Segregation Table for Hazardous Materials:

- Cyanides or cyanide mixtures must not be loaded or stored with acids or acidic materials. The reaction of cyanides with acids releases deadly hydrogen cyanide gas.
- Regardless of the methods of separation employed, Class 8 (corrosive) liquid materials may not be loaded above Class 4 (flammable solid) materials or Class 5 (oxidizing) materials.
- When the §172.101 Table or 49 CFR 172.402 requires a package to bear a subsidiary hazard label, segregation appropriate to the subsidiary hazard must be applied when that segregation is more restrictive than that

required by the primary hazard. However, hazardous materials of the same class may be stowed together without regard to segregation required for any secondary hazard if the materials are not capable of reacting dangerously with each other and causing combustion or dangerous evolution of heat, evolution of flammable, poisonous, or asphyxiant gases, or formation of corrosive or unstable materials.

## SEGREGATION TABLE FOR HAZARDOUS MATERIALS

| Class or Division | Notes | 1.1, 1.2 | 1. 3 | 1. 4 | 1. 5 | 1. 6 | 2. 1 | 2. 2 | 2.3 gas Zone A | 2.3 gas other than Zone A | 3 | 4.1 | 4.2 | 4.3 | 5.1 | 5.2 | 6.1 liquids PG I Zone A | 7 | 8 liquids only |
|---|---|---|---|---|---|---|---|---|---|---|---|---|---|---|---|---|---|---|---|
| Explosives 1.1 and 1.2 | A | * | * | * | * | * | X | X | X | X | X | X | X | X | X | X | X | X | X |
| Explosives 1.3 | | * | * | * | * | * | X | | X | X | X | | X | X | X | X | X | | X |
| Explosives 1.4 | | * | * | * | * | * | O | | O | O | O | | O | | | | O | | O |
| Very insensitive explosives 1.5 | | * | * | * | * | * | X | X | X | X | X | X | X | X | X | X | X | X | X |
| Extremely insensitive explosives 1.6 | | * | * | * | * | * | | | | | | | | | | | | | |
| Flammable gases 2.1 | | X | X | O | X | | | | X | O | | | | | | | O | O | O |
| Non-toxic, nonflammable gases 2.2 | | X | | | X | | | | | | | | | | | | | | |
| Poisonous gas Zone A 2.3 | | X | X | O | X | | X | | | | X | X | X | X | X | X | | | X |
| Poisonous gas other than Zone A 2.3 | | X | X | O | X | | O | | | | O | O | O | O | O | O | | | O |
| Flammable liquids 3 | | X | X | O | X | | | | X | O | | | | | O | | X | | |
| Flammable solids 4.1 | | X | | | X | | | | X | O | | | | | | | X | | O |
| Spontaneously combustible materials 4.2 | | X | X | O | X | | | | X | O | | | | | | | X | | X |
| Dangerous when wet materials 4.3 | | X | X | | X | | | | X | O | | | | | | | X | | O |
| Oxidizers 5.1 | A | X | X | | X | | | | X | O | O | | | | | | X | | O |
| Organic peroxides 5.2 | | X | X | | X | | | | X | O | | | | | | | X | | O |
| Poisonous liquids PG I Zone A 6.1 | | X | X | O | X | | O | | | | X | X | X | X | X | X | | | X |
| Radioactive materials 7 | | X | | | X | | O | | | | | | | | | | | | |
| Corrosive liquids 8 | | X | X | O | X | | O | | X | O | | O | X | O | O | O | X | | |

Class 1 (explosive) materials are broken down into **compatibility groups** for transport. These groups are indicated by letters A–H and J–L, plus N and S. The **classification code** for an explosive consists of the **division number** followed by the compatibility group letter. Compatibility group letters are used to specify the controls for the transportation and transportation related storage of explosives and to prevent an increase in hazard that might result if certain types of explosives were stored or transported together.

Compatibility groups and classification codes for the various types of explosives are set forth in the tables in §173.52. The Classification Code Table sets forth compatibility groups and classification codes for substances and articles described in the first column. Altogether, there are 35 possible classification codes for explosives, because the possible combinations of hazard divisions and compatibility groups are limited.

CLASS 1 (EXPLOSIVES) CLASSIFICATION CODES TABLE

| Description of substances or article to be classified | Compatibility group | Classification code |
|---|---|---|
| Primary explosive substance | A | 1.1A |
| Article containing a primary explosive substance and not containing two or more effective protective features | B | 1.1B<br>1.2B<br>1.4B |
| Propellant explosive substance or other deflagrating explosive substance or article containing such explosive substance | C | 1.1C<br>1.2C<br>1.3C<br>1.4C |
| Secondary detonating explosive substance or black powder or article containing a secondary detonating explosive substance, in each case without means of initiation and without a propelling charge, or article containing a primary explosive substance and containing two or more effective protective features | D | 1.1D<br>1.2D<br>1.4D<br>1.5D |
| Article containing a secondary detonating explosive substance, without means of initiation, with a propelling charge (other than one containing flammable liquid or hypergolic liquid). | E | 1.1E<br>1.2E<br>1.4E |
| Article containing a secondary detonating explosive substance with its means of initiation, with a propelling charge (other than one containing flammable liquid or hypergolic liquid) or without a propelling charge | F | 1.1F<br>1.2F<br>1.3F<br>1.4F |
| Pyrotechnic substance or article containing a pyrotechnic substance, or article containing both an explosive substance and an illuminating, incendiary, tear-producing or smoke-producing substance (other than a water-activated article or one containing white phosphorus, phosphide or flammable liquid or gel or hypergolic liquid) | G | 1.1G<br>1.2G<br>1.3G<br>1.4G |
| Article containing both an explosive substance and white phosphorus | H | 1.2H<br>1.3H |
| Article containing both an explosive substance and flammable liquid or gel | J | 1.1J<br>1.2J<br>1.3J |

## CLASS 1 (EXPLOSIVES) CLASSIFICATION CODES TABLE

| Description of substances or article to be classified | Compatibility Group | Classification Code |
|---|---|---|
| Article containing both an explosive substance and a toxic chemical agent........ | K | 1.2K<br>1.3K |
| Explosive substance or article containing an explosive substance and presenting a special risk (e.g., due to water-activation or presence of hypergolic liquids, phosphides or pyrophoric substances) needing isolation of each type ............ | L | 1.1L<br>1.2L<br>1.3L |
| Articles containing only extremely insensitive detonating substances ........ | N | 1.6N |
| Substance or article so packed or designed that any hazardous effects arising from accidental functioning are limited to the extent that they do not significantly hinder or prohibit fire fighting or other emergency response efforts in the immediate vicinity of the package ........ | S | 1.4S |

Class 1 materials may not be loaded, transported, or stored together, except as allowed in the regulations, and in accordance with the following table:

## COMPATIBILITY TABLE FOR CLASS 1 (EXPLOSIVE) MATERIALS

| Compatibility group | A | B | C | D | E | F | G | H | J | K | L | N | S |
|---|---|---|---|---|---|---|---|---|---|---|---|---|---|
| A | | X | X | X | X | X | X | X | X | X | X | X | X |
| B | X | | X | X | X | X | X | X | X | X | X | X | |
| C | X | X | | 2 | 2 | X | X | X | X | X | X | 3 | |
| D | X | X | 2 | | 2 | X | X | X | X | X | X | 3 | |
| E | X | X | 2 | 2 | | X | X | X | X | X | X | 3 | |
| F | X | X | X | X | X | | X | X | X | X | X | X | |
| G | X | X | X | X | X | X | | X | X | X | X | X | |
| H | X | X | X | X | X | X | X | | X | X | X | X | |
| J | X | X | X | X | X | X | X | X | | X | X | X | |
| K | X | X | X | X | X | X | X | X | X | | X | X | |
| L | X | X | X | X | X | X | X | X | X | X | 1 | X | X |
| N | X | X | 3 | 3 | 3 | X | X | X | X | X | X | | |
| S | X | 4/5 | 4/5 | 4/5 | 4/5 | 4/5 | 4/5 | 4/5 | 4/5 | 4/5 | X | 4/5 | |

Instructions for using the compatibility table for Class 1 materials:

- A **blank** space in the Table indicates that no restrictions apply.
- The letter "**X**" in the Table indicates that explosives of different compatibility groups may not be carried on the same transport vehicle.
- The numbers in the Table mean the following:

  "1" means an explosive from compatibility group L may only be carried on the same vehicle with an identical explosive.

  "2" means any combination of explosives from compatibility groups C, D, or E is assigned to compatibility group E.

"3" means any combination of explosives from compatibility groups C, D, or E with those in compatibility group N is assigned to compatibility group D.

"4" means §177.835(g) applies when transporting detonators.

"5" means Division 1.4S fireworks may not be loaded on the same vehicle with Division 1.1 or 1.2 (Class A explosive) materials.

- Except as excluded below, explosives of the same compatibility group but of different divisions may be transported together provided that the whole shipment is transported as though its entire contents were of the lower numerical division (i.e., Division 1.1 being lower than Division 1.2). For example, a mixed shipment of Division 1.2 (Class A explosive) materials and Division 1.4 (Class C explosive) materials, both of compatibility group D, must be transported as Division 1.2 (Class A explosive) materials.
- When Division 1.5 (blasting agent) materials, compatibility group D, are transported in the same freight container as Division 1.2 (Class A explosive) materials, compatibility group D, the shipment must be transported as Division 1.1 (Class A explosive) materials, compatibility group D.

### 7.3 Vehicle Inspections

Shipments of hazardous materials and hazardous waste should both be thoroughly checked and inspected immediately before leaving the facility for treatment, storage or disposal to confirm compliance with the regulations. The Hazardous Materials Pre-transportation Checklist and the Hazardous Waste Pre-transportation Checklist presented in Figures 7.1 and 7.2 may be used to facilitate these inspections.

Name of Inspector: ______________________ Date: ______________

| Item | Conditions Reviewed | Status | |
|---|---|---|---|
| | | Complete | Incomplete |
| Container Marking | Container Contents (Proper Shipping Name) | | |
| | DOT ID Number | | |
| | Organization Name and Address | | |
| | Emergency Response Information & 24-hour Telephone Number | | |
| Container Label | Appropriate DOT Label(s) on Each Container | | |
| Placard | Appropriate DOT Placard(s) on Truck | | |
| Bill of Lading | | | |
| Condition of Containers: ❑ Good ❑ Poor ❑ Leaking<br>Date Shipment Left Facility:<br>Comments: | | | |

Figure 7.1 Hazardous Material Pre-transportation Checklist

Name of Inspector: ____________________ Date: ____________________

| Item | Conditions Reviewed | Status | |
|---|---|---|---|
| | | Complete | Incomplete |
| Container Marking | Container Contents (Proper Shipping Name) | | |
| | DOT ID Number | | |
| | Begin Accumulation Date | | |
| | U.S. EPA ID Number | | |
| | Warning Mark: "Hazardous Waste—Federal Law Prohibits Improper Disposal. If found, contact nearest police or public safety authority or the U.S. Environmental Protection Agency." | | |
| | Organization Name and Address | | |
| | Manifest Document Number | | |
| Container Label | Appropriate DOT Label(s) on Each Container | | |
| Manifest | Current Manifest (expires 30 September 1994) | | |
| Placard | Appropriate DOT Placard(s) on Each Vehicle | | |
| Condition of Containers: ❑ Good ❑ Poor ❑ Leaking<br>Date Shipment Left Facility:<br>Comments: | | | |

Figure 7.2 Hazardous Waste Pre-transportation Checklist

The following questions should also be answered before the hazardous waste shipment leaves the facility:

| | | |
|---|---|---|
| Does emergency response information accompany the shipment? | ____ Yes | ____ No |
| Have special marks been applied to appropriate containers (RQ, Bung Label, ORM designation, Poison-Inhalation designation, Poison-Inhalation Hazard)? | ____ Yes | ____ No |
| Is the waste restricted from land disposal? | ____ Yes | ____ No |
| If "Yes" above, has the Appropriate Land Disposal Notice or Certification been attached to the manifest? | ____ Yes | ____ No |

## 7.4 Placarding

Placards alert persons to the potential dangers associated with hazardous materials contained within motor vehicles, rail cars, freight containers, cargo tanks and portable tanks. Placards also guide emergency personnel who respond to incidents involving hazardous materials.

Except under very limited circumstances, each highway transport vehicle containing hazardous materials or hazardous waste must be placarded on each end and on each side with the placards specified by the DOT. It is the shipper's responsibility to offer appropriate placard(s) to the transporter. Each shipper must have a supply of appropriate placards which can be offered to transporters hauling hazardous materials and hazardous waste. It is the transporter's responsibility to replace any placards which become dislodged from the vehicle during transportation.

The color, size and design of placards are specified by DOT in 49 CFR 172 Subpart F. The hazard class division number as prescribed by the United Nations may be printed in the lower corner of the placard. All placards must be durable and able to withstand exposure to weather.

### 7.4.1 Placement of Placards

Placards must be visible from the directions in which they face. Placards must be placed on each side and each end of the motor vehicle. They may be placed on the front of the tractor of a tractor-trailer instead of placarding the front of the trailer, or may be placed on both units. Place each placard at least 3" away from any ladders, pipes or other appurtenances which might reduce the effectiveness of the placard. Place placards so dirt or water from vehicle wheels will not be sprayed on the placard. Securely fix each placard on the sides and ends of the vehicle or use a placard holder.

### 7.4.2 Selecting the Proper Placard (Pre-HM-181)

It is the shipper's responsibility to offer the correct placard(s) to the transporter. To determine which placards are required, you must use DOT's placarding rules and tables summarized below.

There are two tables that you must refer to when selecting the proper placards. Table 1 lists those materials, **regardless of the quantity shipped**, which must be placarded as noted.

**Department of Transportation**
**Placard Table 1**

| If The Hazard Class Is | Offer This Placard |
|---|---|
| Class A explosives | EXPLOSIVES A[1] |
| Class B explosives | EXPLOSIVES B[2] |
| Poison A | POISON GAS[1] |
| Flammable solid (DANGEROUS WHEN WET label only) | FLAMMABLE SOLID W[3] |
| Radioactive material | RADIOACTIVE[4,5] |
| Radioactive material ($UF_6$, fissile)† | RADIOACTIVE[4] and CORROSIVE[6] |
| Radioactive material ($UF_6$, low specific activity)‡ | RADIOACTIVE[4,5] and CORROSIVE[6] |

**Notes to Placard Table 1**

1 See 49 CFR 172.510(a)

2 EXPLOSIVES B placard is not required if the transport vehicle contains class A explosives and is placarded EXPLOSIVES A as required.

3 FLAMMABLE SOLID "W" placard is required only when the DANGEROUS WHEN WET label is specified in the Hazardous Materials Table for a material classed as a Flammable solid.

4 Applies only to any quantity of packages bearing the RADIOACTIVE Yellow-III label (refer to 49 CFR 172.403).

5 For exclusive use shipments (as defined in 49 CFR 173.403) of low specific activity radioactive materials transported in accordance with 49 CFR 173.425(b) or (c).

6 CORROSIVE placard is not required to be displayed on transport vehicles for shipments of less than 1,000 pounds gross weight.

† Uranium hexafluoride, fissile (with > 0.1% $U^{235}$)

‡ Uranium hexafluoride, low specific activity (with ≤ 0.1% $U^{235}$)

## Department of Transportation
## Placard Table 2

| If The Hazard Class Is | Offer This Placard |
|---|---|
| Class C explosives | DANGEROUS[1] |
| Blasting Agents | BLASTING AGENTS[9,10] |
| Nonflammable gas | NONFLAMMABLE GAS[8] |
| Nonflammable gas (Chlorine) | CHLORINE[7] |
| Nonflammable gas (Fluorine) | POISON |
| Nonflammable gas (Oxygen, cryogenic liquid) | OXYGEN |
| Flammable gas | FLAMMABLE GAS[8] |
| Combustible liquid | COMBUSTIBLE[3,4] |
| Flammable liquid | FLAMMABLE |
| Flammable solid | FLAMMABLE SOLID[5] |
| Oxidizer | OXIDIZER[9,10] |
| Organic Peroxide | ORGANIC PEROXIDE |
| Poison B | POISON |
| Corrosive material | CORROSIVE[6] |
| Irritating material | DANGEROUS |

**Notes to Placard Table 2**

1 Applies only to a Class C explosive required to be labeled with an EXPLOSIVE C label.

2 [RESERVED]

3 A COMBUSTIBLE placard is required only when a material classed as combustible liquid is transported in a package having a **rated capacity of more than 110 gallons**, a **cargo tank**, or a **tank car**.

4 A FLAMMABLE placard may be used on a cargo tank or portable tank during transportation by highway, rail or water, and on a compartmented tank car containing materials classed as flammable liquid and combustible liquid.

5 Except when offered for transportation by water, a FLAMMABLE placard may be displayed in place of a FLAMMABLE SOLID placard except when a DANGEROUS WHEN WET label is specified for the material on the Hazardous Materials Table.

6 Refer to 49 CFR 173.245(b) for authorized exceptions.

7 A CHLORINE placard is required only for a packages having a **rated capacity of more than 110 gallons**; a NONFLAMMABLE GAS placard is to be displayed for packages having a **rated capacity of 110 gallons or less**.

8 A NONFLAMMABLE GAS placard is not required on a motor vehicle displaying a FLAMMABLE GAS placard or an OXYGEN placard.

9 BLASTING AGENTS, OXIDIZER and DANGEROUS placards need not be displayed if a transport vehicle or freight container also contains Class A or Class B explosives and is placarded EXPLOSIVES A or EXPLOSIVES B as required.

10 Except for shipments by water, OXIDIZER placards need not be displayed if a freight container, motor vehicle, or rail car also contains blasting agents and is placarded BLASTING AGENT as required.

For Placard Table 2, the following guidelines apply:

1. **1,000 pounds** or more of materials classified in Table 2 must be placarded as specified in the table.
2. If two or more Table 2 materials are being transported, the DANGEROUS placard may be displayed instead of separate placards specified in the table. However, when **5,000 pounds** or more of one hazard category is loaded at one facility, the placard specified in the table must be displayed.

### 7.4.3 Selecting the Proper Placard (HM-181)

For most hazardous materials, as of 1 October 1994, new placards and placarding tables must be used (see section 7.4.4 for exceptions). The new placards are intended to align with international shipping standards. Each bulk packaging, freight container, unit load device, transport vehicle or rail car containing hazardous material must be placarded on each side and each end with the placards specified in Tables 1 and 2 below and in accordance with any hazard class specific placarding requirements, including the specifications for the placards named in the tables and described in detail in §§172.519 through 172.558.

**DANGEROUS placard**. A freight container, unit load device, transport vehicle or rail car which contains non-bulk packagings with two or more categories of hazardous materials that require different placards specified in Table 2 may be placarded with DANGEROUS placards instead of the separate placarding specified for each of the materials in Table 2. However, when 2,268 kg (**5,000 pounds**) or more of one category of material is loaded therein at one loading facility, the placard specified in Table 2 of paragraph for that category must be applied.

**Exception for less than 454 kg (1,001 pounds).** Except for transport vehicles and freight containers subject to §172.505, bulk packagings, or transportation by aircraft or vessel, placards for hazardous materials covered by Table 2 are not required on:

- A transport vehicle or freight container which contains less than 454 kg (1,001 pounds) aggregate gross weight of hazardous materials covered by Table 2 of paragraph (e) of this section;
- A rail car loaded with transport vehicles or freight containers, none of which is required to be placarded.

The above exceptions provided do not prohibit the display of placards in the manner prescribed, if not otherwise prohibited (see §172.502), on transport vehicles or freight containers which are not required to be placarded.

**Exception for empty non-bulk packages.** A non-bulk packaging that contains only the residue of a hazardous material covered by Table 2 need not be included in determining placarding requirements.

Placards are specified for hazardous materials in accordance with the following tables:

**Department of Transportation**
**HM-181 Placard Table 1**

| Category of material (Hazard class or division number and additional description, as appropriate) | Placard name | Placard design section reference (§) |
|---|---|---|
| 1.1 | EXPLOSIVES 1.1 | §177.522 |
| 1.2 | EXPLOSIVES 1.2 | §172.522 |
| 1.3 | EXPLOSIVES 1.3 | §172.522 |
| 2.3 | POISON GAS | §172.540 |
| 4.3 | DANGEROUS WHEN WET | §172.548 |
| 6.1 (PG I, inhalation hazard only) | POISON | §172.554 |
| 7 (Radioactive Yellow III label only) | RADIOACTIVE[1] | §172.556 |

[1] RADIOACTIVE placard also required for exclusive use shipments of low specific activity material in accordance with §173.425(b) or (c).

**Any quantity** of material falling within the hazard classes and divisions listed in Table 1 must be placarded.

## Department of Transportation HM-181 Placard Table 2

| Category of material (Hazard class or division number and additional description, as appropriate) | Placard name | Placard design section reference (§) |
|---|---|---|
| 1.4 | EXPLOSIVES 1.4 | §172.523 |
| 1.5 | EXPLOSIVES 1.5 | §172.524 |
| 1.6 | EXPLOSIVES 1.6 | §172.525 |
| 2.1 | FLAMMABLE GAS | §172.532 |
| 2.2 | NON-FLAMMABLE GAS | §172.528 |
| 3 | FLAMMABLE | §172.542 |
| Combustible liquid | COMBUSTIBLE | §172.544 |
| 4.1 | FLAMMABLE SOLID | §172.546 |
| 4.2 | SPONTANEOUSLY COMBUSTIBLE | §172.547 |
| 5.1 | OXIDIZER | §172.550 |
| 5.2 | ORGANIC PEROXIDE | §172.552 |
| 6.1 (PG I or II, other than PG I inhalation hazard) | POISON | §172.554 |
| 6.1 (PG III) | KEEP AWAY FROM FOOD | §172.553 |
| 6.2 | (None) | |
| 8 | CORROSIVE | §172.558 |
| 9 | CLASS 9 | §172.560 |
| ORM-D | (None) | |

For Table 2, the following guidelines apply:

1. **1,000 pounds** or more of materials classified in Table 2 must be placarded as specified in the table.
2. If two or more Table 2 materials are being transported, the DANGEROUS placard may be displayed instead of separate placards specified in the table. However, when **5,000 pounds** or more of one hazard category is loaded at one facility, the placard specified in the table must be displayed.

Under HM-181, placards are not required for the following:

- Infectious substances;
- Hazardous materials classed as ORM-D;
- Hazardous materials authorized by this subchapter to be offered for transportation as Limited Quantities when identified as such on shipping papers in accordance with 49 CFR 172.203(b);
- Hazardous materials which are packaged as small quantities under the provisions of §173.4 of this subchapter; and
- Combustible liquids in non-bulk packagings.

**Placarding for Subsidiary Hazards**

- Each transport vehicle, portable tank, freight container or unit load device that contains a poisonous material subject to the "Poison—Inhalation Hazard" shipping description of 49 CFR 172.203(m)(3) shall be placarded with POISON or POISON GAS placards, as appropriate, on each side and each end, if not so placarded under 49 CFR 172.504.
- In addition to the RADIOACTIVE placard which may be required by 49 CFR 172.504(e), each transport vehicle, portable tank or freight container that contains 454 kg (**1,001 pounds**) or more gross weight of fissile or low specific activity uranium hexafluoride shall be placarded with a CORROSIVE placard on each side and each end.
- Each transport vehicle, portable tank, freight container or unit load device that contains a material which has a subsidiary hazard of being dangerous when wet, as defined in 49 CFR 173.124 of this subchapter, shall be placarded with DANGEROUS WHEN WET placards, on each side and each end, in addition to the placards required by 49 CFR 172.504.
- Hazardous materials that possess secondary hazards may exhibit subsidiary placards that correspond to the primary placards, even when not required by (see also 49 CFR 172.519(b)(4)).

**MARINE POLLUTANT Mark on Vehicles**

A **transport vehicle** or **freight container** that contains either a non-bulk or bulk package subject to the marine pollutant marking requirements of §172.322(a) or (b) must be marked with the MARINE POLLUTANT mark. The mark must appear on each side and each end of the transport vehicle or freight container, and must be visible from the direction it faces. This requirement may be met by the marking displayed on a freight container or portable tank loaded on a motor vehicle or rail car. This mark may be displayed in black lettering on a white square-on-point configuration having the same outside dimensions as a placard. The mark is not required, however, except for transportation by vessel, on a transport vehicle or freight container that bears a specified **label** or **placard**.

### 7.4.4 Selecting Proper Placards (Transitional Provisions)

As noted under compliance deadlines (Chapter I, sect. 1.5), Until **1 October 2001**, placards that conform to specifications for placards in effect on 30 September 1991, or placards specified in the 21 December 1990 final rule may be used, for highway transportation only, in place of the placards specified in subpart F of 49 CFR 172. Because hazard classes are changed, during the transition period, §171.14(c)(2) provides the following table for selecting placards:

PLACARD SUBSTITUTION TABLE

| Hazard class or division No. | Current placard name | Old (30 Sept. 1991) placard name |
|---|---|---|
| Division 1.1 | Explosives 1.1 | Explosives A. |
| Division 1.2 | Explosives 1.2 | Explosives A. |
| Division 1.3 | Explosives 1.3 | Explosives B. |
| Division 1.4 | Explosives 1.4 | Dangerous. |
| Division 1.5 | Explosives 1.5 | Blasting agents. |
| Division 1.6 | Explosives 1.6 | Dangerous. |
| Division 2.1 | Flammable gas | Flammable gas. |
| Division 2.2 | Nonflammable gas | Nonflammable gas. |
| Division 2.3 | Poison gas | Poison gas. |
| Class 3 | Flammable | Flammable. |
| Combustible liquid | Combustible | Combustible. |
| Division 4.1 | Flammable solid | Flammable solid. |
| Division 4.2 | Spontaneously combustible | Flammable solid. |
| Division 4.3 | Dangerous when wet | Flammable solid W. |
| Division 5.1 | Oxidizer | Oxidizer. |
| Division 5.2 | Organic peroxide | Organic peroxide. |
| Division 6.1, PG I and II | Poison | Poison. |
| Division 6.1, PG III | Keep away from food | (none required). |
| Class 7 | Radioactive | Radioactive. |
| Class 8 | Corrosive | Corrosive. |
| Class 9 | Class 9 | (none required.) |

# Chapter 8
# Emergency Response

## 8.1 Emergency Preparedness

### 8.1.1 Emergency Response Information

DOT regulations published in 49 CFR 172, Subpart G, effective **31 December 1990**, require the maintenance of certain types of emergency response information during transportation of hazardous materials and wastes. The following information must be available during transportation and handling and storage incident to transportation:

- the **basic description** and **technical name** of the hazardous material as required by DOT or other applicable regulations
- immediate **hazards** to health
- **risks** of fire or explosion
- immediate **precautions** to be taken in the event of an accident or incident
- immediate **methods** for handling **fires**
- initial **methods** for handling **spills** or leaks in the absence of fire
- preliminary **first aid** measures.

This information must be in English and available for use away from the hazardous material package. The information may be included on or attached to the shipping papers, and may take the form of material safety data sheets for single-component waste streams or comparable materials, or some kind of emergency response guidance manual such as the DOT *Emergency Response Guidebook.* However, if the *Guidebook* is used for emergency response information, the corresponding page of the manual must be clearly referenced on the shipping papers for each material in the shipment. This information must be maintained by the shipper near where the materials are prepared for shipment, and must be maintained by the transporter at all times during transportation so it is available during an emergency. For hazardous wastes, the emergency response information must also be maintained by hazardous waste transfer, storage and handling facilities so it is immediately accessible to facility personnel during an emergency.

### 8.1.2 Oil Release Response Plans

On June 16, 1993, The Research and Special Programs Administration (RSPA) of the DOT published an interim final rule regarding oil spill prevention and response plans. The new section (49 CFR 130) is intended to replace and clarify section 49 CFR 171.5 published on February 2, 1993. This interim final rule implements the Federal Water Pollution Control Act as amended by the Oil Pollution Act of 1990.

The regulations found in 49 CFR 130 apply to all oils. Many of these oils are not hazardous materials under DOT definition (they have a flash point over 200° F) and therefore are not subject to the more stringent regulations under the Hazardous Materials Transportation Act. **The regulation applies to**

**shipments of all oils in excess of 42,000 gallons (1,000 barrels) per package and shipments of petroleum oils in excess of 3,500 gallons per package.** The definitions of oil and petroleum oil are presented below, along with the basic requirements 49 CFR 130.

*Petroleum Oil* - Any oil extracted from geological hydrocarbon deposits, including fractions and derivatives thereof.

*Oil* - Oil of any kind or in any form, including, but not limited to, petroleum, vegetable oil, animal oil, fuel oil, sludge, oil refuse, an oil mixed with waste or other dredged spoil.

**Basic Requirements**

- No person may offer oil subject to part 130 for transportation without providing the carrier with a document indicating the shipment contains oil.
- Packaging used to transport oil must be designed, constructed, maintained, closed, and loaded so that, under conditions normally incident to transportation, there will be no release of oil to the environment.
- As of **September 30, 1993**, no person may transport oil subject to this part unless that person has a written plan that:
  - describes response actions to all potential releases during transportation
  - accounts for maximum potential releases from transported containers
  - identifies who will respond to releases
  - lists the names and addresses of agencies and persons to be contacted in the event of a release.

  Motor carriers and rail carriers must retain this plan on file at their principal place of business and at dispatch locations.
- As of **February 18, 1993**, no person may transport an oil in a quantity greater than 42,000 gallons unless their plan also:
  - is consistent with the National Contingency Plan in 40 CFR 300 and Area Contingency Plans
  - identifies a qualified person with full authority to implement removal actions
  - identifies a means of immediate communication between the above individual and responding agencies and persons providing personnel and equipment
  - includes and ensures arrangements with private personnel (including address and phone number) contracted to respond to a worst-case potential release
  - describes the training, equipment testing, periodic unannounced drills, and response actions of facility personnel to ensure the safety of personnel and the environment

  This plan must be submitted, and resubmitted in the event of significant changes, to the Associate Administrator for Hazardous Materials Safety (for portable tanks), to the Federal Railroad Administrator (for tank cars), or to the Federal Highway Administrator (for cargo tanks) at 400 Seventh Street, SW, Washington, DC 20590-0001.

To summarize, non-petroleum oils shipped in quantities of 42,000 gallons or less are not required to develop response plans or prevention procedures. The requirements of part 130 (outlined above) only apply to non-petroleum oils when they are shipped in quantities in excess of 42,000 gallons.

## 8.2 Hazardous Materials—Transportation-Related Incidents

### 8.2.1 Containing and Controlling the Release

The DOT regulations at 49 CFR 177, Subpart D, detail requirements for the care of hazardous materials that are in transit when an accident occurs. These requirements govern shipments that are aboard disabled vehicles as well as the disposition of containers found broken or leaking during transit. In the case of a vehicle that is disabled during transit (i.e., flat tire, overheated engine, mechanical failure), the regulations state that, "special care shall be taken to guard the vehicle and its load or to take such steps as may be necessary to provide against hazard."

When leaks or spills occur from packages or containers during transportation, the driver should:

- Prevent the leakage from spreading
- Prevent fire hazards
- Repair the packages or containers (if safe and practical to do so)
- If desired, forward the material to destination in a salvage drum in accordance will requirements of 49 CFR 173.3(c).

40 CFR Part 263.31 of the EPA regulations states that "a transporter must clean up any hazardous waste discharge that occurs during transportation or take such action as may be required or approved by federal, state, or local officials so that the hazardous waste discharge no longer presents a hazard to human health or the environment."

### 8.2.2 When to Report Hazardous Materials Incidents

If any of the following incidents occurs as a result of or incident to the transportation of hazardous materials, the **National Response Center must be notified by the carrier** at the earliest practicable moment;

- A person is killed; or
- A person receives injuries requiring his or her hospitalization; or
- Estimated carrier or other property damage exceeds $50,000; or
- An evacuation of the general public occurs lasting one or more hours; or
- One or more major transportation arteries or facilities are closed or shut down for one hour or more; or
- The operational flight pattern or routine of an aircraft is altered; or
- Fire, breakage, spillage, or suspected radioactive contamination occurs involving shipment of radioactive material (see also §§174.45, 175.45, 176.48, and 177.807 of 49 CFR); or

- Fire, breakage, spillage, or suspected contamination occurs involving shipment of etiologic agents; or
- A situation exists of such a nature (e.g., a continuing danger to life exists at the scene of the incident) that, in the judgment of the carrier, it should be reported to the DOT even though it does not meet the above criteria.

Notice involving etiologic agents may be given the Director, Center for Disease Control, U.S. Public Health Service, Atlanta, GA, (404) 633-5313, in place of the notice to the Department or 202-267-2675. Each notice must include the following information:

- Name of reporter.
- Name and address of carrier represented by reporter.
- Phone number where reporter can be contacted.
- Date, time, and location of incident.
- The extent of injuries, if any.
- Classification, name, and quantity of hazardous materials involved, if such information is available.
- Type of incident and nature of hazardous material involvement and whether a continuing danger to life exists at the scene.

In order to expedite the reporting procedure, the checklist presented in the following table may be used.

- ❑ Location and name of person making report
- ❑ Name and address of carrier represented by person making the report
- ❑ Telephone number where person can be reached
- ❑ Date, time, and location of the hazardous materials incident
- ❑ Extent of injuries, if any
- ❑ Classification, proper shipping name, and quantity of hazardous materials involved
- ❑ Type of incident and nature of hazardous materials involved
- ❑ Whether or not a continuing danger to life exists

Telephone Checklist for Hazardous Materials Incident Report

Transporters should also be equipped with spill response equipment such as pallet jacks, dollies, open head drums with fresh adsorbent and (spark-proof) shovels.

### 8.2.3 Written Hazardous Materials Incident Reports.

A written Hazardous Materials Incident Report on DOT Form F 5800.1 (Rev. 6/89) must be submitted to DOT within **30 days** of the date of discovery of an incident for which a telephone report (as described above) is required, each incident that occurs during the course of transportation (including loading, unloading, and temporary storage) in which any of the circumstances described above occurs or there has been an unintentional release of hazardous materials from a package (including a tank) or any quantity of hazardous waste has been discharged during transportation.

For **hazardous waste**, a copy of the hazardous waste manifest for the waste must be attached to the report together with an estimate of the quantity of the waste removed from the scene, the name and address of the facility to which it was taken, and the manner of disposition of any removed waste must be entered Section IX of the report form (Form F 5800.1).

Each carrier making a hazardous materials incident report must send the report to the Information Systems Manager, DHM-63, Research and Special Programs Administration, Department of Transportation, Washington, DC 20590-0001. A **copy** of the report must also be retained, for a period of **two years**, at the carrier's principal place of business, or at other places as authorized and approved in writing by an agency of the Department of Transportation.

Except as provided in the paragraph below, a written report is **not** required for incidents involving the unintentional release of hazardous materials being transported under the following proper shipping names:

- Consumer commodity.
- Battery, electric storage, wet, filled with acid or alkali.
- Paint and paint related material when shipped in packagings of five gallons or less.

The above exceptions to incident reporting do **not** apply to:

- Incidents required to be reported under 49 CFR 171.15(a);
- Incidents involving transportation aboard aircraft; nor
- Incidents involving the transportation of hazardous waste.

A **guidance** document for assisting in the completion of DOT Form F 5800.1 may be obtained from the Office of Hazardous Materials Transportation, DHM-51, U.S. Department of Transportation, Washington, DC 20590-0001.

## 8.3 Non-DOT Incident and Release Reporting Requirements

### On-Site Releases of Hazardous Waste

It is the **responsibility of the facility Emergency Coordinator** (or his designee) to oversee all response actions and ensure the proper notifications are made. Whenever there is an imminent or actual emergency situation, the Emergency Coordinator must:

- immediately **alert** facility personnel who may be adversely affected by the situation. This may be accomplished by activating alarms or communication systems.
- immediately **notify** those emergency response organizations with designated response roles. This notification is to be made to appropriate organizations on-site as well as state or local agencies.

- take all reasonable **safety measures** necessary to ensure that fires, explosions, and releases do not recur or spread to other hazardous waste on-site. If applicable at your facility, these measures must include:
  - collecting and containing released waste,
  - removing or isolating containers, and
  - stopping processes and operations.

  If operations are ceased for a fire, explosion, or release, the Emergency Coordinator must (if appropriate at your facility with respect to a certain situation) **monitor** for ruptures in valves, pipes, or other equipment; leaks; pressure buildup or gas generation.

- immediately **identify** the character, exact source, amount of material released, and a determination of how far the release has spread. This may be done by direct observation of the situation, a review of facility records, and chemical analysis.

  In characterizing the situation, the Emergency Coordinator must **assess** possible hazards to human health and the environment that may result from the release, fire, or explosion. This assessment must consider both direct and indirect effects of the situation. For example, the assessment must consider such questions as:
  - what are the effects of any gases that may be generated?
  - will contaminated water run off the site, and if so, will it harm human health or the environment?
  - how will chemical agents used to control fire or heat-induced explosions affect the environment surrounding the site?

  If the Emergency Coordinator determines that **evacuation** of local areas **off-site** may be necessary, he must immediately notify the appropriate local authorities and arrange for the facility Emergency Coordinator to assist local authorities.

  Additionally, if the Emergency Coordinator determines that a release, fire, or explosion could threaten human health or the environment off-site, the On-Scene Coordinator (for your geographical area) must be notified and provided with the following information:
  - name and telephone number of the individual making the report;
  - name and address of the facility;
  - time and type of situation (e.g., release, fire, explosion);
  - name and quantity of material(s) involved, to the extent known;
  - name and quantity of the material(s) involved, to the extent known; and
  - possible hazards to human health and environment off site.

During the response, the Emergency Coordinator must ensure that no hazardous materials or waste that may be incompatible with the released material is treated, stored, or disposed of in the area affected by the emergency situation, until cleanup procedures are completed. After remediation efforts have been completed, the Emergency Coordinator must ensure that all emergency equipment listed in the contingency plan is cleaned and fit for it's intended use, before operations are resumed. Before operations can begin again in the areas of your facility that were affected, the facility must report to the EPA Regional Administrator (or your state environmental agency if your state has RCRA authorization) on how the Emergency Coordinator met the requirements described in this paragraph.

### 8.3.1 Your Company or Organization's Responsibilities for Hazardous Waste Releases

In addition to reporting to the Regional EPA Administrator as described above, the plant environmental engineer/safety representative must **record** the **time, date,** and **details** of any incident that requires implementing the contingency plan. Within **15 days** after the emergency situation, the facility is required to submit a written report to the Regional Administrator (or state environmental agency) describing the situation. The report must include:

- name, address, and telephone number of the facility owner or operator
- name, address, and main phone number of the facility;
- date, time, and type of emergency situation (e.g., spill, fire, explosion);
- name and quantity of material(s) involved;
- the extent of injuries, if any;
- an assessment of actual or potential hazards to human health or the environment, where this applicable; and
- estimated quantity and disposition of recovered material that resulted from the incident.

### 8.3.2 Initial Response Actions for Hazardous Material and Waste Incidents

In the event of a spill, fire, explosion or other type of release, the **first action** that should be taken by the person discovering the incident is to **alert the facility's Emergency Coordinator**. Based on the type of incident, appropriately trained personnel should be dispatched to respond to the incident.

While waiting for assistance to arrive, there are some general measures you should take to minimize the danger to those near the incident. First, approach the area cautiously; resist the urge to rush onto the scene in hopes of being able to size up the situation rapidly. Next, you should identify the hazards.

Secure the scene. Without entering the immediate hazard area, do what you can to isolate the area and assure the safety of people and the environment in the vicinity of the incident. Keep people away from the scene and the perimeter. Allow room enough to move and remove equipment that may be necessary to respond to the emergency.

Decide on site entry. Any efforts you make to rescue persons, protect property or the environment must be weighed against the possibility that you could become part of the problem. Never allow anyone to enter a hazard area alone, and always ensure that response workers wear appropriate personal protective equipment.

- Alert facility Emergency Coordinator
- Approach area cautiously
- Identify the hazards
- Secure the scene
- Decide on site entry
- Do not allow personnel to enter area alone
- Provide appropriate personal protective equipment

Figure 8.1 Initial Response Actions

### 8.3.3 Responsibilities of the Emergency Coordinator

**Response to Spills**

There are **four steps** that should be taken by those responding to a spill (summarized in Figure 8.2 below). The first involves **identifying the source** of the spill. To do this you must determine where the spill is coming from and the identity of the spilled material. If a liquid is coming from a group of containers, determine which one is leaking. If a spilled material is detected in a waterway, trace the waterway back to determine where the material is entering the water. If a waterway leads you to an underground pipe, trace its path by examining facility diagrams. Be aware that approaching a liquid to determine its source may require you to wear personal protective equipment so you will not be harmed or overcome by the material.

Next, **stop the spill at its source**. This may require turning a drum in such a way that the hole or rupture is above the surface of the liquid, or by closing a valve. The released liquid should then be confined to further prevent it from entering into the environment. Confining a liquid might involve pouring absorbent over the area of the spill, or placing a boom in a stream or shoveling dirt or gravel into a river bed to stop a liquid from flowing further downstream.

Finally, all spill **residues** must be **collected and packaged** and spill response equipment must be decontaminated. Spill residues include absorbent, soil, and all other materials that came into contact with the released liquid, that will not be decontaminated. The residues should be placed in Department of Transportation specification containers.

Decontamination involves cleaning all equipment, clothing, and structures that have come into contact with the released liquid. The level of decontamination necessary depends on the amount of time an article was in contact with the waste, the affinity of the waste for the article, and the hazards of the waste. Decontamination can range from a soapy water wash down of shovels and rubber boots to a kerosene rinse followed by steam cleaning. Use care to contain all rinsate from decontamination activities, as it must also be disposed of. OSHA regulations published in 29 CFR 1910.120

identify additional requirements for responding to incidents involving the release of hazardous waste and hazardous substances.

- Identify the source of the spill
- Stop the spill at its source
- Collect all spill residues and spill response equipment
- Decontaminate all equipment, clothing and structures contacting the released substance.

Figure 8.2 Response to Spills

### Response to Fires and Explosions

In the event of a fire or explosion, the safest response action is to **warn** others nearby, **get away** from the area as quickly as possible, and **alert** the plant fire brigade and local Fire Department, as specified in your Contingency Plan. If you are ever in a situation in which you must contain a small fire until the Fire Department arrives, be aware that fire extinguishers are rated either **A**, **B**, **C**, or **D** (or any combination of those letters) depending on what type of fire they are best suited for.

### Reportable Spills

In general, all spills of any size within your facility should be reported to your facility Emergency Coordinator. It would then be the responsibility of the Emergency Coordinator to determine if the spill or "release" is reportable to government authorities. To determine if a release is reportable, the Emergency Coordinator needs to determine at least 3 things (not all of which are necessary to trigger a report): 1) was the incident classified as a "release;" 2) the quantity of the release; and 3) was the release to the environment.

> A "release," is defined in 40 CFR 355 as any spilling, leaking, pumping, pouring, emitting, emptying, discharging, injecting, escaping, leaching, dumping, or disposing into the environment of any hazardous chemical, extremely hazardous substance, or CERCLA hazardous substance.
>
> The "environment" is defined in 40 CFR 355 as water, air, and land and the interrelationship that exists among and between water, air, and land and all living things.

### 8.3.4 Federal Emergency Response Reporting

Figure 8.3 lists definitions for terms used in this section of the manual.

**Oil** means oil of any kind or in any form (e.g., petroleum, fuel oil, oil refuse, etc.).

**Hazardous Substance** Any material in an amount equal to its reportable quantity. EPA, DOT, and some states have their own lists of hazardous substances and their reportable quantities.

**Hazardous Waste** A waste that either is listed on EPA's (or a state's) list of hazardous wastes; a waste that displays one or more of EPA's (or a state's) characteristics, a mixture of listed waste with a solid waste; or residue derived from the treatment of a listed waste.

**Hazardous Material** a substance or material including a hazardous substance, which has been determined by the Secretary of Transportation to be capable of posing an unreasonable risk to health, safety, and property when transported in commerce, and which has been so designated

**Excess Air Emissions** means any emissions exceeding any permit limits:

- as a direct result of unavoidable upset conditions in the nature of the process,
- because of unavoidable or unforeseeable breakdown or malfunction of any air pollution control equipment or related operating equipment, or
- as a direct result of shutdown or start-up of such equipment for necessary scheduled maintenance.

You should also refer to your facility's air permit if you suspect that a violation has occurred.

**Wastewater Excursions** means any excursions exceeding any permit limits:

- as a direct result of a bypass or upset of any portion of a treatment facility, or
- as a direct result of any activity that may result in noncompliance with permit requirements.

You should also refer to your facility's wastewater discharge permit. (NPDES/SPDES).

**Underground Tank Leak** means any release or reasonable suspicion of a release of a "regulated substance" (e.g., Hazardous Substances and petroleum).

**Waters of the State** includes all surface and groundwaters located within the boundaries of the state, unless otherwise noted.

**Release** as any spilling, leaking, pumping, pouring, emitting, emptying, discharging, injecting, escaping, leaching, dumping, or disposing into the environment of any hazardous chemical, extremely hazardous substance, or CERCLA hazardous substance.

**Environment** water, air, and land and the interrelationship that exists among and between water, air, and land and all living things.

**Note:** Many states have different reporting quantities regarding Hazardous Substances and Hazardous Wastes. If you are unsure of your state's reporting requirements, contact its department of environmental management.

Figure 8.3 Release Reporting Definitions

**Oil**

Report spills into or upon the "navigable waters" (surface waters) of the United States or adjoining shorelines to:

United States Coast Guard (USCG), Washington, DC
National Response Center (NRC)
(24-hour) (800) 424-8802

**Gas**

Report any release of "gas" (natural gas, flammable gas, or gas that is toxic or corrosive) from a pipeline to:

United States Department of Transportation (USDOT), Washington, DC
Office of Pipeline Safety
NRC
(24-hour) (800) 424-8802

**Hazardous Substance**

Report any release, equal to or exceeding the reportable quantity, in any 24-hour period into the environment to:

U.S.C.G., Washington, DC
NRC
(24-hour) (800) 424-8802

Refer to the lists of hazardous substances and their reportable quantities in 40 CFR 355 (appendices A and B), and 40 CFR 302.4.

**Hazardous Waste**

Report any release, fire, or explosion from a facility that could threaten human health or the environment outside the facility, if evacuation of local areas is advisable, or if the release is greater than or equal to the reportable quantity to local authorities, <u>and</u> to:

U.S.C.G., Washington, DC
NRC
(24-hour) (800) 424-8802

Report any release to the environment from a hazardous waste tank system or secondary containment system to the appropriate EPA Regional Administrator, or his designee, within 24 hours of its detection.

Report any "discharge" (accidental or intentional spilling, leaking pouring, emitting, emptying, or dumping of hazardous wastes into or on any land or water) from an air, rail, highway, or water transport to:

USCG, Washington, DC
NRC
(24-hour) (800) 424-8802

| Region | States and Territories | Phone Number |
|---|---|---|
| Region 1 | CT, ME, MA, NH, RI, VT | (617) 223-7265 |
| Region 2 | NJ, NY, Virgin Islands, Puerto Rico | (201) 321-6635 |
| Region 3 | DE, DC, MD, PA, VA, WV | (215) 597-9898 |
| Region 4 | AL, FL, GA, KY, MS, NC, SC, TN | (404) 347-4062 |
| Region 5 | IL, IN, MI, MN, OH, WI | (312) 353-2318 |
| Region 6 | AR, LA, NM, OK, TX | (214) 655-2222 |
| Region 7 | IA, KS, MO, NB | (913) 236-3778 |
| Region 8 | CO, MT, ND, SD, UT, WY | (303) 293-7142 |
| Region 9 | AZ, CA, HI, NV, Guam, America Samoa, and Trust Territories | (415) 974-8131 |
| Region 10 | AK, ID, OR, WA | (206) 442-1263 |

Figure 8.4 USEPA Regions and 24-Hour Emergency Response Numbers

**Hazardous Material**

Report transportation-related (including loading, unloading, and temporary storage) incidents in which, as a direct result, the release of hazardous materials and/or hazardous wastes occurred, to:

USCG, Washington, DC
NRC
(24-hour) (800) 424-8802

**Excess Air Emissions**

Releases of hazardous air pollutants are subject to the same reporting requirements as hazardous wastes above. Hazardous air pollutant releases covered by NESHAP semiannual and special reporting requirements are exempt from this reporting requirement.

**Wastewater Excursions**

Report any excursion (noncompliance) which may endanger health or the environment, within 24 hours of becoming aware of the circumstances, to your facility's appropriate EPA Regional Administrator.

The following must also be reported within 24 hours:

- Any unanticipated "bypass" (the intentional diversion of waste streams from any portion of a treatment facility) which exceeds any effluent limitation in the NPDES permit;
- Any "upset" (an exceptional incident in which there is unintentional and temporary noncompliance with technology based permit effluent limitations because of factors beyond the reasonable control of the permitee) which exceeds any effluent limitation in the NPDES permit;
- Violation of a maximum daily discharge limitation for any pollutants listed in the NPDES permit to be reported within 24 hours.

**Underground Tank Leaks**

Releases and spills from underground storage tanks are to be reported to the appropriate state and local agencies (you do not need to report to federal EPA).

**SARA Title III**

Report any release of an extremely hazardous substance, which is also a hazardous substance equal to or exceeding its reportable quantity in any 24-hour period into the environment, outside the boundaries of the facility, to:

USCG, Washington, DC
NRC
(24-hour) (800) 424-8802

**PCBs**

Under TSCA, spills of 10 lbs. or more by weight of PCB's (any concentration greater than 50 ppm) must be reported to the appropriate EPA Regional Office. Spill cleanup must begin promptly, no later than 24 hours after the spill is discovered (48 hours for a PCB Transformer).

Use the following calculation to determine the pounds of PCBs present in a liquid:

$$\frac{\text{(Gal of Liquid)} \times \text{(Specific Gravity)} \times \text{(8.3 lbs./gal)} \times \text{(ppm of PCBs)}}{1{,}000{,}000} = \text{lbs. of PCBs}$$

For example, your facility released 100 gallons of mineral oil from a transformer. The concentration of PCBs in the mineral oil was 350 ppm. The Material Safety Data Sheet for the mineral oil indicates that it has a specific gravity of 0.9.

$$\text{Pounds of PCBs released} = \frac{100 \text{ Gal} \times .9 \times 8.3 \text{ lbs./gal} \times 350 \text{ ppm}}{1{,}000{,}000}$$

Pounds of PCBs released = .26

Under CERCLA, the Reportable Quantity (RQ) for PCB's is 1 pound. Spills of PCB-containing materials (mixtures and solutions) must be reported when the total volume spilled times the concentration equals or exceeds the RQ. Reports are made to the NRC (1-800-424-8802). Spill containment and cleanup must be timely and effective.

## 3.3 Personal Protective and Safety Equipment

Each employer should identify the hazards posed by the materials and wastes in the workplace and prescribe the personal protective equipment (PPE) that best guards employees against the hazards associated with the handling of those materials. But, it is also the material handler's responsibility to wear the prescribed PPE.

Hazardous material handlers should always wear appropriate personal protective equipment, since contact with hazardous materials can lead to a variety of health disorders. When there is possible exposure to hazardous waste, the degree of hazard will be considered and the personal protective equipment appropriate to safeguard against the hazard posed by the waste should be prescribed.

Protective clothing provides varying degrees of partial and/or total body protection from chemical exposure, fire, minor blasts, or explosions. Such clothing ranges from gloves, aprons, leggings, and sleeve protectors to fully encapsulating suits. Each type of protective clothing has a specific purpose, as well as limitations. Individual components of clothing and respiratory equipment must be assembled into an ensemble to provide the appropriate level of protection.

Gloves are not only meant to protect the hands but the forearms as well. No one type of glove can offer protection from all hazardous materials and wastes. For example, a butyl glove will withstand chemical attack from acetone—a common cleaning solvent—for days but will decompose in less than an hour when exposed to paint thinner. Eventually, after repeated exposure to hazardous materials and wastes, a glove will degrade to the point that it no longer has the ability to offer protection to your hands and forearms.

Like gloves, no one type of boot can provide unlimited protection. If hazardous waste handlers work around solvents, acids, or bases; in a wet or slippery environment; or in an extremely cold environment and standard operating procedures do not call for special protective boots, they should discuss the situation with their shop supervisors or Safety Director to determine the appropriate equipment needed.

Occupational Safety and Health standards require that workers be provided with suitable eye protection. To be effective, the eyewear must be of the appropriate type for the hazard encountered and must be properly fitted. Eye protection should be reasonably comfortable, fit snugly, and not hinder normal movement.

Over half of the injuries to workers wearing eye protection result from objects or chemicals going around or under the protector. Face masks preclude this consequence by protecting the entire face.

Respirators are designed to remove specific contaminants from the air before you inhale them. Filtration is accomplished by filters, cartridges, or canisters that remove airborne contamination. For each type of respirator, there are a variety of air-purifying filters, cartridges, or canisters designed to protect you against specific contaminants. Keep in mind that some contaminants cannot be removed by air-purifying respirators.

In situations where contaminants cannot be removed by an air-purifying respirator, a self-contained breathing apparatus (SCBA) must be used. SCBAs offer protection against most types and levels of airborne contaminants by providing clean air to a face piece from an air cylinder carried on your back. SCBAs must be used in situations:

- that are immediately dangerous to life and health,
- where there is a high concentration of vapors, or

- where there is poor ventilation or an oxygen deficiency, such as when performing maintenance operations in jet fuel storage tanks.

In general, waste paints and solvents require face and hand protection to guard against the dermatitis that could result from contact with the waste. Waste acids and bases can pose a greater threat with just minimal contact. Therefore, in addition to wearing face and hand protection, eye goggles, a splash apron, and safety boots should be worn to guard against the corrosive effects of these types of wastes.

Personal protective equipment is only effective when used properly. Hazardous waste handlers who are required to wear PPE should practice putting on and removing all ensembles at least annually. Additionally, all waste handlers must be trained in the operation and use of their personal protective equipment, including maintenance, decontamination, emergency procedures, and self-rescue in the event of equipment failure.

## 8.4 Emergency Response Training

Personnel may not be exposed or risk exposure to the hazards involved in containing and cleaning up a spill or release involving hazardous materials unless they have been properly trained. During a spill or other transportation incident, personnel should not attempt to clean up the material involved in the incident unless they have proper training, equipment, and support personnel present. The Occupational Safety and Health Administration has specific training and experience requirements for individuals who respond to releases of hazardous materials. These requirements are indicated in Figure 8.5.

It is important for personnel to understand that they should not attempt to respond aggressively to the incident without comprehensive training and on-site support. They should only respond in a defensive fashion (keep themselves and others out of the area).

| Response Level | Job Function | Training Requirements |
|---|---|---|
| First Responder Awareness Level | Witness incident and initiate emergency response | Understanding of hazardous materials, their risks, and how to secure site and notify others. |
| First Responder Operations Level | Respond defensively to protect nearby persons, property, and the environment. | 8 hours of training including awareness level |
| Hazardous Materials Technician | Respond aggressively to stop releases. | 24 hours including operations level topics. |
| Hazardous Materials Specialist | Support technicians, provide information on specific substances and act as liaison with government authorities. | 24 hours equal to technician plus specialized competency. |
| On-Scene Incident Commander | Assumes control of the incident scene. | 24 hours equal to operations level plus competency in commanding incidents and implementing the emergency response plans. |
| Skilled Support Personnel | Temporarily perform immediate emergency support work (e.g., operate back hoe or crane). | Initial pre-entry briefing |
| Specialist Employees | Provide technical advice or assistance to incident commander. | Annual training and competency in their area of specialization. |

Figure 8.5 Training Requirements for Workers who Respond to Releases of Hazardous Substances

# Chapter 9
# Motor Carrier Responsibilities

## 9.1 General Information and Regulations—49 CFR 177

In addition to the applicable requirements in Parts 171, 172, 173, 178 and 180 (and presented in Chapters I through VIII of this book), motor carriers comply with the regulations in Part 177 applicable to the acceptance and transportation of hazardous materials. No person may accept for transportation or transport any shipment of hazardous material that is not in compliance with the prescribed regulations. More specifically, each motor carrier must—

- Comply with and thoroughly instruct employees on the prescribed regulations
- Provide training prescribed in §177.816 and follow special instructions prescribed in §§177.818 and 177.826 when transporting a flammable cryogenic liquid in a cargo tank on a public highway
- Not transport a hazardous material unless it is accompanied by a shipping paper prepared in accordance with §§172.200, 172.201, 172.203, and 172.203
- Not accept a hazardous material offered for transportation unless the shipping paper describing the material includes a shippers' certification that meets the requirements in §172.204
- Maintain an adequate supply of labels specified in Subpart E of Part 172 to replace those that become lost or destroyed
- Not move a transport vehicle containing a hazardous material unless the vehicle is marked and placarded in accordance with Part 172
- Comply with the general requirements for loading and unloading prescribed in 49 CFR 177 Subpart B, §177.834 through §177.844 and Subpart C, §177.848—Segregation and separation chart of hazardous materials
- Comply with the Federal Motor Carrier Safety Regulations prescribed in 49 CFR Parts 390 through 397
- Report those hazardous material incidents described in §§171.15 and §171.16 as required by those sections
- Comply with reporting and record keeping requirements contained in §180.417

  Make available records, equipment, packaging, and containers for examination and inspection by a duly authorized representative of DOT
- Conform to the re-test and inspection requirements set forth in Subpart E of Part 180
- Not transport certain materials which are forbidden or limited for transportation as prescribed in §177.821. Routing and training requirements for radioactive materials are prescribed in §177.825.

Many of these requirements have been discussed in the preceding chapters. In this chapter, we will cover requirements that apply to motor carriers, or that have special application to them.

**Contents of Part 177—Carriage by Public Highway:** It is important that you be familiar with the general organization and subject matter of Part 177 in order to locate the detailed requirements of the regulations. Notice that certain divisions and subdivisions are selected for special treatment. Often, these specialized sections contain requirements similar to those for other divisions, but with specific precautions, response actions, etc., appropriate to the kinds of hazards presented by the materials. The following table lists the contents of Part 177:

**Subpart A—General Information and Regulations**

| Section | |
|---|---|
| 177.800 | Purpose and scope. |
| 177.801 | Unacceptable hazardous materials shipments. |
| 177.802 | Inspection. |
| 177.803 | Export and import shipments by domestic carriers by motor vehicles. |
| 177.804 | Compliance with Federal Motor Carrier Safety Regulations. |
| 177.805 | Canadian shipments and packagings. |
| 177.806 | U.S. Government material. |
| 177.807 | Reporting hazardous materials incidents. |
| 177.808 | Connecting carrier shipments. |
| 177.809 | Carrier's material and supplies. |
| 177.810 | Vehicular tunnels. |
| 177.811 | Astray shipments. |
| 177.812 | Containers required. |
| 177.813 | Inefficient containers. |
| 177.814 | Retention of cargo tank motor vehicle manufacturer's certificate, maintenance and other reports. |
| 177.815 | Lost or destroyed labels. |
| 177.816 | Training. |
| 177.817 | Shipping papers. |
| 177.818 | Special instructions; Division 2.1 (flammable gas) materials that are cryogenic liquids. |
| 177.821 | Hazardous materials forbidden or limited for transportation. |
| 177.822 | Acceptable articles. |
| 177.823 | Marking and placarding motor vehicles. |
| 177.824 | Retesting and inspection of cargo tanks. |
| 177.825 | Routing and training requirements for Class 7 (radioactive) materials. |
| 177.826 | Carrier's registration statement; Division 2.1 (flammable gas) materials that are cryogenic liquids. |

**Subpart B—Loading and unloading**

| | |
|---|---|
| 177.834 | General requirements. |
| 177.835 | Class 1 (explosive) materials. |
| 177.836 | Nonexplosive material. |
| 177.837 | Class 3 (flammable liquid) materials. |
| 177.838 | Class 4 (flammable solid) materials, Class 5 (oxidizing) materials, and Division 4.2 (pyroforic* liquid) materials. |
| 177.839 | Class 8 (corrosive) materials. |
| 177.840 | Class 2 (gases) materials. |
| 177.841 | Division 6.1 (poisonous) and Division 2.3 (poisonous gas) materials. |
| 177.842 | Class 7 (radioactive) material. |
| 177.843 | Contamination of vehicles. |
| 177.844 | Class 9 (miscellaneous hazardous) materials. |

**Subpart C—Segregation and Separation Chart of Hazardous Materials**

| | |
|---|---|
| 177.848 | Segregation of hazardous materials. |

**Subpart D—Vehicles and Shipments in Transit; Accidents**

| | |
|---|---|
| 177.853 | Transportation and delivery of shipments. |
| 177.854 | Disabled vehicles and broken or leaking packages; repairs. |
| 177.855 | Accidents; Class 1 (explosive) materials. |
| 177.856 | Accidents; Class 3 (flammable liquid) materials. |
| 177.857 | Accidents; Class 4 (flammable solid) and Class 5 (oxidizing) materials. |
| 177.858 | Accidents; Class 8 (corrosive) materials. |
| 177.859 | Accidents; Class 2 (gases) materials. |
| 177.860 | Accidents or leakage; Division 6.1 (poisonous) or Division 2.3 (poisonous gas) materials. |
| 177.861 | Accidents; Class 7 (radioactive) materials. |

**Subpart E—Regulations Applying to Hazardous Material on Motor Vehicles Carrying Passengers for Hire**

| | |
|---|---|
| 177.870 | Regulations for passenger-carrying vehicles. |

* In this portion of the regulations, DOT uses the spelling *pyroforic*. The word is usually spelled *pyrophoric*.

## 9.2 Carrier Training Requirements

**Responsibility for training**: A carrier may not transport hazardous materials by motor vehicle unless each of its hazmat employees involved in that transportation is trained as required by §172.816 and subpart H of part 172 (addressed in Chapter II of this manual).

**Specific Training**: In addition to the training requirements applicable to all hazmat employees, no carrier may transport, or cause to be transported, a hazardous material unless each hazmat employee who will operate a motor vehicle has been trained in the applicable requirements of 49 CFR Parts 390 through 397 (Motor Carrier Safety Regulations) and the procedures necessary for the safe operation of that motor vehicle. Driver training must include all of the following subjects:

- Pre-trip safety inspection
- Use of vehicle controls and equipment, including operation of emergency equipment
- Operation of vehicle, including—
  - turning
  - backing
  - braking
  - parking
  - handling
- vehicle characteristics including those that affect vehicle stability, such as—
  - effects of braking and curves
  - effects of speed on vehicle control
  - dangers associated with maneuvering through curves
  - dangers associated with weather or road conditions that a driver may experience (e.g., blizzards, mountainous terrain, high winds)
  - high center of gravity
- Procedures for maneuvering tunnels, bridges, and railroad crossings
- Requirements pertaining to attendance of vehicles, parking, smoking, routing, and incident reporting
- Loading and unloading of materials, including—
  - Compatibility and segregation of cargo in a mixed load
  - Package handling methods
  - Load securement

**Specialized requirements for cargo tanks and portable tanks**: In addition to the specific training requirements outlined above, each person who operates a cargo tank or a vehicle with a portable tank with a capacity of 1,000 gallons or more must receive training applicable to the requirements of the Hazardous Materials Regulations and have the appropriate state-issued commercial driver's license (CDL) required by 49 CFR part 383. Specialized training must include all of the following:

- Operation of emergency control features of the cargo tank or portable tank
- Special vehicle handling characteristics, including—
  - high center of gravity
  - fluid load subject to surge
  - effects of fluid-load surge on braking
  - characteristic differences in stability among baffled, unbaffled, and multi-compartmented tanks
  - effects of partial loads on vehicle stability
- Loading and unloading procedures

- The properties and hazards of the material(s) transported
- Re-test and inspection requirements for cargo tanks

These specialized training requirements may be satisfied by complying with the current requirements for a state CDL with a tank vehicle or hazardous materials endorsement.

Specialized training related to cargo tanks and portable tanks must be conducted at least once **every two years**. Each hazmat employer must create and retain a record of current training for each hazmat employee. The record must include at least the **last two years**. It must be retained for as long as that employee is employed by that employer as a hazmat employee and for **90 days** after termination of employment as a hazmat employee.

### 9.3 Shipping Papers

A carrier may not transport a hazardous material unless it is accompanied by a **shipping paper** (e.g., a bill of lading, waybill, or manifest) prepared in accordance with §§172.200, 172.201, 172.203, and 172.203 (see Chapter VI of this manual). The carrier may not accept a hazardous material offered for transportation unless the shipping paper describing the material includes a **shippers' certification** which meets the requirements in §172.204. The carrier must ensure that the shipping paper is readily available to and recognizable by authorities in the event of an accident or inspection. Specifically, the driver and the carrier must—

- Clearly distinguish the shipping paper by either distinctively tabbing it or by having it appear first among any other papers carried
- Store the shipping paper within the driver's immediate reach and either readily visible to a person entering the driver's compartment or in a holder mounted to the inside of the door
- If the driver is not at the vehicle's controls then the shipping papers must be in a compartment or holder mounted to the inside of the door, or placed in the driver's seat.

The shipping papers are an important source of emergency response information for personnel responding to a possible incident or accident. Adequate emergency response information must accompany each shipment, and can be contained entirely in the shipping papers, or in one or more separate documents in conjunction with the shipping papers. The minimum information required is—

- Description of the hazardous material (check that shipping paper entries and package markings agree)
- Immediate hazards to health
- Risks of fire or explosion
- Immediate precautions to be taken in the event of an incident or accident
- Immediate methods for handling large or small fires
- Initial methods for handling spills or leaks in the absence of fire
- Preliminary first aid measures

Much of this information can be supplied by a Material Safety Data Sheet (MSDS) for the material(s) carried, if appropriate MSDS(s) are available. Another reference frequently used is the DOT *Emergency Response Guidebook* (DOT P 5800). The latest edition is for 1993 (5800.5). Some shippers will indicate on shipping papers the *Emergency Response Guidebook* Guide Number in the format "93 ERG 27." However, the *Guidebook* itself advises emergency responders to first identify the material(s) from the UN or NA number on the placard or shipping paper, or from the shipping name; then select the appropriate Guide number from the number or name index.

### 9.4 Marking, Labeling, and Placarding Requirements

Package labels are supplied by the shipper and should be applied to the package(s) before the load is picked up. However, if labels are missing or destroyed, or if placards are damaged or lost, the carrier is responsible for replacing them. Therefore, the carrier should have a supply of spare labels and placards appropriate to the load.

A carrier may not move a transport vehicle containing a hazardous material unless the vehicle is marked and placarded in accordance with Part 172 or as authorized in §171.12(a). Motor carriers frequently affix and display their own placards for shipments of hazardous materials. However, if the carrier does not have the appropriate placards for a shipment, the placards must be provided by the shipper and displayed by the carrier. Placards are not required in an emergency and one or more of the following conditions applies:

- The vehicle is escorted by a representative of a state or local government
- The carrier has permission from DOT
- Movement of the vehicle is necessary to protect life or property

### 9.5 Loading and Unloading Requirements

Chapter VII covered requirements for loading and unloading hazardous materials onto vehicles that apply to both shippers and transporters. Among these, the following responsibilities are borne by the carrier. If the shipper were performing any of these tasks, he or she would also be required to comply.

- *Packages secured in a vehicle*—tanks, barrels, drums, cylinders, or other packaging not permanently attached, containing Class 3 (flammable liquid), Class 2 (gases). Class 8 (corrosive), Division 6.1 (poisonous), or Class 7 (radioactive) material must be secured against movement within the vehicle during transportation.
- *No hazardous materials on pole trailers.*
- *No smoking while loading or unloading*—for Class 1 (explosive), Class 3 (flammable liquid), Class 4 (flammable solid), Class 5 (oxidizing), or Division 2.1 (flammable gas) materials.
- *Keep fire away during loading and unloading*—for Class 1 (explosive), Class 3 (flammable liquid), Class 4 (flammable solid), Class 5 (oxidizing), or Division 2.1 (flammable gas) materials.

- *Hand brake set while loading and unloading*—and all other reasonable precautions taken to prevent motion of the motor vehicle during loading or unloading.
- *Use of tools in loading and unloading*—no tools likely to damage the effectiveness of the closure of any package or other container, or likely to have an adverse effect on the package or container may be used for loading or unloading of Class 1 (explosive) materials or other dangerous articles.
- *Prevent relative motion between containers*—containers of Class 1 (explosive), Class 3 (flammable liquid), Class 4 (flammable solid), Class 5 (oxidizing), Class 8 (corrosive), Glass 2 (gases), and Division 6.1 (poisonous) materials must be braced to prevent motion relative to the vehicle in transit. Containers with valves or other fittings must be loaded to minimize the likelihood of damage to them during transportation.
- *Prohibited loading combinations*—in any single driven motor vehicle or in any single unit of a combination of motor vehicles, hazardous materials must not be loaded together if prohibited by the loading and storage chart at § 177.848 (section 7.2 in Chapter VII). This should not be interpreted to forbid the carrying of materials essential to safe operation of motor vehicles under the Motor Carrier Safety Regulations (49 CFR 393).

There are further loading and unloading rules that apply more specifically to the transporter:

- *Precautions concerning containers in transit and fueling road units*—the carrier must take reasonable care to prevent excessive rise in temperature of containers and their contents during transit. There must be no tampering with containers or their contents, and no discharge of any contents between the points of origin and billed destination. No container, except a cargo tank, may be discharged before being removed from the vehicle. However, this rule does not prohibit fueling machinery or vehicles used in road construction or maintenance.
- *Attendance requirements.*
  - During loading, a cargo tank must be attended by a qualified person at all times. The person who is responsible for loading the cargo tank is also responsible for ensuring that it is so attended.
  - During unloading, a motor carrier who transports hazardous materials by cargo tank must also ensure that the tank is attended by a qualified person at all times. However, the carrier's obligation to ensure attendance during unloading ceases when the following three conditions are all fulfilled:
    - The carrier's obligation for transporting the materials is completed
    - The cargo tank has been placed on the consignee's premises
    - The motive power has been removed from the cargo tank and removed from the premises.
  - A person "attends" the loading or unloading of a cargo tank when, throughout the process, is awake, has an unobstructed view of the cargo tank, and is within 7.62 meters (25 feet) of the tank.
  - A person is "qualified" when he/she has been made aware of the nature of the hazardous material which is to be loaded or unloaded, has been

instructed on the procedures to be followed in emergencies, is authorized to move the cargo tank, and has the means to do so.
  - A delivery hose, when attached to the cargo tank, is considered a part of the vehicle.

- *Cargo Tanks*—tanks constructed and maintained in compliance with Spec. 106A or 110A (49 CFR 179.300 and 179.301) that are authorized for the shipment of hazardous materials by highway in Part 173 must be carried in accordance with the following requirements:
  - Tanks must be securely chocked or clamped on vehicles to prevent any shifting
  - Equipment suitable for handling a tank must be provided at any point where a tank is to be loaded upon or removed from a vehicle
  - No more than two cargo-carrying vehicles may be in the same combination of vehicles
  - Combination rail freight, highway shipments, and trailer-on-flat-car service must comply with §§174.200 and 174.204, dealing with gases in tank cars and rail cars with fumigated or treated loads
- Specification 56, 57, IM (intermodal) 101, and IM 102 portable tanks, when loaded, may not be stacked on each other, nor placed under other freight during transportation by motor vehicle.

### 9.5 Requirements During Transportation

While a shipment of hazardous material is actually on the road, the provisions of both 49 CFR Part 177—Carriage by Public Highway, and Parts 390–397—Motor Carrier Safety Regulations, apply. In summary, these rules require—

- *Use of cargo heaters when transporting certain hazardous materials* (transportation includes loading, carrying, and unloading)—there are detailed requirements and specifications in 49 CFR 177.834(l) governing whether, when, and what kinds of cargo heaters may be used with Class 1 (explosive) materials (their use is prohibited, and they must be rendered inoperative), Class 3 (flammable liquid) or Division 2.1 (flammable gas) materials.
- *Parking*—a vehicle carrying hazardous materials in general must not park—
  - On or within five feet of the traveled portion of a public street or highway, except for brief periods when necessities of operation require parking and make it impracticable to park in any other place.
- *Parking with explosives*—a vehicle carrying Class A or B explosives must not park—
  - On or within five feet of the traveled portion of a public street or highway
  - On private property (including fueling or eating facilities) without the knowledge and consent of the person in charge, who knows the nature of the hazardous materials on board
  - Within 300 feet of a bridge, tunnel, dwelling, building, or place where people work, congregate, or assemble, except for brief periods when

necessities of operation require parking and make it impracticable to park in any other place.

- *Instructions and documents for explosives*—there are special requirements for motor carriers that transport Class A or B explosive materials:
  - The carrier must supply the driver with a copy of the rules in 49 CFR 397 and a document containing instructions to be followed in case of accident or delay, to include—
    - telephone numbers of persons to contact (including carrier and shipper representatives)
    - nature of the explosives being transported
    - precautions to take in emergencies such as fires, accidents, or leakages
  - The driver must sign a receipt for the document, and the carrier must retain the receipt on file for one year
  - The driver must possess and be familiar with these documents, the shipping papers, and the written route plan.
- *Routing.*
  - Unless there is no practicable alternative, carriers of hazardous materials must operate over routes that do not go through or near heavily populated areas, places where crowds are assembled, tunnels, narrow streets, or alleys ("practicable" does *not* take into account operating convenience)
  - Carriers of Class A or B explosives must have a written route plan
  - Radioactive materials are subject to much stricter routing controls, given at §177.825.
- *Fires*—drivers must not operate a vehicle carrying hazardous materials near an open fire unless they first take precautions to make sure the vehicle can safely pass the fire without stopping, and must not park within 300 feet of an open fire.
- *Smoking*—smoking or carrying a lighted cigarette, cigar, or pipe is prohibited on or within 25 feet of—
  - A vehicle containing explosives, oxidizing materials, or flammable materials
  - An empty vehicle that has been used to transport flammable liquids or gases, and that during that use required marking or placarding under §177.823.
- *Fueling*—when a motor vehicle containing hazardous materials is being fueled—
  - The engine must not be operating
  - A person must be in control of the fueling process at the point where the fuel tank is filled.
- *Tires*—for vehicles with dual tires on any axle, the driver must stop in a safe location (observing the parking restrictions noted above) at least once every two hours or 100 miles of travel, whichever is less, and examine the tires for integrity, inflation, and overheating:

- If a tire is found to be flat, leaking, or improperly inflated, the driver must get the tire repaired, replaced, or properly inflated before driving further, except that the vehicle may be driven to the nearest safe place before the tire is worked on
- If a tire is overheated, the driver must immediately have the tire removed and placed at a safe distance from the vehicle, and must not operate the vehicle until the cause of overheating is corrected.

### 9.6 Reporting Hazardous Materials Incidents

Report those hazardous material incidents described in §§171.15 and §171.16. Reporting requirements for these incidents are presented in Chapter VIII of this manual, section 8.2.

### 9.7 Reporting and Record Keeping—Vehicle Certification Records

Comply with reporting and record keeping requirements contained in §180.417, as outlined below:

- **Owners** of a specification cargo tank must retain the manufacturer's data report or certificate and related papers certifying that the specification cargo tank identified in the documents was manufactured and tested in accordance with the applicable specification. These records must be retained throughout ownership of the specification cargo tank and for one year thereafter. In the event of a change of ownership, the prior owner must retain non-fading photocopies of these documents for at least one year.
- Motor carriers who **use** a specification cargo tank motor vehicle must obtain a copy of the manufacturer's certificate and related papers or an authorized alternative report, and retain the documents. A motor carrier who is not the owner of a cargo tank motor vehicle must retain a copy of the vehicle certification report at its principal place of business (or a DOT-authorized alternative location) for as long as the cargo tank motor vehicle is used by that carrier and for one year afterward. These provisions do not apply to a motor carrier leasing a cargo tank motor vehicle for less than 30 days.
- If you operate DOT Specification cargo tanks that do not have manufacturer's certificates and/or data reports, there are special provisions at §180.417(a)(3) for obtaining or creating certifications.
- Test or Inspection Reporting—each cargo tank that is tested or reinspected as specified in §180.407 must have a written report, in English, that includes the following:
  - Type of test or inspection performed and a listing of all items either tested or inspected (a checklist is acceptable)
  - Owner's and manufacturer's serial numbers
  - DOT Specifications
  - Test date (month and year)
  - Location of defects found and method used to repair each defect
  - Name and address of person performing the test, and the DOT registration number of the facility or the person performing the test

- Disposition statement, such as "Cargo tank returned to service" or "Cargo tank withdrawn from service"
- DOT registration number of the inspector and dated signatures of the inspector and owner.

• Test record retention—the owner and the motor carrier, if not the owner, must each retain a copy of the test and inspection reports until the next test or inspection of the same type is successfully completed. This requirement does not apply to a motor carrier leasing a cargo tank for less than 30 days.

• There are additional requirements at §180.417(c) for Specification MC 330 or MC 331 cargo tank in anhydrous ammonia, liquefied petroleum gas, or any other service that may cause stress corrosion cracking.

### 9.8 Re-test and Inspection Requirements

Cargo tank motor vehicles must conform to the reset and inspection requirements presented in 49 CFR 180 Subpart E—Qualification and Maintenance of Cargo Tanks. These regulations deal with the following subjects:

| Section | |
|---|---|
| 180.401 | Applicability |
| 180.403 | Definitions |
| 180.405 | Qualification of cargo tanks |
| 180.407 | Requirements for test and inspection of cargo tanks |
| 180.409 | Minimum qualifications for inspectors and testers |
| 180.411 | Acceptable results of tests and inspections |
| 180.413 | Repair, modification, stretching, or rebarrelling of cargo tanks |
| 180.415 | Test and inspection markings |
| 180.417 | Reporting and record retention requirements |

### 9.9 Forbidden or Limited Transportation

Carriers transporting a cryogenic liquid that is also a flammable gas in a cargo tank on a public highway must receive specialized training prescribed in §177.816 (see section 9.2 above).

If the material is in a package exceeding 450 liters (119 gallons) water capacity, these carriers must also carry written instructions containing the following information with the required shipping papers:

• General precautions
• Manual venting instructions
• Emergency procedures
• The names and telephone numbers of persons to be contacted in case of emergency or accident.

Further, carriers of cryogenic liquids that are also flammable gases must file a registration statement by certified mail, return receipt requested, with the Associate Administrator for Hazardous Materials Safety, RSPA.

• The registration statement must contain the following information:
  - The carrier's name and principal place of business

- Locations where cargo tanks used to transport Division 2.1 (flammable gas) materials that are cryogenic liquids are domiciled
- The serial number or vehicle identification number of each cargo tank used by the carrier to transport Division 2.1 materials that are cryogenic liquids, and the name of each Division 2.1 material that is a cryogenic liquid transported in each cargo tank.

- The registration statement must be filed—
  - Initially between 1 January and 28 February 1985 (this initial statement was only required to contain information regarding operations that took place during the 90 days prior to the date of the statement)
  - Subsequently, between 1 January and 28 February of each odd numbered year after 1985
  - For equipment obtained or operations begun between the specified two-year filing intervals, the information must be provided on the registration statement filed during the next required filing period.

Certain materials are forbidden or limited for transportation:

- Liquid nitroglycerin, desensitized liquid nitroglycerin or diethylene glycol dinitrate, except as defined in §173.54(e), may not be accepted for transportation or be transported by any common carrier by motor vehicle.
- Any individual container used for the transportation of liquid nitroglycerin, desensitized liquid nitroglycerin or diethylene glycol dinitrate, or any boot used with any of them, that is found in a condition as that would permit leakage, must be discarded and may not afterward be repaired for further use.
- No motor carrier may accept for transportation or transport any hazardous material listed in the Hazardous Materials Table as "forbidden" for transportation by all motor carriers. If there is any prohibition or limitation in regard to the transportation of any particular hazardous material by a particular kind of motor carrier, the material may not be accepted for transportation or be transported by that kind of carrier, unless there is a specific exception in 49 CFR 177 (see §177.870).
- Leaking or damaged packages of Division 1.1 or 1.2 (high explosive) material must not be accepted for transportation or transported. If a package of Division 1.1 or 1.2 (high explosive) material offered for shipment shows excessive dampness, is moldy, shows outward signs of any oily stain, or shows any other indication that absorption of the liquid part of the explosive material is not perfect, or that the amount of the liquid part is greater than the absorbent can carry, the package must always be refused. The shipper must substantiate any claim that a stain is due to contact with material other than the liquid explosive ingredient of the explosive material. In case of doubt the package must be rejected.
- Smokeless powder for small arms in quantities not greater than 45 kg (99 lb) net weight transported in one car or motor vehicle may be classed as a Class 4 (flammable solid) material if the Bureau of Explosives examines it for this classification and the Associate Administrator for Hazardous Materials Safety approves. The maximum quantity in any inside packaging must be no more than 3.6 kg (7.9 pounds) and inside packagings

must be arranged and protected to prevent simultaneous ignition of the contents. The complete package must be a type examined by the Bureau of Explosives and approved by the Director, OHMT. Each outside package must bear a Flammable Solid (4) label.

Special routing and training requirements for radioactive materials are prescribed at §177.825.

# Chapter 10
# Training Exercises and Examinations

## Introduction

As explained in Chapter 2, every hazmat employee must receive both initial and recurrent training that includes—

- General awareness/familiarization training
- Function-specific training
- Safety training
- Testing

To meet these requirements, Environmental Resource Center usually presents the material addressed in this manual in an eight-hour course divided into nine units, or modules:

| No. | Module | Chapter(s) Covered |
|---|---|---|
| 1 | Overview of DOT Hazardous Materials Requirements | 1, 2, 3 (3.1) |
| 2 | How to Use the Hazardous Materials Table | 3 (3.2, 3.3) |
| 3 | Packaging | 4 |
| 4 | Marking and Labeling | 5 |
| 5 | Shipping Papers | 6 |
| 6 | Placarding | 7 (7.4) |
| 7 | Loading, Unloading, and Stowing | 7 (7.1–7.3) |
| 8 | Safe Handling of Hazardous Materials | 7 (general) |
| 9 | Emergency Response | 8 |

On the following pages are nine exercises to build student proficiency and confidence, one for each of the modules listed. These are designed for use during the presentation of the modules. Following each exercise is a multiple-choice test covering the subject matter of the module. At the end of this chapter is a blank answer sheet that may be copied for administration of the test questions, and a keyed answer sheet, with the correct answers filled in.

### 10.1.1 Introduction—Exercise

**What are some tasks that you do that are covered by the DOT Hazardous Materials regulations?**

---

**Draw a line between the material and its definition:**

**Hazardous waste** — **Material with a physical or health hazard to employees. Regulated by OSHA**

**Hazardous chemical** — **Material capable of posing a risk to health and safety when transported. Regulated by DOT.**

**Hazardous material** — **Material that is either "characteristic" or "listed" and its disposal is regulated by EPA.**

**List the name of each DOT hazard class as it is introduced in class:**

| Hazard Class | Name | Hazard Class | Name |
|---|---|---|---|
| 1 | | 6 | |
| 2 | | 7 | |
| 3 | | 8 | |
| 4 | | 9 | |
| 5 | | ORM–D | |

List the names of hazardous materials that you work with:

______________________________________________

______________________________________________

How frequently must you receive hazardous materials training?

______________________________________________

What are the penalties for not complying with the DOT hazardous materials regulations?

______________________________________________

## 10.1.2 Introduction—Test

### Module 1—General Awareness and Familiarization

1. Hazardous materials are defined by the Department of Transportation as:
   a. Any chemical present in the workplace which is a physical hazard or a health hazard
   b. Ignitable, corrosive, or reactive wastes
   c. Materials in an amount greater than or equal to their reportable quantities
   d. Materials capable of posing a risk to health and safety when transported

2. Department of Transportation requires training for:
   a. Operators of any motor vehicle engaged in the transportation of hazardous materials
   b. Personnel who pack, label, or prepare hazardous materials for shipment, as well as personnel who prepare shipping documents, load or unload, or offer hazardous materials for transportation or transport hazardous materials
   c. Employees which respond to spills during transportation
   d. All of the above

3. A material which has a mass explosion hazard that would affect the entire truckload when shipped, is classified by DOT as a:
   a. Division 1.1 Explosive
   b. Division 1.6 Explosive
   c. Pyrophoric Material
   d. Class 3 Flammable Liquid

4. Gasoline has a flash point of -50°F, and has the DOT hazard class:
   a. Combustible liquid
   b. Class 3 flammable liquids
   c. Class 4 flammable solids
   d. Class 5 oxidizers

5. Sodium metal will burn if it gets wet. It is classified by DOT as:
   a. Division 4.1 wetted explosive, self-reactive, readily combustible through friction
   b. Division 4.2 spontaneously combustible
   c. Division 4.3 dangerous when wet
   d. Division 4.5 organometallic

**Module 1—General Awareness and Familiarization (continued)**

6. Caustic soda (sodium hydroxide) can severely damage skin at the site of contact. It is classified by DOT as:

   a. Division 5.2 organic peroxide
   b. Division 5.1 oxidizer
   c. Division 6.1 toxic to humans
   d. Class 8 corrosive

7. Some materials are regulated by the DOT as hazardous when shipped but do not fit into classes 1 through 8. These materials are classified as:

   a. ORM-A irritants
   b. Dangerous goods
   c. Class 9 miscellaneous hazardous materials
   d. Non-hazardous materials

8. Hazmat employees include persons that load, unload, transport, or prepare hazardous materials for shipment. DOT requires Hazmat employees to receive

   a. Rapid promotions
   b. The emergency response guidebook
   c. Training that applies to their jobs
   d. Registration with DOT

9. Which of the following is the minimum penalty for noncompliance?

   a. $25 per day per violation
   b. $250 per day per violation
   c. $2,500 per day per violation
   d. $25,000 per day per violation

### 10.2.1 Hazardous Materials Table—Exercise

**What are some uses for the hazardous materials table?**

---

**Identify what is listed on each column of the hazardous materials table:**

| Column | Use | Column | Use |
|---|---|---|---|
| 1 | | 6 | |
| 2 | | 7 | |
| 3 | | 8 | |
| 4 | | 9 | |
| 5 | | 10 | |

**Draw a line between the Column 1 abbreviation and its meaning.**

| | |
|---|---|
| A | No changes: Hazard class in col. 3 must be used for shipping name in col. 2. |
| W | Air shipments, hazardous wastes, hazardous substances |
| + | Water shipments, hazardous wastes, hazardous substances |
| D | International shipments |
| I | Domestic shipments |

**Circle the most dangerous Packing Group:** PG I PG II PG III

Use the first page of the Hazardous Materials Table and determine the following for **Acetone**:

Hazard Class:

Packing Group

Label:

Special Provisions:

Non-bulk Packaging Regulation: 49 CFR 173.____

The "Basic Description" includes the proper shipping name, hazard class (or division), identification number, and packing group. From the Hazardous Materials Table, determine the Basic Description for **Methanol**.

___

| Material | Density (lb/gal) | RQ (lb) | Amount shipped in one container |
|---|---|---|---|
| Methyl ethyl ketone | 10 | 5000 | 55 gal |
| Potassium cyanide | 12.7 | 10 | 35 gal |

Which of the above shipments is a *hazardous substance*?

## Exercise—Determining the Proper DOT Shipping Name

| Waste Description | 3 55-gallon drums of waste acid bath—weighs 10 lbs./gallon—contains 50% liquid arsenic acid and 50% chromosulfuric acid | |
|---|---|---|
| Is the waste a mixture? Are the concentrations known? | liquid arsenic acid | |
| | chromosulfuric acid | |
| | | |
| Is the mixture listed? | Arsenic acid and chromosulfuric acid mixture? | |
| Is "generic name" listed? | Is "acid bath" listed? | |
| Is one listed but not the other? | Is liquid arsenic acid listed?<br>Is chromosulfuric acid listed? | |
| Is more than one listed? | Is liquid arsenic acid listed?<br>Is chromosulfuric acid listed? | |
| Is there more than one hazard? | Hazard & PG (liquid arsenic acid)? | |
| | Hazard & PG (chromosulfuric acid)? | |
| | | |
| Are the hazards listed together? | "Corrosive liquids, poisonous" or "Poisonous liquids, corrosive" Reference Table 6.3 | |
| Which hazard has precedence? | "*Corrosive* liquids, poisonous" or "*Poisonous* liquids, corrosive" Reference Tables 6.1 and 6.2 | |
| *Proper Shipping name* | Waste __________ __________, __________, n.o.s. | |

| Hazardous Material | Concentrations? | Quantity? | Reportable Quantity? |
|---|---|---|---|
| arsenic acid? | | | |
| chromosulfuric acid? | | | |
| D002 ? (HMT App. A, Table 1) | | | |
| | | | |

| Is the mixture a Hazardous Substance? | See the definition of a Hazardous Substance in Chapter 6 | |
|---|---|---|

Add the hazard class, the UN four digit number, the packing group and add in parentheses the technical names

___ Waste __________ __________, __________ n.o.s., ___, UN ______, *PG* ___
(*Contains* ____________________, ____________________)

## Worksheet—Determining the Proper DOT Shipping Name

| | | |
|---|---|---|
| Waste Description | | |
| Is the waste a mixture? Are the concentrations known? | | |
| | | |
| | | |
| Is the mixture listed? | | |
| Is "generic name" listed? | | |
| Is one listed but not the other? | | |
| Is more than one listed? | | |
| Is there more than one hazard? | | |
| | | |
| | | |
| Are the hazards listed together? | Reference Table 6.3 | |
| Which hazard has precedence? | Reference Tables 6.1 and 6.2 | |
| *Proper Shipping name* | Waste ________ ________, ________, n.o.s. | |

| Hazardous Material | Concentrations? | Quantity? | Reportable Quantity? |
|---|---|---|---|
| | | | |
| | | | |
| | | | |
| | | | |
| Is the mixture a Hazardous Substance? | See the definition of a Hazardous Substance in Chapter 6 | | |

Add the hazard class, the UN four digit number, the packing group and add in parentheses the technical names

____ Waste ________ ________, ________ n.o.s., ___, UN ______, *PG* ___
(*Contains* ________________, ________________)

## 10.2.2 Hazardous Materials Table—Test
### Module 2—Hazardous Materials Table

1. Which of the following are listed on the hazardous materials table:
   a. Shipping names, hazard classes, labels, and identification numbers
   b. Placards, and packaging codes
   c. Compatibility between different hazard classes
   d. LD50, LC50, and other hazard determination data

2. When you find an *A* in column 1 of the hazardous materials table, you know that:
   a. The material is a Class A explosive
   b. The material is classified as a Poison A
   c. The listed basic description must be used for shipments by air or for shipments of hazardous substances and hazardous wastes
   d. The listed basic description must be used for shipments by water or for shipments of hazardous substances and hazardous wastes

3. When column 3 of the table indicates that the hazard class is "Forbidden"
   a. It is illegal to transport, purchase or possess this material
   b. Vehicle placards and container labels are forbidden for this material
   c. The material cannot be transported unless diluted or stabilized
   d. It is forbidden to manufacture this material without an exception from DOT

4. Packing groups identify the degree of danger. A material in packing group III is:
   a. More dangerous than a material in packing group II
   b. Less dangerous than a material in packing group II
   c. Classified as an Explosive, Radioactive, or Etiologic Agent
   d. Not classified as a hazardous material

5. If a material has more than one packing group shown on the hazardous materials table, you should:
   a. Select either of the packing groups
   b. Select the appropriate packing group, based on the properties of the material
   c. Add the packing groups to determine the hazard
   d. Pack the material in more than one container.

## Module 2—Hazardous Materials Table (continued)

6. Column 8B of the hazardous materials table has the requirements for non-bulk packaging. In column 8B, you can find the regulatory references for:
   a. Packaging standards for roll off boxes, tub skids, and similar large containers
   b. Stowing hazardous materials on vessels
   c. Packaging standards for drums
   d. Packaging standards for exceptions

7. Acetic acid has a reportable quantity of 5000 pounds. If a tank truck contained 6000 pounds of acetic acid, the truck contains
   a. A reportable quantity of a hazardous substance
   b. A reportable quantity of a marine pollutant
   c. A reportable quantity of a hazardous chemical
   d. All of the above

### 10.3.1 Packaging—Exercise

To identify a container that meets DOT standards for a hazardous material, you should follow these steps:

1. Find the material to be shipped in column ____ of the HMT.
2. Find the corresponding regulatory reference in column ____ of the table.
3. Read the regulation that you were referred to by column 8 above and then find a container that meets all or any one of the authorized packagings.
4. Make sure that the material that you put in the container is (compatible) or (incompatible) [circle best answer] with the container.

On the first page of the Hazardous Materials Table, find the regulatory reference for bulk containers used to hold acetone.

Identify when the "new" HM-181 containers must be used:

| Type of Material | New Container Compliance Date |
|---|---|
| All Infectious Substances | 1-1-9__ |
| All PIHs | 10-1-9__ |
| All other | 10-1-9__ |

The codes **X**, **Y**, and **Z** identify the packing groups that the container is authorized to hold. Insert on the table the packing groups that containers meeting these standards can hold.

| Performance Standard | Meets Packing Groups: |
|---|---|
| X | |
| Y | |
| Z | |

Write in the meaning of each of the codes on the drum mark below:

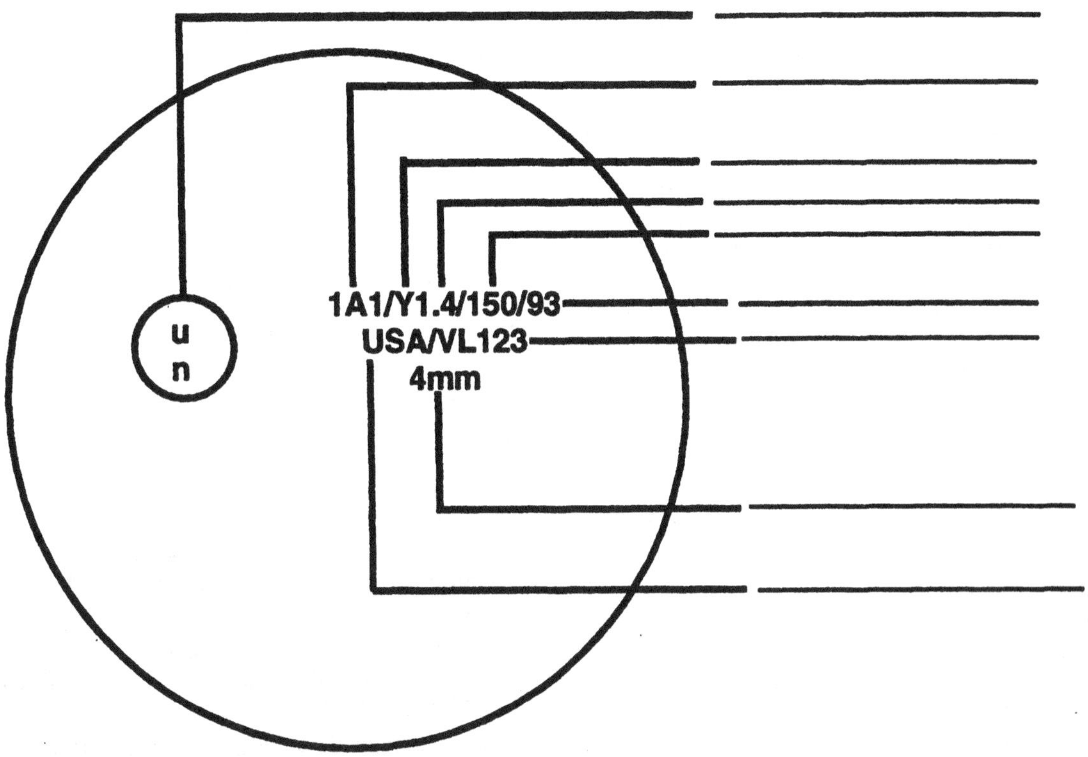


From the following information, select an appropriate non-bulk packaging for a highway shipment of acetone and another highway shipment of solid sodium hydroxide:

| Symbols | Hazardous materials descriptions and proper shipping names | Hazard class or Division | Identification Numbers | Packing group | Label(s) required (if not excepted) | Special provisions | (8) Packaging authorizations (§ 173.***) | | | (9) Quantity limitations | | (10) Vessel stowage requirements | |
|---|---|---|---|---|---|---|---|---|---|---|---|---|---|
| | | | | | | | Exceptions | Non-bulk packaging | Bulk packaging | Passenger aircraft or railcar | Cargo aircraft only | Vessel stowages | Other stowage provisions |
| (1) | (2) | (3) | (4) | (5) | (6) | (7) | (8A) | (8B) | (8C) | (9A) | (9B) | (10A) | (10B) |
| | Acetone | 3 | UN1090 | II | FLAMMABLE LIQUID | T8 | 150 | 202 | 242 | 5 L | 60 L | B | |
| | Sodium hydroxide, solid | 8 | UN1823 | II | CORROSIVE | | 154 | 212 | 240 | 15 kg | 50 kg | A | |
| | Sodium hydroxide solution | 8 | UN1824 | II | CORROSIVE | B2, N34, T8 | 154 | 202 | 242 | 1 L | 30 L | A | |

(See the regulatory extracts on the next two pages.)

49 CFR 173.202 [emphasis added for clarity]

**Non-bulk** packagings for **liquid** hazardous materials in Packing Group II.

(a) When §172.101 of this subchapter specifies that a liquid hazardous material be packaged under this section, only non-bulk packagings prescribed in this section may be used for its transportation. Each packaging must conform to the general packaging requirements of subpart B of Part 173, to the requirements of Part 178 of this subchapter at the Packing Group I or II performance level (unless otherwise excepted), and to the particular requirements of the special provisions of Column 7 of the §172.101 Table.

(b) The following **combination** packagings are authorized:

**Outer** packagings:

Steel drum: 1A1 or 1A2
Aluminum drum: 1B1 or 1B2
Metal drum other than steel or aluminum: 1N1 or 1N2
Plywood drum: 1D
Fiber drum: 1G
Plastic drum: 1H1 or 1H2
Wooden barrel: 2C2
Steel jerrican: 3A1 or 3A2
Plastic jerrican: 3H1 or 3H2
Steel box: 4A1 or 4A2
Aluminum box: 4B1 or 4B2
Natural wood box: 4C1 or 4C2
Plywood box: 4D
Reconstituted wood box: 4F
Fiberboard box: 4G
Expanded plastic box: 4H1
Solid plastic box: 4H2

**Inner** packagings:

Glass or earthenware receptacles
Plastic receptacles
Metal receptacles
Glass ampoules

(c) Except for transportation by passenger aircraft, the following **single** packagings are authorized:

Steel drum: 1A1 or 1A2
Aluminum drum: 1B1 or 1B2
Metal drum other than steel or aluminum: 1N1 or 1N2
Plastic drum: 1H1 or 1H2
Fiber drum: 1G (with liner)
Wooden barrel: 2C1
Steel jerrican: 3A1 or 3A2
Plastic jerrican: 3H1 or 3H2

Plastic receptacle in steel, aluminum, fiber or plastic drum: 6HA1, 6HB1, 6HG1 or 6HH

Plastic receptacle in steel, aluminum, wooden, plywood or fiberboard box: 6HA2, 6HB2, 6HC, 6HD2 or 6HG2

Glass, porcelain or stoneware in steel, aluminum or fiber drum: 6PA1, 6PB1 or 6PG1

Glass, porcelain or stoneware in steel, aluminum, wooden or fiberboard box: 6PA2, 6PB2, 6PC or 6PG2

Glass, porcelain or stoneware in solid or expanded plastic packaging: 6PH1 or 6PH2

Plastic receptacle in plywood drum: 6HD1

Glass, porcelain or stoneware in plywood drum or wickerwork hamper: 6PDl or 6PD2

Cylinders, specification, as prescribed for any compressed gas, except for Specifications 8 and 3HT

[56 FR 66271, 20 Dec 1991]

49 CFR 173.212 [emphasis added for clarity]

**Non-bulk** packagings for **solid** hazardous materials in Packing Group II.

(a) When §172.101 of this subchapter specifies that a solid hazardous material be packaged under this section, only non-bulk packagings prescribed in this section may be used for its transportation. Each package must conform to the general packaging requirements of subpart B of Part 173, to the requirements of Part 178 of this subchapter at the Packing Group I or II performance level, and to the requirements of the special provisions of Column 7 of the §172.101 Table.

(b) The following **combination** packagings are authorized:

**Outer** packagings:

Steel drum: 1A1 or 1A2
Aluminum drum: 1B1 or 1B2

Metal drum other than steel or aluminum: 1N1 or 1N2
Plywood drum: 1D
Fiber drum: 1G
Plastic drum: 1H1 or 1H2
Wooden barrel: 2C2
Steel jerrican: 3A1 or 3A2
Plastic jerrican: 3H1 or 3H2
Steel box: 4A1 or 4A2
Aluminum box: 4B1 or 4B2
Natural wood box: 4C1 or 4C2
Plywood box: 4D
Reconstituted wood box: 4F
Fiberboard box: 4G
Solid plastic box: 4H2

**Inner** packagings:

Glass or earthenware receptacles
Plastic receptacles
Metal receptacles
Glass ampoules

(c) Except for transportation by passenger aircraft, the following **single** packagings are authorized:

Steel drum: 1A1 or 1A2
Aluminum drum: 1B1 or 1B2
Plywood drum: 1D
Plastic drum: 1H1 or 1H2
Fiber drum: 1G
Metal drum other than steel or aluminum: 1N1 or 1N2
Wooden barrel: 2C1 or 2C2
Steel jerrican: 3A1 or 3A2
Plastic jerrican: 3H1 or 3H2
Steel box: 4A1
Steel box with liner: 4A2
Aluminum box: 4B1
Aluminum box with liner: 4B2
Natural wood box: 4C1
Natural wood box, sift proof: 4C2
Plywood box: 4D
Reconstituted wood box: 4F
Fiberboard box: 4G
Expanded plastic box: 4H1
Solid plastic box: 4H2
Bag, woven plastic: 5H1, 5H2 or 5H3
Bag, plastic film: 5H4
Bag, textile: 5L1, 5L2 or 5L3
Bag, paper, multiwall, water resistant: 5M2

Plastic receptacle in steel, aluminum, plywood fiber or plastic drum: 6HA1, 6HB1, 6HD1, 6HG1 or 6HH

Plastic receptacle in steel aluminum, wood, plywood or fiberboard box: 6HA2, 6HB2, 6HC, 6HD2 or 6HG2

Glass, porcelain or stoneware in steel, aluminum, plywood or fiber drum: 6PA1, 6PB1, 6PD1 or 6PG1

Glass, porcelain or stoneware in steel, aluminum, wooden or fiberboard box: 6PA2, 6PB2, 6PC or 6PG2

Glass, porcelain or stoneware in expanded or solid plastic packaging: 6PH1 or 6PH2

[55 FR 52653, 21 Dec 1990]

### 10.3.2 Packaging—Test

### Module 3—Packaging

1. Containers that meet DOT standards must be used for:
   a. Bulk shipments of hazardous materials
   b. Non-bulk shipments of hazardous substances
   c. Shipments of hazardous wastes
   d. Any of the above

2. If more than one container code is listed for a hazardous material, the shipper may:
   a. Use any available container provided it is compatible with the material being shipped
   b. Use a container marked with any code listed
   c. Use a container marked with any code listed provided it is compatible with the material being shipped
   d. Use a container that meets the packing group

3. Packages that meet the pre-HM-181 specifications may be used until October 1, 1996, unless you are shipping:
   a. Radioactive materials
   b. Materials that are poisonous by inhalation
   c. Marine pollutants
   d. Hazardous substances

4. If a package is marked with the performance standard letter Y, it meets packing group(s)
   a. I only
   b. I and II only
   c. I, II and III
   d. II and III only

5. Under the HM-181 standard, the manufacturer's mark on a container designated to contain only solids or inner packages will include:
   a. The words "solids or inner packages only"
   b. The letter "S"
   c. The code IA2
   d. The packing group III

## Module 3—Packaging (continued)

6. A container that has been reconditioned for reuse can be identified by
   a. Its original performance oriented packaging code
   b. Its fresh coat of paint
   c. Its reconditioner's mark
   d. Its minimum thickness mark

7. Prior to shipping a hazardous material, you must ensure that each container is
   a. In good condition and properly sealed
   b. Completely full
   c. Open for inspection
   d. All of the above

### 10.4.1 Marking and Labeling—Exercise

Draw a line from each mark to its definition:

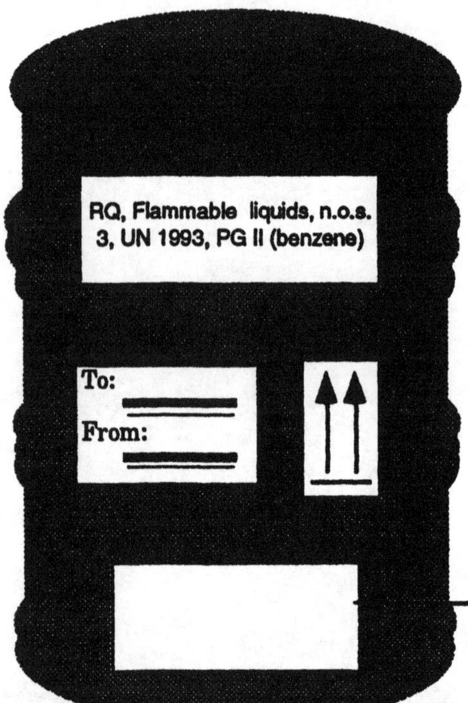


- Proper Shipping Name
- DOT ID Number
- Hazardous Substance Designation
- Orientation Mark
- What else would be marked on container if this was a 2.3 Poison Gas or 6.1 Poison Liquid?

What must be marked on a package if you ship:

a) At elevated temperature ____________________

b) Molten aluminum ____________________

c) A marine pollutant ____________________

What types of containers are labels required on:

________________________________________

Who must ensure that containers and other packages are properly marked and labeled? (shipper) (carrier) [circle best answer]

Use the first page of the Hazardous Materials Table to find the proper label for acetone: ______________________________

**Subsidiary Hazard Table**

| PG | 3 | 4.1 | 4.2 | 4.3 | 5.1 | 6.1 | 8 |
|---|---|---|---|---|---|---|---|
| I | X | *** | *** | X | X | X | X |
| II | X | X | X | X | X | X | X |
| III | * | N | X | X | N | N | ** |

X Required
* Vessel
** Air
*** Impossible
N None

Use the table above to determine the subsidiary label(s) needed for a mixture that is shipped by truck and is known to present the following additional hazards:

Flammable (3) PG I ______________________________

Flammable PG III ______________________________

A ________________ label should be used for flammable liquids with a vapor pressure between 16 and 40 pounds per square inch.

# Marking Requirements

| Marking Requirements for non-bulk packaging (173.301)<br>*Maximum Capacity ≤ 450 L (119 gallons - e.g. a 55 gallon drum)* | |
|---|---|
| Required | • Proper Shipping Name<br>• UN/NA number (except limited quantities or ORM-D materials) |
| Required (unless) | • Consignee/consignor (unless highway only and by one carrier or... from one consignor to one consignee) |
| Required (if) | • Technical Names for n.o.s. material<br>• DOT E____ (exemption number)<br>• Radioactive materials (see 172.310)<br>• Package orientation markings<br>• Inhalation Hazard<br>• Poison (6.1)<br>• ORM-D or ORM-D-Air (172.316)<br>• Explosive Hazardous Materials (172.320)<br>• Hazardous Substances (including HW)<br>  - Name HS/HW<br>  - Waste stream number for HW<br>  - RQ<br>• Bung Label<br>• "Salvage Drum" |

| Marking Requirements for bulk packaging (173.302)<br>*Maximum Capacity ≥ 450 L (119 gallons)* | |
|---|---|
| Required | • UN/NA number (on 4 sides) |
| Required (if) | • DOT E____ (exemption number)<br>• HOT, Molten Aluminum, or Molten Sulfur (on 2 opposing sides) |

| Marking Requirements for Portable Tanks (172.326)<br>*Bulk package designed to be loaded into or on a transport vehicle (e.g. a 500 gallon tote tank)* | |
|---|---|
| Required | • Proper Shipping Name (on 2 opposing sides)<br>• UN/NA number (on 4 sides)<br>• Name of owner or lessee |

| Marking Requirements for Cargo Tanks (172.328)<br>*(Bulk package attached to a motor vehicle (e.g. tank truck)* | |
|---|---|
| Required | • UN/NA number (on 4 sides) (Orange panel/Placard) |
| Required (if) | • Gases—Proper Shipping Name/Appropriate Common Name<br>• QT or NQT (Quenched and tempered steel or other ) |

| Marking Requirements for *tank cars* and multi-unit *tank car* tanks (172.328)<br>*(a type of rail car)* | |
|---|---|
| Required | • Proper Shipping Name/Appropriate Common Name (on each side)<br>• UN/NA number (on 4 sides or 2 sides if multi-unit) |

## DOT LABELING REQUIREMENTS

| Label Requirements (49 CFR 172.400) | |
|---|---|
| Required | • non-bulk packages<br>• bulk packaging (except cargo tanks, portable tanks, tank cars > 640 cubic feet)<br>• portable tanks < 1000 gallons<br>• 106 or 110 multi-unit tank cars (unless placarded)<br>• Overpacks, freight containers or unit load devices < 640 cubic feet (unless placarded) |
| Not Required | • Division 2.1 or 2.2 that is not poisonous and carried by a private carrier, not overpacked and marked in accordance with the Compressed Gas Association<br>• Certain DOD materials<br>• Compressed Gas permanently attached<br>• Freight container, aircraft unit load device or portable tank that is placarded<br>• Overpack or unit load device labels are visible |
| Required (if) | • Cargo Aircraft Only Label<br>• Radioactive Material (172.403)<br>• Explosive materials that are 6.1 PG I or II—labeled "Poison" |
| Subsidiary Label | • Within 6 inches of Primary hazard label |
| Duplicate label (required if) | • Non-bulk package or overpack > 64 cubic feet<br>• Non-bulk packages of radioactive materials<br>• DOT 106 or 110 multi-tank tank cars (each end)<br>• Portable tanks < 1000 gallons<br>• Freight Containers or aircraft unit load devices > 64 cubic feet but < 640 cubic feet |

## 10.4.2 Marking and Labeling—Test

### Module 4—Marking and Labeling

1. Which of the following must be marked by the shipper on packages containing hazardous materials?
   a. Packing code
   b. Reportable quantity of marine pollutants
   c. Name and address of the transporter
   d. Proper shipping name, identification number, and shipper's (or consignee's) name and address

2. Orientation arrows should be marked on a container with inner packages containing liquids to designate:
   a. That they should be stored on the top shelf
   b. The nearest evacuation route
   c. The direction of the openings of the inner packages
   d. The location of the DOT packaging codes and specifications

3. Hazardous substances must be marked with:
   a. The letters HS
   b. The letters RQ
   c. The letters ORM
   d. The letters ORM-D

4. When shipping molten aluminum, each container must be marked:
   a. MOLTEN ALUMINUM
   b. K088
   c. THIS END UP or THIS WAY UP
   d. TYPE A

5. The DOT proper shipping name and identification number must be removed from containers that are:
   a. Empty, which means that they have been cleaned and purged of hazardous materials
   b. Empty, which means that they contain less than 3% of hazardous material or less than 1 inch of hazardous material residue (whichever is less)
   c. Empty, which means that they have been triple rinsed
   d. Empty, which means that each package contains only a residue of hazardous material.

## Module 4 —Marking and Labeling (continued)

6. Who is required to properly label containers prior to shipment:
   a. Shipper
   b. Transporter
   c. Carrier
   d. Vehicle operator

7. A subsidiary label may be required if:
   a. The hazardous material may subside during the shipment
   b. More than one carrier (transporter) will transport the shipment
   c. The package contains more than one hazard class
   d. The package is too small to accommodate a hazard class label.

8. Hazard class labels must be placed
   a. On any surface except the bottom of the container and on the same side as the proper shipping name
   b. On all sides of each container
   c. On non-bulk containers that have been cleaned and purged
   d. On a background of a non-contrasting color

9. If the amount shipped in a container exceeds the limits in column 9A and is within the limits in column 9B of the hazardous materials table, the following label must be applied to the container:
   a. KEEP AWAY FROM FOOD
   b. CAUTION, UNSCREW THIS BUNG SLOWLY
   c. CARGO AIRCRAFT ONLY
   d. Hold at airport

10. A HOT label must be used when shipping
    a. Materials classified as High Order Teratogens
    b. Materials classified as Hazardous Only by Train
    c. Materials which are shipped at elevated temperature
    d. Radioactive materials

### 10.5.1 Shipping Papers—Exercise

The following must be on each hazmat shipping paper:

Name and address of ________________________________.
Address of location to which material is ________________.
______________ ________________ of the hazardous material.

The "basic description" includes the following 4 items. Identify the column of the Hazardous Materials Table where each can be found.

| Item | Column of Hazmat Table |
|---|---|
| Proper shipping name | |
| Hazard class or division number | |
| UN/NA number | |
| Packing Group | |

From the first page of the HMT, determine the basic description for acetone.

______________________________________________________________

When shipping hazardous and non-hazardous materials on the same bill of lading, how should hazardous materials be distinguished?

______________________________________________________________

What letters should be added to a shipping name if the material is a *hazardous substance* (if a certain condition is met)? ____

What words should be added if a material is a Division 2.3 or 6.1 PG I?

______________________________________________

"Ltd. Qty." is added to the basic description when shipping a ______________

______________________.

## 10.5.2 Shipping Papers—Test

### Module 5—Shipping Papers

1. A manifest must be used when shipping—
   a. Hazardous waste
   b. Hazardous materials
   c. Non-hazardous materials
   d. Any of the above

2. Each bill of lading must include the basic description for each hazardous material being shipped. The basic description includes—
   a. Proper shipping name, amount shipped, and hazard class
   b. Proper shipping name, hazard class, and packaging authorizations
   c. Proper shipping name, hazard class, identification number, and packing group (if any)
   d. Technical name, hazard class, and required labels

3. When shipping both hazardous and non-hazardous materials, you should distinguish the hazardous materials by—
   a. Listing the hazardous material first
   b. Enter the hazardous material in a contrasting color
   c. Placing an X in the HM column before the proper shipping name
   d. Any of the above

4. When shipping a hazardous mixture with a generic shipping name such as Flammable liquids, n.o.s. you must add to the shipping name—
   a. Nothing
   b. The flash point of the mixture
   c. The packaging code of the container the material is being shipped in
   d. The technical name(s) for the first two components which are the basis for the hazard of the material.

5. When shipping hazardous materials by air, the following certification should be added to the shipping paper:
   a. Cargo Aircraft Only
   b. Load Last, Unload First
   c. This shipment is within the limitations prescribed for passenger aircraft/cargo aircraft only
   d. Aircraft route-controlled quantity

## Module 5—Shipping Papers (cont.)

6. The shipping paper entry "Residue: Last Contained ______" (name of hazardous material) should be used when shipping:

   a. Packages containing residue only
   b. Rail cars containing residue only
   c. Marine Pollutants
   d. Packages that have been cleaned and purged of hazardous material residue.

7. When shipping hazardous materials in division 4.3, the shipping description must include the entry:

   a. Dangerous when wet
   b. Inhalation hazard
   c. Spontaneously combustible
   d. Wetted explosive

8. The following emergency response information must be included on or with each hazardous material bill of lading:

   a. Twenty-four hour emergency number and emergency response information
   b. The name of a person to contact in the event of an emergency
   c. Twenty four hour National Response Center number
   d. Nearest waterway

9. A hazardous waste manifest must be accompanied by a land disposal notice or certification:

   a. Only when shipping to a land disposal facility
   b. With each shipment of a land disposal restricted waste
   c. With the first shipment of a land disposal restricted waste to each applicable treatment, storage, or disposal facility
   d. When there is no alternative to land disposal

10. If you ship hazardous waste from a state that requires its own state-specific manifest to another state which requires its own state specific manifest, you must use—

    a. Your state's manifest (generator state manifest)
    b. The manifest for the state to which the waste will be transported (consignment state)
    c. The EPA uniform hazardous waste manifest
    d. A DOT straight bill of lading with a hazardous waste certification.

### 10.6.1 Placarding—Exercise

Which of the following must be placarded if they carry hazardous materials (circle all that apply):

| | |
|---|---|
| Bulk Packages | Non-bulk containers |
| Unit Load Devices | 55 gallon drums |
| Rail Cars | Limited quantity containers |
| Motor Vehicles | Airplanes |
| Freight Containers | Compressed gas bottles |

Who must ensure that the vehicle is properly placarded? ______________

Using the placard tables in Chapter 7 (7.4.3), determine the appropriate placards for:

55 gallons of acetone (a flammable liquid) ______________

5000 pounds of sodium hydroxide (a corrosive liquid)______________

1 pound of lithium hydride (dangerous when wet, hazard class 4.3)__________

1500 lbs of an oxidizer plus 1500 lbs of a flammable liquid______________

6000 lbs each of an oxidizer and a flammable liquid______________

Placards must be secured on ____ sides of the vehicle.

When should a "Residue" placard be used?______________

True / False: the Marine Pollutant mark is only to be used when shipping by water.

## 10.6.2 Placarding—Test

### Module 6—Placards

1. When shipping hazardous materials, placards must be placed on:
   a. Bulk packages
   b. Freight containers
   c. Unit load devices
   d. All of the above

2. The _________ should offer appropriate placards for a hazardous materials shipment.
   a. Shipper
   b. Carrier
   c. Consignee (destination)
   d. All of the above

3. If a shipment contains more than one Placard Table I listed material, you must:
   a. Affix the placard for the more dangerous (highest precedence) hazard class
   b. Affix each placard on the vehicle
   c. Affix the placard DANGEROUS
   d. Not allow this shipment because you cannot transport more than one hazard class from Placard Table I on the same shipment.

4. Until a railroad tank car that previously held hazardous materials has been cleaned and purged of vapors or reloaded with another material, you should:
   a. Leave the original placards on the railroad tank car
   b. Apply an orange panel with the hazardous material identification number
   c. Apply a white square on point placard with the hazardous material identification number
   d. Apply hazard class placards for the initial hazard class with the word "Residue" added.

5. Placards are not required for—
   a. Class 6.2 and ORM-D
   b. Limited quantities if identified on the shipping paper
   c. Small quantity packages and combustible liquids in non-bulk packages
   d. All of the above

**Module 6—Placards (continued)**

6. Placards must be placed—
   a. On all four sides of a vehicle
   b. 5 inches from ladders, pipes, or other items that distract from their visibility
   c. Inside the cab of the vehicle
   d. All of the above

### 10.7.1 Loading, Unloading, and Stowing—Exercise

Why should tank car brakes be set and wheels be blocked on all cars during loading?

______________________________________________

When loading flammable liquids, which of the following precautions apply (circle all that apply):

No smoking
Set hand brake
Secure containers against movement within the vehicle
Do not store with poisonous gases or liquids
Motor vehicle engine should be off unless necessary for operation of a pump

During *cargo tank loading*, a __________ person should attend the process by having an _________________ view and be within _____ feet of the shipment.

Using the table on the next page, determine if corrosive liquids (class 8) could be shipped with:

- flammable liquids (class 3) ______________________________
- flammable solids (4.1) ________________________________

## Stowing of Incompatible Materials

DOT prohibits the loading of certain combinations of hazardous materials on the same motor vehicle in order to prevent fires, explosions, and spills caused by reactions of incompatible chemicals. When shipping hazardous materials by highway, the material segregation requirements in the following table must be followed:

To read the chart you must know the hazard class for the material being shipped. If you are shipping a hazardous material with a hazard class which is listed, follow the row on the chart associated with that hazard class and specific material. An "X" in the row indicates that materials may not be loaded, transported, or stored together. An "O" indicates that the materials may not be transported together unless separated by a distance of 1.2 meters (4 feet) in all directions. A "*" indicates the segregation of Class 1 materials is governed by the Compatibility Table for Class 1 materials in 49 CFR 177.848. The two letters in the notes column indicate special prohibitions. An "A" indicates that notwithstanding an "X," ammonium nitrate fertilizer may be loaded or stored with Division 1.1 materials. A "B" indicates that in addition to the requirements designated by the "O" the materials may not be loaded or transported together unless for each class or division the packages are palletized at a minimum height of 10 cm off the floor or separated so that in the event of a leak, there will be no commingling of hazardous materials.

### Segregation Table for Hazardous Materials

| Class or Division | Notes | 1.1, 1.2 | 1.3 | 1.4 | 1.5 | 1.6 | 2.1 | 2.2 | 2.3 gas Zone A | 2.3 gas other than Zone A | 3 | 4.1 | 4.2 | 4.3 | 5.1 | 5.2 | 6.1 liquids PG I Zone A | 7 | 8 liquids only |
|---|---|---|---|---|---|---|---|---|---|---|---|---|---|---|---|---|---|---|---|
| Explosives .... 1.1 and 1.2 | A | • | • | • | • | • | X | X | X | X | X | X | X | X | X | X | X | X | X |
| Explosives ............ 1.3 | | • | • | • | • | • | X | | X | X | X | | X | X | X | X | X | | X |
| Explosives ............ 1.4 | | • | • | • | • | • | O | | O | O | O | | O | | | | O | | O |
| Very insensitive explosives ............ 1.5 | | • | • | • | • | • | X | X | X | X | X | X | X | X | X | X | X | X | X |
| Extremely insensitive explosives ............ 1.6 | | • | • | • | • | • | | | | | | | | | | | | | |
| Flammable gases ....... 2.1 | | X | X | O | X | | | | X | O | | | | | | | O | O | O |
| Non-toxic, nonflammable gases ............ 2.2 | | X | | | X | | | | | | | | | | | | | | |
| Poisonous gas Zone A ............ 2.3 | | X | X | O | X | | X | | | | X | X | X | X | X | X | | | X |
| Poisonous gas other than Zone A ............ 2.3 | | X | X | O | X | | O | | | | O | O | O | O | O | O | | | O |
| Flammable liquids ......... 3 | | X | X | O | X | | | | X | O | | | | | O | | X | | |
| Flammable solids ....... 4.1 | | X | | | X | | | | X | O | | | | | | | X | | O |
| Spontaneously combustible materials ............ 4.2 | | X | X | O | X | | | | X | O | | | | | | | X | | X |
| Dangerous when wet materials ............ 4.3 | | X | X | | X | | | | X | O | | | | | | | X | | O |
| Oxidizers ............ 5.1 | A | X | X | | X | | | | X | O | O | | | | | | X | | O |
| Organic peroxides ...... 5.2 | | X | X | | X | | | | X | O | | | | | | | X | | O |
| Poisonous liquids PG I Zone A ............ 6.1 | | X | X | O | X | | O | | | | X | X | X | X | X | X | | | X |
| Radioactive materials ............ 7 | | X | | | X | | O | | | | | | | | | | | | |
| Corrosive liquids ............ 8 | | X | X | O | X | | O | | X | O | | O | X | O | O | O | X | | |

In addition to the requirements shown on the above table:

- Cyanides or cyanide mixtures must not be loaded or stored with acids or acidic materials. The reaction of cyanides with acids releases deadly hydrogen cyanide gas.
- Regardless of the methods of separation employed, Class 8 (corrosive) liquid materials may not be loaded above Class 4 (flammable solid) materials or Class 5 (oxidizing) materials.

- When the §172.101 Table or 49 CFR 172.402 requires a package to bear a subsidiary hazard label, segregation appropriate to the subsidiary hazard must be applied when that segregation is more restrictive than that required by the primary hazard. However, hazardous materials of the same class may be stowed together without regard to segregation required for any secondary hazard if the materials are not capable of reacting dangerously with each other and causing combustion or dangerous evolution of heat, evolution of flammable, poisonous, or asphyxiant gases, or formation of corrosive or unstable materials.

## General Loading Requirements

- All hazardous wastes that are also DOT flammable liquids, compressed gases, corrosive materials or poisonous materials must be secured against movement within the vehicle during transportation. Brace these containers by using load bars or other devices that will prevent movement. Be careful to avoid damaging valves and fittings on containers during loading and stowage.
- Never load hazardous wastes or hazardous materials into pole trailers.
- Never smoke while loading explosive, flammable and oxidizing materials. Post signs or take other precaution to keep fire away and prevent anyone in the vicinity from smoking.
- Make sure the hand brake of the vehicle is set before loading or unloading all hazardous wastes and materials. The engine must be stopped whenever explosive or flammable materials are loaded unless the engine is needed to operate a pump on a cargo tank.
- Use tools designed for the containers that are loaded into the vehicle. Never use tools or equipment that could damage the containers or the closures on the containers. For example, a fork lift used to lift and move an open-head 55-gallon by the ring on the drum could damage the ring and result in a spill of hazardous waste during transportation.
- Cargo heaters are always prohibited in vehicles hauling explosives. Cargo heaters used in hauling flammable liquids must be catalytic and must meet the design and operational requirements specified in 49 CFR 177.
- Cargo tanks must be attended by a qualified person at all times during loading. "Attended" means throughout the process the person is awake, has an unobstructed view of the tank, and is within 25 feet of the tank or the delivery hose to the tank. A person is "qualified" if he is aware of the nature of the hazardous material that is loaded, has been instructed on the procedures to follow in an emergency, and is authorized to move the tank and is able to do so.
- Follow bonding and grounding procedures when loading flammable liquids into cargo tanks through open filling holes. Bonding and grounding is not required by DOT when a cargo tank is loaded through a vapor tight top or bottom connection so that there is no release of vapor at a point where a spark could occur.
- Load flammable solids and oxidizing materials only into trailers that have an enclosed body. Take precautions to prevent these materials from getting wet during loading and during transportation.

- Never load poisons into a motor vehicle that also contains foodstuffs, feed or any edible material meant for human or animal consumption. Hazardous material should never be shipped with food and animal feed or in tank trucks used for shipping food stuffs. Hazardous material should not be transported in the cab of the vehicle.

### 10.7.2 Loading, Unloading, and Stowing—Test

**Module 7—Loading, Unloading, Stowing**

1. To ensure that liquids do not expand, causing a container to rupture, they may not completely fill a container at:
   a. 100 °C
   b. 75 °C
   c. 55 °C
   d. 5 °C

2. To ensure that the contents of a salvage drum does not rupture, the contents should be:
   a. Cushioned
   b. Filled with water
   c. Crushed
   d. Tightly closed

3. Poisons may not be loaded on the same vehicle with—
   a. Edible materials
   b. Consumer commodities
   c. Class 9
   d. All of the above

4. If there is an X between two rows of materials on the DOT compatibility table, this means that—
   a. They may be loaded or stowed together
   b. They may not be loaded or stowed together
   c. They must be at least four feet apart when loaded on the same vehicle
   d. They must be at least 1.2 meters apart when loaded on the same vehicle.

5. Which of the following rules apply to the loading of hazardous materials on motor vehicles?
   a. Brakes must be securely set, wheels blocked and all other reasonable precautions taken to prevent motion of the motor vehicle during such loading process
   b. A qualified person must attend the unloading of a cargo tank, meaning that throughout the process, he/she is awake, has an unobstructed view of the cargo tank and is within 25 feet of the cargo tank
   c. containers of Class 3, Class 2, Class 8, Division 6.1 or Class 7 hazardous wastes must be secured against movement within a transport vehicle
   d. All of the above

6. ___________ which are likely to damage the effectiveness of the closure of any package or other container, or likely to adversely affect such package or container, must not be used for the loading of any hazardous material into or on a motor vehicle.

   a. Persons
   b. Tools
   c. Companies
   d. Vehicles

7. Except as otherwise provided by DOT regulations, bulk or non-bulk packaging containing the residue of a hazardous material must be offered for transportation and transported in the same manner as:

   a. An empty container
   b. When it previously contained a greater quantity of that hazardous material
   c. A hazardous waste
   d. An unregulated shipment

### 10.8.1 Safe Handling of Hazardous Materials—Exercise

What are some transportation hazards that should be avoided?

______________________________________________

______________________________________________

What could happen if incompatibles mix together?

______________________________________________

What precautions can you take to prevent a drum from bulging?

______________________________________________

Why is it <u>not</u> true that if you cannot smell a chemical, it is safe to breathe?

______________________________________________

What personal protective equipment should be worn when pouring caustic (sodium hydroxide)?

______________________________________________

The DOT emergency response guide for gasoline is Guide 27 (see next page). Using the guide, what precautions would be taken when working with gasoline?

______________________________________________

______________________________________________

## POTENTIAL HAZARDS

**FIRE OR EXPLOSION**

Flammable/combustible material; may be ignited by heat, sparks or flames.
Vapors may travel to a source of ignition and flash back.
Container may explode in heat of fire.
Vapor explosion hazard indoors, outdoors or in sewers.
Runoff to sewer may create fire or explosion hazard.

**HEALTH HAZARDS**

May be poisonous if inhaled or absorbed through skin.
Vapors may cause dizziness or suffocation.
Contact may irritate or burn skin and eyes.
Fire may produce irritating or poisonous gases.
Runoff from fire control or dilution water may cause pollution.

## EMERGENCY ACTION

Keep unnecessary people away; isolate hazard area and deny entry.
Stay upwind; keep out of low areas.
Positive pressure self-contained breathing apparatus (SCBA) and structural firefighters' protective clothing will provide limited protection.
**Isolate for 1/2 mile in all directions if tank, rail car or tank truck is involved in fire.**
**CALL CHEMTREC AT 1-800-424-9300** FOR EMERGENCY ASSISTANCE.
If water pollution occurs, notify the appropriate authorities.

**FIRE**

**Small Fires:** Dry chemical, $CO_2$, water spray or regular foam.
**Large Fires:** Water spray, fog or regular foam.
Move container from fire area if you can do it without risk.
Apply cooling water to sides of containers that are exposed to flames until well after fire is out. Stay away from ends of tanks.
For massive fire in cargo area, use unmanned hose holder or monitor nozzles; if this is impossible, withdraw from area and let fire burn.
Withdraw immediately in case of rising sound from venting safety device or any discoloration of tank due to fire.

**SPILL OR LEAK**

Shut off ignition sources; no flares, smoking or flames in hazard area.
Stop leak if you can do it without risk.
Water spray may reduce vapor; but it may not prevent ignition in closed spaces.
**Small Spills**: Take up with sand or other noncombustible absorbent material and place into containers for later disposal.
**Large Spills**: Dike far ahead of liquid spill for later disposal.

**FIRST AID**

Move victim to fresh air and call emergency medical care; if not breathing, give artificial respiration; if breathing is difficult, give oxygen.
In case of contact with material, immediately flush eyes with running water for at least 15 minutes. Wash skin with soap and water.
Remove and isolate contaminated clothing and shoes at the site.

### 10.8.2 Safe Handling of Hazardous Materials—Test

### Module 8—Safe Handling of Hazardous Materials

1. The best time to prepare for handling a hazardous materials emergency is:
   a. Immediately after you learn about it
   b. After a spill, fire, or explosion
   c. After someone has been injured
   d. Prior to working with hazardous materials

2. Before opening a container of hazardous materials, it is important to—
   a. Read any labels on the container
   b. Remove hazard warning labels from the container
   c. Sniff the bung opening to ensure that you have the proper material
   d. Shake the container to see if there is enough for your application

3. When flammable liquids are transferred from one container to another, the containers must be _______________ to avoid danger from static electricity.
   a. Diluted
   b. Bonded or grounded
   c. Kept cool
   d. Placed in a containment area

4. Smoking a cigarette in a hazardous materials storage area is:
   a. Not dangerous if the containers are sealed
   b. Acceptable if no flammable liquids or solids are present
   c. Acceptable if the area is explosion proof
   d. Not allowed

5. Personal protective equipment should be worn:
   a. Only if it comfortable
   b. Unless a worker feels that he or she is not exposed to hazardous materials
   c. Whenever performing a task in which personal protective equipment is specified by a supervisor, standard operating procedure, job safety analysis, or safety manager.
   d. Whenever performing a task in which there are adequate engineering controls to safely limit worker exposure.

## Module 8—Safe Handling of Hazardous Materials (continued)

6. How can you determine if a glove will protect your hands from exposure to a particular chemical?
   a. Refer to the glove's specification sheet, a material safety data sheet for the chemical, or contact your safety manager for advice
   b. Obtain the thickest and most durable glove available
   c. Test it by blowing in each glove to see if there are any leaks
   d. All of the above

### 10.9.1 Emergency Response—Exercise

List some incidents that would require a report to your plant emergency coordinator.

________________________________________

________________________________________

What is the first thing that you should do if you see a hazardous material spill?

________________________________________

What is the phone number of your on-site emergency coordinator?

________________________________________

What would you report if you were reporting a spill?

________________________________________

List the names of some written materials that you can refer to in the event of an emergency:

________________________________________

### 10.9.2 Emergency Response—Test

**Module 9—Emergency Response**

1. If you detect an unusual odor in a hazardous materials storage area, the first thing you should do is:
   a. Look for a hazardous materials leak or spill
   b. Notify others (such as employees that could be exposed, your supervisor or the plant emergency coordinator)
   c. Contain the material to prevent it from spreading
   d. Stop the source of the leak or spill

2. If a fellow employee falls unconscious while cleaning the inside of a tank car, you should immediately
   a. Go into the tank and rescue him
   b. Look for a hazardous material leak or spill
   c. Look inside the tank car to see if anyone else has fallen unconscious
   d. Report the incident so that trained and properly equipped response personnel will rescue the employee

3. Emergency responders use the four-digit hazardous material identification number to locate product hazard information in the ________________ issued by DOT.
   a. Material safety data sheet
   b. Job safety analysis
   c. Emergency response guide
   d. Regulations

4. The shipping paper for a hazardous materials shipment must have an emergency response telephone number that is monitored:
   a. 8 hours per day, 5 days per week
   b. By the transporter
   c. By the DOT
   d. At all times

5. Who is responsible for filing a written hazardous materials incident report with the DOT following a reported release of hazardous materials which occurs during the course of transportation (including loading, unloading, and temporary storage):
   a. Carrier
   b. Shipper
   c. Person that discovers the incident
   d. Plant manager for the site at which the incident occurs

Name:________________ Course Location: ________________ Date:______

# Hazardous Materials Management
## *Compliance with the New DOT Requirements*
## Test Answer Sheet

### Module 1—Introduction

1. ⓐ ⓑ ⓒ ⓓ
2. ⓐ ⓑ ⓒ ⓓ
3. ⓐ ⓑ ⓒ ⓓ
4. ⓐ ⓑ ⓒ ⓓ
5. ⓐ ⓑ ⓒ ⓓ
6. ⓐ ⓑ ⓒ ⓓ
7. ⓐ ⓑ ⓒ ⓓ
8. ⓐ ⓑ ⓒ ⓓ
9. ⓐ ⓑ ⓒ ⓓ

### Module 2—Hazardous Materials Table

1. ⓐ ⓑ ⓒ ⓓ
2. ⓐ ⓑ ⓒ ⓓ
3. ⓐ ⓑ ⓒ ⓓ
4. ⓐ ⓑ ⓒ ⓓ
5. ⓐ ⓑ ⓒ ⓓ
6. ⓐ ⓑ ⓒ ⓓ
7. ⓐ ⓑ ⓒ ⓓ

### Module 3—Packaging

1. ⓐ ⓑ ⓒ ⓓ
2. ⓐ ⓑ ⓒ ⓓ
3. ⓐ ⓑ ⓒ ⓓ
4. ⓐ ⓑ ⓒ ⓓ
5. ⓐ ⓑ ⓒ ⓓ
6. ⓐ ⓑ ⓒ ⓓ
7. ⓐ ⓑ ⓒ ⓓ

### Module 4—Marking and Labeling

1. ⓐ ⓑ ⓒ ⓓ
2. ⓐ ⓑ ⓒ ⓓ
3. ⓐ ⓑ ⓒ ⓓ
4. ⓐ ⓑ ⓒ ⓓ
5. ⓐ ⓑ ⓒ ⓓ
6. ⓐ ⓑ ⓒ ⓓ
7. ⓐ ⓑ ⓒ ⓓ
8. ⓐ ⓑ ⓒ ⓓ
9. ⓐ ⓑ ⓒ ⓓ
10. ⓐ ⓑ ⓒ ⓓ

### Module 5—Shipping Papers

1. ⓐ ⓑ ⓒ ⓓ
2. ⓐ ⓑ ⓒ ⓓ
3. ⓐ ⓑ ⓒ ⓓ
4. ⓐ ⓑ ⓒ ⓓ
5. ⓐ ⓑ ⓒ ⓓ
6. ⓐ ⓑ ⓒ ⓓ
7. ⓐ ⓑ ⓒ ⓓ
8. ⓐ ⓑ ⓒ ⓓ
9. ⓐ ⓑ ⓒ ⓓ
10. ⓐ ⓑ ⓒ ⓓ

### Module 6—Placards

1. ⓐ ⓑ ⓒ ⓓ
2. ⓐ ⓑ ⓒ ⓓ
3. ⓐ ⓑ ⓒ ⓓ
4. ⓐ ⓑ ⓒ ⓓ
5. ⓐ ⓑ ⓒ ⓓ
6. ⓐ ⓑ ⓒ ⓓ

### Module 7—Loading, Unloading, Stowing

1. ⓐ ⓑ ⓒ ⓓ
2. ⓐ ⓑ ⓒ ⓓ
3. ⓐ ⓑ ⓒ ⓓ
4. ⓐ ⓑ ⓒ ⓓ
5. ⓐ ⓑ ⓒ ⓓ
6. ⓐ ⓑ ⓒ ⓓ
7. ⓐ ⓑ ⓒ ⓓ

### Module 8—Safe Handling of Hazardous Materials

1. ⓐ ⓑ ⓒ ⓓ
2. ⓐ ⓑ ⓒ ⓓ
3. ⓐ ⓑ ⓒ ⓓ
4. ⓐ ⓑ ⓒ ⓓ
5. ⓐ ⓑ ⓒ ⓓ
6. ⓐ ⓑ ⓒ ⓓ

### Module 9—Emergency Response

1. ⓐ ⓑ ⓒ ⓓ
2. ⓐ ⓑ ⓒ ⓓ
3. ⓐ ⓑ ⓒ ⓓ
4. ⓐ ⓑ ⓒ ⓓ
5. ⓐ ⓑ ⓒ ⓓ

# Hazardous Materials Management

## *Compliance with the New DOT Requirements*

## Test Answer Key

**Module 1—Introduction**

1. ⓐ ⓑ ⓒ ●
2. ⓐ ⓑ ⓒ ●
3. ● ⓑ ⓒ ⓓ
4. ⓐ ● ⓒ ⓓ
5. ⓐ ⓑ ● ⓓ
6. ⓐ ⓑ ⓒ ●
7. ⓐ ⓑ ● ⓓ
8. ⓐ ⓑ ● ⓓ
9. ⓐ ● ⓒ ⓓ

**Module 2—Hazardous Materials Table**

1. ● ⓑ ⓒ ⓓ
2. ⓐ ⓑ ● ⓓ
3. ⓐ ⓑ ● ⓓ
4. ⓐ ● ⓒ ⓓ
5. ⓐ ● ⓒ ⓓ
6. ⓐ ⓑ ● ⓓ
7. ● ⓑ ⓒ ⓓ

**Module 3—Packaging**

1. ⓐ ⓑ ⓒ ●
2. ⓐ ⓑ ● ⓓ
3. ⓐ ● ⓒ ⓓ
4. ⓐ ⓑ ⓒ ●
5. ⓐ ● ⓒ ⓓ
6. ⓐ ⓑ ● ⓓ
7. ● ⓑ ⓒ ⓓ

**Module 4—Marking and Labeling**

1. ⓐ ⓑ ⓒ ●
2. ⓐ ⓑ ● ⓓ
3. ⓐ ● ⓒ ⓓ
4. ● ⓑ ⓒ ⓓ
5. ● ⓑ ⓒ ⓓ
6. ● ⓑ ⓒ ⓓ
7. ⓐ ⓑ ● ⓓ
8. ● ⓑ ⓒ ⓓ
9. ⓐ ⓑ ● ⓓ
10. ⓐ ⓑ ● ⓓ

**Module 5—Shipping Papers**

1. ● ⓑ ⓒ ⓓ
2. ⓐ ⓑ ● ⓓ
3. ⓐ ⓑ ⓒ ●
4. ⓐ ⓑ ⓒ ●
5. ⓐ ⓑ ● ⓓ
6. ● ⓑ ⓒ ⓓ
7. ● ⓑ ⓒ ⓓ
8. ● ⓑ ⓒ ⓓ
9. ⓐ ● ⓒ ⓓ
10. ⓐ ● ⓒ ⓓ

**Module 6—Placards**

1. ⓐ ⓑ ⓒ ●
2. ● ⓑ ⓒ ⓓ
3. ⓐ ● ⓒ ⓓ
4. ⓐ ⓑ ⓒ ●
5. ⓐ ⓑ ⓒ ●
6. ● ⓑ ⓒ ⓓ

**Module 7—Loading, Unloading, Stowing**

1. ⓐ ⓑ ● ⓓ
2. ● ⓑ ⓒ ⓓ
3. ● ⓑ ⓒ ⓓ
4. ⓐ ● ⓒ ⓓ
5. ⓐ ⓑ ⓒ ●
6. ⓐ ● ⓒ ⓓ
7. ⓐ ● ⓒ ⓓ

**Module 8—Safe Handling of Hazardous Materials**

1. ⓐ ⓑ ⓒ ●
2. ● ⓑ ⓒ ⓓ
3. ⓐ ● ⓒ ⓓ
4. ⓐ ⓑ ⓒ ●
5. ⓐ ⓑ ● ⓓ
6. ● ⓑ ⓒ ⓓ

**Module 9—Emergency Response**

1. ⓐ ● ⓒ ⓓ
2. ⓐ ⓑ ⓒ ●
3. ⓐ ⓑ ● ⓓ
4. ⓐ ⓑ ⓒ ●
5. ● ⓑ ⓒ ⓓ

# Appendix A
# HM-181 49 CFR Part 172 (1991-1992)

# Appendix A

## 49 CFR PART 172--HAZARDOUS MATERIALS TABLE, SPECIAL PROVISIONS, HAZARDOUS MATERIALS COMMUNICATIONS, EMERGENCY RESPONSE INFORMATION, AND TRAINING REQUIREMENTS

**Subpart A--General**

Sec.
172.1 Purpose and scope.
172.3 Applicability.

**Subpart B--Table of Hazardous Materials and Special Provisions**

Sec.
172.101 Purpose and use of hazardous materials table.
Appendix A to 172.101--List of Hazardous Substances and Reportable Quantities
Appendix B to 172.101--List of Marine Pollutants
172.102 Special provisions.

**Subpart C--Shipping Papers**

Sec.
172.200 Applicability.
172.202 Despcription of hazardous material on shipping papers.
172.203 Additional description requirements.
172.204 Shipper's certification.
172.205 Hazardous waste manifest.

**Subpart D--Marking**

Sec.
172.300 Applicability.
172.301 General marking requirements for non-bulk packagings.
172.302 General marking requirements for bulk packagings.
172.303 Prohibited marking.
172.304 Marking requirements.
172.306 [Removed and Reserved]
172.308 Authorized abbreviations.
172.310 Radioactive materials.
172.312 Liquid hazardous materials in non-bulk packagings.
172.313 Poisonous hazardous materials.
172.316 Packagings containing material classed as ORM-D.
172.320 Explosive hazardous materials.
172.322 Marine pollutants.
172.324 Hazardous substances in non-bulk packagings.
172.325 Elevated temperature materials.
172.326 Portable tanks.
172.328 Cargo tanks.
172.330 Tank cars and multi-unit tank car tanks.
172.331 Bulk packagings other than portable tanks, cargo tanks,

tank cars and multi-unit tank car tanks.

172.332 Identification number markings.

172.334 Identification numbers; prohibited display.

172.336 Identification numbers; special provisions.

172.338 Replacement of identification numbers.

**Subpart E--Labeling**

Sec.

172.400 General labeling requirements.

172.400a Exceptions from labeling.

172.401 Prohibited labeling.

172.402 Additional labeling requirements.

172.403 Radioactive material.

172.404 Labels for mixed and consolidated packaging.

172.405 Authorized label modifications.

172.406 Placement of labels.

172.407 Label specifications.

172.411 EXPLOSIVE 1.1, 1.2, 1.3, 1.4, 1.5 and 1.6 labels, and EXPLOSIVE Subsidiary label.

172.415 NON-FLAMMABLE GAS label.

172.416 POISON GAS label.

172.417 FLAMMABLE GAS label.

172.419 FLAMMABLE LIQUID label.

172.420 FLAMMABLE SOLID label.

172.422 SPONTANEOUSLY COMBUSTABLE label.

172.423 DANGEROUS WHEN WET label.

172.426 OXIDIZER label.

172.427 ORGANIC PEROXIDE label.

172.430 POISON label.

172.431 KEEP AWAY FROM FOOD label.

172.432 INFECTIOUS SUBSTANCE label.

172.436 RADIOACTIVE WHITE-I label.

172.438 RADIOACTIVE YELLOW-II label.

172.440 RADIOACTIVE YELLOW-III label.

172.442 CORROSIVE label.

172.444 [Removed and Reserved]

172.446 CLASS 9 label.

172.448 CARGO AIRCRAFT ONLY label.

172.450 EMPTY label.

**Subpart F--Placarding**

Sec.

172.500 Applicability of placarding requirements.

172.502 Prohibited and permissive placarding.

172.503 Identification number display on placards.

172.504 General placarding requirements.

172.505 Placarding for subsidiary hazards.

172.506 Providing and affixing placards: Highway.

172.507 Special placarding provisions: Highway.

172.508 Providing and affixing placards: Rail.

172.510 Special placarding provisions: Rail.
172.512 Freight containers and aircraft unit load devices.
172.514 Bulk packagings other than tank cars.
172.516 Visibility and display of placards.
172.519 General specifications for placards.
172.521 DANGEROUS placard.
172.522 EXPLOSIVES 1.1, EXPLOSIVES 1.2 and EXPLOSIVES 1.3 placards.
172.523 EXPLOSIVES 1.4 placard.
172.524 EXPLOSIVES 1.5 placard.
172.525 EXPLOSIVES 1.6 placard.
172.526 Standard requirements for the RESIDUE placard.
172.527 Background requirements for certain placards.
172.528 NON-FLAMMABLE GAS placard.
172.530 OXYGEN placard.
172.532 FLAMMABLE GAS placard.
172.536 [Removed and Reserved]
172.540 POISON GAS placard.
172.542 FLAMMABLE placard.
172.544 COMBUSTIBLE placard.
172.546 FLAMMABLE SOLID placard.
172.547 SPONTANEOUSLY COMBUSTIBLE placard.
172.548 DANGEROUS WHEN WET placard.
172.550 OXIDIZER placard.
172.552 ORGANIC PEROXIDE placard.
172.553 KEEP AWAY FROM FOOD placard.
172.554 POISON placard.
172.556 RADIOACTIVE placard.
172.558 CORROSIVE placard.
172.560 CLASS 9 placard.

**Subpart G--Emergency Response Information**

Sec.
172.600 Applicability and general requirements.
172.602 Emergency response information.
172.604 Emergency response telephone number.

**Subpart H--Training**

Sec.
172.700 Purpose and scope.
172.701 Federal-State relationship.
172.702 Applicability and responsibility for training.
172.704 Training requirements.
172.9990A APPENDIX A--OFFICE OF HAZARDOUS MATERIALS TRANSPORTATION COLOR TOLERANCE CHARTS AND TABLES
172.9990B APPENDIX B--[Removed and Reserved]

## 172.9990C APPENDIX C-- DIMENSIONAL SPECIFICATIONS FOR RECOMMENDED PLACARD HOLDER

AUTHORITY: 49 App. U.S.C. 1803, 1804, 1805, 1808; 49 CFR part 1, unless otherwise noted.

## IDENTIFICATION NUMBER CROSS REFERENCE INDEX TO PROPER SHIPPING NAMES IN §172.101

NOTE: This listing is provided for information purposes only. It is compiled and kept up-to-date by RSPA

IDENTIFICATION NUMBER CROSS REFERENCE INDEX TO PROPER SHIPPING NAMES IN §172.101

NOTE: This listing is provided for information purposes only. It is compiled and kept up-to-date by RSPA

IDENTIFICATION NUMBER CROSS REFERENCE INDEX TO PROPER SHIPPING NAMES IN §172.101

NOTE: This listing is provided for information purposes only. It is compiled and kept up-to-date by RSPA

| (1)-Identification Number | (2)-Description |
|---|---|
| UN0004 | Ammonium picrate |
| UN0005 | Cartridges for weapons |
| NA0006 | Explosive pest control devices |
| UN0006 | Cartridges for weapons |
| UN0007 | Cartridges for weapons |
| UN0009 | Ammunition, incendiary |
| UN0010 | Ammunition, incendiary |
| UN0012 | Cartridges for weapons, inert projectile *or* Cartridges, small arms |
| UN0014 | Cartridges for weapons, blank *or* Cartridges, small arms, blank |
| UN0015 | Ammunition, smoke |
| UN0016 | Ammunition, smoke |
| UN0018 | Ammunition, tear-producing |
| UN0019 | Ammunition, tear-producing |
| UN0020 | Ammunition, toxic |
| UN0021 | Ammunition, toxic |
| UN0027 | Black powder *or* Gunpowder |
| UN0028 | Black powder, compressed *or* Gunpowder, compressed *or* Black powder, in pellets *or* Gunpowder, in pellets |
| UN0029 | Detonators, non-electric |
| UN0030 | Detonators, electric |
| UN0033 | Bombs |
| UN0034 | Bombs |
| UN0035 | Bombs |
| UN0037 | Bombs, photo-flash |
| UN0038 | Bombs, photo-flash |
| UN0039 | Bombs, photo-flash |
| UN0042 | Boosters |
| UN0043 | Bursters |
| UN0044 | Primers, cap type |
| UN0048 | Charges, demolition |
| UN0049 | Cartridges, flash |
| UN0050 | Cartridges, flash |
| UN0054 | Cartridges, signal |
| UN0055 | Cases, cartridge, empty with primer |
| UN0056 | Charges, depth |
| UN0059 | Charges, shaped, commercial |
| UN0060 | Charges, supplementary explosive |
| UN0065 | Cord, detonating |
| UN0066 | Cord, igniter |
| UN0070 | Cutters, cable, explosive |
| UN0072 | Cyclotrimethylenetrinitramine, wetted *or* Cyclonite, wetted *or* Hexogen, wetted *or* RDX, wetted |
| UN0073 | Detonators for ammunition |
| UN0074 | Diazodinitrophenol, wetted |
| UN0075 | Diethyleneglycol dinitrate, desensitized |
| UN0076 | Dinitrophenol |
| UN0077 | Dinitrophenolates |
| UN0078 | Dinitroresorcinol |
| UN0079 | Hexanitrodiphenylamine (Dipicrylamine; Hexyl) |
| UN0081 | Explosive, blasting, type A |
| UN0082 | Explosive, blasting, type B |
| UN0083 | Explosive, blasting, type C |
| UN0084 | Explosive, blasting, type D |
| UN0092 | Flares, surface |
| UN0093 | Flares, aerial |
| UN0094 | Flash powder |
| UN0099 | Fracturing devices, explosive |
| UN0101 | Fuse, instantaneous, non-detonating *or* Quickmatch |
| UN0102 | Cord detonating *or* Fuse detonating |
| UN0103 | Fuse, igniter |
| UN0104 | Cord, detonating, mild effect *or* Fuse, detonating, mild effect |
| UN0105 | Fuse, safety |
| UN0106 | Fuzes, detonating |
| UN0107 | Fuzes, detonating |
| UN0110 | Grenades, practice |
| UN0113 | Guanyl nitrosaminoguanylidene hydrazine, wetted |
| UN0114 | Guanyl nitrosaminoguanyltetrazene, wetted *or* Tetrazene, wetted |
| UN0118 | Hexolite |
| UN0121 | Igniters |
| UN0124 | Jet perforating guns, charged |
| UN0129 | Lead azide, wetted |
| UN0130 | Lead styphnate, wetted *or* Lead trinitroresorcinate, wetted |
| UN0131 | Lighters, fuse |
| UN0132 | Deflagrating metal salts of aromatic nitro-derivatives, n.o.s |
| NA0133 | Mannitol hexanitrate (Nitromannite), wetted |
| UN0135 | Mercury fulminate, wetted |
| UN0136 | Mines |
| UN0137 | Mines |
| UN0138 | Mines |
| UN0143 | Nitroglycerin, desensitized |
| UN0144 | Nitroglycerin, solution in alcohol |
| UN0146 | Nitrostarch |
| UN0147 | Nitro urea |
| NA0150 | Pentaerythrite tetranitrate (Pentaerythritol tetranitrate; PETN) wetted |
| UN0151 | Pentolite |
| UN0153 | Trinitroaniline *or* Picramide |
| UN0154 | Trinitrophenol *or* Picric acid |
| UN0155 | Trinitrochlorobenzene *or* Picryl chloride |
| UN0158 | Potassium salts of aromatic nitro-derivatives |
| UN0159 | Powder cake, wetted *or* Powder paste, wetted |
| UN0160 | Powder, smokeless |
| UN0161 | Powder, smokeless |
| UN0167 | Projectiles |
| UN0168 | Projectiles |
| UN0169 | Projectiles |
| UN0171 | Ammunition, illuminating |
| UN0173 | Release devices, explosive |
| UN0174 | Rivets, explosive |
| UN0180 | Rockets |
| UN0181 | Rockets |
| UN0182 | Rockets |
| UN0186 | Rocket motors |
| UN0190 | Samples, explosive |
| UN0191 | Signal devices, hand |
| UN0192 | Signals, railway track, explosive |
| UN0193 | Signals, railway track, explosive |
| UN0194 | Signals, distress |
| UN0195 | Signals, distress |
| UN0196 | Signals, smoke |
| UN0197 | Signals, smoke |
| UN0203 | Sodium salts of aromatic nitro-derivatives, n.o.s. |
| UN0204 | Sounding devices, explosive |
| UN0207 | Tetranitroanilene |
| UN0208 | Trinitrophenylmethylnitramine *or* Tetryl |
| UN0209 | Trinitrotoluene *or* TNT |
| UN0212 | Tracers for ammunition |
| UN0213 | Trinitroanisole |
| UN0214 | Trinitrobenzene |
| UN0215 | Trinitrobenzoic acid |
| UN0216 | Trinitro-meta-creso |
| UN0217 | Trinitronaphthalene |
| UN0218 | Trinitrophenetole |
| UN0219 | Trinitroresorcinol *or* Styphnic acid |
| UN0220 | Urea nitrate |
| UN0221 | Warheads, torpedo |
| UN0222 | Ammonium nitrate |
| UN0224 | Barium azide |
| UN0225 | Boosters with detonator |
| UN0226 | Cyclotetramethylenetetranitramine, wetted *or* HMX, wetted *or* Octogen, wetted |
| UN0234 | Sodium dinitro-o-cresolate |
| UN0235 | Sodium picramate |
| UN0236 | Zirconium picramate |
| UN0237 | Charges, shaped, flexible, linear |
| UN0238 | Rockets, line-throwing |
| UN0240 | Rockets, line-throwing |
| UN0241 | Explosive, blasting, type E |
| UN0242 | Charges, propelling, for cannon |
| UN0243 | Ammunition, incendiary, white phosphorus |
| UN0244 | Ammunition, incendiary, white phosphorus |
| UN0245 | Ammunition smoke, white phosphorus |
| UN0246 | Ammunition, smoke, white phosphorus |
| UN0247 | Ammunition, incendiary |
| UN0248 | Contrivances, water-activated |
| UN0249 | Contrivances, water-activated |
| UN0250 | Rocket motors with hypergolic liquids |
| UN0254 | Ammunition, illuminating |
| UN0255 | Detonators, electric |

## IDENTIFICATION NUMBER CROSS REFERENCE INDEX TO PROPER SHIPPING NAMES IN §172.101

**NOTE:** This listing is provided for information purposes only. It is compiled and kept up-to-date by RSPA

IDENTIFICATION NUMBER CROSS REFERENCE INDEX TO PROPER SHIPPING NAMES IN §172.101

NOTE: This listing is provided for information purposes only. It is compiled and kept up-to-date by RSPA

IDENTIFICATION NUMBER CROSS REFERENCE INDEX TO PROPER SHIPPING NAMES IN §172.101

NOTE: This listing is provided for information purposes only. It is compiled and kept up-to-date by RSPA

| (1)-Identification Number | (2)-Description |
|---|---|
| UN0257 | Fuzes, detonating |
| UN0266 | Octolite *or* Octol |
| UN0267 | Detonators, non-electric |
| UN0268 | Boosters with detonator |
| UN0271 | Charges, propelling, for rocket motors |
| UN0272 | Charges, propelling, for rocket motors |
| NA0273 | Propellant explosive, solid |
| UN0273 | Charges, propelling, for rocket motors |
| NA0274 | Propellant explosive, solid |
| UN0274 | Charges, propelling, for rocket motors |
| UN0275 | Cartridges, power device |
| UN0276 | Cartridges, power device |
| UN0277 | Cartridges, oil well |
| UN0278 | Cartridges, oil well |
| UN0279 | Charges, propelling, for cannon |
| UN0280 | Rocket motors |
| UN0281 | Rocket motors |
| UN0282 | Nitroguanidine *or* Picrite |
| UN0283 | Boosters |
| UN0284 | Grenades |
| UN0285 | Grenades |
| UN0286 | Warheads, rocket |
| UN0287 | Warheads, rocket |
| UN0288 | Charges, shaped, flexible, linear |
| UN0289 | Cord, detonating |
| UN0290 | Cord, detonating *or* Fuse, detonating |
| UN0291 | Bombs |
| UN0292 | Grenades |
| UN0293 | Grenades |
| UN0294 | Mines |
| UN0295 | Rockets |
| UN0296 | Sounding devices, explosive |
| UN0297 | Ammunition, illuminating |
| UN0299 | Bombs, photo-flash |
| UN0300 | Ammunition, incendiary |
| UN0301 | Ammunition, tear-producing |
| UN0303 | Ammunition, smoke |
| UN0305 | Flash powder |
| UN0306 | Tracers for ammunition |
| UN0312 | Cartridges, signal |
| UN0313 | Signals, smoke |
| UN0314 | Igniters |
| UN0315 | Igniters |
| UN0316 | Fuzes, igniting |
| UN0317 | Fuzes, igniting |
| UN0318 | Grenades, practice |
| UN0319 | Primers, tubular |
| UN0320 | Primers, tubular |
| UN0321 | Cartridges for weapons |
| UN0322 | Rocket motors with hypergolic liquids |
| UN0323 | Cartridges, power device |
| UN0324 | Projectiles |
| UN0325 | Igniters |
| UN0326 | Cartridges for weapons, blank |
| UN0327 | Cartridges for weapons, blank *or* Cartridges, small arms, blank |
| UN0328 | Cartridges for weapons, inert projectile *or* Cartridges, small arms |
| UN0329 | Torpedoes |
| UN0330 | Torpedoes |
| NA0331 | Ammonium nitrate-fuel oil mixture (containing only prilled ammonium nitrate and fuel oil) |
| UN0331 | Explosive, blasting, type B *or* Agent blasting, Type B |
| UN0332 | Explosive, blasting, type E *or* Agent blasting, Type E |
| UN0333 | Fireworks |
| UN0334 | Fireworks |
| UN0335 | Fireworks |
| UN0336 | Fireworks |
| UN0337 | Fireworks |
| UN0338 | Cartridges for weapons, blank *or* Cartridges, small arms, blank |
| UN0339 | Cartridges for weapons, inert projectile *or* Cartridges, small arms |
| UN0340 | Nitrocellulose |
| UN0341 | Nitrocellulose |
| UN0342 | Nitrocellulose, wetted |
| UN0344 | Projectiles |
| UN0345 | Projectiles |
| UN0346 | Projectiles |
| UN0347 | Projectiles |
| UN0348 | Cartridges for weapons |
| NA0349 | Grenades, empty primed |
| UN0349 | Articles, explosive, n.o.s. |
| NA0350 | Boosters with detonator |
| UN0350 | Articles, explosive, n.o.s. |
| UN0351 | Articles, explosive, n.o.s. |
| UN0352 | Articles, explosive, n.o.s. |
| UN0353 | Articles, explosive, n.o.s. |
| UN0354 | Articles, explosive, n.o.s. |
| UN0355 | Articles, explosive, n.o.s. |
| UN0356 | Articles, explosive, n.o.s. |
| UN0357 | Substances, explosive, n.o.s. |
| UN0358 | Substances, explosive, n.o.s. |
| UN0359 | Substances, explosive, n.o.s. |
| UN0360 | Detonator assemblies, non-electric |
| UN0361 | Detonator assemblies, non-electric |
| UN0362 | Ammunition, practice |
| UN0363 | Ammunition, proof |
| UN0364 | Detonators for ammunition |
| UN0365 | Detonators for ammunition |
| UN0366 | Detonators for ammunition |
| UN0367 | Fuzes, detonating |
| UN0368 | Fuzes, igniting |
| UN0369 | Warheads, rocket |
| UN0370 | Warheads, rocket |
| UN0371 | Warheads, rocket |
| UN0372 | Grenades, practice |
| UN0373 | Signal devices, hand |
| UN0374 | Sounding devices, explosive |
| UN0375 | Sounding devices, explosive |
| UN0376 | Primers, tubular |
| UN0377 | Primers, cap type |
| UN0378 | Primers, cap type |
| UN0379 | Cases, cartridges, empty with primer |
| UN0380 | Articles, pyrophoric |
| UN0381 | Cartridges, power device |
| UN0382 | Components, explosive train, n.o.s. |
| UN0383 | Components, explosive train, n.o.s. |
| UN0384 | Components, explosive train, n.o.s. |
| UN0385 | 5-Nitrobenzotriazol |
| UN0386 | Trinitrobenzenesulfonic acid |
| UN0387 | Trinitrofluorenone |
| UN0388 | Trinitrotoluene and Trinitrobenzene mixtures *or* Trinitrotoluene *or* TNT and trinitrobenzene mixtures *or* TNT and hexanitrostilbene mixtures and Hexanitrostilbene mixtures |
| UN0389 | Trinitrotoluene mixtures containing Trinitrobenzene and Hexanitrostilbene *or* TNT mixtures containing trinitrobenzene and hexanitrostilbene |
| UN0390 | Tritonal |
| UN0391 | RDX and HMX mixtures, wetted |
| UN0392 | Hexanitrostilbene |
| UN0393 | Hexatonal, cast |
| UN0394 | Trinitroresorcinol, wetted *or* Styphnic acid, wetted |
| UN0395 | Rocket motors, liquid fueled |
| UN0396 | Rocket motors, liquid fueled |
| UN0397 | Rockets, liquid fueled |
| UN0398 | Rockets, liquid fueled |
| UN0399 | Bombs with flammable liquid |
| UN0400 | Bombs with flammable liquid |
| UN0401 | Dipicryl sulfide |
| UN0402 | Ammonium perchlorate |
| UN0403 | Flares, aerial |
| UN0404 | Flares, aerial |
| UN0405 | Cartridges, signal |
| UN0406 | Dinitrosobenzene |
| UN0407 | Tetrazol-1-acetic acid |
| UN0408 | Fuzes, detonating |
| UN0409 | Fuzes, detonating |
| UN0410 | Fuzes, detonating |
| UN0411 | Pentaerythrite tetranitrate; Pentaerythritoltetranitrate (PETN) |
| NA0412 | Explosive pest control devices |
| UN0412 | Cartridges for weapons |
| UN0413 | Cartridges for weapons, blank |
| UN0414 | Charges, propelling, for cannon |
| UN0415 | Charges, propelling, for rocket motors |
| UN0416 | Charges, propelling, for rocket motors |

## IDENTIFICATION NUMBER CROSS REFERENCE INDEX TO PROPER SHIPPING NAMES IN §172.101

**NOTE:** This listing is provided for information purposes only. It is compiled and kept up-to-date by RSPA

| (1)-Identification Number | (2)-Description |
|---|---|
| UN0417 | Cartridges for weapons, inert projectile *or* Cartridges, small arms |
| UN0418 | Flares surface |
| UN0419 | Flares, surface |
| UN0420 | Flares, aerial |
| UN0421 | Flares, aerial |
| UN0424 | Projectiles |
| UN0425 | Projectiles |
| UN0426 | Projectiles |
| UN0427 | Projectiles |
| UN0428 | Articles, pyrotechnic |
| UN0429 | Articles, pyrotechnic |
| UN0430 | Articles, pyrotechnic |
| UN0431 | Articles, pyrotechnic |
| UN0432 | Articles, pyrotechnic |
| UN0433 | Powder cake, wetted *or* Powder paste, wetted |
| UN0434 | Projectiles |
| UN0435 | Projectiles |
| UN0436 | Rockets |
| UN0437 | Rockets |
| UN0438 | Rockets |
| UN0439 | Charges, shaped, commercial |
| UN0440 | Charges, shaped, commercial |
| UN0441 | Charges, shaped, commercial |
| UN0442 | Charges, explosive, commercial |
| UN0443 | Charges, explosive, commercial |
| UN0444 | Charges, explosive, commercial |
| UN0445 | Charges, explosive, commercial |
| UN0446 | Cases, combustible, empty, without primer |
| UN0447 | Cases, combustible, empty, without primer |
| UN0448 | 5-Mercaptotetrazol- 1 -acetic acid |
| UN0449 | Torpedoes, liquid fuelled |
| UN0450 | Torpedoes, liquid fuelled |
| UN0451 | Torpedoes |
| UN0452 | Grenades practice |
| UN0453 | Rockets, line throwing |
| UN0454 | Igniters |
| UN0455 | Detonators, non-electric |
| UN0456 | Detonators, electric |
| UN0457 | Charges, bursting, plastics bonded |
| UN0458 | Charges, bursting, plastics bonded |
| UN0459 | Charges, bursting, plastics bonded |
| UN0460 | Charges, bursting, plastics bonded |
| UN0461 | Components, explosive train, n.o.s. |
| UN0462 | Articles, explosive, n.o.s. |
| UN0463 | Articles, explosive, n.o.s. |
| UN0464 | Articles, explosive, n.o.s. |
| UN0465 | Articles, explosive, n.o.s. |
| UN0466 | Articles, explosive, n.o.s. |
| UN0467 | Articles, explosive, n.o.s. |


## IDENTIFICATION NUMBER CROSS REFERENCE INDEX TO PROPER SHIPPING NAMES IN §172.101

**NOTE:** This listing is provided for information purposes only. It is compiled and kept up-to-date by RSPA


| (1)-Identification Number | (2)-Description |
|---|---|
| UN0468 | Articles, explosive, n.o.s. |
| UN0469 | Articles, explosive, n.o.s. |
| UN0470 | Articles, explosive, n.o.s. |
| UN0471 | Articles, explosive, n.o.s. |
| UN0472 | Articles, explosive, n.o.s. |
| NA0473 | Barium styphnate |
| NA0473 | Lead mononitroresorcinate |
| NA0473 | Nitrosoguanidine |
| UN0473 | Substances, explosive, n.o.s. |
| NA0474 | Propellant explosive, liquid |
| UN0474 | Substances, explosive, n.o.s. |
| UN0475 | Substances, explosive, n.o.s. |
| UN0476 | Substances, explosive, n.o.s. |
| NA0477 | Propellant explosive, liquid |
| UN0477 | Substances, explosive, n.o.s. |
| UN0478 | Substances, explosive, n.o.s. |
| UN0479 | Substances, explosives, n.o.s. |
| UN0480 | Substances, explosives, n.o.s. |
| UN0481 | Substances, explosive, n.o.s. |
| UN0482 | Substances, explosive, very insensitive, n.o.s., *or* Substances, EVI, n.o.s. |
| UN0483 | Cyclotrimethylenetrinitramine, desensitize *or* Cyclonite, desensitized *or* Hexogen, desensitized *or* RDX, desensitized; Hexogen; RDX), desensitized |
| UN0484 | Cyclotetramethylenetetranitramine, desensitized *or* Octogen, desensitized *or* HMX, desensitized |
| UN0485 | Substances, explosive, n.o.s. |
| UN0486 | Articles, explosive, extremely insensitive Articles, EEI |
| UN0487 | Signals, smoke |
| UN0488 | Ammunition, practice |
| UN0489 | Dinitroglycoluril *or* Dingu |
| UN0490 | Nitrotriazolone *or* NTO |
| UN1001 | Acetylene, dissolved |
| UN1002 | Air, compressed |
| UN1003 | Air, refrigerated liquid |
| UN1005 | Ammonia anhydrous liquefied or Ammonia solutions |
| UN1005 | Ammonia, anhydrous, liquefied *or* Ammonia solutions |
| UN1006 | Argon, compressed |
| UN1008 | Boron trifluoride |
| UN1009 | Bromotrifluoromethane |
| UN1010 | Butadienes, inhibited |
| UN1011 | Butane or Butane mixtures |
| UN1012 | Butylene |
| UN1013 | Carbon dioxide |
| UN1014 | Carbon dioxide and oxygen mixtures |
| UN1015 | Carbon dioxide and nitrous oxide mixture |
| UN1016 | Carbon monoxide |
| UN1017 | Chlorine |
| UN1018 | Chlorodifluoromethane |
| UN1020 | Chloropentafluoroethane |
| UN1021 | Chlorotetrafluoroethane |
| UN1022 | Chlorotrifluoromethane |
| UN1023 | Coal gas |


## IDENTIFICATION NUMBER CROSS REFERENCE INDEX TO PROPER SHIPPING NAMES IN §172.101

**NOTE:** This listing is provided for information purposes only. It is compiled and kept up-to-date by RSPA


| (1)-Identification Number | (2)-Description |
|---|---|
| UN1026 | Cyanogen, liquefied |
| UN1027 | Cyclopropane, liquified |
| UN1028 | Dichlorodifluoromethane |
| UN1029 | Dichlorofluoromethane |
| UN1030 | Difluoroethane |
| UN1032 | Dimethylamine, anhydrous |
| UN1033 | Dimethyl ether |
| UN1035 | Ethane, compressed |
| UN1036 | Ethylamine |
| UN1037 | Ethyl chloride |
| UN1038 | Ethylene, refrigerated liquid |
| UN1039 | Ethyl methyl ether |
| UN1040 | Ethylene oxide |
| UN1041 | Carbon dioxide and ethylene oxide mixtures |
| UN1043 | Fertilizer ammoniating solution |
| UN1044 | Fire extinguishers |
| UN1045 | Fluorine, compressed |
| UN1046 | Helium, compressed |
| UN1048 | Hydrogen bromide, anhydrous |
| UN1049 | Hydrogen, compressed |
| UN1050 | Hydrogen chloride, anhydrous |
| UN1051 | Hydrogen cyanide, anhydrous, stabilized |
| UN1052 | Hydrogen fluoride, anhydrous |
| UN1053 | Hydrogen sulfide, liquefied |
| UN1055 | Isobutylene |
| UN1056 | Krypton, compressed |
| UN1057 | Lighters *or* Lighter refills (cigarettes) containing flammable gas |
| UN1058 | Liquefied gases |
| UN1060 | Methylacetylene and propadiene mixtures, stabilized |
| UN1061 | Methylamine, anhydrous |
| UN1062 | Methyl bromide |
| UN1063 | Methyl chloride |
| UN1064 | Methyl mercaptan |
| UN1065 | Neon, compressed |
| UN1066 | Nitrogen, compressed |
| UN1067 | Dinitrogen tetroxide, liquefied |
| UN1069 | Nitrosyl chloride |
| UN1070 | Nitrous oxide, compressed |
| UN1071 | Oil gas |
| UN1072 | Oxygen, compressed |
| UN1073 | Oxygen, refrigerated liquid |
| UN1075 | Petroleum gases, liquefied |
| UN1076 | Phosgene |
| UN1077 | Propylene |
| UN1078 | Refrigerant gases, n.o.s. |
| UN1079 | Sulfur dioxide, liquefied |
| UN1080 | Sulfur hexafluoride |
| UN1081 | Tetrafluoroethylene, inhibited |
| UN1082 | Trifluorochloroethylene, inhibited |
| UN1083 | Trimethylamine, anhydrous |
| UN1085 | Vinyl bromide, inhibited |
| UN1086 | Vinyl chloride, inhibited |
| UN1087 | Vinyl methyl ether, inhibited |
| UN1088 | Acetal |

## IDENTIFICATION NUMBER CROSS REFERENCE INDEX TO PROPER SHIPPING NAMES IN §172.101

NOTE: This listing is provided for information purposes only. It is compiled and kept up-to-date by RSPA

| (1)-Identification Number | (2)-Description |
|---|---|
| UN1089 | Acetaldehyde |
| UN1090 | Acetone |
| UN1091 | Acetone oils |
| UN1092 | Acrolein, inhibited |
| UN1093 | Acrylonitrile, inhibited |
| UN1098 | Allyl alcohol |
| UN1099 | Allyl bromide |
| UN1100 | Allyl chloride |
| UN1104 | Amyl acetates |
| UN1105 | Amyl alcohols |
| UN1106 | Amylamines |
| UN1107 | Amyl chlorides |
| UN1108 | n-Amylene |
| UN1109 | Amyl formates |
| UN1110 | Amyl methyl ketone |
| UN1111 | Amyl mercaptans |
| UN1112 | Amyl nitrate |
| UN1113 | Amyl nitrites |
| UN1114 | Benzene |
| UN1118 | Brake fluid |
| UN1120 | Butanols |
| UN1123 | Butyl acetates |
| UN1125 | n-Butylamine |
| UN1126 | n-Butyl bromide |
| UN1127 | Chlorobutanes |
| UN1128 | n-Butyl formate |
| UN1129 | Butyraldehyde |
| UN1130 | Camphor oil |
| UN1131 | Carbon disulfide |
| UN1133 | Adhesives |
| UN1134 | Chlorobenzene |
| UN1135 | Ethylene chlorohydrin |
| UN1136 | Coal tar distillates, flammable |
| UN1139 | Coating solution |
| UN1143 | Crotonaldehyde, stabilized |
| UN1144 | Crotonylene |
| UN1145 | Cyclohexane |
| UN1146 | Cyclopentane |
| UN1147 | Decahydronaphthalene |
| UN1148 | Diacetone alcohol |
| UN1149 | Dibutyl ethers |
| UN1150 | Dichloroethylene |
| UN1152 | Dichloropentanes |
| UN1153 | Ethylene glycol diethyl ether |
| UN1154 | Diethylamine |
| UN1155 | Diethyl ether or Ethyl ether |
| UN1156 | Diethyl ketone |
| UN1157 | Diisobutyl ketone |
| UN1158 | Diisopropylamine |
| UN1159 | Diisopropyl ether |
| UN1160 | Dimethylamine solution |
| UN1161 | Dimethyl carbonate |
| UN1162 | Dimethyldichlorosilane |
| UN1163 | Dimethylhydrazine, unsymmetrical |
| UN1164 | Dimethyl sulfide |
| UN1165 | Dioxane |
| UN1166 | Dioxolane |
| UN1167 | Divinyl ether, inhibited |
| UN1169 | Extracts, aromatic, liquid |
| UN1170 | Ethanol *or* Ethyl alcohol *or* Ethanol solutions *or* Ethyl alcohol solutions |
| UN1171 | Ethylene glycol monoethyl ether |

IDENTIFICATION NUMBER CROSS REFERENCE INDEX TO PROPER SHIPPING NAMES IN §172.101

NOTE: This listing is provided for information purposes only. It is compiled and kept up-to-date by RSPA

| (1)-Identification Number | (2)-Description |
|---|---|
| UN1172 | Ethylene glycol monoethyl ether acetate |
| UN1173 | Ethyl acetate |
| UN1175 | Ethylbenzene |
| UN1176 | Ethyl borate |
| UN1177 | 2-Ethylbutyl acetate |
| UN1178 | 2-Ethylbutyraldehyde |
| UN1179 | Ethyl butyl ether |
| UN1180 | Ethyl butyrate |
| UN1181 | Ethyl chloroacetate |
| UN1182 | Ethyl chloroformate |
| UN1183 | Ethyldichlorosilane |
| UN1184 | Ethylene dichloride |
| UN1185 | Ethyleneimine, inhibited |
| UN1188 | Ethylene glycol monomethyl ether |
| UN1189 | Ethylene glycol monomethyl ether acetate |
| UN1190 | Ethyl formate |
| UN1191 | Octyl aldehydes |
| UN1192 | Ethyl lactate |
| UN1193 | Ethyl methyl ketone *or* Methyl ethyl ketone |
| UN1194 | Ethyl nitrite solutions |
| UN1195 | Ethyl propionate |
| UN1196 | Ethyltrichlorosilane |
| UN1197 | Extracts, flavoring, liquid |
| UN1198 | Formaldehyde, solutions |
| UN1199 | Furfural |
| UN1201 | Fusel oil |
| UN1202 | Gas oil |
| NA1203 | Gasohol |
| UN1203 | Gasoline |
| UN1204 | Nitroglycerin solution in alcohol |
| UN1206 | Heptanes |
| UN1207 | Hexaldehyde |
| UN1208 | Hexanes |
| UN1210 | Ink |
| UN1212 | Isobutanol *or* Isobutyl alcohol |
| UN1213 | Isobutyl acetate |
| UN1214 | Isobutylamine |
| UN1216 | Isooctenes |
| UN1218 | Isoprene, inhibited |
| UN1219 | Isopropanol *or* Isopropyl alcohol |
| UN1220 | Isopropyl acetate |
| UN1221 | Isopropylamine |
| UN1222 | Isopropyl nitrate |
| UN1223 | Kerosene |
| UN1224 | Ketones, liquid, n.o.s. |
| NA1226 | Lighters for cigars, cigarettes |
| UN1228 | Mercaptans, liquid, n.o.s. *or* Mercaptan mixtures, liquid, n.o.s. |
| UN1229 | Mesityl oxide |
| UN1230 | Methanol, *or* Methyl alcohol |
| UN1231 | Methyl acetate |
| UN1233 | Methylamyl acetate |
| UN1234 | Methylal |
| UN1235 | Methylamine, aqueous solution |
| UN1237 | Methyl butyrate |
| UN1238 | Methyl chloroformate |
| UN1239 | Methylchloromethyl ether |

IDENTIFICATION NUMBER CROSS REFERENCE INDEX TO PROPER SHIPPING NAMES IN §172.101

NOTE: This listing is provided for information purposes only. It is compiled and kept up-to-date by RSPA

| (1)-Identification Number | (2)-Description |
|---|---|
| UN1242 | Methyldichlorosilane |
| UN1243 | Methyl formate |
| UN1244 | Methylhydrazine |
| UN1245 | Methyl isobutyl ketone |
| UN1246 | Methyl isopropenyl ketone, inhibited |
| UN1247 | Methyl methacrylate monomer, inhibited |
| UN1248 | Methyl propionate |
| UN1249 | Methyl propyl ketone |
| UN1250 | Methyltrichlorosilane |
| UN1251 | Methyl vinyl ketone |
| UN1255 | Naphtha |
| UN1256 | Naphtha |
| UN1257 | Natural gasoline |
| UN1259 | Nickel carbonyl |
| UN1261 | Nitromethane |
| UN1262 | Octanes |
| UN1263 | Paint |
| UN1263 | Paint related material |
| UN1264 | Paraldehyde |
| UN1265 | n-Pentanes *or* Isopentane |
| UN1266 | Perfumery products |
| UN1267 | Petroleum crude oil |
| UN1268 | Petroleum distillates, n.o.s. |
| UN1270 | Petroleum oil |
| UN1271 | Petroleum spirit |
| UN1272 | Pine oil |
| UN1274 | n-Propanol *or* Propyl alcohol normal |
| UN1275 | Propionaldehyde |
| UN1276 | n-Propyl acetate |
| UN1277 | Propylamine |
| UN1278 | Propyl chloride |
| UN1279 | Propylene dichloride |
| UN1280 | Propylene oxide |
| UN1281 | Propyl formates |
| UN1282 | Pyridine |
| UN1286 | Rosin oil |
| UN1287 | Rubber solution |
| UN1288 | Shale oil |
| UN1289 | Sodium methylate solutions |
| UN1292 | Tetraethyl silicate |
| UN1293 | Tinctures, medicinal |
| UN1294 | Toluene |
| UN1295 | Trichlorosilane |
| UN1296 | Triethylamine |
| UN1297 | Trimethylamine, aqueous solutions |
| UN1298 | Trimethylchlorosilane |
| UN1299 | Turpentine |
| UN1300 | Turpentine substitute |
| UN1301 | Vinyl acetate, inhibited |
| UN1302 | Vinyl ethyl ether, inhibited |
| UN1303 | Vinylidene chloride, inhibited |
| UN1304 | Vinyl isobutyl ether, inhibited |
| UN1305 | Vinyltrichlorosilane |
| UN1306 | Wood preservatives, liquid |
| UN1307 | Xylenes |
| UN1308 | Zirconium suspended in a liquid |
| UN1309 | Aluminum powder, coated |
| UN1310 | Ammonium picrate, wetted |
| UN1312 | Borneol |
| UN1313 | Calcium resinate |

## IDENTIFICATION NUMBER CROSS REFERENCE INDEX TO PROPER SHIPPING NAMES IN §172.101

NOTE: This listing is provided for information purposes only. It is compiled and kept up-to-date by RSPA

| (1)-Identification Number | (2)-Description |
|---|---|
| UN1314 | Calcium resinate, fused |
| UN1318 | Cobalt resinate, precipitated |
| UN1320 | Dinitrophenol, wetted |
| UN1321 | Dinitrophenolates, wetted |
| UN1322 | Dinitroresorcinol, wetted |
| UN1323 | Ferrocerium |
| UN1324 | Films, nitrocellulose base |
| NA1325 | Fusee |
| NA1325 | Medicines |
| NA1325 | Smokeless powder for small arms |
| UN1325 | Flammable solids, n.o.s. |
| UN1326 | Hafnium powder, wetted |
| UN1328 | Hexamine |
| UN1330 | Manganese resinate |
| UN1331 | Matches, strike anywhere |
| UN1332 | Metaldehyde |
| UN1333 | Cerium |
| UN1334 | Naphthalene, crude *or* refined |
| UN1336 | Nitroguanidine, wetted *or* Picrite, wetted |
| UN1337 | Nitrostarch, wetted |
| UN1338 | Phosphorus, amorphous |
| UN1339 | Phosphorus heptasulfide |
| UN1340 | Phosphorus pentasulfide |
| UN1341 | Phosphorus sesquisulfide |
| UN1343 | Phosphorus trisulfide |
| NA1344 | Picric acid, wet, with not less than 10% water |
| UN1344 | Trinitrophenol, wetted |
| UN1346 | Silicon powder, amorphous |
| UN1347 | Silver picrate, wetted |
| UN1348 | Sodium dinitro-ortho-cresolate, wetted |
| UN1349 | Sodium picramate, wetted |
| UN1350 | Sulfur |
| UN1352 | Titanium powder, wetted |
| UN1353 | Toe puffs |
| UN1354 | Trinitrobenzene, wetted |
| UN1355 | Trinitrobenzoic acid, wetted |
| UN1356 | Trinitrotoluene, wetted |
| UN1357 | Urea nitrate, wetted |
| UN1358 | Zirconium powder, wetted |
| UN1360 | Calcium phosphide |
| NA1361 | Charcoal |
| UN1361 | Carbon |
| UN1362 | Carbon, activated |
| UN1363 | Copra |
| UN1364 | Cotton waste, oily |
| NA1365 | Cotton |
| UN1365 | Cotton, wet |
| UN1366 | Diethylzinc |
| UN1369 | p-Nitrosodimethylaniline |
| UN1370 | Dimethylzinc |
| UN1373 | Fibers *or* Fabrics, animal *or* vegetable, n.o.s. |
| UN1374 | Fish meal, unstablized *or* Fish scrap, unstabilized |
| UN1376 | Iron oxide, spent, *or* Iron sponge, spent |
| UN1378 | Metal catalyst, wetted |
| UN1379 | Paper, unsaturated oil treated |
| UN1380 | Pentaborane |

IDENTIFICATION NUMBER CROSS REFERENCE INDEX TO PROPER SHIPPING NAMES IN §172.101

NOTE: This listing is provided for information purposes only. It is compiled and kept up-to-date by RSPA

| (1)-Identification Number | (2)-Description |
|---|---|
| UN1381 | Phosphorus, white *or* yellow dry *or* under water *or* in solution |
| UN1382 | Potassium sulfide, anhydrous *or* Potassium sulfide |
| UN1383 | Pyrophoric metals, n.o.s. *or* Pyrophoric alloys, n.o.s. |
| UN1384 | Sodium dithionite *or* Sodium hydrosulfite |
| UN1385 | Sodium sulfide, anhydrous *or* Sodium sulfate |
| UN1386 | Seed cake |
| UN1389 | Alkali metal amalgams |
| UN1390 | Alkali metal amides |
| UN1391 | Alkali metal dispersions, *or* Alkaline earth metal dispersions |
| UN1392 | Alkaline earth metal amalgams |
| UN1393 | Alkaline earth metal alloys, n.o.s. |
| UN1394 | Aluminum carbide |
| UN1395 | Aluminum ferrosilicon powder |
| UN1396 | Aluminum powder, uncoated |
| UN1397 | Aluminum phosphide |
| UN1398 | Aluminum silicon powder, uncoated |
| UN1400 | Barium |
| UN1401 | Calcium |
| UN1402 | Calcium carbide |
| UN1403 | Calcium cyanamide |
| UN1404 | Calcium hydride |
| UN1405 | Calcium silicide |
| UN1407 | Cesium *or* Caesium |
| UN1408 | Ferrosilicon |
| UN1409 | Hydrides, metal, n.o.s. |
| UN1410 | Lithium aluminum hydride |
| UN1411 | Lithium aluminum hydride, ethereal |
| UN1413 | Lithium borohydride |
| UN1414 | Lithium hydride |
| UN1415 | Lithium |
| UN1417 | Lithium silicon |
| UN1418 | Magnesium, powder *or* Magnesium alloys, powder |
| UN1419 | Magnesium aluminum phosphide |
| UN1420 | Potassium, metal alloys |
| UN1421 | Alkali metal alloys, liquid, n.o.s. |
| UN1422 | Potassium sodium alloys |
| UN1423 | Rubidium |
| UN1426 | Sodium borohydride |
| UN1427 | Sodium hydride |
| UN1428 | Sodium |
| UN1431 | Sodium methylate |
| UN1432 | Sodium phosphide |
| UN1433 | Stannic phosphide |
| UN1435 | Zinc ashes |
| UN1436 | Zinc powder *or* Zinc dust |
| UN1437 | Zirconium hydride |
| UN1438 | Aluminum nitrate |
| UN1439 | Ammonium dichromate |
| UN1442 | Ammonium perchlorate |

IDENTIFICATION NUMBER CROSS REFERENCE INDEX TO PROPER SHIPPING NAMES IN §172.101

NOTE: This listing is provided for information purposes only. It is compiled and kept up-to-date by RSPA

| (1)-Identification Number | (2)-Description |
|---|---|
| UN1444 | Ammonium persulfate |
| UN1445 | Barium chlorate |
| UN1446 | Barium nitrate |
| UN1447 | Barium perchlorate |
| UN1448 | Barium permanganate |
| UN1449 | Barium peroxide |
| UN1450 | Bromates, inorganic, n.o.s. |
| UN1451 | Cesium nitrate *or* Caesium nitrate |
| UN1452 | Calcium chlorate |
| UN1453 | Calcium chlorite |
| UN1454 | Calcium nitrate |
| UN1455 | Calcium perchlorate |
| UN1456 | Calcium permanganate |
| UN1457 | Calcium peroxide |
| UN1458 | Chlorate and borate mixtures |
| UN1459 | Chlorate and magnesium chloride mixtures |
| UN1461 | Chlorates, inorganic, n.o.s. |
| UN1462 | Chlorites, inorganic, n.o.s. |
| NA1463 | Chromic acid, solid |
| UN1463 | Chromium trioxide, anhydrous |
| UN1465 | Didymium nitrate |
| UN1466 | Ferric nitrate |
| UN1467 | Guanidine nitrate |
| UN1469 | Lead nitrate |
| UN1470 | Lead perchlorate, solid *or* solution |
| UN1471 | Lithium hypochlorite, dry *or* Lithium hypochlorite mixtures, dry |
| UN1472 | Lithium peroxide |
| UN1473 | Magnesium bromate |
| UN1474 | Magnesium nitrate |
| UN1475 | Magnesium perchlorate |
| UN1476 | Magnesium peroxide |
| UN1477 | Nitrates, inorganic, n.o.s. |
| NA1479 | Medicines |
| UN1479 | Oxidizing substances, solid, n.o.s. |
| UN1481 | Perchlorates, inorganic, n.o.s. |
| UN1482 | Permanganates, inorganic, n.o.s. |
| UN1483 | Peroxides, inorganic, n.o.s. |
| UN1484 | Potassium bromate |
| UN1485 | Potassium chlorate |
| UN1486 | Potassium nitrate |
| UN1487 | Potassium nitrate and sodium nitrite mixtures |
| UN1488 | Potassium nitrite |
| UN1489 | Potassium perchlorate, solid *or* solution |
| UN1490 | Potassium permanganate |
| UN1491 | Potassium peroxide |
| UN1492 | Potassium persulfate |
| UN1493 | Silver nitrate |
| UN1494 | Sodium bromate |
| UN1495 | Sodium chlorate |
| UN1496 | Sodium chlorite |
| UN1498 | Sodium nitrate |
| UN1499 | Sodium nitrate and potassium nitrate mixtures |
| UN1500 | Sodium nitrite |
| UN1502 | Sodium perchlorate |

## IDENTIFICATION NUMBER CROSS REFERENCE INDEX TO PROPER SHIPPING NAMES IN §172.10'

**NOTE: This listing is provided for information purposes only. It is compiled and kept up-to-date by RSPA**

| (1)-Identification Number | (2)-Description |
|---|---|
| UN1503 | Sodium permanganate |
| UN1504 | Sodium peroxide |
| UN1505 | Sodium persulfate |
| UN1506 | Strontium chlorate, solid *or* solution |
| UN1507 | Strontium nitrate |
| UN1508 | Strontium perchlorate |
| UN1509 | Strontium peroxide |
| UN1510 | Tetranitromethane |
| UN1511 | Urea hydrogen peroxide |
| UN1512 | Zinc ammonium nitrite |
| UN1513 | Zinc chlorate |
| UN1514 | Zinc nitrate |
| UN1515 | Zinc permanganate |
| UN1516 | Zinc peroxide |
| UN1517 | Zirconium picramate, wetted |
| UN1541 | Acetone cyanohydrin, stabilized |
| UN1544 | Alkaloids, solid, n.o.s. *or* alkaloid salts, solid, n.o.s. |
| UN1545 | Allyl isothiocyanate, stabilized |
| UN1546 | Ammonium arsenate |
| UN1547 | Aniline |
| UN1548 | Aniline hydrochloride |
| NA1549 | Antimony tribromide, solid |
| NA1549 | Antimony tribromide, solution |
| NA1549 | Antimony trifluoride solution |
| NA1549 | Antimony trifluoride, solid |
| UN1549 | Antimony compounds, inorganic, solid, n.o.s. |
| UN1550 | Antimony lactate |
| UN1551 | Antimony potassium tartrate |
| UN1553 | Arsenic acid, liquid |
| UN1554 | Arsenic acid, solid |
| UN1555 | Arsenic bromide |
| NA1556 | Methyldichloroarsine |
| UN1556 | Arsenic compounds, liquid, n.o.s. |
| NA1557 | Arsenic sulfide |
| NA1557 | Arsenic trisulfide |
| UN1557 | Arsenic compounds, solid, n.o.s. |
| UN1558 | Arsenic |
| UN1559 | Arsenic pentoxide |
| UN1560 | Arsenic trichloride |
| UN1561 | Arsenic trioxide |
| UN1562 | Arsenical dust |
| UN1564 | Barium compounds, n.o.s. |
| UN1565 | Barium cyanide |
| UN1566 | Beryllium compounds, n.o.s. |
| UN1567 | Beryllium, powder |
| UN1569 | Bromoacetone |
| UN1570 | Brucine |
| UN1571 | Barium azide, wetted |
| UN1572 | Cacodylic acid |
| UN1573 | Calcium arsenate |
| NA1574 | Calcium arsenite, solid |
| UN1574 | Calcium arsenate and calcium arsenite, mixtures, solid |
| UN1575 | Calcium cyanide |
| UN1577 | Chlorodinitrobenzenes |
| UN1578 | Chloronitrobenzene |
| UN1578 | Chloronitrobenzenes |

## IDENTIFICATION NUMBER CROSS REFERENCE INDEX TO PROPER SHIPPING NAMES IN §172.101

**NOTE: This listing is provided for information purposes only. It is compiled and kept up-to-date by RSPA**

| (1)-Identification Number | (2)-Description |
|---|---|
| UN1579 | 4-Chloro-o-toluidine hydrochloride |
| UN1580 | Chloropicrin |
| UN1581 | Chloropicrin and methyl bromide mixtures |
| UN1582 | Chloropicrin and methyl chloride mixtures |
| UN1583 | Chloropicrin mixtures, n.o.s. |
| UN1584 | Cocculus |
| UN1585 | Copper acetoarsenite |
| UN1586 | Copper arsenite |
| UN1587 | Copper cyanide |
| UN1588 | Cyanides, inorganic, n.o.s. |
| UN1589 | Cyanogen chloride, inhibited |
| UN1590 | Dichloroanilines, solid *or* liquid |
| UN1591 | o-Dichlorobenzene |
| UN1592 | p-Dichlorobenzene |
| UN1593 | Dichloromethane |
| UN1594 | Diethyl sulfate |
| UN1595 | Dimethyl sulfate |
| UN1596 | Dinitroanilines |
| UN1597 | Dinitrobenzenes |
| UN1598 | Dinitro-o-cresol |
| UN1599 | Dinitrophenol solutions |
| UN1600 | Dinitrotoluenes, molten |
| UN1601 | Disinfectants, solid, n.o.s. |
| UN1602 | Dyes, liquid, n.o.s. *or* dye intermediates, liquid, n.o.s. |
| UN1603 | Ethyl bromoacetate |
| UN1604 | Ethylenediamine |
| UN1605 | Ethylene dibromide |
| UN1606 | Ferric arsenate |
| UN1607 | Ferric arsenite |
| UN1608 | Ferrous arsenate |
| UN1610 | Halogenated irritating liquids, n.o s. |
| UN1611 | Hexaethyl tetraphosphate |
| UN1612 | Hexaethyl tetraphosphate and compressed gas mixtures |
| NA1613 | Hydrocyanic acid, aqueous solutions |
| UN1614 | Hydrogen cyanide, anhydrous, stabilized |
| UN1616 | Lead acetate |
| UN1617 | Lead arsenates |
| UN1618 | Lead arsenites |
| UN1620 | Lead cyanide |
| UN1621 | London Purple |
| UN1622 | Magnesium arsenate |
| UN1623 | Mercuric arsenate |
| UN1624 | Mercuric chloride |
| UN1625 | Mercuric nitrate |
| UN1626 | Mercuric potassium cyanide |
| UN1627 | Mercurous nitrate |
| UN1629 | Mercury acetate |
| UN1630 | Mercury ammonium chloride |
| UN1631 | Mercury benzoate |
| UN1634 | Mercury bromides |
| UN1636 | Mercury cyanide |
| UN1637 | Mercury gluconate |
| UN1638 | Mercury iodide |
| UN1639 | Mercury nucleate |

## IDENTIFICATION NUMBER CROSS REFERENCE INDEX TO PROPER SHIPPING NAMES IN §172.101

**NOTE: This listing is provided for information purposes only. It is compiled and kept up-to-date by RSPA**

| (1)-Identification Number | (2)-Description |
|---|---|
| UN1640 | Mercury oleate |
| UN1641 | Mercury oxide |
| UN1642 | Mercury oxycyanide, desensitized |
| UN1643 | Mercury potassium iodide |
| UN1644 | Mercury salicylate |
| UN1645 | Mercury sulfates |
| UN1646 | Mercury thiocyanate |
| UN1647 | Methyl bromide and ethylene dibromide mixtures, liquid |
| UN1648 | Methyl cyanide |
| NA1649 | Tetraethyl lead, liquid |
| UN1649 | Motor fuel anti-knock mixtures |
| UN1650 | beta-Naphthylamine |
| UN1651 | Naphthylthiourea |
| UN1652 | Naphthylurea |
| UN1653 | Nickel cyanide |
| UN1654 | Nicotine |
| UN1655 | Nicotine compounds, solid, n.o.s. *or* Nicotine preparations, solid, n.o.s. |
| UN1656 | Nicotine hydrochloride *or* Nicotine hydrochloride solution |
| UN1657 | Nicotine salicylate |
| UN1658 | Nicotine sulfate |
| UN1659 | Nicotine tartrate |
| UN1660 | Nitric oxide |
| UN1661 | Nitroanilines |
| UN1662 | Nitrobenzene |
| UN1663 | Nitrophenols |
| UN1664 | Nitrotoluenes |
| UN1665 | Nitroxylenes, (o-; m-; p-) |
| UN1669 | Pentachloroethane |
| UN1670 | Perchloromethylmercaptan |
| UN1671 | Phenol, solid |
| UN1672 | Phenylcarbylamine chloride |
| UN1673 | Phenylenediamines |
| UN1674 | Phenylmercuric acetate |
| UN1677 | Potassium arsenate |
| UN1678 | Potassium arsenite |
| UN1679 | Potassium cuprocyanide |
| UN1680 | Potassium cyanide |
| UN1683 | Silver arsenite |
| UN1684 | Silver cyanide |
| UN1685 | Sodium arsenate |
| UN1686 | Sodium arsenite, aqueous solutions |
| UN1687 | Sodium azide |
| UN1688 | Sodium cacodylate |
| UN1689 | Sodium cyanide |
| UN1690 | Sodium fluoride |
| UN1691 | Strontium arsenite |
| UN1692 | Strychnine *or* Strychnine salts |
| NA1693 | Tear gas devices |
| UN1693 | Tear gas substances, n.o.s., liquid |
| UN1693 | Tear gas substances, n.o.s., solid |
| UN1694 | Bromobenzyl cyanides |
| UN1695 | Chloroacetone, stabilized |
| UN1697 | Chloroacetophenone |
| UN1698 | Diphenylamine chloroarsine |

## IDENTIFICATION NUMBER CROSS REFERENCE INDEX TO PROPER SHIPPING NAMES IN §172.101

**NOTE:** This listing is provided for information purposes only. It is compiled and kept up-to-date by RSPA

| (1)-Identification Number | (2)-Description |
|---|---|
| UN1699 | Diphenylchloroarsine, solid *or* liquid |
| UN1700 | Tear gas candles |
| UN1701 | Xylyl bromide |
| UN1702 | Tetrachloroethane |
| UN1703 | Tetraethyl dithiopyrophosphate and gases in solution *or* tetraethyl dithiopyrophospate and gas mixtures |
| UN1703 | Tetraethyl dithiopyrophosphate and gases in solution *or* Tetraethyl dithiopyrophosphate and gases mixtures |
| UN1704 | Tetraethyl dithiopyrophosphate |
| UN1705 | Tetraethyl pyrophosphate and compressed gas mixtures |
| UN1705 | Tetraethyl pyrophosphate and compressed gas mixtures |
| NA1707 | Thallium sulfate, solid |
| UN1707 | Thallium compounds, n.o.s. |
| UN1708 | Toluidines |
| UN1709 | 2,4-Toluylenediamine *or* 2,4-Toluenediamine |
| UN1710 | Trichloroethylene |
| UN1711 | Xylidines, solid *or* solution |
| UN1712 | Zinc arsenate *or* Zinc arsenite *or* Zinc arsenate and Zinc arsenite mixtures |
| UN1713 | Zinc cyanide |
| UN1714 | Zinc phosphide |
| UN1715 | Acetic anhydride |
| UN1716 | Acetyl bromide |
| UN1717 | Acetyl chloride |
| UN1718 | Butyl acid phosphate |
| UN1719 | Caustic alkali liquids, n.o.s. |
| UN1722 | Allyl chloroformate |
| UN1723 | Allyl iodide |
| UN1724 | Allyltrichlorosilane, stabilized |
| UN1725 | Aluminum bromide, anhydrous |
| UN1726 | Aluminum chloride, anhydrous |
| UN1727 | Ammonium hydrogen fluoride, solid |
| UN1728 | Amyltrichlorosilane |
| UN1729 | Anisoyl chloride |
| UN1730 | Antimony pentachloride, liquid |
| UN1731 | Antimony pentachloride, solutions |
| UN1732 | Antimony pentafluoride |
| UN1733 | Antimony trichloride, liquid |
| UN1733 | Antimony trichloride, solid |
| UN1736 | Benzoyl chloride |
| UN1737 | Benzyl bromide |
| UN1738 | Benzyl chloride |
| UN1739 | Benzyl chloroformate |
| UN1740 | Bifluorides, n.o.s. |
| UN1741 | Boron trichloride |
| UN1742 | Boron trifluoride acetic acid complex |

IDENTIFICATION NUMBER CROSS REFERENCE INDEX TO PROPER SHIPPING NAMES IN §172.101

NOTE: This listing is provided for information purposes only. It is compiled and kept up-to-date by RSPA

| (1)-Identification Number | (2)-Description |
|---|---|
| UN1743 | Boron trifluoride propionic acid complex |
| UN1744 | Bromine *or* Bromine solutions |
| UN1745 | Bromine pentafluoride |
| UN1746 | Bromine trifluoride |
| UN1747 | Butyltrichlorosilane |
| UN1748 | Calcium hypochlorite, dry *or* Calcium hypochlorite mixtures, dry |
| UN1749 | Chlorine trifluoride |
| UN1750 | Chloroacetic acid, liquid |
| UN1751 | Chloroacetic acid, solid |
| UN1752 | Chloroacetyl chloride |
| UN1753 | Chlorophenyltrichlorosilane |
| UN1754 | Chlorosulfonic acid |
| UN1755 | Chromic acid solution |
| UN1756 | Chromic fluoride, solid |
| UN1757 | Chromic fluoride, solution |
| UN1758 | Chromium oxychloride |
| NA1759 | Ferrous chloride, solid |
| NA1759 | Medicines |
| UN1759 | Corrosive solids, n.o.s. |
| NA1760 | Chemical kit |
| NA1760 | Compounds, cleaning liquid |
| NA1760 | Compounds, tree *or* weed killing, liquid |
| NA1760 | Ferrous chloride, solution |
| NA1760 | Medicines |
| NA1760 | Titanium sulfate solution |
| UN1760 | Corrosive liquids, n.o.s. |
| UN1761 | Cupriethylenediamine solution |
| UN1762 | Cyclohexenyltrichlorosilane |
| UN1763 | Cyclohexyltrichlorosilane |
| UN1764 | Dichloroacetic acid |
| UN1765 | Dichloroacetyl chloride |
| UN1766 | Dichlorophenyltrichlorosilane |
| UN1767 | Diethyldichlorosilane |
| UN1768 | Difluorophosphoric acid, anhydrous |
| UN1769 | Diphenyldichlorosilane |
| UN1770 | Diphenylmethyl bromide |
| UN1771 | Dodecyltrichlorosilane |
| UN1773 | Ferric chloride |
| UN1774 | Fire extinguisher charges |
| UN1775 | Fluoboric acid |
| UN1776 | Fluorophosphoric acid anhydrous |
| UN1777 | Fluorosulfonic acid |
| UN1778 | Fluorosilicic acid |
| UN1779 | Formic acid |
| UN1780 | Fumaryl chloride |
| UN1781 | Hexadecyltrichlorosilane |
| UN1782 | Hexafluorophosphoric acid |
| UN1783 | Hexamethylenediamine solution |
| UN1784 | Hexyltrichlorosilane |
| UN1786 | Hydrofluoric acid and Sulfuric acid mixtures |
| UN1787 | Hydriodic acid, solution |
| UN1788 | Hydrobromic acid solution |
| UN1789 | Hydrochloric acid, solution |
| UN1790 | Hydrofluoric acid, solution |
| UN1791 | Hypochlorite solutions |
| UN1792 | Iodine monochloride |
| UN1793 | Isopropyl acid phosphate |

IDENTIFICATION NUMBER CROSS REFERENCE INDEX TO PROPER SHIPPING NAMES IN §172.101

NOTE: This listing is provided for information purposes only. It is compiled and kept up-to-date by RSPA

| (1)-Identification Number | (2)-Description |
|---|---|
| UN1794 | Lead sulfate |
| UN1796 | Nitrating acid mixtures |
| UN1798 | Nitrohydrochloric acid |
| UN1799 | Nonyltrichlorosilane |
| UN1800 | Octadecyltrichlorosilane |
| UN1801 | Octyltrichlorosilane |
| UN1802 | Perchloric acid |
| UN1803 | Phenolsulfonic acid, liquid |
| UN1804 | Phenyltrichlorosilane |
| UN1805 | Phosphoric acid |
| UN1806 | Phosphorus pentachloride |
| UN1807 | Phosphorus pentoxide |
| UN1808 | Phosphorus tribromide |
| UN1809 | Phosphorus trichloride |
| UN1810 | Phosphorus oxychloride |
| UN1811 | Potassium bifluoride |
| UN1812 | Potassium fluoride |
| UN1813 | Potassium hydroxide, solid |
| UN1814 | Potassium hydroxide, solution |
| UN1815 | Propionyl chloride |
| UN1816 | Propyltrichlorosilane |
| UN1817 | Pyrosulfuryl chloride |
| UN1818 | Silicon tetrachloride |
| UN1819 | Sodium aluminate, solution |
| UN1821 | Sodium hydrogen sulfate, solid |
| UN1823 | Sodium hydroxide, solid |
| UN1824 | Sodium hydroxide solution |
| UN1825 | Sodium monoxide |
| UN1826 | Nitrating acid mixtures spent |
| UN1826 | Nitrating acid mixtures, spent |
| UN1827 | Stannic chloride, anhydrous |
| UN1828 | Sulfur chlorides |
| NA1829 | Sulfur trioxide, uninhibited |
| UN1829 | Sulfur trioxide, inhibited |
| UN1830 | Sulfuric acid |
| UN1831 | Sulfuric acid, fuming |
| UN1832 | Sulfuric acid, spent |
| UN1833 | Sulfurous acid |
| UN1834 | Sulfuryl chloride |
| UN1835 | Tetramethylammonium hydroxide |
| UN1836 | Thionyl chloride |
| UN1837 | Thiophosphoryl chloride |
| UN1838 | Titanium tetrachloride |
| UN1839 | Trichloroacetic acid |
| UN1840 | Zinc chloride, solution |
| UN1841 | Acetaldehyde ammonia |
| UN1843 | Ammonium dinitro-o-cresolate |
| UN1845 | Carbon dioxide, solid *or* Dry ice |
| UN1846 | Carbon tetrachloride |
| UN1847 | Potassium sulfide, hydrated |
| UN1848 | Propionic acid |
| UN1849 | Sodium sulfide, hydrated |
| UN1854 | Barium alloys, pyrophoric |
| UN1855 | Calcium, pyrophoric *or* Calcium alloys, pyrophoric |
| UN1855 | Hexafluoropropylene |
| UN1859 | Silicon tetrafluoride |
| UN1860 | Vinyl fluoride, inhibited |
| UN1862 | Ethyl crotonate |
| UN1863 | Fuel, aviation, turbine engine |

## IDENTIFICATION NUMBER CROSS REFERENCE INDEX TO PROPER SHIPPING NAMES IN §172.101

NOTE: This listing is provided for information purposes only. It is compiled and kept up-to-date by RSPA

| (1)-Identification Number | (2)-Description |
|---|---|
| UN1864 | Gas drips |
| UN1865 | n-Propyl nitrate |
| UN1866 | Resin solution |
| UN1868 | Decaborane |
| UN1869 | Magnesium *or* Magnesium alloys |
| UN1870 | Potassium borohydride |
| UN1871 | Titanium hydride |
| UN1872 | Lead dioxide |
| UN1873 | Perchloric acid |
| UN1884 | Barium oxide |
| UN1885 | Benzidine |
| UN1886 | Benzylidene chloride |
| UN1887 | Bromochloromethane |
| UN1888 | Chloroform |
| UN1889 | Cyanogen bromide |
| UN1891 | Ethyl bromide |
| UN1892 | Ethyldichloroarsine |
| UN1894 | Phenylmercuric hydroxide |
| UN1895 | Phenylmercuric nitrate |
| UN1897 | Tetrachloroethylene |
| UN1898 | Acetyl iodide |
| UN1902 | Diisooctyl acid phosphate |
| UN1903 | Disinfectants, corrosive liquid, n.o.s. |
| UN1905 | Selenic acid |
| UN1906 | Sludge, acid |
| UN1907 | Soda lime |
| UN1908 | Sodium chlorite solution |
| UN1910 | Calcium oxide |
| NA1911 | Diborane mixtures |
| UN1911 | Diborane |
| UN1912 | Methyl chloride and methylene chloride mixtures |
| UN1913 | Neon, refrigerated liquid |
| UN1914 | Butylpropionate |
| UN1915 | Cyclohexanone |
| UN1916 | 2,2-Dichlorodiethyl ether |
| UN1917 | Ethyl acrylate, inhibited |
| UN1918 | Isopropylbenzene |
| UN1919 | Methyl acrylate, inhibited |
| UN1920 | Nonanes |
| UN1921 | Propyleneimine, inhibited |
| UN1922 | Pyrrolidine |
| UN1923 | Calcium dithionite *or* Calcium hydrosulfite |
| UN1928 | Methyl magnesium bromide, in ethyl ether |
| UN1929 | Potassium dithionite *or* Potassium hydrosulfite |
| UN1931 | Zinc dithionite *or* Zinc hydrosulfite |
| UN1932 | Zirconium scrap |
| UN1935 | Cyanide solutions |
| UN1938 | Bromoacetic acid |
| UN1939 | Phosphorus oxybromide |
| UN1940 | Thioglycolic acid |
| UN1941 | Dibromodifluoromethane |
| UN1942 | Ammonium nitrate |
| UN1944 | Matches, safety |
| UN1945 | Matches, wax, Vesta |
| UN1950 | Aerosols |
| UN1951 | Argon, refrigerated liquid |
| UN1952 | Carbon dioxide and ethylene oxide mixtures |
| UN1953 | Compressed *or* Liquefied gases, flammable, toxic, n.o.s. |


## IDENTIFICATION NUMBER CROSS REFERENCE INDEX TO PROPER SHIPPING NAMES IN §172.101

NOTE: This listing is provided for information purposes only. It is compiled and kept up-to-date by RSPA


| (1)-Identification Number | (2)-Description |
|---|---|
| NA1954 | Refrigerant gases, n.o.s. *or* Dispersant gases, n.o.s. |
| NA1954 | Refrigerating machines |
| UN1954 | Compressed *or* Liquefied gases, flammable, n.o.s. |
| NA1955 | Organic phosphate, Organic phosphate compound, *or* Organic phosphorus compound; mixed with compressed gas |
| UN1955 | Compressed *or* Liquefied gases, toxic, n.o.s. |
| UN1955 | Compressed *or* Liquefied gases, toxic, n.o.s. |
| NA1956 | Accumulators, pressurized, pneumatic *or* hydraulic |
| NA1956 | Hexafluoropropylene oxide |
| UN1956 | Compressed *or* Liquefied gases, n.o.s. |
| UN1957 | Deuterium |
| UN1958 | Dichlorotetrafluoroethane |
| UN1959 | 1,1-Difluoroethylene |
| UN1960 | Engine starting fluid |
| NA1961 | Ethane-Propane mixture, refrigerated liquid |
| UN1961 | Ethane, refrigerated liquid |
| UN1962 | Ethylene, compressed |
| UN1963 | Helium, refrigerated liquid |
| UN1964 | Hydrocarbon gases, compressed, n.o.s. *or* Hydrocarbon gases mixtures, compressed, n.o.s. |
| UN1965 | Hydrocarbon gases, liquefied, n.o.s. *or* Hydrocarbon gases mixtures, liquefied, n.o.s. |
| UN1966 | Hydrogen, refrigerated liquid |
| NA1967 | Parathion and compressed gas mixture |
| UN1967 | Insecticide gases, toxic, n.o.s. |
| UN1968 | Insecticide gases |
| UN1968 | Insecticide gases, n.o.s. |
| UN1969 | Isobutane *or* Isobutane mixtures |
| UN1970 | Krypton, refrigerated liquid |
| UN1971 | Methane, compressed *or* Natural gas, compressed |
| UN1972 | Methane, refrigerated liquid |
| UN1973 | Chlorodifluoromethane and chloropentafluoroethane mixture |
| UN1974 | Chlorodifluorobromomethane |
| UN1975 | Nitric oxide and dinitrogen tetroxide mixtures (Nitric oxide and nitrogen dioxide mixtures) |
| UN1976 | Octafluorocyclobutane |
| UN1977 | Nitrogen, refrigerated liquid |
| UN1978 | Propane |
| UN1979 | Rare gases, mixtures |
| UN1980 | Rare gases and oxygen mixtures |
| UN1981 | Rare gases and nitrogen mixtures |
| UN1982 | Tetrafluoromethane |


## IDENTIFICATION NUMBER CROSS REFERENCE INDEX TO PROPER SHIPPING NAMES IN §172.101

NOTE: This listing is provided for information purposes only. It is compiled and kept up-to-date by RSPA


| (1)-Identification Number | (2)-Description |
|---|---|
| UN1983 | Chlorotrifluoroethane |
| UN1984 | Trifluoromethane |
| NA1986 | Denatured alcohol |
| NA1986 | Propargyl alcohol |
| UN1986 | Alcohols, toxic, n.o.s. |
| NA1987 | Denatured alcohol |
| UN1987 | Alcohols, n.o.s. |
| UN1988 | Aldehydes, toxic, n.o.s. |
| UN1989 | Aldehydes, n.o.s. |
| UN1991 | Chloroprene, inhibited |
| UN1992 | Flammable liquids, poisonous, n.o.s. |
| NA1993 | Combustible liquid, n.o.s. |
| NA1993 | Compounds, cleaning liquid |
| NA1993 | Compounds, tree *or* weed killing, liquid |
| NA1993 | Diesel fuel |
| NA1993 | Fuel oil |
| NA1993 | Medicines |
| NA1993 | Refrigerating machine |
| UN1993 | Flammable liquids, n.o.s. |
| UN1994 | Iron pentacarbonyl |
| NA1999 | Asphalt |
| UN1999 | Tars, liquid |
| UN2000 | Celluloid |
| UN2001 | Cobalt Naphthenates, powder |
| UN2002 | Celluloid, scrap |
| UN2003 | Metal alkyls, n.o.s. |
| UN2004 | Magnesium diamide |
| UN2005 | Magnesium diphenyl |
| UN2006 | Plastics, nitrocellulose based, spontaneously combustible, n.o.s. |
| UN2008 | Zirconium powder, dry |
| UN2009 | Zirconium, dry |
| UN2010 | Magnesium hydride |
| UN2011 | Magnesium phosphide |
| UN2012 | Potassium phosphide |
| UN2013 | Strontium phosphide |
| UN2014 | Hydrogen peroxide, aqueous solutions |
| UN2015 | Hydrogen peroxide, stabilized *or* Hydrogen peroxide aqueous solutions, stabilized |
| UN2016 | Ammunition, toxic, nonexplosive |
| UN2017 | Ammunition, tear-producing, non-explosive |
| UN2018 | Chloroanilines, solid |
| UN2019 | Chloroanilines, liquid |
| UN2020 | Chlorophenols, solid |
| UN2021 | Chlorophenols, liquid |
| UN2022 | Cresylic acid |
| UN2023 | Epichlorohydrin |
| UN2024 | Mercury compounds, liquid, n.o.s. |
| UN2025 | Mercury compounds, solid, n.o.s. |
| UN2026 | Phenylmercuric compounds, n.o.s. |
| UN2027 | Sodium arsenite, solid |
| UN2028 | Bombs, smoke, non-explosive |

**IDENTIFICATION NUMBER CROSS REFERENCE INDEX TO PROPER SHIPPING NAMES IN §172.101**

**NOTE: This listing is provided for information purposes only. It is compiled and kept up-to-date by RSPA**

| (1)-Identification Number | (2)-Description |
|---|---|
| UN2029 | Hydrazine, anhydrous *or* Hydrazine aqueous solutions |
| UN2030 | Hydrazine hydrate *or* Hydrazine aqueous solutions |
| UN2031 | Nitric acid |
| UN2032 | Nitric acid, red fuming |
| UN2033 | Potassium monoxide |
| UN2034 | Hydrogen and Methane mixtures, compressed |
| UN2035 | Trifluoroethane, compressed |
| UN2036 | Xenon |
| UN2037 | Receptacles, small, containing gas |
| UN2038 | Dinitrotoluenes |
| UN2044 | 2,2-Dimethylpropane |
| UN2045 | Isobutyraldehyde *or* isobutyl aldehyde |
| UN2046 | Cymenes |
| UN2047 | Dichloropropene |
| UN2048 | Dicyclopentadiene |
| UN2049 | Diethylbenzene |
| UN2050 | Diisobutylene, isomeric compounds |
| UN2051 | Dimethylethanolamine |
| UN2052 | Dipentene |
| UN2053 | Methyl isobutyl carbinol |
| UN2054 | Morpholine |
| UN2055 | Styrene monomer, inhibited |
| UN2056 | Tetrahydrofuran |
| UN2057 | Tripropylene |
| UN2058 | Valeraldehyde |
| UN2059 | Nitrocellulose, solution, flammable |
| NA2069 | Ammonium nitrate mixed fertilizers |
| UN2071 | Ammonium nitrate fertilizers |
| NA2072 | Ammonium nitrate fertilizers |
| UN2073 | Ammonia solutions |
| UN2074 | Acrylamide |
| UN2075 | Chloral, anhydrous, inhibited |
| UN2076 | Cresols |
| UN2077 | alpha-Naphthylamine |
| UN2078 | Toluene diisocyanate |
| UN2079 | Diethylenetriamine |
| UN2186 | Hydrogen chloride, refrigerated liquid |
| UN2187 | Carbon dioxide, refrigerated liquid |
| UN2188 | Arsine |
| UN2189 | Dichlorosilane |
| UN2190 | Oxygen difluoride |
| UN2191 | Sulfuryl fluoride |
| UN2192 | Germane |
| UN2193 | Hexafluoroethane |
| UN2194 | Selenium hexafluoride |
| UN2195 | Tellurium hexafluoride |
| UN2196 | Tungsten hexafluoride |
| UN2197 | Hydrogen iodide, anhydrous |
| UN2198 | Phosphorus pentafluoride |
| UN2199 | Phosphine |
| UN2200 | Propadiene, inhibited |
| UN2201 | Nitrous oxide, refrigerated liquid |

**IDENTIFICATION NUMBER CROSS REFERENCE INDEX TO PROPER SHIPPING NAMES IN §172.101**

**NOTE: This listing is provided for information purposes only. It is compiled and kept up-to-date by RSPA**

| (1)-Identification Number | (2)-Description |
|---|---|
| UN2202 | Hydrogen selenide, anhydrous |
| UN2203 | Silane |
| UN2204 | Carbonyl sulfide |
| UN2205 | Adiponitrile |
| UN2206 | Isocyanates, n.o.s. *or* Isocyanate solutions, n.o.s. |
| UN2207 | Isocyanates, liquid or solid, n.o.s. *or* Iscyanate solutions, n.o.s. |
| UN2208 | Calcium hypochlorite mixtures, dry |
| UN2209 | Formaldehyde, solutions |
| UN2210 | Maneb *or* Maneb preparations |
| UN2211 | Polystyrene beads, expandable |
| UN2212 | Blue Asbestos |
| UN2213 | Paraformaldehyde |
| UN2214 | Phthalic anhydride |
| NA2215 | Maleic acid |
| UN2215 | Maleic anhydride |
| UN2216 | Fish meal *or* Fish scrap stabilized |
| UN2217 | Seed cake |
| UN2218 | Acrylic acid, inhibited |
| UN2219 | Allyl glycidyl ether |
| UN2222 | Anisole |
| UN2224 | Benzonitrile |
| UN2225 | Benzene sulfonyl chloride |
| UN2226 | Benzotrichloride |
| UN2227 | n-Butyl methacrylate |
| UN2228 | Butylphenols, liquid |
| UN2229 | Butylphenols, solid |
| UN2232 | Chloroacetaldehyde |
| UN2233 | Chloroanisidines |
| UN2234 | Chlorobenzotrifluorides |
| UN2235 | Chlorobenzylchlorides |
| UN2236 | 3-Chloro-4-methylphenylisocyanate |
| UN2237 | Chloronitroanilines |
| UN2238 | Chlorotoluenes |
| UN2239 | Chlorotoluidines |
| UN2240 | Chromosulfuric acid |
| UN2241 | Cycloheptane |
| UN2242 | Cycloheptene |
| UN2243 | Cyclohexyl acetate |
| UN2244 | Cyclopentanol |
| UN2245 | Cyclopentanone |
| UN2246 | Cyclopentene |
| UN2247 | n-Decane |
| UN2248 | Di-n-butylamine |
| UN2249 | Dichlorodimethyl ether, symmetrical |
| UN2250 | Dichlorophenyl isocyanates |
| UN2251 | 2,5 Norbornadiene (Dicycloheptadiene) |
| UN2252 | 1,2-Dimethoxyethane |
| UN2253 | N,N-Dimethylaniline |
| UN2254 | Matches, fusee |
| NA2255 | Polyester resin kits |
| UN2256 | Cyclohexene |
| UN2257 | Potassium |
| UN2258 | 1,2-Propylenediamine |
| UN2259 | Triethylenetetramine |
| UN2260 | Tripropylamine |
| UN2261 | Xylenols |

**IDENTIFICATION NUMBER CROSS REFERENCE INDEX TO PROPER SHIPPING NAMES IN §172.101**

**NOTE: This listing is provided for information purposes only. It is compiled and kept up-to-date by RSPA**

| (1)-Identification Number | (2)-Description |
|---|---|
| UN2262 | Dimethylcarbamoyl chloride |
| UN2263 | Dimethylcyclohexanes |
| UN2264 | Dimethylcyclohexylamine |
| UN2265 | N,N-Dimethylformamide |
| UN2266 | Dimethyl-N-propylamine |
| UN2267 | Dimethyl thiophosphoryl chloride |
| UN2269 | 3,3'-Iminodipropylamine |
| UN2270 | Ethylamine, aqueous solution |
| UN2271 | Ethyl amyl ketone |
| UN2272 | N-Ethylaniline |
| UN2273 | 2-Ethylaniline |
| UN2274 | N-Ethyl-N-benzylaniline |
| UN2275 | 2-Ethylbutanol |
| UN2276 | 2-Ethylhexylamine |
| UN2277 | Ethyl methacrylate |
| UN2278 | n-Heptene |
| UN2279 | Hexachlorobutadiene |
| UN2280 | Hexamethylenediamine, solid |
| UN2281 | Hexamethylene diisocyanate |
| UN2282 | Hexanols |
| UN2283 | Isobutyl methacrylate |
| UN2284 | Isobutyronitrile |
| UN2285 | Isocyanatobenzotrifluorides |
| UN2286 | Pentamethylheptane |
| UN2287 | Isoheptenes |
| UN2288 | Isohexenes |
| UN2289 | Isophoronediamine |
| UN2290 | Isophorone diisocyanate |
| UN2291 | Lead compounds, soluble, n.o.s. |
| UN2293 | 4-Methoxy-4-methylpentan-2-one |
| UN2294 | N-Methylaniline |
| UN2295 | Methyl chloroacetate |
| UN2296 | Methyl cyclohexane |
| UN2297 | Methyl cyclohexanone |
| UN2298 | Methyl cyclopentane |
| UN2299 | Methyl dichloroacetate |
| UN2300 | 2-Methyl-5-ethylpyridine |
| UN2301 | 2-Methylfuran |
| UN2302 | 5-Methylhexan-2-one |
| UN2303 | Isopropenylbenzene |
| UN2304 | Naphthalene, molten |
| UN2305 | Nitrobenzenesulfonic acid |
| UN2306 | Nitrobenzotrifluorides |
| UN2307 | 3-Nitro-4-chlorobenzotrifluoride |
| UN2308 | Nitrosylsulfuric acid |
| UN2309 | Octadiene |
| UN2310 | Pentan-2,4-Dione |
| UN2311 | Phenetidines |
| UN2312 | Phenol, molten |
| UN2313 | Picolines |
| UN2315 | Polychlorinated biphenyls |
| UN2316 | Sodium cuprocyanide, solid |
| UN2317 | Sodium cuprocyanide, solution |
| UN2318 | Sodium hydrosulfide |
| UN2319 | Terpene hydrocarbons, n.o.s. |
| UN2320 | Tetraethylenepentamine |
| UN2321 | Trichlorobenzenes, liquid |

## IDENTIFICATION NUMBER CROSS REFERENCE INDEX TO PROPER SHIPPING NAMES IN §172.101

NOTE: This listing is provided for information purposes only. It is compiled and kept up-to-date by RSPA

| (1)-Identification Number | (2)-Description |
|---|---|
| UN2322 | Trichlorobutene |
| UN2323 | Triethyl phosphite |
| UN2324 | Triisobutylene |
| UN2325 | 1,3,5-Trimethylbenzene |
| UN2326 | Trimethylcyclohexylamine |
| UN2327 | Trimethylhexamethylene-diamines |
| UN2328 | Trimethylhexamethylene diisocyanate |
| UN2329 | Trimethyl phosphite |
| UN2330 | Undecane |
| UN2331 | Zinc chloride, anhydrous |
| UN2332 | Acetaldehyde oxime |
| UN2333 | Allyl acetate |
| UN2334 | Allylamine |
| UN2335 | Allyl ethyl ether |
| UN2336 | Allyl formate |
| UN2337 | Phenyl mercaptan |
| UN2338 | Benzotrifluoride |
| UN2339 | 2-Bromobutane |
| UN2340 | 2-Bromoethyl ethyl ether |
| UN2341 | 1-Bromo-3-methylbutane |
| UN2342 | Bromomethylpropanes |
| UN2343 | 2-Bromopentane |
| UN2344 | 2-Bromopropane |
| UN2345 | 3-Bromopropyne |
| UN2346 | Butanedione |
| UN2347 | Butyl mercaptans |
| UN2348 | Butylacrylate |
| UN2350 | Butyl methyl ether |
| UN2351 | Butyl nitrites |
| UN2352 | Butyl vinyl ether, inhibited |
| UN2353 | Butyryl chloride |
| UN2354 | Chloromethyl ethyl ether |
| UN2356 | 2-Chloropropane |
| UN2357 | Cyclohexylamine |
| UN2358 | Cyclooctatetraene |
| UN2359 | Diallylamine |
| UN2360 | Diallylether |
| UN2361 | Diisobutylamine |
| UN2362 | 1,1-Dichloroethane |
| UN2363 | Ethyl mercaptan |
| UN2364 | n-Propyl benzene |
| UN2366 | Diethyl carbonate |
| UN2367 | alpha-Methylvaleraldehyde |
| UN2368 | alpha-Pinene |
| UN2369 | Ethylene glycol monobutyl ether |
| UN2370 | 1-Hexene |
| UN2371 | Isopentenes |
| UN2372 | 1,2-Di-(dimethylamino) ethane |
| UN2373 | Diethoxymethane |
| UN2374 | 3,3-Diethoxypropene |
| UN2375 | Diethyl sulfide |
| UN2376 | 2,3-Dihydropyran |
| UN2377 | 1,1-Dimethoxyethane |
| UN2378 | 2-Dimethylaminoacetonitrile |
| UN2379 | 1,3-Dimethylbutylamine |
| UN2380 | Dimethyldiethoxysilane |
| UN2381 | Dimethyl disulfide |
| UN2382 | Dimethylhydrazine, symmetrical |
| UN2383 | Dipropylamine |
| UN2384 | Dipropyl ether |

IDENTIFICATION NUMBER CROSS REFERENCE INDEX TO PROPER SHIPPING NAMES IN §172.101

NOTE: This listing is provided for information purposes only. It is compiled and kept up-to-date by RSPA

| (1)-Identification Number | (2)-Description |
|---|---|
| UN2385 | Ethyl isobutyrate |
| UN2386 | 1-Ethylpiperidine |
| UN2387 | Fluorobenzene |
| UN2388 | Fluorotoluenes |
| UN2389 | Furan |
| UN2390 | 2-Iodobutane |
| UN2391 | Iodomethylpropanes |
| UN2392 | Iodopropanes |
| UN2393 | Isobutyl formate |
| UN2394 | Isobutyl propionate |
| UN2395 | Isobutyryl chloride |
| UN2396 | Methacrylaldehyde |
| UN2397 | 3-Methylbutan-2-one |
| UN2398 | Methyl tert butyl ether |
| UN2399 | 1-Methylpiperidine |
| UN2400 | Methyl isovalerate |
| UN2401 | Piperidine |
| UN2402 | Propanethiols |
| UN2403 | Isopropenyl acetate |
| UN2404 | Propionitrile |
| UN2405 | Isopropyl butyrate |
| UN2406 | Isopropyl isobutyrate |
| UN2407 | Isopropyl chloroformate |
| UN2409 | Isopropyl propionate |
| UN2410 | 1,2,3,6-Tetrahydropyridine |
| UN2411 | Butyronitrile |
| UN2412 | Tetrahydrothiophene |
| UN2413 | Tetrapropylorthotitanate |
| UN2414 | Thiophene |
| UN2416 | Trimethyl borate |
| UN2417 | Carbonyl fluoride |
| UN2418 | Sulfur tetrafluoride |
| UN2419 | Bromotrifluoroethylene |
| UN2420 | Hexafluoroacetone |
| UN2421 | Nitrogen trioxide |
| UN2422 | Octafluorobut-2-ene |
| UN2424 | Octafluoropropane |
| UN2426 | Ammonium nitrate, liquid |
| UN2427 | Potassium chlorate, solution |
| UN2428 | Sodium chlorate, solution |
| UN2429 | Calcium chlorate solution |
| UN2430 | Alkyl phenols, solid, n.o.s. |
| UN2431 | Anisidines |
| UN2432 | N,N-Diethyl aniline |
| UN2433 | Chloronitrotoluenes |
| UN2434 | Dibenzyldichlorosilane |
| UN2435 | Ethylphenyldichlorosilane |
| UN2436 | Thioacetic acid |
| UN2437 | Methylphenyldichlorosilane |
| UN2438 | Trimethylacetyl chloride |
| UN2439 | Sodium hydrogen fluoride |
| UN2440 | Stannic chloride, pentahydrate |
| UN2441 | Titanium trichloride, pyrophoric *or* Titanium trichloride mixtures, pyrophoric |
| UN2442 | Trichloroacetyl chloride |
| UN2443 | Vanadium oxytrichloride |
| UN2444 | Vanadium tetrachloride |
| UN2445 | Lithium alkyls |
| UN2446 | Nitrocresols |
| UN2447 | Phosphorus white, molten |
| UN2448 | Sulfur, molten |
| UN2449 | Oxalates |
| UN2451 | Nitrogen trifluoride |
| UN2452 | Ethylacetylene, inhibited |
| UN2453 | Ethyl fluoride |

IDENTIFICATION NUMBER CROSS REFERENCE INDEX TO PROPER SHIPPING NAMES IN §172.101

NOTE: This listing is provided for information purposes only. It is compiled and kept up-to-date by RSPA

| (1)-Identification Number | (2)-Description |
|---|---|
| UN2454 | Methyl fluoride |
| UN2456 | 2-Chloropropene |
| UN2457 | 2,3-Dimethylbutane |
| UN2458 | Hexadienes |
| UN2459 | 2-Methyl-1-butene |
| UN2460 | 2-Methyl-2-butene |
| UN2461 | Methylpentadienes |
| UN2463 | Aluminum hydride |
| UN2464 | Beryllium nitrate |
| UN2465 | Dichloroisocyanuric acid, dry *or* Dichloroisocyanuric acid salts |
| UN2466 | Potassium superoxide |
| UN2467 | Sodium percarbonate |
| NA2468 | (mono-(Trichloro) tetra-(monopotassium dichloro)-penta-s-triazinetrione, dry |
| UN2468 | Trichloroisocyanuric acid, dry |
| UN2469 | Zinc bromate |
| UN2470 | Phenylacetonitrile, liquid |
| UN2471 | Osmium tetroxide |
| UN2473 | Sodium arsanilate |
| UN2474 | Thiophosgene |
| UN2475 | Vanadium trichloride |
| UN2477 | Methyl isothiocyanate |
| UN2478 | Isocyanates, n.o.s. *or* Isocyanate solutions, n.o.s. (flashpoint less than 23 degrees C) |
| UN2480 | Methyl isocyanate |
| UN2481 | Ethyl isocyanate |
| UN2482 | n-Propyl isocyanate |
| UN2483 | Isopropyl isocyanate |
| UN2484 | Tert-Butyl isocyanate |
| UN2485 | n-Butyl isocyanate |
| UN2486 | Isobutyl isocyanate |
| UN2487 | Phenyl isocyanate |
| UN2488 | Cyclohexyl isocyanate |
| UN2489 | Diphenylmethane-4,4'diisocyanate |
| UN2490 | Dichloroisopropyl ether |
| UN2491 | Ethanolamine *or* Ethanolamine solutions |
| UN2493 | Hexamethyleneimine |
| UN2495 | Iodine pentafluoride |
| UN2496 | Propionic anhydride |
| UN2497 | Sodium phenolate, solid |
| UN2498 | 1,2,3,6-Tetrahydrobenzaldehyde |
| UN2501 | Tris-(1-aziridinyl)phosphine oxide, solution |
| UN2502 | Valeryl chloride |
| UN2503 | Zirconium tetrachloride |
| UN2504 | Tetrabromoethane |
| UN2505 | Ammonium fluoride |
| UN2506 | Ammonium hydrogen sulfate |
| UN2507 | Chloroplatinic acid, solid |
| UN2508 | Molybdenum pentachloride |
| UN2509 | Potassium hydrogen sulfate |
| UN2511 | alpha-Chloropropionic acid |
| UN2512 | Aminophenols |
| UN2513 | Bromoacetyl bromide |
| UN2514 | Bromobenzene |
| UN2515 | Bromoform |

## IDENTIFICATION NUMBER CROSS REFERENCE INDEX TO PROPER SHIPPING NAMES IN §172.101

**NOTE:** This listing is provided for information purposes only. It is compiled and kept up-to-date by RSPA

| (1)-Identification Number | (2)-Description |
|---|---|
| UN2516 | Carbon tetrabromide |
| UN2517 | Chlorodifluoroethanes |
| UN2518 | 1,5,9-Cyclododecatriene |
| UN2520 | Cyclooctadienes |
| UN2521 | Diketene, inhibited |
| UN2522 | Dimethylaminoethyl methacrylate |
| UN2524 | Ethyl orthoformate |
| UN2525 | Ethyl oxalate |
| UN2526 | Furfurylamine |
| UN2527 | Isobutyl acrylate |
| UN2528 | Isobutyl isobutyrate |
| UN2529 | Isobutyric acid |
| UN2530 | Isobutyric anhydride |
| UN2531 | Methacrylic acid, inhibited |
| UN2533 | Methyl trichloroacetate |
| UN2534 | Methylchlorosilane |
| UN2535 | Methylmorpholine |
| UN2536 | Methyltetrahydrofuran |
| UN2538 | Nitronaphthalene |
| UN2541 | Terpinolene |
| UN2542 | Tributylamine |
| UN2545 | Hafnium powder, dry |
| UN2546 | Titanium powder, dry |
| UN2547 | Sodium superoxide |
| UN2548 | Chlorine pentafluoride |
| UN2552 | Hexafluoroacetone hydrate |
| UN2553 | Naphtha |
| UN2554 | Methyl allyl chloride |
| UN2555 | Nitrocellulose with water |
| UN2556 | Nitrocellulose with alcohol |
| UN2557 | Nitrocellulose with plasticizing |
| UN2558 | Epibromohydrin |
| UN2560 | 2-Methylpentan-2-ol |
| UN2561 | 3-Methyl-1-butene |
| UN2564 | Trichloroacetic acid, solution |
| UN2565 | Dicyclohexylamine |
| UN2567 | Sodium pentachlorophenate |
| UN2570 | Cadmium compounds |
| UN2571 | Ethylsulfuric acid |
| UN2572 | Phenylhydrazine |
| UN2573 | Thallium chlorate |
| UN2574 | Tricresyl phosphate |
| UN2576 | Phosphorus oxybromide, molten |
| UN2577 | Phenylacetyl chloride |
| UN2578 | Phosphorus trioxide |
| UN2579 | Piperazine |
| UN2580 | Aluminum bromide, solution |
| UN2581 | Aluminum chloride, solution |
| UN2582 | Ferric chloride, solution |
| UN2583 | Alkyl, Aryl *or* Toluene sulfonic acid, solid |
| NA2584 | Dodecylbenzenesulfonic acid |
| UN2584 | Alkyl, Aryl *or* Toluene sulfonic acid, liquid |
| UN2585 | Alkyl, Aryl *or* Toluene sulfonic acid, solid |
| UN2586 | Alkyl, Aryl *or* Toluene sulfonic acid, liquid |
| UN2587 | Benzoquinone |
| UN2588 | Pesticides, solid, toxic, n.o.s. |
| UN2589 | Vinyl chloroacetate |


## IDENTIFICATION NUMBER CROSS REFERENCE INDEX TO PROPER SHIPPING NAMES IN §172.101

**NOTE:** This listing is provided for information purposes only. It is compiled and kept up-to-date by RSPA


| (1)-Identification Number | (2)-Description |
|---|---|
| UN2590 | White asbestos, (chrysotile, actinolite, anthophyllite, tremolite) |
| UN2591 | Xenon, refrigerated liquid |
| UN2599 | Chlorotrifluoromethane and trifluoromethane azeotropic mixture |
| UN2600 | Carbon monoxide and hydrogen mixture |
| UN2601 | Cyclobutane |
| UN2602 | Dichlorodifluoromethane and difluoroethane azeotropic mixture |
| UN2603 | Cycloheptatriene |
| UN2604 | Boron trifluoride diethyl etherate |
| UN2605 | Methoxymethyl isocyanate |
| UN2606 | Methyl orthosilicate |
| UN2607 | Acrolein dimer, stabilized |
| UN2608 | Nitropropanes |
| UN2609 | Triallyl borate |
| UN2610 | Triallylamine |
| UN2611 | Propylene chlorohydrin |
| UN2612 | Methyl propyl ether |
| UN2614 | Methallyl alcohol |
| UN2615 | Ethyl propyl ether |
| UN2616 | Triisopropyl borate |
| UN2617 | Methyl cyclohexanols |
| UN2618 | Vinyl toluene, inhibited |
| UN2619 | Benzyldimethylamine |
| UN2620 | Amyl butyrates |
| UN2621 | Acetyl methyl carbinol |
| UN2622 | Glycidaldehyde |
| UN2623 | Firelighters, solid |
| UN2624 | Magnesium silicide |
| UN2626 | Chloric acid solution |
| UN2627 | Nitrites, inorganic, n.o.s. |
| UN2628 | Potassium fluoroacetate |
| UN2629 | Sodium fluoroacetate |
| NA2630 | Sodium selenite |
| UN2630 | Selenates *or* Selenites |
| UN2642 | Fluoroacetic acid |
| UN2643 | Methyl bromoacetate |
| UN2644 | Methyl iodide |
| UN2645 | Phenacyl bromide |
| UN2646 | Hexachlorocyclopentadiene |
| UN2647 | Malononitrile |
| UN2648 | 1,2-Dibromobutan-3-one |
| UN2649 | 1,3-Dichloroacetone |
| UN2650 | 1,1-Dichloro-1-nitroethane |
| UN2651 | 4,4'-Diaminodiphenyl methane |
| UN2653 | Benzyl iodide |
| UN2655 | Potassium fluorosilicate |
| UN2656 | Quinoline |
| UN2657 | Selenium disulfide |
| UN2658 | Selenium powder |
| UN2659 | Sodium chloroacetate |
| UN2660 | Nitrotoluidines (mono) |
| UN2661 | Hexachloroacetone |
| UN2662 | Hydroquinone, liquid |
| UN2662 | Hydroquinone, solid |
| UN2664 | Dibromomethane |
| UN2666 | Ethyl cyanoacetate |
| UN2667 | Butyltoluenes |
| UN2668 | Chloroacetonitrile |
| UN2669 | Chlorocresols |


## IDENTIFICATION NUMBER CROSS REFERENCE INDEX TO PROPER SHIPPING NAMES IN §172.101

**NOTE:** This listing is provided for information purposes only. It is compiled and kept up-to-date by RSPA


| (1)-Identification Number | (2)-Description |
|---|---|
| UN2670 | Cyanuric chloride |
| UN2671 | Aminopyridines |
| UN2672 | Ammonia solutions |
| UN2673 | 2-Amino-4-chlorophenol |
| UN2674 | Sodium fluorosilicate |
| UN2676 | Stibine |
| UN2677 | Rubidium hydroxide solution |
| UN2678 | Rubidium hydroxide |
| UN2679 | Lithium hydroxide, solution |
| UN2680 | Lithium hydroxide, monohydrate *or* Lithium hydroxide, solid |
| UN2681 | Caesium hydroxide solution |
| UN2682 | Caesium hydroxide, solid |
| UN2683 | Ammonium sulfide solution |
| UN2684 | Diethylaminopropylamine |
| UN2685 | N,N-Diethylethylenediamine |
| UN2686 | Diethylaminoethanol |
| UN2687 | Dicyclohexylammonium nitrite |
| UN2688 | 1-Chloro-3-bromopropane |
| UN2689 | Glycerol alpha-monochlorohydrin |
| UN2690 | N-n-Butyl imidazole |
| UN2691 | Phosphorus pentabromide |
| UN2692 | Boron tribromide |
| UN2693 | Bisulfites, inorganic, aqueous solutions, n.o.s. |
| UN2698 | Tetrahydrophthalic anhydrides |
| UN2699 | Trifluoroacetic acid |
| UN2705 | 1-Pentol |
| UN2707 | Dimethyldioxanes |
| UN2708 | Butoxyl |
| UN2709 | Butyl benzenes |
| UN2710 | Dipropylketone |
| UN2711 | Dibromobenzene |
| UN2713 | Acridine |
| UN2714 | Zinc resinate |
| UN2715 | Aluminum resinate |
| UN2716 | 1,4-Butynediol |
| UN2717 | Camphor |
| UN2719 | Barium bromate |
| UN2720 | Chromium nitrate |
| UN2721 | Copper chlorate |
| UN2722 | Lithium nitrate |
| UN2723 | Magnesium chlorate |
| UN2724 | Manganese nitrate |
| UN2725 | Nickel nitrate |
| UN2726 | Nickel nitrite |
| UN2727 | Thallium nitrate |
| UN2728 | Zirconium nitrate |
| UN2729 | Hexachlorobenzene |
| UN2730 | Nitroanisole |
| UN2732 | Nitrobromobenzenes |
| UN2733 | Alkylamines, n.o.s. *or* Polyalkylamines, n.o.s. |
| UN2734 | Alkylamines, n.o.s. *or* Polyalkylamines, n.o.s. |
| UN2735 | Alkylamines, n.o.s. *or* Polyalkylamines, n.o.s. |
| UN2738 | N-Butylaniline |
| UN2739 | Butyric anhydride |
| UN2740 | n-Propyl chloroformate |
| UN2741 | Barium hypochlorite |
| NA2742 | Isobutyl chloroformate |
| NA2742 | sec-Butyl chloroformate |

## IDENTIFICATION NUMBER CROSS REFERENCE INDEX TO PROPER SHIPPING NAMES IN §172.101

NOTE: This listing is provided for information purposes only. It is compiled and kept up-to-date by RSPA

IDENTIFICATION NUMBER CROSS REFERENCE INDEX TO PROPER SHIPPING NAMES IN §172.101

NOTE: This listing is provided for information purposes only. It is compiled and kept up-to-date by RSPA

IDENTIFICATION NUMBER CROSS REFERENCE INDEX TO PROPER SHIPPING NAMES IN §172.101

NOTE: This listing is provided for information purposes only. It is compiled and kept up-to-date by RSPA

| (1)-Identification Number | (2)-Description |
|---|---|
| UN2742 | Chloroformates, n.o.s. |
| UN2743 | n-Butyl chloroformate |
| UN2744 | Cyclobutylchloroformate |
| UN2745 | Chloromethylchloroformate |
| UN2746 | Phenylchloroformate |
| UN2747 | tert-Butylcyclohexyl-chloroformate |
| UN2748 | 2-Ethylhexylchloroformate |
| UN2749 | Tetramethylsilane |
| UN2750 | 1,3-Dichloropropanol-2 |
| UN2751 | Diethylthiophosphoryl chloride |
| UN2752 | 1,2-Epoxy-3-ethoxypropane |
| UN2753 | N-Ethylbenzyltoluidines solid or liquid |
| UN2754 | N-Ethyltoluidines |
| UN2757 | Carbamate pesticides, solid, toxic, n.o.s. |
| UN2758 | Carbamate pesticides, liquid, flammable toxic, n.o.s. |
| UN2759 | Arsenical pesticides, solid, toxic, n.o.s. |
| UN2760 | Arsenical pesticides, liquid, flammable, toxic, n.o.s. |
| NA2761 | Aldrin |
| NA2761 | Dieldrin |
| UN2761 | Organochlorine pesticides, solid, toxic, n.o.s. |
| NA2762 | Aldrin |
| UN2762 | Organochlorine pesticides liquid, flammable, toxic, n.o.s. |
| UN2763 | Triazine pesticides, solid, toxic, n.o.s. |
| UN2764 | Triazine pesticides, liquid, flammable, toxic, n.o.s. |
| UN2765 | Phenoxy pesticides, solid, toxic, n.o.s. |
| UN2766 | Phenoxy pesticides, liquid, flammable, toxic, n.o.s. |
| UN2767 | Phenyl urea pesticides, solid, toxic, n.o.s. |
| UN2768 | Phenyl urea pesticides, liquid, flammable, toxic, n.o.s. |
| UN2769 | Benzoic derivative pesticides, solid, toxic, n.o.s. |
| UN2770 | Benzoic derivative pesticides, liquid, flammable, toxic, n.o.s. |
| UN2771 | Dithiocarbamate pesticides, solid, toxic, n.o.s. |
| UN2772 | Dithiocarbamate pesticides, liquid, flammable, toxic, n.o.s. |
| UN2773 | Phthalimide derivative pesticides, solid, toxic, n.o.s. |
| UN2774 | Phthalimide derivative pesticides, liquid, flammable, toxic, n.o.s. |
| UN2775 | Copper based pesticides, solid, toxic, n.o.s. |
| UN2776 | Copper based pesticides, liquid, flammable, toxic, n.o.s. |
| UN2777 | Mercury based pesticides, solid, toxic, n.o s. |
| UN2778 | Mercury based pesticides, liquid, flammable, toxic, n.o.s. |
| UN2779 | Substituted nitrophenol pesticides, solid, toxic, n.o.s. |
| UN2780 | Substituted nitrophenol pesticides, liquid, flammable, toxic, n.o.s. |
| UN2781 | Bipyridilium pesticides, solid, toxic, n.o.s. |
| UN2782 | Bipyridilium pesticides, liquid, flammable, toxic, n.o.s. |
| NA2783 | Methyl parathion |
| NA2783 | Parathion |
| NA2783 | Tetraethyl pyrophosphate |
| UN2783 | Organophosphorus pesticides, solid, toxic, n.o.s. |
| UN2784 | Organophosphorus pesticides, liquid, flammable, toxic, n.o.s. |
| UN2785 | Thia-4-pentanal |
| UN2786 | Organotin pesticides, solid, toxic, n.o.s. |
| UN2787 | Organotin pesticides, liquid, flammable, toxic, n.o.s. |
| UN2788 | Organotin compounds, liquid, n.o.s. |
| UN2789 | Acetic acid, glacial *or* Acetic acid solution |
| UN2790 | Acetic acid solution |
| UN2793 | Ferrous metal borings, shavings, turnings *or* cuttings |
| UN2794 | Batteries, wet, filled with acid |
| UN2795 | Batteries, wet, filled with alkali |
| UN2796 | Battery fluid, acid |
| UN2797 | Battery fluid, alkali |
| UN2798 | Phenyl phosphorus dichloride |
| UN2799 | Phenyl phosphorus thiodichloride |
| UN2800 | Batteries, wet, non-spillable |
| UN2801 | Dyes, liquid, n.o.s. *or* dye intermediates, liquid, n.o.s. |
| UN2802 | Copper chloride |
| UN2803 | Gallium |
| UN2805 | Lithium hydride, fused solid |
| UN2806 | Lithium nitride |
| UN2809 | Mercury |
| NA2810 | Compounds, tree *or* weed killing, liquid |
| NA2810 | Medicines |
| UN2810 | Poisonous liquids, n.o s. |
| NA2811 | Medicines |
| NA2811 | Selenium oxide |
| UN2811 | Poisonous solids, n.o.s. |
| UN2812 | Sodium aluminate, solid |
| UN2813 | Substances which in contact with water emit flammable gases, solid, n.o.s. |
| UN2814 | Infectious substances, affecting humans |
| UN2815 | N-Aminoethylpiperazine |
| UN2817 | Ammonium hydrogen fluoride, solution |
| UN2818 | Ammonium polysulfide, solution |
| UN2819 | Amyl acid phosphate |
| UN2820 | Butyric acid |
| UN2821 | Phenol solutions |
| UN2822 | 2-Chloropyridine |
| UN2823 | Crotonic acid |
| UN2826 | Ethyl chlorothioformate |
| UN2829 | Caproic acid |
| UN2830 | Lithium ferrosilicon |
| UN2831 | 1,1,1-Trichloroethane |
| UN2834 | Phosphorous acid, ortho |
| UN2835 | Sodium aluminum hydride |
| UN2837 | Sodium hydrogen sulfate, solution |
| UN2838 | Vinyl butyrate, inhibited |
| UN2839 | Aldol |
| UN2840 | Butyraldoxime |
| UN2841 | Di-n-amylamine |
| UN2842 | Nitroethane |
| UN2844 | Calcium manganese silicon |
| NA2845 | Ethyl phosphonous dichloride, anhydrous |
| NA2845 | Methyl phosphonous dichloride |
| UN2845 | Pyrophoric liquids, n.o.s. |
| UN2846 | Pyrophoric solids, n.o.s. |
| UN2849 | 3-Chloropropanol-1 |
| UN2850 | Propylene tetramer |
| UN2851 | Boron trifluoride dihydrate |
| UN2852 | Dipicryl sulfide, wetted |
| UN2853 | Magnesium fluorosilicate |
| UN2854 | Ammonium fluorosilicate |
| UN2855 | Zinc fluorosilicate |
| UN2856 | Fluorosilicates, n.o.s. |
| UN2857 | Refrigerating machines |
| UN2858 | Zirconium, dry |
| UN2859 | Ammonium metavanadate |
| UN2860 | Vanadium trioxide |
| UN2861 | Ammonium polyvanadate |
| UN2862 | Vanadium pentoxide |
| UN2863 | Sodium ammonium vanadate |
| UN2864 | Potassium metavanadate |
| UN2865 | Hydroxylamine sulfate |
| UN2869 | Titanium trichloride mixtures |
| UN2870 | Aluminum borohydride *or* Aluminum borohydride in devices |
| UN2871 | Antimony powder |
| UN2872 | Dibromochloropropane |
| UN2873 | Dibutylaminoethanol |
| UN2874 | Furfuryl alcohol |
| UN2875 | Hexachlorophene |
| UN2876 | Resorcinol |
| UN2878 | Titanium sponge granules *or* Titanium sponge powders |
| UN2879 | Selenium oxychloride |
| UN2880 | Calcium hypochlorite, hydrated *or* Calcium hypochlorite, hydrated mixtures |
| UN2881 | Metal catalyst, dry |
| UN2900 | Infectious substances, affecting animals |

## IDENTIFICATION NUMBER CROSS REFERENCE INDEX TO PROPER SHIPPING NAMES IN §172.101

**NOTE: This listing is provided for information purposes only. It is compiled and kept up-to-date by RSPA**

IDENTIFICATION NUMBER CROSS REFERENCE INDEX TO PROPER SHIPPING NAMES IN §172.101

NOTE: This listing is provided for information purposes only. It is compiled and kept up-to-date by RSPA

IDENTIFICATION NUMBER CROSS REFERENCE INDEX TO PROPER SHIPPING NAMES IN §172.101

NOTE: This listing is provided for information purposes only. It is compiled and kept up-to-date by RSPA

| (1)-Identification Number | (2)-Description |
|---|---|
| UN2901 | Bromine chloride |
| UN2902 | Pesticides, liquid, toxic, n.o.s. |
| UN2903 | Pesticides, liquid, toxic, flammable, n.o.s. |
| UN2904 | Chlorophenates, liquid |
| UN2905 | Chlorophenates, solid |
| UN2906 | Triisocyanatoisocyanurate of isophoronediisocyanate, solution |
| UN2907 | Isosorbide dinitrate mixture |
| UN2910 | Radioactive material, excepted package-articles manufactured from natural *or* depleted uranium *or* natural thorium |
| UN2910 | Radioactive material, excepted package-empty packaging |
| UN2910 | Radioactive material, excepted package-instruments *or* articles |
| UN2910 | Radioactive material, excepted package-limited quantity of material |
| UN2912 | Radioactive material, low specific activity |
| UN2918 | Radioactive material, fissile, n.o.s. |
| NA2920 | Dichlorobutene |
| UN2920 | Corrosive liquids, flammable, n.o.s. |
| UN2921 | Corrosive solids, flammable, n.o.s. |
| NA2922 | Sodium hydrosulfide, solution |
| UN2922 | Corrosive liquids, poisonous, n.o.s. |
| UN2923 | Corrosive solids, poisonous, n.o.s. |
| UN2924 | Flammable liquids, corrosive, n.o.s. |
| UN2925 | Flammable solids, corrosive, n.o.s. |
| UN2926 | Flammable solids, poisonous, n.o.s. |
| NA2927 | Ethyl phosphonothioic dichloride, anhydrous |
| NA2927 | Ethyl phosphorodichloridate |
| UN2927 | Poisonous liquids, corrosive, n.o.s. |
| UN2928 | Poisonous solids, corrosive, n.o.s. |
| UN2929 | Poisonous liquids, flammable, n o.s. |
| UN2930 | Poisonous solids, flammable, n.o.s. |
| UN2931 | Vanadyl sulfate |
| UN2933 | Methyl-2-chloropropionate |
| UN2934 | Isopropyl-2-chloropropionate |
| UN2935 | Ethyl-2-chloropropionate |
| UN2936 | Thiolactic acid |
| UN2937 | alpha-Methylbenzyl alcohol |
| UN2938 | Methyl benzoate |
| UN2940 | 9-Phosphabicyclononanes *or* Cyclooctadiene phosphines |
| UN2941 | Fluoroanilines |
| UN2942 | 2-Trifluoromethylaniline |
| UN2943 | Tetrahydrofurfurylamine |
| UN2945 | N-Methylbutylamine |
| UN2946 | 2-Amino-5-diethylaminopentane |
| UN2947 | Isopropyl chloroacetate |
| UN2948 | 3-Trifluoromethylaniline |
| UN2949 | Sodium hydrosulfide |
| UN2950 | Magnesium granules, coated |
| UN2951 | Diphenyloxide-4,4'disulfohydrazide |
| UN2952 | Azodiisobutyronitrile |
| UN2953 | 2,2'-Azodi-(2,4 dimethylvaleronitrile) |
| UN2954 | 1,1'-Azodi-(hexahydrobenzonitrile) |
| UN2955 | 2,2'-Azodi-(2,4-dimethyl-4-methoxyvaleronitrile |
| UN2956 | 5-tert-Butyl-2,4,6-trinitro-m-xylene *or* Musk xylene |
| UN2965 | Boron trifluoride dimethyl etherate |
| UN2966 | Thioglycol |
| UN2967 | Sulfamic acid |
| UN2968 | Maneb stabilized *or* Maneb preparations, stabilized |
| UN2969 | Castor beans *or* Castor meal *or* Castor pomace *or* Castor flake |
| UN2970 | Benzene sulfohydrazide |
| UN2971 | Benzene-1,3-disulfohydrazide |
| UN2972 | N,N-Dinitrosopentamethylenetetramine |
| UN2973 | N,N-Dinitroso-N,N'-dimethyl terephthalamide |
| UN2974 | Radioactive material, special form, n.o.s. |
| UN2975 | Thorium metal, pyrophoric |
| UN2976 | Thorium nitrate, solid |
| UN2977 | Uranium hexafluoride, fissile |
| UN2978 | Uranium hexafluoride |
| UN2979 | Uranium metal, pyrophoric |
| UN2980 | Uranyl nitrate hexahydrate solution |
| UN2981 | Uranyl nitrate, solid |
| UN2982 | Radioactive material, n.o.s. |
| UN2983 | Ethylene oxide and propylene oxide mixtures |
| UN2984 | Hydrogen peroxide, aqueous solutions |
| UN2985 | Chlorosilanes, n.o.s. |
| UN2986 | Chlorosilanes, n.o.s. |
| UN2987 | Chlorosilanes, n.o.s. |
| UN2988 | Chlorosilanes, n.o.s. |
| UN2989 | Lead phosphite, dibasic |
| UN2990 | Life-saving appliances, self-inflating |
| UN2991 | Carbamate pesticides, liquid, toxic, flammable, n.o.s. |
| UN2992 | Carbamate pesticides, liquid, toxic, n.o.s. |
| UN2993 | Arsenical pesticides, liquid, toxic, flammable, n.o.s. |
| UN2994 | Arsenical pesticides, liquid, toxic, n.o.s. |
| UN2995 | Organochlorine pesticides, liquid, toxic, flammable, n.o.s. |
| UN2996 | Organochlorine pesticides, liquid, toxic, n.o.s. |
| UN2997 | Triazine pesticides, liquid, toxic, flammable, n.o.s. |
| UN2998 | Triazine pesticides, liquid, toxic, n.o.s. |
| UN2999 | Phenoxy pesticides, liquid, toxic, flammable, n.o.s. |
| UN3000 | Phenoxy pesticides, liquid, toxic, n.o.s. |
| UN3001 | Phenyl urea pesticides, liquid, toxic, flammable, n.o.s. |
| UN3002 | Phenyl urea pesticides, liquid, toxic, n.o.s. |
| UN3003 | Benzoic derivative pesticides, liquid, toxic, flammable, n.o.s. |
| UN3004 | Benzoic derivative pesticides, liquid, toxic, n.o.s. |
| UN3005 | Dithiocarbamate pesticides, liquid, toxic, flammable, n.o.s. |
| UN3006 | Dithiocarbamate pesticides, liquid, toxic, n.o.s. |
| UN3007 | Phthalimide derivative pesticides, liquid, toxic, flammable, n.o.s. |
| UN3008 | Phthalimide derivative pesticides, liquid, toxic, n.o.s. |
| UN3009 | Copper based pesticides, liquid, toxic, flammable, n.o.s. |
| UN3010 | Copper based pesticides, liquid, toxic, n.o.s |
| UN3011 | Mercury based pesticides, liquid, toxic, flammable, n.o.s. |
| UN3012 | Mercury based pesticides, liquid, toxic, n.o.s. |
| UN3013 | Substituted nitrophenol pesticides, liquid, toxic, flammable, n.o.s. |
| UN3014 | Substituted nitrophenol pesticides, liquid, toxic, n.o.s. |
| UN3015 | Bipyridilium pesticides, liquid, toxic, flammable, n.o.s. |
| UN3016 | Bipyridilium pesticides, liquid, toxic, n.o.s. |
| UN3017 | Organophosphorus pesticides, liquid, toxic, flammable, n.o.s. |
| NA3018 | Methyl parathion |
| NA3018 | Tetraethyl pyrophosphate |
| UN3018 | Organophosphorus pesticides, liquid, toxic, n.o.s. |

## IDENTIFICATION NUMBER CROSS REFERENCE INDEX TO PROPER SHIPPING NAMES IN §172.101

**NOTE:** This listing is provided for information purposes only. It is compiled and kept up-to-date by RSPA

| (1)-Identification Number | (2)-Description |
|---|---|
| UN3019 | Organotin pesticides, liquid, toxic, flammable, n.o.s. |
| UN3020 | Organotin pesticides, liquid, toxic, n.o.s. |
| UN3021 | Pesticides, liquid, flammable, toxic, n.o.s. (flashpoint less than 23 degrees C) |
| UN3022 | 1,2-Butylene oxide, stabilized |
| UN3023 | tert-Octylmercaptan |
| UN3024 | Coumarin derivative pesticides, liquid, flammable, toxic, n.o.s. |
| UN3025 | Coumarin derivative pesticides, liquid, toxic, flammable, n.o.s. |
| UN3026 | Coumarin derivative pesticides, liquid, toxic, n.o.s. |
| UN3027 | Coumarin derivative pesticides, solid, toxic, n.o.s. |
| UN3028 | Batteries, dry, containing potassium hydroxide solid |
| UN3030 | 2,2-Azodi (2-methylbutyronitrile) |
| UN3031 | Self-reactive substances |
| UN3032 | Self-reactive substances |
| UN3033 | 3-Chloro-4-diethylaminobenzenediazonium zinc chloride |
| UN3034 | 4-Dipropylaminobenzenediazonium zinc chloride |
| UN3035 | 3-(2-Hydroxyethoxy)-4-pyrrolidin-1-ylbenzenediazonium zinc chloride |
| UN3036 | 2,5-Diethoxy-4-morpholinobenzenediazonium zinc chloride |
| UN3037 | 4-(Benzyl(ethyl)amino)-3-ethoxybenzenediazonium zinc chloride |
| UN3038 | 4-(Benzyl(methyl)amino)3-ethoxybenzenediazonium zinc chloride |
| UN3039 | 4-Dimethylamino-6-(2-dimethylaminoethoxy) toluene-2-diazonium zinc chloride |
| UN3040 | Sodium 2-diazo-1-naphthol-4-sulphonate |
| UN3041 | Sodium 2-diazo-1-naphthol-5-sulphonate |
| UN3042 | 2-Diazo-1-naphthol-4-sulpho-chloride |
| UN3043 | 2-Diazo-1-naphthol-5-sulpho-chloride |
| UN3048 | Aluminum phosphide pesticides |
| UN3049 | Metal alkyl halides, n.o.s. |
| UN3050 | Metal alkyl hydrides, n.o.s. |
| UN3051 | Aluminum alkyls |
| UN3052 | Aluminum alkyl halides |
| UN3053 | Magnesium alkyls |
| UN3054 | Cyclohexyl mercaptan |
| UN3055 | 2-(2-Aminoethoxy) ethanol |
| UN3056 | n-Heptaldehyde |
| UN3057 | Trifluoroacetylchloride |
| UN3064 | Nitroglycerin, solution in alcohol |
| UN3065 | Alcoholic beverages |
| UN3066 | Paint or Paint related material |
| UN3070 | Dichlorodifluoromethane and ethylene oxide mixture |
| UN3071 | Mercaptans, liquid, n.o.s. *or* Mercaptan mixtures, liquid, n.o.s. |
| UN3072 | Life-saving appliances, not self inflating |
| UN3073 | Vinylpyridenes, inhibited |
| UN3076 | Aluminum alkyl hydrides |
| NA3077 | Hazardous waste, solid, n.o.s. |
| NA3077 | Other regulated substances, solid, n.o.s. |
| UN3077 | Environmentally hazardous substances, solid, n.o.s. |
| UN3078 | Cerium |
| UN3079 | Methacrylonitrile, inhibited |
| UN3080 | Isocyanates, n.o.s. *or* Isocyanate solutions, n.o.s. |
| NA3082 | Hazardous waste, liquid, n.o.s. |
| NA3082 | Other regulated substances, liquid, n.o.s. |
| UN3082 | Environmentally hazardous substances, liquid, n.o.s. |
| UN3083 | Perchloryl fluoride |
| UN3084 | Corrosive solids, oxidizing, n.o.s. |
| UN3085 | Oxidizing substances, solid, corrosive, n.o.s. |
| UN3086 | Poisonous solids, oxidizing, n.o.s. |
| UN3087 | Oxidizing substances, solid, poisonous, n.o.s. |
| UN3088 | Self-heating substances, solid, n.o.s. |
| UN3089 | Metal powders, flammable, n.o.s. |
| UN3090 | Lithium battery, liquid cathode |
| UN3090 | Lithium battery, solid cathode |
| UN3091 | Lithium batteries, contained in equipment |
| UN3092 | 1-Methoxy-2-proponal |
| UN3093 | Corrosive liquids, oxidizing, n.o.s. |
| UN3094 | Corrosive liquids, which in contact with water emit flammable gases, n.o.s. |
| UN3095 | Corrosive solids, self heating, n.o.s. |
| UN3096 | Corrosive solids, which in contact with water emit flammable gases, n.o.s. |
| UN3098 | Oxidizing substances, liquid, corrosive, n.o.s. |
| UN3099 | Oxidizing substances, liquid, poisonous, n.o.s. |
| UN3100 | Oxidizing substances, solid, self heating, n.o.s. |
| UN3101 | Organic peroxide type B, liquid |
| UN3102 | Organic peroxide type B, solid |
| UN3103 | Organic peroxide type C, liquid |
| UN3104 | Organic peroxide type C, solid |
| UN3105 | Organic peroxide type D, liquid |
| UN3106 | Organic peroxide type D, solid |
| UN3107 | Organic peroxide type E, liquid |
| UN3108 | Organic peroxide type E, solid |
| UN3109 | Organic peroxide type F, liquid |
| UN3110 | Organic peroxide type F, solid |
| UN3111 | Organic peroxide type B, liquid, temperature controlled |
| UN3112 | Organic peroxide type B, solid, temperature controlled |
| UN3113 | Organic peroxide type C, liquid, temperature controlled |
| UN3114 | Organic peroxide type C, solid, temperature controlled |
| UN3115 | Organic peroxide type D, liquid, temperature controlled |
| UN3116 | Organic peroxide type D, solid, temperature controlled |
| UN3117 | Organic peroxide type E, liquid, temperature controlled |
| UN3118 | Organic peroxide type E, solid, temperature controlled |
| UN3119 | Organic peroxide type F, liquid, temperature controlled |
| UN3120 | Organic peroxide type F, solid, temperature controlled |
| UN3121 | Oxidizing substances, solid, which in contact with water emit flammable gases, n.o.s. |
| UN3122 | Poisonous liquids, oxidizing, n.o.s. |
| UN3123 | Poisonous liquids which in contact water emit flammable gases, n.o.s. |
| UN3124 | Poisonous, solids, self heating, n.o.s. |
| UN3125 | Poisonous, solids, which in contact water emit flammable gases, n.o.s. |

## IDENTIFICATION NUMBER CROSS REFERENCE INDEX TO PROPER SHIPPING NAMES IN §172.101

NOTE: This listing is provided for information purposes only. It is compiled and kept up-to-date by RSPA

| (1)-Identification Number | (2)-Description |
|---|---|
| UN3126 | Self-heating substances, solid, corrosive, n.o.s. |
| UN3127 | Self-heating substances, solid, oxidizing, n.o.s. |
| UN3128 | Self-heating substances, solid, poisonous, n.o.s. |
| UN3129 | Substances which in contact with water emit flammable gases, liquid, corrosive, n.o.s. |
| UN3130 | Substances which in contact with water emit flammable gases, liquid, poisonous, n.o.s. |
| UN3131 | Substances which in contact with water emit flammable gases. solid, corrosive, n.o.s. |
| UN3132 | Substances which in contact with water emit flammable gases, solid, flammable, n.o.s. |
| UN3133 | Substances which in contact with water emit flammable gases, solid, oxidizing, n.o.s. |
| UN3134 | Substances which in contact with water, emit flammable gases, solid, poisonous, n.o.s. |
| UN3135 | Substances which in contact with water, emit flammable gases, solid, self-heating, n.o.s. |
| UN3136 | Trifluoromethane, refrigerated, liquid |
| UN3137 | Oxidizing substances, solid, flammable, n.o.s. |
| UN3138 | Ethylene, acetylene and propylene in mixtures, refrigerated liquid containing at least 71.5 per cent ethylene with not more than 22.5 per cent acetylene and not more than 6 per cent propylene |
| UN3139 | Oxidizing substances, liquid, n.o.s. |
| UN3140 | Alkaloids, liquid |
| UN3141 | Antimony compounds, inorganic, liquid, n.o.s. |
| UN3142 | Disinfectants, liquid, n.o.s |
| UN3143 | Dyes, solid, n.o.s. *or* dye intermediates, solid, n.o.s. |
| UN3144 | Nicotine compounds, liquid, n.o.s. *or* Nicotine preparations, liquid, n.o.s. |
| UN3145 | Alkyl phenols, liquid, n.o.s. |
| UN3146 | Organotin compounds, solid, n.o.s. |
| UN3147 | Dyes, solid, n.o.s. *or* Dye intermediates, solid, n.o.s. |
| UN3148 | Substances which in contact with water emit flammable gases, liquid n.o.s. |
| UN3149 | Hydrogen peroxide and peroxyacetic acid mixtures |
| UN3150 | Devices, small, hydrocarbon gas powered *or* hydrocarbon gas refills for small devices |
| UN3151 | Polyhalogenated biphenyls, liquid *or* Polyhalogenated terphenyls liquid |
| UN3152 | Polyhalogenated biphenyls, solid *or* Polyhalogenated terphenyls, solid |
| UN3153 | Perfluoromethylvinyl ether |
| UN3154 | Perfluoroethylvinyl ether |
| NA9035 | Gas identification set |
| NA9163 | Zirconium sulfate |
| NA9191 | Chlorine dioxide, hydrate, frozen |
| NA9195 | Metal alkyl, solution, n.o.s. |
| NA9202 | Carbon monoxide, refrigerated liquid |
| NA9206 | Methyl phosphonic dichloride |
| NA9263 | Chloropivaloyl chloride |
| NA9264 | 3,5-Dichloro-2,4,6-trifluoropyridine |
| NA9269 | Trimethoxysilane |
| NA9275 | Regulated medical waste |
| NA9302 | Aircraft hydraulic power unit fuel tank |

IDENTIFICATION NUMBER CROSS REFERENCE INDEX TO PROPER SHIPPING NAMES IN §172.101

NOTE: This listing is provided for information purposes only. It is compiled and kept up-to-date by RSPA

### Subpart A — General

#### §172.1 Purpose and scope.

This Part lists and classifies those materials which the Department of Transportation has designated as hazardous materials for purposes of transportation and precribes the requirements for shipping papers, packaging marking, labeling, and transport vehicle placarding applicable to the shipment and transportation of those hazardous materials.

#### §172.3 Applicability.

(a) This Part applies to —

(1) Each person who offers a hazardous material for transportation and

(2) Each carrier by air, highway, rail, or water who transports a hazardous material.

(b) When a person, other than one of those provided for in paragraph (a) of this section, performs a packaging labeling or marking function required by this part, that person shall perform the function in accordance with this part.

### Subpart B—Table of Hazardous Materials and Special Provisions

#### §172.101 Purpose and use of hazardous materials table.

(a) The Hazardous Materials Table (Table) in this section designates the materials listed therein as hazardous materials for the purpose of transportation of those materials. For each listed material, the Table identifies the hazard class or specifies that the material is forbidden in transportation, and gives the proper shipping name or directs the user to the preferred proper shipping name. In addition, the Table specifies or references requirements in this subchapter pertaining to labeling, packaging, quantity limits aboard aircraft and stowage of hazardous materials aboard vessels.

(b) *Column 1: Symbols*. Column 1 of the Table contains five symbols ("+", "A", "D", "I", and "W"), as follows:

(1) The plus (+) fixes the proper shipping name, hazard class and packing group for that entry without regard to whether the material meets the definition of that class or packing group or meets any other hazard class definition. An appropriate alternate proper shipping name and hazard class may be authorized by the Associate Administrator for Hazardous Materials Safety.

(2) The letter "A" restricts the application of requirements of this subchapter to materials offered or intended for transportation by aircraft, unless the material is a hazardous substance or a hazardous waste.

(3) The letter "D" identifies proper shipping names which are appropriate for describing materials for domestic transportation but may be inappropriate for international transportation under the provisions of international regulations (e.g., IMO, ICAO). An alternate proper shipping name may be selected when either domestic or international transportation is involved.

(4) The letter "I" identifies proper shipping names which are appropriate for describing materials in international transportation. An alternate proper shipping name may be selected when only domestic transportation is involved.

(5) The letter "W" restricts the application of requirements of this subchapter to materials offered or intended for transportation by vessel, unless the material is a hazardous substance or a hazardous waste.

(c) *Column 2: Hazardous materials descriptions and proper shipping names*. Column 2 lists the hazardous materials descriptions and proper shipping names of materials designated as hazardous materials. Modification of a proper shipping name may otherwise be required or authorized by this section. Proper shipping names are limited to those shown in Roman type (not italics).

(1) Proper shipping names may be used in the singular or plural and in either capital or lower case letters. Words may be alternatively spelled in the same manner as they appear in the ICAO Technical Instructions or the IMDG Code. For example "aluminum" may be spelled "aluminium" and "sulfur" may be spelled "sulphur". However, the word "inflammable" may not be used in place of the word "flammable".

(2) Punctuation marks and words in italics are not part of the proper shipping name, but may be used in addition to the proper shipping name. The word "or" in italics indicates that terms in the sequence may be used as the proper shipping name, as appropriate.

(3) The abbreviation "n.o.i." or "n.o.i.b.n." may be used interchangeably with "n.o.s."

(4) Except for hazardous wastes, when qualifying words are used as part of the proper shipping name, their sequence in the package markings and shipping paper description is optional. However, the entry in the Table reflects the preferred sequence.

(5) When one entry references another entry by use of the word "see", if both names are in Roman type, either name may be used as the proper shipping name (e.g., Ethyl alcohol, *see* Ethanol).

(6) When a proper shipping name includes a concentration range as part of the shipping descrip-

tion. the actual concentration. if it is within the range stated, may be used in place of the concentration range. For example, an aqueous solution of hydrogen peroxide containing 30 percent peroxide may be described as "Hydrogen peroxide, aqueous solution *with not less than 20 percent but not more than 40 percent hydrogen peroxide*" or "Hydrogen peroxide, aqueous solution *with 30 percent hydrogen peroxide*".

(7) Use of the prefix "mono" is optional in any shipping name, when appropriate. Thus, Iodine monochloride may be used interchangeably with Iodine chloride. In "Glycerol alphamonochlorohydrin" the term "mono" is considered a prefix to the term "chlorohydrin" and may be deleted.

(8) Hazardous substances. The appendix to this section lists materials which are listed or designated as hazardous substances under section 101(14) of the Comprehensive Environmental Response, Compensation, and Liability Act (CERCLA). Proper shipping names for hazardous substances (see appendix) to this section and §171.8 of this subchapter) shall be determined as follows:

(i) If the hazardous substance appears in the Table by technical name, then the technical name is the proper shipping name.

(ii) If the hazardous substance does not appear in the Table and is not a forbidden material, then an appropriate generic, or "n.o.s" , shipping name shall be selected correponding to the hazard class (and packing group, if any) of the material as determined by the defining criteria of this subchapter (see §§173.2 and 173.2a of this subchapter). For example, a hazardous substance which is listed in the appendix but not in the Table and which meets the definition of flammable liquid might be described as "Flammable liquid, n.o.s." or other appropriate shipping name corresponding to the flammable liquid hazard class.

(9) Hazardous wastes. If the word "waste" is not included in the hazardous material description in Column 2 of the Table, the proper shipping name for a hazardous waste (as defined in §171.8 of this subchapter), shall include the word "Waste" preceding the proper shipping name of the material. For example: Waste acetone.

(10) Mixtures and solutions, (i) A mixture or solution not identified specifically by name, comprised of a hazardous material identified in the Table by technical name and non-hazardous material, shall be described using the proper shipping name of the hazardous material and the qualifying word "mixture" or "solution", as appropriate, unless—

(A) Except as provided in §172.101 (i)(4) the packaging specified in Column 8 is inappropriate to the physical state of the material;

(B) The shipping description indicates that the proper shipping name applies only to the pure or technically pure hazardous material;

(C) The hazard class, packing group, or subsidiary hazard of the mixture or solution is different from that specified for the entry;

(D) There is a significant change in the measures to be taken in emergencies;

(ii) If one or more of the conditions specified in paragraph (c)(10)(i) of this section is satisfied then a proper shipping name shall be selected as prescribed in paragraph (c)(12)(ii) of this section.

(E) The material is identified by special provision in Column 7 of the §172.101 Table as a material poisonous by inhalation; however, it no longer meets the definition of poisonous by inhalation or it falls within a different hazard zone than that specified in the special provision; or

(F) The material can be appropriately described by a shipping name that describes its intended application, such as "Coating solution", "Extracts, flavoring" or "Compound, cleaning liquid".

(11) Except for a material subject to or prohibited by §§173.21, 173.51, 173.56(d), or 173.56(e)(1) of this subchapter, a material for which the hazard class is uncertain and must be determined by testing or a material that is a hazardous waste may be assigned a tentative shipping name, hazard class and identification number, based on the shipper's tentative determination according to-

(i) Defining criteria in this subchapter;

(ii) The hazard precedence prescribed in §173.2a of this subchapter; and

(iii) The shipper's knowledge of the material.

(12) Except when the proper shipping name in the Table is preceded by a plus (+)—

(i) If it is specifically determined that a material meets the definition of a hazard class or packing group, other than the class or packing group shown in association with the proper shipping name, or does not meet the defining criteria for a subsidiary hazard shown in Column 6 of the Table, the material shall be described by an appropriate proper shipping name listed in association with the correct hazard class, packing group, or subsidiary hazard for the material.

(ii) Generic or n.o.s. descriptions. If an appropriate technical name is not shown in the Table, selection of a proper shipping name shall be made from the generic or n.o.s. descriptions corresponding to the specific hazard class, packing group, or subsidiary hazard, if any, for the material. The name that most appropriately describes the material shall be used; e.g, an alcohol not listed by its technical name in the Table shall be described as "Alcohol, n.o.s." rather than "Flammable liquid, n.o.s.". Some mixtures may be more appropriately described according to their application, such as "Coating solution" or "Extracts, flavoring, liquid", rather than by an n.o.s. entry, such as "Flammable liquid, n.o.s." It should be noted, however, that an n.o.s. description as a proper shipping name may not provide sufficient information for shipping papers and package marking. Under the provisions of subparts C and D of this part, the technical name of the constituent which makes the product a hazardous material may be required in association with the proper shipping name.

(iii) Multiple hazard materials. If a material meets the definition of more than one hazard class, and is not identified in the Table by a specific description, the hazard class of the material shall be determined by using the precedence specified in §173.2a of this subchapter, and an appropriate shipping description (e.g., "Flammable liquid, corrosive n.o.s.") shall be selected as described in paragraph (c)(12)(ii) of this section. Except for a pesticide, a subsidiary hazard of Division 6.1, Packing Group III, need not be considered for selecting a shipping description.

(iv) If it is specifically determined that a material is not a forbidden material and does not meet the definition of any hazard class, the material is not a hazardous material.

(13) Organic peroxides. Generic proper shipping names for organic peroxides, as listed in Column 2 of the Table, shall be selected based on the technical name of the organic peroxide, in accordance with the provisions of §173.225 of this subchapter.

(d) *Column 3: Hazard class or Division.* Column 3 contains a designation of the hazard class or division corresponding to each proper shipping name, or the word "Forbidden".

(1) A material for which the entry in this column is "Forbidden" may not be offered for transportation or transported. This prohibition does not apply if the material is diluted, stabilized or incorporated in a device and it is classed in accordance with the definitions of hazardous materials contained in part 173 of this subchapter.

(2) When a reevaluation of test data or new data indicates a need to modify the "Forbidden" designation or the hazard class or packing group specified for a material specifically indentified in the Table, this data should be submitted to the Associate Administrator for Hazardous Materials Safety.

(3) A basic description of each hazard class and the section reference for class definitions appear in §173.2 of this subchapter.

(4) Each reference to a Class 3 material is modified to read "Combustible liquid" when that material is reclassified in accordance with §173.150(f) of this subchapter or has a flash point above 60.5°C (141°F) but below 93°C (200°F).

(e) *Column 4: Identification number.* Column 4 lists the identification number assigned to each proper shipping name. Those preceded by the letter "UN" are associated with proper shipping names considered appropriate for international transportation as well as domestic transportation. Those preceded by the letters "NA" are associated with proper shipping names not recognized for international transportation, except to and from Canada. Identification numbers in the "NA9000" series are associated with proper shipping names not appropriately covered by international hazardous materials (dangerous goods) transportation standards, or not appropriately addressed by the international transportation standards for emergency response information purposes, except for transportation between the United States and Canada.

(f) *Column 5: Packing group.* Column 5 specifies one or more packing groups assigned to a material corresponding to the proper shipping name and hazard class for that material. Classes 2 and 7 materials and ORM-D materials do not have packing groups, Packing Groups I, II, and III indicate the degree of danger presented by the material is either great, medium or minor, respectively. If more than one packing group is indicated for an entry, the packing group for the hazardous material is determined using the criteria for assignment of packing groups specified in subpart D of part 173. When a reevaluation of test data or new data indicates a need to modify the specified packing group(s), the data should be submitted to the Associate Administrator for Hazardous Materials Safety. Each reference in this column to a material which is a hazardous waste or a hazardous substance, and whose proper shipping name is preceded in Column 1 of the Table by the letter "A" or "W", is modified to read "III" on those occasions when the material is offered for transportation or transported by a mode in which its transportation is not otherwise subject to requirements of this subchapter.

(g) *Column 6: Labels.* Column 6 specifies the hazard warning label(s) required for a package filled with a material conforming to the associated hazard class and proper shipping name, unless the package is otherwise excepted from labeling by provisions in subpart D of part 172, or part 173 of this subchapter. The first label shown for each entry is indicative of the primary hazard of the material, additional labels are indicative of subsidiary hazards. Provisions in §172.402 of this part may require that a label other than that specified in Column 6 be affixed to the package in addition to that specified in Column 6.

(h) *Column 7: Special provisions.* Column 7 specifies codes for special provisions applicable to hazardous materials. When Column 7 refers to a special provision for a hazardous material, the meaning and requirements of that special provision are as set forth in §172.102 of this subpart.

(i) *Column 8: Packaging authorizations.* Columns 8A, 8B and 8C specify the applicable sections for exceptions, non-bulk packaging requirements and bulk-packaging requirements, respectively, in part 173 of this subchapter. Columns 8A, 8B and 8C are completed in a manner which indicates that "§173." precedes the designated numerical entry. For example, the entry "202" in Column 8B associated with the proper shipping name "Gasoline" indicates that for this material conformance to non-bulk packaging requirements prescribed in §173.202 of this subchapter is required. When packaging requirements are specified, they are in addition to the standard requirements for all packagings prescribed in §173.24 of this subchapter and any other applicable requirements in subparts A and B of part 173 of this subchapter.

(1) Exceptions. Column 8A contains exceptions from some of the requirements of this subchapter. The referenced exceptions are in addition to those specified in subpart A of part 173 and elsewhere in this subchapter. A "None" in this column means no packaging exceptions are authorized, except as may be provided by special provisions in Column 7.

(2) Non-bulk packagings. Column 8B references the section in part 173 of this subchapter which prescribes packaging requirements for non-bulk packagings. A "None" in this column means non-bulk packagings are not authorized, except as may be provided by special provisions in Column 7. Each reference in this column to a material which is a hazardous waste or a hazardous substance, and whose proper shipping name is preceded in Column 1 of the Table by the letter "A" or "W", is modified to include "§173.203" or "§173.213", as appropriate for liquids and solids, respectively, on those occasions when the material is offered for transportation or transported by a mode in which its transportation is not otherwise subject to the requirements of this subchapter.

(3) Bulk packaging. Column 8C specifies the section in part 173 of this subchapter which prescribes packaging requirements for bulk packagings, subject to the limitations, requirements and additional authorizations of Column 7. A "None" in this column means bulk packagings are not authorized, except as may be provided by special provisions in Column 7. Additional authorizations and limitations for use of IM portable tanks are set forth in Column 7. For each reference in this column to a material which is a hazardous waste or a hazardous substance, and whose proper shipping name is preceded in Column 1 of the Table by the letter "A" or "W" and which is offered for transportation or transported by a mode in which its transportation is not otherwise subject to the requirements of this subchapter.

(i) The column reference is §§173.240 or 173.241, as appropriate.

(ii) For a solid material, the exception provided in Special provision B54 is applicable.

(iii) For a Class 9 material which meets the definition of an elevated temperature material, the column reference is §173.247.

(4) For a hazardous material which is specifically named in the Table and whose packaging sections specify packagings not applicable to the form of the material (e.g., packaging specified is for solid material and the material is being offered for transportation in a liquid form) the following table should be used to determine the appropriate packaging section:

| Packaging section reference for solid materials | Corresponding packaging section for liquid materials |
|---|---|
| §173.211 | §173.201 |
| §173.212 | §173.202 |
| §173.213 | §173.203 |
| §173.240 | §173.241 |
| §173.242 | §173.243 |

(j) *Column 9: Quantity limitations.* Columns 9A and 9B specify the maximum quantities that may be offered for transportation in one package by passenger-carrying aircraft or passenger-carrying rail car (Column 9A) or by cargo aircraft only (Column 9B), subject to the following:

(1) "Forbidden" means the material may not be offered for transportation or transported in the applicable mode of transport.

(2) The quantity limitation is "net" except where otherwise specified, such as for "Consumer commodity" which specifies "30 kg gross."

(3) When articles or devices are specifically listed by name, the net quantity limitation applies to the entire article or device (less packaging and packaging materials) rather than only to its hazardous components.

(4) A package offered or intended for transportation by aircraft and which is filled with a material forbidden on passenger-carrying aircraft but permitted on cargo aircraft only, or which exceeds the maximum net quantity authorized on passenger-carrying aircraft, shall be labelled with the CARGO AIRCRAFT ONLY label specified in §172.448 of this part.

(k) *Column 10: Vessel stowage requirements.* Column 10A [Vessel stowage] specifies the authorized stowage locations on board cargo and passenger vessels. Column 10B [Other provisions] specifies codes for stowage requirements for specific hazardous materials. The meaning of each code in Column 10B is set forth in §176.84 of this subchapter. Section 176.63 of this subchapter sets forth the physical requirements for each of the authorized locations listed in Column 10A. (For bulk transportation by vessel, see 46 CFR parts

30 to 40, 70, 98,148, 151, 153 and 154.) The authorized stowage locations specified in Column 10A are defined as follows:

(1) Stowage category "A" means the material may be stowed "on deck" or "under deck" on a cargo vessel and a passenger vessel.

(2) Stowage category "B" means the material may be stowed "on deck" or "under deck" on a cargo vessel, but must be stowed "on deck" on a passenger vessel.

(3) Stowage category "C" means the material must be stowed "on deck" on a cargo vessel and on a passenger vessel.

(4) Stowage category "D" means the material must be stowed "on deck" on a cargo vessel, but is prohibited on a passenger vessel.

(5) Stowage category "E" means the material may be stowed "on deck" or "under deck" on a cargo vessel, but is prohibited on a passenger vessel.

(l) *Changes to the Table*. (1) Unless specifically stated otherwise in a rule document published in the FEDERAL REGISTER amending the Table—

(i) Such a change does not apply to the shipment of any package filled prior to the effective date of the amendment; and

(ii) Stocks of preprinted shipping papers and package marking may be continued in use, in the manner previously authorized, until depleted or for a one-year period, subsequent to the effective date of the amendment, whichever is less.

(2) Except as otherwise provided in this section any alteration of a shipping description or associated entry which is listed in the §172.101 Table must receive prior written approval from the Associate Administrator for Hazardous Materials Safety.

## §172.101 HAZARDOUS MATERIALS TABLE

| Symbols | Hazardous materials descriptions and proper shipping names | Hazard class or Division | Identification Numbers | Packing group | Label(s) required (if not excepted) | Special provisions | (8) Packaging authorizations (§173.***) | | | (9) Quantity limitations | | (10) Vessel stowage requirements | |
|---|---|---|---|---|---|---|---|---|---|---|---|---|---|
| | | | | | | | Exceptions | Non-bulk packaging | Bulk packaging | Passenger aircraft or railcar | Cargo aircraft only | Vessel stowages | Other stowage provisions |
| (1) | (2) | (3) | (4) | (5) | (6) | (7) | (8A) | (8B) | (8C) | (9A) | (9B) | (10A) | (10B) |
| | *Accellerene, see* p-Nitrosodimethylaniline | | | | | | | | | | | | |
| | *Accumulators, electric; see* Batteries, wet *etc.* | | | | | | | | | | | | |
| D | Accumulators, pressurized, pneumatic *or* hydraulic *(containing non-flammable gas)* | 2.2 | NA1956 | | NONFLAMMABLE GAS | | 306 | 306 | None | No limit | No limit | A | |
| | Acetal | 3 | UN1088 | II | FLAMMABLE LIQUID | T7 | 150 | 202 | 242 | 5 L | 60 L | E | |
| | Acetaldehyde | 3 | UN1089 | I | FLAMMABLE LIQUID | A3, B16, T20, T26, T29 | None | 201 | 243 | Forbidden | 30 L | E | |
| A | Acetaldehyde ammonia | 9 | UN1841 | III | CLASS 9 | | 155 | 204 | 241 | 200 kg | 200 kg | A | 34 |
| | Acetaldehyde oxime | 3 | UN2332 | II | FLAMMABLE LIQUID | T8 | 150 | 202 | 242 | 5 L | 60 L | A | |
| | Acetic acid, glacial *or* Acetic acid solution, *more than 80 per cent acid, by mass* | 8 | UN2789 | II | CORROSIVE | A3, A6, A7, A10 B2, T8 | 154 | 202 | 242 | 1 L | 30 L | A | 12, 21, 48 |
| | Acetic acid solution, *more than 10 per cent but not more than 80 per cent acid, by mass* | 8 | UN2790 | II | CORROSIVE | A3, A6, A7, A10 B2, T8 | 154 | 202 | 242 | 1 L | 30 L | A | 112 |
| | Acetic anhydride | 8 | UN1715 | II | CORROSIVE | A3, A6, A7, A10, B2, T8 | 154 | 202 | 242 | 1 L | 30 L | A | 40 |
| | Acetone | 3 | UN1090 | II | FLAMMABLE LIQUID | T8 | 150 | 202 | 242 | 5 L | 60 L | B | |
| | Acetone cyanohydrin, stabilized | 6.1 | UN1541 | I | POISON | 2, A3, B9, B14, B32, B74, B76, B77, N34, T38, T43, T45 | None | 227 | 244 | Forbidden | 30 L | D | 25, 40, 49 |
| | Acetone oils | 3 | UN1091 | II | FLAMMABLE LIQUID | T7, T30 | 150 | 202 | 242 | 5 L | 60 L | B | |
| | *Acetonitrile, see* Methyl cyanide | | | | | | | | | | | | |
| | *Acetyl acetone peroxide with more than 9% by mass active oxygen* | Forbidden | | | | | | | | | | | |
| | *Acetyl benzoyl peroxide, solid, or more than 40% in solution* | Forbidden | | | | | | | | | | | |
| | Acetyl bromide | 8 | UN1716 | II | CORROSIVE | B2, T12, T26 | 154 | 202 | 242 | 1 L | 30 L | C | 8, 40 |
| | Acetyl chloride | 3 | UN1717 | II | FLAMMABLE LIQUID, CORROSIVE | A3, A6, A7, N34, T18, T26 | None | 202 | 243 | 1 L | 5 L | B | 40 |
| | *Acetyl cyclohexanesulfonyl peroxide, more than 82 per cent wetted with less than 12 per cent water* | Forbidden | | | | | | | | | | | |
| | Acetyl iodide | 8 | UN1898 | II | CORROSIVE | B2, T9 | 154 | 202 | 242 | 1 L | 30 L | C | 8, 40 |
| | Acetyl methyl carbinol | 3 | UN2621 | III | FLAMMABLE LIQUID | B1, T1 | 150 | 203 | 242 | 60 L | 220 L | A | |
| | *Acetyl peroxide, solid, or more than 25 percent in solution* | Forbidden | | | | | | | | | | | |
| | Acetylene, dissolved | 2.1 | UN1001 | | FLAMMABLE GAS | | None | 303 | None | Forbidden | 15 kg | D | 25, 40, 57 |
| | *Acetylene (liquefied)* | Forbidden | | | | | | | | | | | |
| | *Acetylene silver nitrate* | Forbidden | | | | | | | | | | | |
| | *Acetylene tetrabromide, see* Tetrabromoethane | | | | | | | | | | | | |
| | *Acid butyl phosphate, see* Butyl acid phosphate | | | | | | | | | | | | |
| | Acid, sludge, *see* Sludge acid | | | | | | | | | | | | |
| | Acridine | 6.1 | UN2713 | III | KEEP AWAY FROM FOOD | | 153 | 213 | 240 | 100 kg | 200 kg | A | |
| | Acrolein dimer, stabilized | 3 | UN2607 | III | FLAMMABLE LIQUID | B1, T1 | 150 | 203 | 242 | 60 L | 220 L | A | 40 |
| | Acrolein, inhibited | 6.1 | UN1092 | I | POISON, FLAMMABLE LIQUID | 1, B9, B12, B14, B30, B42, B72, B77, T38, T43, T44 | None | 226 | 244 | Forbidden | Forbidden | D | 40 |
| | Acrylamide | 6.1 | UN2074 | III | KEEP AWAY FROM FOOD | T8 | 153 | 213 | 240 | 100 kg | 200 kg | A | 12 |
| | Acrylic acid, inhibited | 8 | UN2218 | II | CORROSIVE | B2, T8 | 154 | 202 | 242 | 1 L | 30 L | C | 8, 12, 21, 25, 40 |
| | Acrylonitrile, inhibited | 3 | UN1093 | I | FLAMMABLE LIQUID, POISON | B9, T18, T26 | None | 201 | 243 | Forbidden | 30 L | E | 40 |
| | *Actuating cartridge, explosive, see* Cartridges, power device | | | | | | | | | | | | |
| | Adhesives, *containing a flammable liquid* | 3 | UN1133 | II | FLAMMABLE LIQUID | B52, T7, T30 | 150 | 173 | 242 | 5 L | 60 L | B | |
| | | | | III | FLAMMABLE LIQUID | B1, B52, T7, T30 | 150 | 173 | 242 | 60 L | 220 L | A | |
| | Adiponitrile | 6.1 | UN2205 | III | KEEP AWAY FROM FOOD | T1 | 153 | 203 | 241 | 60 L | 220 L | A | 26 |
| | Aerosols, *corrosive, Packing Group II or III, (each not exceeding 1 L capacity)* | 2.2 | UN1950 | | NONFLAMMABLE GAS, CORROSIVE | A34 | 306 | None | None | 75 kg | 150 kg | A | 40, 48, 85 |
| | Aerosols, *flammable, (each not exceeding 1 L capacity)* | 2.1 | UN1950 | | FLAMMABLE GAS | N82 | 306 | None | None | 75 kg | 150 kg | A | 40, 48, 85 |
| | Aerosols, *non-flammable, (each not exceeding 1 L capacity)* | 2.2 | UN1950 | | NONFLAMMABLE GAS | | 306, 307 | None | None | 75 kg | 150 kg | A | 48, 85 |

## §172.101 HAZARDOUS MATERIALS TABLE, Continued

| Symbols | Hazardous materials descriptions and proper shipping names | Hazard class or Division | Identification Numbers | Packing group | Label(s) required (if not excepted) | Special provisions | (8) Packaging authorizations (§173.***) | | | (9) Quantity limitations | | (10) Vessel stowage requirements | |
|---|---|---|---|---|---|---|---|---|---|---|---|---|---|
| | | | | | | | Exceptions | Non-bulk packaging | Bulk packaging | Passenger aircraft or railcar | Cargo aircraft only | Vessel stowages | Other stowage provisions |
| (1) | (2) | (3) | (4) | (5) | (6) | (7) | (8A) | (8B) | (8C) | (9A) | (9B) | (10A) | (10B) |
| | Aerosols, *poison, each not exceeding 1 L capacity* | 2.2 | UN1950 | | NONFLAMMABLE GAS | | 306 | None | None | Forbidden | Forbidden | A | 40, 48, 85 |
| D | Air bag inflators *or* Air bag modules *for supplemental restraint systems. See also* Articles, pyrotechnic *for technical purposes (UN0430, UN0431, UN0432)* | 4.1 | NA1325 | III | FLAMMABLE SOLID | | 166 | 166 | 166 | 25 kg | 100 kg | A | |
| | Air, compressed | 2.2 | UN1002 | | NONFLAMMABLE GAS | | 306 | 302 | 302 | 75 kg | 150 kg | A | |
| | Air, refrigerated liquid *(cryogenic liquid)* | 2.2 | UN1003 | | NONFLAMMABLE GAS, OXIDIZER | | 320 | 316 | 318, 319 | Forbidden | 150 kg | A | 51 |
| | Air, refrigerated liquid, *(cryogenic liquid)non-pressurized* | 2.2 | UN1003 | | NONFLAMMABLE GAS, OXIDIZER | | 320 | 316 | 318, 319 | Forbidden | Forbidden | A | 85 |
| | Aircraft evacuation slides, *see* Life saving appliances *etc* | | | | | | | | | | | | |
| | Aircraft hydraulic power unit fuel tank *(containing a mixture of anhydrous hydrazine and monomethyl hydrazine) (M86 fuel)* | 3 | UN3165 | I | FLAMMABLE LIQUID, POISON, CORROSIVE | | None | 172 | None | Forbidden | 30 L | E | |
| | Aircraft survival kits, *see* Life saving appliances *etc* | | | | | | | | | | | | |
| | Alcoholic beverages | 3 | UN3065 | II | FLAMMABLE LIQUID | B1, T1 | 150 | 202 | 242 | 5 L | 60 L | A | |
| | | | | III | FLAMMABLE LIQUID | B1, N11, T1 | 150 | 203 | 242 | 60 L | 220 L | A | |
| | Alcohols, n.o.s. | 3 | UN1987 | I | FLAMMABLE LIQUID | T8, T31 | None | 201 | 243 | 1 L | 30 L | E | |
| | | | | II | FLAMMABLE LIQUID | T8, T31 | 150 | 202 | 242 | 5 L | 60 L | B | |
| | | | | III | FLAMMABLE LIQUID | B1, T7, T30 | 150 | 203 | 242 | 60 L | 220 L | A | |
| | Alcohols, toxic, n.o.s. | 3 | UN1986 | I | FLAMMABLE LIQUID, POISON | T8, T31 | None | 201 | 243 | Forbidden | 30 L | E | 40 |
| | | | | II | FLAMMABLE LIQUID, POISON | T8, T31 | None | 202 | 243 | 1 L | 60 L | B | 40 |
| | Aldehydes, n.o.s. | 3 | UN1989 | I | FLAMMABLE LIQUID | T8, T31 | None | 201 | 243 | 1 L | 30 L | E | |
| | | | | II | FLAMMABLE LIQUID | T8, T31 | 150 | 202 | 242 | 5 L | 60 L | B | |
| | | | | III | FLAMMABLE LIQUID | B1, T7, T30 | 150 | 203 | 242 | 60 L | 220 L | A | |
| | Aldehydes, toxic, n.o.s. | 3 | UN1988 | I | FLAMMABLE LIQUID, POISON | T8, T31 | None | 201 | 243 | Forbidden | 30 L | E | 40 |
| | | | | II | FLAMMABLE LIQUID, POISON | T8, T31 | None | 202 | 243 | 1 L | 60 L | B | 40 |
| | Aldol | 6.1 | UN2839 | II | POISON | T8 | None | 202 | 243 | 5 L | 60 L | A | 48 |
| D | Aldrin, *liquid* | 6.1 | NA2762 | II | POISON | | None | 202 | 243 | 5 L | 60 L | B | |
| D | Aldrin, *solid* | 6.1 | NA2761 | II | POISON | | None | 212 | 242 | 25 kg | 100 kg | A | 40 |
| | Alkali metal alloys, liquid, n.o.s. | 4.3 | UN1421 | I | DANGEROUS WHEN WET | A2, A3, B48, N34 | None | 201 | 244 | Forbidden | 1 L | D | |
| | Alkali metal amalgams | 4.3 | UN1389 | I | DANGEROUS WHEN WET | A2, A3, N34 | None | 201 | 244 | Forbidden | 1 L | D | |
| | Alkali metal amides | 4.3 | UN1390 | II | DANGEROUS WHEN WET | A6, A7, A8, A19, A20 | None | 212 | 241 | 15 kg | 50 kg | E | 40 |
| | Alkali metal dispersions, *or* Alkaline earth metal dispersions | 4.3 | UN1391 | I | DANGEROUS WHEN WET | A2, A3 | None | 201 | 244 | Forbidden | 1 L | D | |
| | *Alkaline corrosive liquids, n.o.s., see* Caustic alkali liquids, n.o.s. | | | | | | | | | | | | |
| | Alkaline earth metal alloys, n.o.s. | 4.3 | UN1393 | II | DANGEROUS WHEN WET | A19 | None | 212 | 241 | 15 kg | 50 kg | E | |
| | Alkaline earth metal amalgams | 4.3 | UN1392 | I | DANGEROUS WHEN WET | A19, N34, N40 | None | 211 | 242 | Forbidden | 15 kg | D | |
| | Alkaloids, liquid, *poisonous*, n.o.s., *or* Alkaloid salts, liquid, *poisonous*, n.o.s. | 6.1 | UN3140 | I | POISON | A4, T42 | None | 201 | 243 | 1 L | 30 L | A | |
| | | | | II | POISON | T14 | None | 202 | 243 | 5 L | 60 L | A | |
| | | | | III | KEEP AWAY FROM FOOD | T7 | 153 | 203 | 241 | 60 L | 220 L | A | |
| | Alkaloids, solid, n.o.s. *or* alkaloid salts, solid, n.o.s. *poisonous* | 6.1 | UN1544 | I | POISON | | None | 211 | 242 | 5 kg | 50 kg | A | |
| | | | | II | POISON | | None | 212 | 242 | 25 kg | 100 kg | A | |
| | | | | III | KEEP AWAY FROM FOOD | | 153 | 213 | 240 | 100 kg | 200 kg | A | |
| | Alkyl, Aryl *or* Toluene sulfonic acid, liquid, *with more than 5 per cent free sulfuric acid* | 8 | UN2584 | II | CORROSIVE | B2, T8, T27 | 154 | 202 | 242 | 1 L | 30 L | B | 9 |
| | Alkyl, Aryl *or* Toluene sulfonic acid, liquid, *with not more than 5 per cent free sulfuric acid* | 8 | UN2586 | III | CORROSIVE | T8 | 154 | 203 | 241 | 5 L | 60 L | B | 9 |
| | Alkyl, Aryl *or* Toluene sulfonic acid, solid, *with more than 5 per cent free sulfuric acid* | 8 | UN2583 | II | CORROSIVE | | 154 | 212 | 240 | 15 kg | 50 kg | A | |

## §172.101 HAZARDOUS MATERIALS TABLE, Continued

| Symbols | Hazardous materials descriptions and proper shipping names | Hazard class or Division | Identification Numbers | Packing group | Label(s) required (if not excepted) | Special provisions | (8) Packaging authorizations (§173.***) | | | (9) Quantity limitations | | (10) Vessel stowage requirements | |
|---|---|---|---|---|---|---|---|---|---|---|---|---|---|
| | | | | | | | Exceptions | Non-bulk packaging | Bulk packaging | Passenger aircraft or railcar | Cargo aircraft only | Vessel stowages | Other stowage provisions |
| (1) | (2) | (3) | (4) | (5) | (6) | (7) | (8A) | (8B) | (8C) | (9A) | (9B) | (10A) | (10B) |
| | Alkyl, Aryl *or* Toluene sulfonic acid, solid, *with not more than 5 per cent free sulfuric acid* | 8 | UN2585 | III | CORROSIVE | | 154 | 213 | 240 | 25 kg | 100 kg | A | |
| | Alkyl phenols, liquid, n.o.s. *(including C2-C8 homologues)* | 6.1 | UN3145 | III | KEEP AWAY FROM FOOD | T7 | 153 | 203 | 241 | 60 L | 220 L | A | |
| | Alkyl phenols, solid, n.o.s. *(including C2-C8 homologues)* | 6.1 | UN2430 | III | KEEP AWAY FROM FOOD | | 153 | 213 | 240 | 100 kg | 200 kg | A | |
| | Alkylamines, n.o.s. *or* Polyalkylamines, n.o.s., *corrosive* | 8 | UN2735 | I | CORROSIVE | A3, A6, B10, N34, T42 | None | 201 | 242 | 0.5 L | 2.5 L | A | |
| | | | | II | CORROSIVE | B2, T8 | 154 | 202 | 242 | 1 L | 30 L | A | |
| | | | | III | CORROSIVE | T8 | 154 | 203 | 241 | 5 L | 60 L | A | |
| | Alkylamines, n.o.s. *or* Polyalkylamines, n.o.s. *corrosive, flammable* | 8 | UN2734 | I | CORROSIVE, FLAMMABLE LIQUID | A3, A6, N34, T8, T31 | None | 201 | 243 | 0.5 L | 2.5 L | A | 48 |
| | | | | II | CORROSIVE, FLAMMABLE LIQUID | T8, T31 | None | 202 | 243 | 1 L | 30 L | A | 48 |
| | Alkylamines, n.o.s. *or* Polyalkylamines, n.o.s. *flammable, corrosive* | 3 | UN2733 | I | FLAMMABLE LIQUID, CORROSIVE | T42 | None | 201 | 243 | 0.5 L | 2.5 L | D | 40 |
| | | | | II | FLAMMABLE LIQUID, CORROSIVE | T8, T31 | None | 202 | 243 | 1 L | 5 L | B | 40 |
| | | | | III | FLAMMABLE LIQUID, CORROSIVE | B1, T8, T31 | None | 203 | 242 | 5 L | 60 L | A | 40 |
| | *Allethrin, see* Pesticides, liquid, toxic, n.o.s. | | | | | | | | | | | | |
| | Allyl acetate | 3 | UN2333 | II | FLAMMABLE LIQUID, POISON | T8 | None | 202 | 243 | 1 L | 60 L | E | 40 |
| | Allyl alcohol | 6.1 | UN1098 | I | POISON, FLAMMABLE LIQUID | 2, B9, B14, B32, B74, B77, T38, T43, T45 | None | 227 | 244 | Forbidden | Forbidden | E | 40 |
| | Allyl bromide | 3 | UN1099 | I | FLAMMABLE LIQUID, POISON | T18 | None | 201 | 243 | Forbidden | 30 L | B | 40 |
| | Allyl chloride | 3 | UN1100 | I | FLAMMABLE LIQUID, POISON | T18, T26 | None | 201 | 243 | Forbidden | 30 L | E | 40 |
| | *Allyl chlorocarbonate, see* Allyl chloroformate | | | | | | | | | | | | |
| + | Allyl chloroformate | 8 | UN1722 | I | CORROSIVE, POISON | 1, A3, B9, B14, B30, B72, N41, T38, T43, T44 | None | 226 | 244 | Forbidden | 2.5 L | D | 21, 40, 48, 95 |
| | Allyl ethyl ether | 3 | UN2335 | II | FLAMMABLE LIQUID, POISON | T8 | None | 202 | 243 | 1 L | 60 L | E | 40 |
| | Allyl formate | 3 | UN2336 | I | FLAMMABLE LIQUID, POISON | T18, T26 | None | 201 | 243 | Forbidden | 30 L | E | 40 |
| | Allyl glycidyl ether | 3 | UN2219 | III | FLAMMABLE LIQUID, POISON | B1, T7 | 150 | 203 | 242 | 60 L | 220 L | A | |
| | Allyl iodide | 3 | UN1723 | II | FLAMMABLE LIQUID, CORROSIVE | A3, A6, N34, T18 | None | 201 | 243 | 0.5 L | 2.5 L | B | 40 |
| + | Allyl isothiocyanate, stabilized | 6.1 | UN1545 | II | POISON | 2, A3, A7, B9, B14, B32, B74, T38, T43, T45 | None | 227 | 244 | Forbidden | 60 L | D | 40 |
| | Allylamine | 6.1 | UN2334 | I | POISON, FLAMMABLE LIQUID | 2, B9, B14, B32, B74, T38, T43, T45 | None | 227 | 244 | Forbidden | Forbidden | E | 40 |
| | Allyltrichlorosilane, stabilized | 8 | UN1724 | II | CORROSIVE, FLAMMABLE LIQUID | A7, B2, B6, N34, T8, T26 | None | 202 | 242 | Forbidden | 30 L | C | 12, 21, 40, 48 |
| | Aluminum alkyl halides | 4.2 | UN3052 | I | SPONTANEOUSLY COMBUSTIBLE | B9, B11, T28, T29, T40 | None | 181 | 244 | Forbidden | Forbidden | D | |
| | Aluminum alkyl hydrides | 4.2 | UN3076 | I | SPONTANEOUSLY COMBUSTIBLE | B9, B11, T28, T29, T40 | None | 181 | 244 | Forbidden | Forbidden | D | |
| | Aluminum alkyls | 4.2 | UN3051 | I | SPONTANEOUSLY COMBUSTIBLE | B9, B11, T28, T29, T40 | None | 181 | 244 | Forbidden | Forbidden | D | |
| | Aluminum borohydride *or* Aluminum borohydride in devices | 4.2 | UN2870 | I | SPONTANEOUSLY COMBUSTIBLE, DANGEROUS WHEN WET | B11 | None | 181 | 244 | Forbidden | Forbidden | D | |
| | Aluminum bromide, anhydrous | 8 | UN1725 | II | CORROSIVE | | 154 | 212 | 240 | 15 kg | 50 kg | A | 40 |
| | Aluminum bromide, solution | 8 | UN2580 | III | CORROSIVE | T8 | 154 | 203 | 241 | 5 L | 60 L | A | |
| | Aluminum carbide | 4.3 | UN1394 | II | DANGEROUS WHEN WET | A20, N41 | None | 212 | 242 | 15 kg | 50 kg | A | |
| | Aluminum chloride, anhydrous | 8 | UN1726 | II | CORROSIVE | | 154 | 212 | 240 | 15 kg | 50 kg | A | 40 |
| | Aluminum chloride, solution | 8 | UN2581 | III | CORROSIVE | T8 | 154 | 203 | 241 | 5 L | 60 L | A | |
| | *Aluminum dross, wet or hot* | Forbidden | | | | | | | | | | | |
| | Aluminum ferrosilicon powder | 4.3 | UN1395 | II | DANGEROUS WHEN WET, POISON | A19 | None | 212 | 242 | 15 kg | 50 kg | A | 40, 85, 103 |
| | Aluminum hydride | 4.3 | UN2463 | I | DANGEROUS WHEN WET | A19, N40 | None | 211 | 242 | Forbidden | 15 kg | E | |

| Symbols | Hazardous materials descriptions and proper shipping names | Hazard class or Division | Identification Numbers | Packing group | Label(s) required (if not excepted) | Special provisions | (8) Packaging authorizations (§173.***) | | | (9) Quantity limitations | | (10) Vessel stowage requirements | |
|---|---|---|---|---|---|---|---|---|---|---|---|---|---|
| | | | | | | | Exceptions | Non-bulk packaging | Bulk packaging | Passenger aircraft or railcar | Cargo aircraft only | Vessel stowages | Other stowage provisions |
| (1) | (2) | (3) | (4) | (5) | (6) | (7) | (8A) | (8B) | (8C) | (9A) | (9B) | (10A) | (10B) |
| D | Aluminum, molten | 9 | NA9260 | III | CLASS 9 | | None | None | 247 | Forbidden | Forbidden | D | |
| | Aluminum nitrate | 5.1 | UN1438 | III | OXIDIZER | A1, A29 | 152 | 213 | 240 | 25 kg | 100 kg | A | |
| | *Aluminum phosphate solution, see* Corrosive liquids, n.o.s. | | | | | | | | | | | | |
| | Aluminum phosphide | 4.3 | UN1397 | I | DANGEROUS WHEN WET, POISON | A8, A19, N40 | None | 211 | 242 | Forbidden | 15 kg | E | 40, 85 |
| | Aluminum phosphide pesticides | 6.1 | UN3048 | I | POISON | A8 | None | 211 | 242 | Forbidden | 15 kg | E | 40, 85 |
| | Aluminum powder, coated | 4.1 | UN1309 | II | FLAMMABLE SOLID | | 151 | 212 | 240 | 15 kg | 50 kg | A | 13, 39, 101 |
| | | | | III | FLAMMABLE SOLID | | 151 | 213 | 240 | 25 kg | 100 kg | A | 13, 39, 101 |
| | Aluminum powder, uncoated | 4.3 | UN1396 | II | DANGEROUS WHEN WET | A19, A20 | None | 212 | 242 | 15 kg | 50 kg | A | 39 |
| | Aluminum resinate | 4.1 | UN2715 | III | FLAMMABLE SOLID | | 151 | 213 | 240 | 25 kg | 100 kg | A | |
| | Aluminum silicon powder, uncoated | 4.3 | UN1398 | III | DANGEROUS WHEN WET | A1, A19 | None | 213 | 241 | 25 kg | 100 kg | A | 40, 85, 103 |
| | *Amatols, see* Explosives, blasting, type B | | | | | | | | | | | | |
| | 2-Amino-4-chlorophenol | 6.1 | UN2673 | II | POISON | | None | 212 | 242 | 25 kg | 100 kg | A | |
| | 2-Amino-5-diethylaminopentane | 6.1 | UN2946 | III | KEEP AWAY FROM FOOD | T1 | 153 | 203 | 240 | 60 L | 220 L | A | |
| | 2-(2-Aminoethoxy) ethanol | 8 | UN3055 | III | CORROSIVE | T2 | 154 | 203 | 241 | 5 L | 60 L | A | |
| | N-Aminoethylpiperazine | 8 | UN2815 | III | CORROSIVE | T7 | 154 | 203 | 241 | 5 L | 60 L | A | 12, 48 |
| | Aminophenols *(o-, m-, p-)* | 6.1 | UN2512 | III | KEEP AWAY FROM FOOD | T1 | 153 | 213 | 240 | 100 kg | 200 kg | A | |
| | *Aminopropyldiethanolamine, see* Alkylamines, n.o.s. | | | | | | | | | | | | |
| | *n-Aminopropylmorpholine, see* Alkylamines, n.o.s. | | | | | | | | | | | | |
| | Aminopyridines *(o-, m-, p-)* | 6.1 | UN2671 | II | POISON | T7 | None | 212 | 242 | 25 kg | 100 kg | B | 12, 40, 48 |
| I | Ammonia, anhydrous, liquefied *or* Ammonia solutions *relative density less than 0.880 at 15 degrees C in water, with more than 50 per cent ammonia* | 2.3 | UN1005 | | POISON GAS | 4 | None | 304 | 314, 315 | Forbidden | 25 kg | D | 40, 57 |
| D | Ammonia, anhydrous, liquefied *or* Ammonia solutions, *relative density less than 0.880 at 15 degrees C in water, with more than 50 percent ammonia* | 2.2 | UN1005 | | NONFLAMMABLE GAS | 13 | None | 304 | 314, 315 | Forbidden | 25 kg | D | 40, 57 |
| | Ammonia solutions, *relative density between 0.880 and 0.957 at 15 degrees C in water, with more than 10 per cent but not more than 35 percent ammonia* | 8 | UN2672 | III | CORROSIVE | T14 | 154 | 203 | 241 | 5 L | 60 L | A | 40, 85 |
| | Ammonia solutions, *relative density less than 0.880 at 15 degrees C in water, with more than 35 per cent but not more than 50 per cent ammonia* | 2.2 | UN2073 | | NONFLAMMABLE GAS | | 306 | 304 | 314, 315 | Forbidden | 150 kg | E | 40, 57 |
| | Ammonium arsenate | 6.1 | UN1546 | II | POISON | | None | 212 | 242 | 25 kg | 100 kg | A | |
| | *Ammonium azide* | Forbidden | | | | | | | | | | | |
| | *Ammonium bifluoride, solid, see* Ammonium hydrogen fluoride, solid | | | | | | | | | | | | |
| | *Ammonium bifluoride solution, see* Ammonium hydrogen fluoride, solution | | | | | | | | | | | | |
| | *Ammonium bromate* | Forbidden | | | | | | | | | | | |
| | *Ammonium chlorate* | Forbidden | | | | | | | | | | | |
| | Ammonium dichromate | 5.1 | UN1439 | II | OXIDIZER | | 152 | 212 | 242 | 5 kg | 25 kg | A | |
| | Ammonium dinitro-o-cresolate | 6.1 | UN1843 | II | POISON | T8 | None | 212 | 242 | 25 kg | 100 kg | B | 36, 65, 66, 77 |
| | Ammonium fluoride | 6.1 | UN2505 | III | KEEP AWAY FROM FOOD | | 153 | 213 | 240 | 100 kg | 200 kg | A | 26 |
| | Ammonium fluorosilicate | 6.1 | UN2854 | III | KEEP AWAY FROM FOOD | | 153 | 213 | 240 | 100 kg | 200 kg | A | 26 |
| | *Ammonium fulminate* | Forbidden | | | | | | | | | | | |
| | Ammonium hydrogen fluoride, solid | 8 | UN1727 | II | CORROSIVE | N34 | 154 | 212 | 240 | 15 kg | 50 kg | A | 25, 26, 40 |
| | Ammonium hydrogen fluoride, solution | 8 | UN2817 | II | CORROSIVE, POISON | N34, T15 | None | 202 | 243 | 1 L | 30 L | B | 40, 95 |
| | Ammonium hydrogen sulfate | 8 | UN2506 | II | CORROSIVE | | 154 | 212 | 240 | 15 kg | 50 kg | A | 40 |
| | *Ammonium hydrosulfide, solution, see* Ammonium sulfide solution | | | | | | | | | | | | |
| D | Ammonium hydroxide, *see* Ammonia solutions, *etc* | | | | | | | | | | | | |
| | Ammonium metavanadate | 6.1 | UN2859 | II | POISON | | None | 212 | 242 | 25 kg | 100 kg | A | |
| D | Ammonium nitrate fertilizers | 5.1 | NA2072 | III | OXIDIZER | 7 | 152 | 213 | 240 | 25 kg | 100 kg | B | 48, 59, 60, 117 |

## §172.101 HAZARDOUS MATERIALS TABLE, Continued

| Symbols | Hazardous materials descriptions and proper shipping names | Hazard class or Division | Identification Numbers | Packing group | Label(s) required (if not excepted) | Special provisions | (8) Packaging authorizations (§173.***) | | | (9) Quantity limitations | | (10) Vessel stowage requirements | |
|---|---|---|---|---|---|---|---|---|---|---|---|---|---|
| | | | | | | | Exceptions | Non-bulk packaging | Bulk packaging | Passenger aircraft or railcar | Cargo aircraft only | Vessel stowages | Other stowage provisions |
| (1) | (2) | (3) | (4) | (5) | (6) | (7) | (8A) | (8B) | (8C) | (9A) | (9B) | (10A) | (10B) |
| AW | Ammonium nitrate fertilizers: *uniform non-segregating mixtures of nitrogen/ phosphate or nitrogen/potash types or complete fertilizers of nitrogen/ phosphate/ potash type, with not more than 70 per cent ammonium nitrate and not more than 0.4 per cent total added combustible material or with not more than 45 per cent ammonium nitrate with unrestricted combustible material* | 9 | UN2071 | III | CLASS 9 | | 155 | 213 | 240 | 200 kg | 200 kg | A | |
| D | Ammonium nitrate-fuel oil mixture (containing only prilled ammonium nitrate and fuel oil) | 1.5D | NA0331 | II | EXPLOSIVE 1.5D | | None | 62 | None | Forbidden | Forbidden | B | 1E, 5E, 19E |
| | Ammonium nitrate, liquid *(hot concentrated solution)* | 5.1 | UN2426 | | OXIDIZER | B5, T25 | None | None | 243 | Forbidden | Forbidden | D | 59, 60 |
| D | Ammonium nitrate mixed fertilizers | 5.1 | NA2069 | III | OXIDIZER | 10 | 152 | 213 | 240 | 25 kg | 100 kg | B | 48, 59, 60, 117 |
| | Ammonium nitrate, *with more than 0.2 per cent combustible substances, including any organic substance calculated as carbon, to the exclusion of any other added substance* | 1.1D | UN0222 | II | EXPLOSIVE 1.1D | | None | 62 | None | Forbidden | Forbidden | B | 1E, 5E, 19E |
| | Ammonium nitrate, *with not more than 0.2 per cent of combustible substances, including any organic substance calculated as carbon, to the exclusion of any other added substance* | 5.1 | UN1942 | III | OXIDIZER | A1, A29 | 152 | 213 | 240 | 25 kg | 100 kg | A | 48, 59, 60 116 |
| | *Ammonium nitrite* | Forbidden | | | | | | | | | | | |
| | Ammonium perchlorate | 5.1 | UN1442 | II | OXIDIZER | A9 | 152 | 212 | 242 | 5 kg | 25 kg | E | 58, 69, 106 |
| | Ammonium perchlorate | 1.1D | UN0402 | II | EXPLOSIVE 1.1D | 107 | None | 62 | None | Forbidden | Forbidden | B | 1E, 5E, 19E |
| | *Ammonium permanganate, see* Permanganates, inorganic, n.o.s. | | | | | | | | | | | | |
| | Ammonium persulfate | 5.1 | UN1444 | III | OXIDIZER | A1, A29 | 152 | 213 | 240 | 25 kg | 100 kg | A | |
| | Ammonium picrate, *dry or wetted with less than 10 per cent water, by mass* | 1.1D | UN0004 | II | EXPLOSIVE 1.1D | | None | 62 | None | Forbidden | Forbidden | B | 1E, 5E, 19E |
| | Ammonium picrate, wetted *with not less than 10 per cent water, by mass* | 4.1 | UN1310 | I | FLAMMABLE SOLID | A2, N41 | None | 211 | None | 0.5 kg | 0.5 kg | D | 28, 36 |
| | Ammonium polysulfide, solution | 8 | UN2818 | II | CORROSIVE, POISON | T14 | None | 202 | 243 | 1 L | 30 L | B | 12, 26, 40, 48, 95 |
| | Ammonium polyvanadate | 6.1 | UN2861 | II | POISON | | None | 212 | 242 | 25 kg | 100 kg | A | |
| | *Ammonium silicofluoride, see* Ammonium fluorosilicate | | | | | | | | | | | | |
| | Ammonium sulfide solution | 8 | UN2683 | II | CORROSIVE, POISON, FLAMMABLE LIQUID | T14 | None | 202 | 243 | 1 L | 30 L | B | 12, 22, 26, 48, 100 |
| | *Ammunition, blank, see* Cartridges for weapons, blank | | | | | | | | | | | | |
| | Ammunition, illuminating *with or without burster, expelling charge or propelling charge* | 1.2G | UN0171 | II | EXPLOSIVE 1.2G | | 63(b) | 62 | None | Forbidden | Forbidden | B | |
| | Ammunition, illuminating *with or without burster, expelling charge or propelling charge* | 1.3G | UN0254 | II | EXPLOSIVE 1.3G | | 63(b) | 62 | None | Forbidden | Forbidden | B | |
| | Ammunition, illuminating *with or without burster, expelling charge or propelling charge* | 1.4G | UN0297 | II | EXPLOSIVE 1.4G | | 63(b) | 62 | None | Forbidden | 75 kg | A | 24E |
| | Ammunition, incendiary *liquid or gel, with burster, expelling charge or propelling charge* | 1.3J | UN0247 | II | EXPLOSIVE 1.3J | | 63(b) | 62 | None | Forbidden | Forbidden | E | 7E, 13E, 23E |
| | *Ammunition, incendiary (water-activated contrivances) with burster, expelling charge or propelling charge, see* Contrivances, water-activated, *etc.* | | | | | | | | | | | | |
| | Ammunition, incendiary, white phosphorus, *with burster, expelling charge or propelling charge* | 1.2H | UN0243 | II | EXPLOSIVE 1.2H | | 63(b) | 62 | None | Forbidden | Forbidden | E | 8E, 14E, 15E |
| | Ammunition, incendiary, white phosphorus, *with burster, expelling charge or propelling charge* | 1.3H | UN0244 | II | EXPLOSIVE 1.3H | | 63(b) | 62 | None | Forbidden | Forbidden | E | 8E, 14E, 15E |
| | Ammunition, incendiary *with or without burster, expelling charge, or propelling charge* | 1.2G | UN0009 | II | EXPLOSIVE 1.2G | | 63(b) | 62 | None | Forbidden | Forbidden | B | |
| | Ammunition, incendiary *with or without burster, expelling charge, or propelling charge* | 1.3G | UN0010 | II | EXPLOSIVE 1.3G | | 63(b) | 62 | None | Forbidden | Forbidden | B | |
| | Ammunition, incendiary *with or without burster, expelling charge or propelling charge* | 1.4G | UN0300 | II | EXPLOSIVE 1.4G | | 63(b) | 62 | None | Forbidden | 75 kg | A | 24E |
| | Ammunition, practice | 1.4G | UN0362 | II | EXPLOSIVE 1.4G | | 63(b) | 62 | None | Forbidden | 75 kg | A | 24E |
| | Ammunition, practice | 1.3G | UN0488 | II | EXPLOSIVE 1.3G | | 63(b) | 62 | None | Forbidden | Forbidden | B | 1E, 7E, 25E |

## §172.101 HAZARDOUS MATERIALS TABLE, Continued

| Symbols | Hazardous materials descriptions and proper shipping names | Hazard class or Division | Identification Numbers | Packing group | Label(s) required (if not excepted) | Special provisions | (8) Packaging authorizations (§173.***) | | | (9) Quantity limitations | | (10) Vessel stowage requirements | |
|---|---|---|---|---|---|---|---|---|---|---|---|---|---|
| | | | | | | | Exceptions | Non-bulk packaging | Bulk packaging | Passenger aircraft or railcar | Cargo aircraft only | Vessel stowages | Other stowage provisions |
| (1) | (2) | (3) | (4) | (5) | (6) | (7) | (8A) | (8B) | (8C) | (9A) | (9B) | (10A) | (10B) |
| | Ammunition, proof | 1.4G | UN0363 | II | EXPLOSIVE 1.4G | | 63(b) | 62 | None | Forbidden | 75 kg | A | 24E |
| | *Ammunition, rocket, see* Warheads, rocket *etc.* | | | | | | | | | | | | |
| | *Ammunition, SA (small arms), see* Cartridges for weapons, *etc.* | | | | | | | | | | | | |
| | *Ammunition, smoke (water-activated contrivances), white phosphorus, with burster, expelling charge or propelling charge, see* Contrivances, water-activated, *etc. (UN 0248)* | | | | | | | | | | | | |
| | *Ammunition, smoke (water-activated contrivances), without white phosphorus or phosphides, with burster, expelling charge or propelling charge, see* Contrivances, water-activated, *etc. (UN 0249)* | | | | | | | | | | | | |
| | Ammunition, smoke, white phosphorus *with burster, expelling charge, or propelling charge* | 1.3H | UN0246 | II | EXPLOSIVE 1.3H | | 63(b) | 62 | None | Forbidden | Forbidden | E | 8E, 14E, 15E, 17E |
| | Ammunition smoke, white phosphorus *with burster,expelling charge, or propelling charge* | 1.2H | UN0245 | II | EXPLOSIVE 1.2H | | 63(b) | 62 | None | Forbidden | Forbidden | E | 8E, 14E, 15E, 17E |
| | Ammunition, smoke *with or without burster, expelling charge or propelling charge* | 1.3G | UN0016 | II | EXPLOSIVE 1.3G, CORROSIVE | | 63(b) | 62 | None | Forbidden | Forbidden | E | 20E |
| | Ammunition, smoke *with or without burster, expelling charge or propelling charge* | 1.4G | UN0303 | II | EXPLOSIVE 1.4G, CORROSIVE | | 63(b) | 62 | None | Forbidden | 75 kg | E | 20E |
| | Ammunition, smoke *with or without burster, expelling charge or propelling charge* | 1.2G | UN0015 | II | EXPLOSIVE 1.2G, CORROSIVE | | 63(b) | 62 | None | Forbidden | Forbidden | E | 20E |
| | *Ammunition, sporting, see* Cartridges for weapons, *etc. (UN 0012; UN 0328; UN 0339)* | | | | | | | | | | | | |
| | Ammunition, tear-producing, non-explosive, *without burster or expelling charge, non-fuzed* | 6.1 | UN2017 | II | POISON, CORROSIVE | | None | 212 | None | Forbidden | 50 kg | E | 13, 40 |
| | Ammunition, tear-producing *with burster, expelling charge or propelling charge* | 1.2G | UN0018 | II | EXPLOSIVE 1.2G, CORROSIVE, POISON | | 63(b) | 62 | None | Forbidden | Forbidden | E | 20E |
| | Ammunition, tear-producing *with burster, expelling charge or propelling charge* | 1.3G | UN0019 | II | EXPLOSIVE 1.3G, CORROSIVE, POISON | | 63(b) | 62 | None | Forbidden | Forbidden | E | 20E |
| | Ammunition, tear-producing *with burster, expelling charge or propelling charge* | 1.4G | UN0301 | II | EXPLOSIVE 1.4G, CORROSIVE, POISON | | 63(b) | 62 | None | Forbidden | 75 kg | E | 20E |
| | Ammunition, toxic, nonexplosive, *without burster or expelling charge, non-fuzed* | 6.1 | UN2016 | II | POISON | | None | 212 | None | Forbidden | 100 kg | E | 13, 40 |
| | *Ammunition, toxic (water-activated contrivances), with burster, expelling charge or propelling charge, see* Contrivances, water-activated, *etc.* | | | | | | | | | | | | |
| | Ammunition, toxic *with burster, expelling charge, or propelling charge* | 1.2K | UN0020 | II | EXPLOSIVE 1.2K, POISON | | 63(b) | 62 | None | Forbidden | Forbidden | E | 2E, 8E, 11E, 17E |
| | Ammunition, toxic *with burster, expelling charge, or propelling charge* | 1.3K | UN0021 | II | EXPLOSIVE 1.3K, POISON | | 63(b) | 62 | None | Forbidden | Forbidden | E | 2E, 8E, 11E, 17E |
| | Amyl acetates | 3 | UN1104 | III | FLAMMABLE LIQUID | B1, T1 | 150 | 203 | 241 | 60 L | 220 L | A | |
| | Amyl acid phosphate | 8 | UN2819 | III | CORROSIVE | T7 | 154 | 203 | 241 | 5 L | 60 L | A | |
| | Amyl alcohols | 3 | UN1105 | II | FLAMMABLE LIQUID | T1 | 150 | 202 | 242 | 5 L | 60 L | B | |
| | | | | III | FLAMMABLE LIQUID | B1, B3, T1 | 150 | 203 | 242 | 60 L | 220 L | A | |
| | Amyl butyrates | 3 | UN2620 | III | FLAMMABLE LIQUID | B1, T1 | 150 | 203 | 242 | 60 L | 220 L | A | |
| | Amyl chlorides | 3 | UN1107 | II | FLAMMABLE LIQUID | T1 | 150 | 202 | 242 | 5 L | 60 L | B | |
| | Amyl formates | 3 | UN1109 | III | FLAMMABLE LIQUID | B1, T1 | 150 | 203 | 242 | 60 L | 220 L | A | |
| | Amyl mercaptans | 3 | UN1111 | II | FLAMMABLE LIQUID | A3, T8 | None | 202 | 242 | 5 L | 60 L | B | 95, 102 |
| | Amyl methyl ketone | 3 | UN1110 | III | FLAMMABLE LIQUID | B1, T1 | 150 | 203 | 242 | 60 L | 220 L | A | |
| | Amyl nitrate | 3 | UN1112 | III | FLAMMABLE LIQUID | B1, T1 | 150 | 202 | 242 | 5 L | 60 L | A | 40 |
| | Amyl nitrites | 3 | UN1113 | II | FLAMMABLE LIQUID | T8 | 150 | 202 | 242 | 5 L | 60 L | E | 40 |
| | Amylamines | 3 | UN1106 | II | FLAMMABLE LIQUID | T1 | 150 | 202 | 242 | 5 L | 60 L | B | |
| | n-Amylene | 3 | UN1108 | I | FLAMMABLE LIQUID | T14 | 150 | 201 | 243 | 1 L | 30 L | E | |
| | Amyltrichlorosilane | 8 | UN1728 | II | CORROSIVE | A7, B2, B6, N34, T8, T26 | None | 202 | 242 | Forbidden | 30 L | C | 40 |
| | *Anhydrous, ammonia see Ammonia, anhydrous, liquefied* | | | | | | | | | | | | |
| | *Anhydrous hydrofluoric acid, see* Hydrogen fluoride, anhydrous | | | | | | | | | | | | |

## §172.101 HAZARDOUS MATERIALS TABLE, Continued

| Symbols | Hazardous materials descriptions and proper shipping names | Hazard class or Division | Identification Numbers | Packing group | Label(s) required (if not excepted) | Special provisions | (8) Packaging authorizations (§173.***) | | | (9) Quantity limitations | | (10) Vessel stowage requirements | |
|---|---|---|---|---|---|---|---|---|---|---|---|---|---|
| | | | | | | | Exceptions | Non-bulk packaging | Bulk packaging | Passenger aircraft or railcar | Cargo aircraft only | Vessel stowages | Other stowage provisions |
| (1) | (2) | (3) | (4) | (5) | (6) | (7) | (8A) | (8B) | (8C) | (9A) | (9B) | (10A) | (10B) |
| + | Aniline | 6.1 | UN1547 | II | POISON | T8 | None | 202 | 243 | 5 L | 60 L | A | 40 |
| | Aniline hydrochloride | 6.1 | UN1548 | III | KEEP AWAY FROM FOOD | | 153 | 213 | 240 | 100 kg | 200 kg | A | |
| | *Aniline oil, see* Aniline | | | | | | | | | | | | |
| | Anisidines | 6.1 | UN2431 | III | KEEP AWAY FROM FOOD | T1 | 153 | 203 | 241 | 60 L | 220 L | A | |
| | Anisole | 3 | UN2222 | III | FLAMMABLE LIQUID | B1, T1 | 150 | 203 | 242 | 60 L | 220 L | A | |
| | Anisoyl chloride | 8 | UN1729 | II | CORROSIVE | B2, T8 | 154 | 202 | 242 | 1 L | 30 L | C | 8, 40 |
| | *Anti-freeze, liquid, see* Flammable liquids, n.o.s. | | | | | | | | | | | | |
| | *Antimonous chloride, see* Antimony trichloride | | | | | | | | | | | | |
| | Antimony compounds, inorganic, liquid, n.o.s. | 6.1 | UN3141 | I | POISON | A4, T42 | None | 201 | 243 | 1 L | 30 L | A | |
| | | | | II | POISON | T14 | None | 202 | 243 | 5 L | 60 L | A | |
| | | | | III | KEEP AWAY FROM FOOD | T7 | 153 | 203 | 241 | 60 L | 220 L | A | |
| | Antimony compounds, inorganic, solid, n.o.s. | 6.1 | UN1549 | I | POISON | | None | 211 | 242 | 5 kg | 50 kg | A | |
| | | | | II | POISON | | None | 212 | 242 | 25 kg | 100 kg | A | |
| | | | | III | KEEP AWAY FROM FOOD | | 153 | 213 | 240 | 100 kg | 200 kg | A | |
| | Antimony lactate | 6.1 | UN1550 | III | KEEP AWAY FROM FOOD | | 153 | 213 | 24[illegible] | 100 kg | 200 kg | A | |
| | Antimony pentachloride, liquid | 8 | UN1730 | II | CORROSIVE | B2, T8, T26 | None | 202 | 242 | 1 L | 30 L | C | 8, 40 |
| | Antimony pentachloride, solutions | 8 | UN1731 | II | CORROSIVE | B2, T8, T27 | 154 | 202 | 242 | 1 L | 30 L | C | 8, 40 |
| | Antimony pentafluoride | 8 | UN1732 | II | CORROSIVE, POISON | A3, A6, A7, A10, N3, T12, T26 | None | 202 | 243 | Forbidden | 30 L | D | 40, 95 |
| | Antimony potassium tartrate | 6.1 | UN1551 | III | KEEP AWAY FROM FOOD | | 153 | 213 | 240 | 100 kg | 200 kg | A | |
| | Antimony powder | 6.1 | UN2871 | III | KEEP AWAY FROM FOOD | | 153 | 213 | 240 | 100 kg | 200 kg | A | |
| | *Antimony sulfide and a chlorate, mixtures of* | Forbidden | | | | | | | | | | | |
| | *Antimony sulfide, solid, see* Antimony compounds, inorganic, n.o.s. | | | | | | | | | | | | |
| D | Antimony tribromide, solid | 8 | NA1549 | II | CORROSIVE | | 154 | 212 | 240 | 25 kg | 100 kg | A | 13 |
| D | Antimony tribromide, solution | 8 | NA1549 | II | CORROSIVE | B2 | 154 | 202 | 242 | 1 L | 30 L | C | 13 |
| | Antimony trichloride, liquid | 8 | UN1733 | II | CORROSIVE | B2 | 154 | 202 | 242 | 1 L | 30 L | C | 40 |
| | Antimony trichloride, solid | 8 | UN1733 | II | CORROSIVE | | 154 | 212 | 240 | 15 kg | 50 kg | A | 40 |
| D | Antimony trifluoride, solid | 8 | NA1549 | II | CORROSIVE | | 154 | 212 | 240 | 25 kg | 25 kg | A | 13 |
| D | Antimony trifluoride solution | 8 | NA1549 | II | CORROSIVE | B2 | 154 | 202 | 242 | 1 L | 30 L | C | 13 |
| | *Aqua ammonia, see* Ammonia solution, *etc.* | | | | | | | | | | | | |
| | Argon, compressed | 2.2 | UN1006 | | NONFLAMMABLE GAS | | 306 | 302 | 314, 315 | 75 kg | 150 kg | A | |
| | Argon, refrigerated liquid *(cryogenic liquid)* | 2.2 | UN1951 | | NONFLAMMABLE GAS | | 320 | 316 | 318 | 50 kg | 500 kg | B | |
| | Arsenic | 6.1 | UN1558 | II | POISON | | None | 212 | 242 | 25 kg | 100 kg | A | |
| | Arsenic acid, liquid | 6.1 | UN1553 | I | POISON | T18, T27 | None | 201 | 243 | 1 L | 30 L | B | |
| | Arsenic acid, solid | 6.1 | UN1554 | II | POISON | | None | 212 | 242 | 25 kg | 100 kg | A | |
| | Arsenic bromide | 6.1 | UN1555 | II | POISON | | None | 212 | 242 | 25 kg | 100 kg | A | 12, 40 |
| | *Arsenic chloride, see* Arsenic trichloride | | | | | | | | | | | | |
| | Arsenic compounds, liquid, n.o.s. *including arsenates n.o.s.; arsenites, n.o.s.; arsenic sulfides, n.o.s.; and organic compounds of arsenic, n.o.s.* | 6.1 | UN1556 | I | POISON | | None | 201 | 243 | 1 L | 30 L | B | 40 |
| | | | | II | POISON | | None | 202 | 243 | 5 L | 60 L | B | 40 |
| | | | | III | KEEP AWAY FROM FOOD | | 153 | 203 | 241 | 60 L | 220 L | B | 40 |
| | Arsenic compounds, solid, n.o.s. *including arsenates, n.o.s.; arsenites, n.o.s.; arsenic sulfides, n.o.s.; and organic compounds of arsenic, n.o.s.* | 6.1 | UN1557 | I | POISON | | None | 211 | 242 | 5 kg | 50 kg | A | |
| | | | | II | POISON | | None | 212 | 242 | 25 kg | 100 kg | A | |
| | | | | III | KEEP AWAY FROM FOOD | | 153 | 213 | 240 | 100 kg | 200 kg | A | |
| | Arsenic pentoxide | 6.1 | UN1559 | II | POISON | | None | 212 | 242 | 25 kg | 100 kg | A | |
| D | Arsenic sulfide | 6.1 | NA1557 | II | POISON | | None | 212 | 242 | 25 kg | 100 kg | A | |
| | *Arsenic sulfide and a chlorate, mixtures of* | Forbidden | | | | | | | | | | | |
| | Arsenic trichloride | 6.1 | UN1560 | I | POISON | 2, B9, B14, B32, B74, T38, T43, T45 | None | 227 | 244 | 1L | 30L | B | 40 |
| | Arsenic trioxide | 6.1 | UN1561 | II | POISON | | None | 212 | 242 | 25 kg | 100 kg | A | |
| D | Arsenic trisulfide | 6.1 | NA1557 | II | POISON | | None | 212 | 242 | 25 kg | 100 kg | A | |
| | *Arsenic, white, solid, see* Arsenic trioxide | | | | | | | | | | | | |
| | Arsenical dust | 6.1 | UN1562 | II | POISON | | None | 212 | 242 | 25 kg | 100 kg | A | |
| | Arsenical pesticides, liquid, flammable, toxic, n.o.s., *flash point less than 23 degrees C.* | 3 | UN2760 | I | FLAMMABLE LIQUID, POISON | | None | 201 | 243 | Forbidden | 30 L | B | 40 |
| | | | | II | FLAMMABLE LIQUID, POISON | | None | 202 | 243 | 1 L | 60 L | B | 40 |

## §172.101 HAZARDOUS MATERIALS TABLE, Continued

| Symbols | Hazardous materials descriptions and proper shipping names | Hazard class or Division | Identification Numbers | Packing group | Label(s) required (if not excepted) | Special provisions | (8) Packaging authorizations (§173.***) | | | (9) Quantity limitations | | (10) Vessel stowage requirements | |
|---|---|---|---|---|---|---|---|---|---|---|---|---|---|
| | | | | | | | Exceptions | Non-bulk packaging | Bulk packaging | Passenger aircraft or railcar | Cargo aircraft only | Vessel stowages | Other stowage provisions |
| (1) | (2) | (3) | (4) | (5) | (6) | (7) | (8A) | (8B) | (8C) | (9A) | (9B) | (10A) | (10B) |
| | Arsenical pesticides, liquid, toxic, flammable, n.o.s. *flashpoint not less than 23 degrees C.* | 6.1 | UN2993 | I | POISON, FLAMMABLE LIQUID | T42 | None | 201 | 243 | 1 L | 30 L | B | 40 |
| | | | | II | POISON, FLAMMABLE LIQUID | T14 | None | 202 | 243 | 5 L | 60 L | B | 40 |
| | | | | III | KEEP AWAY FROM FOOD, FLAMMABLE LIQUID | B1, T14 | 153 | 203 | 241 | 60 L | 220 L | A | 40 |
| | Arsenical pesticides, liquid, toxic, n.o.s. | 6.1 | UN2994 | I | POISON | T42 | None | 201 | 243 | 1 L | 30 L | B | 40 |
| | | | | II | POISON | T14 | None | 202 | 243 | 5 L | 60 L | B | 40 |
| | | | | III | KEEP AWAY FROM FOOD | T14 | 153 | 203 | 241 | 60 L | 220 L | A | 40 |
| | Arsenical pesticides, solid, toxic, n.o.s. | 6.1 | UN2759 | I | POISON | | None | 211 | 242 | 5 kg | 50 kg | A | 40 |
| | | | | II | POISON | | None | 212 | 242 | 25 kg | 100 kg | A | 40 |
| | | | | III | KEEP AWAY FROM FOOD | | 153 | 213 | 240 | 100 kg | 200 kg | A | 40 |
| | *Arsenious acid, solid, see* Arsenic trioxide | | | | | | | | | | | | |
| | *Arsenious and mercuric iodide solution, see* Arsenic compounds, liquid, n.o.s. | | | | | | | | | | | | |
| | Arsine | 2.3 | UN2188 | | POISON GAS, FLAMMABLE GAS | 1 | None | 192 | 245 | Forbidden | Forbidden | D | 40 |
| | Articles, explosive, extremely insensitive *or* Articles, EEI | 1.6N | UN0486 | II | EXPLOSIVE 1.6N | 101 | None | 62 | None | Forbidden | Forbidden | B | |
| | Articles, explosive, n.o.s. | 1.4S | UN0349 | II | EXPLOSIVE 1.4S | 101 | None | 62 | None | 25 kg | 100 kg | A | 9E |
| | Articles, explosive, n.o.s. | 1.4B | UN0350 | II | EXPLOSIVE 1.4B | 101 | None | 62 | None | Forbidden | Forbidden | A | 24E |
| | Articles, explosive, n.o.s. | 1.4C | UN0351 | II | EXPLOSIVE 1.4C | 101 | None | 62 | None | Forbidden | Forbidden | A | 24E |
| | Articles, explosive, n.o.s. | 1.4D | UN0352 | II | EXPLOSIVE 1.4D | 101 | None | 62 | None | Forbidden | Forbidden | A | 24E |
| | Articles, explosive, n.o.s. | 1.4G | UN0353 | II | EXPLOSIVE 1.4G | 101 | None | 62 | None | Forbidden | Forbidden | A | 24E |
| | Articles, explosive, n.o.s. | 1.1L | UN0354 | II | EXPLOSIVE 1.1L | 101 | None | 62 | None | Forbidden | Forbidden | E | 2E, 8E, 17E |
| | Articles, explosive, n.o.s. | 1.2L | UN0355 | II | EXPLOSIVE 1.2L | 101 | None | 62 | None | Forbidden | Forbidden | E | 2E, 8E, 17E |
| | Articles, explosive, n.o.s. | 1.3L | UN0356 | II | EXPLOSIVE 1.3L | 101 | None | 62 | None | Forbidden | Forbidden | E | 2E, 8E, 17E |
| | Articles, explosive, n.o.s. | 1.1C | UN0462 | II | EXPLOSIVE 1.1C | 101 | None | 62 | None | Forbidden | Forbidden | B | |
| | Articles, explosive, n.o.s. | 1.1D | UN0463 | II | EXPLOSIVE 1.1D | 101 | None | 62 | None | Forbidden | Forbidden | B | |
| | Articles, explosive, n.o.s. | 1.1E | UN0464 | II | EXPLOSIVE 1.1E | 101 | None | 62 | None | Forbidden | Forbidden | B | |
| | Articles, explosive, n.o.s. | 1.1F | UN0465 | II | EXPLOSIVE 1.1F | 101 | None | 62 | None | Forbidden | Forbidden | E | |
| | Articles, explosive, n.o.s. | 1.2C | UN0466 | II | EXPLOSIVE 1.2C | 101 | None | 62 | None | Forbidden | Forbidden | B | |
| | Articles, explosive, n.o.s. | 1.2D | UN0467 | II | EXPLOSIVE 1.2D | 101 | None | 62 | None | Forbidden | Forbidden | B | |
| | Articles, explosive, n.o.s. | 1.2E | UN0468 | II | EXPLOSIVE 1.2E | 101 | None | 62 | None | Forbidden | Forbidden | B | |
| | Articles, explosive, n.o.s. | 1.2F | UN0469 | II | EXPLOSIVE 1.2F | 101 | None | 62 | None | Forbidden | Forbidden | E | |
| | Articles, explosive, n.o.s. | 1.3C | UN0470 | II | EXPLOSIVE 1.3C | 101 | None | 62 | None | Forbidden | Forbidden | B | |
| | Articles, explosive, n.o.s. | 1.4E | UN0471 | II | EXPLOSIVE 1.4E | 101 | None | 62 | None | Forbidden | 75kg | A | 24E |
| | Articles, explosive, n.o.s. | 1.4F | UN0472 | II | EXPLOSIVE 1.4F | 101 | None | 62 | None | Forbidden | Forbidden | E | |
| | Articles, pyrophoric | 1.2L | UN0380 | II | EXPLOSIVE 1.2L | | None | 62 | None | Forbidden | Forbidden | E | 2E, 8E, 17E |
| | Articles, pyrotechnic *for technical purposes* | 1.1G | UN0428 | II | EXPLOSIVE 1.1G | | None | 62 | None | Forbidden | Forbidden | B | |
| | Articles, pyrotechnic *for technical purposes* | 1.2G | UN0429 | II | EXPLOSIVE 1.2G | | None | 62 | None | Forbidden | Forbidden | B | |
| | Articles, pyrotechnic *for technical purposes* | 1.3G | UN0430 | II | EXPLOSIVE 1.3G | | None | 62 | None | Forbidden | Forbidden | B | |
| | Articles, pyrotechnic *for technical purposes* | 1.4G | UN0431 | II | EXPLOSIVE 1.4G | | None | 62 | None | Forbidden | 75 kg | A | 24E |
| | Articles, pyrotechnic *for technical purposes* | 1.4S | UN0432 | II | EXPLOSIVE 1.4S | | None | 62 | None | 25 kg | 100 kg | A | 9E |
| D | Asbestos | 9 | NA2212 | III | CLASS 9 | | 155 | 216 | 240 | 200 kg | 200 kg | A | 34, 40 |
| | *Ascaridole (organic peroxide)* | Forbidden | | | | | | | | | | | |
| D | Asphalt, *at or above its flashpoint* | 3 | NA1999 | III | FLAMMABLE LIQUID | B1 | 150 | 203 | 247 | Forbidden | Forbidden | D | |
| D | Asphalt, cut back, *see* Tars, liquid, *etc.* | | | | | | | | | | | | |
| | *Automobile, motorcycle, tractor, or other self-propelled vehicle, engine, or other mechanical apparatus, see* Vehicles, self-propelled | | | | | | | | | | | | |
| | *Azaurolic acid (salt of) (dry)* | Forbidden | | | | | | | | | | | |
| | *Azido guanidine picrate (dry)* | Forbidden | | | | | | | | | | | |
| | *5-Azido-1-hydroxy tetrazole* | Forbidden | | | | | | | | | | | |
| | *Azido hydroxy tetrazole (mercury and silver salts)* | Forbidden | | | | | | | | | | | |
| | *3-Azido-1,2-Propylene glycol dinitrate* | Forbidden | | | | | | | | | | | |

## §172.101 HAZARDOUS MATERIALS TABLE, Continued

| Symbols | Hazardous materials descriptions and proper shipping names | Hazard class or Division | Identification Numbers | Packing group | Label(s) required (if not excepted) | Special provisions | (8) Packaging authorizations (§173.***) | | | (9) Quantity limitations | | (10) Vessel stowage requirements | |
|---|---|---|---|---|---|---|---|---|---|---|---|---|---|
| | | | | | | | Exceptions | Non-bulk packaging | Bulk packaging | Passenger aircraft or railcar | Cargo aircraft only | Vessel stowages | Other stowage provisions |
| (1) | (2) | (3) | (4) | (5) | (6) | (7) | (8A) | (8B) | (8C) | (9A) | (9B) | (10A) | (10B) |
| | *Azidodithiocarbonic acid* | Forbidden | | | | | | | | | | | |
| | *Azidoethyl nitrate* | Forbidden | | | | | | | | | | | |
| | *1-Aziridinyl phosphine oxide-(tris), see* Tris-(1-aziridinyl) phosphine oxide, solution | | | | | | | | | | | | |
| | 2,2'-Azodi-(2,4-dimethyl-4-methoxyvaleronitrile | 4.1 | UN2955 | II | FLAMMABLE SOLID | | None | 224 | None | Forbidden | Forbidden | D | 2 |
| | 2,2'-Azodi-(2,4 dimethylvaleronitrile) | 4.1 | UN2953 | II | FLAMMABLE SOLID | | None | 224 | None | Forbidden | Forbidden | D | 2 |
| | 1,1'-Azodi-(hexahydrobenzonitrile) | 4.1 | UN2954 | II | FLAMMABLE SOLID | | None | 224 | None | 15 kg | 50 kg | B | 12, 61, 85 |
| | 2,2'-Azodi (2-methyl-butyronitrile) | 4.1 | UN3030 | II | FLAMMABLE SOLID | | None | 224 | None | Forbidden | Forbidden | D | 2, 61 |
| | Azodiisobutyronitrile | 4.1 | UN2952 | II | FLAMMABLE SOLID, EXPLOSIVES | 41, 53 | None | 224 | None | Forbidden | Forbidden | D | 2 |
| | *Azotetrazole (dry)* | Forbidden | | | | | | | | | | | |
| | Barium | 4.3 | UN1400 | II | DANGEROUS WHEN WET | A19 | None | 212 | 241 | 15 kg | 50 kg | E | |
| | Barium alloys, pyrophoric | 4.2 | UN1854 | I | SPONTANEOUSLY COMBUSTIBLE | | None | 181 | None | Forbidden | Forbidden | D | |
| | Barium azide, *dry or wetted with less than 50 per cent water, by mass* | 1.1A | UN0224 | II | EXPLOSIVE 1.1A, Poison | 111, 117 | None | 62 | None | Forbidden | Forbidden | E | 2E, 6E |
| | Barium azide, wetted *with not less than 50 per cent water, by mass* | 4.1 | UN1571 | I | FLAMMABLE SOLID, POISON | A2 | None | 182 | None | Forbidden | 0.5 kg | D | 28, 36 |
| | Barium bromate | 5.1 | UN2719 | II | OXIDIZER, POISON | | None | 212 | 242 | 5 kg | 25 kg | A | 56, 58, 106 |
| | Barium chlorate | 5.1 | UN1445 | II | OXIDIZER, POISON | A9, N34, T8 | None | 212 | 242 | 5 kg | 25 kg | A | 56, 58, 106 |
| | Barium compounds, n.o.s. | 6.1 | UN1564 | I<br>II<br>III | POISON<br>POISON<br>KEEP AWAY FROM FOOD | | None<br>None<br>153 | 211<br>212<br>213 | 242<br>242<br>240 | 5 kg<br>25 kg<br>100 kg | 50 kg<br>100 kg<br>200 kg | A<br>A<br>A | |
| | Barium cyanide | 6.1 | UN1565 | I | POISON | N74, N75 | None | 211 | 242 | 5 kg | 50 kg | A | 26, 40 |
| | Barium hypochlorite *with more than 22 per cent available chlorine* | 5.1 | UN2741 | II | OXIDIZER | A7, A9, N34 | 152 | 212 | None | 5 kg | 25 kg | B | 56, 58, 106 |
| | Barium nitrate | 5.1 | UN1446 | II | OXIDIZER, POISON | | None | 212 | 242 | 5 kg | 25 kg | A | |
| | Barium oxide | 6.1 | UN1884 | III | KEEP AWAY FROM FOOD | | 153 | 213 | 240 | 100 kg | 200 kg | A | |
| | Barium perchlorate | 5.1 | UN1447 | II | OXIDIZER, POISON | T8 | None | 212 | 242 | 5 kg | 25 kg | A | 56, 58, 106 |
| | Barium permanganate | 5.1 | UN1448 | II | OXIDIZER, POISON | | None | 212 | 242 | 5 kg | 25 kg | D | 56, 58, 69, 106, 107 |
| | Barium peroxide | 5.1 | UN1449 | II | OXIDIZER, POISON | | None | 212 | 242 | 5 kg | 25 kg | A | 13, 75, 106 |
| | Barium selenate, *see* Selenates *or* Selenites | | | | | | | | | | | | |
| | Barium selenite, *see* Selenates *or* Selenites | | | | | | | | | | | | |
| D | Barium styphnate | 1.1A | NA0473 | II | EXPLOSIVE 1.1A | 111, 117 | None | 62 | None | Forbidden | Forbidden | E | 2E, 6E |
| | Batteries, dry, containing potassium hydroxide solid, *electric, storage* | 8 | UN3028 | III | CORROSIVE | | None | 213 | None | 25 kg gross | 230 kg gross | A | |
| | Batteries, wet, filled with acid, *electric storage* | 8 | UN2794 | III | CORROSIVE | | 159 | 159 | None | 25 kg gross | No limit | A | |
| | Batteries, wet, filled with alkali, *electric storage* | 8 | UN2795 | III | CORROSIVE | | 159 | 159 | None | 25 kg gross | No limit | A | |
| AW | Batteries, wet, non-spillable, *electric storage* | 8 | UN2800 | III | CORROSIVE | | 159 | 159 | None | No Limit | No Limit | A | |
| | Battery fluid, acid | 8 | UN2796 | II | CORROSIVE | A3, A7, B2, B15, N6, N34, T9, T27 | 154 | 202 | 242 | 1 L | 30 L | B | |
| | Battery fluid, alkali | 8 | UN2797 | II | CORROSIVE | B2, N6, T8 | 154 | 202 | 242 | 1 L | 30 L | A | |
| | *Battery lithium type, see* Lithium batteries *etc.* | | | | | | | | | | | | |
| | *Battery, wet, filled with acid or alkali with automobile (or named self-propelled vehicle or mechanical equipment containing internal combustion engine) see* Vehicles, self-propelled *etc.* | | | | | | | | | | | | |
| | *Battery, wet, with wheelchair, see* Wheelchair, electric | | | | | | | | | | | | |
| | Benzene | 3 | UN1114 | II | FLAMMABLE LIQUID | T8 | 150 | 202 | 242 | 5 L | 60 L | B | 40 |
| | *Benzene diazonium chloride (dry)* | Forbidden | | | | | | | | | | | |
| | *Benzene diazonium nitrate (dry)* | Forbidden | | | | | | | | | | | |

# §172.101 HAZARDOUS MATERIALS TABLE, Continued

| Symbols | Hazardous materials descriptions and proper shipping names | Hazard class or Division | Identification Numbers | Packing group | Label(s) required (if not excepted) | Special provisions | (8) Packaging authorizations (§173.***) | | | (9) Quantity limitations | | (10) Vessel stowage requirements | |
|---|---|---|---|---|---|---|---|---|---|---|---|---|---|
| | | | | | | | Exceptions | Non-bulk packaging | Bulk packaging | Passenger aircraft or railcar | Cargo aircraft only | Vessel stowages | Other stowage provisions |
| (1) | (2) | (3) | (4) | (5) | (6) | (7) | (8A) | (8B) | (8C) | (9A) | (9B) | (10A) | (10B) |
| | Benzene-1,3-disulfohydrazide, *not more than 52 per cent as a paste* | 4.1 | UN2971 | II | FLAMMABLE SOLID | | None | 224 | None | 15 kg | 50 kg | B | 12, 61, 85 |
| | *Benzene phosphorus dichloride, see* Phenyl phosphorus dichloride | | | | | | | | | | | | |
| | *Benzene phosphorus thiodichloride, see* Phenyl phosphorus thiodichloride | | | | | | | | | | | | |
| | Benzene sulfohydrazide | 4.1 | UN2970 | II | FLAMMABLE SOLID | | None | 224 | None | 15 kg | 50 kg | B | 12, 61, 85 |
| | Benzene sulfonyl chloride | 8 | UN2225 | III | CORROSIVE | T8 | 154 | 203 | 241 | 5 L | 60 L | A | 40 |
| | *Benzene triozonide* | Forbidden | | | | | | | | | | | |
| | *Benzenethiol, see* Phenyl mercaptan | | | | | | | | | | | | |
| | Benzidine | 6.1 | UN1885 | II | POISON | | None | 212 | 242 | 25 kg | 100 kg | A | |
| | Benzoic derivative pesticides, liquid, flammable, toxic, n.o.s., *flash point less than 23 degrees C.* | 3 | UN2770 | I | FLAMMABLE LIQUID, POISON | | None | 201 | 243 | Forbidden | 30 L | E | |
| | | | | II | FLAMMABLE LIQUID, POISON | | None | 202 | 243 | 1 L | 60 L | B | |
| | Benzoic derivative pesticides, liquid, toxic, flammable, n.o.s. *flashpoint not less than 23 degrees C.* | 6.1 | UN3003 | I | POISON, FLAMMABLE LIQUID | T42 | None | 201 | 243 | 1 L | 30 L | B | 40 |
| | | | | II | POISON, FLAMMABLE LIQUID | T14 | None | 202 | 243 | 5 L | 60 L | B | 40 |
| | | | | III | KEEP AWAY FROM FOOD, FLAMMABLE LIQUID | T14 | 153 | 203 | 241 | 60 L | 220 L | A | 40 |
| | Benzoic derivative pesticides, liquid, toxic, n.o.s. | 6.1 | UN3004 | I | POISON | T42 | None | 201 | 243 | 1 L | 30 L | B | 40 |
| | | | | II | POISON | T14 | None | 202 | 243 | 5 L | 60 L | B | 40 |
| | | | | III | KEEP AWAY FROM FOOD | T14 | 153 | 203 | 241 | 60 L | 220 L | A | 40 |
| | Benzoic derivative pesticides, solid, toxic, n.o.s. | 6.1 | UN2769 | I | POISON | | None | 211 | 242 | 5 kg | 50 kg | A | 40 |
| | | | | II | POISON | | None | 212 | 242 | 25 kg | 100 kg | A | |
| | | | | III | KEEP AWAY FROM FOOD | | 153 | 213 | 240 | 100 kg | 200 kg | A | |
| | *Benzol, see* Benzene | | | | | | | | | | | | |
| | Benzonitrile | 6.1 | UN2224 | II | POISON | T14 | None | 202 | 243 | 5 L | 60 L | A | 26, 40 |
| | Benzoquinone | 6.1 | UN2587 | II | POISON | | None | 212 | 242 | 25 kg | 100 kg | A | |
| | Benzotrichloride | 8 | UN2226 | II | CORROSIVE | B2, T15 | 154 | 202 | 242 | 1 L | 30 L | A | 40 |
| | Benzotrifluoride | 3 | UN2338 | II | FLAMMABLE LIQUID | T2 | 150 | 202 | 242 | 5 L | 60 L | B | 40 |
| | *Benzoxidiazoles (dry)* | Forbidden | | | | | | | | | | | |
| | *Benzoyl azide* | Forbidden | | | | | | | | | | | |
| | Benzoyl chloride | 8 | UN1736 | II | CORROSIVE | B2, T9, T26 | 154 | 202 | 242 | 1 L | 30 L | C | 8, 40 |
| | Benzyl bromide | 6.1 | UN1737 | II | POISON, CORROSIVE | A3, A7, N33, N34, T12, T26 | None | 202 | 243 | 1 L | 30 L | D | 13, 40 |
| | Benzyl chloride | 6.1 | UN1738 | II | POISON, CORROSIVE | A3, A7, B41, B70, N33, N43, T12, T26 | None | 202 | 243 | 1 L | 30 L | D | 13, 40 |
| | Benzyl chloride *unstabilized* | 6.1 | UN1738 | II | POISON, CORROSIVE | A3, A7, B8, B11, N33, N34, N43, T12, T26 | None | 202 | 243 | 1 L | 30 L | D | 13, 20 |
| | Benzyl chloroformate | 8 | UN1739 | I | CORROSIVE | A3, A6, B4, N41, T18, T26 | None | 201 | 242 | Forbidden | 2.5 L | D | 40 |
| | 4-(Benzyl(ethyl)amino)-3-ethoxybenzenediazonium zinc chloride | 4.1 | UN3037 | II | FLAMMABLE SOLID | | None | 224 | None | Forbidden | Forbidden | D | 2 |
| | Benzyl iodide | 6.1 | UN2653 | II | POISON | T8 | None | 202 | 243 | 5 L | 60 L | B | 12, 40, 48 |
| | 4-(Benzyl(methyl)amino)3-ethoxybenzenediazonium zinc chloride | 4.1 | UN3038 | II | FLAMMABLE SOLID | | None | 224 | None | Forbidden | Forbidden | D | 2 |
| | Benzyldimethylamine | 8 | UN2619 | II | CORROSIVE | B2, T1 | 154 | 202 | 242 | 1 L | 30 L | A | 12, 21, 40, 48 |
| | Benzylidene chloride | 6.1 | UN1886 | II | POISON | T8 | None | 202 | 243 | 5 L | 60 L | D | 40 |
| | Beryllium compounds, n.o.s. | 6.1 | UN1566 | II | POISON | | None | 212 | 242 | 25 kg | 100 kg | A | |
| | Beryllium nitrate | 5.1 | UN2464 | II | OXIDIZER, POISON | | None | 212 | 242 | 5 kg | 25 kg | A | |
| | Beryllium, powder | 6.1 | UN1567 | II | POISON, FLAMMABLE SOLID | | None | 212 | 242 | 15 kg | 50 kg | A | |
| | Bifluorides, n.o.s. *solid* | 8 | UN1740 | II | CORROSIVE | N3, N34 | None | 212 | 240 | 15 kg | 50 kg | A | 25, 26, 40 |
| | Bifluorides, n.o.s. *solutions* | 8 | UN1740 | II | CORROSIVE | N3, N34 | None | 202 | 243 | 1 L | 30 L | A | 25, 26, 40 |
| | *Biphenyl triozonide* | Forbidden | | | | | | | | | | | |
| | Bipyridilium pesticides, liquid, flammable, toxic, n.o.s., *flash point less than 23 degrees C.* | 3 | UN2782 | I | FLAMMABLE LIQUID, POISON | | None | 201 | 243 | Forbidden | 30 L | E | |
| | | | | II | FLAMMABLE LIQUID, POISON | | None | 202 | 243 | 1 L | 60 L | B | |

## §172.101 HAZARDOUS MATERIALS TABLE, Continued

| Symbols | Hazardous materials descriptions and proper shipping names | Hazard class or Division | Identification Numbers | Packing group | Label(s) required (if not excepted) | Special provisions | (8) Packaging authorizations (§173.***) | | | (9) Quantity limitations | | (10) Vessel stowage requirements | |
|---|---|---|---|---|---|---|---|---|---|---|---|---|---|
| | | | | | | | Exceptions | Non-bulk packaging | Bulk packaging | Passenger aircraft or railcar | Cargo aircraft only | Vessel stowages | Other stowage provisions |
| (1) | (2) | (3) | (4) | (5) | (6) | (7) | (8A) | (8B) | (8C) | (9A) | (9B) | (10A) | (10B) |
| | Bipyridilium pesticides, liquid, toxic, flammable, n.o.s. *flashpoint not less than 23 degrees C.* | 6.1 | UN3015 | I | POISON, FLAMMABLE LIQUID | T42 | None | 201 | 243 | 1 L | 30 L | B | 21, 40, 48 |
| | | | | II | POISON, FLAMMABLE LIQUID | T14 | None | 202 | 243 | 5 L | 60 L | B | 21, 40, 48 |
| | | | | III | KEEP AWAY FROM FOOD, FLAMMABLE LIQUID | B1, T14 | 153 | 203 | 242 | 60 L | 220 L | A | 21, 40, 48 |
| | Bipyridilium pesticides, liquid, toxic, n.o.s. | 6.1 | UN3016 | I | POISON | T42 | None | 201 | 243 | 1 L | 30 L | B | 40 |
| | | | | II | POISON | T14 | None | 202 | 243 | 5 L | 60 L | B | 40 |
| | | | | III | KEEP AWAY FROM FOOD | T14 | 153 | 203 | 241 | 60 L | 220 L | A | 40 |
| | Bipyridilium pesticides, solid, toxic, n.o.s. | 6.1 | UN2781 | I | POISON | | None | 211 | 242 | 5 kg | 50 kg | A | 40 |
| | | | | II | POISON | | None | 212 | 242 | 25 kg | 100 kg | A | 40 |
| | | | | III | KEEP AWAY FROM FOOD | | 153 | 213 | 240 | 100 kg | 200 kg | A | 40 |
| | *bis (Aminopropyl) piperazine, see* Corrosive liquid, n.o.s. | | | | | | | | | | | | |
| | Bisulfites, inorganic, aqueous solutions, n.o.s. | 8 | UN2693 | III | CORROSIVE | T8 | 154 | 203 | 241 | 1 L | 30 L | A | 8, 26, 40 |
| | Black powder, compressed *or* Gunpowder, compressed *or* Black powder, in pellets *or* Gunpowder, in pellets | 1.1D | UN0028 | II | EXPLOSIVE 1.1D | | None | 62 | None | Forbidden | Forbidden | B | 1E, 5E |
| | Black powder *or* Gunpowder, *granular or as a meal* | 1.1D | UN0027 | II | EXPLOSIVE 1.1D | | None | 62 | None | Forbidden | Forbidden | B | 10E, 26E |
| | *Blasting agent, n.o.s., see* Explosives, blasting *etc.* | | | | | | | | | | | | |
| | *Blasting cap assemblies, see* Detonator assemblies, non-electric, *for blasting* | | | | | | | | | | | | |
| | *Blasting caps, electric, see* Detonators, electric *for blasting* | | | | | | | | | | | | |
| | *Blasting caps, non-electric, see* Detonators, non-electric, *for blasting* | | | | | | | | | | | | |
| | *Bleaching powder, see* Calcium hypochlorite mixtures, *etc.* | | | | | | | | | | | | |
| I | Blue asbestos *(Crocidolite) or* Brown asbestos *(amosite, mysorite)* | 9 | UN2212 | II | CLASS 9 | | 155 | 216 | 240 | Forbidden | Forbidden | A | 34, 40 |
| | Bombs, photo-flash | 1.1F | UN0037 | II | EXPLOSIVE 1.1F | | 63(b) | 62 | None | Forbidden | Forbidden | E | |
| | Bombs, photo-flash | 1.1D | UN0038 | II | EXPLOSIVE 1.1D | | 63(b) | 62 | None | Forbidden | Forbidden | B | |
| | Bombs, photo-flash | 1.2G | UN0039 | II | EXPLOSIVE 1.2G | | 63(b) | 62 | None | Forbidden | Forbidden | B | |
| | Bombs, photo-flash | 1.3G | UN0299 | II | EXPLOSIVE 1.3G | | 63(b) | 62 | None | Forbidden | Forbidden | B | |
| | Bombs, smoke, non-explosive, *with corrosive liquid, without initiating device* | 8 | UN2028 | II | CORROSIVE | | None | 160 | None | Forbidden | 50 kg | E | 40 |
| | Bombs, *with bursting charge* | 1.1F | UN0033 | II | EXPLOSIVE 1.1F | | 63(b) | 62 | None | Forbidden | Forbidden | E | |
| | Bombs, *with bursting charge* | 1.1D | UN0034 | II | EXPLOSIVE 1.1D | | 63(b) | 62 | None | Forbidden | Forbidden | B | 3E, 7E |
| | Bombs, *with bursting charge* | 1.2D | UN0035 | II | EXPLOSIVE 1.2D | | 63(b) | 62 | None | Forbidden | Forbidden | B | 3E, 7E |
| | Bombs, *with bursting charge* | 1.2F | UN0291 | II | EXPLOSIVE 1.2F | | 63(b) | 62 | None | Forbidden | Forbidden | E | |
| | Bombs with flammable liquid, *with bursting charge* | 1.1J | UN0399 | II | EXPLOSIVE 1.1J | | 63(b) | 62 | None | Forbidden | Forbidden | E | 7E, 16E, 23E |
| | Bombs with flammable liquid, *with bursting charge* | 1.2J | UN0400 | II | EXPLOSIVE 1.2J | | 63(b) | 62 | None | Forbidden | Forbidden | E | 7E, 16E, 23E |
| | Boosters with detonator | 1.1B | UN0225 | II | EXPLOSIVE 1.1B | | None | 62 | None | Forbidden | Forbidden | B | 2E, 6E |
| | Boosters with detonator | 1.2B | UN0268 | II | EXPLOSIVE 1.2B | | None | 62 | None | Forbidden | Forbidden | E | 1E, 7E |
| D | Boosters with detonator | 1.4B | NA0350 | II | EXPLOSIVE 1.4B | 115 | None | 62 | None | Forbidden | 75 kg | A | 24E |
| | Boosters, *without detonator* | 1.1D | UN0042 | II | EXPLOSIVE 1.1D | | None | 62 | None | Forbidden | Forbidden | B | |
| | Boosters, *without detonator* | 1.2D | UN0283 | II | EXPLOSIVE 1.2D | | None | 62 | None | Forbidden | Forbidden | B | |
| | *Borate and chlorate mixtures, see* Chlorate and borate mixtures | | | | | | | | | | | | |
| | Borneol | 4.1 | UN1312 | III | FLAMMABLE SOLID | A1 | None | 213 | 240 | 25 kg | 100 kg | A | |
| + | Boron tribromide | 8 | UN2692 | I | CORROSIVE, POISON | 2, A3, A7, B9, B14, B32, B74, N34, T38, T43, T45 | None | 227 | 244 | Forbidden | 2.5 L | C | 12 |
| | Boron trichloride | 2.3 | UN1741 | | POISON GAS, CORROSIVE | 1, B9, B14 | None | 304 | 314 | Forbidden | Forbidden | E | 25, 40 |
| | Boron trifluoride | 2.3 | UN1008 | | POISON GAS | 2, B9, B14 | None | 302 | 314, 315 | Forbidden | Forbidden | D | 40 |
| | Boron trifluoride acetic acid complex | 8 | UN1742 | II | CORROSIVE | B2, B6, T9, T27 | 154 | 202 | 242 | 1 L | 30 L | A | |
| | Boron trifluoride diethyl etherate | 8 | UN2604 | I | CORROSIVE, FLAMMABLE LIQUID | A19, T8, T26 | None | 202 | 243 | 0.5 L | 2.5 L | D | 12, 21, 40, 48 |
| | Boron trifluoride dihydrate | 8 | UN2851 | II | CORROSIVE | T9, T27 | 154 | 212 | 240 | 15 kg | 50 kg | B | 12, 40, 48 |
| | Boron trifluoride dimethyl etherate | 4.3 | UN2965 | II | DANGEROUS WHEN WET, CORROSIVE, FLAMMABLE LIQUID | A19, T8, T26 | None | 202 | 243 | 1 L | 5 L | D | 21, 28, 40, 49, 100 |
| | Boron trifluoride propionic acid complex | 8 | UN1743 | II | CORROSIVE | B2, T9, T27 | 154 | 202 | 242 | 1 L | 30 L | A | |
| | *Box toe gum, see* Nitrocellulose *etc.* | | | | | | | | | | | | |

## §172.101 HAZARDOUS MATERIALS TABLE, Continued

| Symbols | Hazardous materials descriptions and proper shipping names | Hazard class or Division | Identification Numbers | Packing group | Label(s) required (if not excepted) | Special provisions | (8) Packaging authorizations (§173.***) | | | (9) Quantity limitations | | (10) Vessel stowage requirements | |
|---|---|---|---|---|---|---|---|---|---|---|---|---|---|
| | | | | | | | Exceptions | Non-bulk packaging | Bulk packaging | Passenger aircraft or railcar | Cargo aircraft only | Vessel stowages | Other stowage provisions |
| (1) | (2) | (3) | (4) | (5) | (6) | (7) | (8A) | (8B) | (8C) | (9A) | (9B) | (10A) | (10B) |
| | Brake fluid, *hydraulic* | 3 | UN1118 | II | FLAMMABLE LIQUID | T7, T30 | 150 | 202 | 242 | 5 L | 60 L | B | |
| | | | | III | FLAMMABLE LIQUID | B1, T7, T30 | 150 | 203 | 242 | 60 L | 220 L | A | |
| | Bromates, inorganic, n.o.s. | 5.1 | UN1450 | II | OXIDIZER | | 152 | 212 | 242 | 5 kg | 25 kg | A | 56, 58, 106 |
| | *Bromine azide* | Forbidden | | | | | | | | | | | |
| + | Bromine *or* Bromine solutions | 8 | UN1744 | I | CORROSIVE, POISON | 1, A3, A6, B9, B12, B31, B64, B73, B85, N34, N43, T18, T41 | None | 226 | 249 | Forbidden | 2.5 L | D | 12, 40, 66, 74, 89, 90, 95 |
| | Bromine chloride | 2.3 | UN2901 | | POISON GAS, CORROSIVE, OXIDIZER | 2, B9, B12, B14 | None | 304 | 314, 315 | Forbidden | Forbidden | D | 40, 89, 90 |
| + | Bromine pentafluoride | 5.1 | UN1745 | I | OXIDIZER, POISON, CORROSIVE | 1, B9, B14, B30, B72, T38, T43, T44 | None | 228 | 244 | Forbidden | Forbidden | D | 25, 40, 66, 90 |
| + | Bromine trifluoride | 5.1 | UN1746 | I | OXIDIZER, POISON, CORROSIVE | 2, B9, B14, B32, B74, T38, T43, T45 | None | 228 | 244 | Forbidden | Forbidden | D | 25, 40, 66, 90 |
| | *4-Bromo-1,2-dinitrobenzene* | Forbidden | | | | | | | | | | | |
| | *4-Bromo-1,2-dinitrobenzene (unstable at 59 deg C.)* | Forbidden | | | | | | | | | | | |
| | 1-Bromo-3-methylbutane | 3 | UN2341 | III | FLAMMABLE LIQUID | B1, T7, T30 | 150 | 203 | 242 | 60 L | 220 L | A | |
| | *1-Bromo-3-nitrobenzene (unstable at 56 deg C)* | Forbidden | | | | | | | | | | | |
| | Bromoacetic acid, *solid* | 8 | UN1938 | II | CORROSIVE | A7, N34, T9 | 154 | 212 | 240 | 15 kg | 50 kg | A | |
| | Bromoacetic acid, *solution* | 8 | UN1938 | II | CORROSIVE | B2, T9 | 154 | 202 | 242 | 1 L | 30 L | A | 40 |
| + | Bromoacetone | 6.1 | UN1569 | II | POISON | 2 | None | 193 | 245 | Forbidden | Forbidden | D | 40 |
| | Bromoacetyl bromide | 8 | UN2513 | II | CORROSIVE | B2, T9, T26 | 154 | 202 | 242 | 1 L | 30 L | C | 8 |
| | Bromobenzene | 3 | UN2514 | III | FLAMMABLE LIQUID | B1, T1 | 150 | 203 | 242 | 60 L | 220 L | A | |
| | Bromobenzyl cyanides, *liquid* | 6.1 | UN1694 | I | POISON | T18 | None | 201 | 243 | Forbidden | 30 L | D | 12, 40 |
| | Bromobenzyl cyanides, *solid* | 6.1 | UN1694 | I | POISON | T18 | None | 211 | 241 | Forbidden | 50 kg | D | 12, 40 |
| | 2-Bromobutane | 3 | UN2339 | II | FLAMMABLE LIQUID | B1, T1 | 150 | 202 | 242 | 5 L | 60 L | B | 40 |
| | Bromochloromethane | 6.1 | UN1887 | III | KEEP AWAY FROM FOOD | T7 | 153 | 203 | 241 | 60 L | 220 L | A | |
| | 2-Bromoethyl ethyl ether | 3 | UN2340 | II | FLAMMABLE LIQUID | T7 | 150 | 202 | 242 | 5 L | 60 L | B | 40 |
| | Bromoform | 6.1 | UN2515 | III | KEEP AWAY FROM FOOD | T7 | 153 | 203 | 241 | 60 L | 220 L | A | 12, 40, 48 |
| | Bromomethylpropanes | 3 | UN2342 | II | FLAMMABLE LIQUID | T7, T30 | 150 | 202 | 242 | 5 L | 60 L | B | |
| | 2-Bromopentane | 3 | UN2343 | II | FLAMMABLE LIQUID | T1 | 150 | 202 | 242 | 5 L | 60 L | B | |
| | 2-Bromopropane | 3 | UN2344 | II | FLAMMABLE LIQUID | T7 | 150 | 202 | 242 | 5 L | 60 L | B | 40 |
| | 3-Bromopropyne | 3 | UN2345 | II | FLAMMABLE LIQUID | T8 | 150 | 202 | 242 | 5 L | 60 L | D | 40 |
| | *Bromosilane* | Forbidden | | | | | | | | | | | |
| | *Bromotoluene-alpha, see* Benzyl bromide | | | | | | | | | | | | |
| | Bromotrifluoroethylene | 2.1 | UN2419 | | FLAMMABLE GAS | | None | 304 | 314, 315 | Forbidden | 150 kg | B | 40 |
| | Bromotrifluoromethane, *R1381* | 2.2 | UN1009 | | NONFLAMMABLE GAS | | 306 | 304 | 314, 315 | 75 kg | 150 kg | A | |
| | Brucine | 6.1 | UN1570 | I | POISON | | None | 212 | 242 | 5 kg | 50 kg | A | |
| | Bursters, *explosive* | 1.1D | UN0043 | II | EXPLOSIVE 1.1D | | None | 62 | None | Forbidden | Forbidden | B | |
| | Butadienes, inhibited | 2.1 | UN1010 | | FLAMMABLE GAS | | 306 | 304 | 314, 315 | Forbidden | 150 kg | B | 40 |
| | Butane *or* Butane mixtures *see also* Petroleum gases, liquefied | 2.1 | UN1011 | | FLAMMABLE GAS | 19 | 306 | 304 | 314, 315 | Forbidden | 150 kg | E | 40 |
| | *Butane, butane mixtures and mixtures having similar properties in cartridges each not exceeding 500 grams, see* Receptacles, etc. | | | | | | | | | | | | |
| | Butanedione | 3 | UN2346 | II | FLAMMABLE LIQUID | T1 | 150 | 202 | 242 | 5 L | 60 L | B | |
| | *1,2,4-Butanetriol trinitrate* | Forbidden | | | | | | | | | | | |
| | Butanols | 3 | UN1120 | II | FLAMMABLE LIQUID | T1 | 150 | 202 | 242 | 5 L | 60 L | B | |
| | | | | III | FLAMMABLE LIQUID | B1, T1 | 150 | 203 | 242 | 60 L | 220 L | A | |
| | *tert-Butoxycarbonyl azide* | Forbidden | | | | | | | | | | | |

## §172.101 HAZARDOUS MATERIALS TABLE, Continued

| Symbols | Hazardous materials descriptions and proper shipping names | Hazard class or Division | Identification Numbers | Packing group | Label(s) required (if not excepted) | Special provisions | (8) Packaging authorizations (§173.***) | | | (9) Quantity limitations | | (10) Vessel stowage requirements | |
|---|---|---|---|---|---|---|---|---|---|---|---|---|---|
| | | | | | | | Exceptions | Non-bulk packaging | Bulk packaging | Passenger aircraft or railcar | Cargo aircraft only | Vessel stowages | Other stowage provisions |
| (1) | (2) | (3) | (4) | (5) | (6) | (7) | (8A) | (8B) | (8C) | (9A) | (9B) | (10A) | (10B) |
| | Butoxyl | 3 | UN2708 | III | FLAMMABLE LIQUID | B1, T1 | 150 | 203 | 241 | 60 L | 220 L | A | |
| | Butyl acetates | 3 | UN1123 | II | FLAMMABLE LIQUID | T1 | 150 | 202 | 242 | 5 L | 60 L | B | |
| | | | | III | FLAMMABLE LIQUID | B1, T1 | 150 | 202 | 241 | 60 L | 220 L | A | |
| | Butyl acid phosphate | 8 | UN1718 | III | CORROSIVE | T7 | 154 | 203 | 241 | 5 L | 60 L | A | |
| | *Butyl alcohols, see* Butanols | | | | | | | | | | | | |
| | Butyl benzenes | 3 | UN2709 | III | FLAMMABLE LIQUID | B1, T1 | 150 | 203 | 242 | 60 L | 220 L | A | |
| | n-Butyl bromide | 3 | UN1126 | II | FLAMMABLE LIQUID | T1 | 150 | 202 | 242 | 5 L | 60 L | B | |
| | *n-Butyl chloride, see* Chlorobutanes | | | | | | | | | | | | |
| | n-Butyl chloroformate | 6.1 | UN2743 | I | POISON, CORROSIVE | 2, B9, B14, B32, B74, T38, T43, T45 | None | 227 | 244 | 1 L | 30 L | A | 12, 13, 22, 25, 40, 48, 100 |
| D | sec-Butyl chloroformate | 6.1 | NA2742 | I | POISON, FLAMMABLE LIQUID, CORROSIVE | 2, B9, B14, B32, B74, T38, T43, T45 | None | 227 | 244 | 1 L | 30 L | A | 12, 13, 22, 25, 40, 48, 100 |
| | *Butyl ethers, see* Dibutyl ethers | | | | | | | | | | | | |
| | *Butyl ethyl ether, see* Ethyl butyl ether | | | | | | | | | | | | |
| | n-Butyl formate | 3 | UN1128 | II | FLAMMABLE LIQUID | T1 | 150 | 202 | 242 | 5 L | 60 L | B | |
| | *tert-Butyl hydroperoxide, more than 90 per cent with water* | Forbidden | | | | | | | | | | | |
| | N-n-Butyl imidazole | 6.1 | UN2690 | II | POISON | T8 | None | 202 | 243 | 5 L | 60 L | A | 40 |
| + | Tert-Butyl isocyanate | 3 | UN2484 | I | FLAMMABLE LIQUID, POISON | 1, A7, B9, B14, B30, B72, T38, T43, T44 | None | 226 | 244 | Forbidden | 30 L | D | 40 |
| + | n-Butyl isocyanate | 3 | UN2485 | II | FLAMMABLE LIQUID, POISON | 1, A7, B9, B14, B30, B72, B77, T38, T43, T44 | None | 226 | 244 | Forbidden | 60 L | D | 40 |
| | Butyl mercaptans | 3 | UN2347 | II | FLAMMABLE LIQUID | A3, T8 | 150 | 202 | 242 | 5 L | 60 L | B | 26, 95, 102 |
| | n-Butyl methacrylate | 3 | UN2227 | III | FLAMMABLE LIQUID | B1, T1 | 150 | 203 | 242 | 60 L | 220 L | A | |
| | Butyl methyl ether | 3 | UN2350 | II | FLAMMABLE LIQUID | T8 | 150 | 202 | 242 | 5 L | 60 L | B | |
| | Butyl nitrites | 3 | UN2351 | I | FLAMMABLE LIQUID | T8 | 150 | 202 | 242 | 1 L | 30 L | E | 40 |
| | | | | II | FLAMMABLE LIQUID | T8 | 150 | 202 | 242 | 5 L | 60 L | B | 40 |
| | | | | III | FLAMMABLE LIQUID | B1, T8 | 150 | 202 | 242 | 60 L | 220 L | A | 40 |
| | *tert-Butyl peroxyacetate, more than 76 per cent in solution* | Forbidden | | | | | | | | | | | |
| | *n-Butyl peroxydicarbonate, more than 52 percent in solution* | Forbidden | | | | | | | | | | | |
| | *tert-Butyl peroxyisobutyrate, more than 77 per cent in solution* | Forbidden | | | | | | | | | | | |
| | *Butyl phosphoric acid, see* Butyl acid phosphate | | | | | | | | | | | | |
| | 5-tert-Butyl-2,4,6-trinitro-m-xylene *or* Musk xylene | 4.1 | UN2956 | III | FLAMMABLE SOLID. | | None | 214 | None | Forbidden | Forbidden | D | 12 |
| | Butyl vinyl ether, inhibited | 3 | UN2352 | II | FLAMMABLE LIQUID | T7 | 150 | 202 | 242 | 5 L | 60 L | B | 40 |
| | Butylacrylate | 3 | UN2348 | III | FLAMMABLE LIQUID | B1, T8, T31 | 150 | 202 | 241 | 60 L | 220 L | A | |
| | n-Butylamine | 3 | UN1125 | II | FLAMMABLE LIQUID | T8 | 150 | 202 | 242 | 5 L | 60 L | B | 40 |
| | N-Butylaniline | 6.1 | UN2738 | II | POISON | T8 | None | 202 | 243 | 5 L | 60 L | A | 40 |
| | tert-Butylcyclohexylchloroformate | 6.1 | UN2747 | III | KEEP AWAY FROM FOOD | T8 | 153 | 203 | 241 | 60 L | 220 L | A | 12, 13, 25, 48 |
| | Butylene *see also* Petroleum gases, liquefied | 2.1 | UN1012 | | FLAMMABLE GAS | 19 | None | 304 | 314, 315 | Forbidden | 150 kg | E | 40 |
| | 1,2-Butylene oxide, stabilized | 3 | UN3022 | II | FLAMMABLE LIQUID | T8 | 150 | 202 | 242 | 5 L | 60 L | B | 49 |
| | Butylphenols, liquid | 6.1 | UN2228 | III | KEEP AWAY FROM FOOD | T7 | 153 | 203 | 241 | 60 L | 220 L | A | |
| | Butylphenols, solid | 6.1 | UN2229 | III | KEEP AWAY FROM FOOD | T7 | 153 | 213 | 240 | 100 kg | 200 kg | A | |
| | Butylpropionate | 3 | UN1914 | III | FLAMMABLE LIQUID | B1, T1 | 150 | 203 | 242 | 60 L | 220 L | A | |
| | Butyltoluenes | 6.1 | UN2667 | III | KEEP AWAY FROM FOOD | T2 | 153 | 203 | 241 | 60 L | 220 L | A | |
| | Butyltrichlorosilane | 8 | UN1747 | II | CORROSIVE | A7, B2, B6, N34, T8, T26 | None | 202 | 242 | Forbidden | 30 L | C | 12, 21, 40, 48 |
| | 1,4-Butynediol | 6.1 | UN2716 | III | KEEP AWAY FROM FOOD | A1 | None | 213 | 240 | 100 kg | 200 kg | A | 61, 70 |

| Symbols | Hazardous materials descriptions and proper shipping names | Hazard class or Division | Identification Numbers | Packing group | Label(s) required (if not excepted) | Special provisions | (8) Packaging authorizations (§173.***) | | | (9) Quantity limitations | | (10) Vessel stowage requirements | |
|---|---|---|---|---|---|---|---|---|---|---|---|---|---|
| | | | | | | | Exceptions | Non-bulk packaging | Bulk packaging | Passenger aircraft or railcar | Cargo aircraft only | Vessel stowages | Other stowage provisions |
| (1) | (2) | (3) | (4) | (5) | (6) | (7) | (8A) | (8B) | (8C) | (9A) | (9B) | (10A) | (10B) |
| | Butyraldehyde | 3 | UN1129 | II | FLAMMABLE LIQUID | T8 | 150 | 202 | 242 | 5 L | 60 L | B | |
| | Butyraldoxime | 3 | UN2840 | III | FLAMMABLE LIQUID | B1, T1 | 150 | 203 | 242 | 60 L | 220 L | A | |
| | Butyric acid | 8 | UN2820 | III | CORROSIVE | T1 | 154 | 203 | 241 | 5 L | 60 L | A | 12, 48 |
| | Butyric anhydride | 8 | UN2739 | III | CORROSIVE | T2 | 154 | 203 | 241 | 5 L | 60 L | A | |
| | Butyronitrile | 3 | UN2411 | II | FLAMMABLE LIQUID, POISON | T14 | None | 202 | 243 | 1 L | 60 L | E | 40 |
| | Butyryl chloride | 3 | UN2353 | II | FLAMMABLE LIQUID, CORROSIVE | T9, T26 | None | 202 | 243 | 1 L | 5 L | C | 40 |
| | *Cabazide* | Forbidden | | | | | | | | | | | |
| | Cacodylic acid | 6.1 | UN1572 | II | POISON | | None | 212 | 242 | 25 kg | 100 kg | E | 26 |
| | Cadmium compounds | 6.1 | UN2570 | I | POISON | | None | 211 | 242 | 5 kg | 50 kg | A | |
| | | | | II | POISON | | None | 212 | 242 | 25 kg | 100 kg | A | |
| | | | | III | KEEP AWAY FROM FOOD | | 153 | 213 | 240 | 100 kg | 200 kg | A | |
| | Caesium hydroxide, solid | 8 | UN2682 | II | CORROSIVE | | 154 | 212 | 240 | 15 kg | 50 kg | A | |
| | Caesium hydroxide solution | 8 | UN2681 | II | CORROSIVE | B2, T8 | 154 | 202 | 242 | 1 L | 30 L | A | |
| | Calcium | 4.3 | UN1401 | II | DANGEROUS WHEN WET | | None | 212 | 241 | 15 kg | 50 kg | E | |
| | Calcium arsenate | 6.1 | UN1573 | II | POISON | | None | 212 | 242 | 25 kg | 100 kg | A | |
| | Calcium arsenate and calcium arsenite, mixtures, solid | 6.1 | UN1574 | II | POISON | | None | 212 | 242 | 25 kg | 100 kg | A | |
| D | Calcium arsenite, solid | 6.1 | NA1574 | II | POISON | | None | 212 | 242 | 25 kg | 100 kg | A | |
| | *Calcium bisulfite solution, see* Bisulfites, inorganic, aqueous solutions, n.o.s. | | | | | | | | | | | | |
| | Calcium carbide | 4.3 | UN1402 | II | DANGEROUS WHEN WET | A1, A8, B55, N34 | None | 212 | 241 | 15 kg | 50 kg | B | |
| | Calcium chlorate | 5.1 | UN1452 | II | OXIDIZER | N34 | 152 | 212 | 242 | 5 kg | 25 kg | A | 56, 58, 106 |
| | Calcium chlorate solution | 5.1 | UN2429 | II | OXIDIZER | A2, N41, T8 | 152 | 202 | 242 | 1 L | 5 L | B | 56, 58, 106 |
| | Calcium chlorite | 5.1 | UN1453 | II | OXIDIZER | A9, N34 | 152 | 212 | 242 | 5 kg | 25 kg | A | 56, 58, 106 |
| | Calcium cyanamide *with more than 0.1 per cent of calcium carbide* | 4.3 | UN1403 | III | DANGEROUS WHEN WET | A1, A19 | None | 213 | 241 | 25 kg | 100 kg | A | |
| | Calcium cyanide | 6.1 | UN1575 | I | POISON | N79, N80 | None | 211 | 242 | 5 kg | 50 kg | A | 26, 40 |
| | Calcium dithionite *or* Calcium hydrosulfite | 4.2 | UN1923 | II | SPONTANEOUSLY COMBUSTIBLE | A19, A20 | None | 212 | 241 | 15 kg | 50 kg | E | 13 |
| | Calcium hydride | 4.3 | UN1404 | I | DANGEROUS WHEN WET | A19, N40 | None | 211 | 242 | Forbidden | 15 kg | E | |
| | Calcium hydrosulfite, *see* Calcium dithionite | | | | | | | | | | | | |
| | Calcium hypochlorite, dry *or* Calcium hypochlorite mixtures dry *with more than 39 per cent available chlorine (8.8 per cent available oxygen)* | 5.1 | UN1748 | II | OXIDIZER | A7, A9, N34 | 152 | 212 | None | 5 kg | 25 kg | D | 48, 56, 58, 69, 106, 118 |
| | Calcium hypochlorite, hydrated *or* Calcium hypochlorite, hydrated mixtures, *with not less than 5.5 per cent but not more than 10 per cent water* | 5.1 | UN2880 | II | OXIDIZER | | 152 | 212 | 240 | 5 kg | 25 kg | A | 50, 56, 58, 69, 106 |
| | Calcium hypochlorite mixtures, dry, *with more than 10 per cent but not more than 39 per cent available chlorine* | 5.1 | UN2208 | III | OXIDIZER | A1, A29, N34 | 152 | 213 | 240 | 25 kg | 100 kg | A | 56, 58, 69, 106 |
| | Calcium manganese silicon | 4.3 | UN2844 | III | DANGEROUS WHEN WET | A1, A19 | None | 213 | 241 | 25 kg | 100 kg | A | 85, 103 |
| | Calcium nitrate | 5.1 | UN1454 | III | OXIDIZER | T2 | 152 | 213 | 240 | 25 kg | 100 kg | A | |
| A | Calcium oxide | 8 | UN1910 | III | CORROSIVE | | 154 | 213 | 240 | 25 kg | 100 kg | A | |
| | Calcium perchlorate | 5.1 | UN1455 | II | OXIDIZER | T8 | 152 | 212 | 242 | 5 kg | 25 kg | A | 56, 58, 106 |
| | Calcium permanganate | 5.1 | UN1456 | II | OXIDIZER | | 152 | 212 | 242 | 5 kg | 25 kg | D | 56, 58, 69, 106, 107 |
| | Calcium peroxide | 5.1 | UN1457 | II | OXIDIZER | | 152 | 212 | 242 | 5 kg | 25 kg | A | 13, 75, 106 |
| | Calcium phosphide | 4.3 | UN1360 | I | DANGEROUS WHEN WET, POISON | A8, A19, N40 | None | 211 | 242 | Forbidden | 15 kg | E | 40, 85 |
| | Calcium, pyrophoric *or* Calcium alloys, pyrophoric | 4.2 | UN1855 | I | SPONTANEOUSLY COMBUSTIBLE | | None | 187 | None | Forbidden | Forbidden | D | |
| | Calcium resinate | 4.1 | UN1313 | III | FLAMMABLE SOLID | A1, A19 | None | 213 | 240 | 25 kg | 100 kg | A | |
| | Calcium resinate, fused | 4.1 | UN1314 | III | FLAMMABLE SOLID | A1, A19 | None | 213 | 240 | 25 kg | 100 kg | A | |
| | Calcium selenate, *see* Selenates *or* Selenites | | | | | | | | | | | | |
| | Calcium silicide | 4.3 | UN1405 | II | DANGEROUS WHEN WET | A19 | None | 212 | 241 | 15 kg | 50 kg | B | 85, 103 |
| | | | | III | DANGEROUS WHEN WET | A1, A19 | None | 213 | 241 | 25 kg | 100 kg | B | 85, 103 |

## §172.101 HAZARDOUS MATERIALS TABLE, Continued

| Symbols | Hazardous materials descriptions and proper shipping names | Hazard class or Division | Identification Numbers | Packing group | Label(s) required (if not excepted) | Special provisions | (8) Packaging authorizations (§173.***) Exceptions | (8) Non-bulk packaging | (8) Bulk packaging | (9) Quantity limitations Passenger aircraft or railcar | (9) Cargo aircraft only | (10) Vessel stowage requirements Vessel stowages | (10) Other stowage provisions |
|---|---|---|---|---|---|---|---|---|---|---|---|---|---|
| (1) | (2) | (3) | (4) | (5) | (6) | (7) | (8A) | (8B) | (8C) | (9A) | (9B) | (10A) | (10B) |
| | Camphor oil | 3 | UN1130 | III | FLAMMABLE LIQUID | B1, T1 | 150 | 203 | 242 | 60 L | 220 L | A | |
| | Camphor, *synthetic* | 4.1 | UN2717 | III | FLAMMABLE SOLID | A1 | None | 213 | 240 | 25 kg | 100 kg | A | |
| | *Cannon primers, see* Primers, tubular | | | | | | | | | | | | |
| | Caproic acid | 8 | UN2829 | III | CORROSIVE | T1 | 154 | 203 | 241 | 5 L | 60 L | A | |
| | *Caps, blasting, see* Detonators, *etc.* | | | | | | | | | | | | |
| | Carbamate pesticides, liquid, flammable, toxic, n.o.s., *flash point less than 23 degrees C.* | 3 | UN2758 | I | FLAMMABLE LIQUID, POISON | | None | 201 | 243 | Forbidden | 30 L | B | 40 |
| | | | | II | FLAMMABLE LIQUID, POISON | | None | 202 | 243 | 1 L | 60 L | B | 40 |
| | Carbamate pesticides, liquid, toxic, flammable, n.o.s. *flash point not less than 23 degrees C.* | 6.1 | UN2991 | I | POISON, FLAMMABLE LIQUID | T42 | None | 201 | 243 | 1 L | 30 L | B | 40 |
| | | | | II | POISON, FLAMMABLE LIQUID | T14 | None | 202 | 243 | 5 L | 60 L | B | 40 |
| | | | | III | KEEP AWAY FROM FOOD, FLAMMABLE LIQUID | B1, T14 | 153 | 203 | 242 | 60 L | 220 L | A | 40 |
| | Carbamate pesticides, liquid, toxic, n.o.s. | 6.1 | UN2992 | I | POISON | T42 | None | 201 | 243 | 1 L | 30 L | B | 40 |
| | | | | II | POISON | T14 | None | 202 | 243 | 5 L | 60 L | B | 40 |
| | | | | III | KEEP AWAY FROM FOOD | T14 | 153 | 203 | 241 | 60 L | 220 L | A | 40 |
| | Carbamate pesticides, solid, toxic, n.o.s. | 6.1 | UN2757 | I | POISON | | None | 211 | 242 | 5 kg | 50 kg | A | 40 |
| | | | | II | POISON | | None | 212 | 242 | 25 kg | 100 kg | A | 40 |
| | | | | III | KEEP AWAY FROM FOOD | | 153 | 213 | 240 | 100 kg | 200 kg | A | 40 |
| | *Carbolic acid, see* Phenol, solid *or* Phenol, molten | | | | | | | | | | | | |
| | *Carbolic acid solutions, see* Phenol solutions | | | | | | | | | | | | |
| I | Carbon, activated | 4.2 | UN1362 | III | SPONTANEOUSLY COMBUSTIBLE | | None | 213 | 241 | 0.5 kg | 0.5 kg | A | 12 |
| I | Carbon, *animal or vegetable origin* | 4.2 | UN1361 | III | SPONTANEOUSLY COMBUSTIBLE | | None | 213 | 241 | Forbidden | Forbidden | A | 12 |
| | *Carbon bisulfide, see* Carbon disulfide | | | | | | | | | | | | |
| | Carbon dioxide | 2.2 | UN1013 | | NONFLAMMABLE GAS | | 306 | 302, 304 | 302, 314, 315 | 75 kg | 150 kg | A | |
| | Carbon dioxide and ethylene oxide mixtures *with more than 6 per cent but not more than 25 percent ethylene oxide* | 2.1 | UN1041 | | FLAMMABLE GAS | | None | 304 | 314, 315 | Forbidden | 25 kg | D | 40 |
| | Carbon dioxide and ethylene oxide mixtures *with more than 25 percent ethylene oxide* | 2.3 | UN1041 | | POISON GAS, FLAMMABLE GAS | 6, B9, B14 | None | 304 | 314, 315 | Forbidden | 25 kg | D | 40 |
| | Carbon dioxide and ethylene oxide mixtures *with not more than 6 per cent ethylene oxide* | 2.2 | UN1952 | | NONFLAMMABLE GAS | | None | 304 | 314, 315 | 75 kg | 150 kg | B | |
| | Carbon dioxide and nitrous oxide mixtures | 2.2 | UN1015 | | NONFLAMMABLE GAS | | 306 | None | 314, 315 | 75 kg | 150 kg | A | |
| | Carbon dioxide and oxygen mixtures | 2.2 | UN1014 | | NONFLAMMABLE GAS | | 306 | None | 244 | 75 kg | 150 kg | A | |
| | Carbon dioxide, refrigerated liquid | 2.2 | UN2187 | | NONFLAMMABLE GAS | B43 | 306 | None | 314, 315 | 50 kg | 500 kg | B | |
| AW | Carbon dioxide, solid *or* Dry ice | 9 | UN1845 | III | None | B12 | 217 | 217 | 240 | 200 kg | 200 kg | C | 40 |
| | Carbon disulfide | 3 | UN1131 | I | FLAMMABLE LIQUID, POISON | B16, T18, T26, T29 | None | 201 | 243 | Forbidden | Forbidden | D | 18, 40, 115 |
| | Carbon monoxide | 2.3 | UN1016 | | POISON GAS, FLAMMABLE GAS | 4 | None | 302 | None | Forbidden | 25 kg | D | 40 |
| | Carbon monoxide and hydrogen mixture | 2.1 | UN2600 | | FLAMMABLE GAS | 5 | None | 302 | 302 | Forbidden | Forbidden | D | 40 |
| D | Carbon monoxide, refrigerated liquid *(cryogenic liquid)* | 2.3 | NA9202 | | POISON GAS, FLAMMABLE GAS | 4 | None | 316 | 318 | Forbidden | Forbidden | D | |
| | Carbon tetrabromide | 6.1 | UN2516 | III | KEEP AWAY FROM FOOD | | 153 | 213 | 240 | 100 kg | 200 kg | A | 25 |
| | Carbon tetrachloride | 6.1 | UN1846 | II | POISON | N36, T8 | None | 202 | 243 | 5 L | 60 L | A | 40 |
| | *Carbonyl chloride, see* Phosgene | | | | | | | | | | | | |
| | Carbonyl fluoride | 2.3 | UN2417 | | POISON GAS | 2 | None | 302 | None | Forbidden | Forbidden | D | 40 |
| | Carbonyl sulfide | 2.3 | UN2204 | | POISON GAS, FLAMMABLE GAS | 2, B9, B14 | None | 304 | 314, 315 | Forbidden | 25 kg | D | 40 |
| | *Cartridge cases, empty primed, see* Cases, cartridge, empty, with primer | | | | | | | | | | | | |
| | *Cartridges, actuating, for aircraft ejector seat catapult, fire extinguisher, canopy removal or apparatus, see* Cartridges, power device | | | | | | | | | | | | |
| | *Cartridges, explosive, see* Charges, demolition | | | | | | | | | | | | |
| | Cartridges, flash | 1.1G | UN0049 | II | EXPLOSIVE 1.1G | | None | 62 | None | Forbidden | Forbidden | B | |
| | Cartridges, flash | 1.3G | UN0050 | II | EXPLOSIVE 1.3G | | None | 62 | None | Forbidden | 75 kg | B | |
| | Cartridges for weapons, blank | 1.1C | UN0326 | II | EXPLOSIVE 1.1C | | None | 62 | None | Forbidden | Forbidden | B | |

## §172.101 HAZARDOUS MATERIALS TABLE, Continued

| Symbols | Hazardous materials descriptions and proper shipping names | Hazard class or Division | Identification Numbers | Packing group | Label(s) required (if not excepted) | Special provisions | (8) Packaging authorizations (§173.***) | | | (9) Quantity limitations | | (10) Vessel stowage requirements | |
|---|---|---|---|---|---|---|---|---|---|---|---|---|---|
| | | | | | | | Exceptions | Non-bulk packaging | Bulk packaging | Passenger aircraft or railcar | Cargo aircraft only | Vessel stowages | Other stowage provisions |
| (1) | (2) | (3) | (4) | (5) | (6) | (7) | (8A) | (8B) | (8C) | (9A) | (9B) | (10A) | (10B) |
| | Cartridges for weapons, blank | 1.2C | UN0413 | II | EXPLOSIVE 1.2C | | None | 62 | None | Forbidden | Forbidden | B | |
| | Cartridges for weapons, blank *or* Cartridges, small arms, blank | 1.3C | UN0327 | II | EXPLOSIVE 1.3C | | None | 62 | None | Forbidden | Forbidden | B | |
| | Cartridges for weapons, blank *or* Cartridges, small arms, blank | 1.4C | UN0338 | II | EXPLOSIVE 1.4C | | None | 62 | None | Forbidden | 75 kg | A | 24E |
| | Cartridges for weapons, blank *or* Cartridges, small arms, blank | 1.4S | UN0014 | II | None | 112 | None | 62 | None | 25 kg | 100 kg | A | 9E |
| | Cartridges for weapons, inert projectile *or* Cartridges, small arms | 1.2C | UN0328 | II | EXPLOSIVE 1.2C | | None | 62 | None | Forbidden | Forbidden | B | |
| | Cartridges for weapons, inert projectile *or* Cartridges, small arms | 1.4C | UN0339 | II | EXPLOSIVE 1.4C | | None | 62 | None | Forbidden | 75 kg | B | |
| | Cartridges for weapons, inert projectile *or* Cartridges, small arms | 1.3C | UN0417 | II | EXPLOSIVE 1.3C | | None | 62 | None | Forbidden | Forbidden | B | |
| | Cartridges for weapons, inert projectile *or* Cartridges, small arms, *other than blank* | 1.4S | UN0012 | II | None | 112 | None | 62 | None | 25 kg | 100 kg | A | 9E |
| | Cartridges for weapons, *with bursting charge* | 1.1F | UN0005 | II | EXPLOSIVE 1.1F | | None | 62 | None | Forbidden | Forbidden | E | |
| | Cartridges for weapons, *with bursting charge* | 1.2F | UN0007 | II | EXPLOSIVE 1.2F | | None | 62 | None | Forbidden | Forbidden | E | |
| | Cartridges for weapons, *with bursting charge* | 1.4F | UN0348 | II | EXPLOSIVE 1.4F | | None | 62 | None | Forbidden | Forbidden | E | |
| | Cartridges for weapons, *with bursting charge* | 1.4E | UN0412 | II | EXPLOSIVE 1.4E | | None | 62 | None | Forbidden | 75 kg | A | 24E |
| | Cartridges for weapons, *with bursting charge* | 1.1E | UN0006 | II | EXPLOSIVE 1.1E | | None | 62 | None | Forbidden | Forbidden | B | |
| | Cartridges for weapons, *with bursting charge* | 1.2E | UN0321 | II | EXPLOSIVE 1.2E | | None | 62 | None | Forbidden | Forbidden | B | |
| | Cartridges, oil well | 1.3C | UN0277 | II | EXPLOSIVE 1.3C | | None | 62 | None | Forbidden | Forbidden | B | |
| | Cartridges, oil well | 1.4C | UN0278 | II | EXPLOSIVE 1.4C | | None | 62 | None | Forbidden | 75 kg | A | 24E |
| | Cartridges, power device | 1.3C | UN0275 | II | EXPLOSIVE 1.3C | | None | 62 | None | Forbidden | 75 kg | B | |
| | Cartridges, power device | 1.4C | UN0276 | II | EXPLOSIVE 1.4C | 110 | None | 62 | None | Forbidden | 75 kg | A | 24E |
| | Cartridges, power device | 1.2C | UN0381 | II | EXPLOSIVE 1.2C | | None | 62 | None | Forbidden | Forbidden | B | |
| | Cartridges, power device | 1.4S | UN0323 | II | EXPLOSIVE 1.4S | 110, 112 | None | 62 | None | 25 kg | 100 kg | A | 9E |
| | Cartridges, safety, blank, *see* Cartridges for weapons, blank *(UN 0014)* | | | | | | | | | | | | |
| | Cartridges, safety, *see* Cartridges for weapons, *other than blank or* Cartridges, power device *(UN 0323)* | | | | | | | | | | | | |
| | Cartridges, signal | 1.3G | UN0054 | II | EXPLOSIVE 1.3G | | None | 62 | None | Forbidden | 75 kg | B | |
| | Cartridges, signal | 1.4G | UN0312 | II | EXPLOSIVE 1.4G | | None | 62 | None | Forbidden | 75 kg | A | 24E |
| | Cartridges, signal | 1.4S | UN0405 | II | EXPLOSIVE 1.4S | | None | 62 | None | 25 kg | 100 kg | A | 9E |
| D | Cartridges, small arms | ORM-D | None | | None | 112 | 230 | None | None | 30 kg gross | 30 kg gross | A | |
| | *Cartridges, sporting, see* Cartridges for weapons, *other than blank* | | | | | | | | | | | | |
| | *Cartridges, starter, jet engine, see* Cartridges, power device | | | | | | | | | | | | |
| I | Cases, cartridge, empty with primer | 1.4S | UN0055 | II | EXPLOSIVE 1.4S | | None | 62 | None | 25 kg | 100 kg | A | 9E |
| | Cases, cartridges, empty with primer | 1.4C | UN0379 | II | EXPLOSIVE 1.4C | | None | 62 | None | Forbidden | 75 kg | A | 24E |
| | Cases, combustible, empty, without primer | 1.3C | UN0447 | II | EXPLOSIVE 1.3C | | None | 62 | None | Forbidden | Forbidden | B | |
| | Cases, combustible, empty, without primer | 1.4C | UN0446 | II | EXPLOSIVE 1.4C | | None | 62 | None | Forbidden | 75 kg | A | 24E |
| | *Casinghead gasoline see* Natural gasoline | | | | | | | | | | | | |
| AW | Castor beans *or* Castor meal *or* Castor pomace *or* Castor flake | 9 | UN2969 | II | None | | 155 | 204 | 240 | No limit | No limit | E | 34, 40, 44, 120 |
| | Caustic alkali liquids, n.o.s. | 8 | UN1719 | I | CORROSIVE | A7, B10, T42 | None | 201 | 242 | 0.5 L | 2.5 L | A | |
| | | | | II | CORROSIVE | B2, T14 | 154 | 202 | 242 | 1 L | 30 L | A | |
| | | | | III | CORROSIVE | T7 | 154 | 203 | 241 | 5 L | 60 L | A | |
| | *Caustic potash, see* Potassium hydroxide *etc.* | | | | | | | | | | | | |
| | *Caustic soda, (etc.) see* Sodium hydroxide *etc.* | | | | | | | | | | | | |
| | Celluloid, *in block, rods, rolls, sheets, tubes, etc., except scrap* | 4.1 | UN2000 | III | FLAMMABLE SOLID | | None | 213 | 240 | 25 kg | 100 kg | A | |
| | Celluloid, scrap | 4.2 | UN2002 | III | SPONTANEOUSLY COMBUSTIBLE | | None | 213 | 241 | Forbidden | Forbidden | D | |
| | *Cement, see* Adhesives *containing flammable liquid* | | | | | | | | | | | | |
| | Cerium, *slabs, ingots, or rods* | 4.1 | UN1333 | II | FLAMMABLE SOLID | N34 | None | 212 | 240 | 15 kg | 50 kg | A | 74, 91 |
| | Cerium, *turnings or gritty powder* | 4.3 | UN3078 | II | DANGEROUS WHEN WET | A1 | None | 213 | 242 | 15 kg | 50 kg | E | |
| | Cesium *or* Caesium | 4.3 | UN1407 | I | DANGEROUS WHEN WET | A19, N34, N40 | None | 211 | 242 | Forbidden | 15 kg | D | |
| | Cesium nitrate *or* Caesium nitrate | 5.1 | UN1451 | III | OXIDIZER | A1, A29 | 152 | 213 | 240 | 25 kg | 100 kg | A | |
| D | Charcoal *briquettes, shell, screenings, wood, etc.* | 4.2 | NA1361 | III | SPONTANEOUSLY COMBUSTIBLE | | 151 | 213 | 240 | 25 kg | 100 kg | A | 12 |
| | Charges, bursting, plastics bonded | 1.1D | UN0457 | II | EXPLOSIVE 1.1D | | None | 62 | None | Forbidden | Forbidden | B | |
| | Charges, bursting, plastics bonded | 1.2D | UN0458 | II | EXPLOSIVE 1.2D | | None | 62 | None | Forbidden | Forbidden | B | |
| | Charges, bursting, plastics bonded | 1.4D | UN0459 | II | EXPLOSIVE 1.4D | | None | 62 | None | Forbidden | 75kg | A | 24E |

## §172.101 HAZARDOUS MATERIALS TABLE, Continued

| Symbols | Hazardous materials descriptions and proper shipping names | Hazard class or Division | Identification Numbers | Packing group | Label(s) required (if not excepted) | Special provisions | (8) Packaging authorizations (§173.***) | | | (9) Quantity limitations | | (10) Vessel stowage requirements | |
|---|---|---|---|---|---|---|---|---|---|---|---|---|---|
| | | | | | | | Exceptions | Non-bulk packaging | Bulk packaging | Passenger aircraft or railcar | Cargo aircraft only | Vessel stowages | Other stowage provisions |
| (1) | (2) | (3) | (4) | (5) | (6) | (7) | (8A) | (8B) | (8C) | (9A) | (9B) | (10A) | (10B) |
| + | Chloropicrin | 6.1 | UN1580 | I | POISON | 2, B7, B9, B14, B32, B46, B74, T38, T43, T45 | None | 227 | 244 | Forbidden | Forbidden | D | 40 |
| | Chloropicrin and methyl bromide mixtures | 2.3 | UN1581 | | POISON GAS | 2, B9, B14 | None | 193 | 314, 315 | Forbidden | Forbidden | D | 25, 40 |
| | Chloropicrin and methyl chloride mixtures | 2.3 | UN1582 | | POISON GAS | 2 | None | 193 | 245 | Forbidden | Forbidden | D | 25, 40 |
| | *Chloropicrin mixture, flammable (pressure not exceeding 14.7 psia at 115 degrees F flash point below 100 deg F) see* Poisonous liquids, flammable, n.o.s. | | | | | | | | | | | | |
| | Chloropicrin mixtures, n.o.s. | 6.1 | UN1583 | I | POISON | 5 | None | 201 | 243 | Forbidden | Forbidden | C | 40 |
| | | | | II | POISON | | None | 202 | 243 | Forbidden | Forbidden | C | 40 |
| | | | | III | KEEP AWAY FROM FOOD | | 153 | 203 | 241 | Forbidden | Forbidden | C | 40 |
| D | Chloropivaloyl chloride | 6.1 | NA9263 | I | POISON, CORROSIVE | 2, B9, B14, B32, B74, T38, T43, T45 | None | 227 | 244 | Forbidden | Forbidden | B | 40 |
| | Chloroplatinic acid, solid | 8 | UN2507 | III | CORROSIVE | | 154 | 213 | 240 | 25 kg | 100 kg | A | |
| | Chloroprene, inhibited | 3 | UN1991 | I | FLAMMABLE LIQUID, POISON | B57, T15 | None | 201 | 243 | Forbidden | 30 L | D | 40 |
| | *Chloroprene, uninhibited* | Forbidden | | | | | | | | | | | |
| | 2-Chloropropane | 3 | UN2356 | I | FLAMMABLE LIQUID | N36, T14 | 150 | 201 | 243 | 1 L | 30 L | E | |
| | 3-Chloropropanol-1 | 6.1 | UN2849 | III | KEEP AWAY FROM FOOD | T8 | 153 | 203 | 241 | 60 L | 220 L | A | |
| | 2-Chloropropene | 3 | UN2456 | I | FLAMMABLE LIQUID | A3, N36, T20 | 150 | 201 | 243 | 1 L | 30 L | E | |
| | alpha-Chloropropionic acid | 8 | UN2511 | III | CORROSIVE | T8 | 154 | 203 | 241 | 5 L | 60 L | A | 8 |
| | 2-Chloropyridine | 6.1 | UN2822 | II | POISON | T14 | None | 202 | 243 | 5 L | 60 L | A | 40 |
| | Chlorosilanes, n.o.s. | 8 | UN2987 | II | CORROSIVE | B2 | 154 | 202 | 242 | 1 L | 30 L | C | 40 |
| | Chlorosilanes, n.o.s., *flash point not less than 23 degrees C.* | 8 | UN2986 | II | CORROSIVE, FLAMMABLE LIQUID | | None | 202 | 243 | 1 L | 30 L | C | 23, 40 |
| | Chlorosilanes, n.o.s. *flashpoint less than 23 degrees C.* | 3 | UN2985 | II | FLAMMABLE LIQUID, CORROSIVE | T18, T26 | None | 201 | 243 | 1 L | 5 L | B | 40 |
| | Chlorosilanes, n.o.s., *which in contact with water emit flammable gases* | 4.3 | UN2988 | I | DANGEROUS WHEN WET, FLAMMABLE LIQUID, CORROSIVE | A2 | None | 201 | 244 | Forbidden | 1 L | D | 21, 28, 40, 49, 100 |
| + | Chlorosulfonic acid *(with or without sulfur trioxide)* | 8 | UN1754 | I | CORROSIVE, POISON | 2, A3, A6, A10, B10, B9, B14, B32, B74, T38, T43, T45 | None | 227 | 244 | 0.5 L | 2.5 L | C | 8, 40 |
| | Chlorotetrafluoroethane, *R124* | 2.2 | UN1021 | | NONFLAMMABLE GAS | | 306 | 304 | 314, 315 | 75 kg | 150 kg | A | |
| | Chlorotoluenes | 3 | UN2238 | III | FLAMMABLE LIQUID | B1, T1 | 150 | 203 | 242 | 60 L | 220 L | A | |
| | Chlorotoluidines *liquid* | 6.1 | UN2239 | III | KEEP AWAY FROM FOOD | T7 | 153 | 203 | 241 | 60 L | 220 L | A | |
| | Chlorotoluidines *solid* | 6.1 | UN2239 | III | KEEP AWAY FROM FOOD | | 153 | 213 | 240 | 100 kg | 200 kg | A | |
| | Chlorotrifluoroethane, *R133a* | 2.2 | UN1983 | | NONFLAMMABLE GAS | | 306 | 304 | 314, 315 | 75 kg | 150 kg | A | |
| | Chlorotrifluoromethane and trifluoromethane azeotropic mixture *with approximately 60 per cent chlorotrifluoromethane, R503* | 2.2 | UN2599 | | NONFLAMMABLE GAS | | 306 | 304 | 314, 315 | 75 kg | 150 kg | A | |
| | Chlorotrifluoromethane, *R13* | 2.2 | UN1022 | | NONFLAMMABLE GAS | | 306 | 304 | 314, 315 | 75 kg | 150 kg | A | |
| D | Chromic acid, solid | 5.1 | NA1463 | II | OXIDIZER, CORROSIVE | | None | 212 | 242 | 5 kg | 25 kg | A | |
| | Chromic acid solution | 8 | UN1755 | II | CORROSIVE | B2, T9, T27 | 154 | 202 | 242 | 1 L | 30 L | C | 40 |
| | *Chromic anhydride, see* Chromium trioxide, anhydrous | | | | | | | | | | | | |
| | Chromic fluoride, solid | 8 | UN1756 | II | CORROSIVE | | 154 | 212 | 240 | 15 kg | 50 kg | A | 26 |
| | Chromic fluoride, solution | 8 | UN1757 | II | CORROSIVE | B2, T8 | 154 | 202 | 242 | 1 L | 30 L | A | |
| | Chromium nitrate | 5.1 | UN2720 | III | OXIDIZER | A1, A29 | 152 | 213 | 240 | 25 kg | 100 kg | A | |
| | Chromium oxychloride | 8 | UN1758 | I | CORROSIVE | A3, A6, A7, B10, N34, T12, T26 | None | 201 | 242 | 0.5 L | 2.5 L | C | 8, 40, 66, 74, 89, 90 |
| | Chromium trioxide, anhydrous | 5.1 | UN1463 | II | OXIDIZER, CORROSIVE | | None | 212 | 242 | 5 kg | 25 kg | A | |
| | Chromosulfuric acid | 8 | UN2240 | I | CORROSIVE | A3, A6, A7, B4, B6, N34, T12, T27 | None | 201 | 242 | 0.5 L | 2.5 L | B | 8, 9, 40, 66, 74, 89, 90 |
| | *Chromyl chloride, see* Chromium oxychloride | | | | | | | | | | | | |
| | *Cigar and cigarette lighters, charged with fuel, see* Lighters for cigars, cigarettes, etc. | | | | | | | | | | | | |
| | *Coal briquettes, hot* | Forbidden | | | | | | | | | | | |
| | Coal gas | 2.3 | UN1023 | | POISON GAS, FLAMMABLE GAS | 3 | None | 302 | 314, 315 | Forbidden | 25 kg | D | 40 |

| Symbols | Hazardous materials descriptions and proper shipping names | Hazard class or Division | Identification Numbers | Packing group | Label(s) required (if not excepted) | Special provisions | (8) Packaging authorizations (§173.***) | | | (9) Quantity limitations | | (10) Vessel stowage requirements | |
|---|---|---|---|---|---|---|---|---|---|---|---|---|---|
| | | | | | | | Exceptions | Non-bulk packaging | Bulk packaging | Passenger aircraft or railcar | Cargo aircraft only | Vessel stowages | Other stowage provisions |
| (1) | (2) | (3) | (4) | (5) | (6) | (7) | (8A) | (8B) | (8C) | (9A) | (9B) | (10A) | (10B) |
| | Coal tar distillates, flammable | 3 | UN1136 | II | FLAMMABLE LIQUID | T8, T31 | 150 | 202 | 242 | 5 L | 60 L | B | |
| | | | | III | FLAMMABLE LIQUID | B1, T7, T30 | 150 | 203 | 242 | 60 L | 220 L | A | |
| | *Coal tar dye, corrosive, liquid, n.o.s. see* Dyes, liquid *or* solid, n.o.s. *or* Dye intermediates, liquid *or* solid, n.o.s., *corrosive* | | | | | | | | | | | | |
| | Coating solution | 3 | UN1139 | II | FLAMMABLE LIQUID | T7, T30 | 150 | 202 | 242 | 5 L | 60 L | B | |
| | | | | III | FLAMMABLE LIQUID | B1, T7, T30 | 150 | 203 | 242 | 60 L | 220 L | A | |
| | Cobalt naphthenates, powder | 4.1 | UN2001 | III | FLAMMABLE SOLID | A19 | 151 | 213 | 240 | 25 kg | 100 kg | A | |
| | Cobalt resinate, precipitated | 4.1 | UN1318 | III | FLAMMABLE SOLID | A1, A19 | 151 | 213 | 240 | 25 kg | 100 kg | A | |
| | Cocculus | 6.1 | UN1584 | II | POISON | | None | 212 | 242 | 25 kg | 100 kg | A | |
| | *Coke, hot* | Forbidden | | | | | | | | | | | |
| | *Collodion, see* Nitrocellulose *etc.* | | | | | | | | | | | | |
| D | Combustible liquid, n.o.s. | Combustible liquid | NA1993 | III | None | T1 | 150 | 203 | 241 | 60 L | 220 L | A | |
| D | Compounds, cleaning liquid | 8 | NA1760 | I | CORROSIVE | A7, B10, T42 | None | 201 | 242 | 0.5 L | 2.5 L | B | 40 |
| | | | | II | CORROSIVE | B2, N37, T14 | 154 | 202 | 242 | 1 L | 30 L | B | 40 |
| | | | | III | CORROSIVE | N37, T7 | 154 | 203 | 241 | 5 L | 60 L | A | 40 |
| D | Compounds, cleaning liquid | 3 | NA1993 | I | FLAMMABLE LIQUID | T42 | 150 | 201 | 243 | 1 L | 30 L | E | |
| | | | | II | FLAMMABLE LIQUID | T8, T31 | 150 | 202 | 242 | 5 L | 60 L | B | |
| | | | | III | FLAMMABLE LIQUID | B1, B52, T7, T30 | 150 | 203 | 242 | 60 L | 220 L | A | |
| D | Compounds, tree *or* weed killing, liquid | 8 | NA1760 | I | CORROSIVE | A7, B10, T42 | None | 201 | 242 | 0.5 L | 2.5 L | B | 40 |
| | | | | II | CORROSIVE | B2, N37, T14 | 154 | 202 | 242 | 1 L | 30 L | B | 40 |
| | | | | III | CORROSIVE | N37, T7 | 154 | 203 | 241 | 5 L | 60 L | A | 40 |
| D | Compounds, tree *or* weed killing, liquid | 3 | NA1993 | I | FLAMMABLE LIQUID | T42 | 150 | 201 | 243 | 1 L | 30 L | E | |
| | | | | II | FLAMMABLE LIQUID | T8, T31 | 150 | 202 | 242 | 5 L | 60 L | B | |
| | | | | III | FLAMMABLE LIQUID | B1, B52, T7, T30 | 150 | 203 | 242 | 60 L | 220 L | A | |
| D | Compounds, tree *or* weed killing, liquid | 6.1 | NA2810 | I | POISON | | 153 | 201 | 243 | 1 L | 30 L | B | 40 |
| | | | | II | POISON | | 153 | 202 | 243 | 5 L | 60 L | B | 40 |
| | | | | III | KEEP AWAY FROM FOOD | | 153 | 203 | 241 | 60 L | 220 L | A | 40 |
| | Components, explosive train, n.o.s. | 1.2B | UN0382 | II | EXPLOSIVE 1.2B | 101 | None | 62 | None | Forbidden | Forbidden | B | 1E, 6E |
| | Components, explosive train, n.o.s. | 1.4B | UN0383 | II | EXPLOSIVE 1.4B | 101 | None | 62 | None | Forbidden | 75 kg | A | 24E |
| | Components, explosive train, n.o.s. | 1.4S | UN0384 | II | EXPLOSIVE 1.4S | 101 | None | 62 | None | 25 kg | 100 kg | A | 9E |
| | Components, explosive train, n.o.s. | 1.1B | UN0461 | II | EXPLOSIVE 1.1B | 101 | None | 62 | None | Forbidden | Forbidden | B | 1E, 6E |
| | *Composition B, see* Hexolite, *etc.* | | | | | | | | | | | | |
| | Compressed *or* Liquefied gases, flammable, n.o.s. | 2.1 | UN1954 | | FLAMMABLE GAS | | 306 | 302, 304, 305 | 314, 315 | Forbidden | 150 kg | | 40, 105 |
| | Compressed *or* Liquefied gases, flammable, toxic, n.o.s. *Inhalation hazard Zone A* | 2.3 | UN1953 | | POISON GAS, FLAMMABLE GAS | 1 | None | 192 | 245 | Forbidden | Forbidden | D | 40, 95 |
| | Compressed *or* Liquefied gases, flammable, toxic, n.o.s. *Inhalation Hazard Zone B* | 2.3 | UN1953 | | POISON GAS, FLAMMABLE GAS | 2, B9, B14 | None | 302, 304, 305 | 314, 315 | Forbidden | Forbidden | D | 40 |
| | Compressed *or* Liquefied gases, flammable, toxic, n.o.s. *Inhalation Hazard Zone C* | 2.3 | UN1953 | | POISON GAS, FLAMMABLE GAS | 3, B14 | None | 302, 304, 305 | 314, 315 | Forbidden | Forbidden | D | 40 |
| | Compressed *or* Liquefied gases, flammable, toxic, n.o.s. *Inhalation Hazard Zone D* | 2.3 | UN1953 | | POISON GAS, FLAMMABLE GAS | 4 | None | 302, 304, 305 | 314, 315 | Forbidden | Forbidden | D | 40 |
| | Compressed *or* Liquefied gases, n.o.s. | 2.2 | UN1956 | | NONFLAMMABLE GAS | | 306, 307 | 302, 304, 305 | 314, 315 | 75 kg | 150 kg | A | |
| | Compressed *or* Liquefied gases, toxic, n.o.s. *Inhalation Hazard Zone A* | 2.3 | UN1955 | | POISON GAS | 1 | None | 192 | 245 | Forbidden | Forbidden | D | 40 |
| | Compressed *or* Liquefied gases, toxic, n.o.s. *Inhalation Hazard Zone B* | 2.3 | UN1955 | | POISON GAS | 2, B9, B14 | None | 302, 304, 305 | 314, 315 | Forbidden | Forbidden | D | 40 |
| | Compressed *or* Liquefied gases, toxic, n.o.s. *Inhalation Hazard Zone C* | 2.3 | UN1955 | | POISON GAS | 3, B14 | None | 302, 304, 305 | 314, 315 | Forbidden | Forbidden | D | 40 |
| | Compressed *or* Liquefied gases, toxic, n.o.s. *Inhalation Hazard Zone D* | 2.3 | UN1955 | | POISON GAS | 4 | None | 302, 304, 305 | 314, 315 | Forbidden | Forbidden | D | 40 |
| D | Consumer commodity | ORM-D | None | | None | | 156, 306 | 156, 306 | None | 30 kg gross | 30 kg gross | A | |
| | Contrivances, water-activated, *with burster, expelling charge or propelling charge* | 1.2L | UN0248 | II | EXPLOSIVE 1.2L | 101 | None | 62 | None | Forbidden | Forbidden | E | 2E, 8E, 11E, 17E |
| | Contrivances, water-activated, *with burster, expelling charge or propelling charge* | 1.3L | UN0249 | II | EXPLOSIVE 1.3L | 101 | None | 62 | None | Forbidden | Forbidden | E | 2E, 8E, 11E, 17E |
| | Copper acetoarsenite | 6.1 | UN1585 | II | POISON | | None | 212 | 242 | 25 kg | 100 kg | A | |
| | *Copper acetylide* | Forbidden | | | | | | | | | | | |
| | *Copper amine azide* | Forbidden | | | | | | | | | | | |
| | Copper arsenite | 6.1 | UN1586 | II | POISON | | None | 212 | 242 | 25 kg | 100 kg | A | |

## §172.101 HAZARDOUS MATERIALS TABLE, Continued

| Symbols | Hazardous materials descriptions and proper shipping names | Hazard class or Division | Identification Numbers | Packing group | Label(s) required (if not excepted) | Special provisions | (8) Packaging authorizations (§173.***) | | | (9) Quantity limitations | | (10) Vessel stowage requirements | |
|---|---|---|---|---|---|---|---|---|---|---|---|---|---|
| | | | | | | | Exceptions | Non-bulk packaging | Bulk packaging | Passenger aircraft or railcar | Cargo aircraft only | Vessel stowages | Other stowage provisions |
| (1) | (2) | (3) | (4) | (5) | (6) | (7) | (8A) | (8B) | (8C) | (9A) | (9B) | (10A) | (10B) |
| | Copper based pesticides, liquid, flammable, toxic n.o.s., *flash point less than 23 degrees C.* | 3 | UN2776 | I | FLAMMABLE LIQUID, POISON | | None | 201 | 243 | Forbidden | 30 L | B | 40 |
| | | | | II | FLAMMABLE LIQUID, POISON | | None | 202 | 243 | 1 L | 60 L | B | 40 |
| | Copper based pesticides, liquid, toxic, flammable, n.o.s. *flashpoint not less than 23 degrees C.* | 6.1 | UN3009 | I | POISON, FLAMMABLE LIQUID | T42 | None | 201 | 243 | 1 L | 30 L | B | 40 |
| | | | | II | POISON, FLAMMABLE LIQUID | T14 | None | 202 | 243 | 5 L | 60 L | B | 40 |
| | | | | III | KEEP AWAY FROM FOOD, FLAMMABLE LIQUID | B1, T14 | 153 | 203 | 241 | 60 L | 220 L | A | 40 |
| | Copper based pesticides, liquid, toxic, n.o.s. | 6.1 | UN3010 | I | POISON | T42 | None | 201 | 243 | 1 L | 30 L | B | 40 |
| | | | | II | POISON | T14 | None | 202 | 243 | 5 L | 60 L | B | 40 |
| | | | | III | KEEP AWAY FROM FOOD | T14 | 153 | 203 | 241 | 60 L | 220 L | A | 40 |
| | Copper based pesticides, solid, toxic, n.o.s. | 6.1 | UN2775 | I | POISON | | None | 211 | 242 | 5 kg | 50 kg | A | 40 |
| | | | | II | POISON | | None | 212 | 242 | 25 kg | 100 kg | A | 40 |
| | | | | III | KEEP AWAY FROM FOOD | | 153 | 213 | 24[illegible] | 100 kg | 200 kg | A | 40 |
| | Copper chlorate | 5.1 | UN2721 | II | OXIDIZER | A1 | 152 | 212 | 242 | 5 kg | 25 kg | A | 56, 58, 106 |
| | Copper chloride | 8 | UN2802 | III | CORROSIVE | | 154 | 213 | 240 | 25 kg | 100 kg | A | |
| | Copper cyanide | 6.1 | UN1587 | II | POISON | | None | 204 | 242 | 25 kg | 100 kg | A | 26 |
| | Copper selenate, *see* Selenates *or* Selenites | | | | | | | | | | | | |
| | Copper selenite, *see* Selenates *or* Selenites | | | | | | | | | | | | |
| | *Copper tetramine nitrate* | Forbidden | | | | | | | | | | | |
| AW | Copra | 4.2 | UN1363 | III | SPONTANEOUSLY COMBUSTIBLE | | None | 213 | 241 | Forbidden | Forbidden | A | 13, 19, 48, 119 |
| | Cord, detonating, *flexible* | 1.1D | UN0065 | II | EXPLOSIVE 1.1D | 102 | 63(a) | 62 | None | Forbidden | Forbidden | B | |
| | Cord, detonating, *flexible* | 1.4D | UN0289 | II | EXPLOSIVE 1.4D | | None | 62 | None | Forbidden | 75 kg | A | 24E |
| | Cord detonating *or* Fuse detonating *metal clad* | 1.2D | UN0102 | II | EXPLOSIVE 1.2D | | None | 62 | None | Forbidden | Forbidden | B | |
| | Cord, detonating *or* Fuse, detonating *metal clad* | 1.1D | UN0290 | II | EXPLOSIVE 1.1D | | None | 62 | None | Forbidden | Forbidden | B | |
| | Cord, detonating, mild effect *or* Fuse, detonating, mild effect *metal clad* | 1.4D | UN0104 | II | EXPLOSIVE 1.4D | | None | 62 | None | Forbidden | 75 kg | A | 24E |
| | Cord, igniter | 1.4G | UN0066 | II | EXPLOSIVE 1.4G | | None | 62 | None | Forbidden | 75 kg | A | 24E |
| | *Cordeau detonant fuse, see* Cord, detonating, *etc.*; Cord, detonating, *flexible* | | | | | | | | | | | | |
| | *Cordite, see* Powder, smokeless | | | | | | | | | | | | |
| | Corrosive liquids, flammable, n.o.s. | 8 | UN2920 | I | CORROSIVE, FLAMMABLE LIQUID | B10, T42 | None | 201 | 243 | 0.5 L | 2.5 L | C | 12, 21, 25, 40, 48 |
| | | | | II | CORROSIVE, FLAMMABLE LIQUID | B2, T15, T26 | None | 202 | 243 | 1 L | 30 L | C | 12, 21, 25, 40, 48 |
| | Corrosive liquids, n.o.s. | 8 | UN1760 | I | CORROSIVE | A7, B10, T42 | None | 201 | 243 | 0.5 L | 2.5 L | B | 40 |
| | | | | II | CORROSIVE | B2, T14 | 154 | 202 | 242 | 1 L | 30 L | B | 40 |
| | | | | III | CORROSIVE | T7 | 154 | 203 | 241 | 5 L | 60 L | A | 40 |
| | Corrosive liquids, oxidizing, n.o.s. | 8 | UN3093 | I | CORROSIVE, OXIDIZER | | None | 201 | 243 | Forbidden | 2.5 L | C | 89 |
| | | | | II | CORROSIVE, OXIDIZER | | None | 202 | 243 | 1 L | 30 L | C | 89 |
| | Corrosive liquids, poisonous, n.o.s. | 8 | UN2922 | I | CORROSIVE, POISON | A7, B10 | None | 201 | 243 | 0.5 L | 2.5 L | B | 40, 95 |
| | | | | II | CORROSIVE, POISON | B3 | None | 202 | 243 | 1 L | 30 L | B | 40, 95 |
| | Corrosive liquids, which in contact with water emit flammable gases, n.o.s. | 8 | UN3094 | I | CORROSIVE, DANGEROUS WHEN WET | | None | 201 | 243 | Forbidden | 1 L | E | |
| | | | | II | CORROSIVE, DANGEROUS WHEN WET | | None | 202 | 243 | 1 L | 5 L | E | |
| | Corrosive solids, flammable, n.o.s. | 8 | UN2921 | I | CORROSIVE, FLAMMABLE SOLID | | None | 211 | 242 | 1 kg | 25 kg | B | 12, 24, 25, 48 |
| | | | | II | CORROSIVE, FLAMMABLE SOLID | | None | 212 | 242 | 15 kg | 50 kg | B | 12, 24, 25, 48 |
| | Corrosive solids, n.o.s. | 8 | UN1759 | I | CORROSIVE | | None | 211 | 242 | 1 kg | 25 kg | B | |
| | | | | II | CORROSIVE | | 154 | 212 | 240 | 15 kg | 50 kg | A | |
| | | | | III | CORROSIVE | | 154 | 213 | 240 | 25 kg | 100 kg | A | |
| | Corrosive solids, oxidizing, n.o.s. | 8 | UN3084 | I | CORROSIVE, OXIDIZER | | None | 211 | 240 | 1 kg | 25 kg | C | 89 |
| | | | | II | CORROSIVE, OXIDIZER | | None | 212 | 240 | 15 kg | 50 kg | C | 89 |

| Symbols | Hazardous materials descriptions and proper shipping names | Hazard class or Division | Identification Numbers | Packing group | Label(s) required (if not excepted) | Special provisions | (8) Packaging authorizations (§173.***) | | | (9) Quantity limitations | | (10) Vessel stowage requirements | |
|---|---|---|---|---|---|---|---|---|---|---|---|---|---|
| | | | | | | | Exceptions | Non-bulk packaging | Bulk packaging | Passenger aircraft or railcar | Cargo aircraft only | Vessel stowages | Other stowage provisions |
| (1) | (2) | (3) | (4) | (5) | (6) | (7) | (8A) | (8B) | (8C) | (9A) | (9B) | (10A) | (10B) |
| | Corrosive solids, poisonous, n.o.s. | 8 | UN2923 | I | CORROSIVE, POISON | | None | 211 | 242 | 1 kg | 25 kg | B | 40, 95 |
| | | | | II | CORROSIVE, POISON | | None | 212 | 240 | 15 kg | 50 kg | B | 40, 95 |
| | Corrosive solids, self heating, n.o.s. | 8 | UN3095 | I | CORROSIVE, SPONTANEOUSLY COMBUSTIBLE | | None | 211 | 243 | 1 kg | 25 kg | C | |
| | | | | II | CORROSIVE, SPONTANEOUSLY COMBUSTIBLE | | None | 212 | 242 | 15 kg | 50 kg | C | |
| | Corrosive solids, which in contact with water emit flammable gases, n.o.s. | 8 | UN3096 | I | CORROSIVE, DANGEROUS WHEN WET | | None | 211 | 243 | 1 kg | 25 kg | E | |
| | | | | II | CORROSIVE, DANGEROUS WHEN WET | | None | 212 | 242 | 15 kg | 50 kg | E | |
| DW | Cotton | 9 | NA1365 | | | W41 | None | None | None | No limit | No limit | A | |
| I | Cotton waste, oily | 4.2 | UN1364 | III | SPONTANEOUSLY COMBUSTIBLE | N9 | None | 213 | None | Forbidden | Forbidden | A | 54 |
| AIW | Cotton, wet | 4.2 | UN1365 | | | | None | 204 | 241 | Forbidden | Forbidden | A | |
| | Coumarin derivative pesticides, liquid, flammable, toxic, n.o.s., *flashpoint less than 23 deg C.* | 3 | UN3024 | I | FLAMMABLE LIQUID, POISON | | None | 201 | 243 | Forbidden | 30 L | B | 40 |
| | | | | II | FLAMMABLE LIQUID, POISON | | None | 202 | 243 | 1 L | 60 L | B | 40 |
| | Coumarin derivative pesticides, liquid, toxic, flammable, n.o.s. *flashpoint not less than 23 degrees C.* | 6.1 | UN3025 | I | POISON, FLAMMABLE LIQUID | | None | 201 | 243 | 1 L | 30 L | B | 40 |
| | | | | II | POISON, FLAMMABLE LIQUID | | None | 202 | 243 | 5 L | 60 L | B | 40 |
| | | | | III | KEEP AWAY FROM FOOD, FLAMMABLE LIQUID | B1 | 153 | 203 | 241 | 60 L | 220 L | A | 40 |
| | Coumarin derivative pesticides, liquid, toxic, n.o.s. | 6.1 | UN3026 | I | POISON | | None | 201 | 243 | 1 L | 30 L | B | 40 |
| | | | | II | POISON | | None | 202 | 243 | 5 L | 60 L | B | 40 |
| | | | | III | KEEP AWAY FROM FOOD | | 153 | 203 | 241 | 60 L | 220 L | A | 40 |
| | Coumarin derivative pesticides, solid, toxic, n.o.s. | 6.1 | UN3027 | I | POISON | | None | 211 | 242 | 5 kg | 50 kg | A | 40 |
| | | | | II | POISON | | None | 212 | 242 | 25 kg | 100 kg | A | 40 |
| | | | | III | KEEP AWAY FROM FOOD | | 153 | 213 | 240 | 100 kg | 200 kg | A | 40 |
| | Cresols *(o-; m-; p-)* | 6.1 | UN2076 | II | POISON | T8 | None | 202 | 243 | 5 L | 60 L | A | |
| | Cresylic acid | 6.1 | UN2022 | II | POISON | T8 | None | 202 | 243 | 5 L | 60 L | A | |
| + | Crotonaldehyde, stabilized | 3 | UN1143 | II | FLAMMABLE LIQUID, POISON | 2, B9, B14, B32, B74, B77, T38, T43, T45 | None | 227 | 244 | Forbidden | 60 L | B | 40 |
| | Crotonic acid *liquid* | 8 | UN2823 | III | CORROSIVE | | 154 | 203 | 240 | 5 L | 60 L | A | 12, 48 |
| | Crotonic acid, *solid* | 8 | UN2823 | III | CORROSIVE | | 154 | 213 | 240 | 25 kg | 100 kg | A | 12, 48 |
| | Crotonylene | 3 | UN1144 | I | FLAMMABLE LIQUID | T20 | 150 | 201 | 243 | 1 L | 30 L | E | |
| | Cupriethylenediamine solution | 8 | UN1761 | II | CORROSIVE, POISON | T8, T26 | 154 | 202 | 243 | 1 L | 30 L | A | 95 |
| | Cutters, cable, explosive | 1.4S | UN0070 | II | EXPLOSIVE 1.4S | | None | 62 | None | 25 kg | 100 kg | A | 9E |
| | *Cyanide or cyanide mixtures, dry, see* Cyanides, inorganic, n.o.s. | | | | | | | | | | | | |
| | Cyanide solutions | 6.1 | UN1935 | I | POISON | B37, T18, T26 | None | 201 | 243 | 1 L | 30 L | B | 40, 52 |
| | | | | II | POISON | T18, T26 | None | 202 | 243 | 5 L | 60 L | A | 40, 52 |
| | | | | III | KEEP AWAY FROM FOOD | T18, T26 | 153 | 203 | 241 | 60 L | 220 L | A | 40, 52 |
| | Cyanides, inorganic, n.o.s. | 6.1 | UN1588 | I | POISON | N74, N75 | None | 211 | 242 | 5 kg | 50 kg | A | 52 |
| | | | | II | POISON | N74, N75 | None | 212 | 242 | 25 kg | 100 kg | A | 52 |
| | | | | III | KEEP AWAY FROM FOOD | N74, N75 | 153 | 203 | 241 | 100 kg | 200 kg | A | 52 |
| | Cyanogen bromide | 6.1 | UN1889 | I | POISON, CORROSIVE | 1, A6, A8, B9, B14, B30, B72, T38, T43, T44 | None | 226 | 244 | Forbidden | 25 kg | D | 40 |
| | Cyanogen chloride, inhibited | 2.3 | UN1589 | | POISON GAS, FLAMMABLE GAS | 1 | None | 192 | 245 | Forbidden | Forbidden | D | 40 |
| | Cyanogen, liquefied | 2.3 | UN1026 | | POISON GAS, FLAMMABLE GAS | 2 | None | 192 | 245 | Forbidden | Forbidden | D | 40 |
| | Cyanuric chloride | 8 | UN2670 | III | CORROSIVE | | None | 213 | 240 | 25 kg | 100 kg | A | 12, 40, 48 |
| | *Cyanuric triazide* | Forbidden | | | | | | | | | | | |
| | Cyclobutane | 2.1 | UN2601 | | FLAMMABLE GAS | | 306 | 304 | 314, 315 | Forbidden | 150 kg | B | 40 |
| | Cyclobutylchloroformate | 6.1 | UN2744 | II | POISON, CORROSIVE | T18 | None | 202 | 243 | 1 L | 30 L | A | 12, 13, 22, 25, 40, 48, 100 |

## §172.101 HAZARDOUS MATERIALS TABLE, Continued

| Symbols | Hazardous materials descriptions and proper shipping names | Hazard class or Division | Identification Numbers | Packing group | Label(s) required (if not excepted) | Special provisions | (8) Packaging authorizations (§173.***) | | | (9) Quantity limitations | | (10) Vessel stowage requirements | |
|---|---|---|---|---|---|---|---|---|---|---|---|---|---|
| | | | | | | | Exceptions | Non-bulk packaging | Bulk packaging | Passenger aircraft or railcar | Cargo aircraft only | Vessel stowages | Other stowage provisions |
| (1) | (2) | (3) | (4) | (5) | (6) | (7) | (8A) | (8B) | (8C) | (9A) | (9B) | (10A) | (10B) |
| | 1,5,9-Cyclododecatriene | 6.1 | UN2518 | III | KEEP AWAY FROM FOOD | T7 | 153 | 203 | 241 | 60 L | 220 L | A | 40 |
| | Cycloheptane | 3 | UN2241 | II | FLAMMABLE LIQUID | T1 | 150 | 202 | 242 | 5 L | 60 L | B | 40 |
| | Cycloheptatriene | 3 | UN2603 | II | FLAMMABLE LIQUID, POISON | T14 | None | 202 | 243 | 1 L | 60 L | E | 40 |
| | Cycloheptene | 3 | UN2242 | II | FLAMMABLE LIQUID | T7 | 150 | 202 | 241 | 5 L | 60 L | B | |
| | Cyclohexane | 3 | UN1145 | II | FLAMMABLE LIQUID | T8 | 150 | 202 | 242 | 5 L | 60 L | E | |
| | Cyclohexanone | 3 | UN1915 | III | FLAMMABLE LIQUID | B1, T1 | 150 | 203 | 242 | 60 L | 220 L | A | |
| | Cyclohexene | 3 | UN2256 | II | FLAMMABLE LIQUID | T7 | 150 | 202 | 242 | 5 L | 60 L | E | |
| | Cyclohexenyltrichlorosilane | 8 | UN1762 | II | CORROSIVE | A7, B2, N34, T8, T26 | None | 202 | 242 | Forbidden | 30 L | C | 40 |
| | Cyclohexyl acetate | 3 | UN2243 | III | FLAMMABLE LIQUID | B1, T1 | 150 | 203 | 242 | 60 L | 220 L | A | |
| + | Cyclohexyl isocyanate | 6.1 | UN2488 | II | POISON | 2, B9, B14, B32, B74, B77, T38, T43, T45 | None | 227 | 244 | 5 L | 60 L | D | 40 |
| | Cyclohexyl mercaptan | 3 | UN3054 | III | FLAMMABLE LIQUID | B1, T1 | 150 | 203 | 242 | 60 L | 220 L | A | 40, 95, 102 |
| | Cyclohexylamine | 8 | UN2357 | II | CORROSIVE, FLAMMABLE LIQUID | T8, T26 | None | 202 | 243 | 1 L | 30 L | A | 12, 21, 40, 48 |
| | Cyclohexyltrichlorosilane | 8 | UN1763 | II | CORROSIVE | A7, B2, N34, T8 T26 | None | 202 | 242 | Forbidden | 30 L | C | 40 |
| | Cyclonite and cyclotetramethylenetetranitramine mixtures, wetted *or* desensitized *see* RDX and HMX mixtures, wetted *or* desensitized *etc* ed | | | | | | | | | | | | |
| | Cyclonite and HMX mixtures, wetted *or* desensitized *see* RDX and HMX mixtures, wetted *or* desensitized *etc.* | | | | | | | | | | | | |
| | Cyclonite and octogen mixtures, wetted *or* desensitized *see* RDX and HMX mixtures, wetted *or* desensitized *etc* | | | | | | | | | | | | |
| | Cyclonite, *see* Cyclotrimethylenetrinitramine, *etc.* | | | | | | | | | | | | |
| | Cyclooctadiene phosphines, *see* 9-Phosphabicyclononanes | | | | | | | | | | | | |
| | Cyclooctadienes | 3 | UN2520 | III | FLAMMABLE LIQUID | B1, T1 | 150 | 203 | 242 | 60 L | 220 L | A | |
| | Cyclooctatetraene | 3 | UN2358 | II | FLAMMABLE LIQUID | T8 | 150 | 202 | 242 | 5 L | 60 L | B | |
| | Cyclopentane | 3 | UN1146 | II | FLAMMABLE LIQUID | T14 | 150 | 202 | 242 | 5 L | 60 L | E | |
| | *Cyclopentane, methyl, see* Methyl cyclopentane | | | | | | | | | | | | |
| | Cyclopentanol | 3 | UN2244 | III | FLAMMABLE LIQUID | B1, T1 | 150 | 203 | 242 | 60 L | 220 L | A | |
| | Cyclopentanone | 3 | UN2245 | III | FLAMMABLE LIQUID | B1, T1 | 150 | 203 | 242 | 60 L | 220 L | A | |
| | Cyclopentene | 3 | UN2246 | II | FLAMMABLE LIQUID | T13 | 150 | 202 | 242 | 5 L | 60 L | E | |
| | Cyclopropane, liquified | 2.1 | UN1027 | | FLAMMABLE GAS | | 306 | 304 | 314, 315 | Forbidden | 150 kg | E | 40 |
| | *Cyclotetramethylene tetranitramine (dry or unphlegmatized) (HMX)* | Forbidden | | | | | | | | | | | |
| | Cyclotetramethylenetetranitramine, desensitized *or* Octogen, desensitized *or* HMX, desensitized | 1.1D | UN0484 | II | EXPLOSIVE 1.1D | | None | 62 | None | Forbidden | Forbidden | B | 1E, 5E |
| | Cyclotetramethylenetetranitramine, wetted *or* HMX, wetted *or* Octogen, wetted *with not less than 15 per cent water, by mass* | 1.1D | UN0226 | II | EXPLOSIVE 1.1D | | None | 62 | None | Forbidden | Forbidden | B | 1E, 5E |
| | Cyclotrimethylenenitramine and octogen, mixtures, wetted *or* desensitized *see* RDX and HMX mixtures, wetted *or* desensitized *etc.* | | | | | | | | | | | | |
| | Cyclotrimethylenetrinitramine and cyclotetramethylenetetranitramine mixtures, wetted *or* desensitized *see* RDX and HMX mixtures, wetted *or* desensitized *etc.* | | | | | | | | | | | | |
| | Cyclotrimethylenetrinitramine and HMX mixtures, wetted *or* desensitized *see* RDX and HMX mixtures, wetted *or* desensitized *etc.* | | | | | | | | | | | | |

| Symbols | Hazardous materials descriptions and proper shipping names | Hazard class or Division | Identification Numbers | Packing group | Label(s) required (if not excepted) | Special provisions | (8) Packaging authorizations (§173.***) | | | (9) Quantity limitations | | (10) Vessel stowage requirements | |
|---|---|---|---|---|---|---|---|---|---|---|---|---|---|
| | | | | | | | Exceptions | Non-bulk packaging | Bulk packaging | Passenger aircraft or railcar | Cargo aircraft only | Vessel stowages | Other stowage provisions |
| (1) | (2) | (3) | (4) | (5) | (6) | (7) | (8A) | (8B) | (8C) | (9A) | (9B) | (10A) | (10B) |
| | Cyclotrimethylenetrinitramine, desensitized *or* Cyclonite, desensitized *or* Hexogen, desensitized *or* RDX, desensitizede; Hexogen; RDX), desensitized | 1.1D | UN0483 | II | EXPLOSIVE 1.1D | | None | 62 | None | Forbidden | Forbidden | B | 1E, 5E |
| | Cyclotrimethylenetrinitramine, wetted *or* Cyclonite, wetted *or* Hexogen, wetted *or* RDX, wetted *with not less than 15 percent water by mass* | 1.1D | UN0072 | II | EXPLOSIVE 1.1D | | None | 62 | None | Forbidden | Forbidden | B | 1E, 5E |
| | Cymenes | 3 | UN2046 | III | FLAMMABLE LIQUID | B1, T1 | 150 | 203 | 242 | 60 L | 220 L | A | |
| | Decaborane | 4.1 | UN1868 | II | FLAMMABLE SOLID, POISON | A19, A20 | None | 212 | None | Forbidden | 50 kg | A | |
| | Decahydronaphthalene | 3 | UN1147 | III | FLAMMABLE LIQUID | B1, T1 | 150 | 203 | 242 | 60 L | 220 L | A | |
| | n-Decane | 3 | UN2247 | III | FLAMMABLE LIQUID | B1, T1 | 150 | 203 | 242 | 60 L | 220 L | A | |
| | Deflagrating metal salts of aromatic nitroderivatives, n.o.s | 1.3C | UN0132 | II | EXPLOSIVE 1.3C | | None | 62 | None | Forbidden | Forbidden | B | 1E, 5E |
| | *Delay electric igniter, see* Igniters | | | | | | | | | | | | |
| D | Denatured alcohol | 3 | NA1987 | II | FLAMMABLE LIQUID | T8, T31 | 150 | 202 | 242 | 5 L | 60 L | B | |
| | | | | III | FLAMMABLE LIQUID | B1, T7, T30 | 150 | 203 | 242 | 60 L | 220 L | A | |
| D | Denatured alcohol | 3 | NA1986 | I | FLAMMABLE LIQUID, POISON | T8, T31 | None | 201 | 243 | Forbidden | 30 L | E | 40 |
| | | | | II | FLAMMABLE LIQUID, POISON | T8, T31 | 150 | 202 | 243 | 1 L | 60 L | E | 40 |
| | *Depth charges, see* Charges, depth | | | | | | | | | | | | |
| | *Detonating relays, see* Detonators, *etc.* | | | | | | | | | | | | |
| | Detonator assemblies, non-electric *for blasting* | 1.1B | UN0360 | II | EXPLOSIVE 1.1B | | None | 62 | None | Forbidden | Forbidden | B | 1E, 2E, 6E |
| | Detonator assemblies, non-electric, *for blasting* | 1.4B | UN0361 | II | EXPLOSIVE 1.4B | 103 | 63(f), 63(g) | 62 | None | Forbidden | 75 kg | A | 1E, 2E, 6E, 24E |
| | Detonators, electric, *for blasting* | 1.1B | UN0030 | II | EXPLOSIVE 1.1B | | 63(f), 63(g) | 62 | None | Forbidden | Forbidden | B | |
| | Detonators, electric, *for blasting* | 1.4B | UN0255 | II | EXPLOSIVE 1.4B | 103 | 63(f), 63(g) | 62 | None | Forbidden | 75 kg | A | 24E |
| | Detonators, electric *for blasting* | 1.4S | UN0456 | II | EXPLOSIVE 1.4S | 104 | 63(f), 63(g) | 62 | None | 25 kg | 100 kg | A | 9E |
| | Detonators for ammunition | 1.1B | UN0073 | II | EXPLOSIVE 1.1B | | None | 62 | None | Forbidden | Forbidden | B | 2E, 6E, 25E |
| | Detonators for ammunition | 1.2B | UN0364 | II | EXPLOSIVE 1.2B | | None | 62 | None | Forbidden | Forbidden | B | 2E, 6E |
| | Detonators for ammunition | 1.4B | UN0365 | II | EXPLOSIVE 1.4B | 103 | None | 62 | None | Forbidden | 75 kg | A | 24E |
| | Detonators for ammunition | 1.4S | UN0366 | II | EXPLOSIVE 1.4S | 104 | None | 62 | None | 25 kg | 100 kg | A | 9E |
| | Detonators, non-electric, *for blasting* | 1.1B | UN0029 | II | EXPLOSIVE 1.1B | | None | 62 | None | Forbidden | Forbidden | B | 1E, 2E, 6E |
| | Detonators, non-electric, *for blasting* | 1.4B | UN0267 | II | EXPLOSIVE 1.4B | 103 | 63(f), 63(g) | 62 | None | Forbidden | 75 kg | B | |
| | Detonators, non-electric *for blasting* | 1.4S | UN0455 | II | EXPLOSIVE 1.4S | 104 | None | 62 | None | 25 kg | 100 kg | A | 9E |
| | Deuterium | 2.1 | UN1957 | | FLAMMABLE GAS | | 306 | 302 | None | Forbidden | 150 kg | E | 40 |
| | Devices, small, hydrocarbon gas powered *or* hydrocarbon gas refills for small devices *with release device* | 2.1 | UN3150 | | FLAMMABLE GAS | | 306 | 304 | | Forbidden | 150 kg | B | 40 |
| | Di-n-amylamine | 6.1 | UN2841 | III | KEEP AWAY FROM FOOD | T8 | 153 | 203 | 241 | 60 L | 220 L | A | |
| | *Di-n-butyl peroxydicarbonate, more than 52 per cent in solution* | Forbidden | | | | | | | | | | | |
| | Di-n-butylamine | 8 | UN2248 | II | CORROSIVE, FLAMMABLE LIQUID | T8 | None | 202 | 243 | 1 L | 30 L | A | 12, 21, 48 |
| | *2,2-Di-(tert-butylperoxy) butane, more than 55 per cent in solution* | Forbidden | | | | | | | | | | | |
| | *Di-(tert-butylperoxy) phthalate, more than 55 per cent in solution* | Forbidden | | | | | | | | | | | |
| | *2,2-Di-(4,4-di-tert-butylperoxycyclohexyl) propane, more than 42 per cent with inert solid* | Forbidden | | | | | | | | | | | |
| | *Di-2,4-dichlorobenzoyl peroxide, more than 75 per cent with water* | Forbidden | | | | | | | | | | | |
| | 1,2-Di-(dimethylamino) ethane | 3 | UN2372 | II | FLAMMABLE LIQUID | T8 | 150 | 202 | 242 | 5 L | 60 L | B | |
| | *Di-2-ethylhexyl phosphoric acid, see* Diisooctyl acid phosphate | | | | | | | | | | | | |
| | *Di-(1-hydroxytetrazole) (dry)* | Forbidden | | | | | | | | | | | |
| | *Di-(1-naphthoyl) peroxide* | Forbidden | | | | | | | | | | | |
| | *α,α'-Di-(nitroxy) methylether* | Forbidden | | | | | | | | | | | |
| | *Di-(beta-nitroxyethyl) ammonium nitrate* | Forbidden | | | | | | | | | | | |

## §172.101 HAZARDOUS MATERIALS TABLE, Continued

| Symbols | Hazardous materials descriptions and proper shipping names | Hazard class or Division | Identification Numbers | Packing group | Label(s) required (if not excepted) | Special provisions | (8) Packaging authorizations (§173.***) | | | (9) Quantity limitations | | (10) Vessel stowage requirements | |
|---|---|---|---|---|---|---|---|---|---|---|---|---|---|
| | | | | | | | Exceptions | Non-bulk packaging | Bulk packaging | Passenger aircraft or railcar | Cargo aircraft only | Vessel stowages | Other stowage provisions |
| (1) | (2) | (3) | (4) | (5) | (6) | (7) | (8A) | (8B) | (8C) | (9A) | (9B) | (10A) | (10B) |
| | Diacetone alcohol | 3 | UN1148 | III | FLAMMABLE LIQUID | B1, T1 | 150 | 203 | 242 | 60 L | 220 L | A | |
| | *Diacetone alcohol peroxides, more than 57% in solution with more than 9% hydrogen peroxide, less than 26% diacetone alcohol and less than 9% water; total active oxygen content more than 9% by mass* | Forbidden | | | | | | | | | | | |
| | *Diacetyl, see* Butanedione | | | | | | | | | | | | |
| | *Diacetyl peroxide, solid, or more than 25% in solution* | Forbidden | | | | | | | | | | | |
| | Diallylamine | 3 | UN2359 | II | FLAMMABLE LIQUID | T8 | 150 | 202 | 242 | 5 L | 60 L | B | 40 |
| | Diallylether | 3 | UN2360 | II | FLAMMABLE LIQUID, POISON | N12, T8 | None | 202 | 243 | 1 L | 60 L | E | 40 |
| | 4,4'-Diaminodiphenyl methane | 6.1 | UN2651 | III | KEEP AWAY FROM FOOD | | 153 | 213 | 240 | 100 kg | 200 kg | A | |
| | *p-Diazidobenzene* | Forbidden | | | | | | | | | | | |
| | *1,2-Diazidoethane* | Forbidden | | | | | | | | | | | |
| | 2-Diazo-1-naphthol-4-sulpho-chloride | 4.1 | UN3042 | II | FLAMMABLE SOLID, EXPLOSIVE | 53 | None | 224 | None | Forbidden | Forbidden | D | |
| | 2-Diazo-1-naphthol-5-sulpho-chloride | 4.1 | UN3043 | II | FLAMMABLE SOLID, EXPLOSIVE | 53 | None | 224 | None | Forbidden | Forbidden | D | |
| | *1,1'-Diazoaminonaphthalene* | Forbidden | | | | | | | | | | | |
| | *Diazoaminotetrazole (dry)* | Forbidden | | | | | | | | | | | |
| | *Diazodinitrophenol (dry)* | Forbidden | | | | | | | | | | | |
| | Diazodinitrophenol, wetted *with not less than 40 per cent water or mixture of alcohol and water, by mass* | 1.1A | UN0074 | II | EXPLOSIVE 1.1A | 111, 117 | None | 62 | None | Forbidden | Forbidden | E | 2E, 6E |
| | *Diazodiphenylmethane* | Forbidden | | | | | | | | | | | |
| | *Diazonium nitrates (dry)* | Forbidden | | | | | | | | | | | |
| | *Diazonium perchlorates (dry)* | Forbidden | | | | | | | | | | | |
| | *1,3-Diazopropane* | Forbidden | | | | | | | | | | | |
| | *Dibenzyl peroxydicarbonate, more than 87 per cent with water* | Forbidden | | | | | | | | | | | |
| | Dibenzyldichlorosilane | 8 | UN2434 | II | CORROSIVE | B2, T8, T26 | 154 | 202 | 242 | 1 L | 30 L | C | 40 |
| | Diborane | 2.3 | UN1911 | | POISON GAS, FLAMMABLE GAS | 1 | None | 302 | None | Forbidden | Forbidden | D | 40, 57 |
| | Diborane mixtures | 2.1 | NA1911 | | FLAMMABLE GAS | 5 | None | 302 | 245 | Forbidden | Forbidden | D | 40, 57 |
| | *Dibromoacetylene* | Forbidden | | | | | | | | | | | |
| | Dibromobenzene | 3 | UN2711 | III | FLAMMABLE LIQUID | B1, T1 | 150 | 203 | 242 | 60 L | 220 L | A | |
| | 1,2-Dibromobutan-3-one | 6.1 | UN2648 | II | POISON | | None | 202 | 243 | 5 L | 60 L | B | 40 |
| | Dibromochloropropane | 6.1 | UN2872 | III | KEEP AWAY FROM FOOD | T7 | 153 | 203 | 241 | 60 L | 220 L | A | |
| A | Dibromodifluoromethane, *R1282* | 9 | UN1941 | III | None | T22 | 155 | 203 | 241 | 100 L | 220 L | A | 25 |
| | *1,2-Dibromoethane, see* Ethylene dibromide | | | | | | | | | | | | |
| | Dibromomethane | 6.1 | UN2664 | III | KEEP AWAY FROM FOOD | T7 | 153 | 203 | 241 | 60 L | 220 L | A | |
| | Dibutyl ethers | 3 | UN1149 | III | FLAMMABLE LIQUID | B1, T1 | 150 | 203 | 242 | 60 L | 220 L | A | |
| | Dibutylaminoethanol | 6.1 | UN2873 | III | KEEP AWAY FROM FOOD | T1 | 153 | 203 | 241 | 60 L | 220 L | A | |
| | *N,N'-Dichlorazodicarbonamidine (salts of) (dry)* | Forbidden | | | | | | | | | | | |
| | 1,1-Dichloro-1-nitroethane | 6.1 | UN2650 | II | POISON | T8 | None | 202 | 243 | 5 L | 60 L | A | 12, 40, 48 |
| D | 3,5-Dichloro-2,4,6-trifluoropyridine | 6.1 | NA9264 | I | POISON | 2, B9, B14, B32, B74, T38, T43, T45 | None | 227 | 244 | Forbidden | Forbidden | A | 40, 95 |
| | Dichloroacetic acid | 8 | UN1764 | II | CORROSIVE | A3, A6, A7, B2, N34, T9, T27 | 154 | 202 | 242 | 1 L | 30 L | A | |
| | 1,3-Dichloroacetone | 6.1 | UN2649 | II | POISON | | None | 212 | 243 | 25 kg | 100 kg | B | 12, 40, 48 |
| | Dichloroacetyl chloride | 8 | UN1765 | II | CORROSIVE | A3, A6, A7, B2, B6, N34, T8, T26 | 154 | 202 | 242 | 1 L | 30 L | D | 40 |
| | *Dichloroacetylene* | Forbidden | | | | | | | | | | | |
| | Dichloroanilines, solid *or* liquid | 6.1 | UN1590 | II | POISON | T14 | None | 202 | 242 | 5 L | 60 L | A | 40 |
| | o-Dichlorobenzene | 6.1 | UN1591 | III | KEEP AWAY FROM FOOD | T7 | 153 | 203 | 241 | 60 L | 220 L | A | |
| | p-Dichlorobenzene | 6.1 | UN1592 | III | KEEP AWAY FROM FOOD | | 153 | 213 | 240 | 100 kg | 200 kg | A | |

## §172.101 HAZARDOUS MATERIALS TABLE, Continued

| Symbols | Hazardous materials descriptions and proper shipping names | Hazard class or Division | Identification Numbers | Packing group | Label(s) required (if not excepted) | Special provisions | (8) Packaging authorizations (§173.***) | | | (9) Quantity limitations | | (10) Vessel stowage requirements | |
|---|---|---|---|---|---|---|---|---|---|---|---|---|---|
| | | | | | | | Exceptions | Non-bulk packaging | Bulk packaging | Passenger aircraft or railcar | Cargo aircraft only | Vessel stowages | Other stowage provisions |
| (1) | (2) | (3) | (4) | (5) | (6) | (7) | (8A) | (8B) | (8C) | (9A) | (9B) | (10A) | (10B) |
| D | Dichlorobutene | 8 | NA2920 | I | CORROSIVE, FLAMMABLE LIQUID | | None | 201 | 243 | 0.5 L | 2.5 L | C | 12, 21, 25, 40, 48 |
| | 2,2-Dichlorodiethyl ether | 6.1 | UN1916 | II | POISON, FLAMMABLE LIQUID | N33, N34, T8 | None | 202 | 243 | 5 L | 60 L | A | |
| | Dichlorodifluoromethane and difluoroethane azeotropic mixture *with approximately 74 per cent dichlorodifluoromethane, R500* | 2.2 | UN2602 | | NONFLAMMABLE GAS | | 306 | 304 | 314, 315 | 75 kg | 150 kg | A | |
| | Dichlorodifluoromethane and ethylene oxide mixture, *with not more than 12% ethylene oxide* | 2.2 | UN3070 | | NONFLAMMABLE GAS | | None | 304 | 314, 315 | Forbidden | 25 kg | D | 40 |
| | Dichlorodifluoromethane, *R12* | 2.2 | UN1028 | | NONFLAMMABLE GAS | | 306 | 304 | 314, 315 | 75 kg | 150 kg | A | |
| | Dichlorodimethyl ether, symmetrical | 6.1 | UN2249 | I | POISON | T25 | None | 201 | 243 | Forbidden | Forbidden | D | 40, 105 |
| | 1,1-Dichloroethane | 3 | UN2362 | II | FLAMMABLE LIQUID | T7 | 150 | 202 | 242 | 5 L | 60 L | B | 40 |
| | *1,2-Dichloroethane, see* Ethylene dichloride | | | | | | | | | | | | |
| | *Dichloroethyl sulfide* | Forbidden | | | | | | | | | | | |
| | Dichloroethylene | 3 | UN1150 | II | FLAMMABLE LIQUID | T14 | 150 | 202 | 242 | 5 L | 60 L | B | |
| | Dichlorofluoromethane, *R21* | 2.2 | UN1029 | | NONFLAMMABLE GAS | | 306 | 304 | 314, 315 | 75 kg | 150 kg | A | |
| | Dichloroisocyanuric acid, dry *or* Dichloroisocyanuric acid salts | 5.1 | UN2465 | II | OXIDIZER | 28 | 152 | 212 | 240 | 5 kg | 25 kg | A | 13 |
| | Dichloroisopropyl ether | 6.1 | UN2490 | II | POISON | T8 | None | 202 | 243 | 5 L | 60 L | B | |
| | Dichloromethane | 6.1 | UN1593 | III | KEEP AWAY FROM FOOD | N36, T13 | 153 | 203 | 241 | 60 L | 220 L | A | |
| | Dichloropentanes | 3 | UN1152 | III | FLAMMABLE LIQUID | B1, T1 | 150 | 202 | 241 | 60 L | 220 L | A | |
| | Dichlorophenyl isocyanates | 6.1 | UN2250 | II | POISON | | None | 212 | 242 | 25 kg | 100 kg | A | 25, 40, 48 |
| | Dichlorophenyltrichlorosilane | 8 | UN1766 | II | CORROSIVE | A7, B2, B6, N34, T8, T26 | None | 202 | 242 | Forbidden | 30 L | C | 40 |
| | *Dichloropropane, see* Propylene dichloride | | | | | | | | | | | | |
| | 1,3-Dichloropropanol-2 | 6.1 | UN2750 | II | POISON | T8 | None | 202 | 243 | 5 L | 60 L | A | 12, 40, 48 |
| | Dichloropropene | 3 | UN2047 | II | FLAMMABLE LIQUID | T8 | 150 | 202 | 242 | 5 L | 60 L | A | |
| | *Dichloropropene and propylene dichloride mixture, see* Propylene dichloride | | | | | | | | | | | | |
| | Dichlorosilane | 2.3 | UN2189 | | POISON GAS, FLAMMABLE GAS | 2, B9, B14 | None | 304 | 314, 315 | Forbidden | Forbidden | D | 40 |
| | Dichlorotetrafluoroethane, *R114* | 2.2 | UN1958 | | NONFLAMMABLE GAS | | 306 | 304 | 314, 315 | 75 kg | 150 kg | A | |
| | *Dichlorovinylchloroarsine* | Forbidden | | | | | | | | | | | |
| | Dicycloheptadiene, *see* Norbornadiene | | | | | | | | | | | | |
| | Dicyclohexylamine | 8 | UN2565 | III | CORROSIVE | T8 | 154 | 203 | 241 | 5 L | 60 L | A | |
| | Dicyclohexylammonium nitrite | 4.1 | UN2687 | III | FLAMMABLE SOLID | | 153 | 213 | 240 | 25 kg | 100 kg | A | 48 |
| | Dicyclopentadiene | 3 | UN2048 | III | FLAMMABLE LIQUID | B1, T1 | 150 | 203 | 242 | 60 L | 220 L | A | |
| | Didymium nitrate | 5.1 | UN1465 | III | OXIDIZER | A1 | 152 | 213 | 240 | 25 kg | 100 kg | A | |
| D | Dieldrin | 6.1 | NA2761 | II | POISON | | None | 212 | 242 | 0.5 kg | 5 kg | A | 40 |
| D | Diesel fuel | 3 | NA1993 | III | None | B1 | 150 | 203 | 241 | 60 L | 220 L | A | |
| | *Diethanol nitrosamine dinitrate (dry)* | Forbidden | | | | | | | | | | | |
| | 2,5-Diethoxy-4-morpholinobenzenediazonium zinc chloride | 4.1 | UN3036 | II | FLAMMABLE SOLID | | None | 224 | None | 15 kg | 50 kg | C | 2 |
| | Diethoxymethane | 3 | UN2373 | II | FLAMMABLE LIQUID | T8 | 150 | 202 | 242 | 5 L | 60 L | B | |
| | 3,3-Diethoxypropene | 3 | UN2374 | II | FLAMMABLE LIQUID | T1 | 150 | 202 | 242 | 5 L | 60 L | B | |
| | N,N-Diethyl aniline | 6.1 | UN2432 | III | KEEP AWAY FROM FOOD | T2 | 153 | 203 | 241 | 60 L | 220 L | A | |
| | Diethyl carbonate | 3 | UN2366 | III | FLAMMABLE LIQUID | B1, T1 | 150 | 203 | 242 | 60 L | 220 L | A | |
| | *Diethyl cellosolve, see* Ethylene glycol diethyl ether | | | | | | | | | | | | |
| | Diethyl ether *or* Ethyl ether | 3 | UN1155 | I | FLAMMABLE LIQUID | T21 | 150 | 201 | 243 | 1 L | 30 L | E | 40 |
| | Diethyl ketone | 3 | UN1156 | II | FLAMMABLE LIQUID | T1 | 150 | 202 | 242 | 5 L | 60 L | B | |
| | *Diethyl peroxydicarbonate, more than 27 per cent in solution* | Forbidden | | | | | | | | | | | |
| | Diethyl sulfate | 6.1 | UN1594 | II | POISON | T14 | None | 202 | 243 | 5 L | 60 L | C | |
| | Diethyl sulfide | 3 | UN2375 | II | FLAMMABLE LIQUID | T14 | None | 202 | 243 | 1 L | 60 L | E | |

| Symbols | Hazardous materials descriptions and proper shipping names | Hazard class or Division | Identification Numbers | Packing group | Label(s) required (if not excepted) | Special provisions | (8) Packaging authorizations (§173.***) Exceptions | Non-bulk packaging | Bulk packaging | (9) Quantity limitations Passenger aircraft or railcar | Cargo aircraft only | (10) Vessel stowage requirements Vessel stowages | Other stowage provisions |
|---|---|---|---|---|---|---|---|---|---|---|---|---|---|
| (1) | (2) | (3) | (4) | (5) | (6) | (7) | (8A) | (8B) | (8C) | (9A) | (9B) | (10A) | (10B) |
| | Diethylamine | 3 | UN1154 | II | FLAMMABLE LIQUID | N34, T8 | 150 | 202 | 242 | 5 L | 60 L | E | |
| | Diethylaminoethanol | 3 | UN2686 | III | FLAMMABLE LIQUID | B1, T1 | 150 | 203 | 242 | 60 L | 220 L | A | |
| | Diethylaminopropylamine | 8 | UN2684 | III | CORROSIVE, FLAMMABLE LIQUID | B1, B2, T8 | 154 | 203 | 243 | 5 L | 60 L | A | 12, 21, 48 |
| | Diethylbenzene | 3 | UN2049 | III | FLAMMABLE LIQUID | B1, T1 | 150 | 203 | 242 | 60 L | 220 L | A | |
| | Diethyldichlorosilane | 8 | UN1767 | II | CORROSIVE, FLAMMABLE LIQUID | A7, B6, N34, T8, T26 | None | 202 | 243 | Forbidden | 30 L | C | 21, 40 |
| | *Diethylene glycol dinitrate* | Forbidden | | | | | | | | | | | |
| | Diethyleneglycol dinitrate, desensitized *with not less than 25 percent non-volatile water-insoluble phlegmatizer, by mass* | 1.1D | UN0075 | II | EXPLOSIVE 1.1D | | None | 62 | None | Forbidden | Forbidden | B | 1E, 4E, 21E |
| | Diethylenetriamine | 8 | UN2079 | II | CORROSIVE | B2, T8 | 154 | 202 | 242 | 1 L | 30 L | A | 40 |
| | N,N-Diethylethylenediamine | 8 | UN2685 | II | CORROSIVE, FLAMMABLE LIQUID | T8 | None | 202 | 243 | 1 L | 30 L | A | 12, 21, 48 |
| | *Diethylgold bromide* | Forbidden | | | | | | | | | | | |
| | Diethylthiophosphoryl chloride | 8 | UN2751 | II | CORROSIVE | B2, T8 | None | 202 | 242 | 15 kg | 50 kg | D | 12, 40 |
| | Diethylzinc | 4.2 | UN1366 | I | SPONTANEOUSLY COMBUSTIBLE | B11, T28, T40 | None | 181 | 244 | Forbidden | Forbidden | D | 18 |
| | Difluorochloroethanes, *see* Chlorodifluoroethanes | | | | | | | | | | | | |
| | Difluoroethane, *R152a* | 2.1 | UN1030 | | FLAMMABLE GAS | | 306 | 304 | 314, 315 | Forbidden | 150 kg | B | 40 |
| | 1,1-Difluoroethylene, *R1132a* | 2.1 | UN1959 | | FLAMMABLE GAS | | 306 | 304 | None | Forbidden | 150 kg | E | 40 |
| | Difluorophosphoric acid, anhydrous | 8 | UN1768 | II | CORROSIVE | A6, A7, B2, N5, N34, T9, T27 | None | 202 | 242 | 1 L | 30 L | A | 40 |
| | 2,3-Dihydropyran | 3 | UN2376 | II | FLAMMABLE LIQUID | T7 | 150 | 202 | 242 | 5 L | 60 L | B | |
| | *1,8-Dihydroxy-2,4,5,7-tetranitroanthraquinone (chrysamminic acid)* | Forbidden | | | | | | | | | | | |
| | *Diiodoacetylene* | Forbidden | | | | | | | | | | | |
| | Diisobutyl ketone | 3 | UN1157 | III | FLAMMABLE LIQUID | B1, T1 | 150 | 203 | 242 | 60 L | 220 L | A | |
| | Diisobutylamine | 3 | UN2361 | III | FLAMMABLE LIQUID | B1, T1 | 150 | 203 | 242 | 60 L | 220 L | A | |
| | Diisobutylene, isomeric compounds | 3 | UN2050 | II | FLAMMABLE LIQUID | T1 | 150 | 202 | 242 | 5 L | 60 L | B | |
| | Diisooctyl acid phosphate | 8 | UN1902 | III | CORROSIVE | T7 | 154 | 203 | 241 | 5 L | 60 L | A | |
| | Diisopropyl ether | 3 | UN1159 | II | FLAMMABLE LIQUID | T8 | 150 | 202 | 242 | 5 L | 60 L | E | 40 |
| | Diisopropylamine | 3 | UN1158 | II | FLAMMABLE LIQUID | T8 | 150 | 202 | 242 | 5 L | 60 L | B | |
| | *Diisopropylbenzene hydroperoxide, more than 72 percent in solution* | Forbidden | | | | | | | | | | | |
| + | Diketene, inhibited | 3 | UN2521 | III | FLAMMABLE LIQUID, POISON | 2, B9, B14, B32, B74, T38, T43, T45 | None | 227 | 244 | 60 L | 220 L | A | 49 |
| | 1,1-Dimethoxyethane | 3 | UN2377 | II | FLAMMABLE LIQUID | T13 | 150 | 202 | 242 | 5 L | 60 L | B | |
| | 1,2-Dimethoxyethane | 3 | UN2252 | II | FLAMMABLE LIQUID | T1 | 150 | 202 | 242 | 5 L | 60 L | B | |
| | Dimethyl carbonate | 3 | UN1161 | II | FLAMMABLE LIQUID | T8 | 150 | 202 | 242 | 5 L | 60 L | B | |
| | *Dimethyl chlorothiophosphate, see* Dimethyl thiophosphoryl chloride | | | | | | | | | | | | |
| | *2,5-Dimethyl-2,5-dihydroperoxy hexane, more than 82 per cent with water* | Forbidden | | | | | | | | | | | |
| | Dimethyl disulfide | 3 | UN2381 | II | FLAMMABLE LIQUID | T8 | 150 | 202 | 242 | 5 L | 60 L | B | 40 |
| | Dimethyl ether | 2.1 | UN1033 | | FLAMMABLE GAS | | 306 | 304 | 314, 315 | Forbidden | 150 kg | B | 40 |
| | Dimethyl-N-propylamine | 3 | UN2266 | II | FLAMMABLE LIQUID, CORROSIVE | T14, T26 | None | 202 | 243 | 1 L | 5 L | B | 40 |
| | Dimethyl sulfate | 6.1 | UN1595 | I | POISON, CORROSIVE | 2, B9, B14, B32, B74, B77, T38, T43, T45 | None | 227 | 244 | Forbidden | 2.5 L | D | 40 |
| | Dimethyl sulfide | 3 | UN1164 | II | FLAMMABLE LIQUID | T14 | None | 201 | 243 | 1 L | 30 L | E | 40 |
| | Dimethyl thiophosphoryl chloride | 8 | UN2267 | III | CORROSIVE | T7 | None | 203 | 241 | 5 L | 60 L | B | 8 |
| | Dimethylamine, anhydrous | 2.1 | UN1032 | | FLAMMABLE GAS | | None | 304 | 314, 315 | Forbidden | Forbidden | D | 40 |
| | Dimethylamine solution | 3 | UN1160 | II | FLAMMABLE LIQUID | T8 | 150 | 202 | 242 | 5 L | 60 L | B | |

| Symbols | Hazardous materials descriptions and proper shipping names | Hazard class or Division | Identification Numbers | Packing group | Label(s) required (if not excepted) | Special provisions | (8) Packaging authorizations (§173.***) | | | (9) Quantity limitations | | (10) Vessel stowage requirements | |
|---|---|---|---|---|---|---|---|---|---|---|---|---|---|
| | | | | | | | Exceptions | Non-bulk packaging | Bulk packaging | Passenger aircraft or railcar | Cargo aircraft only | Vessel stowages | Other stowage provisions |
| (1) | (2) | (3) | (4) | (5) | (6) | (7) | (8A) | (8B) | (8C) | (9A) | (9B) | (10A) | (10B) |
| | 4-Dimethylamino-6-(2-dimethylaminoethoxy) toluene-2-diazonium zinc chloride | 4.1 | UN3039 | II | FLAMMABLE SOLID | | None | 224 | None | Forbidden | Forbidden | D | 2 |
| | 2-Dimethylaminoacetonitrile | 3 | UN2378 | II | FLAMMABLE LIQUID, POISON | T8 | None | 202 | 243 | 1 L | 60 L | A | 26, 40 |
| | Dimethylaminoethyl methacrylate | 6.1 | UN2522 | II | POISON | T8 | None | 202 | 243 | 5 L | 60 L | B | 40 |
| | N,N-Dimethylaniline | 6.1 | UN2253 | II | POISON | T8 | None | 202 | 243 | 5 L | 60 L | A | |
| | 2,3-Dimethylbutane | 3 | UN2457 | II | FLAMMABLE LIQUID | T13 | 150 | 202 | 242 | 5 L | 60 L | E | |
| | 1,3-Dimethylbutylamine | 3 | UN2379 | II | FLAMMABLE LIQUID | T8 | 150 | 202 | 242 | 5 L | 60 L | B | |
| | Dimethylcarbamoyl chloride | 8 | UN2262 | II | CORROSIVE | B2, T8 | 154 | 202 | 242 | 1 L | 30 L | A | 40 |
| | Dimethylcyclohexanes | 3 | UN2263 | II | FLAMMABLE LIQUID | T1 | 150 | 202 | 242 | 5 L | 60 L | B | |
| | Dimethylcyclohexylamine | 8 | UN2264 | II | CORROSIVE | B2, T8 | 154 | 202 | 242 | 1 L | 30 L | A | 12, 21, 40, 48 |
| | Dimethyldichlorosilane | 3 | UN1162 | I | FLAMMABLE LIQUID, CORROSIVE | B77, T15, T26 | None | 201 | 243 | Forbidden | Forbidden | B | 40 |
| | Dimethyldiethoxysilane | 3 | UN2380 | II | FLAMMABLE LIQUID | T8 | 150 | 202 | 242 | 5 L | 60 L | B | |
| | Dimethyldioxanes | 3 | UN2707 | II | FLAMMABLE LIQUID | T8, T31 | 150 | 202 | 242 | 5 L | 60 L | B | |
| | | | | III | FLAMMABLE LIQUID | B1, T7, T30 | 150 | 203 | 242 | 60 L | 220 L | A | |
| | Dimethylethanolamine | 3 | UN2051 | III | FLAMMABLE LIQUID | B1, T1 | 150 | 203 | 242 | 60 L | 220 L | A | |
| | N,N-Dimethylformamide | 3 | UN2265 | III | FLAMMABLE LIQUID | B1, T1 | 150 | 203 | 242 | 60 L | 220 L | A | |
| | *Dimethylhexane dihydroperoxide (dry)* | Forbidden | | | | | | | | | | | |
| + | Dimethylhydrazine, symmetrical | 3 | UN2382 | I | FLAMMABLE LIQUID, POISON | 2, A7, B9, B14, B32, B74, B77, T38, T43, T45 | None | 227 | 244 | Forbidden | 30 L | E | 40 |
| | Dimethylhydrazine, unsymmetrical | 6.1 | UN1163 | I | POISON, FLAMMABLE LIQUID, CORROSIVE | 2, A7, B9, B14, B32, B74, B79, T38, T43, T45 | None | 227 | 244 | Forbidden | Forbidden | D | 21, 40, 49, 61, 100 |
| | 2,2-Dimethylpropane *other than pentane and isopentane* | 2.1 | UN2044 | | FLAMMABLE GAS | | 306 | 304 | 314, 315 | Forbidden | 150 kg | E | 40 |
| | Dimethylzinc | 4.2 | UN1370 | I | SPONTANEOUSLY COMBUSTIBLE | B11, B16, T28, T29, T40 | None | 181 | 244 | Forbidden | Forbidden | D | 18 |
| | Dinitro-o-cresol, *solid* | 6.1 | UN1598 | II | POISON | T14 | None | 212 | 242 | 25 kg | 100 kg | A | |
| | Dinitro-o-cresol, *solution* | 6.1 | UN1598 | II | POISON | T14 | None | 202 | 243 | 5 L | 60 L | A | |
| | *1,3-Dinitro-5,5-dimethyl hydantoin* | Forbidden | | | | | | | | | | | |
| | *Dinitro-7,8-dimethylglycoluril (dry)* | Forbidden | | | | | | | | | | | |
| | *1,3-Dinitro-4,5-dinitrosobenzene* | Forbidden | | | | | | | | | | | |
| | *1,4-Dinitro-1,1,4,4-tetramethylolbutanetetranitrate (dry)* | Forbidden | | | | | | | | | | | |
| | *2,4-Dinitro-1,3,5-trimethylbenzene* | Forbidden | | | | | | | | | | | |
| | Dinitroanilines | 6.1 | UN1596 | II | POISON | T14 | None | 212 | 242 | 25 kg | 100 kg | A | 91 |
| | Dinitrobenzenes, *liquid* | 6.1 | UN1597 | II | POISON | 11, T14 | None | 202 | 243 | 5 L | 60 L | A | 91 |
| | Dinitrobenzenes, *solid* | 6.1 | UN1597 | II | POISON | 11 | None | 212 | 242 | 25 kg | 100 kg | A | 91 |
| | *Dinitrochlorobenzene, see* Chlorodinitrobenzene | | | | | | | | | | | | |
| | *1,2-Dinitroethane* | Forbidden | | | | | | | | | | | |
| | *1,1-Dinitroethane (dry)* | Forbidden | | | | | | | | | | | |
| | Dinitrogen tetroxide, liquefied | 2.3 | UN1067 | | POISON GAS, OXIDIZER | 1, B7, B12, B14, B45, B46, B61, B66, B67, B77 | None | 336 | 314 | Forbidden | Forbidden | D | 40, 43 |
| | Dinitroglycoluril *or* Dingu | 1.1D | UN0489 | II | EXPLOSIVE 1.1D | | None | 62 | None | Forbidden | Forbidden | B | 1E, 5E, 25E |
| | *Dinitromethane* | Forbidden | | | | | | | | | | | |
| | Dinitrophenol, *dry or wetted with less than 15 per cent water, by mass* | 1.1D | UN0076 | II | EXPLOSIVE 1.1D, POISON | | None | 62 | None | Forbidden | Forbidden | B | 1E, 5E |
| | Dinitrophenol solutions | 6.1 | UN1599 | II | POISON | T8 | None | 202 | 243 | 5 L | 60 L | A | 36 |
| | Dinitrophenol, wetted *with not less than 15 per cent water, by mass* | 4.1 | UN1320 | I | FLAMMABLE SOLID, POISON | A8, A19, A20, N41 | None | 211 | None | 1 kg | 15 kg | E | 28, 36 |
| | Dinitrophenolates *alkali metals, dry or wetted with less than 15 per cent water, by mass* | 1.3C | UN0077 | II | EXPLOSIVE 1.3C, POISON | | None | 62 | None | Forbidden | Forbidden | B | 1E, 5E |

| Symbols | Hazardous materials descriptions and proper shipping names | Hazard class or Division | Identification Numbers | Packing group | Label(s) required (if not excepted) | Special provisions | (8) Packaging authorizations (§173.***) | | | (9) Quantity limitations | | (10) Vessel stowage requirements | |
|---|---|---|---|---|---|---|---|---|---|---|---|---|---|
| | | | | | | | Exceptions | Non-bulk packaging | Bulk packaging | Passenger aircraft or railcar | Cargo aircraft only | Vessel stowages | Other stowage provisions |
| (1) | (2) | (3) | (4) | (5) | (6) | (7) | (8A) | (8B) | (8C) | (9A) | (9B) | (10A) | (10B) |
| | Dinitrophenolates, wetted *with not less than 15 per cent water, by mass* | 4.1 | UN1321 | I | FLAMMABLE SOLID, POISON | A8, A19, A20, N41 | None | 211 | None | 1 kg | 15 kg | E | 28, 36 |
| | *Dinitropropylene glycol* | Forbidden | | | | | | | | | | | |
| | Dinitroresorcinol, *dry or wetted with less than 15 per cent water, by mass* | 1.1D | UN0078 | II | EXPLOSIVE 1.1D | | None | 62 | None | Forbidden | Forbidden | B | 1E, 5E |
| | *4,6-Dinitroresorcinol (heavy metal salts of) (dry)* | Forbidden | | | | | | | | | | | |
| | *2,4-Dinitroresorcinol (heavy metal salts of) (dry)* | Forbidden | | | | | | | | | | | |
| | Dinitroresorcinol, wetted *with not less than 15 per cent water, by mass* | 4.1 | UN1322 | I | FLAMMABLE SOLID | A8, A19, A20, N41 | None | 211 | None | 1 kg | 15 kg | E | 28, 36 |
| | *3,5-Dinitrosalicylic acid (lead salt) (dry)* | Forbidden | | | | | | | | | | | |
| | N,N'-Dinitroso-N,N'-dimethyl terephthalamide *not more than 72% as a paste* | 4.1 | UN2973 | II | FLAMMABLE SOLID, EXPLOSIVE | 41, 53 | None | 224 | None | Forbidden | Forbidden | D | 12, 61 |
| | Dinitrosobenzene | 1.3C | UN0406 | II | EXPLOSIVE 1.3C | | None | 62 | None | Forbidden | Forbidden | B | 1E, 5E |
| | *Dinitrosobenzylamidine and salts of (dry)* | Forbidden | | | | | | | | | | | |
| | N,N'-Dinitrosopentamethylenetetramine *not more than 82% with phlegmatizer* | 4.1 | UN2972 | II | FLAMMABLE SOLID, EXPLOSIVE | 41, 53 | None | 224 | None | Forbidden | Forbidden | D | 12, 61 |
| | *2,2-Dinitrostilbene* | Forbidden | | | | | | | | | | | |
| | Dinitrotoluenes, *liquid* | 6.1 | UN2038 | II | POISON | T8 | None | 202 | 243 | 5 L | 60 L | A | |
| | Dinitrotoluenes, molten | 6.1 | UN1600 | II | POISON | T14 | None | 202 | 243 | Forbidden | Forbidden | C | |
| | Dinitrotoluenes, *solid* | 6.1 | UN2038 | II | POISON | T8 | None | 212 | 242 | 25 kg | 100 kg | A | |
| | *1,9-Dinitroxy pentamethylene-2,4, 6,8-tetramine (dry)* | Forbidden | | | | | | | | | | | |
| | Dioxane | 3 | UN1165 | II | FLAMMABLE LIQUID | T8 | 150 | 202 | 242 | 5 L | 60 L | B | |
| | Dioxolane | 3 | UN1166 | II | FLAMMABLE LIQUID | T8 | 150 | 202 | 242 | 5 L | 60 L | B | 40 |
| | Dipentene | 3 | UN2052 | III | FLAMMABLE LIQUID | B1, T1 | 150 | 203 | 242 | 60 L | 220 L | A | |
| | Diphenylamine chloroarsine | 6.1 | UN1698 | I | POISON | | None | 201 | None | Forbidden | Forbidden | D | 40 |
| | Diphenylchloroarsine, solid *or liquid* | 6.1 | UN1699 | I | POISON | A8, B14, B32, N33, N34 | None | 211 | 244 | Forbidden | 15 kg | D | 40 |
| | Diphenyldichlorosilane | 8 | UN1769 | II | CORROSIVE | A7, B2, N34, T8, T26 | None | 202 | 244 | Forbidden | 30 L | C | 40 |
| | Diphenylmethane-4,4'diisocyanate | 6.1 | UN2489 | III | KEEP AWAY FROM FOOD | T8 | 153 | 203 | 241 | 60 L | 220 L | A | 48, 108 |
| | Diphenylmethyl bromide | 8 | UN1770 | II | CORROSIVE | | 154 | 212 | 240 | 15 kg | 50 kg | D | 40 |
| | Diphenyloxide-4,4'disulfohydrazide | 4.1 | UN2951 | II | FLAMMABLE SOLID | | None | 224 | None | 15 kg | 50 kg | B | 12, 61, 85 |
| | Dipicryl sulfide, *dry or wetted with less than 10 per cent water, by mass* | 1.1D | UN0401 | II | EXPLOSIVE 1.1D | | None | 62 | None | Forbidden | Forbidden | B | 1E, 5E, 19E |
| | Dipicryl sulfide, wetted *with not less than 10 per cent water, by mass* | 4.1 | UN2852 | I | FLAMMABLE SOLID | A2, N41 | None | 211 | None | Forbidden | 0.5 kg | D | 28, 36 |
| | Dipicrylamine, *see* Hexanitrodiphenylamine | | | | | | | | | | | | |
| | *Dipropionyl peroxide, more than 28 per cent in solution* | Forbidden | | | | | | | | | | | |
| | Dipropyl ether | 3 | UN2384 | II | FLAMMABLE LIQUID | T1 | 150 | 202 | 242 | 5 L | 60 L | B | |
| | Dipropylamine | 3 | UN2383 | II | FLAMMABLE LIQUID | T8 | 150 | 202 | 242 | 5 L | 60 L | B | |
| | 4-Dipropylaminobenzenediazonium zinc chloride | 4.1 | UN3034 | II | FLAMMABLE SOLID | | None | 224 | None | 15 kg | 50 kg | C | |
| | Dipropylketone | 3 | UN2710 | III | FLAMMABLE LIQUID | B1, T1 | 150 | 203 | 242 | 60 L | 220 L | A | |
| | Disinfectants, corrosive liquid, n.o.s. | 8 | UN1903 | II | CORROSIVE | B2 | 154 | 202 | 242 | 1 L | 30 L | A | |
| | | | | III | CORROSIVE | | 154 | 203 | 241 | 5 L | 60 L | A | 40 |
| | Disinfectants, liquid, n.o.s *poisonous* | 6.1 | UN3142 | I | POISON | A4, T42 | None | 201 | 243 | 1 L | 30 L | A | 40 |
| | | | | II | POISON | T14 | None | 202 | 243 | 5 L | 60 L | A | 40 |
| | | | | III | KEEP AWAY FROM FOOD | T7 | 153 | 203 | 241 | 60 L | 220 L | A | 40 |
| | Disinfectants, solid, n.o.s. *poisonous* | 6.1 | UN1601 | II | POISON | | None | 212 | 242 | 25 kg | 100 kg | A | 40 |
| | | | | III | KEEP AWAY FROM FOOD | | 153 | 213 | 240 | 100 kg | 200 kg | A | 40 |
| | Dispersant gases, n.o.s. *see* Refrigerant gases, n.o.s. | | | | | | | | | | | | |
| | Dithiocarbamate pesticides, liquid, flammable, toxic, n.o.s., *flash point less than 23 degrees C.* | 3 | UN2772 | I | FLAMMABLE LIQUID, POISON | | None | 201 | 243 | Forbidden | 30 L | B | 40 |
| | | | | II | FLAMMABLE LIQUID, POISON | | None | 202 | 243 | 1 L | 60 L | B | 40 |

## §172.101 HAZARDOUS MATERIALS TABLE, Continued

| Symbols | Hazardous materials descriptions and proper shipping names | Hazard class or Division | Identification Numbers | Packing group | Label(s) required (if not excepted) | Special provisions | (8) Packaging authorizations (§173.***) | | | (9) Quantity limitations | | (10) Vessel stowage requirements | |
|---|---|---|---|---|---|---|---|---|---|---|---|---|---|
| | | | | | | | Exceptions | Non-bulk packaging | Bulk packaging | Passenger aircraft or railcar | Cargo aircraft only | Vessel stowages | Other stowage provisions |
| (1) | (2) | (3) | (4) | (5) | (6) | (7) | (8A) | (8B) | (8C) | (9A) | (9B) | (10A) | (10B) |
| | Dithiocarbamate pesticides, liquid, toxic, flammable, n.o.s. *flashpoint not less than 23 degrees C.* | 6.1 | UN3005 | I | POISON, FLAMMABLE LIQUID | T42 | None | 201 | 243 | 1 L | 30 L | B | 40 |
| | | | | II | POISON, FLAMMABLE LIQUID | T14 | None | 202 | 243 | 5 L | 60 L | B | 40 |
| | | | | III | KEEP AWAY FROM FOOD, FLAMMABLE LIQUID | T14 | 153 | 203 | 241 | 60 L | 220 L | A | 40 |
| | Dithiocarbamate pesticides, liquid, toxic, n.o.s. | 6.1 | UN3006 | I | POISON | T42 | None | 201 | 243 | 1 L | 30 L | B | 40 |
| | | | | II | POISON | T14 | None | 202 | 243 | 5 L | 60 L | B | 40 |
| | | | | III | KEEP AWAY FROM FOOD | T14 | 153 | 203 | 241 | 60 L | 220 L | A | 40 |
| | Dithiocarbamate pesticides, solid, toxic, n.o.s. | 6.1 | UN2771 | I | POISON | | None | 211 | 242 | 5 kg | 50 kg | A | 40 |
| | | | | II | POISON | | None | 212 | 242 | 25 kg | 100 kg | A | 40 |
| | | | | III | KEEP AWAY FROM FOOD | | 153 | 213 | 240 | 100 kg | 200 kg | A | 40 |
| | Divinyl ether, inhibited | 3 | UN1167 | I | FLAMMABLE LIQUID | T14 | None | 202 | 241 | 5 L | 60 L | E | 40 |
| D | Dodecylbenzenesulfonic acid | 8 | NA2584 | II | CORROSIVE | B2 | 154 | 202 | 242 | 1 L | 30 L | B | 9 |
| | Dodecyltrichlorosilane | 8 | UN1771 | II | CORROSIVE | A7, B2, B6, N34, T8, T26 | None | 202 | 242 | Forbidden | 30 L | C | 40 |
| | Dry ice, *see* Carbon dioxide, solid | | | | | | | | | | | | |
| | Dyes, liquid, n.o.s. *or* dye intermediates, liquid, n.o.s. *corrosive* | 8 | UN2801 | II | CORROSIVE | 11, B2, T14 | 154 | 202 | 242 | 1 L | 30 L | A | 23 |
| | | | | III | CORROSIVE | 11, T7 | 154 | 203 | 241 | 5 L | 60 L | A | 23 |
| | Dyes, liquid, n.o.s. *or* dye intermediates, liquid, n.o.s. *poisonous* | 6.1 | UN1602 | II | POISON | | None | 202 | 243 | 5 L | 60 L | A | |
| | | | | III | KEEP AWAY FROM FOOD | | 153 | 203 | 241 | 60 L | 220 | A | |
| | Dyes, solid, n.o.s. *or* Dye intermediates, solid, n.o.s. *corrosive* | 8 | UN3147 | II | CORROSIVE | | 154 | 212 | 240 | 15 kg | 50 kg | A | |
| | | | | III | CORROSIVE | | 154 | 213 | 240 | 25 kg | 100 kg | A | |
| | Dyes, solid, n.o.s. *or* dye intermediates, solid, n.o.s. *poisonous* | 6.1 | UN3143 | I | POISON | A5 | None | 211 | 242 | 5kg | 50kg | A | |
| | | | | II | POISON | | None | 212 | 242 | 25 kg | 100 kg | A | |
| | | | | III | KEEP AWAY FROM FOOD | | 153 | 213 | 240 | 100 kg | 200 kg | A | |
| | *Dynamite, see* Explosive, blasting, type A | | | | | | | | | | | | |
| | *Electrolyte (acid or alkali) for batteries, see* Battery fluid, acid *or* Battery fluid, alkali | | | | | | | | | | | | |
| D | Elevated temperature material, liquid, n.o.s. *(at or above 100°C (212°F) and below its flash point)* | 9 | NA9259 | III | Class 9 | | None | None | 247 | Forbidden | Forbidden | D | M4 |
| D | *Elevated temperature material at or above its flash point, see* Flammable liquid, elevated temperature material, n.o.s. *(see §173.247(d)(2))* | | | | | | | | | | | | |
| D | *Elevated temperature material, solid, n.o.s. (see §173.247(h)(4))* | | | | | | | | | | | | |
| | Engine starting fluid, *with flammable gas* | 2.1 | UN1960 | | FLAMMABLE GAS | | None | 304 | None | Forbidden | 150 kg | D | 40 |
| | Engines, internal combustion, *see* Vehicle, self-propelled | | | | | | | | | | | | |
| | Environmentally hazardous substances, liquid, n.o.s. | 9 | UN3082 | III | CLASS 9 | 8, T1, N50 | 155 | 203 | 241 | None | None | A | |
| | Environmentally hazardous substances, solid, n.o.s. | 9 | UN3077 | III | CLASS 9 | 8, B54, N50 | 155 | 213 | 240 | None | None | A | |
| | Epibromohydrin | 6.1 | UN2558 | I | POISON | T18, T26 | None | 201 | 243 | Forbidden | Forbidden | D | 40 |
| | Epichlorohydrin | 6.1 | UN2023 | II | POISON | T14 | None | 202 | 243 | 5 L | 60 L | A | 40 |
| | 1,2-Epoxy-3-ethoxypropane | 3 | UN2752 | III | FLAMMABLE LIQUID | B1, T1 | 150 | 203 | 242 | 60 L | 220 L | A | |
| | *Etching acid, liquid, n.o.s., see* Hydrofluoric acid, solution *etc.* | | | | | | | | | | | | |
| | Ethane, compressed | 2.1 | UN1035 | | FLAMMABLE GAS | | 306 | 304 | 302 | Forbidden | 150 kg | E | 40 |
| D | Ethane-Propane mixture, refrigerated liquid | 2.1 | NA1961 | | FLAMMABLE GAS | | None | 316 | 314, 315 | Forbidden | Forbidden | D | 40 |
| | Ethane, refrigerated liquid | 2.1 | UN1961 | | FLAMMABLE GAS | | None | None | 315 | Forbidden | Forbidden | D | 40 |
| | *Ethanol amine dinitrate* | Forbidden | | | | | | | | | | | |
| | Ethanol *or* Ethyl alcohol *or* Ethanol solutions *or* Ethyl alcohol solutions | 3 | UN1170 | II | FLAMMABLE LIQUID | T1 | 150 | 202 | 242 | 5 L | 60 L | A | |
| | | | | III | FLAMMABLE LIQUID | B1, T1 | 150 | 203 | 242 | 60 L | 220 L | A | |
| | Ethanolamine *or* Ethanolamine solutions | 8 | UN2491 | III | CORROSIVE | T7 | 154 | 203 | 241 | 5 L | 60 L | A | |
| | *Ether, see* Diethyl ether | | | | | | | | | | | | |
| | Ethyl acetate | 3 | UN1173 | II | FLAMMABLE LIQUID | T2 | 150 | 202 | 242 | 5 L | 60 L | B | |

## §172.101 HAZARDOUS MATERIALS TABLE, Continued

| Symbols | Hazardous materials descriptions and proper shipping names | Hazard class or Division | Identification Numbers | Packing group | Label(s) required (if not excepted) | Special provisions | (8) Packaging authorizations (§173.***) | | | (9) Quantity limitations | | (10) Vessel stowage requirements | |
|---|---|---|---|---|---|---|---|---|---|---|---|---|---|
| | | | | | | | Exceptions | Non-bulk packaging | Bulk packaging | Passenger aircraft or railcar | Cargo aircraft only | Vessel stowages | Other stowage provisions |
| (1) | (2) | (3) | (4) | (5) | (6) | (7) | (8A) | (8B) | (8C) | (9A) | (9B) | (10A) | (10B) |
| | Ethyl acrylate, inhibited | 3 | UN1917 | II | FLAMMABLE LIQUID | T8 | 150 | 202 | 242 | 5 L | 60 L | B | 40 |
| | Ethyl alcohol, *see* Ethanol | | | | | | | | | | | | |
| | *Ethyl aldehyde, see* Acetaldehyde | | | | | | | | | | | | |
| | Ethyl amyl ketone | 3 | UN2271 | III | FLAMMABLE LIQUID | B1, T1 | 150 | 203 | 242 | 60 L | 220 L | A | |
| | N-Ethyl-N-benzylaniline | 6.1 | UN2274 | III | KEEP AWAY FROM FOOD | T2 | 153 | 203 | 241 | 60 L | 220 L | A | |
| | Ethyl borate | 3 | UN1176 | II | FLAMMABLE LIQUID | T8 | 150 | 202 | 242 | 5 L | 60 L | B | |
| | Ethyl bromide | 6.1 | UN1891 | II | POISON | T17 | None | 202 | 243 | 5 L | 60 L | B | 40 |
| | Ethyl bromoacetate | 6.1 | UN1603 | II | POISON | T14 | None | 202 | 243 | Forbidden | Forbidden | D | 40 |
| | Ethyl butyl ether | 3 | UN1179 | II | FLAMMABLE LIQUID | B1, T1 | 150 | 202 | 242 | 5 L | 60 L | B | |
| | Ethyl butyrate | 3 | UN1180 | III | FLAMMABLE LIQUID | B1, T1 | 150 | 202 | 242 | 5 L | 60 L | A | |
| | Ethyl chloride | 2.1 | UN1037 | | FLAMMABLE GAS | B63, B77 | None | 322 | 314, 315 | Forbidden | 150 kg | B | 40 |
| | Ethyl chloroacetate | 6.1 | UN1181 | II | POISON | T14 | None | 202 | 243 | 5 L | 60 L | A | |
| | Ethyl chloroformate | 6.1 | UN1182 | I | POISON, FLAMMABLE LIQUID, CORROSIVE | 2, A3, A6, A7, B9, B14, B32, B74, N34, T38, T43, T45 | None | 227 | 244 | Forbidden | Forbidden | D | 21, 40, 100 |
| | Ethyl-2-chloropropionate | 3 | UN2935 | III | FLAMMABLE LIQUID | B1, T1 | 150 | 203 | 242 | 60 L | 220 L | A | |
| + | Ethyl chlorothioformate | 8 | UN2826 | II | CORROSIVE, POISON | 2, B9, B14, B32, B74, T38, T43, T45 | None | 227 | 244 | 1 L | 30 L | A | 12, 21, 40, 48 |
| | Ethyl crotonate | 3 | UN1862 | II | FLAMMABLE LIQUID | T1 | 150 | 202 | 242 | 5 L | 60 L | B | |
| | Ethyl cyanoacetate | 6.1 | UN2666 | III | KEEP AWAY FROM FOOD | T8 | 153 | 203 | 241 | 60 L | 220 L | A | 26 |
| | Ethyl ether, *see* Diethyl ether | | | | | | | | | | | | |
| | Ethyl fluoride | 2.1 | UN2453 | | FLAMMABLE GAS | | 306 | 304 | 314, 315 | 75 kg | 150 kg | E | 40 |
| | Ethyl formate | 3 | UN1190 | II | FLAMMABLE LIQUID | T8 | 150 | 202 | 242 | 5 L | 60 L | E | |
| | *Ethyl hydroperoxide* | Forbidden | | | | | | | | | | | |
| | Ethyl isobutyrate | 3 | UN2385 | II | FLAMMABLE LIQUID | T1 | 150 | 202 | 242 | 5 L | 60 L | B | |
| + | Ethyl isocyanate | 3 | UN2481 | I | FLAMMABLE LIQUID, POISON | 1, A7, B9, B14, B30, B72, T38, T43, T44 | None | 226 | 244 | Forbidden | 30 L | D | 40 |
| | Ethyl lactate | 3 | UN1192 | III | FLAMMABLE LIQUID | B1, T1 | 150 | 203 | 242 | 60 L | 220 L | A | |
| | Ethyl mercaptan | 3 | UN2363 | I | FLAMMABLE LIQUID | T21 | None | 201 | 243 | Forbidden | 30 L | E | 95, 102 |
| | Ethyl methacrylate | 3 | UN2277 | II | FLAMMABLE LIQUID | T1 | 150 | 202 | 242 | 5 L | 60 L | B | |
| | Ethyl methyl ether | 2.1 | UN1039 | | FLAMMABLE GAS | B63 | None | 324 | 314, 315 | Forbidden | 150 kg | B | 40 |
| | Ethyl methyl ketone *or* Methyl ethyl ketone | 3 | UN1193 | II | FLAMMABLE LIQUID | T8 | 150 | 202 | 242 | 5 L | 60 L | B | |
| | Ethyl nitrite solutions | 3 | UN1194 | I | FLAMMABLE LIQUID, POISON | | None | 201 | None | Forbidden | Forbidden | E | 40, 105 |
| | Ethyl orthoformate | 3 | UN2524 | III | FLAMMABLE LIQUID | B1, T7 | 150 | 203 | 242 | 60 L | 220 L | A | |
| | Ethyl oxalate | 6.1 | UN2525 | III | KEEP AWAY FROM FOOD | T1 | 153 | 203 | 241 | 60 L | 220 L | A | |
| | *Ethyl perchlorate* | Forbidden | | | | | | | | | | | |
| D | Ethyl phosphonothioic dichloride, anhydrous | 6.1 | NA2927 | I | POISON, CORROSIVE | 2, B9, B14, B32, B74, T38, T43, T45 | None | 227 | 244 | Forbidden | Forbidden | D | 20, 40, 95 |
| D | Ethyl phosphonous dichloride, anhydrous *pyrophoric liquid* | 6.1 | NA2845 | I | POISON, SPONTANEOUSLY COMBUSTIBLE | 2, B9, B14, B32, B74, T38, T43, T45 | None | 227 | 244 | Forbidden | Forbidden | D | 18 |
| D | Ethyl phosphorodichloridate | 6.1 | NA2927 | I | POISON, CORROSIVE | 2, B9, B14, B32, B74, T38, T43, T45 | None | 227 | 244 | Forbidden | Forbidden | D | 20, 40, 95 |
| | Ethyl propionate | 3 | UN1195 | II | FLAMMABLE LIQUID | T1 | 150 | 202 | 242 | 5 L | 60 L | B | |
| | Ethyl propyl ether | 3 | UN2615 | II | FLAMMABLE LIQUID | T8 | 150 | 202 | 242 | 5 L | 60 L | E | |
| | *Ethyl silicate, see* Tetraethyl silicate | | | | | | | | | | | | |
| | Ethylacetylene, inhibited | 2.1 | UN2452 | | FLAMMABLE GAS | | None | 304 | 314, 315 | Forbidden | 150 kg | B | 40 |
| | Ethylamine | 2.1 | UN1036 | | FLAMMABLE GAS | B77 | None | 321 | 314, 315 | Forbidden | Forbidden | D | 40 |
| | Ethylamine, aqueous solution *with not less than 50 per cent but not more than 70 per cent ethylamine* | 3 | UN2270 | II | FLAMMABLE LIQUID | T14 | 150 | 202 | 242 | 5 L | 60 L | B | 40 |
| | N-Ethylaniline | 6.1 | UN2272 | III | KEEP AWAY FROM FOOD | T2 | 153 | 203 | 241 | 60 L | 220 L | A | |

## §172.101 HAZARDOUS MATERIALS TABLE, Continued

| Symbols | Hazardous materials descriptions and proper shipping names | Hazard class or Division | Identification Numbers | Packing group | Label(s) required (if not excepted) | Special provisions | (8) Packaging authorizations (§173.***) | | | (9) Quantity limitations | | (10) Vessel stowage requirements | |
|---|---|---|---|---|---|---|---|---|---|---|---|---|---|
| | | | | | | | Exceptions | Non-bulk packaging | Bulk packaging | Passenger aircraft or railcar | Cargo aircraft only | Vessel stowages | Other stowage provisions |
| (1) | (2) | (3) | (4) | (5) | (6) | (7) | (8A) | (8B) | (8C) | (9A) | (9B) | (10A) | (10B) |
| | 2-Ethylaniline | 6.1 | UN2273 | III | KEEP AWAY FROM FOOD | T2 | 153 | 203 | 241 | 60 L | 220 L | A | |
| | Ethylbenzene | 3 | UN1175 | II | FLAMMABLE LIQUID | T1 | 150 | 202 | 242 | 5 L | 60 L | B | |
| | N-Ethylbenzyltoluidines solid *or* liquid | 6.1 | UN2753 | III | KEEP AWAY FROM FOOD | T14 | 153 | 203 | 241 | 60 L | 220 L | A | |
| | 2-Ethylbutanol | 3 | UN2275 | III | FLAMMABLE LIQUID | B1, T1 | 150 | 203 | 242 | 60 L | 220 L | A | |
| | 2-Ethylbutyl acetate | 3 | UN1177 | III | FLAMMABLE LIQUID | B1, T1 | 150 | 203 | 242 | 60 L | 220 L | A | |
| | 2-Ethylbutyraldehyde | 3 | UN1178 | II | FLAMMABLE LIQUID | B1, T1 | 150 | 202 | 242 | 5 L | 60 L | B | |
| | Ethyldichloroarsine | 6.1 | UN1892 | I | POISON | 2, B9, B14, B32, B74, T38, T43, T45 | None | 227 | 244 | Forbidden | Forbidden | D | 40 |
| | Ethyldichlorosilane | 4.3 | UN1183 | I | DANGEROUS WHEN WET, CORROSIVE, FLAMMABLE LIQUID | A2, A3, A7, N34, T18, T26 | None | 201 | 244 | Forbidden | 1 L | D | 21, 28, 40, 49, 100 |
| | Ethylene, acetylene and propylene in mixtures, refrigerated liquid containing at least 71.5 per cent ethylene with not more than 22.5 per cent acetylene and not more than 6 per cent propylene | 2.1 | UN3138 | | FLAMMABLE GAS | | None | 304 | 314, 315 | Forbidden | Forbidden | D | 40 |
| + | Ethylene chlorohydrin | 6.1 | UN1135 | II | POISON | 2, B9, B14, B32, B74, T38, T43, T45 | None | 227 | 244 | Forbidden | 60 L | A | 40 |
| | Ethylene, compressed | 2.1 | UN1962 | | FLAMMABLE GAS | | 306 | 304 | 302 | Forbidden | 150 kg | E | 40 |
| | *Ethylene diamine diperchlorate* | Forbidden | | | | | | | | | | | |
| + | Ethylene dibromide | 6.1 | UN1605 | II | POISON | 2, B9, B14, B32, B74, B77, T38, T43, T45 | None | 227 | 244 | 5 L | 60 L | A | 40 |
| | *Ethylene dibromide and methyl bromide liquid mixtures, see* Methyl bromide and ethylene dibromide, liquid mixtures | | | | | | | | | | | | |
| | Ethylene dichloride | 3 | UN1184 | II | FLAMMABLE LIQUID, POISON | T14 | None | 202 | 243 | 1 L | 60 L | B | 40 |
| | Ethylene glycol diethyl ether | 3 | UN1153 | III | FLAMMABLE LIQUID | B1, T1 | 150 | 203 | 242 | 60 L | 220 L | A | |
| | *Ethylene glycol dinitrate* | Forbidden | | | | | | | | | | | |
| | Ethylene glycol monobutyl ether | 6.1 | UN2369 | III | KEEP AWAY FROM FOOD | T1 | 153 | 203 | 241 | 60 L | 220 L | A | |
| | Ethylene glycol monoethyl ether | 3 | UN1171 | III | FLAMMABLE LIQUID | B1, T1 | 150 | 203 | 242 | 60 L | 220 L | A | |
| | Ethylene glycol monoethyl ether acetate | 3 | UN1172 | III | FLAMMABLE LIQUID | B1, T1 | 150 | 203 | 242 | 60 L | 220 L | A | |
| | Ethylene glycol monomethyl ether | 3 | UN1188 | III | FLAMMABLE LIQUID | B1, T1 | 150 | 203 | 242 | 60 L | 220 L | A | |
| | Ethylene glycol monomethyl ether acetate | 3 | UN1189 | III | FLAMMABLE LIQUID | B1, T1 | 150 | 203 | 242 | 60 L | 220 L | A | |
| | *Ethylene oxide and carbon dioxide mixtures, see* Carbon dioxide and ethylene oxide mixtures, *etc.* | | | | | | | | | | | | |
| | Ethylene oxide and propylene oxide mixtures, *not more than 30 per cent ethylene oxide* | 3 | UN2983 | I | FLAMMABLE LIQUID, POISON | 5, A11, N4, N34, T24, T29 | None | 201 | 243 | Forbidden | 30 L | E | 40 |
| | Ethylene oxide, *pure or with nitrogen* | 2.3 | UN1040 | | POISON GAS, FLAMMABLE GAS | 3 | None | 323 | 323 | Forbidden | 25 kg | D | 40 |
| | Ethylene, refrigerated liquid *(cryogenic liquid)* | 2.1 | UN1038 | | FLAMMABLE GAS | | None | 316 | 318, 319 | Forbidden | Forbidden | D | 40 |
| | Ethylenediamine | 8 | UN1604 | II | CORROSIVE, FLAMMABLE LIQUID | T14 | 154 | 202 | 243 | 1 L | 30 L | A | 12, 21, 40, 48 |
| | Ethyleneimine, inhibited | 6.1 | UN1185 | I | POISON, FLAMMABLE LIQUID | 1, B9, B14, B30, B72, B77, N25, N32, T38, T43, T44 | None | 226 | 244 | Forbidden | Forbidden | D | 40 |
| | *Ethylhexaldehyde, see* Octyl aldehydes *etc* | | | | | | | | | | | | |
| | 2-Ethylhexylamine | 8 | UN2276 | III | CORROSIVE | T2 | 154 | 203 | 241 | 5 L | 60 L | A | 12, 21, 40, 48 |
| | 2-Ethylhexylchloroformate | 6.1 | UN2748 | II | POISON, CORROSIVE | T42 | None | 202 | 243 | 1 L | 30 L | A | 12, 13, 25, 40, 48, 95 |
| | Ethylphenyldichlorosilane | 8 | UN2435 | II | CORROSIVE | A7, B2, N34, T8, T26 | None | 202 | 242 | Forbidden | 30 L | C | |
| | 1-Ethylpiperidine | 3 | UN2386 | II | FLAMMABLE LIQUID | T8 | 150 | 202 | 242 | 5 L | 60 L | B | |
| | Ethylsulfuric acid | 8 | UN2571 | II | CORROSIVE | B2, T9, T27 | 154 | 202 | 242 | 1 L | 30 L | C | 14 |
| | N-Ethyltoluidines | 6.1 | UN2754 | II | POISON | T14 | None | 202 | 243 | 5 L | 60 L | A | 48 |
| | Ethyltrichlorosilane | 3 | UN1196 | II | FLAMMABLE LIQUID, CORROSIVE | A7, N34, T15, T26 | None | 201 | 243 | Forbidden | 2.5 L | B | 40 |

## §172.101 HAZARDOUS MATERIALS TABLE, Continued

| Symbols | Hazardous materials descriptions and proper shipping names | Hazard class or Division | Identification Numbers | Packing group | Label(s) required (if not excepted) | Special provisions | (8) Packaging authorizations (§173.***) | | | (9) Quantity limitations | | (10) Vessel stowage requirements | |
|---|---|---|---|---|---|---|---|---|---|---|---|---|---|
| | | | | | | | Exceptions | Non-bulk packaging | Bulk packaging | Passenger aircraft or railcar | Cargo aircraft only | Vessel stowages | Other stowage provisions |
| (1) | (2) | (3) | (4) | (5) | (6) | (7) | (8A) | (8B) | (8C) | (9A) | (9B) | (10A) | (10B) |
| | *Etiologic agent, see* Infectious substances, *etc.* | | | | | | | | | | | | |
| | *Explosive articles, see* Articles, explosive, n.o.s. *etc.* | | | | | | | | | | | | |
| | Explosive, blasting, type A | 1.1D | UN0081 | II | EXPLOSIVE 1.1D | | None | 62 | None | Forbidden | Forbidden | B | 1E, 5E, 19E, 21E |
| | Explosive, blasting, type B | 1.1D | UN0082 | II | EXPLOSIVE 1.1D | | None | 62 | None | Forbidden | Forbidden | B | 1E, 5E, 19E |
| | Explosive, blasting, type B *or* Agent blasting, Type B | 1.5D | UN0331 | II | EXPLOSIVE 1.5D | 105, 106 | None | 62 | None | Forbidden | Forbidden | B | 1E, 5E, 19E |
| | Explosive, blasting, type C | 1.1D | UN0083 | II | EXPLOSIVE 1.1D | | None | 62 | None | Forbidden | Forbidden | B | 1E, 5E, 22E |
| | Explosive, blasting, type D | 1.1D | UN0084 | II | EXPLOSIVE 1.1D | | None | 62 | None | Forbidden | Forbidden | B | 1E, 5E |
| | Explosive, blasting, type E | 1.1D | UN0241 | II | EXPLOSIVE 1.1D | | None | 62 | None | Forbidden | Forbidden | B | 1E, 5E, 19E |
| | Explosive, blasting, type E *or* Agent blasting, Type E | 1.5D | UN0332 | II | EXPLOSIVE 1.5D | 105, 106 | None | 62 | None | Forbidden | Forbidden | B | 1E, 5E |
| | *Explosive, forbidden. See Sec. 173.51* | Forbidden | | | | | | | | | | | |
| D | Explosive pest control devices | 1.1E | NA0006 | II | EXPLOSIVE 1.1E | | None | 62 | None | Forbidden | Forbidden | E | |
| D | Explosive pest control devices | 1.4E | NA0412 | II | EXPLOSIVE 1.4E | | None | 62 | None | Forbidden | 75 kg | A | 24E |
| | *Explosive substances, see* Substances, explosive, n.o.s. *etc* | | | | | | | | | | | | |
| | *Explosives, slurry, see* Explosive, blasting, type E | | | | | | | | | | | | |
| | *Explosives, water gels, see* Explosive, blasting, type E | | | | | | | | | | | | |
| | Extracts, aromatic, liquid | 3 | UN1169 | II | FLAMMABLE LIQUID | T7, T30 | 150 | 202 | 242 | 5 L | 60 L | B | |
| | | | | III | FLAMMABLE LIQUID | B1, T7, T30 | 150 | 203 | 242 | 60 L | 220 L | A | |
| | Extracts, flavoring, liquid | 3 | UN1197 | II | FLAMMABLE LIQUID | T7, T30 | 150 | 202 | 242 | 5 L | 60 L | B | |
| | | | | III | FLAMMABLE LIQUID | B1, T7, T30 | 150 | 203 | 242 | 60 L | 220 L | A | |
| | *Fabric with animal or vegetable oil, see* Fibers or fabrics, *etc.* | | | | | | | | | | | | |
| | Ferric arsenate | 6.1 | UN1606 | II | POISON | | None | 212 | 242 | 25 kg | 100 kg | A | |
| | Ferric arsenite | 6.1 | UN1607 | II | POISON | | None | 212 | 242 | 25 kg | 100 kg | A | |
| | Ferric chloride | 8 | UN1773 | III | CORROSIVE | | 154 | 213 | 240 | 25 kg | 100 kg | A | |
| | Ferric chloride, solution | 8 | UN2582 | III | CORROSIVE | B15, T8 | 154 | 203 | 241 | 5 L | 60 L | A | |
| | Ferric nitrate | 5.1 | UN1466 | III | OXIDIZER | A1, A29 | 152 | 213 | 240 | 25 kg | 100 kg | A | |
| | Ferrocerium | 4.1 | UN1323 | II | FLAMMABLE SOLID | A19 | 151 | 212 | 240 | 15 kg | 50 kg | A | |
| | Ferrosilicon, *with 30 percent or more but less than 90 percent silicon* | 4.3 | UN1408 | III | DANGEROUS WHEN WET | A1, A19 | None | 213 | 240 | 25 kg | 100 kg | A | 13, 40, 85, 103 |
| | Ferrous arsenate | 6.1 | UN1608 | II | POISON | | None | 212 | 242 | 25 kg | 100 kg | A | |
| D | Ferrous chloride, solid | 8 | NA1759 | II | CORROSIVE | | 154 | 212 | 240 | 15 kg | 50 kg | A | |
| D | Ferrous chloride, solution | 8 | NA1760 | II | CORROSIVE | B3 | 154 | 202 | 242 | 1 L | 30 L | B | 40 |
| | Ferrous metal borings, shavings, turnings *or* cuttings *in a form liable to self-heating* | 4.2 | UN2793 | III | SPONTANEOUSLY COMBUSTIBLE | A1, A19 | None | 213 | 241 | 25 kg | 100 kg | A | |
| | Fertilizer ammoniating solution *with free ammonia* | 2.2 | UN1043 | | NONFLAMMABLE GAS | | 306 | 304 | 314, 315 | Forbidden | 150 kg | E | 40 |
| A/W | Fibers *or* Fabrics, animal *or* vegetable, n.o.s *with animal or vegetable oil* | 4.2 | UN1373 | III | SPONTANEOUSLY COMBUSTIBLE | | None | 213 | 241 | Forbidden | Forbidden | A | |
| | *Films, nitrocellulose base, from which gelatine has been removed, film scrap, see* Celluloid scrap | | | | | | | | | | | | |
| | Films, nitrocellulose base, *gelatine coated (except scrap)* | 4.1 | UN1324 | III | FLAMMABLE SOLID | | None | 183 | None | 25 kg | 100 kg | D | 91 |
| | Fire extinguisher charges, *corrosive liquid* | 8 | UN1774 | II | CORROSIVE | N41 | 154 | 202 | None | 1 L | 30 L | A | |
| | *Fire extinguisher charges, expelling, explosive, see* Cartridges, power device | | | | | | | | | | | | |
| | Fire extinguishers *containing compressed or liquefied gas* | 2.2 | UN1044 | | NONFLAMMABLE GAS | | 306 | 306 | None | 75 kg | 150 kg | A | |
| | Firelighters, solid *with flammable liquid* | 4.1 | UN2623 | II | FLAMMABLE SOLID | A19 | None | 212 | None | 15 kg | 50 kg | A | 119 |
| | | | | III | FLAMMABLE SOLID | A1, A19 | None | 213 | None | 25 kg | 100 kg | A | 119 |
| | Fireworks | 1.1G | UN0333 | II | EXPLOSIVE 1.1G | 108 | None | 62 | None | Forbidden | Forbidden | B | |
| | Fireworks | 1.2G | UN0334 | II | EXPLOSIVE 1.2G | 108 | None | 62 | None | Forbidden | Forbidden | B | |
| | Fireworks | 1.3G | UN0335 | II | EXPLOSIVE 1.3G | 108 | None | 62 | None | Forbidden | Forbidden | B | |
| | Fireworks | 1.4G | UN0336 | II | EXPLOSIVE 1.4G | 108 | None | 62 | None | Forbidden | 75 kg | A | 24E |
| | Fireworks | 1.4S | UN0337 | II | EXPLOSIVE 1.4S | 108 | None | 62 | None | 25 kg | 100 kg | A | 9E |
| W | Fish meal, stabilized *or* Fish scrap, stabilized | 9 | UN2216 | III | None | | 155 | 218 | 218 | No limit | No limit | A | 88, 120 |
| | Fish meal, unstabilized *or* Fish scrap, unstabilized | 4.2 | UN1374 | II | SPONTANEOUSLY COMBUSTIBLE | A1, A19 | None | 212 | 241 | 15 kg | 50 kg | A | 119, 120 |

## §172.101 HAZARDOUS MATERIALS TABLE, Continued

| Symbols | Hazardous materials descriptions and proper shipping names | Hazard class or Division | Identification Numbers | Packing group | Label(s) required (if not excepted) | Special provisions | (8) Packaging authorizations (§173.***) | | | (9) Quantity limitations | | (10) Vessel stowage requirements | |
|---|---|---|---|---|---|---|---|---|---|---|---|---|---|
| | | | | | | | Exceptions | Non-bulk packaging | Bulk packaging | Passenger aircraft or railcar | Cargo aircraft only | Vessel stowages | Other stowage provisions |
| (1) | (2) | (3) | (4) | (5) | (6) | (7) | (8A) | (8B) | (8C) | (9A) | (9B) | (10A) | (10B) |
| | *Fissile radioactive materials, see* Radioactive material, fissile, n.o.s | | | | | | | | | | | | |
| | *Flammable compressed gas, see* Compressed or Liquefied gas, flammable, *etc* | | | | | | | | | | | | |
| | *Flammable compressed gas (small receptacles not fitted with a dispersion device, not refillable), see* Receptacles, *etc* | | | | | | | | | | | | |
| | Flammable gas in lighters, *see* Lighters for cigars *or* cigarettes, *with flammable gas* | | | | | | | | | | | | |
| | Flammable liquid, elevated temperature material, n.o.s. | 3 | NA9276 | III | FLAMMABLE LIQUID | | None | None | 247 | Forbidden | Forbidden | D | M4 |
| | Flammable liquids, corrosive, n.o.s. | 3 | UN2924 | I | FLAMMABLE LIQUID, CORROSIVE | T42 | None | 201 | 243 | 0.5 L | 2.5 L | E | 40 |
| | | | | II | FLAMMABLE LIQUID, CORROSIVE | T15, T26 | None | 202 | 243 | 1 L | 5 L | B | 40 |
| | | | | III | FLAMMABLE LIQUID, CORROSIVE | B1, T15, T26 | 150 | 203 | 242 | 5 L | 60 L | A | 40 |
| | Flammable liquids, n.o.s. | 3 | UN1993 | I | FLAMMABLE LIQUID | T42 | 150 | 201 | 243 | 1 L | 30 L | E | |
| | | | | II | FLAMMABLE LIQUID | T8, T31 | 150 | 202 | 242 | 5 L | 60 L | B | |
| | | | | III | FLAMMABLE LIQUID | B1, B52, T7, T30 | 150 | 203 | 242 | 60 L | 220 L | A | |
| | Flammable liquids, poisonous, n.o.s. | 3 | UN1992 | I | FLAMMABLE LIQUID, POISON | T42 | None | 201 | 243 | Forbidden | 30 L | E | 40 |
| | | | | II | FLAMMABLE LIQUID, POISON | T18 | None | 202 | 243 | 1 L | 60 L | B | 40 |
| | Flammable solids, corrosive, n.o.s | 4.1 | UN2925 | II | FLAMMABLE SOLID, CORROSIVE | | None | 212 | 242 | 15 kg | 50 kg | D | 40 |
| | | | | III | FLAMMABLE SOLID, CORROSIVE | A1 | 151 | 213 | 242 | 25 kg | 100 kg | D | 40 |
| | Flammable solids, n.o.s | 4.1 | UN1325 | II | FLAMMABLE SOLID | | 151 | 212 | 240 | 15 kg | 50 kg | B | |
| | | | | III | FLAMMABLE SOLID | | 151 | 213 | 240 | 25 kg | 100 kg | B | |
| | Flammable solids, poisonous, n.o.s | 4.1 | UN2926 | II | FLAMMABLE SOLID, POISON | | None | 212 | 242 | 15 kg | 50 kg | B | 40 |
| | | | | III | FLAMMABLE SOLID, KEEP AWAY FROM FOOD | A1 | 151 | 213 | 242 | 25 kg | 100 kg | B | 40 |
| | Flares, aerial | 1.3G | UN0093 | II | EXPLOSIVE 1.3G | | None | 62 | None | Forbidden | 75 kg | B | |
| | Flares, aerial | 1.4G | UN0403 | II | EXPLOSIVE 1.4G | | None | 62 | None | Forbidden | 75 kg | A | 24E |
| | Flares, aerial | 1.4S | UN0404 | II | EXPLOSIVE 1.4S | | None | 62 | None | 25 kg | 100 kg | A | 9E |
| | Flares, aerial | 1.1G | UN0420 | II | EXPLOSIVE 1.1G | | None | 62 | None | Forbidden | Forbidden | B | |
| | Flares, aerial | 1.2G | UN0421 | II | EXPLOSIVE 1.2G | | None | 62 | None | Forbidden | Forbidden | B | |
| | *Flares, airplane, see* Flares, aerial | | | | | | | | | | | | |
| | *Flares, signal, see* Cartridges, signal | | | | | | | | | | | | |
| | Flares surface | 1.1G | UN0418 | II | EXPLOSIVE 1.1G | | None | 62 | None | Forbidden | Forbidden | B | |
| | Flares, surface | 1.2G | UN0419 | II | EXPLOSIVE 1.2G | | None | 62 | None | Forbidden | Forbidden | B | |
| | Flares, surface | 1.3G | UN0092 | II | EXPLOSIVE 1.3G | | None | 62 | None | Forbidden | 75 kg | B | |
| | *Flares, water-activated, see* Contrivances, water-activated, *etc* | | | | | | | | | | | | |
| | Flash powder | 1.1G | UN0094 | II | EXPLOSIVE 1.1G | | None | 62 | None | Forbidden | Forbidden | B | 1E, 5E |
| | Flash powder | 1.3G | UN0305 | II | EXPLOSIVE 1.3G | | None | 62 | None | Forbidden | Forbidden | B | 1E, 5E |
| | *Flue dusts, poisonous, see* Arsenical dust | | | | | | | | | | | | |
| | Fluoboric acid | 8 | UN1775 | II | CORROSIVE | A6, A7, B2, B15, N3, N34, T15, T27 | 154 | 202 | 242 | 1 L | 30 L | A | |
| | *Fluoric acid, see* Hydrofluoric acid, solution, *etc* | | | | | | | | | | | | |
| | Fluorine, compressed | 2.3 | UN1045 | | POISON GAS, OXIDIZER | 1 | None | 302 | None | Forbidden | Forbidden | D | 40 |
| | Fluoroacetic acid | 6.1 | UN2642 | I | POISON | | None | 211 | 242 | 1 kg | 15 kg | E | |
| | Fluoroanilines | 6.1 | UN2941 | III | KEEP AWAY FROM FOOD | T8 | 153 | 203 | 241 | 60 L | 220 L | A | |
| | Fluorobenzene | 3 | UN2387 | II | FLAMMABLE LIQUID | T8 | 150 | 202 | 242 | 5 L | 60 L | B | |
| | Fluorophosphoric acid anhydrous | 8 | UN1776 | II | CORROSIVE | A6, A7, B2, N3, N34, T9, T27 | None | 202 | 242 | 1 L | 30 L | A | |
| | Fluorosilicates, n.o.s | 6.1 | UN2856 | III | KEEP AWAY FROM FOOD | | 153 | 213 | 240 | 100 kg | 200 kg | A | 26 |
| | Fluorosilicic acid | 8 | UN1778 | II | CORROSIVE | A6, A7, B2, B15, N3, N34, T12, T27 | None | 202 | 242 | 1 L | 30 L | A | |
| | Fluorosulfonic acid | 8 | UN1777 | I | CORROSIVE | A3, A6, A7, A10, B6, B10, B41, N3, T9, T27 | None | 201 | 242 | 0.5 L | 2.5 L | D | 40 |

## §172.101 HAZARDOUS MATERIALS TABLE, Continued

| Symbols | Hazardous materials descriptions and proper shipping names | Hazard class or Division | Identification Numbers | Packing group | Label(s) required (if not excepted) | Special provisions | (8) Packaging authorizations (§173 ***) | | | (9) Quantity limitations | | (10) Vessel stowage requirements | |
|---|---|---|---|---|---|---|---|---|---|---|---|---|---|
| | | | | | | | Exceptions | Non-bulk packaging | Bulk packaging | Passenger aircraft or railcar | Cargo aircraft only | Vessel stowages | Other stowage provisions |
| (1) | (2) | (3) | (4) | (5) | (6) | (7) | (8A) | (8B) | (8C) | (9A) | (9B) | (10A) | (10B) |
| | Fluorotoluenes | 3 | UN2388 | II | FLAMMABLE LIQUID | T8 | 150 | 202 | 242 | 5 L | 60 L | B | 40 |
| | *Forbidden materials See 173.21* | Forbidden | | | | | | | | | | | |
| | Formaldehyde, solutions | 9 | UN2209 | III | None | T1 | 155 | 204 | 240 | 100 L | 220 L | A | |
| | Formaldehyde, solutions, *flammable* | 3 | UN1198 | III | FLAMMABLE LIQUID | B1 T8 | 150 | 203 | 242 | 60 L | 220 L | A | 40 |
| | *Formalin see* Formaldehyde, solutions | | | | | | | | | | | | |
| | Formic acid | 8 | UN1779 | II | CORROSIVE | B2 B12, B28, T8 | 154 | 202 | 242 | 1 L | 30 L | A | 40 |
| | Fracturing devices, explosive, *without detonators for oil wells* | 1.1D | UN0099 | II | EXPLOSIVE 1.1D | | None | 62 | None | Forbidden | Forbidden | B | |
| | Fuel, aviation, turbine engine | 3 | UN1863 | I | FLAMMABLE LIQUID | T7 | 150 | 201 | 243 | 1 L | 30 L | E | |
| | | | | II | FLAMMABLE LIQUID | T1 | 150 | 202 | 242 | 5 L | 60 L | B | |
| | | | | III | FLAMMABLE LIQUID | B1, T1 | 150 | 203 | 242 | 60 L | 220 L | A | |
| D | Fuel oil *(No. 1, 2, 4, 5, or 6)* | 3 | NA1993 | III | FLAMMABLE LIQUID | B1 | 150 | 203 | 241 | 60 L | 220 L | A | |
| | *Fulminate of mercury (dry)* | Forbidden | | | | | | | | | | | |
| | *Fulminate of mercury, wet, see* Mercury fulminate, *etc* | | | | | | | | | | | | |
| | *Fulminating gold* | Forbidden | | | | | | | | | | | |
| | *Fulminating mercury* | Forbidden | | | | | | | | | | | |
| | *Fulminating platinum* | Forbidden | | | | | | | | | | | |
| | *Fulminating silver* | Forbidden | | | | | | | | | | | |
| | *Fulminic acid* | Forbidden | | | | | | | | | | | |
| | Fumaryl chloride | 8 | UN1780 | II | CORROSIVE | B2, T8, T26 | 154 | 202 | 242 | 1 L | 30 L | C | 8, 40 |
| | Furan | 3 | UN2389 | I | FLAMMABLE LIQUID | T18 | None | 201 | 242 | 1 L | 30 L | E | 40 |
| | Furfural | 3 | UN1199 | III | FLAMMABLE LIQUID | B1, T1 | 150 | 203 | 241 | 60 L | 220 L | A | |
| | Furfuryl alcohol | 6.1 | UN2874 | III | KEEP AWAY FROM FOOD | T2 | 153 | 203 | 240 | 60 L | 220 L | A | 26, 74 |
| | Furfurylamine | 3 | UN2526 | III | FLAMMABLE LIQUID | B1, T1 | 150 | 203 | 241 | 60 L | 220 L | A | 40 |
| | Fuse, detonating, *metal clad, see* Cord, detonating, *metal clad* | | | | | | | | | | | | |
| | Fuse, detonating, mild effect, *metal clad, see* Cord, detonating, mild effect, *metal clad* | | | | | | | | | | | | |
| | Fuse, igniter *tubular metal clad* | 1.4G | UN0103 | II | EXPLOSIVE 1.4G | | None | 62 | None | Forbidden | 75 kg | A | 24E |
| | Fuse, instantaneous, non-detonating *or* Quickmatch | 1.3G | UN0101 | II | EXPLOSIVE 1.3G | | None | 62 | None | Forbidden | Forbidden | B | |
| | Fuse, safety | 1.4S | UN0105 | II | EXPLOSIVE 1.4S | | None | 62 | None | 25 kg | 100 kg | A | 9E |
| D | Fusee *(railway or highway)* | 4.1 | NA1325 | II | FLAMMABLE SOLID | | None | 184 | None | 15 kg | 50 kg | B | |
| | Fusel oil | 3 | UN1201 | II | FLAMMABLE LIQUID | T1 | 150 | 202 | 242 | 5 L | 60 L | B | |
| | | | | III | FLAMMABLE LIQUID | B1, T1 | 150 | 203 | 242 | 60 L | 220 L | A | |
| | *Fuses, tracer, see* Tracers for ammunition | | | | | | | | | | | | |
| | *Fuzes, combination, percussion and time, see* Fuzes, detonating *(UN 0257, UN 0367)*; Fuzes, igniting (UN 0317, UN 0368) | | | | | | | | | | | | |
| | Fuzes, detonating | 1.1B | UN0106 | II | EXPLOSIVE 1.1B | | None | 62 | None | Forbidden | Forbidden | B | 2E, 6E |
| | Fuzes, detonating | 1.2B | UN0107 | II | EXPLOSIVE 1.2B | | None | 62 | None | Forbidden | Forbidden | B | 2E, 6E |
| | Fuzes, detonating | 1.4B | UN0257 | II | EXPLOSIVE 1.4B | 116 | None | 62 | None | Forbidden | 75 kg | A | 24E |
| | Fuzes, detonating | 1.4S | UN0367 | II | EXPLOSIVE 1.4S | 116 | None | 62 | None | 25 kg | 100 kg | A | 9E |
| | Fuzes, detonating, *with protective features* | 1.1D | UN0408 | II | EXPLOSIVE 1.1D | | None | 62 | None | Forbidden | Forbidden | B | |
| | Fuzes, detonating, *with protective features* | 1.2D | UN0409 | II | EXPLOSIVE 1.2D | | None | 62 | None | Forbidden | Forbidden | B | |
| | Fuzes, detonating, *with protective features* | 1.4D | UN0410 | II | EXPLOSIVE 1.4D | 116 | None | 62 | None | Forbidden | 75 kg | A | 24E |
| | Fuzes, igniting | 1.3G | UN0316 | II | EXPLOSIVE 1.3G | | None | 62 | None | Forbidden | Forbidden | B | |
| | Fuzes, igniting | 1.4G | UN0317 | II | EXPLOSIVE 1.4G | | None | 62 | None | Forbidden | 75 kg | B | |
| | Fuzes, igniting | 1.4S | UN0368 | II | EXPLOSIVE 1.4S | | None | 62 | None | 25 kg | 100 kg | A | 9E |
| | *Galactsan trinitrate* | Forbidden | | | | | | | | | | | |
| | Gallium | 8 | UN2803 | III | CORROSIVE | | None | 162 | 240 | 20 kg | 20 kg | B | 48 |
| | Gas drips, *hydrocarbon* | 3 | UN1864 | II | FLAMMABLE LIQUID | T7, T30 | 150 | 202 | 242 | 5 L | 60 L | B | |
| | Gas generator assemblies (aircraft), *containing a non-flammable non-toxic gas and a propellant cartridge* | 2.2 | | | NONFLAMMABLE GAS | | None | 335 | None | 75 kg | 150 kg | A | |

| Symbols | Hazardous materials descriptions and proper shipping names | Hazard class or Division | Identification Numbers | Packing group | Label(s) required (if not excepted) | Special provisions | (8) Packaging authorizations (§173 ***) | | | (9) Quantity limitations | | (10) Vessel stowage requirements | |
|---|---|---|---|---|---|---|---|---|---|---|---|---|---|
| | | | | | | | Exceptions | Non-bulk packaging | Bulk packaging | Passenger aircraft or railcar | Cargo aircraft only | Vessel stowages | Other stowage provisions |
| (1) | (2) | (3) | (4) | (5) | (6) | (7) | (8A) | (8B) | (8C) | (9A) | (9B) | (10A) | (10B) |
| D | Gas identification set | 2.3 | NA9035 | | POISON GAS | 6 | None | 194 | None | Forbidden | Forbidden | D | |
| | Gas oil | 3 | UN1202 | III | FLAMMABLE LIQUID | B1, T7, T30 | 150 | 203 | 242 | 60 L | 220 L | A | |
| D | Gasohol (gasoline mixed with ethyl alcohol, max. 20% alcohol) | 3 | NA1203 | II | FLAMMABLE LIQUID | | 150 | 202 | 242 | 5 L | 60 L | E | |
| | Gasoline | 3 | UN1203 | II | FLAMMABLE LIQUID | B33, T8 | 150 | 202 | 242 | 5 L | 60 L | E | |
| | Gasoline, casinghead, see Natural gasoline | | | | | | | | | | | | |
| | Gelatine, blasting, see Explosive, blasting, type A | | | | | | | | | | | | |
| | Gelatine dynamites, see Explosive, blasting, type A | | | | | | | | | | | | |
| | Germane | 2.3 | UN2192 | | POISON GAS FLAMMABLE GAS | 1 | None | 192 | 245 | Forbidden | Forbidden | D | 40 |
| | Glycerol-1,3-dinitrate | Forbidden | | | | | | | | | | | |
| | Glycerol gluconate trinitrate | Forbidden | | | | | | | | | | | |
| | Glycerol lactate trinitrate | Forbidden | | | | | | | | | | | |
| | Glycerol alpha-monochlorohydrin | 6.1 | UN2689 | III | KEEP AWAY FROM FOOD | T2 | 153 | 203 | 241 | 60 L | 220 L | A | |
| | Glyceryl trinitrate, see Nitroglycerin, etc. | | | | | | | | | | | | |
| | Glycidaldehyde | 3 | UN2622 | II | FLAMMABLE LIQUID, POISON | T8 | 150 | 202 | 243 | 1 L | 60 L | A | 40 |
| D | Grenades, empty primed | 1.4S | NA0349 | II | None | None | None | 62 | None | 25 kg | 100 kg | A | 9E |
| | Grenades, hand or rifle, with bursting charge | 1.1D | UN0284 | II | EXPLOSIVE 1.1D | | 63(b) | 62 | None | Forbidden | Forbidden | B | |
| | Grenades, hand or rifle, with bursting charge | 1.2D | UN0285 | II | EXPLOSIVE 1.2D | | 63(b) | 62 | None | Forbidden | Forbidden | B | |
| | Grenades, hand or rifle, with bursting charge | 1.1F | UN0292 | II | EXPLOSIVE 1.1F | | 63(b) | 62 | None | Forbidden | Forbidden | E | |
| | Grenades, hand or rifle, with bursting charge | 1.2F | UN0293 | II | EXPLOSIVE 1.2F | | 63(b) | 62 | None | Forbidden | Forbidden | E | |
| | Grenades, illuminating, see Ammunition, illuminating, etc. | | | | | | | | | | | | |
| | Grenades practice Hand or rifle | 1.4G | UN0452 | II | EXPLOSIVE 1.4G | | 63(b) | 62 | None | Forbidden | 75 kg | B | |
| | Grenades, practice, hand or rifle | 1.4S | UN0110 | II | EXPLOSIVE 1.4S | | 63(b) | 62 | None | 25 kg | 100 kg | A | 9E |
| | Grenades, practice, hand or rifle | 1.3G | UN0318 | II | EXPLOSIVE 1.3G | | 63(b) | 62 | None | Forbidden | Forbidden | B | |
| | Grenades, practice, hand or rifle | 1.2G | UN0372 | II | EXPLOSIVE 1.2G | | 63(b) | 62 | None | Forbidden | Forbidden | B | |
| | Grenades, smoke, see Ammunition, smoke, etc. | | | | | | | | | | | | |
| | Guanidine nitrate | 5.1 | UN1467 | III | OXIDIZER | A1 | 152 | 213 | 240 | 25 kg | 100 kg | A | 73 |
| | Guanyl nitrosaminoguanylidene hydrazine (dry) | Forbidden | | | | | | | | | | | |
| | Guanyl nitrosaminoguanylidene hydrazine, wetted with not less than 30 per cent water, by mass | 1.1A | UN0113 | II | EXPLOSIVE 1.1A | 111, 117 | None | 62 | None | Forbidden | Forbidden | E | 2E, 6E |
| | Guanyl nitrosaminoguanyltetrazene (dry) | Forbidden | | | | | | | | | | | |
| | Guanyl nitrosaminoguanyltetrazene, wetted or Tetrazene, wetted with not less than 30 per cent water or mixture of alcohol and water, by mass | 1.1A | UN0114 | II | EXPLOSIVE 1.1A | 111, 117 | None | 62 | None | Forbidden | Forbidden | E | 2E, 6E |
| | Gunpowder, compressed or Gunpowder in pellets, see Black powder (UN 0028) | | | | | | | | | | | | |
| | Gunpowder, granular or as a meal, see Black powder (UN 0027) | | | | | | | | | | | | |
| | Hafnium powder, dry | 4.2 | UN2545 | I | SPONTANEOUSLY COMBUSTIBLE | | None | 211 | 242 | Forbidden | Forbidden | D | |
| | | | | II | SPONTANEOUSLY COMBUSTIBLE | A19, A20, N34 | None | 212 | 241 | 15 kg | 50 kg | D | |
| | | | | III | SPONTANEOUSLY COMBUSTIBLE | | None | 213 | 241 | 25 kg | 100 kg | D | |
| | Hafnium powder, wetted with not less than 25 per cent water (a visible excess of water must be present) (a) mechanically produced, particle size less than 53 microns; (b) chemically produced, particle size less than 840 microns | 4.1 | UN1326 | II | FLAMMABLE SOLID | A6, A19, A20, N34 | None | 212 | 241 | 15 kg | 50 kg | E | |
| | Halogenated irritating liquids, n.o.s. | 6.1 | UN1610 | I | POISON | T42 | None | 201 | 243 | Forbidden | Forbidden | D | 40 |
| | | | | II | POISON | T14 | None | 202 | 243 | 5 L | 60 L | D | 40 |
| | | | | III | KEEP AWAY FROM FOOD | T14 | 153 | 202 | 241 | 60 L | 220 L | D | 40 |
| | Hand signal device, see Signal devices, hand | | | | | | | | | | | | |
| | Hazardous substances, liquid or solid, n.o.s., see Environmentally hazardous substances, etc. | | | | | | | | | | | | |
| D | Hazardous waste, liquid, n.o.s. | 9 | NA3082 | III | CLASS 9 | | 155 | 203 | 241 | No limit | No limit | A | |
| D | Hazardous waste, solid, n.o.s. | 9 | NA3077 | III | CLASS 9 | B54 | 155 | 213 | 240 | No limit | No limit | A | |
| | Helium, compressed | 2.2 | UN1046 | | NONFLAMMABLE GAS | | 306 | 302 | 302, 314 | 75 kg | 150 kg | A | 85 |

## §172.101 HAZARDOUS MATERIALS TABLE, Continued

| Symbols | Hazardous materials descriptions and proper shipping names | Hazard class or Division | Identification Numbers | Packing group | Label(s) required (if not excepted) | Special provisions | (8) Packaging authorizations (§173.***) | | | (9) Quantity limitations | | (10) Vessel stowage requirements | |
|---|---|---|---|---|---|---|---|---|---|---|---|---|---|
| | | | | | | | Exceptions | Non-bulk packaging | Bulk packaging | Passenger aircraft or railcar | Cargo aircraft only | Vessel stowages | Other stowage provisions |
| (1) | (2) | (3) | (4) | (5) | (6) | (7) | (8A) | (8B) | (8C) | (9A) | (9B) | (10A) | (10B) |
| | *Helium-oxygen mixture, see* Rare gases and oxygen mixtures | | | | | | | | | | | | |
| | Helium, refrigerated liquid *(cryogenic liquid)* | 2.2 | UN1963 | | NONFLAMMABLE GAS | | 320 | 316 | 318 | 50 kg | 500 kg | B | |
| | n-Heptaldehyde | 3 | UN3056 | III | FLAMMABLE LIQUID | B1, T1 | 150 | 203 | 242 | 60 L | 220 L | A | |
| | Heptanes | 3 | UN1206 | II | FLAMMABLE LIQUID | T2 | 150 | 202 | 242 | 5 L | 60 L | B | |
| | n-Heptene | 3 | UN2278 | II | FLAMMABLE LIQUID | T8 | 150 | 202 | 242 | 5 L | 60 L | B | |
| | Hexachloroacetone | 6.1 | UN2661 | III | KEEP AWAY FROM FOOD | T8 | 153 | 203 | 241 | 60 L | 220 L | B | 12, 40 |
| | Hexachlorobenzene | 6.1 | UN2729 | III | KEEP AWAY FROM FOOD | | 153 | 203 | 241 | 60 L | 220 L | A | |
| | Hexachlorobutadiene | 6.1 | UN2279 | III | KEEP AWAY FROM FOOD | T7 | 153 | 203 | 241 | 60 L | 220 L | A | |
| | Hexachlorocyclopentadiene | 6.1 | UN2646 | I | POISON | 2, B9, B14, B32, B74, B77, T38, T43, T45 | None | 227 | 244 | Forbidden | Forbidden | D | 40 |
| | Hexachlorophene | 6.1 | UN2875 | III | KEEP AWAY FROM FOOD | | 153 | 213 | 240 | 100 kg | 200 kg | A | |
| | Hexadecyltrichlorosilane | 8 | UN1781 | II | CORROSIVE | A7, B2, B6, N34, T8 | None | 202 | 242 | Forbidden | 30 L | C | 40 |
| | Hexadienes | 3 | UN2458 | II | FLAMMABLE LIQUID | T7 | None | 202 | 242 | 5 L | 60 L | B | |
| | Hexaethyl tetraphosphate and compressed gas mixtures | 2.3 | UN1612 | | POISON GAS | 3 | None | 334 | None | Forbidden | Forbidden | D | 40 |
| | Hexaethyl tetraphosphate *liquid* | 6.1 | UN1611 | I | POISON | A4 | None | 201 | 243 | 1 L | 30 L | E | 40 |
| | | | | II | POISON | N76 | None | 202 | 243 | 5 L | 60 L | E | 40 |
| | | | | III | KEEP AWAY FROM FOOD | N77 | None | 203 | 241 | 60 L | 220 L | E | 40 |
| | Hexaethyl tetraphosphate, *solid* | 6.1 | UN1611 | I | POISON | | None | 211 | 242 | Forbidden | 15 kg | E | 95 |
| | | | | II | POISON | N76 | None | 212 | 242 | 25 kg | 100 kg | E | 95 |
| | | | | III | KEEP AWAY FROM FOOD | N77 | 153 | 213 | 240 | 100 kg | 200 kg | E | 34 |
| | Hexafluoroacetone | 2.3 | UN2420 | | POISON GAS | 2, B9, B14 | None | 304 | 314, 315 | Forbidden | 25 kg | D | 40 |
| | Hexafluoroacetone hydrate | 6.1 | UN2552 | II | POISON | T14 | None | 202 | 243 | 5 L | 60 L | B | 40 |
| | Hexafluoroethane, *R116* | 2.2 | UN2193 | | NONFLAMMABLE GAS | | 306 | 304 | 314, 315 | 75 kg | 150 kg | A | |
| | Hexafluorophosphoric acid | 8 | UN1782 | II | CORROSIVE | A6, A7, B2, N3, N34, T9, T27 | None | 202 | 242 | 1 L | 30 L | A | |
| D | Hexafluoropropylene oxide | 2.2 | NA1956 | | NONFLAMMABLE GAS | | 306 | 304 | 314, 315 | 75 kg | 150 kg | A | |
| | Hexafluoropropylene, *R1216* | 2.2 | UN1858 | | NONFLAMMABLE GAS | | 306 | 304 | 314, 315 | 75 kg | 150 kg | A | |
| | Hexaldehyde | 3 | UN1207 | III | FLAMMABLE LIQUID | B1, T1 | 150 | 203 | 242 | 60 L | 220 L | A | |
| | Hexamethylene diisocyanate | 6.1 | UN2281 | II | POISON | T14 | None | 202 | 243 | 5 L | 60 L | B | 13, 40 |
| | *Hexamethylene triperoxide diamine (dry)* | Forbidden | | | | | | | | | | | |
| | Hexamethylenediamine, solid | 8 | UN2280 | III | CORROSIVE | | 154 | 213 | 240 | 25 kg | 100 kg | A | 12, 48 |
| | Hexamethylenediamine solution | 8 | UN1783 | II | CORROSIVE | T8 | None | 202 | 243 | 1 L | 30 L | A | |
| | Hexamethyleneimine | 3 | UN2493 | II | FLAMMABLE LIQUID, CORROSIVE | T8 | None | 202 | 243 | 1 L | 5 L | B | 40 |
| | *Hexamethylol benzene hexanitrate* | Forbidden | | | | | | | | | | | |
| | Hexamine | 4.1 | UN1328 | III | FLAMMABLE SOLID | A1 | 151 | 213 | 240 | 25 kg | 100 kg | A | |
| | Hexanes | 3 | UN1208 | II | FLAMMABLE LIQUID | T8 | 150 | 202 | 242 | 5 L | 60 L | E | |
| | *2,2',4,4',6,6'-Hexanitro-3,3'-dihydroxyazobenzene (dry)* | Forbidden | | | | | | | | | | | |
| | *Hexanitroazoxy benzene* | Forbidden | | | | | | | | | | | |
| | *N,N'-(hexanitrodiphenyl) ethylene dinitramine (dry)* | Forbidden | | | | | | | | | | | |
| | *Hexanitrodiphenyl urea* | Forbidden | | | | | | | | | | | |
| | *2,2',3',4,4',6-Hexanitrodiphenylamine* | Forbidden | | | | | | | | | | | |
| | Hexanitrodiphenylamine (Dipicrylamine; Hexyl) | 1.1D | UN0079 | II | EXPLOSIVE 1.1D | | None | 62 | None | Forbidden | Forbidden | B | 1E, 5E |
| | *2,3',4,4',6,6'-Hexanitrodiphenylether* | Forbidden | | | | | | | | | | | |
| | *Hexanitroethane* | Forbidden | | | | | | | | | | | |
| | *Hexanitrooxanilide* | Forbidden | | | | | | | | | | | |

| Symbols | Hazardous materials descriptions and proper shipping names | Hazard class or Division | Identification Numbers | Packing group | Label(s) required (if not excepted) | Special provisions | (8) Packaging authorizations (§173.***) | | | (9) Quantity limitations | | (10) Vessel stowage requirements | |
|---|---|---|---|---|---|---|---|---|---|---|---|---|---|
| | | | | | | | Exceptions | Non-bulk packaging | Bulk packaging | Passenger aircraft or railcar | Cargo aircraft only | Vessel stowages | Other stowage provisions |
| (1) | (2) | (3) | (4) | (5) | (6) | (7) | (8A) | (8B) | (8C) | (9A) | (9B) | (10A) | (10B) |
| | Hexanitrostilbene | 1 1D | UN0392 | II | EXPLOSIVE 1 1D | | None | 62 | None | Forbidden | Forbidden | B | 1E, 5E |
| | *Hexanoic acid, see* Corrosive liquids, n.o.s | | | | | | | | | | | | |
| | Hexanols | 3 | UN2282 | III | FLAMMABLE LIQUID | B1, T1 | 150 | 203 | 242 | 60 L | 220 L | A | |
| | Hexatonal, cast | 1 1D | UN0393 | II | EXPLOSIVE 1 1D | | None | 62 | None | Forbidden | Forbidden | B | 1E, 5E |
| | 1-Hexene | 3 | UN2370 | II | FLAMMABLE LIQUID | T8 | 150 | 202 | 242 | 5 L | 60 L | E | |
| | Hexogen and cyclotetramethylenetetranitramine mixtures, wetted *or* desensitized *see* RDX and HMX mixtures, wetted *or* desensitized *etc.* | | | | | | | | | | | | |
| | Hexogen and HMX mixtures, wetted *or* desensitized *see* RDX and HMX mixtures, wetted *or* desensitized *etc* | | | | | | | | | | | | |
| | Hexogen and octogen mixtures, wetted *or* desensitized *see* RDX and HMX mixtures, wetted *or* desensitized *etc* | | | | | | | | | | | | |
| | Hexogen, *see* Cyclotrimethylenetrinitramine *etc* | | | | | | | | | | | | |
| | Hexolite *dry or wetted with less than 15 per cent water, by mass* | 1 1D | UN0118 | II | EXPLOSIVE 1 1D | | None | 62 | None | Forbidden | Forbidden | B | 1E 5E |
| | Hexyl, *see* Hexanitrodiphenylamine | | | | | | | | | | | | |
| | Hexyltrichlorosilane | 8 | UN1784 | II | CORROSIVE | A7, B2, B6 N34, T8, T26 | None | 202 | 242 | Forbidden | 30 L | C | 40 |
| | *High explosives, see individual explosives entries* | | | | | | | | | | | | |
| | HMX, *see* Cyclotetramethylene-tetranitramine, *etc* | | | | | | | | | | | | |
| | Hydrazine, anhydrous *or* Hydrazine aqueous solutions *with more than 64 per cent hydrazine, by mass* | 3 | UN2029 | I | FLAMMABLE LIQUID, POISON, CORROSIVE | A3, A6, A7, A10, B16, B53, T25 | None | 201 | 243 | Forbidden | 2 5 L | D | 21, 40, 42, 100 |
| | *Hydrazine azide* | Forbidden | | | | | | | | | | | |
| | *Hydrazine chlorate* | Forbidden | | | | | | | | | | | |
| | *Hydrazine dicarbonic acid diazide* | Forbidden | | | | | | | | | | | |
| | Hydrazine hydrate *or* Hydrazine aqueous solutions, *with not more than 64 per cent hydrazine, by mass* | 8 | UN2030 | II | CORROSIVE, POISON | B16, B53, T15 | None | 202 | 243 | Forbidden | 30 L | D | 40, 42, 82 |
| | *Hydrazine perchlorate* | Forbidden | | | | | | | | | | | |
| | *Hydrazine selenate* | Forbidden | | | | | | | | | | | |
| | Hydrides, metal, n.o.s. | 4.3 | UN1409 | I | DANGEROUS WHEN WET | A19, N34, N40 | None | 211 | 242 | Forbidden | 15 kg | D | |
| | *Hydriodic acid, anhydrous, see* Hydrogen iodide, anhydrous | | | | | | | | | | | | |
| | Hydriodic acid, solution | 8 | UN1787 | II | CORROSIVE | A3, A6, B2, N41, T9, T27 | 154 | 202 | 242 | 1 L | 30 L | C | 8 |
| | *Hydrobromic acid, anhydrous, see* Hydrogen bromide, anhydrous | | | | | | | | | | | | |
| | Hydrobromic acid solution *more than 49% hydrobromic acid* | 8 | UN1788 | II | CORROSIVE | B2, B15, N41, T9, T27 | 154 | 202 | 242 | Forbidden | Forbidden | C | 8 |
| | Hydrobromic acid solution, *not more than 49% hydrobromic acid* | 8 | UN1788 | II | CORROSIVE | A3, A6, B2, B15, N41, T9, T27 | 154 | 202 | 242 | 1 L | 30 L | C | 8 |
| | Hydrocarbon gases, compressed, n.o.s. *or* Hydrocarbon gases mixtures, compressed, n.o.s. | 2.1 | UN1964 | | FLAMMABLE GAS | | 306 | 302 | 314, 315 | Forbidden | 150 kg | E | 40 |
| | Hydrocarbon gases, liquefied, n.o.s. *or* Hydrocarbon gases mixtures, liquefied, n.o.s. | 2.1 | UN1965 | | FLAMMABLE GAS | | 306 | 304 | 314, 315 | Forbidden | 150 kg | E | 40 |
| | *Hydrochloric acid, anhydrous, see* Hydrogen chloride, anhydrous | | | | | | | | | | | | |
| | Hydrochloric acid, solution | 8 | UN1789 | II | CORROSIVE | A3, A6, B3, B15, N41, T9, T27 | 154 | 202 | 242 | 1 L | 30 L | C | 8 |
| | *Hydrocyanic acid, anhydrous, see* Hydrogen cyanide *etc* | | | | | | | | | | | | |
| D | Hydrocyanic acid, aqueous solutions *less than 5% HCN* | 6.1 | NA1613 | II | POISON | B12, T18, T26 | None | 195 | 243 | Forbidden | 5 L | D | 40 |
| | Hydrocyanic acid, aqueous solutions *not more than 20% hydrocyanic acid* | 6 1 | UN1613 | I | POISON | 2, B12, B61, B65, B77, B82 | None | 195 | 244 | Forbidden | Forbidden | D | 40 |
| | *Hydrocyanic acid, liquefied, see* Hydrogen cyanide, *etc.* | | | | | | | | | | | | |
| | *Hydrocyanic acid (prussic), unstabilized* | Forbidden | | | | | | | | | | | |

| Symbols | Hazardous materials descriptions and proper shipping names | Hazard class or Division | Identification Numbers | Packing group | Label(s) required (if not excepted) | Special provisions | (8) Packaging authorizations (§173.***) | | | (9) Quantity limitations | | (10) Vessel stowage requirements | |
|---|---|---|---|---|---|---|---|---|---|---|---|---|---|
| | | | | | | | Exceptions | Non-bulk packaging | Bulk packaging | Passenger aircraft or railcar | Cargo aircraft only | Vessel stowages | Other stowage provisions |
| (1) | (2) | (3) | (4) | (5) | (6) | (7) | (8A) | (8B) | (8C) | (9A) | (9B) | (10A) | (10B) |
| | Hydrofluoric acid and Sulfuric acid mixtures | 8 | UN1786 | I | CORROSIVE, POISON | A6, A7, B15, B23, N5, N34, T18, T27 | None | 201 | 243 | Forbidden | 2.5 L | D | 40, 95 |
| | *Hydrofluoric acid, anhydrous, see* Hydrogen fluoride, anhydrous | | | | | | | | | | | | |
| | Hydrofluoric acid, solution, *more than 60 per cent strength* | 8 | UN1790 | I | CORROSIVE, POISON | A6, A7, B4, B12, B15, B23, N5, N34, T18, T27 | None | 201 | 243 | 0.5 L | 2.5 L | D | 12, 40, 48, 95 |
| | Hydrofluoric acid, solution, *not more than 60 per cent strength* | 8 | UN1790 | II | CORROSIVE, POISON | A6, A7, B12, B15 N5 N34, T18, T27 | None | 202 | 243 | 1 L | 30 L | D | 12, 40, 48, 95 |
| | *Hydrofluoroboric acid, see* Fluoboric acid | | | | | | | | | | | | |
| | *Hydrofluorosilicic acid, see* Fluorosilicic acid | | | | | | | | | | | | |
| | Hydrogen and Methane mixtures, compressed | 2.1 | UN2034 | | FLAMMABLE GAS | | 306 | 302 | 302, 314, 315 | Forbidden | 150 kg | E | 40 |
| | Hydrogen bromide, anhydrous | 2.3 | UN1048 | | POISON GAS, CORROSIVE | 3, B14 | None | 304 | 314, 315 | Forbidden | 25 kg | D | 40 |
| | Hydrogen chloride, anhydrous | 2.3 | UN1050 | | POISON GAS, CORROSIVE | 3 | None | 304 | None | Forbidden | Forbidden | D | 40 |
| | Hydrogen chloride, refrigerated liquid | 2.3 | UN2186 | | POISON GAS, CORROSIVE | 3, B6, B43 | None | None | 314, 315 | Forbidden | Forbidden | B | 40 |
| | Hydrogen, compressed | 2.1 | UN1049 | | FLAMMABLE GAS | | 306 | 302 | 302, 314 | Forbidden | 150 kg | E | 40, 57 |
| | Hydrogen cyanide, anhydrous, stabilized | 6.1 | UN1051 | I | POISON, FLAMMABLE LIQUID | 1, B12, B35, B61, B65, B77 B82 | None | 195 | 244 | Forbidden | Forbidden | D | 40 |
| | Hydrogen cyanide, anhydrous, stabilized, *absorbed in a porous inert material* | 6.1 | UN1614 | I | POISON | 5 | None | 195 | None | Forbidden | Forbidden | D | 25, 40 |
| | Hydrogen fluoride, anhydrous | 8 | UN1052 | I | CORROSIVE, POISON | 3, B7, B12, B46, B71, B77, T24, T27 | None | 163 | 243 | Forbidden | Forbidden | D | 40, 95 |
| | Hydrogen iodide, anhydrous | 2.2 | UN2197 | | NONFLAMMABLE GAS, CORROSIVE | | 306 | 304 | 314, 315 | Forbidden | Forbidden | D | 40 |
| | *Hydrogen iodide solution, see* Hydriodic acid, solution | | | | | | | | | | | | |
| | Hydrogen peroxide and peroxyacetic acid mixtures, *with acids, water and not more than 5 per cent peroxyacetic acid, stabilized* | 5.1 | UN3149 | II | OXIDIZER, CORROSIVE | A2, A3, A6, B12, B53, T14 | None | 202 | 243 | 1 L | 5 L | D | 25, 66, 75, 106 |
| | Hydrogen peroxide, aqueous solutions *with more than 40 per cent but not more than 60 per cent hydrogen peroxide (stabilized as necessary)* | 5.1 | UN2014 | II | OXIDIZER, CORROSIVE | 12, A3, A6, B12, B53, B80, B81, B85, T14, T37 | None | 202 | 243 | Forbidden | Forbidden | D | 25, 66, 75, 106 |
| | Hydrogen peroxide, aqueous solutions *with not less than 8 per cent but less than 20 per cent hydrogen peroxide (stabilized as necessary)* | 5.1 | UN2984 | III | OXIDIZER | 17, A1, T8, T37 | 152 | 203 | 241 | 2.5 L | 30 L | B | 25, 75, 106 |
| | Hydrogen peroxide, aqueous solutions *with not less than 20 per cent but not more than 40 per cent hydrogen peroxide (stabilized as necessary)* | 5.1 | UN2014 | II | OXIDIZER, CORROSIVE | A2, A3, A6, B12 B53, T14, T37 | None | 202 | 243 | 1 L | 5 L | D | 25, 66, 75, 106 |
| | Hydrogen peroxide, stabilized *or* Hydrogen peroxide aqueous solutions, stabilized *with more than 60 per cent hydrogen peroxide* | 5.1 | UN2015 | I | OXIDIZER, CORROSIVE | 12, A3, A6, B12, B53, B80, B81, B85, T15, T37 | None | 201 | 243 | Forbidden | Forbidden | D | 25, 66, 75, 106 |
| | Hydrogen, refrigerated liquid *(cryogenic liquid)* | 2.1 | UN1966 | | FLAMMABLE GAS | | None | 316 | 318, 319 | Forbidden | Forbidden | D | 40 |
| | Hydrogen selenide, anhydrous | 2.3 | UN2202 | | POISON GAS, FLAMMABLE GAS | 1 | None | 192 | 245 | Forbidden | Forbidden | D | 40 |
| | *Hydrogen sulfate, see* Sulfuric acid | | | | | | | | | | | | |
| | Hydrogen sulfide, liquefied | 2.3 | UN1053 | | POISON GAS, FLAMMABLE GAS | 2, B9, B14 | None | 304 | 314, 315 | Forbidden | Forbidden | D | 40 |
| | Hydroquinone, liquid | 6.1 | UN2662 | III | KEEP AWAY FROM FOOD | T8 | 153 | 213 | 240 | 100 kg | 200 kg | A | |
| | Hydroquinone, solid | 6.1 | UN2662 | III | KEEP AWAY FROM FOOD | | 153 | 213 | 240 | 100 kg | 200 kg | AA | |
| | *Hydrosilicofluoric acid, see* Fluorosilicic acid | | | | | | | | | | | | |
| | 3-(2-Hydroxyethoxy)-4-pyrrolidin-1-ylbenzenediazonium zinc chloride | 4.1 | UN3035 | II | FLAMMABLE SOLID | | None | 224 | 240 | Forbidden | Forbidden | D | 2 |
| | *Hydroxyl amine iodide* | Forbidden | | | | | | | | | | | |
| | Hydroxylamine sulfate | 8 | UN2865 | III | CORROSIVE | | 154 | 213 | 240 | 25 kg | 100 kg | A | |
| | Hypochlorite solutions *with more than 5 per cent but less than 16 per cent available chlorine* | 8 | UN1791 | III | CORROSIVE | N34, T7 | 154 | 203 | 241 | 5 L | 60 L | B | 26 |
| | Hypochlorite solutions *with 16 per cent or more available chlorine* | 8 | UN1791 | II | CORROSIVE | A7, B2, B15, N34, T7 | 154 | 202 | 242 | 1 L | 30 L | B | 26 |
| | *Hyponitrous acid* | Forbidden | | | | | | | | | | | |
| | *Igniter fuse, metal clad, see* Fuse, igniter, tubular, metal clad | | | | | | | | | | | | |
| | Igniters | 1.1G | UN0121 | II | EXPLOSIVE 1.1G | | None | 62 | None | Forbidden | Forbidden | B | |
| | Igniters | 1.2G | UN0314 | II | EXPLOSIVE 1.2G | | None | 62 | None | Forbidden | Forbidden | B | |
| | Igniters | 1.3G | UN0315 | II | EXPLOSIVE 1.3G | | None | 62 | None | Forbidden | Forbidden | A | |
| | Igniters | 1.4G | UN0325 | II | EXPLOSIVE 1.4G | | None | 62 | None | Forbidden | 75 kg | A | 24E |

## §172.101 HAZARDOUS MATERIALS TABLE, Continued

| Symbols (1) | Hazardous materials descriptions and proper shipping names (2) | Hazard class or Division (3) | Identification Numbers (4) | Packing group (5) | Label(s) required (if not excepted) (6) | Special provisions (7) | (8) Packaging authorizations (§173.***) Exceptions (8A) | Non-bulk packaging (8B) | Bulk packaging (8C) | (9) Quantity limitations Passenger aircraft or railcar (9A) | Cargo aircraft only (9B) | (10) Vessel stowage requirements Vessel stowages (10A) | Other stowage provisions (10B) |
|---|---|---|---|---|---|---|---|---|---|---|---|---|---|
| | Igniters | 1 4S | UN0454 | II | EXPLOSIVE 1 4S | | None | 62 | None | 25 kg | 100 kg | A | 9E |
| | 3,3'-Iminodipropylamine | 8 | UN2269 | III | CORROSIVE | T8 | 154 | 203 | 241 | 5 L | 60 L | A | |
| | Infectious substances, affecting animals *only* | 6.2 | UN2900 | | INFECTIOUS SUBSTANCE | | 196 | 196 | None | 50 mL or 50 g | 4 L or 4 kg | B | |
| | Infectious substances, affecting humans | 6 2 | UN2814 | | INFECTIOUS SUBSTANCE | | 196 | 196 | None | 50 mL or 50 g | 4 L or 4 kg | B | |
| | *Inflammable, see Flammable* | | | | | | | | | | | | |
| | *Initiating explosives (dry)* | Forbidden | | | | | | | | | | | |
| | Ink, *printer's, flammable* | 3 | UN1210 | II | FLAMMABLE LIQUID | T7, T30, | 150 | 173 | 242 | 5 L | 60 L | B | |
| | | | | III | FLAMMABLE LIQUID | B1, T7, T30 | 150 | 173 | 242 | 60 L | 220 L | A | |
| | *Inositol hexanitrate (dry)* | Forbidden | | | | | | | | | | | |
| | Insecticide gases *flammable* n.o.s | 2 1 | UN1968 | | FLAMMABLE GAS | | 306 | 304 | 314, 315 | 75 kg | 150 kg | D | 40 |
| | Insecticide gases, n.o.s | 2 2 | UN1968 | | NONFLAMMABLE GAS | | 306 | 304 | 314, 315 | 75 kg | 150 kg | A | 105 |
| | Insecticide gases toxic, n.o.s | 2 3 | UN1967 | | POISON GAS | 3 | None | 193, 334 | 245 | Forbidden | Forbidden | D | 40 |
| | *Inulin trinitrate (dry)* | Forbidden | | | | | | | | | | | |
| | *Iodine azide (dry)* | Forbidden | | | | | | | | | | | |
| | Iodine monochloride | 8 | UN1792 | II | CORROSIVE | B6, N41, T8, T26 | None | 202 | 242 | Forbidden | 50 kg | D | 40, 66 74 89, 90 |
| | Iodine pentafluoride | 5 1 | UN2495 | I | OXIDIZER, POISON | | None | 205 | 243 | Forbidden | 2.5 L | D | 25, 40, 66, 90 |
| | 2-Iodobutane | 3 | UN2390 | II | FLAMMABLE LIQUID | T8 | 150 | 202 | 242 | 5 L | 60 L | B | |
| | Iodomethylpropanes | 3 | UN2391 | II | FLAMMABLE LIQUID | T8 | 150 | 202 | 242 | 5 L | 60 L | B | |
| | Iodopropanes | 3 | UN2392 | III | FLAMMABLE LIQUID | B1, T8 | 150 | 203 | 241 | 60 L | 220 L | A | |
| | *Iodoxy compounds (dry)* | Forbidden | | | | | | | | | | | |
| | *Iridium nitratopentamine iridium nitrate* | Forbidden | | | | | | | | | | | |
| | *Iron chloride, see* Ferric chloride | | | | | | | | | | | | |
| | Iron oxide, spent, *or* Iron sponge, spent *obtained from coal gas purification* | 4.2 | UN1376 | III | SPONTANEOUSLY COMBUSTIBLE | B18 | None | 213 | 240 | Forbidden | Forbidden | E | |
| | Iron pentacarbonyl | 6.1 | UN1994 | I | POISON, FLAMMABLE LIQUID | 1, B9, B14, B30, B72, B77, T38, T43, T44 | None | 192 | 244 | Forbidden | Forbidden | D | 40 |
| | *Iron sesquichloride, see* Ferric chloride | | | | | | | | | | | | |
| | *Irritating material, see* Tear gas substances, etc | | | | | | | | | | | | |
| | Isobutane *or* Isobutane mixtures *see also* Petroleum gases, liquefied | 2.1 | UN1969 | | FLAMMABLE GAS | 19 | 306 | 304 | 314, 315 | Forbidden | 150 kg | E | 40 |
| | Isobutanol *or* isobutyl alcohol | 3 | UN1212 | III | FLAMMABLE LIQUID | B1, T1 | 150 | 203 | 241 | 60 L | 220 L | A | |
| | Isobutyl acetate | 3 | UN1213 | II | FLAMMABLE LIQUID | T1 | 150 | 202 | 242 | 5 L | 60 L | B | |
| | Isobutyl acrylate | 3 | UN2527 | III | FLAMMABLE LIQUID | B1, T1 | 150 | 203 | 242 | 60 L | 220 L | A | |
| | Isobutyl alcohol, see Isobutanol | | | | | | | | | | | | |
| | Isobutyl aldehyde, see Isobutyraldehyde | | | | | | | | | | | | |
| D | Isobutyl chloroformate | 6.1 | NA2742 | I | POISON, FLAMMABLE LIQUID, CORROSIVE | 2, B9, B14, B32, B74, T38, T43, T45 | None | 227 | 244 | 1 L | 30 L | A | 12, 13, 22, 25, 40, 48, 100 |
| | Isobutyl formate | 3 | UN2393 | II | FLAMMABLE LIQUID | T1 | 150 | 202 | 242 | 5 L | 60 L | B | |
| | Isobutyl isobutyrate | 3 | UN2528 | III | FLAMMABLE LIQUID | B1, T1 | 150 | 203 | 242 | 60 L | 220 L | A | |
| + | Isobutyl isocyanate | 3 | UN2486 | II | FLAMMABLE LIQUID, POISON | 1, B9, B14, B30, B72, T38, T43, T44 | None | 226 | 244 | 1 L | 60 L | D | 40 |
| | Isobutyl methacrylate | 3 | UN2283 | III | FLAMMABLE LIQUID | B1, T1 | 150 | 203 | 242 | 60 L | 220 L | A | |
| | Isobutyl propionate | 3 | UN2394 | III | FLAMMABLE LIQUID | T1 | 150 | 203 | 242 | 60 L | 220 L | B | |
| | Isobutylamine | 3 | UN1214 | II | FLAMMABLE LIQUID | T8 | 150 | 202 | 242 | 5 L | 60 L | B | 40 |
| | Isobutylene *see also* Petroleum gases, liquefied | 2 1 | UN1055 | | FLAMMABLE GAS | | 306 | 304 | 314, 315 | Forbidden | 150 kg | E | 40 |
| | Isobutyraldehyde *or* isobutyl aldehyde | 3 | UN2045 | II | FLAMMABLE LIQUID | T8 | 150 | 202 | 242 | 5 L | 60 L | E | 40 |
| | Isobutyric acid | 3 | UN2529 | III | FLAMMABLE LIQUID | B1, T1 | 150 | 203 | 242 | 60 L | 220 L | A | |
| | Isobutyric anhydride | 3 | UN2530 | III | FLAMMABLE LIQUID | B1, T1 | 150 | 203 | 242 | 60 L | 220 L | A | |

## §172.101 HAZARDOUS MATERIALS TABLE, Continued

| Symbols | Hazardous materials descriptions and proper shipping names | Hazard class or Division | Identification Numbers | Packing group | Label(s) required (if not excepted) | Special provisions | (8) Packaging authorizations (§173.***) | | | (9) Quantity limitations | | (10) Vessel stowage requirements | |
|---|---|---|---|---|---|---|---|---|---|---|---|---|---|
| | | | | | | | Exceptions | Non-bulk packaging | Bulk packaging | Passenger aircraft or railcar | Cargo aircraft only | Vessel stowages | Other stowage provisions |
| (1) | (2) | (3) | (4) | (5) | (6) | (7) | (8A) | (8B) | (8C) | (9A) | (9B) | (10A) | (10B) |
| | Isobutyronitrile | 3 | UN2284 | II | FLAMMABLE LIQUID POISON | T17 | None | 202 | 243 | 1 L | 60 L | E | 40 |
| | Isobutyryl chloride | 3 | UN2395 | II | FLAMMABLE LIQUID CORROSIVE | T9, T26 | None | 202 | 243 | 1 L | 5 L | C | 40 |
| | Isocyanates liquid or solid n.o.s. or Isocyanate solutions n.o.s. *boiling point not less than 300 degrees C* | 6.1 | UN2207 | III | KEEP AWAY FROM FOOD | | 153 | 203 | 241 | 60 L | 220 L | A | 48 |
| | Isocyanates n.o.s. or Isocyanate solutions n.o.s., *flash point more than 61 degrees C and boiling point less than 300 degrees C* | 6.1 | UN2206 | II | POISON | T15 | None | 202 | 243 | 5 L | 60 L | D | 25, 40, 48 |
| | Isocyanates n.o.s. or Isocyanate solutions n.o.s. *flash point not less than 23 degrees C but not more than 61 degrees C and boiling point less than 300 degrees C* | 6.1 | UN3080 | II | POISON, FLAMMABLE LIQUID | T15 | None | 202 | 243 | 5 L | 60 L | D | 25, 40, 48 |
| | Isocyanates, n.o.s. or Isocyanate solutions, n.o.s. (flashpoint less than 23 degrees C) | 3 | UN2478 | II | FLAMMABLE LIQUID, POISON | 5, A3, A7 | None | 202 | 243 | 1 L | 60 L | D | 40 |
| | Isocyanatobenzotrifluorides | 6.1 | UN2285 | II | POISON | 5, T14 | None | 202 | 243 | 5 L | 60 L | B | 25, 40, 48 |
| | Isoheptenes | 3 | UN2287 | II | FLAMMABLE LIQUID | T7 | 150 | 202 | 242 | 5 L | 60 L | B | |
| | Isohexenes | 3 | UN2288 | II | FLAMMABLE LIQUID | T7 | 150 | 202 | 242 | 5 L | 60 L | E | |
| | *Isooctane, see* Octanes | | | | | | | | | | | | |
| | Isooctenes | 3 | UN1216 | II | FLAMMABLE LIQUID | T8 | 150 | 202 | 242 | 5 L | 60 L | B | |
| | Isopentane, *see* n-Pentane | | | | | | | | | | | | |
| | *Isopentanoic acid, see* Corrosive liquids, n.o.s. | | | | | | | | | | | | |
| | Isopentenes | 3 | UN2371 | I | FLAMMABLE LIQUID | T20 | 150 | 201 | 243 | 1 L | 30 L | E | |
| | Isophorone diisocyanate | 6.1 | UN2290 | III | KEEP AWAY FROM FOOD | T7 | 153 | 203 | 241 | 60 L | 220 L | B | 40 |
| | Isophoronediamine | 8 | UN2289 | III | CORROSIVE | T8 | 154 | 203 | 241 | 5 L | 60 L | A | 8 |
| | Isoprene, inhibited | 3 | UN1218 | I | FLAMMABLE LIQUID | T20 | 150 | 201 | 243 | 1 L | 30 L | E | |
| | Isopropanol or Isopropyl alcohol | 3 | UN1219 | II | FLAMMABLE LIQUID | T1 | 150 | 202 | 242 | 5 L | 60 L | B | |
| | Isopropenyl acetate | 3 | UN2403 | II | FLAMMABLE LIQUID | T1 | 150 | 202 | 242 | 5 L | 60 L | B | |
| | Isopropenylbenzene | 3 | UN2303 | III | FLAMMABLE LIQUID | B1, T1 | 150 | 203 | 242 | 60 L | 220 L | A | |
| | Isopropyl acetate | 3 | UN1220 | II | FLAMMABLE LIQUID | T1 | 150 | 202 | 242 | 5 L | 60 L | B | |
| | Isopropyl acid phosphate | 8 | UN1793 | III | CORROSIVE | T7 | 154 | 213 | 240 | 25 kg | 100 kg | A | |
| | Isopropyl alcohol, *see* Isopropanol | | | | | | | | | | | | |
| | Isopropyl butyrate | 3 | UN2405 | III | FLAMMABLE LIQUID | B1, T1 | 150 | 203 | 242 | 60 L | 220 L | A | |
| | Isopropyl chloroacetate | 3 | UN2947 | III | FLAMMABLE LIQUID | B1, T1 | 150 | 203 | 242 | 60 L | 220 L | A | |
| + | Isopropyl chloroformate | 3 | UN2407 | I | FLAMMABLE LIQUID, CORROSIVE, POISON | 2, B9, B14, B32, B74, B77, T38, T43, T45 | None | 227 | 244 | Forbidden | Forbidden | B | 40 |
| | Isopropyl 2-chloropropionate | 3 | UN2934 | III | FLAMMABLE LIQUID | B1, T1 | 150 | 203 | 242 | 60 L | 220 L | A | |
| | Isopropyl isobutyrate | 3 | UN2406 | II | FLAMMABLE LIQUID | T1 | 150 | 202 | 242 | 5 L | 60 L | B | |
| + | Isopropyl isocyanate | 3 | UN2483 | I | FLAMMABLE LIQUID, POISON | 1, B9, B14, B30, B72, T38, T43, T44 | None | 226 | 244 | Forbidden | 30 L | D | 40 |
| | *Isopropyl mercaptan, see* Propanethiols | | | | | | | | | | | | |
| | Isopropyl nitrate | 3 | UN1222 | II | FLAMMABLE LIQUID | T25 | 150 | 202 | None | 5 L | 60 L | D | |
| | *Isopropyl phosphoric acid, see* Isopropyl acid phosphate | | | | | | | | | | | | |
| | Isopropyl propionate | 3 | UN2409 | II | FLAMMABLE LIQUID | T1 | 150 | 202 | 242 | 5 L | 60 L | B | |
| | Isopropylamine | 3 | UN1221 | I | FLAMMABLE LIQUID | T20 | None | 201 | 243 | 1 L | 30 L | E | |
| | Isopropylbenzene | 3 | UN1918 | III | FLAMMABLE LIQUID | B1, T1 | 150 | 203 | 242 | 60 L | 220 L | A | |
| | *Isopropylcumyl hydroperoxide, more than 72 per cent in solution* | Forbidden | | | | | | | | | | | |
| | Isosorbide dinitrate mixture *with not less than 60 per cent lactose, mannose, starch or calcium hydrogen phosphate* | 4.1 | UN2907 | II | FLAMMABLE SOLID | | None | 212 | None | 15 kg | 50 kg | E | |
| | *Isothiocyanic acid* | Forbidden | | | | | | | | | | | |

## §172.101 HAZARDOUS MATERIALS TABLE, Continued

| Symbols | Hazardous materials descriptions and proper shipping names | Hazard class or Division | Identification Numbers | Packing group | Label(s) required (if not excepted) | Special provisions | (8) Packaging authorizations (§173.***) | | | (9) Quantity limitations | | (10) Vessel stowage requirements | |
|---|---|---|---|---|---|---|---|---|---|---|---|---|---|
| | | | | | | | Exceptions | Non-bulk packaging | Bulk packaging | Passenger aircraft or railcar | Cargo aircraft only | Vessel stowages | Other stowage provisions |
| (1) | (2) | (3) | (4) | (5) | (6) | (7) | (8A) | (8B) | (8C) | (9A) | (9B) | (10A) | (10B) |
| | *Jet fuel, see* Fuel aviation, turbine engine | | | | | | | | | | | | |
| | Jet perforating guns, charged oil well, *without detonator* | 1 1D | UN0124 | II | EXPLOSIVE 1 1D | | None | 62 | None | Forbidden | Forbidden | B | |
| D | Jet perforating guns, charged *oil well, without detonator* | 1 4D | NA0124 | II | EXPLOSIVE 1 4D | 114 | None | 62 | None | Forbidden | Forbidden | A | 24E |
| | *Jet perforators, see* Charges, shaped, commercial *etc* | | | | | | | | | | | | |
| | *Jet tappers, without detonator, see* Charges, shaped commercial, *etc* | | | | | | | | | | | | |
| | *Jet thrust igniters, for rocket motors or Jato, see* Igniters | | | | | | | | | | | | |
| | Jet thrust unit (Jato), *see* Rocket motors | | | | | | | | | | | | |
| | Kerosene | 3 | UN1223 | III | FLAMMABLE LIQUID | B1 T1 | 150 | 203 | 242 | 60 L | 220 L | A | |
| | Ketones, liquid, n o s | 3 | UN1224 | I | FLAMMABLE LIQUID | T8, T31 | None | 201 | 243 | 1 L | 30 L | E | |
| | | | | II | FLAMMABLE LIQUID | T8, T31 | 150 | 202 | 242 | 5 L | 60 L | B | |
| | | | | III | FLAMMABLE LIQUID | B1, T7, T30 | 150 | 203 | 242 | 60 L | 220 L | A | |
| | Krypton, compressed | 2 2 | UN1056 | | NONFLAMMABLE GAS | | 306 | 302 | None | 75 kg | 150 kg | A | |
| | Krypton, refrigerated liquid *(cryogenic liquid)* | 2.2 | UN1970 | | NONFLAMMABLE GAS | | 320 | None | None | 50 kg | 500 kg | B | |
| | *Lacquer base or lacquer chips nitrocellulose, dry, see* Nitrocellulose, *etc (UN 2557)* | | | | | | | | | | | | |
| | *Lacquer base or lacquer chips, plastic, wet with alcohol or solvent, see* Nitrocellulose *(UN 2059, UN 2060, UN 2555, UN2556) or* Paint *etc (UN1263)* | | | | | | | | | | | | |
| | Lead acetate | 6.1 | UN1616 | III | KEEP AWAY FROM FOOD | | 153 | 213 | 240 | 100 kg | 200 kg | A | |
| | Lead arsenates | 6.1 | UN1617 | II | POISON | | None | 212 | 242 | 25 kg | 100 kg | A | |
| | Lead arsenites | 6.1 | UN1618 | II | POISON | | None | 212 | 242 | 25 kg | 100 kg | A | |
| | *Lead azide (dry)* | Forbidden | | | | | | | | | | | |
| | Lead azide, wetted *with not less than 20 per cent water or mixture of alcohol and water, by mass* | 1.1A | UN0129 | II | EXPLOSIVE 1.1A | 111, 117 | None | 62 | None | Forbidden | Forbidden | E | 2E, 6E |
| | Lead compounds, soluble, n.o.s. | 6.1 | UN2291 | III | KEEP AWAY FROM FOOD | | 153 | 213 | 240 | 100 kg | 200 kg | A | |
| | Lead cyanide | 6.1 | UN1620 | II | POISON | | None | 212 | 242 | 25 kg | 100 kg | A | 26 |
| | Lead dioxide | 5.1 | UN1872 | III | OXIDIZER | A1 | 152 | 213 | 240 | 25 kg | 100 kg | A | 34 |
| | *Lead dross, see* Lead sulphate, *with more than 3% free acid* | | | | | | | | | | | | |
| D | Lead mononitroresorcinate | 1 1A | NA0473 | II | EXPLOSIVE 1.1A | 111, 117 | None | 62 | None | Forbidden | Forbidden | E | 2E, 6E |
| | Lead nitrate | 5.1 | UN1469 | II | OXIDIZER, POISON | | None | 212 | 242 | 5 kg | 25 kg | A | |
| | *Lead nitroresorcinate (dry)* | Forbidden | | | | | | | | | | | |
| | Lead perchlorate, solid *or* solution | 5.1 | UN1470 | II | OXIDIZER, POISON | T8 | None | 212 | 242 | 5 kg | 25 kg | A | 56, 58, 106 |
| | *Lead peroxide, see* Lead dioxide | | | | | | | | | | | | |
| | Lead phosphite, dibasic | 4.1 | UN2989 | II | FLAMMABLE SOLID | | None | 212 | 240 | 5 kg | 25 kg | B | 34 |
| | | | | III | FLAMMABLE SOLID | | 151 | 213 | 240 | 15 kg | 50 kg | B | 34 |
| | *Lead picrate (dry)* | Forbidden | | | | | | | | | | | |
| | *Lead styphnate (dry)* | Forbidden | | | | | | | | | | | |
| | Lead styphnate, wetted *or* Lead trinitroresorcinate, wetted *with not less than 20 per cent water or mixture of alcohol and water, by mass* | 1 1A | UN0130 | II | EXPLOSIVE 1.1A | 111, 117 | None | 62 | None | Forbidden | Forbidden | E | 2E, 6E |
| | Lead sulfate *with more than 3 per cent free acid* | 8 | UN1794 | II | CORROSIVE | | 154 | 212 | 240 | 15 kg | 50 kg | A | |
| | Lead trinitroresorcinate, *see* Lead styphnate, *etc* | | | | | | | | | | | | |
| | Life-saving appliances, not self inflating containing dangerous goods as equipment | 9 | UN3072 | | None | | None | 219 | None | No limit | No limit | A | 120 |
| | Life-saving appliances, self inflating | 9 | UN2990 | | None | | None | 219 | None | No limit | No limit | A | 120 |
| | *Lighter replacement cartridges containing liquefied petroleum gases (and similar devices, each not exceeding 65 grams), see* Lighters for cigars, cigarettes, etc *with flammable gas* | | | | | | | | | | | | |
| D | Lighters for cigars, cigarettes, etc., *with lighter fluids* | 3 | NA1226 | II | FLAMMABLE LIQUID | N10 | None | 21 | None | Forbidden | Forbidden | B | |

## §172.101 HAZARDOUS MATERIALS TABLE, Continued

| Symbols | Hazardous materials descriptions and proper shipping names | Hazard class or Division | Identification Numbers | Packing group | Label(s) required (if not excepted) | Special provisions | (8) Packaging authorizations (§173.***) | | | (9) Quantity limitations | | (10) Vessel stowage requirements | |
|---|---|---|---|---|---|---|---|---|---|---|---|---|---|
| | | | | | | | Exceptions | Non-bulk packaging | Bulk packaging | Passenger aircraft or railcar | Cargo aircraft only | Vessel stowages | Other stowage provisions |
| (1) | (2) | (3) | (4) | (5) | (6) | (7) | (8A) | (8B) | (8C) | (9A) | (9B) | (10A) | (10B) |
| | Lighters, fuse | 1.4S | UN0131 | II | EXPLOSIVE 1.4S | | None | 62 | None | 25 kg | 100 kg | A | 9E |
| | Lighters or Lighter refills (cigarettes) containing flammable gas | 2.1 | UN1057 | | FLAMMABLE GAS | N10 | None | 21, 308 | None | 1 kg | 15 kg | B | 40 |
| | *Lime, unslaked, see* Calcium oxide | | | | | | | | | | | | |
| | Liquefied gases, *non-flammable charged with nitrogen, carbon dioxide or air* | 2.2 | UN1058 | | NONFLAMMABLE GAS | | 306 | 304 | None | 75 kg | 150 kg | A | |
| | *Liquefied hydrocarbon gas, see* Hydrocarbon gases, liquefied, n.o.s., *etc* | | | | | | | | | | | | |
| | *Liquefied natural gas, see* Methane, *etc. (UN 1972)* | | | | | | | | | | | | |
| | Liquefied petroleum gas *see* Petroleum gases, liquefied | | | | | | | | | | | | |
| | Lithium | 4.3 | UN1415 | II | DANGEROUS WHEN WET | A7, A19, N45 | None | 212 | 244 | Forbidden | 50 kg | E | |
| | *Lithium acetylide ethylenediamine complex, see* Substances which in contact with water emit flammable gases | | | | | | | | | | | | |
| | Lithium alkyls | 4.2 | UN2445 | I | SPONTANEOUSLY COMBUSTIBLE | B11, T28, T40 | None | 181 | 244 | Forbidden | Forbidden | D | |
| | Lithium aluminum hydride | 4.3 | UN1410 | I | DANGEROUS WHEN WET | A19, N40 | None | 211 | 242 | Forbidden | 15 kg | E | |
| | Lithium aluminum hydride, ethereal | 4.3 | UN1411 | I | DANGEROUS WHEN WET, FLAMMABLE LIQUID | A2, A3, A11, N34 | None | 201 | 244 | Forbidden | 1 L | D | 40 |
| | Lithium batteries, contained in equipment | 9 | UN3091 | II | CLASS 9 | 18, A12 | None | 185 | None | See A12 | See A12 | A | |
| | Lithium battery, liquid cathode | 9 | UN3090 | II | CLASS 9 | A12 | None | 185 | None | See A12 | 0.5 kg | A | |
| | Lithium battery, solid cathode | 9 | UN3090 | II | CLASS 9 | A12 | None | 185 | None | See A12 | 0.5 kg | A | |
| | Lithium borohydride | 4.3 | UN1413 | I | DANGEROUS WHEN WET | A19, N40 | None | 211 | 242 | Forbidden | 15 kg | E | |
| | Lithium ferrosilicon | 4.3 | UN2830 | II | DANGEROUS WHEN WET | A19 | None | 212 | 241 | 15 kg | 50 kg | E | 40, 85, 103 |
| | Lithium hydride | 4.3 | UN1414 | I | DANGEROUS WHEN WET | A19, N40 | None | 211 | 242 | Forbidden | 15 kg | E | |
| | Lithium hydride, fused solid | 4.3 | UN2805 | II | DANGEROUS WHEN WET | A8, A19, A20 | None | 212 | 241 | 15 kg | 50 kg | E | |
| | Lithium hydroxide, monohydrate *or* Lithium hydroxide, solid | 8 | UN2680 | II | CORROSIVE | | 154 | 212 | 240 | 15 kg | 50 kg | A | |
| | Lithium hydroxide, solution | 8 | UN2679 | II | CORROSIVE | B2, T8 | 154 | 202 | 242 | 1 L | 30 L | A | 96 |
| | Lithium hypochlorite, dry *or* Lithium hypochlorite mixtures, dry | 5.1 | UN1471 | II | OXIDIZER | A9, N34 | 152 | 212 | 240 | 5 kg | 25 kg | A | 13, 48, 56, 58, 69, 106, 116 |
| | *Lithium in cartridges, see* Lithium | | | | | | | | | | | | |
| | Lithium nitrate | 5.1 | UN2722 | III | OXIDIZER | A1 | 152 | 213 | 240 | 25 kg | 100 kg | A | |
| | Lithium nitride | 4.3 | UN2806 | I | DANGEROUS WHEN WET | A19, N40 | None | 211 | 242 | Forbidden | 15 kg | E | |
| | Lithium peroxide | 5.1 | UN1472 | II | OXIDIZER | A9, N34 | 152 | 212 | None | 5 kg | 25 kg | A | 13, 75, 106 |
| | Lithium silicon | 4.3 | UN1417 | II | DANGEROUS WHEN WET | A19, A20 | None | 212 | 241 | 15 kg | 50 kg | A | 85, 103 |
| | *LNG, see* Methane *etc. (UN 1972)* | | | | | | | | | | | | |
| | London Purple | 6.1 | UN1621 | II | POISON | | None | 212 | 242 | 25 kg | 100 kg | A | |
| | *LPG, see* Petroleum gases, liquefied | | | | | | | | | | | | |
| | *Lye, see* Sodium hydroxide, solutions | | | | | | | | | | | | |
| | Magnesium alkyls | 4.2 | UN3053 | I | SPONTANEOUSLY COMBUSTIBLE | B11, T28, T29, T40 | None | 181 | 244 | Forbidden | Forbidden | D | 18 |
| | Magnesium aluminum phosphide | 4.3 | UN1419 | I | DANGEROUS WHEN WET, POISON | A19, N34, N40 | None | 211 | 242 | Forbidden | 15 kg | E | 40, 85 |
| + | Magnesium arsenate | 6.1 | UN1622 | II | POISON | | None | 212 | 242 | 25 kg | 100 kg | A | |
| | *Magnesium bisulfite solution, see* Bisulfites inorganic aqueous solutions, n.o.s | | | | | | | | | | | | |
| | Magnesium bromate | 5.1 | UN1473 | II | OXIDIZER | A1 | 152 | 212 | 242 | 5 kg | 25 kg | A | 56, 58, 106 |
| | Magnesium chlorate | 5.1 | UN2723 | II | OXIDIZER | | 152 | 212 | 242 | 5 kg | 25 kg | A | 56, 58, 106 |
| | Magnesium diamide | 4.2 | UN2004 | II | SPONTANEOUSLY COMBUSTIBLE | A8, A19, A20 | None | 212 | 241 | 15 kg | 50 kg | C | |
| | Magnesium diphenyl | 4.2 | UN2005 | I | SPONTANEOUSLY COMBUSTIBLE | | None | 187 | 244 | Forbidden | Forbidden | C | |
| | *Magnesium dross, wet or hot* | Forbidden | | | | | | | | | | | |
| | Magnesium fluorosilicate | 6.1 | UN2853 | III | KEEP AWAY FROM FOOD | | 153 | 213 | 240 | 100 kg | 200 kg | A | 26 |
| | Magnesium granules, *coated particle size not less than 149 microns* | 4.3 | UN2950 | III | DANGEROUS WHEN WET | A1, A19 | None | 213 | 240 | 25 kg | 100 kg | A | |
| | Magnesium hydride | 4.3 | UN2010 | I | DANGEROUS WHEN WET | A19, N40 | None | 211 | 242 | Forbidden | 15 kg | E | |

## §172.101 HAZARDOUS MATERIALS TABLE, Continued

| Symbols | Hazardous materials descriptions and proper shipping names | Hazard class or Division | Identification Numbers | Packing group | Label(s) required (if not excepted) | Special provisions | (8) Packaging authorizations (§173.***) | | | (9) Quantity limitations | | (10) Vessel stowage requirements | |
|---|---|---|---|---|---|---|---|---|---|---|---|---|---|
| | | | | | | | Exceptions | Non-bulk packaging | Bulk packaging | Passenger aircraft or railcar | Cargo aircraft only | Vessel stowages | Other stowage provisions |
| (1) | (2) | (3) | (4) | (5) | (6) | (7) | (8A) | (8B) | (8C) | (9A) | (9B) | (10A) | (10B) |
| | Magnesium *or* Magnesium alloys *with more than 50 per cent magnesium in pellets, turnings or ribbons* | 4.1 | UN1869 | III | FLAMMABLE SOLID | A1 | 151 | 213 | 240 | 25 kg | 100 kg | A | 39 |
| | Magnesium nitrate | 5 1 | UN1474 | III | OXIDIZER | A1, T2 | 152 | 213 | 240 | 25 kg | 100 kg | A | |
| | Magnesium perchlorate | 5 1 | UN1475 | II | OXIDIZER | T8 | 152 | 212 | 242 | 5 kg | 25 kg | A | 56, 58, 106 |
| | Magnesium peroxide | 5 1 | UN1476 | II | OXIDIZER | | 152 | 212 | 242 | 5 kg | 25 kg | A | 13, 75, 106 |
| | Magnesium phosphide | 4 3 | UN2011 | I | DANGEROUS WHEN WET, POISON | A19, N40 | None | 211 | None | Forbidden | 15 kg | E | 40, 85 |
| | Magnesium, powder *or* Magnesium alloys, powder | 4 3 | UN1418 | II | DANGEROUS WHEN WET, SPONTANEOUSLY COMBUSTIBLE | A19, B56 | None | 212 | 241 | 15 kg | 50 kg | A | 39 |
| | *Magnesium scrap, see* Magnesium, *etc (UN 1869)* | | | | | | | | | | | | |
| | Magnesium silicide | 4 3 | UN2624 | II | DANGEROUS WHEN WET | A19, A20 | None | 212 | 241 | 15 kg | 50 kg | B | 85, 103 |
| | *Magnetized material, see section 173 21* | | | | | | | | | | | | |
| D | *Maleic acid* | 8 | NA2215 | III | CORROSIVE | | 154 | 213 | 240 | 25 kg | 100 kg | A | |
| | *Maleic anhydride* | 8 | UN2215 | III | CORROSIVE | T7 | 154 | 213 | 240 | 25 kg | 100 kg | A | |
| | *Malononitrile* | 6 1 | UN2647 | II | POISON | | None | 212 | 242 | 25 kg | 100 kg | A | 12 |
| | *Maneb or Maneb preparations with not less than 60 per cent maneb* | 4 2 | UN2210 | III | SPONTANEOUSLY COMBUSTIBLE, DANGEROUS WHEN WET | A1, A19 | None | 213 | 242 | 25 kg | 100 kg | A | 34 |
| | Maneb stabilized *or* Maneb preparations, *stabilized against self-heating* | 4 3 | UN2968 | III | DANGEROUS WHEN WET | A1, A19 | None | 213 | 242 | 25 kg | 100 kg | B | 34 |
| | Manganese nitrate | 5 1 | UN2724 | III | OXIDIZER | A1 | 152 | 213 | 240 | 25 kg | 100 kg | A | |
| | Manganese resinate | 4 1 | UN1330 | III | FLAMMABLE SOLID | A1 | 151 | 213 | 240 | 25 kg | 100 kg | A | |
| | *Manniton tetranitrate* | Forbidden | | | | | | | | | | | |
| | *Mannitol hexanitrate (dry)* | Forbidden | | | | | | | | | | | |
| D | Mannitol hexanitrate (Nitromannite), wetted *with not less than 40 percent water, by mass or mixture of alcohol and water* | 1.1A | NA0133 | II | EXPLOSIVE 1.1A | 111 | None | 62 | None | Forbidden | Forbidden | E | 1E, 5E |
| | *Marine pollutants, liquid or solid, n.o.s., see* Environmentally hazardous substances, liquid *or* solid n.o.s. | | | | | | | | | | | | |
| | *Matches, block, see* Matches, *strike anywhere* | | | | | | | | | | | | |
| | *Matches, fusee* | 4 1 | UN2254 | III | FLAMMABLE SOLID | | 186 | 186 | None | Forbidden | Forbidden | A | |
| | Matches, safety *(book, card or strike on box)* | 4 1 | UN1944 | III | FLAMMABLE SOLID | | 186 | 186 | None | 25 kg | 100 kg | A | |
| | Matches, strike anywhere | 4 1 | UN1331 | III | FLAMMABLE SOLID | | 186 | 186 | None | Forbidden | Forbidden | B | |
| | Matches, wax, Vesta | 4 1 | UN1945 | III | FLAMMABLE SOLID | | 186 | 186 | None | 25 kg | 100 kg | B | |
| | *Matting acid, see* Sulfuric acid | | | | | | | | | | | | |
| D | Medicines, *corrosive, liquid*, n o s | 8 | NA1760 | II | CORROSIVE | B3 | 154 | 202 | 242 | 1 L | 30 L | B | 40 |
| | | | | III | CORROSIVE | | 154 | 203 | 241 | 5 L | 60 L | A | 40 |
| D | Medicines, *corrosive, solid, n.o.s* | 8 | NA1759 | II | CORROSIVE | | 154 | 212 | 240 | 15 kg | 50 kg | A | |
| | | | | III | CORROSIVE | | 154 | 213 | 240 | 25 kg | 100 kg | A | |
| D | Medicines, *flammable, liquid, n o.s* | 3 | NA1993 | I | FLAMMABLE LIQUID | | 150 | 201 | 243 | 1 L | 30 L | E | |
| | | | | II | FLAMMABLE LIQUID | | 150 | 202 | 242 | 5 L | 60 L | B | |
| | | | | III | FLAMMABLE LIQUID | B1 | 150 | 203 | 242 | 60 L | 220 L | A | |
| D | Medicines, *flammable, solid, n.o.s* | 4 1 | NA1325 | II | FLAMMABLE SOLID | | 151 | 212 | 240 | 15 kg | 50 kg | B | |
| D | Medicines, *oxidizing substance, solid n.o.s* | 5 1 | NA1479 | II | OXIDIZER | | 152 | 212 | 242 | 5 kg | 25 kg | B | 56, 58, 69, 106 |
| D | Medicines, *poisonous, liquid, n.o s* | 6 1 | NA2810 | I | POISON | | 153 | 201 | 243 | 1 L | 30 L | B | 40 |
| | | | | II | POISON | | 153 | 202 | 243 | 5 L | 60 L | B | 40 |
| | | | | III | KEEP AWAY FROM FOOD | | 153 | 203 | 241 | 60 L | 220 L | A | 40 |
| D | Medicines, *poisonous, solid, n.o.s.* | 6 1 | NA2811 | I | POISON | | 153 | 211 | 242 | 5 kg | 50 kg | B | |
| | | | | II | POISON | | 153 | 212 | 242 | 25 kg | 100 kg | B | |
| | | | | III | KEEP AWAY FROM FOOD | | 153 | 213 | 240 | 100 kg | 200 kg | A | |
| | *Memtetrahydrophthalic anhydride, see* Corrosive liquids, n.o.s. | | | | | | | | | | | | |
| | Mercaptans, liquid, n.o.s. *or* Mercaptan mixtures, liquid, n.o.s. | 3 | UN1228 | II | FLAMMABLE LIQUID, POISON | T14 | None | 202 | 243 | Forbidden | 60 L | B | 40, 95, 102 |
| | Mercaptans, liquid n.o.s. *or* Mercaptan mixtures, liquid, *n.o.s., flash point not less than 23 degrees C* | 6 1 | UN3071 | II | POISON, FLAMMABLE LIQUID | T14 | None | 202 | 243 | 5 L | 60 L | C | 40, 121 |
| | 5-Mercaptotetrazol-1-acetic acidd | 1.4C | UN0448 | II | EXPLOSIVE 1.4C | | None | 62 | None | Forbidden | 75 kg | E | 1E, 5E |

## §172.101 HAZARDOUS MATERIALS TABLE, Continued

| Symbols | Hazardous materials descriptions and proper shipping names | Hazard class or Division | Identification Numbers | Packing group | Label(s) required (if not excepted) | Special provisions | (8) Packaging authorizations (§173.***) | | | (9) Quantity limitations | | (10) Vessel stowage requirements | |
|---|---|---|---|---|---|---|---|---|---|---|---|---|---|
| | | | | | | | Exceptions | Non-bulk packaging | Bulk packaging | Passenger aircraft or railcar | Cargo aircraft only | Vessel stowages | Other stowage provisions |
| (1) | (2) | (3) | (4) | (5) | (6) | (7) | (8A) | (8B) | (8C) | (9A) | (9B) | (10A) | (10B) |
| | Mercuric arsenate | 6.1 | UN1623 | II | POISON | | None | 212 | 242 | 25 kg | 100 kg | A | |
| | Mercuric chloride | 6.1 | UN1624 | II | POISON | | None | 212 | 242 | 25 kg | 100 kg | A | |
| | *Mercuric compounds, see* Mercury compounds, etc | | | | | | | | | | | | |
| | Mercuric nitrate | 6.1 | UN1625 | II | POISON | N73 | None | 212 | 242 | 25 kg | 100 kg | A | |
| + | Mercuric potassium cyanide | 6.1 | UN1626 | I | POISON | N74, N75 | None | 211 | 242 | 5 kg | 50 kg | A | 26 |
| | *Mercuric sulfocyanate, see* Mercury thiocyanate | | | | | | | | | | | | |
| | *Mercurol see* Mercury nucleate | | | | | | | | | | | | |
| | *Mercurous azide* | Forbidden | | | | | | | | | | | |
| | *Mercurous compounds, see* Mercury compounds, etc | | | | | | | | | | | | |
| | Mercurous nitrate | 6.1 | UN1627 | II | POISON | | None | 212 | 242 | 25 kg | 100 kg | A | |
| A W | Mercury | 8 | UN2809 | III | CORROSIVE | | 164 | 164 | 240 | 35 kg | 35 kg | B | 40, 97 |
| | Mercury acetate | 6.1 | UN1629 | II | POISON | | None | 212 | 242 | 25 kg | 100 kg | A | |
| | *Mercury acetylide* | Forbidden | | | | | | | | | | | |
| | Mercury ammonium chloride | 6.1 | UN1630 | II | POISON | | None | 212 | 242 | 25 kg | 100 kg | A | |
| | Mercury based pesticides, liquid, flammable, toxic, n.o.s *flash point less than 23 deg C* | 3 | UN2778 | I | FLAMMABLE LIQUID, POISON | | None | 201 | 243 | Forbidden | 30 L | B | 40 |
| | | | | II | FLAMMABLE LIQUID, POISON | | None | 202 | 243 | 1 L | 60 L | B | 40 |
| | Mercury based pesticides, liquid, toxic, flammable, n.o.s *flashpoint not less than 23 degrees C* | 6.1 | UN3011 | I | POISON, FLAMMABLE LIQUID | T42 | None | 201 | 243 | 1 L | 30 L | B | 40 |
| | | | | II | POISON, FLAMMABLE LIQUID | T14 | None | 202 | 243 | 5 L | 60 L | B | 40 |
| | | | | III | KEEP AWAY FROM FOOD, FLAMMABLE LIQUID | T14 | 153 | 202 | 241 | 60 L | 220 L | A | 40 |
| | Mercury based pesticides, liquid, toxic, n.o.s | 6.1 | UN3012 | I | POISON | T42 | None | 201 | 243 | 1 L | 30 L | B | 40 |
| | | | | II | POISON | T14 | None | 202 | 243 | 5 L | 60 L | B | 40 |
| | | | | III | KEEP AWAY FROM FOOD | T14 | 153 | 203 | 241 | 60 L | 220 L | A | 40 |
| | Mercury based pesticides, solid, toxic, n.o.s. | 6.1 | UN2777 | I | POISON | | None | 211 | 242 | 5 kg | 50 kg | A | 40 |
| | | | | II | POSION | | None | 212 | 242 | 25 kg | 100 kg | A | 40 |
| | | | | III | KEEP AWAY FROM FOOD | | 153 | 213 | 240 | 100 kg | 200 kg | A | 40 |
| | Mercury benzoate | 6.1 | UN1631 | II | POISON | | None | 212 | 242 | 25 kg | 100 kg | A | |
| | Mercury bromides | 6.1 | UN1634 | II | POISON | | None | 212 | 242 | 25 kg | 100 kg | A | |
| | Mercury compounds, liquid, n.o.s | 6.1 | UN2024 | I | POISON | | None | 201 | 243 | 1 L | 30 L | B | 40 |
| | | | | II | POISON | | None | 202 | 243 | 5 L | 60 L | B | 40 |
| | | | | III | KEEP AWAY FROM FOOD | | 153 | 203 | 241 | 60 L | 220 L | B | 40 |
| | Mercury compounds, solid, n.o.s | 6.1 | UN2025 | I | POISON | | None | 211 | 242 | 5 kg | 50 kg | A | |
| | | | | II | POISON | | None | 212 | 242 | 25 kg | 100 kg | A | |
| | | | | III | KEEP AWAY FROM FOOD | | 153 | 213 | 240 | 100 kg | 200 kg | A | |
| A | Mercury *contained in manufactured articles* | 8 | UN2809 | I | CORROSIVE | | None | 164 | None | No limit | No limit | B | 40 |
| | Mercury cyanide | 6.1 | UN1636 | II | POISON | N74, N75 | None | 212 | 242 | 25 kg | 100 kg | A | 26 |
| | Mercury fulminate, wetted *with not less than 20 per cent water, or mixture of alcohol and water, by mass* | 1.1A | UN0135 | II | EXPLOSIVE 1.1A | 111, 117 | None | 62 | None | Forbidden | Forbidden | E | 2E, 6E |
| | Mercury gluconate | 6.1 | UN1637 | II | POISON | | None | 212 | 242 | 25 kg | 100 kg | A | |
| | Mercury iodide | 6.1 | UN1638 | II | POISON | | None | 212 | 242 | 25 kg | 100 kg | A | |
| | *Mercury iodide aquabasic ammonobasic (Iodide of Millon's base)* | Forbidden | | | | | | | | | | | |
| | Mercury iodide, solution | 6.1 | UN1638 | II | POISON | | None | 202 | 243 | 5 L | 60 L | A | |
| | *Mercury nitride* | Forbidden | | | | | | | | | | | |
| | Mercury nucleate | 6.1 | UN1639 | II | POISON | | None | 212 | 242 | 25 kg | 100 kg | A | |
| | Mercury oleate | 6.1 | UN1640 | II | POISON | | None | 212 | 242 | 25 kg | 100 kg | A | |
| | Mercury oxide | 6.1 | UN1641 | II | POISON | | None | 212 | 242 | 25 kg | 100 kg | A | |
| | *Mercury oxycyanide* | Forbidden | | | | | | | | | | | |
| | Mercury oxycyanide, desensitized | 6.1 | UN1642 | II | POISON | | None | 212 | 242 | 25 kg | 100 kg | A | 26, 91 |
| | Mercury potassium iodide | 6.1 | UN1643 | II | POISON | | None | 212 | 242 | 25 kg | 100 kg | A | |
| | Mercury salicylate | 6.1 | UN1644 | II | POISON | | None | 212 | 242 | 25 kg | 100 kg | A | |
| + | Mercury sulfates | 6.1 | UN1645 | II | POISON | | None | 212 | 242 | 25 kg | 100 kg | A | |
| | Mercury thiocyanate | 6.1 | UN1646 | II | POISON | | None | 212 | 242 | 25 kg | 100 kg | A | |
| | Mesityl oxide | 3 | UN1229 | III | FLAMMABLE LIQUID | B1, T1 | None | 203 | 242 | 60 L | 220 L | A | |
| | Metal alkyl halides, n.o.s. | 4.2 | UN3049 | I | SPONTANEOUSLY COMBUSTIBLE | B9, B11, T28, T29, T40 | None | 181 | 244 | Forbidden | Forbidden | D | |

## §172.101 HAZARDOUS MATERIALS TABLE, Continued

| Symbols | Hazardous materials descriptions and proper shipping names | Hazard class or Division | Identification Numbers | Packing group | Label(s) required (if not excepted) | Special provisions | (8) Packaging authorizations (§173.***) | | | (9) Quantity limitations | | (10) Vessel stowage requirements | |
|---|---|---|---|---|---|---|---|---|---|---|---|---|---|
| | | | | | | | Exceptions | Non-bulk packaging | Bulk packaging | Passenger aircraft or railcar | Cargo aircraft only | Vessel stowages | Other stowage provisions |
| (1) | (2) | (3) | (4) | (5) | (6) | (7) | (8A) | (8B) | (8C) | (9A) | (9B) | (10A) | (10B) |
| | Metal alkyl hydrides, n.o.s | 4.2 | UN3050 | I | SPONTANEOUSLY COMBUSTIBLE | B9, B11, T28, T29, T40 | None | 181 | 244 | Forbidden | Forbidden | D | |
| D | Metal alkyl, solution, n.o.s | 3 | NA9195 | II | FLAMMABLE LIQUID | | 150 | 202 | 242 | 1 L | 4 L | B | |
| | Metal alkyls, n.o.s | 4.2 | UN2003 | I | SPONTANEOUSLY COMBUSTIBLE | B11, T42 | None | 181 | 244 | Forbidden | Forbidden | D | |
| | Metal catalyst, dry | 4.2 | UN2881 | I | SPONTANEOUSLY COMBUSTIBLE | N34 | None | 187 | None | Forbidden | Forbidden | B | |
| | Metal catalyst, wetted *with not less than 40 per cent water or other suitable liquid, by mass, finely divided activated or spent* | 4.2 | UN1378 | II | SPONTANEOUSLY COMBUSTIBLE | A2, A8, N34 | None | 212 | None | Forbidden | 50 kg | C | |
| | Metal powders, flammable, n.o.s | 4.1 | UN3089 | II | FLAMMABLE SOLID | | 151 | 212 | 240 | 15 kg | 50 kg | B | |
| | | | | III | FLAMMABLE SOLID | | 151 | 212 | 240 | 25 kg | 100 kg | B | |
| | *Metal salts of methyl nitramine (dry)* | Forbidden | | | | | | | | | | | |
| | Metaldehyde | 4.1 | UN1332 | III | FLAMMABLE SOLID | A1 | 151 | 213 | 240 | 25 kg | 100 kg | A | |
| | Methacrylaldehyde | 3 | UN2396 | II | FLAMMABLE LIQUID, POISON | T8 | None | 202 | 243 | 1 L | 60 L | E | 40 |
| | Methacrylic acid, inhibited | 8 | UN2531 | III | CORROSIVE | T8 | 154 | 203 | 241 | 5 L | 60 L | A | 8, 12, 48 |
| | Methacrylonitrile, inhibited | 3 | UN3079 | I | FLAMMABLE LIQUID, POISON | | None | 201 | 243 | Forbidden | 30 L | D | 12, 40, 48 |
| | Methallyl alcohol | 3 | UN2614 | III | FLAMMABLE LIQUID | B1, T1 | 150 | 203 | 242 | 60 L | 220 L | A | |
| | *Methane and hydrogen mixtures, see* Hydrogen and methane, mixtures, etc | | | | | | | | | | | | |
| | Methane, compressed *or* Natural gas, compressed *(with high methane content)* | 2.1 | UN1971 | | FLAMMABLE GAS | | 306 | 302 | 302 | Forbidden | 150 kg | E | 40 |
| | Methane, refrigerated liquid *(cryogenic liquid)or* Natural gas, refrigerated liquid *(cryogenic liquid)(with high methane content)* | 2.1 | UN1972 | | FLAMMABLE GAS | | None | None | 318 | Forbidden | Forbidden | D | 40 |
| | Methanol, *or* Methyl alcohol | 3 | UN1230 | II | FLAMMABLE LIQUID, POISON | T8 | None | 202 | 242 | 1 L | 60 L | B | 40 |
| | *Methazoic acid* | Forbidden | | | | | | | | | | | |
| | 4-Methoxy-4-methylpentan-2-one | 3 | UN2293 | III | FLAMMABLE LIQUID | B1, T1 | 150 | 203 | 242 | 60 L | 220 L | A | |
| | 1-Methoxy-2-propanol | 3 | UN3092 | III | FLAMMABLE LIQUID | B1, T1 | 150 | 203 | 242 | 60 L | 220 L | A | 44 |
| + | Methoxymethyl isocyanate | 3 | UN2605 | I | FLAMMABLE LIQUID, POISON | 1, B9, B14, B30, B72, T38, T43, T44 | None | 226 | 244 | Forbidden | 30 L | D | 40 |
| | Methyl acetate | 3 | UN1231 | II | FLAMMABLE LIQUID | T8 | 150 | 202 | 242 | 5 L | 60 L | B | |
| | Methyl acrylate, inhibited | 3 | UN1919 | II | FLAMMABLE LIQUID | T8 | 150 | 202 | 242 | 5 L | 60 L | B | |
| | Methyl alcohol, *see* Methanol | | | | | | | | | | | | |
| | Methyl allyl chloride | 3 | UN2554 | II | FLAMMABLE LIQUID | T8 | 150 | 202 | 242 | 5 L | 60 L | E | |
| | *Methyl amyl ketone, see* Amyl methyl ketone | | | | | | | | | | | | |
| | Methyl benzoate | 6.1 | UN2938 | III | KEEP AWAY FROM FOOD | T1 | 153 | 203 | 240 | 60 L | 220 L | A | |
| | Methyl bromide | 2.3 | UN1062 | | POISON GAS | 3, B14 | None | 193 | 314, 315 | Forbidden | 25 kg | D | 40 |
| | *Methyl bromide and chloropicrin mixtures with more than 2 per cent chloropicrin, see* Chloropicrin and methyl bromide mixtures | | | | | | | | | | | | |
| | *Methyl bromide and chloropicrin mixtures with not more than 2 per cent chloropicrin, see* Methyl bromide | | | | | | | | | | | | |
| | Methyl bromide and ethylene dibromide mixtures, liquid | 6.1 | UN1647 | I | POISON | 2, B9, B14, B32, B74, N65, T38, T43, T45 | None | 227 | 244 | Forbidden | 30 L | C | 40 |
| | Methyl bromoacetate | 6.1 | UN2643 | II | POISON | T8 | None | 202 | 243 | 5 L | 60 L | D | 40 |
| | 3-Methyl-1-butene | 3 | UN2561 | I | FLAMMABLE LIQUID | T20 | None | 201 | 243 | 1 L | 30 L | E | |
| | 2-Methyl-1-butene | 3 | UN2459 | I | FLAMMABLE LIQUID | T14 | None | 201 | 243 | 1 L | 30 L | E | |
| | 2-Methyl-2-butene | 3 | UN2460 | II | FLAMMABLE LIQUID | T14 | None | 202 | 242 | 5 L | 60 L | E | |
| | Methyl tert butyl ether | 3 | UN2398 | II | FLAMMABLE LIQUID | T14 | 150 | 202 | 242 | 5 L | 60 L | E | |
| | Methyl butyrate | 3 | UN1237 | II | FLAMMABLE LIQUID | T1 | 150 | 202 | 242 | 5 L | 60 L | B | |
| | Methyl chloride | 2.1 | UN1063 | | FLAMMABLE GAS | | 306 | 304 | 314, 315 | Forbidden | 25 kg | D | 40 |
| | *Methyl chloride and chloropicrin mixtures, see* Chloropicrin and methyl chloride mixtures | | | | | | | | | | | | |

## §172.101 HAZARDOUS MATERIALS TABLE, Continued

| Symbols | Hazardous materials descriptions and proper shipping names | Hazard class or Division | Identification Numbers | Packing group | Label(s) required (if not excepted) | Special provisions | (8) Packaging authorizations (§173 ***) | | | (9) Quantity limitations | | (10) Vessel stowage requirements | |
|---|---|---|---|---|---|---|---|---|---|---|---|---|---|
| | | | | | | | Exceptions | Non-bulk packaging | Bulk packaging | Passenger aircraft or railcar | Cargo aircraft only | Vessel stowages | Other stowage provisions |
| (1) | (2) | (3) | (4) | (5) | (6) | (7) | (8A) | (8B) | (8C) | (9A) | (9B) | (10A) | (10B) |
| | Methyl chloride and methylene chloride mixtures | 2.1 | UN1912 | | FLAMMABLE GAS | | 306 | 304 | 314, 315 | Forbidden | 150 kg | D | 40 |
| | Methyl chloroacetate | 6.1 | UN2295 | II | POISON | T11 | None | 202 | 243 | 5 L | 60 L | C | |
| | *Methyl chlorocarbonate*, see Methyl chloroformate | | | | | | | | | | | | |
| | *Methyl chloroform*, see 1,1,1-Trichloroethane | | | | | | | | | | | | |
| | Methyl chloroformate | 6.1 | UN1238 | I | POISON, FLAMMABLE LIQUID, CORROSIVE | 1, A3, A6, A7, B9, B14, B30, B72, N34, T38, T43, T44 | None | 226 | 244 | Forbidden | Forbidden | D | 21, 40, 100 |
| | Methyl-2-chloropropionate | 3 | UN2933 | III | FLAMMABLE LIQUID | B1, T7 | 150 | 203 | 242 | 60 L | 220 L | A | |
| | Methyl cyanide | 3 | UN1648 | II | FLAMMABLE LIQUID, POISON | T14 | None | 202 | 243 | 1 L | 60 L | B | 40 |
| | Methyl cyclohexane | 3 | UN2296 | II | FLAMMABLE LIQUID | B1, T1 | 150 | 202 | 242 | 5 L | 60 L | B | |
| | Methyl cyclohexanols, *flash point not more than 60.5 degrees C* | 3 | UN2617 | III | FLAMMABLE LIQUID | B1, T2 | 150 | 203 | 242 | 60 L | 220 L | A | |
| | Methyl cyclohexanone | 3 | UN2297 | III | FLAMMABLE LIQUID | B1, T1 | 150 | 203 | 242 | 60 L | 220 L | A | |
| | Methyl cyclopentane | 3 | UN2298 | II | FLAMMABLE LIQUID | T8 | 150 | 202 | 242 | 5 L | 60 L | B | |
| | Methyl dichloroacetate | 6.1 | UN2299 | III | KEEP AWAY FROM FOOD | T1 | 153 | 203 | 241 | 60 L | 220 L | A | |
| | *Methyl ethyl ether*, see Ethyl methyl ether | | | | | | | | | | | | |
| | Methyl ethyl ketone, see Ethyl methyl ketone | | | | | | | | | | | | |
| | *Methyl ethyl ketone peroxide, in solution with more than 9 per cent by mass active oxygen* | Forbidden | | | | | | | | | | | |
| | 2-Methyl-5-ethylpyridine | 6.1 | UN2300 | III | KEEP AWAY FROM FOOD | T7 | 153 | 203 | 241 | 60 L | 220 L | A | |
| | Methyl fluoride | 2.1 | UN2454 | | FLAMMABLE GAS | | 306 | 304 | 314, 315 | Forbidden | 150 kg | E | 40 |
| | Methyl formate | 3 | UN1243 | I | FLAMMABLE LIQUID | T20 | 150 | 201 | 243 | 1 L | 30 L | E | |
| + | Methyl iodide | 6.1 | UN2644 | II | POISON | 2, B9, B14, B32, B74, T38, T43, T45 | None | 227 | 244 | 5 L | 60 L | A | 12, 40, 48 |
| | Methyl isobutyl carbinol | 3 | UN2053 | III | FLAMMABLE LIQUID | B1, T1 | 150 | 203 | 242 | 60 L | 220 L | A | |
| | Methyl isobutyl ketone | 3 | UN1245 | II | FLAMMABLE LIQUID | T1 | 150 | 202 | 242 | 5 L | 60 L | B | |
| | *Methyl isobutyl ketone peroxide, in solution with more than 9 per cent by mass active oxygen* | Forbidden | | | | | | | | | | | |
| | Methyl isocyanate | 6.1 | UN2480 | I | POISON, FLAMMABLE LIQUID | 1, A7, B9, B14, B30, B72, T38, T43, T44 | None | 226 | 244 | Forbidden | Forbidden | D | 26, 40 |
| | Methyl isopropenyl ketone, inhibited | 3 | UN1246 | II | FLAMMABLE LIQUID | T7 | 150 | 202 | 242 | 5 L | 60 L | B | |
| + | Methyl isothiocyanate | 3 | UN2477 | II | FLAMMABLE LIQUID, POISON | 1, B9, B14, B30, B72, T38, T43, T44 | None | 226 | 244 | Forbidden | 60 L | A | |
| | Methyl isovalerate | 3 | UN2400 | II | FLAMMABLE LIQUID | T1 | 150 | 202 | 242 | 5 L | 60 L | B | |
| | Methyl magnesium bromide, in ethyl ether | 4.3 | UN1928 | I | DANGEROUS WHEN WET, FLAMMABLE LIQUID | | None | 201 | 243 | Forbidden | 1 L | D | |
| | Methyl mercaptan | 2.3 | UN1064 | | POISON GAS, FLAMMABLE GAS | 2, B7, B9, B14 | None | 304 | 314, 315 | Forbidden | 25 kg | D | 40 |
| | *Methyl mercaptopropionaldehyde*, see Thia-4-pentanal | | | | | | | | | | | | |
| | Methyl methacrylate monomer, inhibited | 3 | UN1247 | II | FLAMMABLE LIQUID | T8 | 150 | 202 | 242 | 5 L | 60 L | B | 40 |
| | *Methyl nitramine (dry)* | Forbidden | | | | | | | | | | | |
| | *Methyl nitrate* | Forbidden | | | | | | | | | | | |
| | *Methyl nitrite* | Forbidden | | | | | | | | | | | |
| | *Methyl norbornene dicarboxylic anhydride*, see Corrosive liquids, n.o.s | | | | | | | | | | | | |
| + | Methyl orthosilicate | 3 | UN2606 | I | FLAMMABLE LIQUID, POISON | 2, B9, B14, B32, B74, T38, T43, T45 | None | 227 | 244 | Forbidden | 30L | E | 40 |
| D | Methyl parathion *liquid* | 6.1 | NA3018 | II | POISON | N76, T14 | None | 202 | 243 | Forbidden | 1 L | A | 40 |
| D | Methyl parathion *solid* | 6.1 | NA2783 | II | POISON | N77 | None | 212 | 242 | 25 kg | 100 kg | A | 40 |
| D | Methyl phosphonic dichloride | 6.1 | NA9206 | I | POISON, CORROSIVE | 2, A3, B9, B14, B32, B74, N34, N43, T38, T43, T45 | None | 227 | 244 | Forbidden | Forbidden | C | |

## §172.101 HAZARDOUS MATERIALS TABLE, Continued

| Symbols | Hazardous materials descriptions and proper shipping names | Hazard class or Division | Identification Numbers | Packing group | Label(s) required (if not excepted) | Special provisions | (8) Packaging authorizations (§173.***) | | | (9) Quantity limitations | | (10) Vessel stowage requirements | |
|---|---|---|---|---|---|---|---|---|---|---|---|---|---|
| | | | | | | | Exceptions | Non-bulk packaging | Bulk packaging | Passenger aircraft or railcar | Cargo aircraft only | Vessel stowages | Other stowage provisions |
| (1) | (2) | (3) | (4) | (5) | (6) | (7) | (8A) | (8B) | (8C) | (9A) | (9B) | (10A) | (10B) |
| | *Methyl phosphonothioic dichloride anhydrous, see* Corrosive liquid, n.o.s | | | | | | | | | | | | |
| D | Methyl phosphonous dichloride, *pyrophoric liquid* | 6.1 | NA2845 | I | POISON, SPONTANEOUSLY COMBUSTIBLE | 2, B9, B14, B16, B32, B74, T38, T43, T45 | None | 227 | 244 | Forbidden | Forbidden | D | 18 |
| | *Methyl picric acid (heavy metal salts of)* | Forbidden | | | | | | | | | | | |
| | Methyl propionate | 3 | UN1248 | II | FLAMMABLE LIQUID | T2 | 150 | 202 | 242 | 5 L | 60 L | B | |
| | Methyl propyl ether | 3 | UN2612 | II | FLAMMABLE LIQUID | T14 | 150 | 202 | 242 | 5 L | 60 L | E | 40 |
| | Methyl propyl ketone | 3 | UN1249 | II | FLAMMABLE LIQUID | T1 | 150 | 202 | 242 | 5 L | 60 L | B | |
| | *Methyl sulfate, see* Dimethyl sulfate | | | | | | | | | | | | |
| | *Methyl sulfide, see* Dimethyl sulfide | | | | | | | | | | | | |
| | Methyl trichloroacetate | 6.1 | UN2533 | III | KEEP AWAY FROM FOOD | T1 | 153 | 203 | 241 | 60 L | 220 L | A | |
| | *Methyl trimethylol methane trinitrate* | Forbidden | | | | | | | | | | | |
| | Methyl vinyl ketone | 3 | UN1251 | II | FLAMMABLE LIQUID | T8 | 150 | 202 | 242 | 5 L | 60 L | B | |
| | Methylacetylene and propadiene mixtures stabilized | 2.1 | UN1060 | | FLAMMABLE GAS | | 306 | 304 | 314, 315 | Forbidden | 150 kg | B | 40 |
| | Methylal | 3 | UN1234 | II | FLAMMABLE LIQUID | T14 | None | 202 | 242 | 5L | 60 L | E | |
| | Methylamine, anhydrous | 2.3 | UN1061 | | POISON GAS, FLAMMABLE GAS | 3, B14 | None | 304 | 314, 315 | Forbidden | 150 kg | B | 40 |
| | Methylamine, aqueous solution | 3 | UN1235 | II | FLAMMABLE LIQUID, CORROSIVE | B1, T8 | 150 | 202 | 242 | 5 L | 60 L | E | 41 |
| | *Methylamine dinitramine and dry salts thereof* | Forbidden | | | | | | | | | | | |
| | *Methylamine nitroform* | Forbidden | | | | | | | | | | | |
| | *Methylamine perchlorate (dry)* | Forbidden | | | | | | | | | | | |
| | Methylamyl acetate | 3 | UN1233 | III | FLAMMABLE LIQUID | B1, T1 | 150 | 203 | 242 | 60 L | 220 L | A | |
| | N-Methylaniline | 6.1 | UN2294 | III | KEEP AWAY FROM FOOD | T7 | 153 | 203 | 241 | 60 L | 220 L | A | |
| | alpha-Methylbenzyl alcohol | 6.1 | UN2937 | III | KEEP AWAY FROM FOOD | T1 | 153 | 203 | 241 | 60 L | 220 L | A | |
| | 3-Methylbutan-2-one | 3 | UN2397 | II | FLAMMABLE LIQUID | T1 | 150 | 202 | 242 | 5 L | 60 L | B | |
| | N-Methylbutylamine | 3 | UN2945 | II | FLAMMABLE LIQUID | T8 | 150 | 202 | 242 | 5 L | 60 L | B | 40 |
| | Methylchloromethyl ether | 6.1 | UN1239 | I | POISON, FLAMMABLE LIQUID | 1, B9, B14, B30, B72, T38, T43, T44 | None | 226 | 244 | Forbidden | Forbidden | D | 40 |
| | Methylchlorosilane | 2.3 | UN2534 | | POISON GAS, FLAMMABLE GAS | 2, A2, A3, A7, B9, B14, N34 | None | 226 | 314, 315 | Forbidden | 25 kg | D | 40 |
| D | Methyldichloroarsine | 6.1 | NA1556 | I | POISON | 2 | None | 192 | None | Forbidden | Forbidden | D | 40, 95 |
| | Methyldichlorosilane | 4.3 | UN1242 | I | DANGEROUS WHEN WET, CORROSIVE, FLAMMABLE LIQUID | A2, A3, A7, B6, B77, N34, T16, T26 | None | 201 | 243 | Forbidden | 1 L | D | 21, 28, 40, 49, 100 |
| | *Methylene chloride, see* Dichloromethane | | | | | | | | | | | | |
| | *Methylene glycol dinitrate* | Forbidden | | | | | | | | | | | |
| | 2-Methylfuran | 3 | UN2301 | II | FLAMMABLE LIQUID | T7 | 150 | 202 | 242 | 5 L | 60 L | E | |
| | *a-Methylglucoside tetranitrate* | Forbidden | | | | | | | | | | | |
| | *a-Methylglycerol trinitrate* | Forbidden | | | | | | | | | | | |
| | 5-Methylhexan-2-one | 3 | UN2302 | III | FLAMMABLE LIQUID | B1, T1 | 150 | 203 | 242 | 60 L | 220 L | A | |
| | Methylhydrazine | 6.1 | UN1244 | I | POISON, FLAMMABLE LIQUID, CORROSIVE | 1, B9, B14, B30, B72, B77, N34, T38, T43, T44 | None | 226 | 244 | Forbidden | Forbidden | D | 21, 49, 100 |
| | Methylmorpholine | 3 | UN2535 | II | FLAMMABLE LIQUID, CORROSIVE | B6, T8 | None | 202 | 243 | 1 L | 5 L | B | 40 |
| | Methylpentadienes | 3 | UN2461 | II | FLAMMABLE LIQUID | T7 | 150 | 202 | 241 | 5 L | 60 L | E | |
| | 2-Methylpentan-2-ol | 3 | UN2560 | III | FLAMMABLE LIQUID | B1, T1 | 150 | 203 | 242 | 60 L | 220 L | A | |

## §172.101 HAZARDOUS MATERIALS TABLE, Continued

| Symbols | Hazardous materials descriptions and proper shipping names | Hazard class or Division | Identification Numbers | Packing group | Label(s) required (if not excepted) | Special provisions | (8) Packaging authorizations (§173.***) | | | (9) Quantity limitations | | (10) Vessel stowage requirements | |
|---|---|---|---|---|---|---|---|---|---|---|---|---|---|
| | | | | | | | Exceptions | Non-bulk packaging | Bulk packaging | Passenger aircraft or railcar | Cargo aircraft only | Vessel stowages | Other stowage provisions |
| (1) | (2) | (3) | (4) | (5) | (6) | (7) | (8A) | (8B) | (8C) | (9A) | (9B) | (10A) | (10B) |
| | *Methylpentanes, see* Hexanes | | | | | | | | | | | | |
| | Methylphenyldichlorosilane | 8 | UN2437 | II | CORROSIVE | T8, T26 | 154 | 202 | 242 | 1 L | 30 L | C | 40 |
| | 1-Methylpiperidine | 3 | UN2399 | II | FLAMMABLE LIQUID | T8 | 150 | 202 | 242 | 5 L | 60 L | B | |
| | Methyltetrahydrofuran | 3 | UN2536 | II | FLAMMABLE LIQUID | T7 | 150 | 202 | 242 | 5 L | 60 L | B | |
| | Methyltrichlorosilane | 3 | UN1250 | I | FLAMMABLE LIQUID, CORROSIVE | A7, B6, B77, N34, T14, T26 | None | 201 | 243 | Forbidden | 2.5 L | B | 40 |
| | alpha-Methylvaleraldehyde | 3 | UN2367 | II | FLAMMABLE LIQUID | B1, T1 | 150 | 203 | 242 | 60 L | 220 L | B | |
| | *Mine rescue equipment containing carbon dioxide, see* Carbon dioxide | | | | | | | | | | | | |
| | Mines *with bursting charge* | 1.1F | UN0136 | II | EXPLOSIVE 1.1F | | 63(b) | 62 | None | Forbidden | Forbidden | E | |
| | Mines *with bursting charge* | 1.1D | UN0137 | II | EXPLOSIVE 1.1D | | 63(b) | 62 | None | Forbidden | Forbidden | B | 3E, 7E |
| | Mines *with bursting charge* | 1.2D | UN0138 | II | EXPLOSIVE 1.2D | | 63(b) | 62 | None | Forbidden | Forbidden | B | 3E, 7E |
| | Mines *with bursting charge* | 1.2F | UN0294 | II | EXPLOSIVE 1.2F | | 63(b) | 62 | None | Forbidden | Forbidden | E | |
| | *Mixed acid, see* Nitrating acid, mixtures etc | | | | | | | | | | | | |
| | Molybdenum pentachloride | 8 | UN2508 | III | CORROSIVE | T8, T26 | 154 | 213 | 240 | 25 kg | 100 kg | C | 40 |
| | *Monochloroacetone (unstabilized)* | Forbidden | | | | | | | | | | | |
| | *Monochloroethylene, see* Vinyl chloride, inhibited | | | | | | | | | | | | |
| | *Monoethanolamine, see* Ethanolamine, solutions | | | | | | | | | | | | |
| | *Monoethylamine, see* Ethylamine | | | | | | | | | | | | |
| | Morpholine | 3 | UN2054 | III | FLAMMABLE LIQUID | B1, T1 | 150 | 203 | 242 | 60 L | 220 L | A | |
| | *Morpholine, aqueous, mixture, see* Corrosive liquids, n.o.s. | | | | | | | | | | | | |
| | Motor fuel anti-knock compounds see Motor fuel anti-knock mixtures | | | | | | | | | | | | |
| + | Motor fuel anti-knock mixtures | 6.1 | UN1649 | I | POISON, FLAMMABLE LIQUID | 14, B9, B12, B90, T26, T39 | None | 201 | 244 | Forbidden | 30 L | E | 40 |
| | Motor spirit, see Gasoline | | | | | | | | | | | | |
| | *Motor vehicle, see* Vehicles, self-propelled | | | | | | | | | | | | |
| | *Motorcycles, see* Vehicles, self-propelled | | | | | | | | | | | | |
| | *Muriatic acid, see* Hydrochloric acid solution | | | | | | | | | | | | |
| | Musk xylene, *see* 5-tert-Butyl2,4,6-trinitro-m-xylene | | | | | | | | | | | | |
| | Naphtha | 3 | UN2553 | II | FLAMMABLE LIQUID | T8, T31 | 150 | 202 | 242 | 5 L | 60 L | B | |
| | | | | III | FLAMMABLE LIQUID | B1, T7, T30 | 150 | 203 | 242 | 60 L | 220 L | A | |
| | Naphtha, *petroleum* | 3 | UN1255 | I | FLAMMABLE LIQUID | T8 | 150 | 201 | 243 | 1 L | 30 L | E | |
| | | | | II | FLAMMABLE LIQUID | T8 | 150 | 202 | 242 | 5 L | 60 L | B | |
| | Naphtha, *solvent* | 3 | UN1256 | II | FLAMMABLE LIQUID | T8, T31 | 150 | 202 | 242 | 5 L | 60 L | B | |
| | | | | III | FLAMMABLE LIQUID | B1, T7, T30 | 150 | 203 | 242 | 60 L | 220 L | A | |
| | Naphthalene, crude *or* refined | 4.1 | UN1334 | III | FLAMMABLE SOLID | A1 | 151 | 213 | 240 | 25 kg | 100 kg | A | |
| | *Naphthalene diozonide* | Forbidden | | | | | | | | | | | |
| | Naphthalene, molten | 4.1 | UN2304 | III | FLAMMABLE SOLID | A1, T8 | 151 | 213 | 241 | 25 kg | 100 kg | C | |
| | *Naphthyl amineperchlorate* | Forbidden | | | | | | | | | | | |
| | alpha-Naphthylamine | 6.1 | UN2077 | III | KEEP AWAY FROM FOOD | T7 | 153 | 213 | 240 | 100 kg | 200 kg | A | |
| | beta-Naphthylamine | 6.1 | UN1650 | II | POISON | T12, T26 | None | 212 | 242 | 25 kg | 100 kg | A | |
| | Naphthylthiourea | 6.1 | UN1651 | II | POISON | | None | 212 | 242 | 25 kg | 100 kg | A | |
| | Naphthylurea | 6.1 | UN1652 | II | POISON | | None | 212 | 242 | 25 kg | 100 kg | A | |
| | *Natural gases (with high methane content), see* Methane, *etc (UN 1971, UN 1972)* | | | | | | | | | | | | |
| | Natural gasoline | 3 | UN1257 | I | FLAMMABLE LIQUID | T8 | 150 | 201 | 243 | 1L | 30L | E | |
| | | | | II | FLAMMABLE LIQUID | T8 | 150 | 202 | 242 | 5 L | 60 L | E | |
| | *Neohexane, see* Hexanes | | | | | | | | | | | | |
| | Neon, compressed | 2.2 | UN1065 | | NONFLAMMABLE GAS | | 306 | 302 | 302 | 75 kg | 150 kg | A | |
| | Neon, refrigerated liquid *(cryogenic liquid)* | 2.2 | UN1913 | | NONFLAMMABLE GAS | | 320 | 316 | None | 50 kg | 500 kg | B | |

## §172.101 HAZARDOUS MATERIALS TABLE, Continued

| Symbols (1) | Hazardous materials descriptions and proper shipping names (2) | Hazard class or Division (3) | Identification Numbers (4) | Packing group (5) | Label(s) required (if not excepted) (6) | Special provisions (7) | (8) Packaging authorizations (§173.***) Exceptions (8A) | Non-bulk packaging (8B) | Bulk packaging (8C) | (9) Quantity limitations Passenger aircraft or railcar (9A) | Cargo aircraft only (9B) | (10) Vessel stowage requirements Vessel stowages (10A) | Other stowage provisions (10B) |
|---|---|---|---|---|---|---|---|---|---|---|---|---|---|
| | *New explosive or explosive device, see sections 173.51 and 173.56* | | | | | | | | | | | | |
| | Nickel carbonyl | 6.1 | UN1259 | I | POISON, FLAMMABLE LIQUID | 1 | None | 198 | None | Forbidden | Forbidden | D | 18, 40 |
| | Nickel cyanide | 6.1 | UN1653 | II | POISON | N74, N75 | None | 212 | 242 | 25 kg | 100 kg | A | 26 |
| | Nickel nitrate | 5.1 | UN2725 | III | OXIDIZER | A1 | 152 | 213 | 240 | 25 kg | 100 kg | A | |
| | Nickel nitrite | 5.1 | UN2726 | III | OXIDIZER | A1 | 152 | 213 | 240 | 25 kg | 100 kg | A | 56, 58 |
| | *Nickel picrate* | Forbidden | | | | | | | | | | | |
| | Nicotine | 6.1 | UN1654 | II | POISON | | None | 202 | 243 | 5 L | 60 L | A | |
| | Nicotine compounds, liquid, n.o.s. *or* Nicotine preparations, liquid, n.o.s. | 6.1 | UN3144 | I | POISON | A4, T42 | None | 201 | 243 | 1 L | 30 L | B | 40 |
| | | | | II | POISON | T14 | None | 202 | 243 | 5 L | 60 L | B | 40 |
| | | | | III | KEEP AWAY FROM FOOD | T7 | 153 | 203 | 241 | 60 L | 220 L | B | 40 |
| | Nicotine compounds, solid, *n.o.s or* Nicotine preparations, solid, n.o.s | 6.1 | UN1655 | I | POISON | | None | 211 | 242 | 5 kg | 50 kg | B | |
| | | | | II | POISON | | None | 212 | 242 | 25 kg | 100 kg | A | |
| | | | | III | KEEP AWAY FROM FOOD | | 153 | 213 | 240 | 100 kg | 200 kg | A | |
| | Nicotine hydrochloride *or* Nicotine hydrochloride solution | 6.1 | UN1656 | II | POISON | | None | 202 | 243 | 5 L | 60 L | A | |
| | Nicotine salicylate | 6.1 | UN1657 | II | POISON | | None | 212 | 242 | 25 kg | 100 kg | A | |
| | Nicotine sulfate, *solid* | 6.1 | UN1658 | II | POISON | | None | 212 | 242 | 25 kg | 100 kg | A | |
| | Nicotine sulfate, *solution* | 6.1 | UN1658 | II | POISON | T14 | None | 202 | 243 | 5 L | 60 L | A | |
| | Nicotine tartrate | 6.1 | UN1659 | II | POISON | | None | 212 | 242 | 25 kg | 100 kg | A | |
| | *Nitrated paper (unstable)* | Forbidden | | | | | | | | | | | |
| | Nitrates, inorganic, n.o.s | 5.1 | UN1477 | II | OXIDIZER | | 152 | 212 | 240 | 5 kg | 25 kg | A | 46 |
| | *Nitrates of diazonium compounds* | Forbidden | | | | | | | | | | | |
| | Nitrating acid mixtures spent *with not more than 50 per cent nitric acid* | 8 | UN1826 | II | CORROSIVE | B2, T12, T27 | None | 158 | 242 | Forbidden | 30 L | D | 40 |
| | Nitrating acid mixtures, spent *with 50 per cent or more nitric acid* | 8 | UN1826 | I | CORROSIVE, OXIDIZER | T12, T27 | None | 158 | 243 | Forbidden | 2.5 L | D | 40, 66, 89 |
| | Nitrating acid mixtures *with not more than 50 per cent nitric acid* | 8 | UN1796 | II | CORROSIVE | B2, T12, T27 | None | 158 | 242 | Forbidden | 30 L | D | 40 |
| | Nitrating acid mixtures *with 50 per cent or more nitric acid* | 8 | UN1796 | I | CORROSIVE, OXIDIZER | T12, T27 | None | 158 | 243 | Forbidden | 2.5 L | D | 40, 66, 89 |
| | Nitric acid *other than red fuming with more than 70 percent nitric acid* | 8 | UN2031 | I | CORROSIVE | B12, B47, B53, T9, T27 | None | 158 | 243 | Forbidden | 2.5 L | D | 110, 111 |
| | Nitric acid *other than red fuming, with not more than 70 percent nitric acid* | 8 | UN2031 | II | CORROSIVE | B2, B12, B47, B53, T9, T27 | None | 158 | 242 | Forbidden | 30 L | D | 110, 111 |
| + | Nitric acid, red fuming | 8 | UN2032 | I | CORROSIVE, OXIDIZER, POISON | 2, B9, B32, B74, T38, T43, T45 | None | 227 | 244 | Forbidden | Forbidden | D | 40, 66, 74, 89, 90, 95 |
| | Nitric oxide | 2.3 | UN1660 | | POISON GAS | 2, B12, B37, B46, B50, B60, B77 | None | 337 | None | Forbidden | Forbidden | D | 40 |
| | Nitric oxide and dinitrogen tetroxide mixtures (Nitric oxide and nitrogen dioxide mixtures) | 2.3 | UN1975 | | POISON GAS, OXIDIZER | 2, B7, B9, B12, B14, B45, B46, B61, B66, B67, B77 | None | 337 | None | Forbidden | Forbidden | D | 40 |
| | Nitrites, inorganic, n.o.s | 5.1 | UN2627 | II | OXIDIZER | A33 | 152 | 212 | None | 5 kg | 25 kg | A | 46, 56, 58 |
| | 3-Nitro-4-chlorobenzotrifluoride | 6.1 | UN2307 | II | POISON | T8 | None | 202 | 243 | 5 L | 60 L | A | 40 |
| | *6-Nitro-4-diazotoluene-3-sulfonic acid (dry)* | Forbidden | | | | | | | | | | | |
| | *Nitro isobutane triol trinitrate* | Forbidden | | | | | | | | | | | |
| | *N-Nitro-N-methylglycolamide nitrate* | Forbidden | | | | | | | | | | | |
| | *2-Nitro-2-methylpropanol nitrate* | Forbidden | | | | | | | | | | | |
| | Nitro urea | 1.1D | UN0147 | II | EXPLOSIVE 1.1D | | None | 62 | None | Forbidden | Forbidden | B | 1E, 5E |
| | *N-Nitroaniline* | Forbidden | | | | | | | | | | | |
| | Nitroanilines *(o-; m-; p-;)* | 6.1 | UN1661 | II | POISON | T14 | None | 212 | 242 | 25 kg | 100 kg | A | |
| | Nitroanisole | 6.1 | UN2730 | III | KEEP AWAY FROM FOOD | T8 | 153 | 213 | 240 | 100 kg | 200 kg | A | |
| | Nitrobenzene | 6.1 | UN1662 | II | POISON | T14 | None | 202 | 243 | 5 L | 60 L | A | 40 |
| | *m-Nitrobenzene diazonium perchlorate* | Forbidden | | | | | | | | | | | |
| | Nitrobenzenesulfonic acid | 8 | UN2305 | II | CORROSIVE | | 154 | 212 | 240 | 1 L | 30 L | A | |
| | *Nitrobenzol, see* Nitrobenzene | | | | | | | | | | | | |
| | 5-Nitrobenzotriazol | 1.1D | UN0385 | II | EXPLOSIVE 1.1D | | None | 62 | None | Forbidden | Forbidden | B | 1E, 5E, 19E |
| | Nitrobenzotrifluorides | 6.1 | UN2306 | II | POISON | T8 | None | 202 | 243 | 5 L | 60 L | A | 40 |
| | Nitrobromobenzenes *liquid* | 6.1 | UN2732 | III | KEEP AWAY FROM FOOD | T8, T38 | 153 | 203 | 241 | 60 L | 220 L | A | 48 |

## §172.101 HAZARDOUS MATERIALS TABLE, Continued

| Symbols | Hazardous materials descriptions and proper shipping names | Hazard class or Division | Identification Numbers | Packing group | Label(s) required (if not excepted) | Special provisions | (8) Packaging authorizations (§173.***) | | | (9) Quantity limitations | | (10) Vessel stowage requirements | |
|---|---|---|---|---|---|---|---|---|---|---|---|---|---|
| | | | | | | | Exceptions | Non-bulk packaging | Bulk packaging | Passenger aircraft or railcar | Cargo aircraft only | Vessel stowages | Other stowage provisions |
| (1) | (2) | (3) | (4) | (5) | (6) | (7) | (8A) | (8B) | (8C) | (9A) | (9B) | (10A) | (10B) |
| | Nitrobromobenzenes solid | 6.1 | UN2732 | III | KEEP AWAY FROM FOOD | | 153 | 213 | 240 | 100 kg | 200 kg | A | 48 |
| | Nitrocellulose, dry or wetted with less than 25 per cent water (or alcohol), by mass | 1.1D | UN0340 | II | EXPLOSIVE 1.1D | | None | 62 | None | Forbidden | Forbidden | B | 4E, 27E |
| | Nitrocellulose, plasticized with not less than 18 per cent plasticizing substance, by mass | 1.3C | UN0343 | II | EXPLOSIVE 1.3C | | None | 62 | None | Forbidden | Forbidden | B | 1E, 5E |
| | Nitrocellulose, solution, flammable with not more than 12.6 per cent nitrogen, by mass, and not more than 55 per cent nitrocellulose | 3 | UN2059 | II | FLAMMABLE LIQUID | T8, T31 | 150 | 202 | 242 | 5 L | 60 L | B | |
| | | | | III | FLAMMABLE LIQUID | B1, T7, T30 | 150 | 203 | 242 | 60 L | 220 L | A | |
| | Nitrocellulose, unmodified or plasticized with less than 18 per cent plasticizing substance, by mass | 1.1D | UN0341 | II | EXPLOSIVE 1.1D | | None | 62 | None | Forbidden | Forbidden | B | 4E, 27E |
| | Nitrocellulose, wetted with not less than 25 per cent alcohol, by mass | 1.3C | UN0342 | II | EXPLOSIVE 1.3C | | None | 62 | None | Forbidden | Forbidden | B | 1E, 5E |
| | Nitrocellulose with alcohol not less than 25 per cent alcohol by mass, and not more than 12.6 per cent nitrogen, by dry mass | 4.1 | UN2556 | II | FLAMMABLE SOLID | | 151 | 212 | None | 1 kg | 15 kg | D | 28 |
| | Nitrocellulose with plasticizing not less than 18 per cent, plasticizing substance, by mass, and not more than 12.6 per cent nitrogen, by dry mass | 4.1 | UN2557 | II | FLAMMABLE SOLID | | 151 | 212 | None | 1 kg | 15 kg | D | 28 |
| | Nitrocellulose with water not less than 25 per cent water, by mass | 4.1 | UN2555 | II | FLAMMABLE SOLID | | 151 | 212 | None | 15 kg | 50 kg | E | 28 |
| | Nitrochlorobenzene, see Chloronitrobenzenes etc | | | | | | | | | | | | |
| | Nitrocresols | 6.1 | UN2446 | III | KEEP AWAY FROM FOODS | | 153 | 213 | 240 | 100 kg | 200 kg | A | |
| | Nitroethane | 3 | UN2842 | III | FLAMMABLE LIQUID | B1, T8 | 150 | 203 | 242 | 60 L | 220 L | A | |
| | Nitroethyl nitrate | Forbidden | | | | | | | | | | | |
| | Nitroethylene polymer | Forbidden | | | | | | | | | | | |
| | Nitrogen, compressed | 2.2 | UN1066 | | NONFLAMMABLE GAS | | 306 | 302 | 314, 315 | 75 kg | 150 kg | A | |
| | Nitrogen dioxide, liquefied see Dinitrogen tetroxide, liquefied | | | | | | | | | | | | |
| | Nitrogen fertilizer solution, see Fertilizer ammoniating solution etc. | | | | | | | | | | | | |
| | Nitrogen, mixtures with rare gases, see Rare gases and nitrogen mixtures | | | | | | | | | | | | |
| | Nitrogen peroxide, see Dinitrogen tetroxide, liquefied | | | | | | | | | | | | |
| | Nitrogen, refrigerated liquid (cryogenic liquid) | 2.2 | UN1977 | | NONFLAMMABLE GAS | | 320 | 316 | 318 | 50 kg | 500 kg | A | |
| | Nitrogen tetroxide and nitric oxide mixtures, see Nitric oxide and nitrogen tetroxide mixtures | | | | | | | | | | | | |
| | Nitrogen tetroxide, see Dinitrogen tetroxide, liquefied | | | | | | | | | | | | |
| | Nitrogen trichloride | Forbidden | | | | | | | | | | | |
| | Nitrogen trifluoride | 2.2 | UN2451 | | NONFLAMMABLE GAS, OXIDIZER | | None | 302 | None | Forbidden | 25 kg | D | 40 |
| | Nitrogen triiodide | Forbidden | | | | | | | | | | | |
| | Nitrogen triiodide monoamine | Forbidden | | | | | | | | | | | |
| | Nitrogen trioxide | 2.3 | UN2421 | | POISON GAS, OXIDIZER | 1 | None | 336 | 245 | Forbidden | Forbidden | D | 40 |
| | Nitroglycerin, desensitized with not less than 40 per cent non-volatile water insoluble phlegmatizer, by mass | 1.1D | UN0143 | II | EXPLOSIVE 1.1D, POISON | | None | 62 | None | Forbidden | Forbidden | B | 1E, 4E, 21E |
| | Nitroglycerin, liquid, not desensitized | Forbidden | | | | | | | | | | | |
| | Nitroglycerin, solution in alcohol, with more than 1% but not more than 5% nitroglycerin | 3 | UN3064 | II | FLAMMABLE LIQUID | N8 | None | 202 | None | Forbidden | 5 L | E | |
| | Nitroglycerin, solution in alcohol, with more than 1 percent but not more than 10 percent nitroglycerin | 1.1D | UN0144 | II | EXPLOSIVE 1.1D | 20 | None | 62 | None | Forbidden | Forbidden | B | 1E, 4E, 21E |
| | Nitroglycerin solution in alcohol with not more than 1 per cent nitroglycerin | 3 | UN1204 | II | FLAMMABLE LIQUID | N34, T25 | None | 202 | None | 5 L | 60 L | B | |
| | Nitroguanidine nitrate | Forbidden | | | | | | | | | | | |

## §172.101 HAZARDOUS MATERIALS TABLE, Continued

| Symbols | Hazardous materials descriptions and proper shipping names | Hazard class or Division | Identification Numbers | Packing group | Label(s) required (if not excepted) | Special provisions | (8) Packaging authorizations (§173.***) | | | (9) Quantity limitations | | (10) Vessel stowage requirements | |
|---|---|---|---|---|---|---|---|---|---|---|---|---|---|
| | | | | | | | Exceptions | Non-bulk packaging | Bulk packaging | Passenger aircraft or railcar | Cargo aircraft only | Vessel stowages | Other stowage provisions |
| (1) | (2) | (3) | (4) | (5) | (6) | (7) | (8A) | (8B) | (8C) | (9A) | (9B) | (10A) | (10B) |
| | Nitroguanidine or Picrite, *dry or wetted with less than 20 per cent water, by mass* | 1.1D | UN0282 | II | EXPLOSIVE 1.1D | | None | 62 | None | Forbidden | Forbidden | B | 1E, 5E |
| | Nitroguanidine, wetted or Picrite, wetted *with not less than 20 per cent water, by mass* | 4.1 | UN1336 | I | FLAMMABLE SOLID | A8, A19, A20, N41 | None | 211 | None | 1 kg | 15 kg | E | 28 |
| | *1-Nitrohydantoin* | Forbidden | | | | | | | | | | | |
| | Nitrohydrochloric acid | 8 | UN1798 | I | CORROSIVE | A3, B10, N41, T18, T27 | None | 201 | 242 | Forbidden | 2.5 L | D | 40, 66, 74, 89, 90 |
| | *Nitromannite (dry)* | Forbidden | | | | | | | | | | | |
| | Nitromannite, wetted, *see* Mannitol hexanitrate, etc | | | | | | | | | | | | |
| | Nitromethane | 3 | UN1261 | II | FLAMMABLE LIQUID | T25 | 150 | 202 | None | Forbidden | 60 L | A | |
| | *Nitromuriatic acid, see* Nitrohydrochloric acid | | | | | | | | | | | | |
| | Nitronaphthalene | 4.1 | UN2538 | III | FLAMMABLE SOLID | A1 | 151 | 213 | 240 | 25 kg | 100 kg | A | |
| | Nitrophenols (o-; m-; p-;) | 6.1 | UN1663 | III | KEEP AWAY FROM FOOD | T8, T38 | 153 | 213 | 240 | 100 kg | 200 kg | A | |
| | *m-Nitrophenyldinitro methane* | Forbidden | | | | | | | | | | | |
| | Nitropropanes | 3 | UN2608 | III | FLAMMABLE LIQUID | B1, T1 | 150 | 203 | 242 | 60 L | 220 L | A | |
| | p-Nitrosodimethylaniline | 4.2 | UN1369 | II | SPONTANEOUSLY COMBUSTIBLE | A19, A20, N34 | None | 212 | 241 | 15 kg | 50 kg | D | 34 |
| D | Nitrosoguanidine | 1.1A | NA0473 | II | EXPLOSIVE 1.1A | 111, 117 | None | 62 | None | Forbidden | Forbidden | E | 2E, 6E |
| | Nitrostarch, *dry or wetted with less than 20 per cent water, by mass* | 1.1D | UN0146 | II | EXPLOSIVE 1.1D | | None | 62 | None | Forbidden | Forbidden | B | 1E, 5E |
| | Nitrostarch, wetted *with not less than 20 per cent water, by mass* | 4.1 | UN1337 | I | FLAMMABLE SOLID | A8, A19, A20, N41 | None | 211 | None | 1 kg | 15 kg | D | 28 |
| | *Nitrosugars (dry)* | Forbidden | | | | | | | | | | | |
| | Nitrosyl chloride | 2.3 | UN1069 | | POISON GAS, CORROSIVE | 3, B14 | None | 304 | 314, 315 | Forbidden | Forbidden | D | 40 |
| | Nitrosylsulfuric acid | 8 | UN2308 | II | CORROSIVE | A3, A6, A7, B2, N34, T9, T27 | 154 | 202 | 242 | 1 L | 30 L | D | 40, 66, 74, 89, 90 |
| | Nitrotoluenes, *liquid o-; m-; p-;* | 6.1 | UN1664 | II | POISON | T14 | None | 202 | 243 | 5 L | 60 L | A | |
| | Nitrotoluenes, *solid m-, or p-* | 6.1 | UN1664 | II | POISON | T14 | None | 212 | 242 | 25 kg | 100 kg | A | |
| | Nitrotoluidines (mono) | 6.1 | UN2660 | III | KEEP AWAY FROM FOOD | | 153 | 213 | 240 | 100 kg | 200 kg | A | |
| | Nitrotriazolone or NTO | 1.1D | UN0490 | II | EXPLOSIVE 1.1D | | None | 62 | | Forbidden | Forbidden | B | 1E, 5E, 25E |
| | *Nitrous oxide and carbon dioxide mixtures, see* Carbon dioxide and nitrous oxide mixtures | | | | | | | | | | | | |
| | Nitrous oxide, compressed | 2.2 | UN1070 | | NONFLAMMABLE GAS | | 306 | 304 | 314, 315 | 75 kg | 150 kg | A | 40 |
| | Nitrous oxide, refrigerated liquid | 2.2 | UN2201 | | NONFLAMMABLE GAS | B6 | None | None | 314, 315 | 75 kg | 150 kg | B | 40 |
| | Nitroxylenes, (o-; m-; p-) | 6.1 | UN1665 | II | POISON | T14 | None | 202 | 243 | 5 L | 60 L | A | |
| | *Nitroxylol, see* Nitroxylenes | | | | | | | | | | | | |
| | Nonanes | 3 | UN1920 | III | FLAMMABLE LIQUID | B1, T1 | 150 | 203 | 242 | 60 L | 220 L | A | |
| | *Nonflammable gas, n.o.s., see* Compressed or Liquefied gases, etc. *(UN 1955, UN 1956)* | | | | | | | | | | | | |
| | *Nonliquefied gases, see* Compressed gases, *etc.* | | | | | | | | | | | | |
| | *Nonliquefied hydrocarbon gas, see* Hydrocarbon gases, compressed, n.o.s. | | | | | | | | | | | | |
| | Nonyltrichlorosilane | 8 | UN1799 | II | CORROSIVE | A7, B2, B6, N34, T8, T26 | None | 202 | 242 | Forbidden | 30 L | C | 40 |
| | 2,5 Norbornadiene (Dicycloheptadiene) | 3 | UN2251 | II | FLAMMABLE LIQUID | | 150 | 202 | 241 | 5 L | 60 L | D | |
| | *Nordhausen acid, see* Sulfuric acid, fuming *etc.* | | | | | | | | | | | | |
| | Octadecyltrichlorosilane | 8 | UN1800 | II | CORROSIVE | A7, B2, B6 N34, T8 | None | 202 | 242 | Forbidden | 30 L | C | 40 |
| | Octadiene | 3 | UN2309 | II | FLAMMABLE LIQUID | B1, T1 | 150 | 202 | 242 | 5 L | 60 L | B | |
| | *1,7-Octadine-3,5-diyne-1,8-dimethoxy-9-octadecynoic acid* | Forbidden | | | | | | | | | | | |
| | Octafluorobut-2-ene | 2.2 | UN2422 | | NONFLAMMABLE GAS | | None | 304 | 314, 315 | 75 kg | 150 kg | A | |
| | Octafluorocyclobutane, *RC318* | 2.2 | UN1976 | | NONFLAMMABLE GAS | | None | 304 | 314, 315 | 75 kg | 150 kg | A | |
| | Octafluoropropane, *R218* | 2.2 | UN2424 | | NONFLAMMABLE GAS | | None | 304 | 314, 315 | 75 kg | 150 kg | A | |
| | Octanes | 3 | UN1262 | II | FLAMMABLE LIQUID | T1 | 150 | 202 | 242 | 5 L | 60 L | B | |

| Symbols | Hazardous materials descriptions and proper shipping names | Hazard class or Division | Identification Numbers | Packing group | Label(s) required (if not excepted) | Special provisions | (8) Packaging authorizations (§173.***) | | | (9) Quantity limitations | | (10) Vessel stowage requirements | |
|---|---|---|---|---|---|---|---|---|---|---|---|---|---|
| | | | | | | | Exceptions | Non-bulk packaging | Bulk packaging | Passenger aircraft or railcar | Cargo aircraft only | Vessel stowages | Other stowage provisions |
| (1) | (2) | (3) | (4) | (5) | (6) | (7) | (8A) | (8B) | (8C) | (9A) | (9B) | (10A) | (10B) |
| | Octogen, *see* Cyclotetramethylene tetranitramine, etc | | | | | | | | | | | | |
| | Octolite or Octol, *dry or wetted with less than 15 per cent water, by mass* | 1.1D | UN0266 | II | EXPLOSIVE 1.1D | | None | 62 | | Forbidden | Forbidden | B | 1E, 5E |
| | Octyl aldehydes, *flammable* | 3 | UN1191 | III | FLAMMABLE LIQUID | B1, T1 | 150 | 203 | 242 | 60 L | 220 L | A | |
| • | tert-Octylmercaptan | 6.1 | UN3023 | II | POISON, FLAMMABLE LIQUID | 2, B9, B14, B32, B74, T38, T43, T45 | None | 227 | 244 | Forbidden | 60 L | B | 40 |
| | Octyltrichlorosilane | 8 | UN1801 | II | CORROSIVE | A7, B2, B6, N34, T8, T26 | None | 202 | 242 | Forbidden | 30 L | C | 40 |
| D | Oil, n.o.s., *with a flashpoint not less than 93 C (200°F)* | 9 | NA9277 | III | Class 9 | | 155 | 203 | 241 | | | A | |
| | Oil gas | 2.1 | UN1071 | | FLAMMABLE GAS | | None | 304 | 314, 315 | Forbidden | 150 kg | D | 40 |
| | *Oleum, see* Sulfuric acid, fuming | | | | | | | | | | | | |
| | Organic peroxide type A, liquid or solid | Forbidden | | | | | | | | | | | |
| | Organic peroxide type B, liquid | 5.2 | UN3101 | II | ORGANIC PEROXIDE EXPLOSIVE | 53 | None | 225 | None | Forbidden | Forbidden | D | 12, 40 |
| | Organic peroxide type B, liquid, temperature controlled | 5.2 | UN3111 | II | ORGANIC PEROXIDE EXPLOSIVE | 53 | None | 225 | None | Forbidden | Forbidden | D | 2, 40 |
| | Organic peroxide type B, solid | 5.2 | UN3102 | II | ORGANIC PEROXIDE EXPLOSIVE | 53 | None | 225 | None | Forbidden | Forbidden | D | 12, 40 |
| | Organic peroxide type B, solid, temperature controlled | 5.2 | UN3112 | II | ORGANIC PEROXIDE EXPLOSIVE | 53 | None | 225 | None | Forbidden | Forbidden | D | 2, 40 |
| | Organic peroxide type C, liquid | 5.2 | UN3103 | II | ORGANIC PEROXIDE | | 152 | 225 | None | 5 L | 10 L | D | 12, 40 |
| | Organic peroxide type C, liquid, temperature controlled | 5.2 | UN3113 | II | ORGANIC PEROXIDE | | None | 225 | None | Forbidden | Forbidden | D | 2, 40 |
| | Organic peroxide type C, solid | 5.2 | UN3104 | II | ORGANIC PEROXIDE | | 152 | 225 | None | 5 kg | 10 kg | D | 12, 40 |
| | Organic peroxide type C, solid, temperature controlled | 5.2 | UN3114 | II | ORGANIC PEROXIDE | | None | 225 | None | Forbidden | Forbidden | D | 2, 40 |
| | Organic peroxide type D, liquid | 5.2 | UN3105 | II | ORGANIC PEROXIDE | | 152 | 225 | None | 5 L | 10 L | D | 12, 40 |
| | Organic peroxide type D, liquid, temperature controlled | 5.2 | UN3115 | II | ORGANIC PEROXIDE | | None | 225 | None | Forbidden | Forbidden | D | 2, 40 |
| | Organic peroxide type D, solid | 5.2 | UN3106 | II | ORGANIC PEROXIDE | | 152 | 225 | None | 5 kg | 10 kg | D | 12, 40 |
| | Organic peroxide type D, solid, temperature controlled | 5.2 | UN3116 | II | ORGANIC PEROXIDE | | None | 225 | None | Forbidden | Forbidden | D | 2, 40 |
| | Organic peroxide type E, liquid | 5.2 | UN3107 | II | ORGANIC PEROXIDE | | 152 | 225 | None | 10 L | 25 L | D | 12, 40 |
| | Organic peroxide type E, liquid, temperature controlled | 5.2 | UN3117 | II | ORGANIC PEROXIDE | | None | 225 | None | Forbidden | Forbidden | D | 2, 40 |
| | Organic peroxide type E, solid | 5.2 | UN3108 | II | ORGANIC PEROXIDE | | 152 | 225 | None | 10 kg | 25 kg | D | 12, 40 |
| | Organic peroxide type E, solid, temperature controlled | 5.2 | UN3118 | II | ORGANIC PEROXIDE | | None | 225 | None | Forbidden | Forbidden | D | 2, 40 |
| | Organic peroxide type F, liquid | 5.2 | UN3109 | II | ORGANIC PEROXIDE | | 152 | 225 | 225 | 10 L | 25 L | D | 12, 40 |
| | Organic peroxide type F, liquid, temperature controlled | 5.2 | UN3119 | II | ORGANIC PEROXIDE | | None | 225 | None | Forbidden | Forbidden | D | 2, 40 |
| | Organic peroxide type F, solid | 5.2 | UN3110 | II | ORGANIC PEROXIDE | T42 | 152 | 225 | None | 10 kg | 25 kg | D | 12, 40 |
| | Organic peroxide type F, solid, temperature controlled | 5.2 | UN3120 | II | ORGANIC PEROXIDE | | None | 225 | None | Forbidden | Forbidden | D | 2, 40 |
| D | Organic phosphate, Organic phosphate compound, *or* Organic phosphorus compound; mixed with compressed gas | 2.3 | NA1955 | | POISON GAS | 3 | None | 334 | None | Forbidden | Forbidden | D | 40 |
| | Organochlorine pesticides liquid, flammable, toxic, n.o.s. *flash point less than 23 degrees C* | 3 | UN2762 | I | FLAMMABLE LIQUID, POISON | | None | 201 | 243 | Forbidden | 30 L | B | |
| | | | | II | FLAMMABLE LIQUID, POISON | | None | 202 | 243 | 1 L | 60 L | B | |
| | Organochlorine pesticides, liquid, toxic, flammable, n.o.s. *flashpoint not less than 23 degrees C* | 6.1 | UN2995 | I | POISON, FLAMMABLE LIQUID | T42 | None | 201 | 243 | 1 L | 30 L | B | 40 |
| | | | | II | POISON, FLAMMABLE LIQUID | T14 | None | 202 | 243 | 5 L | 60 L | B | 40 |
| | | | | III | KEEP AWAY FROM FOOD, FLAMMABLE LIQUID | B1, T14 | 153 | 203 | 242 | 60 L | 220 L | A | 40 |
| | Organochlorine pesticides, liquid, toxic, n.o.s. | 6.1 | UN2996 | I | POISON | T42 | None | 201 | 243 | 1 L | 30 L | B | 40 |
| | | | | II | POISON | T14 | None | 202 | 242 | 5 L | 60 L | B | 40 |
| | | | | III | KEEP AWAY FROM FOOD | T14 | 153 | 203 | 241 | 60 L | 220 L | A | 40 |

## §172.101 HAZARDOUS MATERIALS TABLE, Continued

| Symbols | Hazardous materials descriptions and proper shipping names | Hazard class or Division | Identification Numbers | Packing group | Label(s) required (if not excepted) | Special provisions | (8) Packaging authorizations (§173.***) | | | (9) Quantity limitations | | (10) Vessel stowage requirements | |
|---|---|---|---|---|---|---|---|---|---|---|---|---|---|
| | | | | | | | Exceptions | Non-bulk packaging | Bulk packaging | Passenger aircraft or railcar | Cargo aircraft only | Vessel stowages | Other stowage provisions |
| (1) | (2) | (3) | (4) | (5) | (6) | (7) | (8A) | (8B) | (8C) | (9A) | (9B) | (10A) | (10B) |
| | Organochlorine pesticides, solid toxic n.o.s. | 6.1 | UN2761 | I | POISON | | None | 211 | 242 | 5 kg | 50 kg | A | 40 |
| | | | | II | POISON | | None | 212 | 242 | 25 kg | 100 kg | A | 40 |
| | | | | III | KEEP AWAY FROM FOOD | | 153 | 213 | 240 | 100 kg | 200 kg | A | 40 |
| | Organophosphorus pesticides, liquid, flammable, toxic, n.o.s., *flash point less than 23 degrees C*. | 3 | UN2784 | I | FLAMMABLE LIQUID, POISON | | None | 201 | 243 | Forbidden | 30 L | B | |
| | | | | II | FLAMMABLE LIQUID, POISON | | None | 202 | 243 | 1 L | 60 L | B | |
| | Organophosphorus pesticides, liquid, toxic, flammable, n.o.s. *flashpoint not less than 23 degrees C*. | 6 1 | UN3017 | I | POISON, FLAMMABLE LIQUID | N76, T42 | None | 201 | 243 | 1 L | 30 L | B | 40 |
| | | | | II | POISON, FLAMMABLE LIQUID | N76, T14 | None | 202 | 243 | 5 L | 60 L | B | 40 |
| | | | | III | KEEP AWAY FROM FOOD, FLAMMABLE LIQUID | B1, N76, T14 | 153 | 203 | 242 | 60 L | 220 L | A | 40 |
| | Organophosphorus pesticides, liquid, toxic, n.o.s. | 6 1 | UN3018 | I | POISON | N76, T42 | None | 201 | 243 | 1 L | 30 L | B | 40 |
| | | | | II | POISON | N76, T14 | None | 202 | 243 | 5 L | 60 L | B | 40 |
| | | | | III | KEEP AWAY FROM FOOD | N76, T14 | 153 | 203 | 241 | 60 L | 220 L | A | 40 |
| | Organophosphorus pesticides, solid, toxic, n.o.s. | 6 1 | UN2783 | I | POISON | N77 | None | 211 | 242 | 5 kg | 50 kg | A | 40 |
| | | | | II | POISON | N77 | None | 212 | 242 | 25 kg | 100 kg | A | 40 |
| | | | | III | KEEP AWAY FROM FOOD | N77 | 153 | 213 | 240 | 100 kg | 200 kg | A | 40 |
| | Organotin compounds, liquid, n.o.s | 6 1 | UN2788 | I | POISON | A3, N33, N34, T42 | None | 201 | 243 | 1 L | 30 L | B | 40 |
| | | | | II | POISON | A3, N33, N34, T14 | None | 202 | 243 | 5 L | 60 L | A | 40 |
| | | | | III | KEEP AWAY FROM FOOD | T14 | 153 | 203 | 241 | 60 L | 220 L | A | 40 |
| | Organotin compounds, solid, n.o.s. | 6 1 | UN3146 | I | POISON | A5 | None | 211 | 242 | 5 kg | 50 kg | B | 40 |
| | | | | II | POISON | | None | 212 | 242 | 25 kg | 100 kg | A | 40 |
| | | | | III | KEEP AWAY FROM FOOD | | 153 | 213 | 240 | 100 kg | 200 kg | A | 40 |
| | Organotin pesticides, liquid, flammable, toxic, n.o.s., *flash point less than 23deg C* | 3 | UN2787 | I | FLAMMABLE LIQUID, POISON | | None | 201 | 243 | Forbidden | 30 L | B | |
| | | | | II | FLAMMABLE LIQUID, POISON | | None | 202 | 243 | 1 L | 60 L | B | |
| | Organotin pesticides, liquid, toxic, flammable, n.o.s. *flashpoint not less than 23 degrees C* | 6.1 | UN3019 | I | POISON, FLAMMABLE LIQUID | T42 | None | 201 | 243 | 1 L | 30 L | B | 40 |
| | | | | II | POISON, FLAMMABLE LIQUID | T14 | None | 202 | 243 | 5 L | 60 L | B | 40 |
| | | | | III | KEEP AWAY FROM FOOD, FLAMMABLE LIQUID | B1, T14 | 153 | 203 | 242 | 60 L | 220 L | A | 40 |
| | Organotin pesticides, liquid, toxic, n.o.s. | 6 1 | UN3020 | I | POISON | T42 | None | 201 | 243 | 1 L | 30 L | B | 40 |
| | | | | II | POISON | T14 | None | 202 | 243 | 5 L | 60 L | B | 40 |
| | | | | III | KEEP AWAY FROM FOOD | T14 | 153 | 203 | 241 | 60 L | 220 L | A | 40 |
| | Organotin pesticides, solid, toxic, n.o.s | 6 1 | UN2786 | I | POISON | | None | 211 | 242 | 5 kg | 50 kg | A | 40 |
| | | | | II | POISON | | None | 212 | 242 | 25 kg | 100 kg | A | 40 |
| | | | | III | KEEP AWAY FROM FOOD | | 153 | 213 | 240 | 100 kg | 200 kg | A | 40 |
| | *Orthonitroaniline, see* Nitroanilines *etc.* | | | | | | | | | | | | |
| | Osmium tetroxide | 6 1 | UN2471 | I | POISON | A8, N33, N34 | None | 211 | 242 | 5 kg | 50 kg | B | 40 |
| AD | Other regulated substances, liquid, n.o.s | 9 | NA3082 | III | CLASS 9 | | 155 | 203 | 241 | No limit | No limit | A | |
| AD | Other regulated substances, solid, n.o.s. | 9 | NA3077 | III | CLASS 9 | | 155 | 213 | 240 | No limit | No limit | A | |
| | Oxalates, *water soluble* | 6 1 | UN2449 | III | KEEP AWAY FROM FOOD | | 153 | 213 | 240 | 100 kg | 200 kg | A | |
| | Oxidizing substances, liquid, corrosive, n.o.s. | 5.1 | UN3098 | I | OXIDIZER, CORROSIVE | | None | 201 | 244 | Forbidden | 2.5 L | B | 34, 56, 58, 69, 106 |
| | | | | II | OXIDIZER, CORROSIVE | | None | 202 | 243 | 1 L | 5 L | D | 34, 56, 58, 69, 106 |
| | | | | III | OXIDIZER, CORROSIVE | | 152 | 203 | 242 | 2.5 L | 30 L | B | 34, 56, 58, 69, 106 |
| | Oxidizing substances, liquid, n.o.s | 5 1 | UN3139 | II | OXIDIZER | A2 | 152 | 202 | 242 | 1 L | 5 L | B | 56, 58, 69, 106 |
| | Oxidizing substances, liquid, poisonous, n.o.s | 5 1 | UN3099 | I | OXIDIZER, POISON | | None | 201 | 244 | Forbidden | 2.5 L | D | 56, 58, 95, 106 |
| | | | | II | OXIDIZER, POISON | | None | 202 | 243 | 1 L | 5 L | B | 56, 58, 95, 106 |
| | | | | III | OXIDIZER, KEEP AWAY FROM FOOD | | 152 | 203 | 242 | 2.5L | 30L | B | 56, 58, 95, 106 |

## §172.101 HAZARDOUS MATERIALS TABLE, Continued

| Symbols (1) | Hazardous materials descriptions and proper shipping names (2) | Hazard class or Division (3) | Identification Numbers (4) | Packing group (5) | Label(s) required (if not excepted) (6) | Special provisions (7) | (8) Packaging authorizations (§173 ***) Exceptions (8A) | Non-bulk packaging (8B) | Bulk packaging (8C) | (9) Quantity limitations Passenger aircraft or railcar (9A) | Cargo aircraft only (9B) | (10) Vessel stowage requirements Vessel stowages (10A) | Other stowage provisions (10B) |
|---|---|---|---|---|---|---|---|---|---|---|---|---|---|
| | Oxidizing substances, solid, corrosive, n.o.s | 5.1 | UN3085 | I | OXIDIZER, CORROSIVE | | None | 211 | 242 | 1 kg | 15 kg | D | 13, 34, 56, 58, 69, 106 |
| | | | | II | OXIDIZER, CORROSIVE | | None | 212 | 242 | 5 kg | 25 kg | B | 13, 34, 56, 58, 69, 106 |
| | | | | III | OXIDIZER, CORROSIVE | | 152 | 213 | 240 | 25 kg | 100 kg | B | 13, 34, 56, 58, 69, 106 |
| | Oxidizing substances, solid, flammable, n.o.s | 5.1 | UN3137 | | OXIDIZER, FLAMMABLE SOLID | | None | 214 | 214 | Forbidden | Forbidden | | |
| | Oxidizing substances, solid, n.o.s | 5.1 | UN1479 | I | OXIDIZER | | 152 | 211 | 242 | 1 kg | 15 kg | B | 56, 58, 69, 106 |
| | | | | II | OXIDIZER | | 152 | 212 | 240 | 5 kg | 25 kg | D | 56, 58, 69, 106 |
| | | | | III | OXIDIZER | | 152 | 213 | 240 | 25 kg | 100 kg | B | 56, 58, 69, 106 |
| | Oxidizing substances, solid, poisonous, n.o.s | 5.1 | UN3087 | I | OXIDIZER, POISON | | None | 211 | 242 | 1 kg | 15 kg | D | 56, 58, 69, 95, 106 |
| | | | | II | OXIDIZER, POISON | | None | 212 | 242 | 5 kg | 25 kg | B | 56, 58, 69, 95, 106 |
| | | | | III | OXIDIZER, KEEP AWAY FROM FOOD | | 152 | 213 | 240 | 25 kg | 100 kg | B | 56, 58, 69, 95, 106 |
| | Oxidizing substances, solid, self heating, n.o.s | 5.1 | UN3100 | | OXIDIZER, SPONTANEOUSLY COMBUSTIBLE | | None | 214 | 214 | Forbidden | Forbidden | | |
| | Oxidizing substances, solid, which in contact with water emit flammable gases, n.o.s | 5.1 | UN3121 | | OXIDIZER, DANGEROUS WHEN WET | | None | 214 | 214 | Forbidden | Forbidden | | |
| | *Oxygen and carbon dioxide mixtures, see* Carbon dioxide and oxygen mixtures | | | | | | | | | | | | |
| | Oxygen, compressed | 2.2 | UN1072 | | NONFLAMMABLE GAS, OXIDIZER | | 306 | 302 | 314, 315 | 75 kg | 150 kg | A | |
| | Oxygen difluoride | 2.3 | UN2190 | | POISON GAS, OXIDIZER | 1 | None | 304 | None | Forbidden | Forbidden | D | 13, 40 |
| | *Oxygen, mixtures with rare gases, see* Rare gases and oxygen mixtures | | | | | | | | | | | | |
| | Oxygen, refrigerated liquid *(cryogenic liquid)* | 2.2 | UN1073 | | NONFLAMMABLE GAS, OXIDIZER | | 320 | 316 | 318 | Forbidden | Forbidden | D | |
| | Paint *including paint, lacquer, enamel, stain, shellac solutions, varnish, polish, liquid filler, and liquid lacquer base* | 3 | UN1263 | II | FLAMMABLE LIQUID | B52, T7, T30 | 150 | 173 | 242 | 5 L | 60 L | B | |
| | | | | III | FLAMMABLE LIQUID | B1, B52, T7, T30 | 150 | 173 | 242 | 60 L | 220 L | A | |
| | Paint or Paint related material | 8 | UN3066 | II | CORROSIVE | B2, N71, T7 | 154 | 202 | 242 | 1 L | 30 L | A | |
| | | | | III | CORROSIVE | B52, N71, T7 | 154 | 203 | 241 | 5 L | 60 L | A | |
| | Paint related material *including paint thinning, drying, removing, or reducing compound* | 3 | UN1263 | II | FLAMMABLE LIQUID | B52, T7, T30 | 150 | 173 | 242 | 5 L | 60 L | B | |
| | | | | III | FLAMMABLE LIQUID | B1, B52, T7, T30 | 150 | 173 | 242 | 60 L | 220 L | A | |
| | Paper, unsaturated oil treated *incompletely dried (including carbon paper)* | 4.2 | UN1379 | III | SPONTANEOUSLY COMBUSTIBLE | | None | 213 | 241 | Forbidden | Forbidden | A | |
| | Paraformaldehyde | 4.1 | UN2213 | III | NONE | A1 | 151 | 213 | 240 | 25 kg | 100 kg | A | |
| | Paraldehyde | 3 | UN1264 | III | FLAMMABLE LIQUID | B1, T1 | 150 | 203 | 242 | 60 L | 220 L | A | |
| | *Paranitroaniline, solid, see* Nitroanilines etc | | | | | | | | | | | | |
| D | Parathion | 6.1 | NA2783 | I | POISON | T42 | None | 201 | 243 | Forbidden | 1 L | A | 40 |
| | | | | II | POISON | T14 | None | 202 | 243 | Forbidden | 5 L | A | 40 |
| D | Parathion and compressed gas mixture | 2.3 | NA1967 | | POISON GAS | 3 | None | 334 | 245 | Forbidden | Forbidden | E | 40 |
| | *Paris green, solid, see* Copper acetoarsenite | | | | | | | | | | | | |
| AW | *PCB, see* Polychlorinated biphenyls | | | | | | | | | | | | |
| + | Pentaborane | 4.2 | UN1380 | I | SPONTANEOUSLY COMBUSTIBLE, POISON | 1 | None | 205 | 245 | Forbidden | Forbidden | D | |
| | Pentachloroethane | 6.1 | UN1669 | II | POISON | T14 | None | 202 | 243 | 5 L | 60 L | A | 40 |
| | *Pentaerythrite tetranitrate (dry)* | Forbidden | | | | | | | | | | | |
| D | Pentaerythrite tetranitrate or Pentaerythritol tetranitrate or PETN, *wetted with not less than 25 per cent water, by mass, or* Pentaerythrite tetranitrate or Pentaerythritol tetranitrate or PETN, *desensitized with not less than 15 per cent phlegmatizer by mass* | 1.1A | NA0150 | II | EXPLOSIVE 1.1A | 111, 117 | None | 62 | None | Forbidden | Forbidden | B | 1E, 5E |
| | Pentaerythrite tetranitrate or Pentaerythritol tetranitrate or PETN, *with not less than 7 per cent wax by mass* | 1.1D | UN0411 | II | EXPLOSIVE 1.1D | | None | 62 | None | Forbidden | Forbidden | B | 1E, 5E |
| | *Pentaerythritol tetranitrate, see* Pentaerythrite tetranitrate, etc | | | | | | | | | | | | |
| | Pentamethylheptane | 3 | UN2286 | III | FLAMMABLE LIQUID | B1, T1 | 150 | 203 | 242 | 60 L | 220 L | A | |
| | Pentan-2,4-Dione | 3 | UN2310 | III | FLAMMABLE LIQUID | B1, T1 | 150 | 203 | 242 | 60 L | 220 L | A | |

## §172.101 HAZARDOUS MATERIALS TABLE, Continued

| Symbols | Hazardous materials descriptions and proper shipping names | Hazard class or Division | Identification Numbers | Packing group | Label(s) required (if not excepted) | Special provisions | (8) Packaging authorizations (§173 ***) Exceptions | Non-bulk packaging | Bulk packaging | (9) Quantity limitations Passenger aircraft or railcar | Cargo aircraft only | (10) Vessel stowage requirements Vessel stowages | Other stowage provisions |
|---|---|---|---|---|---|---|---|---|---|---|---|---|---|
| (1) | (2) | (3) | (4) | (5) | (6) | (7) | (8A) | (8B) | (8C) | (9A) | (9B) | (10A) | (10B) |
| | n-Pentanes or isopentane | 3 | UN1265 | I | FLAMMABLE LIQUID | T20 | 150 | 201 | 243 | 1 L | 30 L | E | |
| | *Pentanitroaniline (dry)* | Forbidden | | | | | | | | | | | |
| | 1-Pentol | 8 | UN2705 | II | CORROSIVE | B2, T8 | 154 | 202 | 242 | 1 L | 30 L | B | 38 |
| | Pentolite, *dry or wetted with less than 15 per cent water, by mass* | 1.1D | UN0151 | II | EXPLOSIVE 1 1D | | None | 62 | None | Forbidden | Forbidden | B | 1E, 5E |
| | Perchlorates, inorganic, n o s | 5 1 | UN1481 | II | OXIDIZER | | 152 | 212 | 242 | 5 kg | 25 kg | A | 46, 56 |
| | *Perchloric acid more than 72 per cent acid by mass* | Forbidden | | | | | | | | | | | |
| | Perchloric acid *more than 50 per cent but not more than 72 per cent acid, by mass* | 5.1 | UN1873 | I | OXIDIZER, CORROSIVE | A2, A3, N41, T9, T27 | None | 201 | 243 | Forbidden | 2.5 L | D | 66 |
| | Perchloric acid *not more than 50 per cent acid by mass* | 8 | UN1802 | II | CORROSIVE, OXIDIZER | N41, T9 | None | 202 | 243 | Forbidden | 30 L | C | 66, 89 |
| | *Perchloroethylene, see Tetrachloroethylene* | | | | | | | | | | | | |
| | Perchloromethylmercaptan | 6 1 | UN1670 | I | POISON | 2, A3, A7, B9, B14, B32, B74, N34, T38, T43, T45 | None | 227 | 244 | Forbidden | 30 L | D | 40 |
| | Perchloryl fluoride | 2.3 | UN3083 | | POISON GAS, OXIDIZER | 3, B12, B14 | None | 302 | 314, 315 | Forbidden | Forbidden | D | 40, 43, 95 |
| | *Percussion caps, see Primers, cap type* | | | | | | | | | | | | |
| | *Perfluoro-2-butene, see Octafluorobut-2-ene* | | | | | | | | | | | | |
| | Perfluoroethylvinyl ether | 2.1 | UN3154 | | FLAMMABLE GAS | | 306 | 302, 304, 305 | 314, 315 | Forbidden | 150 kg | E | 40 |
| | Perfluoromethylvinyl ether | 2 1 | UN3153 | | FLAMMABLE GAS | | 306 | 302, 304, 305 | 314, 315 | Forbidden | 150 kg | E | 40 |
| | Perfumery products *with flammable solvents* | 3 | UN1266 | II | FLAMMABLE LIQUID | T7, T30 | 150 | 202 | 242 | 15 L | 60 L | B | |
| | | | | III | FLAMMABLE LIQUID | B1, T7, T30 | 150 | 203 | 242 | 60 L | 220 L | A | |
| | Permanganates, inorganic, n o s *(except ammonium permanganate, the transport of which is prohibited except when approved by the Associate Administrator for Hazardous Materials Safety y the competent authorities)* | 5 1 | UN1482 | II | OXIDIZER | A30 | 152 | 212 | 242 | 5 kg | 25 kg | D | 56, 58, 69, 106, 107 |
| | Peroxides, inorganic, n.o.s | 5 1 | UN1483 | II | OXIDIZER | A7, A20, N34 | None | 212 | 242 | 5 kg | 25 kg | A | 13, 75, 106 |
| | *Peroxyacetic acid, more than 43 per cent and with more than 6 per cent hydrogen peroxide* | Forbidden | | | | | | | | | | | |
| | Pesticides, liquid, flammable, toxic, n.o.s (flashpoint less than 23 degrees C.) | 3 | UN3021 | I | FLAMMABLE LIQUID, POISON | B5 | None | 201 | 243 | Forbidden | 30 L | B | |
| | | | | II | FLAMMABLE LIQUID, POISON | | None | 202 | 243 | 1 L | 60 L | B | |
| | Pesticides, liquid, toxic, flammable, n.o.s *flashpoint not less than 23 degrees C* | 6.1 | UN2903 | I | POISON, FLAMMABLE LIQUID | T42 | None | 201 | 243 | 1 L | 30 L | B | 40 |
| | | | | II | POISON, FLAMMABLE LIQUID | T14 | None | 202 | 243 | 5 L | 60 L | B | 40 |
| | | | | III | KEEP AWAY FROM FOOD, FLAMMABLE LIQUID | B1, T14 | 153 | 203 | 242 | 60 L | 220 L | A | 40 |
| | Pesticides, liquid, toxic, n o s | 6 1 | UN2902 | I | POISON | T42 | None | 201 | 243 | 1 L | 30 L | B | 40 |
| | | | | II | POISON | T14 | None | 202 | 243 | 5 L | 60 L | B | 40 |
| | | | | III | KEEP AWAY FROM FOOD | T14 | 153 | 203 | 241 | 60 L | 220 L | A | 40 |
| | Pesticides, solid, toxic, n.o.s | 6 1 | UN2588 | I | POISON | | None | 211 | 242 | 5 kg | 50 kg | A | 40 |
| | | | | II | POISON | | None | 212 | 242 | 25 kg | 100 kg | A | 40 |
| | | | | III | KEEP AWAY FROM FOOD | | 153 | 213 | 240 | 100 kg | 200 kg | A | 40 |
| | PETN, see Pentaerythrite tetranitrate | | | | | | | | | | | | |
| | PETN/TNT, see Pentolite, etc | | | | | | | | | | | | |
| | Petrol, see Gasoline | | | | | | | | | | | | |
| | Petroleum crude oil | 3 | UN1267 | I | FLAMMABLE LIQUID | T8, T31 | None | 201 | 243 | 1 L | 30 L | E | |
| | | | | II | FLAMMABLE LIQUID | T8, T31 | 150 | 202 | 242 | 5 L | 60 L | B | |
| | | | | III | FLAMMABLE LIQUID | B1, T7, T30 | 150 | 203 | 242 | 60 L | 220 L | A | |
| | Petroleum distillates, n o s | 3 | UN1268 | I | FLAMMABLE LIQUID | T8, T31 | 150 | 201 | 243 | 1 L | 30 L | E | |
| | | | | II | FLAMMABLE LIQUID | T8, T31 | 150 | 202 | 242 | 5 L | 60 L | B | |
| | | | | III | FLAMMABLE LIQUID | B1, T7, T30 | 150 | 203 | 242 | 60 L | 220 L | A | |
| | *Petroleum ether, see Petroleum spirit* | | | | | | | | | | | | |
| | Petroleum gases, liquefied *or* Liquefied petroleum gas | 2 1 | UN1075 | | FLAMMABLE GAS | | 306 | 304 | 314, 315 | Forbidden | 150 kg | E | 40 |

## §172.101 HAZARDOUS MATERIALS TABLE, Continued

| Symbols | Hazardous materials descriptions and proper shipping names | Hazard class or Division | Identification Numbers | Packing group | Label(s) required (if not excepted) | Special provisions | (8) Packaging authorizations (§173.***) | | | (9) Quantity limitations | | (10) Vessel stowage requirements | |
|---|---|---|---|---|---|---|---|---|---|---|---|---|---|
| | | | | | | | Exceptions | Non-bulk packaging | Bulk packaging | Passenger aircraft or railcar | Cargo aircraft only | Vessel stowages | Other stowage provisions |
| (1) | (2) | (3) | (4) | (5) | (6) | (7) | (8A) | (8B) | (8C) | (9A) | (9B) | (10A) | (10B) |
| | *Petroleum naphtha, see* Naphtha, petroleum | | | | | | | | | | | | |
| | Petroleum oil | 3 | UN1270 | I | FLAMMABLE LIQUID | T8, T31 | None | 201 | 243 | 1 L | 30 L | E | |
| | | | | II | FLAMMABLE LIQUID | T8, T31 | 150 | 202 | 242 | 5 L | 60 L | B | |
| | | | | III | FLAMMABLE LIQUID | B1, T7, T30 | 150 | 203 | 242 | 60 L | 220 L | A | |
| | Petroleum spirit | 3 | UN1271 | I | FLAMMABLE LIQUID | T8 | 150 | 201 | 243 | 1 L | 30 L | E | |
| | | | | II | FLAMMABLE LIQUID | B1, T8 | 150 | 202 | 242 | 5 L | 60 L | B | |
| | Phenacyl bromide | 6.1 | UN2645 | II | POISON | | None | 212 | 242 | 25 kg | 100 kg | B | 40, 48 |
| | Phenetidines | 6.1 | UN2311 | III | KEEP AWAY FROM FOOD | T7 | 153 | 203 | 241 | 60 L | 220 L | A | |
| | Phenol, molten | 6.1 | UN2312 | II | POISON | B14, T8 | None | 202 | 243 | Forbidden | Forbidden | B | 40 |
| + | Phenol, solid | 6.1 | UN1671 | II | POISON | N78, T14 | None | 212 | 240 | 25 kg | 100 kg | A | |
| | Phenol solutions | 6.1 | UN2821 | II | POISON | T14 | None | 202 | 243 | 5 L | 60 L | A | |
| | Phenolsulfonic acid, liquid | 8 | UN1803 | II | CORROSIVE | B2, N41, T8 | 154 | 202 | 242 | 1 L | 30 L | C | 14 |
| | Phenoxy pesticides, liquid, flammable, toxic n.o.s., *flash point less than 23 degrees C* | 3 | UN2766 | I | FLAMMABLE LIQUID, POISON | | None | 201 | 243 | Forbidden | 30 L | B | |
| | | | | II | FLAMMABLE LIQUID, POISON | | None | 202 | 243 | 1 L | 60 L | B | |
| | Phenoxy pesticides, liquid, toxic, flammable, n.o.s *flashpoint not less than 23 degrees C* | 6.1 | UN2999 | I | POISON, FLAMMABLE LIQUID | T42 | None | 201 | 243 | 1 L | 30 L | B | 40 |
| | | | | II | POISON, FLAMMABLE LIQUID | T14 | None | 202 | 243 | 5 L | 60 L | B | 40 |
| | | | | III | KEEP AWAY FROM FOOD, FLAMMABLE LIQUID | B1, T14 | 153 | 203 | 242 | 60 L | 220 L | A | 40 |
| | Phenoxy pesticides, liquid, toxic, n.o.s | 6.1 | UN3000 | I | POISON | T42 | None | 201 | 243 | 1 L | 30 L | B | 40 |
| | | | | II | POISON | T14 | None | 202 | 243 | 5 L | 60 L | B | 40 |
| | | | | III | KEEP AWAY FROM FOOD | T14 | 153 | 203 | 241 | 60 L | 220 L | A | 40 |
| | Phenoxy pesticides, solid, toxic, n.o.s. | 6.1 | UN2765 | I | POISON | | None | 211 | 242 | 5 kg | 50 kg | A | 40 |
| | | | | II | POISON | | None | 212 | 242 | 25 kg | 100 kg | A | 40 |
| | | | | III | KEEP AWAY FROM FOOD | | 153 | 213 | 240 | 100 kg | 200 kg | A | 40 |
| + | Phenyl isocyanate | 6.1 | UN2487 | II | POISON | 2, A3, B9, B14, B32, B74, B77, N33, N34, T38, T43, T45 | None | 227 | 244 | 5 L | 60 L | D | 40 |
| + | Phenyl mercaptan | 6.1 | UN2337 | II | POISON, FLAMMABLE LIQUID | 2, B9, B14, B32, B74, B77, T38, T43, T45 | None | 227 | 244 | 5 L | 60 L | B | 26, 40 |
| | Phenyl phosphorus dichloride | 8 | UN2798 | II | CORROSIVE | B2, B15, T8, T26 | 154 | 202 | 242 | Forbidden | 30 L | B | 40 |
| | Phenyl phosphorus thiodichloride | 8 | UN2799 | II | CORROSIVE | B2, B15, T8, T26 | 154 | 202 | 242 | Forbidden | 30 L | B | 40 |
| | Phenyl urea pesticides, liquid, flammable toxic, n.o.s., *flash point less than 23 degrees C.* | 3 | UN2768 | I | FLAMMABLE LIQUID, POISON | | None | 201 | 243 | Forbidden | 30 L | B | |
| | | | | II | FLAMMABLE LIQUID, POISON | | None | 202 | 243 | 1 L | 60 L | B | |
| | Phenyl urea pesticides, liquid, toxic, flammable, n.o.s., *flash point not less than 23 degrees C* | 6.1 | UN3001 | I | POISON, FLAMMABLE LIQUID | T42 | None | 201 | 243 | 1 L | 30 L | B | 40 |
| | | | | II | POISON, FLAMMABLE LIQUID | T14 | None | 202 | 243 | 5 L | 60 L | B | 40 |
| | | | | III | KEEP AWAY FROM FOOD, FLAMMABLE LIQUID | B1, T14 | 153 | 203 | 242 | 60 L | 220 L | A | 40 |
| | Phenyl urea pesticides, liquid, toxic, n.o.s. | 6.1 | UN3002 | I | POISON | T42 | None | 201 | 243 | 1 L | 30 L | B | 40 |
| | | | | II | POISON | T14 | None | 202 | 243 | 5 L | 60 L | B | 40 |
| | | | | III | KEEP AWAY FROM FOOD | T14 | 153 | 203 | 241 | 60 L | 220 L | A | 40 |
| | Phenyl urea pesticides, solid, toxic, n.o.s. | 6.1 | UN2767 | I | POISON | | None | 211 | 242 | 5 kg | 50 kg | A | 40 |
| | | | | II | POISON | | None | 212 | 242 | 25 kg | 100 kg | A | 40 |
| | | | | III | KEEP AWAY FROM FOOD | | 153 | 213 | 240 | 100 kg | 200 kg | A | 40 |
| | Phenylacetonitrile, liquid | 6.1 | UN2470 | III | KEEP AWAY FROM FOOD | T8 | 153 | 203 | 241 | 60 L | 220 L | A | 26 |
| | Phenylacetyl chloride | 8 | UN2577 | II | CORROSIVE | B2, T8, T26 | 154 | 202 | 242 | 1 L | 30 L | C | 40 |
| | Phenylcarbylamine chloride | 6.1 | UN1672 | I | POISON | 2, B9, B14, B32, B74, T38, T43, T45 | None | 227 | 244 | Forbidden | Forbidden | D | 40 |
| | Phenylchloroformate | 6.1 | UN2746 | II | POISON, CORROSIVE | T12 | None | 202 | 243 | 1 L | 30 L | A | 12, 13, 25, 40, 48 |
| | *m-Phenylene diaminediperchlorate (dry)* | Forbidden | | | | | | | | | | | |
| | Phenylenediamines (o-, m-, p-) | 6.1 | UN1673 | III | KEEP AWAY FROM FOOD | | 153 | 213 | 240 | 100 kg | 200 kg | A | 108 |

## §172.101 HAZARDOUS MATERIALS TABLE, Continued

| Symbols (1) | Hazardous materials descriptions and proper shipping names (2) | Hazard class or Division (3) | Identification Numbers (4) | Packing group (5) | Label(s) required (if not excepted) (6) | Special provisions (7) | (8) Packaging authorizations (§173.***) Exceptions (8A) | Non-bulk packaging (8B) | Bulk packaging (8C) | (9) Quantity limitations Passenger aircraft or railcar (9A) | Cargo aircraft only (9B) | (10) Vessel stowage requirements Vessel stowages (10A) | Other stowage provisions (10B) |
|---|---|---|---|---|---|---|---|---|---|---|---|---|---|
| | Phenylhydrazine | 6.1 | UN2572 | II | POISON | T8 | None | 202 | 243 | 5 L | 60 L | A | 40 |
| | Phenylmercuric acetate | 6.1 | UN1674 | II | POISON | | None | 212 | 242 | 25 kg | 100 kg | A | |
| | Phenylmercuric compounds, n.o.s. | 6.1 | UN2026 | I | POISON | | None | 211 | 242 | 5 kg | 50 kg | A | |
| | | | | II | POISON | | None | 212 | 242 | 25 kg | 100 kg | A | |
| | | | | III | KEEP AWAY FROM FOOD | | 153 | 213 | 240 | 100 kg | 200 kg | A | |
| | Phenylmercuric hydroxide | 6.1 | UN1894 | II | POISON | | None | 212 | 242 | 25 kg | 100 kg | A | |
| | Phenylmercuric nitrate | 6.1 | UN1895 | II | POISON | | None | 212 | 242 | 25 kg | 100 kg | A | |
| | Phenyltrichlorosilane | 8 | UN1804 | II | CORROSIVE | A7, B6, N34, T8 | None | 202 | 242 | Forbidden | 30 L | C | 40 |
| | Phosgene | 2.3 | UN1076 | | POISON GAS, CORROSIVE | 1, B7, B46 | None | 192 | 314 | Forbidden | Forbidden | D | 40 |
| | 9-Phosphabicyclononanes or Cyclooctadiene phosphines | 4.2 | UN2940 | II | SPONTANEOUSLY COMBUSTIBLE | A19 | None | 212 | 241 | 15 kg | 50 kg | A | |
| | Phosphine | 2.3 | UN2199 | | POISON GAS, FLAMMABLE GAS | 1 | None | 192 | 245 | Forbidden | Forbidden | D | 40 |
| | Phosphoric acid | 8 | UN1805 | III | CORROSIVE | A7, N34, T7 | 154 | 203 | 241 | 5 L | 60 L | A | |
| | *Phosphoric acid triethyleneimine, see* Tri-(1-aziridinyl) phosphine oxide, solution | | | | | | | | | | | | |
| | *Phosphoric anhydride, see* Phosphorus pentoxide | | | | | | | | | | | | |
| | Phosphorous acid, ortho | 8 | UN2834 | III | CORROSIVE | T7 | 154 | 213 | 240 | 25 kg | 100 kg | A | 48 |
| | Phosphorus, amorphous | 4.1 | UN1338 | III | FLAMMABLE SOLID | A1, A19, B1, B9, B12, B26 | None | 213 | 243 | 25 kg | 100 kg | A | 74 |
| | *Phosphorus bromide, see* Phosphorus tribromide | | | | | | | | | | | | |
| | *Phosphorus chloride, see* Phosphorus trichloride | | | | | | | | | | | | |
| | Phosphorus heptasulfide, *free from yellow or white phosphorus* | 4.1 | UN1339 | II | FLAMMABLE SOLID | A20, N34 | None | 212 | 240 | 15 kg | 50 kg | B | 74 |
| | Phosphorus oxybromide | 8 | UN1939 | II | CORROSIVE | B8, N41, N43 | None | 212 | 240 | Forbidden | 50 kg | C | 12, 40, 48 |
| | Phosphorus oxybromide, molten | 8 | UN2576 | II | CORROSIVE | B2, B8, N41, N43, T8, T27 | None | 202 | 242 | Forbidden | Forbidden | C | 40 |
| + | Phosphorus oxychloride | 8 | UN1810 | II | CORROSIVE, POISON | 2, A7, B9, B14, B32, B74, B77, N34, T38, T43, T45 | None | 227 | 244 | Forbidden | 30 L | C | 8, 40 |
| | Phosphorus pentabromide | 8 | UN2691 | II | CORROSIVE | A7, N34 | 154 | 202 | 240 | Forbidden | 50 kg | B | 12, 40, 48 |
| | Phosphorus pentachloride | 8 | UN1806 | II | CORROSIVE | A7, N34 | None | 202 | 240 | Forbidden | 50 kg | C | 40 |
| | Phosphorus pentafluoride | 2.3 | UN2198 | | POISON GAS | 1 | None | 302 | None | Forbidden | Forbidden | D | 40 |
| | Phosphorus pentasulfide, *free from yellow or white phosphorus* | 4.3 | UN1340 | II | DANGEROUS WHEN WET | A20, B59 | None | 212 | 242 | 15 kg | 50 kg | B | 74 |
| | Phosphorus pentoxide | 8 | UN1807 | II | CORROSIVE | A7, N34 | 154 | 212 | 240 | 15 kg | 50 kg | A | 10 |
| | Phosphorus sesquisulfide, *free from yellow or white phosphorous* | 4.1 | UN1341 | II | FLAMMABLE SOLID | A20, N34 | None | 212 | 240 | 15 kg | 50 kg | B | 74 |
| | Phosphorus tribromide | 8 | UN1808 | II | CORROSIVE | A3, A6, A7, B2, B25, N34, N43, T8 | None | 202 | 242 | Forbidden | 30 L | C | 8, 40 |
| + | Phosphorus trichloride | 8 | UN1809 | II | CORROSIVE, POISON | 2, A3, A7, B9, B14, B15, B32, B74, B77, N34, T38, T43, T45 | None | 227 | 244 | Forbidden | 30L | C | 8, 40 |
| | Phosphorus trioxide | 8 | UN2578 | III | CORROSIVE | | 154 | 213 | 241 | 25 kg | 100 kg | A | 12, 48 |
| | Phosphorus trisulfide, *free from yellow or white phosphorus* | 4.1 | UN1343 | II | FLAMMABLE SOLID | A20, N34 | None | 212 | 240 | 15 kg | 50 kg | B | 74 |
| | Phosphorus white, molten | 4.2 | UN2447 | I | SPONTANEOUSLY COMBUSTIBLE, POISON | B9, B12, B26, N34, T15, T26, T29 | None | 188 | 243 | Forbidden | Forbidden | D | |
| | *Phosphorus (white or red) and a chlorate, mixtures of* | Forbidden | | | | | | | | | | | |
| | Phosphorus, white or yellow dry or under water or in solution | 4.2 | UN1381 | I | SPONTANEOUSLY COMBUSTIBLE, POISON | B9, B12, B26, N34, T15, T26, T33 | None | 188 | 243 | Forbidden | Forbidden | E | |
| | *Phosphoryl chloride, see* Phosphorus oxychloride | | | | | | | | | | | | |
| | Phthalic anhydride *with more than .05 per cent maleic anhydride* | 8 | UN2214 | III | CORROSIVE | T7 | 154 | 213 | 240 | 25 kg | 100 kg | A | |
| | Phthalimide derivative pesticides, liquid, flammable, toxic, n.o.s., *flash point less than 23 degrees C.* | 3 | UN2774 | I | FLAMMABLE LIQUID, POISON | | None | 201 | 243 | Forbidden | 30 L | B | |
| | | | | II | FLAMMABLE LIQUID, POISON | | None | 202 | 243 | 1 L | 60 L | B | |
| | Phthalimide derivative pesticides, liquid, toxic, flammable, n.o.s. *flashpoint not less than 23 degrees C* | 6.1 | UN3007 | I | POISON, FLAMMABLE LIQUID | T42 | None | 201 | 243 | 1 L | 30 L | B | 40 |
| | | | | II | POISON, FLAMMABLE LIQUID | T14 | None | 202 | 243 | 5 L | 60 L | B | 40 |
| | | | | III | KEEP AWAY FROM FOOD, FLAMMABLE LIQUID | T14 | 153 | 203 | 242 | 60 L | 220 L | A | 40 |

# §172.101 HAZARDOUS MATERIALS TABLE, Continued

| Symbols | Hazardous materials descriptions and proper shipping names | Hazard class or Division | Identification Numbers | Packing group | Label(s) required (if not excepted) | Special provisions | (8) Packaging authorizations (§173.***) | | | (9) Quantity limitations | | (10) Vessel stowage requirements | |
|---|---|---|---|---|---|---|---|---|---|---|---|---|---|
| | | | | | | | Exceptions | Non-bulk packaging | Bulk packaging | Passenger aircraft or railcar | Cargo aircraft only | Vessel stowages | Other stowage provisions |
| (1) | (2) | (3) | (4) | (5) | (6) | (7) | (8A) | (8B) | (8C) | (9A) | (9B) | (10A) | (10B) |
| | Phthalimide derivative pesticides, liquid, toxic, n.o.s. | 6.1 | UN3008 | I | POISON | T42 | None | 201 | 243 | 1 L | 30 L | B | 40 |
| | | | | II | POISON | T14 | None | 202 | 243 | 5 L | 60 L | B | 40 |
| | | | | III | KEEP AWAY FROM FOOD | T14 | 153 | 203 | 241 | 60 L | 220 L | A | 40 |
| | Phthalimide derivative pesticides, solid, toxic, n.o.s. | 6.1 | UN2773 | I | POISON | | None | 211 | 242 | 5 kg | 50 kg | A | 40 |
| | | | | II | POISON | | None | 212 | 242 | 25 kg | 100 kg | A | 40 |
| | | | | III | KEEP AWAY FROM FOOD | | 153 | 213 | 240 | 100 kg | 200 kg | A | 40 |
| | Picolines | 3 | UN2313 | III | FLAMMABLE LIQUID | B1, T8 | 150 | 202 | 242 | 5 L | 60 L | A | 40 |
| | Picric acid, see Trinitrophenol, etc | | | | | | | | | | | | |
| D | Picric acid, wet, with not less than 10% water | 4.1 | NA1344 | I | FLAMMABLE SOLID | A19, A20, N41 | None | 211 | None | Forbidden | Forbidden | D | |
| | Picrite, see Nitroguanidine, etc | | | | | | | | | | | | |
| | Picryl chloride, see Trinitrochlorobenzene | | | | | | | | | | | | |
| | Pine oil | 3 | UN1272 | III | FLAMMABLE LIQUID | B1, T1 | 150 | 203 | 242 | 60 L | 220 L | A | |
| | alpha-Pinene | 3 | UN2368 | III | FLAMMABLE LIQUID | B1, T1 | 150 | 203 | 242 | 60 L | 220 L | A | |
| | Piperazine | 8 | UN2579 | III | CORROSIVE | T7 | 154 | 213 | 240 | 25 kg | 100 kg | A | 12, 48 |
| | Piperidine | 3 | UN2401 | II | FLAMMABLE LIQUID | T2 | 150 | 202 | 242 | 5 L | 60 L | B | |
| | *Pivaloyl chloride*, see Trimethyl acetyl chloride | | | | | | | | | | | | |
| A | Plastic molding material *in dough, sheet or extruded rope form* | 9 | | III | CLASS 9 | | 155 | 213 | None | 100 kg | 200 kg | A | |
| | *Plastic solvent, n.o.s.*, see Flammable liquids, n.o.s. | | | | | | | | | | | | |
| | Plastics, nitrocellulose based, spontaneously combustible, n.o.s. | 4.2 | UN2006 | III | SPONTANEOUSLY COMBUSTIBLE | | None | 213 | None | Forbidden | Forbidden | C | |
| | *Poisonous gases, n.o.s.*, see Compressed or liquefied gases, flammable or toxic, n.o.s. | | | | | | | | | | | | |
| | Poisonous liquids, corrosive, n.o.s. | 6.1 | UN2927 | I | POISON, CORROSIVE | T42 | None | 201 | 243 | 0.5 L | 2.5 L | B | 40 |
| | | | | II | POISON, CORROSIVE | | None | 202 | 243 | 1 L | 30 L | B | 40 |
| | Poisonous liquids, corrosive, n.o.s., *inhalation hazard, Packing Group I, Zone A* | 6.1 | UN2927 | I | POISON, CORROSIVE | 1, B9, B14, B30, B72, T38, T43, T44 | None | 226 | 244 | Forbidden | Forbidden | D | 20, 40, 95 |
| | Poisonous liquids, corrosive, n.o.s., *inhalation hazard, Packing Group I, Zone B* | 6.1 | UN2927 | I | POISON, CORROSIVE | 2, B9, B14, B32, B74, T38, T43, T45 | None | 227 | 244 | Forbidden | Forbidden | D | 20, 40, 95 |
| | Poisonous liquids, flammable, n.o.s. | 6.1 | UN2929 | I | POISON, FLAMMABLE LIQUID | T42 | None | 201 | 243 | 1 L | 30 L | B | 40 |
| | | | | II | POISON, FLAMMABLE LIQUID | T15 | None | 202 | 243 | 5 L | 60 L | B | 40 |
| | Poisonous liquids, flammable, n.o.s., *inhalation hazard, Packing Group I, Zone A* | 6.1 | UN2929 | I | POISON, FLAMMABLE LIQUID | 1, B9, B14, B30, B72, T38, T43, T44 | None | 226 | 244 | Forbidden | Forbidden | D | 20, 40, 95 |
| | Poisonous liquids, flammable, n.o.s., *inhalation hazard, Packing Group I, Zone B* | 6.1 | UN2929 | I | POISON, FLAMMABLE LIQUID | 2, B9, B14, B32, B74, T38, T43, T45 | None | 227 | 244 | Forbidden | Forbidden | D | 20, 40, 95 |
| | Poisonous liquids, n.o.s. | 6.1 | UN2810 | I | POISON | T42 | None | 201 | 243 | 1 L | 30 L | B | 40 |
| | | | | II | POISON | T14 | None | 202 | 243 | 5 L | 60 L | B | 40 |
| | | | | III | KEEP AWAY FROM FOOD | T7 | 153 | 203 | 241 | 60 L | 220 L | A | 40 |
| | Poisonous liquids, n.o.s. *inhalation hazard, Packing Group I, Zone A* | 6.1 | UN2810 | I | POISON | 1, B9, B14, B30, B72, T38, T43, T44 | None | 226 | 244 | Forbidden | Forbidden | D | 20, 40, 95 |
| | Poisonous liquids, n.o.s. *inhalation hazard, Packing Group I, Zone B* | 6.1 | UN2810 | I | POISON | 2, B9, B14, B32, B74, T38, T43, T45 | None | 227 | 244 | Forbidden | Forbidden | D | 20, 40, 95 |
| | Poisonous liquids, oxidizing, n.o.s. | 6.1 | UN3122 | I | POISON, OXIDIZER | A4 | None | 201 | 243 | Forbidden | 2.5 L | C | 89 |
| | | | | II | POISON, OXIDIZER | | None | 202 | 243 | 1 L | 5 L | C | 89 |
| | Poisonous liquids, oxidizing, n.o.s. *inhalation hazard, packing group I, Zone A* | 6.1 | UN3122 | I | POISON, OXIDIZER | 1, B9, B14, B30, B72, T38, T43, T44 | None | 226 | 244 | Forbidden | 2.5 L | C | 89 |
| | Poisonous liquids, oxidizing, n.o.s. *Inhalation Hazard, Packing Group I, Zone B* | 6.1 | UN3122 | I | POISON, OXIDIZER | 2, B9, B14, B32, T38, T43, T45 | None | 227 | 244 | Forbidden | Forbidden | C | 89 |
| | Poisonous liquids which in contact with water emit flammable gases, n.o.s. | 6.1 | UN3123 | I | POISON, DANGEROUS WHEN WET | A4 | None | 201 | 243 | Forbidden | 1 L | E | |
| | | | | II | POISON, DANGEROUS WHEN WET | | None | 202 | 243 | 1 L | 5 L | E | |
| | Poisonous solids, corrosive, n.o.s. | 6.1 | UN2928 | I | POISON, CORROSIVE | | None | 211 | 242 | 1 kg | 25 kg | B | 40 |
| | | | | II | POISON, CORROSIVE | | None | 212 | 242 | 15 kg | 50 kg | B | 40 |

## §172.101 HAZARDOUS MATERIALS TABLE, Continued

| Symbols | Hazardous materials descriptions and proper shipping names | Hazard class or Division | Identification Numbers | Packing group | Label(s) required (if not excepted) | Special provisions | (8) Packaging authorizations (§173.***) | | | (9) Quantity limitations | | (10) Vessel stowage requirements | |
|---|---|---|---|---|---|---|---|---|---|---|---|---|---|
| | | | | | | | Exceptions | Non-bulk packaging | Bulk packaging | Passenger aircraft or railcar | Cargo aircraft only | Vessel stowages | Other stowage provisions |
| (1) | (2) | (3) | (4) | (5) | (6) | (7) | (8A) | (8B) | (8C) | (9A) | (9B) | (10A) | (10B) |
| | Poisonous solids, flammable, n.o.s. | 6.1 | UN2930 | I | POISON, FLAMMABLE SOLID | | None | 211 | 242 | 1 kg | 15 kg | B | |
| | | | | II | POISON, FLAMMABLE SOLID | | None | 212 | 242 | 15 kg | 50 kg | B | |
| | Poisonous solids, n.o.s. | 6.1 | UN2811 | I | POISON | | None | 211 | 242 | 5 kg | 50 kg | B | |
| | | | | II | POISON | | None | 212 | 242 | 25 kg | 100 kg | B | |
| | | | | III | KEEP AWAY FROM FOOD | | 153 | 213 | 240 | 100 kg | 200 kg | A | |
| | Poisonous solids, oxidizing, n.o.s. | 6.1 | UN3086 | I | POISON, OXIDIZER | | None | 211 | 242 | 1 kg | 15 kg | C | 89 |
| | | | | II | POISON, OXIDIZER | | None | 212 | 242 | 15 kg | 50 kg | C | 89 |
| | Poisonous, solids, self heating, n.o.s. | 6.1 | UN3124 | I | POISON, SPONTANEOUSLY COMBUSTIBLE | A5 | None | 211 | 241 | 5 kg | 15 kg | C | |
| | | | | II | POISON, SPONTANEOUSLY COMBUSTIBLE | | None | 212 | 242 | 15 kg | 50 kg | C | |
| | Poisonous, solids, which in contact with water emit flammable gases, n.o.s. | 6.1 | UN3125 | I | POISON, DANGEROUS WHEN WET | A5 | None | 211 | 241 | 5 kg | 15 kg | E | |
| | | | | II | POISON, DANGEROUS WHEN WET | | None | 212 | 242 | 15 kg | 50 kg | E | |
| | Polyalkylamines, n.o.s., see Alkylamines, etc. | | | | | | | | | | | | |
| AW | Polychlorinated biphenyls | 9 | UN2315 | II | CLASS 9 | 9, N81 | 155 | 202 | 241 | 100 L | 220 L | A | 34 |
| D | Polyester resin kits | 5.2 | NA2255 | II | ORGANIC PEROXIDE | | None | 225 | 246 | 5 kg | 5 kg | D | |
| | Polyhalogenated biphenyls, liquid *or* Polyhalogenated terphenyls liquid | 9 | UN3151 | II | CLASS 9 | | 155 | 204 | 241 | 100 L | 220 L | A | 34 |
| | Polyhalogenated biphenyls, solid *or* Polyhalogenated terphenyls, solid | 9 | UN3152 | II | CLASS 9 | | 155 | 204 | 241 | 100 kg | 200 kg | A | 34 |
| | Polystyrene beads, expandable, *evolving flammable vapor.* | 9 | UN2211 | III | None | | 221 | 221 | 240 | 100 kg | 200 kg | A | 85, 87 |
| | Potassium | 4.3 | UN2257 | II | DANGEROUS WHEN WET | A19, A20, B27, N6, N34, T15, T26 | None | 212 | 244 | Forbidden | 50 kg | D | |
| | Potassium arsenate | 6.1 | UN1677 | II | POISON | | None | 212 | 242 | 25 kg | 100 kg | A | |
| | Potassium arsenite | 6.1 | UN1678 | II | POISON | | None | 212 | 242 | 25 kg | 100 kg | A | |
| | Potassium bifluoride, *solid* | 8 | UN1811 | II | CORROSIVE, POISON | N3, N34, T8 | 154 | 212 | 242 | 15 kg | 50 kg | A | 25, 26, 40, 95 |
| | Potassium bifluoride, *solution* | 8 | UN1811 | II | CORROSIVE, POISON | N3, N34, T8 | 154 | 202 | 242 | 1 L | 30 L | A | 26, 40, 95 |
| | Potassium bisulfite solution, see Bisulfites, inorganic, aqueous solutions, n.o.s. | | | | | | | | | | | | |
| | Potassium borohydride | 4.3 | UN1870 | I | DANGEROUS WHEN WET | A19, N40 | None | 211 | 242 | Forbidden | 15 kg | E | |
| | Potassium bromate | 5.1 | UN1484 | II | OXIDIZER | | 152 | 212 | 242 | 5 kg | 25 kg | A | 56, 58, 106 |
| | *Potassium carbonyl* | Forbidden | | | | | | | | | | | |
| | Potassium chlorate | 5.1 | UN1485 | II | OXIDIZER | A9, N34 | 152 | 212 | 242 | 5 kg | 25 kg | A | 56, 58, 106 |
| | *Potassium chlorate mixed with mineral oil, see* Explosive, blasting, type C | | | | | | | | | | | | |
| | Potassium chlorate, solution | 5.1 | UN2427 | II | OXIDIZER | A2, T8 | 152 | 202 | 241 | 1 L | 5 L | B | 56, 58, 106 |
| | Potassium cuprocyanide | 6.1 | UN1679 | II | POISON | | None | 212 | 242 | 25 kg | 100 kg | A | 26 |
| | Potassium cyanide | 6.1 | UN1680 | I | POISON | B69, B77, N74, N75, T18, T26 | None | 211 | 242 | 5 kg | 50 kg | B | 52 |
| | *Potassium dichloro isocyanurate or Potassium dichloro-s-triazinetrione, see* Dichloroisocyanuric acid, dry *or* Dichloroisocyanuric acid salts *etc.* | | | | | | | | | | | | |
| | Potassium dithionite *or* Potassium hydrosulfite | 4.2 | UN1929 | II | SPONTANEOUSLY COMBUSTIBLE | A8, A19, A20 | None | 212 | 241 | 15 kg | 50 kg | E | 13 |
| | Potassium fluoride | 6.1 | UN1812 | III | KEEP AWAY FROM FOOD | T8 | 153 | 213 | 240 | 100 kg | 200 kg | A | 26 |
| | Potassium fluoroacetate | 6.1 | UN2628 | I | POISON | | None | 211 | 242 | 5 kg | 50 kg | E | |
| | Potassium fluorosilicate | 6.1 | UN2655 | III | KEEP AWAY FROM FOOD | | 153 | 213 | 240 | 100 kg | 200 kg | A | 26 |
| | *Potassium hydrate, see* Potassium hydroxide, solid | | | | | | | | | | | | |
| | *Potassium hydrogen fluoride, see* Potassium bifluoride | | | | | | | | | | | | |
| | *Potassium hydrogen fluoride solution, see* Corrosive liquid, n.o.s. | | | | | | | | | | | | |

| Symbols | Hazardous materials descriptions and proper shipping names | Hazard class or Division | Identification Numbers | Packing group | Label(s) required (if not excepted) | Special provisions | (8) Packaging authorizations (§173 ***) | | | (9) Quantity limitations | | (10) Vessel stowage requirements | |
|---|---|---|---|---|---|---|---|---|---|---|---|---|---|
| | | | | | | | Exceptions | Non-bulk packaging | Bulk packaging | Passenger aircraft or railcar | Cargo aircraft only | Vessel stowages | Other stowage provisions |
| (1) | (2) | (3) | (4) | (5) | (6) | (7) | (8A) | (8B) | (8C) | (9A) | (9B) | (10A) | (10B) |
| | Potassium hydrogen sulfate | 8 | UN2509 | II | CORROSIVE | A7, N34 | 154 | 212 | 240 | 15 kg | 50 kg | A | |
| | Potassium hydrosulfite, *see* Potassium dithionite | | | | | | | | | | | | |
| | *Potassium hydroxide, liquid, see* Potassium hydroxide solution | | | | | | | | | | | | |
| | Potassium hydroxide, solid | 8 | UN1813 | II | CORROSIVE | | 154 | 212 | 240 | 15 kg | 50 kg | A | |
| | Potassium hydroxide, solution | 8 | UN1814 | II | CORROSIVE | B2, T8 | 154 | 202 | 242 | 1 L | 30 L | A | |
| | *Potassium hypochlorite, solution, see* Hypochlorite solutions, *etc.* | | | | | | | | | | | | |
| | Potassium, metal alloys | 4.3 | UN1420 | I | DANGEROUS WHEN WET | A19, A20, B27 | None | 212 | 244 | Forbidden | 50 kg | D | |
| | *Potassium metal, liquid alloy, see* Alkali metal alloys, liquid | | | | | | | | | | | | |
| | Potassium metavanadate | 6.1 | UN2864 | II | POISON | | None | 212 | 242 | 25 kg | 100 kg | A | |
| | Potassium monoxide | 8 | UN2033 | II | CORROSIVE | | 154 | 212 | 240 | 15 kg | 50 kg | A | |
| | Potassium nitrate | 5.1 | UN1486 | III | OXIDIZER | A1, A29 | 152 | 212 | 240 | 25 kg | 100 kg | A | |
| | Potassium nitrate and sodium nitrite mixtures | 5.1 | UN1487 | II | OXIDIZER | B12, B78 | 152 | 212 | 240 | 5 kg | 25 kg | A | 56, 58 |
| | Potassium nitrite | 5.1 | UN1488 | II | OXIDIZER | | 152 | 212 | 242 | 5 kg | 25 kg | A | 56, 58 |
| | Potassium perchlorate, solid *or* solution | 5.1 | UN1489 | II | OXIDIZER | T8 | 152 | 212 | 242 | 5 kg | 25 kg | A | 56, 58, 106 |
| | Potassium permanganate | 5.1 | UN1490 | II | OXIDIZER | B12 | 152 | 212 | 242 | 5 kg | 25 kg | D | 56, 58, 69, 106, 107 |
| | Potassium peroxide | 5.1 | UN1491 | I | OXIDIZER | A20, N34 | None | 211 | None | Forbidden | 15 kg | B | 13, 75, 106 |
| | Potassium persulfate | 5.1 | UN1492 | III | OXIDIZER | A1, A29 | 152 | 213 | 240 | 25 kg | 100 kg | A | |
| | Potassium phosphide | 4.3 | UN2012 | I | DANGEROUS WHEN WET, POISON | A19, N40 | None | 211 | None | Forbidden | 15 kg | E | 40, 85 |
| | Potassium salts of aromatic nitro-derivatives, *explosive* | 1.3C | UN0158 | II | EXPLOSIVE 1.3C | | None | 62 | None | Forbidden | Forbidden | B | 1E, 5E |
| | Potassium selenate, *see* Selenates *or* Selenites | | | | | | | | | | | | |
| | Potassium selenite, *see* Selenates *or* Selenites | | | | | | | | | | | | |
| | Potassium sodium alloys | 4.3 | UN1422 | I | DANGEROUS WHEN WET | A19, B27, N34, N40, T15, T26 | None | 201 | 244 | Forbidden | 15 kg | D | |
| | Potassium sulfide, anhydrous *or* Potassium sulfide *with less than 30 per cent water of crystallization* | 4.2 | UN1382 | II | SPONTANEOUSLY COMBUSTIBLE | A19, A20, B16, N34 | None | 212 | 241 | 15 kg | 50 kg | A | |
| | Potassium sulfide, hydrated *with not less than 30 per cent water of crystallization* | 8 | UN1847 | II | CORROSIVE | | 154 | 212 | 240 | 15 kg | 50 kg | A | 26 |
| | Potassium superoxide | 5.1 | UN2466 | I | OXIDIZER | A20 | None | 211 | None | Forbidden | 15 kg | B | 13, 75, 106 |
| | Powder cake, wetted *or* Powder paste, wetted *with not less than 25 per cent water, by mass* | 1.3C | UN0159 | II | EXPLOSIVE 1.3C | | None | 62 | None | Forbidden | Forbidden | B | 1E, 5E |
| | Powder cake, wetted *or* Powder paste, wetted *with not less than 17 percent alcohol by mass* | 1.1C | UN0433 | II | EXPLOSIVE 1.1C | | None | 62 | None | Forbidden | Forbidden | B | 1E, 5E |
| | Powder paste, *see* Powder cake, *etc* | | | | | | | | | | | | |
| | Powder, smokeless | 1.1C | UN0160 | II | EXPLOSIVE 1.1C | | None | 62 | None | Forbidden | Forbidden | B | 10E, 26E |
| | Powder, smokeless | 1.3C | UN0161 | II | EXPLOSIVE 1.3C | | None | 62 | None | Forbidden | Forbidden | B | 10E, 26E |
| | *Power device, explosive, see* Cartridges, power device | | | | | | | | | | | | |
| | Primers, cap type | 1.4S | UN0044 | II | None | | None | 62 | None | 25 kg | 100 kg | A | 9E |
| | Primers, cap type | 1.1B | UN0377 | II | EXPLOSIVE 1.1B | | None | 62 | None | Forbidden | Forbidden | B | 2E, 6E |
| | Primers, cap type | 1.4B | UN0378 | II | EXPLOSIVE 1.4B | | None | 62 | None | Forbidden | 75 kg | B | |
| | *Primers, small arms, see* Primers, cap type | | | | | | | | | | | | |
| | Primers, tubular | 1.3G | UN0319 | II | EXPLOSIVE 1.3G | | None | 62 | None | Forbidden | Forbidden | B | |
| | Primers, tubular | 1.4G | UN0320 | II | EXPLOSIVE 1.4G | | None | 62 | None | Forbidden | 75 kg | A | 24E |
| | Primers, tubular | 1.4S | UN0376 | II | None | | None | 62 | None | 25 kg | 100 kg | A | 9E |
| | *Projectiles, illuminating, see* Ammunition, illuminating, *etc.* | | | | | | | | | | | | |
| | Projectiles, *inert with tracer* | 1.4S | UN0345 | II | EXPLOSIVE 1.4S | | 63(b) | 62 | None | 25 kg | 100 kg | A | 3E, 7E, 9E |
| | Projectiles, *inert, with tracer* | 1.3G | UN0424 | II | EXPLOSIVE 1.3G | | 63(b) | 62 | None | Forbidden | Forbidden | B | 3E, 7E |
| | Projectiles, *inert, with tracer* | 1.4G | UN0425 | II | EXPLOSIVE 1.4G | | 63(b) | 62 | None | Forbidden | 75 kg | A | 3E, 7E, 24E |
| | Projectiles, *with burster or expelling charge* | 1.2D | UN0346 | II | EXPLOSIVE 1.2D | | 63(b) | 62 | None | Forbidden | Forbidden | B | 3E, 7E |
| | Projectiles, *with burster or expelling charge* | 1.4D | UN0347 | II | EXPLOSIVE 1.4D | | 63(b) | 62 | None | Forbidden | 75 kg | A | 3E, 7E, 24E |
| | Projectiles, *with burster or expelling charge* | 1.2F | UN0426 | II | EXPLOSIVE 1.2F | | 63(b) | 62 | None | Forbidden | Forbidden | E | |
| | Projectiles, *with burster or expelling charge* | 1.4F | UN0427 | II | EXPLOSIVE 1.4F | | 63(b) | 62 | None | Forbidden | Forbidden | E | |
| | Projectiles, *with burster or expelling charge* | 1.2G | UN0434 | II | EXPLOSIVE 1.2G | | 63(b) | 62 | None | Forbidden | Forbidden | B | 3E, 7E |
| | Projectiles, *with burster or expelling charge* | 1.4G | UN0435 | II | EXPLOSIVE 1.4G | | 63(b) | 62 | None | Forbidden | 75 kg | A | 3E, 7E, 24E |
| | Projectiles, *with bursting charge* | 1.1F | UN0167 | II | EXPLOSIVE 1.1F | | 63(b) | 62 | None | Forbidden | Forbidden | E | |
| | Projectiles, *with bursting charge* | 1.1D | UN0168 | II | EXPLOSIVE 1.1D | | 63(b) | 62 | None | Forbidden | Forbidden | B | 3E, 7E |
| | Projectiles, *with bursting charge* | 1.2D | UN0169 | II | EXPLOSIVE 1.2D | | 63(b) | 62 | None | Forbidden | Forbidden | B | 3E, 7E |

| Symbols | Hazardous materials descriptions and proper shipping names | Hazard class or Division | Identification Numbers | Packing group | Label(s) required (if not excepted) | Special provisions | (8) Packaging authorizations (§173.***) | | | (9) Quantity limitations | | (10) Vessel stowage requirements | |
|---|---|---|---|---|---|---|---|---|---|---|---|---|---|
| | | | | | | | Exceptions | Non-bulk packaging | Bulk packaging | Passenger aircraft or railcar | Cargo aircraft only | Vessel stowages | Other stowage provisions |
| (1) | (2) | (3) | (4) | (5) | (6) | (7) | (8A) | (8B) | (8C) | (9A) | (9B) | (10A) | (10B) |
| | Projectiles, *with bursting charge* | 1.2F | UN0324 | II | EXPLOSIVE 1.2F | | 63(b) | 62 | None | Forbidden | Forbidden | E | |
| | Projectiles, *with bursting charge* | 1.4D | UN0344 | II | EXPLOSIVE 1.4D | | 63(b) | 62 | None | Forbidden | 75 kg | A | 3E, 7E, 24E |
| | Propadiene, inhibited | 2.1 | UN2200 | | FLAMMABLE GAS | | None | 304 | 314, 315 | Forbidden | 150 kg | B | 40 |
| | *Propadiene mixed with methyl acetylene, see* Methyl acetylene and propadiene mixtures, stabilized | | | | | | | | | | | | |
| | Propane *see also* Petroleum gases, liquefied | 2.1 | UN1978 | | FLAMMABLE GAS | 19 | 306 | 304 | 314, 315 | Forbidden | 150 kg | E | 40 |
| | Propanethiols | 3 | UN2402 | II | FLAMMABLE LIQUID | T8 | 150 | 202 | 242 | 5 L | 60 L | E | 95, 102 |
| | n-Propanol *or* Propyl alcohol normal | 3 | UN1274 | II | FLAMMABLE LIQUID | B1, T1 | 150 | 202 | 242 | 5 L | 60 L | B | |
| D | Propargyl alcohol | 3 | NA1986 | II | FLAMMABLE LIQUID, POISON | | None | 202 | 243 | Forbidden | 1 L | B | 40 |
| D | Propellant explosive, liquid | 1.1C | NA0474 | II | EXPLOSIVE 1.1C | | None | 62 | None | Forbidden | Forbidden | B | 1E, 5E |
| D | Propellant explosive, liquid | 1.3C | NA0477 | II | EXPLOSIVE 1.3C | | None | 62 | None | Forbidden | Forbidden | B | 1E, 5E |
| D | Propellant explosive, solid | 1.1C | NA0273 | II | EXPLOSIVE 1.1C | | None | 62 | None | Forbidden | 5 kg | B | |
| D | Propellant explosive, solid | 1.3C | NA0274 | II | EXPLOSIVE 1.3C | | None | 62 | None | Forbidden | 5 kg | B | |
| | Propionaldehyde | 3 | UN1275 | II | FLAMMABLE LIQUID | T14 | 150 | 202 | 242 | 5 L | 60 L | E | |
| | Propionic acid | 8 | UN1848 | III | CORROSIVE | T7 | 154 | 203 | 241 | 5 L | 60 L | A | 22, 48 |
| | Propionic anhydride | 8 | UN2496 | III | CORROSIVE | T2 | 154 | 203 | 241 | 5 L | 60 L | A | 8 |
| | Propionitrile | 3 | UN2404 | II | FLAMMABLE LIQUID, POISON | T14 | None | 201 | 243 | Forbidden | 60 L | E | 40 |
| | Propionyl chloride | 3 | UN1815 | II | FLAMMABLE LIQUID, CORROSIVE | T8, T26 | None | 202 | 243 | 1 L | 5 L | B | 40 |
| | n-Propyl acetate | 3 | UN1276 | II | FLAMMABLE LIQUID | T1 | 150 | 202 | 242 | 5 L | 60 L | B | |
| | Propyl alcohol, *see* Propanol | | | | | | | | | | | | |
| | n-Propyl benzene | 3 | UN2364 | III | FLAMMABLE LIQUID | B1, T1 | 150 | 203 | 242 | 60 L | 220 L | A | |
| | Propyl chloride | 3 | UN1278 | II | FLAMMABLE LIQUID | N34, T14 | None | 202 | 242 | Forbidden | 60 L | E | |
| | n-Propyl chloroformate | 6.1 | UN2740 | I | POISON, FLAMMABLE LIQUID, CORROSIVE | 2, A3, A6, A7, B9, B14, B32, B74, B77, N34, T38, T43, T45 | None | 227 | 244 | Forbidden | 2.5 L | B | 21, 40, 48, 100 |
| | Propyl formates | 3 | UN1281 | II | FLAMMABLE LIQUID | T8 | 150 | 202 | 242 | 5 L | 60 L | B | |
| + | n-Propyl isocyanate | 3 | UN2482 | I | FLAMMABLE LIQUID, POISON | 1, A7, B9, B14, B30, B72, T38, T43, T44 | None | 226 | 244 | Forbidden | 30 L | D | 40 |
| | *Propyl mercaptan, see* Propanethiols | | | | | | | | | | | | |
| | n-Propyl nitrate | 3 | UN1865 | II | FLAMMABLE LIQUID | T25 | 150 | 202 | None | 5 L | 60 L | D | |
| | Propylamine | 3 | UN1277 | II | FLAMMABLE LIQUID, CORROSIVE | N34, T14 | None | 202 | 242 | 5 L | 60 L | E | 40 |
| | Propylene *see also* Petroleum gases, liquefied | 2.1 | UN1077 | | FLAMMABLE GAS | | 306 | 304 | 314, 315 | Forbidden | 150 kg | E | 40 |
| | Propylene chlorohydrin | 6.1 | UN2611 | II | POISON | T9 | None | 202 | 243 | 5 L | 60 L | A | 12, 40, 48 |
| | Propylene dichloride | 3 | UN1279 | II | FLAMMABLE LIQUID | N36, T1 | 150 | 202 | 242 | 5 L | 60 L | B | |
| | Propylene oxide | 3 | UN1280 | I | FLAMMABLE LIQUID | A3, N34, T20, T29 | None | 201 | 243 | 1 L | 30 L | E | 40 |
| | Propylene tetramer | 3 | UN2850 | III | FLAMMABLE LIQUID | B1, T1 | 150 | 203 | 242 | 60 L | 220 L | A | |
| | 1,2-Propylenediamine | 8 | UN2258 | II | CORROSIVE | A3, A6, N34, T8 | None | 202 | 243 | 1 L | 30 L | A | 12, 20, 40, 48 |
| | Propyleneimine, inhibited | 3 | UN1921 | I | FLAMMABLE LIQUID | A3, N34, T25 | None | 201 | 243 | 1 L | 30 L | B | 40 |
| | Propyltrichlorosilane | 8 | UN1816 | II | CORROSIVE | A7, B2, B6, N34, T8, T26 | None | 202 | 242 | Forbidden | 30 L | C | 12, 21, 40, 48 |
| | *Prussic acid, see* Hydrogen cyanide | | | | | | | | | | | | |
| | Pyridine | 3 | UN1282 | II | FLAMMABLE LIQUID, POISON | T8 | None | 202 | 243 | 1 L | 60 L | B | 21, 100 |
| | *Pyridine perchlorate* | Forbidden | | | | | | | | | | | |
| | Pyrophoric liquids, n.o.s | 4.2 | UN2845 | I | SPONTANEOUSLY COMBUSTIBLE | B11, T42 | None | 181 | 244 | Forbidden | Forbidden | D | 18 |
| | Pyrophoric metals, n.o.s., *or* Pyrophoric alloys, n.o.s. | 4.2 | UN1383 | I | SPONTANEOUSLY COMBUSTIBLE | B11 | None | 187 | 242 | Forbidden | Forbidden | D | |
| | Pyrophoric solids, n.o.s. | 4.2 | UN2846 | I | SPONTANEOUSLY COMBUSTIBLE | | None | 187 | 242 | Forbidden | Forbidden | D | |
| | Pyrosulfuryl chloride | 8 | UN1817 | II | CORROSIVE | B2, T9, T27 | 154 | 202 | 242 | 1 L | 30 L | C | 8, 40 |
| | *Pyroxylin solution or solvent, see* Nitrocellulose | | | | | | | | | | | | |
| | Pyrrolidine | 3 | UN1922 | II | FLAMMABLE LIQUID | T1 | 150 | 202 | 242 | 5 L | 60 L | B | 40 |

## §172.101 HAZARDOUS MATERIALS TABLE, Continued

| Symbols | Hazardous materials descriptions and proper shipping names | Hazard class or Division | Identification Numbers | Packing group | Label(s) required (if not excepted) | Special provisions | (8) Packaging authorizations (§173.***) | | | (9) Quantity limitations | | (10) Vessel stowage requirements | |
|---|---|---|---|---|---|---|---|---|---|---|---|---|---|
| | | | | | | | Exceptions | Non-bulk packaging | Bulk packaging | Passenger aircraft or railcar | Cargo aircraft only | Vessel stowages | Other stowage provisions |
| (1) | (2) | (3) | (4) | (5) | (6) | (7) | (8A) | (8B) | (8C) | (9A) | (9B) | (10A) | (10B) |
| | *Quebrachitol pentanitrate* | Forbidden | | | | | | | | | | | |
| | *Quicklime, see* Calcium oxide | | | | | | | | | | | | |
| | Quinoline | 6.1 | UN2656 | III | KEEP AWAY FROM FOOD | T8 | 153 | 203 | 241 | 60 L | 220 L | A | 12, 48 |
| | *R 114, see* Dichlorotetrafluoroethane | | | | | | | | | | | | |
| | *R 115, see* Chloropentafluoroethane | | | | | | | | | | | | |
| | *R 116, see* Hexafluoroethane | | | | | | | | | | | | |
| | *R 124, see* Chlorotetrafluoroethane | | | | | | | | | | | | |
| | *R 133a, see* Chlorotrifluoroethane | | | | | | | | | | | | |
| | *R 152a, see* Difluoroethane | | | | | | | | | | | | |
| | *R 500, see* Dichlorodifluoromethane and difluoroethane, *etc.* | | | | | | | | | | | | |
| | *R 502, see* Chlorodifluoromethane and chloropentafluoroethane mixture, *etc.* | | | | | | | | | | | | |
| | *R 503, see* Chlorotrifluoromethane and trifluoromethane, *etc.* | | | | | | | | | | | | |
| | *R 12, see* Dichlorodifluoromethane | | | | | | | | | | | | |
| | *R 1281, see* Chlorodifluorobromomethane | | | | | | | | | | | | |
| | *R 13, see* Chlorotrifluoromethane | | | | | | | | | | | | |
| | *R 1381, see* Bromotrifluoromethane | | | | | | | | | | | | |
| | *R 14, see* Tetrafluoromethane | | | | | | | | | | | | |
| | *R 21, see* Dichlorofluoromethane | | | | | | | | | | | | |
| | *R 22, see* Chlorodifluoromethane | | | | | | | | | | | | |
| | Radioactive material, excepted package-articles manufactured from natural or depleted uranium or natural thorium | 7 | UN2910 | | None | | 421-1, 424 | 421-1, 424 | 421-1, 424 | | | A | |
| | Radioactive material, excepted package-empty packaging | 7 | UN2910 | | EMPTY | | 427 | 427 | 427 | | | A | |
| | Radioactive material, excepted package-instruments or articles | 7 | UN2910 | | None | | 421-1, 422 | 421-1, 422 | 421-1, 422 | | | A | |
| | Radioactive material, excepted package-limited quantity of material | 7 | UN2910 | | None | | 421, 421-1 | 421, 421-1 | 421, 421-1 | | | A | |
| | Radioactive material, fissile, n.o.s. | 7 | UN2918 | | RADIOACTIVE | | 453 | 417 | 417 | | | A | 40, 95 |
| | Radioactive material, low specific activity LSA, n.o.s. | 7 | UN2912 | | RADIOACTIVE | | 421, 422, 424 | 425 | 425 | | | A | |
| | Radioactive material, n.o.s. | 7 | UN2982 | | RADIOACTIVE | | 421, 422, 424 | 415, 416 | 415, 416 | | | A | 40, 95 |
| | Radioactive material, special form, n.o.s. | 7 | UN2974 | | RADIOACTIVE | | 421, 422 | 415, 416 | 415, 416 | | | A | |
| | *Railway torpedo, see* Signals, railway track, explosive | | | | | | | | | | | | |
| | Rare gases and nitrogen mixtures | 2.2 | UN1981 | | NONFLAMMABLE GAS | | 306 | 302 | None | 75 kg | 150 kg | A | |
| | Rare gases and oxygen mixtures | 2.2 | UN1980 | | NONFLAMMABLE GAS | | 306 | 302 | None | 75 kg | 150 kg | A | |
| | Rare gases, mixtures | 2.2 | UN1979 | | NONFLAMMABLE GAS | | 306 | 302 | None | 75 kg | 150 kg | A | |
| | *RC 318, see* Octafluorocyclobutane | | | | | | | | | | | | |
| | RDX and cyclotetramethylenetetranitramine, wetted or desensitized *see* RDX and HMX mixtures, wetted or desensitized | | | | | | | | | | | | |
| | RDX and HMX mixtures, wetted *with not less than 15 percent water by mass* or RDX and HMX mixtures, desensitized *with not less than 10 percent phlegmatizer by mass* | 1.1D | UN0391 | II | EXPLOSIVE 1.1D | | None | 62 | None | Forbidden | Forbidden | B | 1E, 5E |
| | RDX and Octogen mixtures, wetted or desensitized see RDX and HMX mixtures, wetted or desensitized etc. | | | | | | | | | | | | |
| | RDX and Octogen mixtures, wetted or desensitized see RDX and HMX mixtures, wetted or desensitized etc | | | | | | | | | | | | |
| | *RDX, see* Cyclotrimethylene trinitramine, *etc.* | | | | | | | | | | | | |
| | Receptacles, small, containing gas *flammable, without release device, not refillable and not exceeding 1 L capacity* | 2.1 | UN2037 | | FLAMMABLE GAS | | 306 | 304 | None | 1 kg | 15 kg | B | 40 |
| | Receptacles, small, containing gas *nonflammable, without release device, not refillable and not exceeding 1 L capacity* | 2.2 | UN2037 | | NONFLAMMABLE GAS | | 306 | 304 | None | 1 kg | 15 kg | B | 40 |
| | *Red phosphorus, see* Phosphorus, amorphous | | | | | | | | | | | | |
| | Refrigerant gases, n.o.s. | 2.2 | UN1078 | | NONFLAMMABLE GAS | | 306 | 304 | 314, 315 | 75 kg | 150 kg | A | |
| D | Refrigerant gases, n.o.s. or Dispersant gases, n.o.s. | 2.1 | NA1954 | | FLAMMABLE GAS | | 306 | 304 | 314, 315 | Forbidden | 150 kg | C | 40 |
| D | Refrigerating machine | 3 | NA1993 | III | FLAMMABLE LIQUID | | 174 | 174 | None | 10 L | 10 L | A | |
| D | Refrigerating machines, containing *flammable, non-poisonous, liquefied gas* | 2.1 | NA1954 | | FLAMMABLE GAS | | 306 | 306 | 306 | Forbidden | 25 kg | C | 40 |
| | Refrigerating machines, containing *non-flammable, non-poisonous, liquefied gas* | 2.2 | UN2857 | | NONFLAMMABLE GAS | | 306, 307 | 306 | 306, 307 | Forbidden | 450 kg | A | |

## §172.101 HAZARDOUS MATERIALS TABLE, Continued

| Symbols | Hazardous materials descriptions and proper shipping names | Hazard class or Division | Identification Numbers | Packing group | Label(s) required (if not excepted) | Special provisions | (8) Packaging authorizations (§173.***) | | | (9) Quantity limitations | | (10) Vessel stowage requirements | |
|---|---|---|---|---|---|---|---|---|---|---|---|---|---|
| | | | | | | | Exceptions | Non-bulk packaging | Bulk packaging | Passenger aircraft or railcar | Cargo aircraft only | Vessel stowages | Other stowage provisions |
| (1) | (2) | (3) | (4) | (5) | (6) | (7) | (8A) | (8B) | (8C) | (9A) | (9B) | (10A) | (10B) |
| D | Regulated medical waste | 6.2 | NA9275 | II | INFECTIOUS SUBSTANCE | | 197 | 197 | None | Forbidden | Forbidden | E | |
| | Release devices, explosive | 1.4S | UN0173 | II | EXPLOSIVE 1.4S | | None | 62 | None | 25 kg | 100 kg | A | 24E |
| | Resin solution, *flammable* | 3 | UN1866 | II | FLAMMABLE LIQUID | B52, T7, T30 | 150 | 202 | 242 | 5 L | 60 L | B | |
| | | | | III | FLAMMABLE LIQUID | B1, B52, T7, T30 | 150 | 203 | 242 | 60 L | 220 L | A | |
| | Resorcinol | 6.1 | UN2876 | III | KEEP AWAY FROM FOOD | | 153 | 213 | 240 | 100 kg | 200 kg | A | |
| | *Rifle grenade, see* Grenades, *hand or rifle, etc.* | | | | | | | | | | | | |
| | *Rifle powder, see* Powder, smokeless *(UN 0160)* | | | | | | | | | | | | |
| | Rivets, explosive | 1.4S | UN0174 | II | EXPLOSIVE 1.4S | | None | 62 | None | 25 kg | 100 kg | A | 9E |
| | *Road asphalt or tar liquid, see* Tars, liquid, *etc.* | | | | | | | | | | | | |
| | Rocket motors | 1.3C | UN0186 | II | EXPLOSIVE 1.3C | 109 | None | 62 | None | Forbidden | 220 kg | B | 9E |
| | Rocket motors | 1.1C | UN0280 | II | EXPLOSIVE 1.1C | 109 | None | 62 | None | Forbidden | Forbidden | B | |
| | Rocket motors | 1.2C | UN0281 | II | EXPLOSIVE 1.2C | 109 | None | 62 | None | Forbidden | Forbidden | B | |
| | Rocket motors, liquid fueled | 1.2J | UN0395 | II | EXPLOSIVE 1.2J | 109 | None | 62 | None | Forbidden | Forbidden | E | 7E, 16E, 23E |
| | Rocket motors, liquid fueled | 1.3J | UN0396 | II | EXPLOSIVE 1.3J | 109 | None | 62 | None | Forbidden | Forbidden | E | 7E, 16E, 23E |
| | Rocket motors with hypergolic liquids *with or without an expelling charge* | 1.3L | UN0250 | II | EXPLOSIVE 1.3L | 109 | None | 62 | None | Forbidden | Forbidden | E | 2E, 8E, 11E, 17E |
| | Rocket motors with hypergolic liquids *with or without an expelling charge* | 1.2L | UN0322 | II | EXPLOSIVE 1.2L | 109 | None | 62 | None | Forbidden | Forbidden | E | 2E, 8E, 11E, 17E |
| | Rockets, line throwing | 1.4G | UN0453 | II | EXPLOSIVE 1.4G | | None | 62 | None | Forbidden | 75 kg | B | |
| | Rockets, line-throwing | 1.2G | UN0238 | II | EXPLOSIVE 1.2G | | None | 62 | None | Forbidden | Forbidden | B | |
| | Rockets, line-throwing | 1.3G | UN0240 | II | EXPLOSIVE 1.3G | | None | 62 | None | Forbidden | 75 kg | B | |
| | Rockets, liquid fueled *with bursting charge* | 1.2J | UN0398 | II | EXPLOSIVE 1.2J | | None | 62 | None | Forbidden | Forbidden | E | 7E, 16E, 23E |
| | Rockets, liquid fueled *with bursting charge* | 1.1J | UN0397 | II | EXPLOSIVE 1.1J | | None | 62 | None | Forbidden | Forbidden | E | 7E, 16E, 23E |
| | Rockets, *with bursting charge* | 1.1F | UN0180 | II | EXPLOSIVE 1.1F | | None | 62 | None | Forbidden | Forbidden | E | |
| | Rockets, *with bursting charge* | 1.1E | UN0181 | II | EXPLOSIVE 1.1E | | None | 62 | None | Forbidden | Forbidden | B | |
| | Rockets, *with bursting charge* | 1.2E | UN0182 | II | EXPLOSIVE 1.2E | | None | 62 | None | Forbidden | Forbidden | B | |
| | Rockets, *with bursting charge* | 1.2F | UN0295 | II | EXPLOSIVE 1.2F | | None | 62 | None | Forbidden | Forbidden | E | |
| | Rockets, *with expelling charge* | 1.2C | UN0436 | II | EXPLOSIVE 1.2C | | None | 62 | None | Forbidden | Forbidden | B | |
| | Rockets, *with expelling charge* | 1.3C | UN0437 | II | EXPLOSIVE 1.3C | | None | 62 | None | Forbidden | Forbidden | B | |
| | Rockets, *with expelling charge* | 1.4C | UN0438 | II | EXPLOSIVE 1.4C | | None | 62 | None | Forbidden | 75 kg | A | 24E |
| | Rockets, *with inert head* | 1.3C | UN0183 | II | EXPLOSIVE 1.3C | | None | 62 | None | Forbidden | Forbidden | B | |
| | Rosin oil | 3 | UN1286 | II | FLAMMABLE LIQUID | T7 | 150 | 202 | 242 | 5 L | 60 L | B | |
| | | | | III | FLAMMABLE LIQUID | B1, T1 | 150 | 203 | 242 | 60 L | 220 L | A | |
| | Rubber solution | 3 | UN1287 | II | FLAMMABLE LIQUID | T7, T30 | 150 | 202 | 242 | 5 L | 60 L | B | |
| | | | | III | FLAMMABLE LIQUID | B1, T7, T30 | 150 | 203 | 242 | 60 L | 220 L | A | |
| | Rubidium | 4.3 | UN1423 | I | DANGEROUS WHEN WET | 22, A7, A19, N34, N40, N45 | None | 211 | 242 | Forbidden | 15 kg | D | |
| | Rubidium hydroxide | 8 | UN2678 | II | CORROSIVE | T8 | 154 | 212 | 240 | 15 kg | 50 kg | A | |
| | Rubidium hydroxide solution | 8 | UN2677 | II | CORROSIVE | B2, T8 | 154 | 202 | 242 | 1 L | 30 L | A | |
| | *Safety fuse, see* Fuse, safety | | | | | | | | | | | | |
| | Samples, explosive, *other than initiating explosives* | | UN0190 | | | 113 | | 62 | None | Forbidden | Forbidden | E | 12E |
| | Sand acid, *see* Fluorosilicic acid | | | | | | | | | | | | |
| | Seed cake, *containing vegetable oil solvent extractions and expelled seeds, containing not more than 10% of oil and when the amount of moisture is higher than 11%, not more than 20% of oil and moisture combined.* | 4.2 | UN1386 | III | None | N7 | None | 213 | 241 | Forbidden | Forbidden | A | 13 |
| I | Seed cake *with more than 1.5 percent oil and not more than 11 percent moisture* | 4.2 | UN1386 | III | None | N7 | None | 213 | 241 | Forbidden | Forbidden | E | 13 |
| I | Seed cake *with not more than 1.5 percent oil and not more than 11 percent moisture* | 4.2 | UN2217 | III | None | N7 | None | 213 | 241 | Forbidden | Forbidden | A | 13 |
| | Selenates *or* Selenites | 6.1 | UN2630 | I | POISON | | None | 211 | 242 | 5 kg | 50 kg | E | |
| | Selenic acid | 8 | UN1905 | I | CORROSIVE | N34 | None | 211 | 242 | Forbidden | 25 kg | A | |
| | Selenites, *see* Selenates *or* Selenites | | | | | | | | | | | | |
| | Selenium disulfide | 6.1 | UN2657 | II | POISON | | None | 212 | 242 | 25 kg | 100 kg | A | |
| | Selenium hexafluoride | 2.3 | UN2194 | | POISON GAS | 1 | None | 302 | None | Forbidden | Forbidden | D | 40 |
| | *Selenium nitride* | Forbidden | | | | | | | | | | | |
| D | Selenium oxide | 6.1 | NA2811 | I | POISON | | None | 211 | 242 | 5 kg | 50 kg | B | |
| | Selenium oxychloride | 8 | UN2879 | I | CORROSIVE, POISON | A3, A6, A7, N34, T12, T27 | None | 201 | 243 | 0.5 L | 2.5 L | E | 40, 95 |

## §172.101 HAZARDOUS MATERIALS TABLE, Continued

| Symbols | Hazardous materials descriptions and proper shipping names | Hazard class or Division | Identification Numbers | Packing group | Label(s) required (if not excepted) | Special provisions | (8) Packaging authorizations (§173 ***) | | | (9) Quantity limitations | | (10) Vessel stowage requirements | |
|---|---|---|---|---|---|---|---|---|---|---|---|---|---|
| | | | | | | | Exceptions | Non-bulk packaging | Bulk packaging | Passenger aircraft or railcar | Cargo aircraft only | Vessel stowages | Other stowage provisions |
| (1) | (2) | (3) | (4) | (5) | (6) | (7) | (8A) | (8B) | (8C) | (9A) | (9B) | (10A) | (10B) |
| | Selenium powder | 6.1 | UN2658 | III | KEEP AWAY FROM FOOD | | 153 | 213 | 240 | 100 kg | 200 kg | A | |
| | Self-heating substances, solid, corrosive, n.o.s. | 4.2 | UN3126 | II | SPONTANEOUSLY COMBUSTIBLE, CORROSIVE | | None | 212 | 242 | 15 kg | 50 kg | C | |
| | | | | III | SPONTANEOUSLY COMBUSTIBLE, CORROSIVE | | None | 213 | 242 | 25 kg | 100 kg | C | |
| | Self-heating substances, solid, n.o.s | 4.2 | UN3088 | II | SPONTANEOUSLY COMBUSTIBLE | | None | 212 | 241 | 15 kg | 50 kg | C | |
| | | | | III | SPONTANEOUSLY COMBUSTIBLE | | None | 213 | 241 | 25 kg | 100 kg | C | |
| | Self-heating substances, solid, oxidizing, n.o.s. | 4.2 | UN3127 | | SPONTANEOUSLY COMBUSTIBLE, OXIDIZER | | None | 214 | 214 | Forbidden | Forbidden | | |
| | Self-heating substances, solid, poisonous, n.o.s | 4.2 | UN3128 | II | SPONTANEOUSLY COMBUSTIBLE, POISON | | None | 212 | 242 | 15 kg | 50 kg | C | |
| | | | | III | SPONTANEOUSLY COMBUSTIBLE, KEEP AWAY FROM FOOD | | None | 213 | 242 | 25 kg | 100 kg | C | |
| | *Self-propelled vehicle, see* Vehicles, self-propelled | | | | | | | | | | | | |
| | Self-reactive substances *(aliphatic azocompounds, aromatic sulphohydrazides, N-nitroso compounds, diazonium salts)* Sample, n.o.s | 4.1 | UN3031 | II | FLAMMABLE SOLID | | None | 214 | None | Forbidden | Forbidden | D | |
| | Self-reactive substances *(aliphatic azocompounds, aromatic sulphohydrazides, N-nitroso compounds, diazonium salts)*, Trial quantities, n.o.s | 4.1 | UN3032 | II | FLAMMABLE SOLID | | None | 214 | None | Forbidden | Forbidden | D | |
| | Shale oil | 3 | UN1288 | I | FLAMMABLE LIQUID | T7 | None | 201 | 243 | 1 L | 30 L | B | |
| | | | | II | FLAMMABLE LIQUID | T7, T30 | 150 | 202 | 242 | 5 L | 60 L | B | |
| | | | | III | FLAMMABLE LIQUID | B1, T7, T30 | 150 | 203 | 242 | 60 L | 220 L | A | |
| | *Shaped charges, commercial, see* Charges, shaped, commercial *etc.* | | | | | | | | | | | | |
| | Signal devices, hand | 1.4G | UN0191 | II | EXPLOSIVE 1.4G | | None | 62 | None | Forbidden | 75 kg | A | 24E |
| | Signal devices, hand | 1.4S | UN0373 | II | EXPLOSIVE 1.4S | | None | 62 | None | 25 kg | 100 kg | A | 9E |
| | Signals, distress.ship | 1.1G | UN0194 | II | EXPLOSIVE 1.1G | | None | 62 | None | Forbidden | Forbidden | B | |
| | Signals, distress, ship | 1.3G. | UN0195 | II | EXPLOSIVE 1.3G | | None | 62 | None | Forbidden | 75 kg | B | |
| | *Signals, highway, see* Signal devices, hand; Fireworks, type D | | | | | | | | | | | | |
| | Signals, railway track, explosive | 1.1G | UN0192 | II | EXPLOSIVE 1.1G | | None | 62 | None | Forbidden | Forbidden | B | |
| | Signals, railway track, explosive | 1.4S | UN0193 | II | EXPLOSIVE 1.4S | | None | 62 | None | 25 kg | 100 kg | A | 9E |
| | *Signals, ship distress, wateractivated, see* Contrivances, water-activated, *etc.* | | | | | | | | | | | | |
| | Signals, smoke | 1.1G | UN0196 | II | EXPLOSIVE 1.1G | | None | 62 | None | Forbidden | Forbidden | B | |
| | Signals, smoke | 1.2G | UN0313 | II | EXPLOSIVE 1.2G | | None | 62 | None | Forbidden | Forbidden | B | |
| | Signals, smoke | 1.4G | UN0197 | II | EXPLOSIVE 1.4G | | None | 62 | None | Forbidden | 75 kg | A | 24E |
| | Signals, smoke | 1.3G | UN0487 | II | EXPLOSIVE 1.3G | | None | 62 | None | Forbidden | Forbidden | B | 1E, 7E, 25E |
| | Silane | 2.1 | UN2203 | | FLAMMABLE GAS | | None | 302 | None | Forbidden | Forbidden | E | 40, 57, 104 |
| | *Silicofluoric acid, see* Fluorosilicic acid | | | | | | | | | | | | |
| | *Silicon chloride, see* Silicon tetrachloride | | | | | | | | | | | | |
| | Silicon powder, amorphous | 4.1 | UN1346 | III | FLAMMABLE SOLID | A1 | None | 213 | 240 | 25 kg | 100 kg | A | |
| | Silicon tetrachloride | 8 | UN1818 | II | CORROSIVE | A3, A6, B2, B6, T18, T26, T29 | 154 | 202 | 242 | 1 L | 30 L | C | 8, 40 |
| | Silicon tetrafluoride | 2.3 | UN1859 | | POISON GAS, CORROSIVE | 4 | None | 302 | None | Forbidden | 25 kg | D | 40 |
| | *Silver acetylide (dry)* | Forbidden | | | | | | | | | | | |
| | Silver arsenite | 6.1 | UN1683 | II | POISON | | None | 212 | 242 | 25 kg | 100 kg | A | |
| | *Silver azide (dry)* | Forbidden | | | | | | | | | | | |
| | *Silver chlorite (dry)* | Forbidden | | | | | | | | | | | |
| | Silver cyanide | 6.1 | UN1684 | II | POISON | | None | 212 | 242 | 25 kg | 100 kg | A | 26, 40 |
| | *Silver fulminate (dry)* | Forbidden | | | | | | | | | | | |
| | Silver nitrate | 5.1 | UN1493 | II | OXIDIZER | | 152 | 212 | 242 | 5 kg | 25 kg | A | |
| | *Silver oxalate (dry)* | Forbidden | | | | | | | | | | | |
| | *Silver picrate (dry)* | Forbidden | | | | | | | | | | | |

## §172.101 HAZARDOUS MATERIALS TABLE, Continued

| Symbols | Hazardous materials descriptions and proper shipping names | Hazard class or Division | Identification Numbers | Packing group | Label(s) required (if not excepted) | Special provisions | (8) Packaging authorizations (§173 ***) | | | (9) Quantity limitations | | (10) Vessel stowage requirements | |
|---|---|---|---|---|---|---|---|---|---|---|---|---|---|
| | | | | | | | Exceptions | Non-bulk packaging | Bulk packaging | Passenger aircraft or railcar | Cargo aircraft only | Vessel stowages | Other stowage provisions |
| (1) | (2) | (3) | (4) | (5) | (6) | (7) | (8A) | (8B) | (8C) | (9A) | (9B) | (10A) | (10B) |
| | Silver picrate, wetted *with not less than 30 per cent water, by mass* | 4.1 | UN1347 | I | FLAMMABLE SOLID | | None | 211 | None | Forbidden | Forbidden | D | 28, 36 |
| | Sludge, acid | 8 | UN1906 | II | CORROSIVE | A3, A7, B2, N34, T9, T27 | None | 202 | 242 | Forbidden | 30 L | C | 14 |
| D | Smokeless powder for small arms (*100 pounds or less*) | 4.1 | NA1325 | I | FLAMMABLE SOLID | 16 | None | 171 | None | Forbidden | Forbidden | A | |
| | Soda lime *with more than 4 per cent sodium hydroxide* | 8 | UN1907 | III | CORROSIVE | | 154 | 213 | 240 | 25 kg | 100 kg | A | |
| | Sodium | 4.3 | UN1428 | II | DANGEROUS WHEN WET | A7, A8, A19, A20, B9, B28, B48, B68, N34, T15, T29, T46 | None | 212 | 244 | Forbidden | 50 kg | D | |
| | Sodium aluminate, solid | 8 | UN2812 | III | CORROSIVE | | 154 | 213 | 240 | 25 kg | 100 kg | A | |
| | Sodium aluminate, solution | 8 | UN1819 | II | CORROSIVE | B2, T8 | 154 | 202 | 242 | 1 L | 30 L | A | |
| | Sodium aluminum hydride | 4.3 | UN2835 | II | DANGEROUS WHEN WET | A8, A19, A20 | None | 212 | 242 | Forbidden | 50 kg | E | |
| | Sodium ammonium vanadate | 6.1 | UN2863 | II | POISON | | None | 212 | 242 | 25 kg | 100 kg | A | |
| | Sodium arsanilate | 6.1 | UN2473 | III | KEEP AWAY FROM FOOD | | 153 | 213 | 240 | 100 kg | 200 kg | A | |
| | Sodium arsenate | 6.1 | UN1685 | II | POISON | | None | 212 | 240 | 25 kg | 100 kg | A | |
| | Sodium arsenite, aqueous solutions | 6.1 | UN1686 | II | POISON | T15 | None | 202 | 243 | 5 L | 60 L | A | |
| | | | | III | KEEP AWAY FROM FOOD | T15 | 153 | 203 | 241 | 60 L | 220 L | A | |
| | Sodium arsenite, solid | 6.1 | UN2027 | II | POISON | | None | 212 | 242 | 25 kg | 100 kg | A | |
| | Sodium azide | 6.1 | UN1687 | II | POISON | B28 | None | 212 | 242 | 25 kg | 100 kg | A | 36, 52, 91 |
| | *Sodium bifluoride, see* Sodium hydrogen fluoride | | | | | | | | | | | | |
| | *Sodium bisulfate, solution, see* Sodium hydrogen sulfate, solution | | | | | | | | | | | | |
| | *Sodium bisulfite, solution, see* Bisulfites inorganic, aqueous solutions, n.o.s. | | | | | | | | | | | | |
| | Sodium borohydride | 4.3 | UN1426 | I | DANGEROUS WHEN WET | N40 | None | 211 | 242 | Forbidden | 15 kg | E | |
| | Sodium bromate | 5.1 | UN1494 | II | OXIDIZER | | 152 | 212 | 242 | 5 kg | 25 kg | A | 56, 58, 106 |
| | Sodium cacodylate | 6.1 | UN1688 | II | POISON | | None | 212 | 242 | 25 kg | 100 kg | A | 26 |
| | Sodium chlorate | 5.1 | UN1495 | II | OXIDIZER | A9, N34, T8 | 152 | 212 | 240 | 5 kg | 25 kg | A | 56, 58, 106 |
| | *Sodium chlorate mixed with dinitrotoluene, see* Explosive blasting, type C | | | | | | | | | | | | |
| | Sodium chlorate, solution | 5.1 | UN2428 | II | OXIDIZER | A2, B6, T8 | 152 | 202 | 241 | 1 L | 5 L | B | 56, 58, 106 |
| | Sodium chlorite | 5.1 | UN1496 | II | OXIDIZER | A9, N34, T8 | None | 212 | 242 | 5 kg | 25 kg | A | 56, 58, 106 |
| | Sodium chlorite solution *with more than 5 per cent available chlorine* | 8 | UN1908 | II | CORROSIVE | A3, A6, A7, B2, N34, T8 | 154 | 202 | 242 | 1 L | 30 L | B | 26 |
| | Sodium chloroacetate | 6.1 | UN2659 | III | KEEP AWAY FROM FOOD | | 153 | 213 | 240 | 100 kg | 200 kg | A | |
| | Sodium cuprocyanide, solid | 6.1 | UN2316 | I | POISON | | None | 211 | 242 | 5 kg | 50 kg | A | 26 |
| | Sodium cuprocyanide, solution | 6.1 | UN2317 | I | POISON | T8, T26 | None | 201 | 243 | 1 L | 30 L | B | 26, 40 |
| | Sodium cyanide | 6.1 | UN1689 | I | POISON | B69, B77, N74, N75, T42 | None | 211 | 242 | 5 kg | 50 kg | B | 52 |
| | Sodium 2-diazo-1-naphthol-4-sulphonate | 4.1 | UN3040 | II | FLAMMABLE SOLID | | None | 224 | None | 15 kg | 50 kg | C | 61 |
| | Sodium 2-diazo-1-naphthol-5-sulphonate | 4.1 | UN3041 | II | FLAMMABLE SOLID | | None | 224 | None | 15 kg | 50 kg | C | 61 |
| | *Sodium dichloroisocyanurate or Sodium dichloro-s-triazinetrione, see* Dichloroisocyanuric acid etc | | | | | | | | | | | | |
| | Sodium dinitro-o-cresolate, *dry or wetted with less than 15 per cent water, by mass* | 1.3C | UN0234 | II | EXPLOSIVE 1.3C | | None | 62 | None | Forbidden | Forbidden | B | 1E, 5E |
| | Sodium dinitro-ortho-cresolate, wetted *with not less than 15 per cent water, by mass* | 4.1 | UN1348 | I | FLAMMABLE SOLID, POISON | A8, A19, A20, N41 | None | 211 | None | 1 kg | 15 kg | E | 28, 36 |
| | Sodium dithionite *or* Sodium hydrosulfite | 4.2 | UN1384 | II | SPONTANEOUSLY COMBUSTIBLE | A19, A20 | None | 212 | 241 | 15 kg | 50 kg | E | 13 |
| | Sodium fluoride | 6.1 | UN1690 | III | KEEP AWAY FROM FOOD | T8 | 153 | 213 | 240 | 100 kg | 200 kg | A | 26 |
| | Sodium fluoroacetate | 6.1 | UN2629 | I | POISON | | None | 211 | 242 | 5 kg | 50 kg | E | |
| | Sodium fluorosilicate | 6.1 | UN2674 | III | KEEP AWAY FROM FOOD | | 153 | 213 | 240 | 100 kg | 200 kg | A | 26 |
| | *Sodium hydrate, see* Sodium hydroxide, solid | | | | | | | | | | | | |
| | Sodium hydride | 4.3 | UN1427 | I | DANGEROUS WHEN WET | A19, N40 | None | 211 | 242 | Forbidden | 15 kg | E | |
| | Sodium hydrogen fluoride | 8 | UN2439 | II | CORROSIVE | N3, N34 | 154 | 212 | 240 | 15 kg | 50 kg | A | 12, 25, 26, 40, 48 |
| | Sodium hydrogen fluoride, *solution* | 8 | UN2439 | II | CORROSIVE | N3, N34 | 154 | 202 | 242 | 1 L | 30 L | A | 12, 25, 26, 40 |
| | Sodium hydrogen sulfate, solution | 8 | UN2837 | II | CORROSIVE | A7, B2, N34, T8, T26 | 154 | 202 | 242 | 1 L | 30 L | A | |

## §172.101 HAZARDOUS MATERIALS TABLE, Continued

| Symbols | Hazardous materials descriptions and proper shipping names | Hazard class or Division | Identification Numbers | Packing group | Label(s) required (if not excepted) | Special provisions | (8) Packaging authorizations (§173 ***) | | | (9) Quantity limitations | | (10) Vessel stowage requirements | |
|---|---|---|---|---|---|---|---|---|---|---|---|---|---|
| | | | | | | | Exceptions | Non-bulk packaging | Bulk packaging | Passenger aircraft or railcar | Cargo aircraft only | Vessel stowages | Other stowage provisions |
| (1) | (2) | (3) | (4) | (5) | (6) | (7) | (8A) | (8B) | (8C) | (9A) | (9B) | (10A) | (10B) |
| D | Sodium hydrosulfide, solution | 8 | NA2922 | II | CORROSIVE, POISON | B2 | 154 | 202 | 242 | 1 L | 30 L | B | 40, 95 |
| | Sodium hydrosulfide, *with less than 25 per cent water of crystallization* | 4.2 | UN2318 | II | SPONTANEOUSLY COMBUSTIBLE | A7, A19, A20 | None | 212 | 241 | 15 kg | 50 kg | A | |
| | Sodium hydrosulfide *with not less than 25 per cent water of crystallization* | 8 | UN2949 | II | CORROSIVE | A7 | 154 | 212 | 240 | 15 kg | 50 kg | A | 26 |
| | Sodium hydrosulfite, *see* Sodium dithionite | | | | | | | | | | | | |
| | Sodium hydroxide, solid | 8 | UN1823 | II | CORROSIVE | | 154 | 212 | 240 | 15 kg | 50 kg | A | |
| | Sodium hydroxide solution | 8 | UN1824 | II | CORROSIVE | B2, N34, T8 | 154 | 202 | 242 | 1 L | 30 L | A | |
| | *Sodium hypochlorite, solution, see* Hypochlorite solutions *etc.* | | | | | | | | | | | | |
| | *Sodium metal, liquid alloy, see* Alkali metal alloys, liquid, n.o.s. | | | | | | | | | | | | |
| | Sodium methylate | 4.2 | UN1431 | II | SPONTANEOUSLY COMBUSTIBLE, CORROSIVE | A19 | None | 211 | 242 | 15kg | 50 kg | B | |
| | Sodium methylate solutions *in alcohol* | 3 | UN1289 | II | FLAMMABLE LIQUID | T8, T31 | 150 | 202 | 242 | 5 L | 60 L | B | |
| | | | | III | FLAMMABLE LIQUID | B1, T7, T30 | 150 | 202 | 242 | 60 L | 220 L | A | |
| | Sodium monoxide | 8 | UN1825 | II | CORROSIVE | | 154 | 212 | 240 | 15 kg | 50 kg | A | |
| | Sodium nitrate | 5.1 | UN1498 | III | OXIDIZER | A1, A29 | 152 | 213 | 240 | 25 kg | 100 kg | A | |
| | Sodium nitrate and potassium nitrate mixtures | 5.1 | UN1499 | III | OXIDIZER | A1, A29 | 152 | 213 | 240 | 25 kg | 100 kg | A | |
| | Sodium nitrite | 5.1 | UN1500 | III | OXIDIZER | A1, A29 | 152 | 213 | 240 | 25 kg | 100 kg | A | 56, 58 |
| | Sodium pentachlorophenate | 6.1 | UN2567 | II | POISON | | None | 212 | 242 | 25 kg | 100 kg | A | |
| | Sodium percarbonate | 5.1 | UN2467 | III | OXIDIZER | 27, A1, A29 | 152 | 213 | 240 | 25 kg | 100 kg | A | 13 |
| | Sodium perchlorate | 5.1 | UN1502 | II | OXIDIZER | T8 | 152 | 212 | 242 | 5 kg | 25 kg | A | 56, 58, 106 |
| | Sodium permanganate | 5.1 | UN1503 | II | OXIDIZER | | 152 | 212 | 242 | 5 kg | 25 kg | D | 56, 58, 69, 106, 107 |
| | Sodium peroxide | 5.1 | UN1504 | I | OXIDIZER | A20, N34 | None | 211 | None | Forbidden | 15 kg | B | 13, 75, 106 |
| | Sodium persulfate | 5.1 | UN1505 | III | OXIDIZER | A1 | 152 | 213 | 240 | 25 kg | 100 kg | A | |
| | Sodium phenolate, solid | 8 | UN2497 | III | CORROSIVE | | 154 | 213 | 240 | 25 kg | 100 kg | A | |
| | Sodium phosphide | 4.3 | UN1432 | I | DANGEROUS WHEN WET, POISON | A19, N40 | None | 211 | None | Forbidden | 15 kg | E | 40, 85 |
| | Sodium picramate, *dry or wetted with less than 20 per cent water, by mass* | 1.3C | UN0235 | II | EXPLOSIVE 1.3C | | None | 62 | None | Forbidden | Forbidden | B | 1E, 5E |
| | Sodium picramate, wetted *with not less than 20 per cent water, by mass* | 4.1 | UN1349 | I | FLAMMABLE SOLID | A8, A19, N41 | None | 211 | None | Forbidden | 15 kg | E | 28, 36 |
| | *Sodium picryl peroxide* | Forbidden | | | | | | | | | | | |
| | *Sodium potassium alloys, see* Potassium sodium alloys | | | | | | | | | | | | |
| | Sodium salts of aromatic nitro-derivatives, n.o.s. *explosive* | 1.3C | UN0203 | II | EXPLOSIVE 1.3C | | None | 62 | None | Forbidden | Forbidden | B | 1E, 5E |
| | *Sodium selenate, see* Selenates *or* Selenites | | | | | | | | | | | | |
| D | Sodium selenite | 6.1 | NA2630 | II | POISON | | None | 212 | 242 | 25 kg | 100 kg | E | |
| | Sodium sulfide, anhydrous *or* Sodium sulfide *with less than 30 per cent water of crystallization* | 4.2 | UN1385 | II | SPONTANEOUSLY COMBUSTIBLE | A19, A20, N34 | None | 212 | 241 | 15 kg | 50 kg | A | |
| | Sodium sulfide, hydrated *with at least 30 per cent water* | 8 | UN1849 | II | CORROSIVE | T8 | 154 | 212 | 240 | 15 kg | 50 kg | A | 26 |
| | Sodium superoxide | 5.1 | UN2547 | I | OXIDIZER | A20, N34 | None | 211 | None | Forbidden | 15 kg | E | 13, 75, 106 |
| | *Sodium tetranitride* | Forbidden | | | | | | | | | | | |
| | Sounding devices, explosive | 1.2F | UN0204 | II | EXPLOSIVE 1.2F | | None | 62 | None | Forbidden | Forbidden | E | |
| | Sounding devices, explosive | 1.1F | UN0296 | II | EXPLOSIVE 1.1F | | None | 62 | None | Forbidden | Forbidden | E | |
| | Sounding devices, explosive | 1.1D | UN0374 | II | EXPLOSIVE 1.1D | | None | 62 | None | Forbidden | Forbidden | B | |
| | Sounding devices, explosive | 1.2D | UN0375 | II | EXPLOSIVE 1.2D | | None | 62 | None | Forbidden | Forbidden | B | |
| | *Spirits of salt, see* Hydrochloric acid | | | | | | | | | | | | |
| | *Squibs, see* Igniters *etc* | | | | | | | | | | | | |
| | Stannic chloride, anhydrous | 8 | UN1827 | II | CORROSIVE | B2, T8, T26 | 154 | 202 | 242 | 1 L | 30 L | C | 8 |
| | Stannic chloride, pentahydrate | 8 | UN2440 | III | CORROSIVE | | 154 | 213 | 240 | 25 kg | 100 kg | A | |
| | Stannic phosphide | 4.3 | UN1433 | I | DANGEROUS WHEN WET, POISON | A19, N40 | None | 211 | 242 | Forbidden | 15 kg | E | 40, 85 |
| | Steel swarf *see* Ferrous metal borings, *etc* | | | | | | | | | | | | |
| | Stibine | 2.3 | UN2676 | | POISON GAS, FLAMMABLE GAS | 1 | None | 304 | None | Forbidden | Forbidden | D | 40 |
| | *Storage batteries, wet, see* Batteries, *wet etc* | | | | | | | | | | | | |
| | Strontium arsenite | 6.1 | UN1691 | II | POISON | | None | 212 | 242 | 25 kg | 100 kg | A | |
| | Strontium chlorate, solid *or* solution | 5.1 | UN1506 | II | OXIDIZER | A1, A9, N34, T8 | 152 | 212 | 242 | 5 kg | 25 kg | A | 56, 58, 106 |

# §172.101 HAZARDOUS MATERIALS TABLE, Continued

| Symbols | Hazardous materials descriptions and proper shipping names | Hazard class or Division | Identification Numbers | Packing group | Label(s) required (if not excepted) | Special provisions | (8) Packaging authorizations (§173.***) | | | (9) Quantity limitations | | (10) Vessel stowage requirements | |
|---|---|---|---|---|---|---|---|---|---|---|---|---|---|
| | | | | | | | Exceptions | Non-bulk packaging | Bulk packaging | Passenger aircraft or railcar | Cargo aircraft only | Vessel stowages | Other stowage provisions |
| (1) | (2) | (3) | (4) | (5) | (6) | (7) | (8A) | (8B) | (8C) | (9A) | (9B) | (10A) | (10B) |
| | Strontium nitrate | 5.1 | UN1507 | III | OXIDIZER | A1, A29 | 152 | 213 | 240 | 25 kg | 100 kg | A | |
| | Strontium perchlorate | 5.1 | UN1508 | II | OXIDIZER | T8 | 152 | 212 | 242 | 5 kg | 25 kg | A | 56, 58, 106 |
| | Strontium peroxide | 5.1 | UN1509 | II | OXIDIZER | | 152 | 212 | 242 | 5 kg | 25 kg | A | 13, 75, 106 |
| | Strontium phosphide | 4.3 | UN2013 | I | DANGEROUS WHEN WET, POISON | A19, N40 | None | 211 | None | Forbidden | 15 kg | E | 40, 85 |
| | Strychnine or Strychnine salts | 6.1 | UN1692 | I | POISON | | None | 211 | 242 | 5 kg | 50 kg | A | 40 |
| | Styphnic acid, *see* Trinitroresorcinol, *etc* | | | | | | | | | | | | |
| | Styrene monomer, inhibited | 3 | UN2055 | III | FLAMMABLE LIQUID | B1, T1 | 150 | 203 | 242 | 60 L | 220 L | A | |
| | Substances, explosive, n.o.s. | 1.1L | UN0357 | II | EXPLOSIVE 1.1L | 101 | None | 62 | None | Forbidden | Forbidden | E | 2E, 8E, 11E, 17E |
| | Substances, explosive, n.o.s. | 1.2L | UN0358 | II | EXPLOSIVE 1.2L | 101 | None | 62 | None | Forbidden | Forbidden | E | 2E, 8E, 11E, 17E |
| | Substances, explosive, n.o.s. | 1.3L | UN0359 | II | EXPLOSIVE 1.3L | 101 | None | 62 | None | Forbidden | Forbidden | E | 2E, 8E, 11E, 17E |
| | Substances, explosive, n.o.s. | 1.1A | UN0473 | II | EXPLOSIVE 1.1A | 101, 111 | None | 62 | None | Forbidden | Forbidden | E | 2E, 6E |
| | Substances, explosive, n.o.s. | 1.1C | UN0474 | II | EXPLOSIVE 1.1C | 101 | None | 62 | None | Forbidden | Forbidden | B | 1E, 5E |
| | Substances, explosive, n.o.s. | 1.1D | UN0475 | II | EXPLOSIVE 1.1D | 101 | None | 62 | None | Forbidden | Forbidden | B | 1E, 5E |
| | Substances, explosive, n.o.s. | 1.1G | UN0476 | II | EXPLOSIVE 1.1G | 101 | None | 62 | None | Forbidden | Forbidden | B | 1E, 8E |
| | Substances, explosive, n.o.s. | 1.3C | UN0477 | II | EXPLOSIVE 1.3C | 101 | None | 62 | None | Forbidden | Forbidden | B | 1E, 5E |
| | Substances, explosive, n.o.s. | 1.3G | UN0478 | II | EXPLOSIVE 1.3G | 101 | None | 62 | None | Forbidden | Forbidden | E | 1E, 8E |
| | Substances, explosive, n.o.s. | 1.4G | UN0485 | II | EXPLOSIVE 1.4G | 101 | None | 62 | None | Forbidden | 75 kg | E | 1E, 5E |
| | Substances, explosive, n.o.s. | 1.4S | UN0481 | II | EXPLOSIVE 1.4S | 101 | None | 62 | None | 25 kg | 75 kg | A | 1E, 5E |
| | Substances, explosive, very insensitive, n.o.s., *or* Substances, EVI, n.o.s. | 1.5D | UN0482 | II | EXPLOSIVE 1.5D | 101 | None | 62 | None | Forbidden | Forbidden | B | 1E, 5E |
| | Substances, explosives, n.o.s. | 1.4C | UN0479 | II | EXPLOSIVE 1.4C | 101 | None | 62 | None | Forbidden | 75 kg | A | 1E, 5E |
| | Substances, explosives, n.o.s. | 1.4D | UN0480 | II | EXPLOSIVE 1.4D | 101 | None | 62 | None | Forbidden | 75 kg | A | 1E, 5E |
| | Substances which in contact with water emit flammable gases, liquid, corrosive, n.o.s. | 4.3 | UN3129 | I | DANGEROUS WHEN WET, CORROSIVE | | None | 201 | 243 | Forbidden | 1 L | E | |
| | | | | II | DANGEROUS WHEN WET, CORROSIVE | | None | 202 | 243 | 1 L | 5 L | E | |
| | | | | III | DANGEROUS WHEN WET, CORROSIVE | | None | 203 | 242 | 5 L | 60 L | E | |
| | Substances which in contact with water emit flammable gases, liquid n.o.s. | 4.3 | UN3148 | I | DANGEROUS WHEN WET | | None | 201 | 244 | Forbidden | 1 L | E | 40 |
| | | | | II | DANGEROUS WHEN WET | | None | 202 | 243 | 1 L | 5 L | E | 40 |
| | | | | III | DANGEROUS WHEN WET | | None | 203 | 243 | 5 L | 60 L | E | 40 |
| | Substances which in contact with water emit flammable gases, liquid, poisonous, n.o.s. | 4.3 | UN3130 | I | DANGEROUS WHEN WET, POISON | A4 | None | 201 | 243 | Forbidden | 1 L | E | |
| | | | | II | DANGEROUS WHEN WET, POISON | | None | 202 | 243 | 1 L | 5 L | E | |
| | | | | III | DANGEROUS WHEN WET, KEEP AWAY FROM FOOD | | None | 203 | 242 | 5 L | 60 L | E | |
| | Substances which in contact with water emit flammable gases solid, corrosive, n.o.s. | 4.3 | UN3131 | I | DANGEROUS WHEN WET, CORROSIVE | N40 | None | 211 | 242 | Forbidden | 15 kg | E | |
| | | | | II | DANGEROUS WHEN WET, CORROSIVE | | None | 212 | 242 | 15 kg | 50 kg | E | |
| | | | | III | DANGEROUS WHEN WET, CORROSIVE | | None | 213 | 241 | 25 kg | 100 kg | E | |
| | Substances which in contact with water emit flammable gases solid, flammable n.o.s. | 4.3 | UN3132 | I | DANGEROUS WHEN WET, FLAMMABLE SOLID | N40 | None | 211 | 242 | Forbidden | 15 kg | E | |
| | | | | II | DANGEROUS WHEN WET, FLAMMABLE SOLID | | None | 212 | 242 | 15 kg | 50 kg | E | |
| | | | | III | DANGEROUS WHEN WET, FLAMMABLE SOLID | | None | 213 | 241 | 25 kg | 100 kg | E | |
| I | Substances which in contact with water emit flammable gases, solid, n.o.s. | 4.3 | UN2813 | I | DANGEROUS WHEN WET | N40 | None | 211 | 242 | Forbidden | 15kg | E | 40 |
| | | | | II | DANGEROUS WHEN WET | | None | 212 | 242 | 15kg | 50kg | E | 40 |
| | | | | III | DANGEROUS WHEN WET | | None | 213 | 241 | 25kg | 100kg | E | 40 |

## §172.101 HAZARDOUS MATERIALS TABLE, Continued

| Symbols | Hazardous materials descriptions and proper shipping names | Hazard class or Division | Identification Numbers | Packing group | Label(s) required (if not excepted) | Special provisions | (8) Packaging authorizations (§173.***) | | | (9) Quantity limitations | | (10) Vessel stowage requirements | |
|---|---|---|---|---|---|---|---|---|---|---|---|---|---|
| | | | | | | | Exceptions | Non-bulk packaging | Bulk packaging | Passenger aircraft or railcar | Cargo aircraft only | Vessel stowages | Other stowage provisions |
| (1) | (2) | (3) | (4) | (5) | (6) | (7) | (8A) | (8B) | (8C) | (9A) | (9B) | (10A) | (10B) |
| | Substances which in contact with water emit flammable gases, solid, oxidizing, n.o.s. | 4.3 | UN3133 | | DANGEROUS WHEN WET, OXIDIZER | | None | 214 | 214 | Forbidden | Forbidden | | |
| | Substances which in contact with water, emit flammable gases, solid, poisonous, n.o.s. | 4.3 | UN3134 | I | DANGEROUS WHEN WET, POISON | A5, N40 | None | 211 | 242 | Forbidden | 15kg | E | |
| | | | | II | DANGEROUS WHEN WET, POISON | | None | 212 | 242 | 15kg | 50kg | E | |
| | | | | III | DANGEROUS WHEN WET, KEEP AWAY FROM FOOD | | None | 213 | 241 | 25kg | 100kg | E | |
| | Substances which in contact with water emit flammable gases, solid, self-heating, n.o.s | 4.3 | UN3135 | I | DANGEROUS WHEN WET SPONTANEOUSLY COMBUSTIBLE | N40 | None | 211 | 242 | Forbidden | 15 kg | E | |
| | | | | II | DANGEROUS WHEN WET, SPONTANEOUSLY COMBUSTIBLE | | None | 212 | 242 | 15 kg | 50 kg | E | |
| | | | | III | DANGEROUS WHEN WET, SPONTANEOUSLY COMBUSTIBLE | | None | 213 | 241 | 25 kg | 100 kg | E | |
| | Substituted nitrophenol pesticides, liquid, flammable, toxic, n.o.s., *flash point less than 23 degrees C.* | 3 | UN2780 | I | FLAMMABLE LIQUID, POISON | | None | 201 | 243 | Forbidden | 30 L | B | |
| | | | | II | FLAMMABLE LIQUID, POISON | | None | 202 | 243 | 1 L | 60 L | B | |
| | Substituted nitrophenol pesticides, liquid, toxic, flammable, n.o.s., *flashpoint not less than 23 degrees C* | 6.1 | UN3013 | I | POISON, FLAMMABLE LIQUID | T42 | None | 201 | 243 | 1 L | 30 L | B | 40 |
| | | | | II | POISON, FLAMMABLE LIQUID | T14 | None | 202 | 243 | 5 L | 60 L | B | 40 |
| | | | | III | KEEP AWAY FROM FOOD, FLAMMABLE LIQUID | B1, T14 | 153 | 203 | 242 | 60 L | 220 L | A | 40 |
| | Substituted nitrophenol pesticides, liquid, toxic, n.o.s | 6.1 | UN3014 | I | POISON | T42 | None | 201 | 243 | 1 L | 30 L | B | 40 |
| | | | | II | POISON | T14 | None | 202 | 243 | 5 L | 60 L | B | 40 |
| | | | | III | KEEP AWAY FROM FOOD | T14 | 153 | 203 | 241 | 60 L | 220 L | A | 40 |
| | Substituted nitrophenol pesticides, solid, toxic, n.o.s. | 6.1 | UN2779 | I | POISON | | None | 211 | 242 | 5 kg | 50 kg | A | 40 |
| | | | | II | POISON | | None | 212 | 242 | 25 kg | 100 kg | A | 40 |
| | | | | III | KEEP AWAY FROM FOOD | | 153 | 213 | 240 | 100 kg | 200 kg | A | 40 |
| | *Sucrose octanitrate (dry)* | Forbidden | | | | | | | | | | | |
| | Sulfamic acid | 8 | UN2967 | III | CORROSIVE | | 154 | 213 | 240 | 25 kg | 100 kg | A | |
| I | Sulfur | 4.1 | UN1350 | III | FLAMMABLE SOLID | A1, N20 | 151 | 213 | 240 | 25 kg | 100 kg | A | 19, 74 |
| D | Sulfur | 9 | NA1350 | III | CLASS 9 | A1, N20 | 151 | 213 | 240 | 25 kg | 100 kg | A | 19, 74 |
| | *Sulfur and chlorate, loose mixtures of* | Forbidden | | | | | | | | | | | |
| | Sulfur chlorides | 8 | UN1828 | I | CORROSIVE | 5, A3, B10, B77, N34 | None | 201 | 243 | Forbidden | 2.5 L | C | 8, 40 |
| | *Sulfur dichloride, see* Sulfur chlorides | | | | | | | | | | | | |
| | Sulfur dioxide, liquefied | 2.3 | UN1079 | | POISON GAS | 3, B14 | None | 304 | 314, 315 | Forbidden | 25 kg | D | 40 |
| | *Sulfur dioxide solution, see* Sulfurous acid | | | | | | | | | | | | |
| | Sulfur hexafluoride | 2.2 | UN1080 | | NONFLAMMABLE GAS | | 306 | 304 | 314, 315 | 75 kg | 150 kg | A | |
| I | Sulfur, molten | 4.1 | UN2448 | III | FLAMMABLE SOLID | B12, B85, T9 | None | 213 | 241 | Forbidden | Forbidden | C | 61 |
| D | Sulfur, molten | 9 | NA2448 | III | CLASS 9 | B12, B85 T9 | None | 213 | 241 | Forbidden | Forbidden | C | 61 |
| | Sulfur tetrafluoride | 2.3 | UN2418 | | POISON GAS | 1 | None | 302 | 245 | Forbidden | 25 kg | D | 40 |
| + | Sulfur trioxide inhibited | 8 | UN1829 | I | CORROSIVE, POISON | 2, A7, B9, B12, B14, B32, B49, B74, B77, N34, T38, T43, T45 | None | 227 | 244 | Forbidden | 25 kg | A | 10, 40 |
| +,D | Sulfur trioxide, uninhibited | 8 | NA1829 | I | CORROSIVE, POISON | 2, A7 B9, B12, B14, B32, B49, B74, B77, N34, T38, T43 T45 | None | 227 | 244 | Forbidden | 25 kg | A | 10 40 |
| | *Sulfuretted hydrogen, see* Hydrogen sulfide, liquefied | | | | | | | | | | | | |
| | Sulfuric acid | 8 | UN1830 | II | CORROSIVE | A3, A7, B3, B83, B84, N34, T9, T27 | 154 | 202 | 242 | 1 L | 30 L | C | 14 |

## §172.101 HAZARDOUS MATERIALS TABLE, Continued

| Symbols | Hazardous materials descriptions and proper shipping names | Hazard class or Division | Identification Numbers | Packing group | Label(s) required (if not excepted) | Special provisions | (8) Packaging authorizations (§173 ***) | | | (9) Quantity limitations | | (10) Vessel stowage requirements | |
|---|---|---|---|---|---|---|---|---|---|---|---|---|---|
| | | | | | | | Exceptions | Non-bulk packaging | Bulk packaging | Passenger aircraft or railcar | Cargo aircraft only | Vessel stowages | Other stowage provisions |
| (1) | (2) | (3) | (4) | (5) | (6) | (7) | (8A) | (8B) | (8C) | (9A) | (9B) | (10A) | (10B) |
| + | Sulfuric acid, fuming *greater than or equal to 30 percent free sulfur trioxide* | 8 | UN1831 | I | CORROSIVE | 2, A3, A6, A7, B9, B14, B32, B74, B77, B84, N34, T38, T43, T45 | None | 227 | 244 | Forbidden | 2.5 L | C | 14, 40 |
| | Sulfuric acid, fuming *less than 30 percent free sulfur trioxide* | 8 | UN1831 | I | CORROSIVE | A3, A7, B84, N34, T18, T27 | None | 201 | 243 | Forbidden | 2.5 L | C | 14, 40 |
| | Sulfuric acid, spent | 8 | UN1832 | II | CORROSIVE | A3, A7, B2, B83, B84, N34, T9, T27 | None | 202 | 242 | Forbidden | 30 L | C | 14 |
| | Sulfuric and hydrofluoric acid mixtures, *see* Hydrofluoric and sulfuric acid mixtures | | | | | | | | | | | | |
| | *Sulfuric anhydride, see* Sulfur trioxide, inhibited | | | | | | | | | | | | |
| | Sulfurous acid | 8 | UN1833 | II | CORROSIVE | B3, T8 | 154 | 202 | 242 | 1 L | 30 L | B | 40 |
| + | Sulfuryl chloride | 8 | UN1834 | I | CORROSIVE, POISON | 1, A3, B6, B9, B10, B14, B30, B74, B77, N34, T38, T43, T44 | None | 226 | 244 | 0.5 L | 2.5 L | C | 8, 40 |
| | Sulfuryl fluoride | 2.3 | UN2191 | | POISON GAS | 4 | None | 304 | 314, 315 | Forbidden | 25 kg | D | 40 |
| | Tars, liquid *including road asphalt and oils bitumen and cut backs* | 3 | UN1999 | II | FLAMMABLE LIQUID | B13, T7, T30 | 150 | 202 | 242 | 5 L | 60 L | B | |
| | | | | III | FLAMMABLE LIQUID | B1, B13, T7, T30 | 150 | 203 | 242 | 60 L | 220 L | A | |
| | Tear gas candles | 6.1 | UN1700 | II | POISON, FLAMMABLE SOLID | | None | 340 | None | Forbidden | 50 kg | D | 40 |
| | *Tear gas cartridges, see* Ammunition, tear-producing, etc | | | | | | | | | | | | |
| D | Tear gas devices *with more than 2 per cent tear gas substances, by mass* | 6.1 | NA1693 | I | POISON | | None | 340 | None | Forbidden | Forbidden | D | 40, 95 |
| | | | | II | POISON | | None | 340 | None | Forbidden | Forbidden | D | 40, 95 |
| | *Tear gas devices, with not more than 2 per cent tear gas substances by mass, see* Aerosols, etc | | | | | | | | | | | | |
| | *Tear gas grenades, see* Tear gas candles | | | | | | | | | | | | |
| | Tear gas substances, n.o.s., liquid | 6.1 | UN1693 | I | POISON | | None | 201 | None | Forbidden | Forbidden | D | 40 |
| | | | | II | POISON | | None | 202 | None | Forbidden | 5 L | D | 40 |
| | Tear gas substances, n.o.s., solid | 6.1 | UN1693 | I | POISON | | None | 211 | None | Forbidden | Forbidden | D | 40 |
| | | | | II | POISON | | None | 212 | None | Forbidden | 25 kg | D | 40 |
| | Tellurium hexafluoride | 2.3 | UN2195 | | POISON GAS | 1 | None | 302 | None | Forbidden | Forbidden | D | 40 |
| | Terpene hydrocarbons, n.o.s. | 3 | UN2319 | III | FLAMMABLE LIQUID | B1, T1 | 150 | 203 | 242 | 60 L | 220 L | A | |
| | Terpinolene | 3 | UN2541 | III | FLAMMABLE LIQUID | B1, T1 | 150 | 203 | 242 | 60 L | 220 L | A | |
| | *Tetraazido benzene quinone* | Forbidden | | | | | | | | | | | |
| | Tetrabromoethane | 6.1 | UN2504 | III | KEEP AWAY FROM FOOD | T7 | 153 | 203 | 241 | 60 L | 220 L | A | |
| | Tetrachloroethane | 6.1 | UN1702 | II | POISON | N36, T14 | None | 202 | 243 | 5 L | 60 L | A | 40 |
| | Tetrachloroethylene | 6.1 | UN1897 | III | KEEP AWAY FROM FOOD | N36, T1 | 153 | 203 | 241 | 60 L | 220 L | A | 40 |
| | Tetraethyl dithiopyrophosphate | 6.1 | UN1704 | II | POISON | | None | 212 | 242 | 25 kg | 100 kg | D | 40 |
| | Tetraethyl dithiopyrophosphate and gases in solution *or* tetraethyl dithiopyrophosphate and gas mixtures *LC50 over 200 but not greater than 5000 ppm* | 2.3 | UN1703 | | POISON GAS | 6 | None | 334 | 245 | Forbidden | Forbidden | D | 40 |
| | Tetraethyl dithiopyrophosphate and gases in solution *or* Tetraethyl dithiopyrophosphate and gases mixtures *LC50 less than or equal to 200 ppm* | 2.3 | UN1703 | | POISON GAS | 1 | None | 334 | 245 | Forbidden | Forbidden | D | 40, 85, 95 |
| D | Tetraethyl lead, liquid | 6.1 | NA1649 | I | POISON, FLAMMABLE LIQUID | | None | 201 | None | Forbidden | Forbidden | E | 40 |
| | Tetraethyl pyrophosphate and compressed gas mixtures *LC50 less than or equal to 200 ppm* | 2.3 | UN1705 | | POISON GAS | 1 | None | 334 | 245 | Forbidden | Forbidden | D | 40, 85, 95 |
| | Tetraethyl pyrophosphate and compressed gas mixtures) *LC50 over 200 ppm but not greater than 5000 ppm* | 2.3 | UN1705 | | POISON GAS | 6, B9, B14 | None | 334 | 314, 315 | Forbidden | Forbidden | D | 40 |
| D | Tetraethyl pyrophosphate, *liquid* | 6.1 | NA3018 | I | POISON | | None | 201 | 243 | Forbidden | 1 L | A | 40 |
| D | Tetraethyl pyrophosphate *solid* | 6.1 | NA2783 | I | POISON | N77 | None | 211 | 242 | Forbidden | 50 kg | A | 40 |
| | Tetraethyl silicate | 3 | UN1292 | III | FLAMMABLE LIQUID | B1, T1 | 150 | 203 | 241 | 60 L | 220 L | A | |
| | *Tetraethylammonium perchlorate (dry)* | Forbidden | | | | | | | | | | | |
| | Tetraethylenepentamine | 8 | UN2320 | III | CORROSIVE | T2 | 154 | 203 | 241 | 5 L | 60 L | A | 8 |
| | Tetrafluoroethylene, inhibited | 2.1 | UN1081 | | FLAMMABLE GAS | | 306 | 304 | None | Forbidden | 150 kg | E | 40 |
| | Tetrafluoromethane, *R14* | 2.2 | UN1982 | | NONFLAMMABLE GAS | | None | 302 | None | 75 kg | 150 kg | A | |
| | 1,2,3,6-Tetrahydrobenzaldehyde | 3 | UN2498 | III | FLAMMABLE LIQUID | B1, T1 | 150 | 203 | 242 | 60 L | 220 L | A | |

## §172.101 HAZARDOUS MATERIALS TABLE, Continued

| Symbols | Hazardous materials descriptions and proper shipping names | Hazard class or Division | Identification Numbers | Packing group | Label(s) required (if not excepted) | Special provisions | (8) Packaging authorizations (§173.***) | | | (9) Quantity limitations | | (10) Vessel stowage requirements | |
|---|---|---|---|---|---|---|---|---|---|---|---|---|---|
| | | | | | | | Exceptions | Non bulk packaging | Bulk packaging | Passenger aircraft or railcar | Cargo aircraft only | Vessel stowages | Other stowage provisions |
| (1) | (2) | (3) | (4) | (5) | (6) | (7) | (8A) | (8B) | (8C) | (9A) | (9B) | (10A) | (10B) |
| | Tetrahydrofuran | 3 | UN2056 | II | FLAMMABLE LIQUID | T8 | None | 202 | 242 | 5 L | 60 L | B | |
| | Tetrahydrofurfurylamine | 3 | UN2943 | III | FLAMMABLE LIQUID | B1, T1 | 150 | 203 | 242 | 60 L | 220 L | A | |
| | Tetrahydrophthalic anhydrides *with more than 0.05 percent of maleic anhydride* | 8 | UN2698 | III | CORROSIVE | | 154 | 213 | 240 | 25 kg | 100 kg | A | |
| | 1,2,3,6-Tetrahydropyridine | 3 | UN2410 | II | FLAMMABLE LIQUID | T8 | 150 | 202 | 242 | 5 L | 60 L | B | |
| | Tetrahydrothiophene | 3 | UN2412 | II | FLAMMABLE LIQUID | T7 | 150 | 202 | 242 | 5 L | 60 L | B | |
| | Tetramethylammonium hydroxide | 8 | UN1835 | II | CORROSIVE | B2, T8 | 154 | 202 | 242 | 1 L | 30 L | A | |
| | *Tetramethylene diperoxide dicarbamide* | Forbidden | | | | | | | | | | | |
| | Tetramethylsilane | 3 | UN2749 | I | FLAMMABLE LIQUID | T21, T26 | None | 201 | 243 | Forbidden | 30 L | D | |
| | *Tetranitro diglycerin* | Forbidden | | | | | | | | | | | |
| | Tetranitroaniline | 1.1D | UN0207 | II | EXPLOSIVE 1.1D | | None | 62 | None | Forbidden | Forbidden | B | 1E, 5E, 19E |
| + | Tetranitromethane | 5.1 | UN1510 | I | OXIDIZER, POISON | 2, B9, B14, B32, B74, T38, T43, T45 | None | 227 | None | Forbidden | Forbidden | D | 40, 66, 106 |
| | *2,3,4,6-Tetranitrophenol* | Forbidden | | | | | | | | | | | |
| | *2,3,4,6-Tetranitrophenyl methyl nitramine* | Forbidden | | | | | | | | | | | |
| | *2,3,4,6-Tetranitrophenylnitramine* | Forbidden | | | | | | | | | | | |
| | *Tetranitroresorcinol (dry)* | Forbidden | | | | | | | | | | | |
| | *2,3,5,6-Tetranitroso-1,4-dinitrobenzene* | Forbidden | | | | | | | | | | | |
| | *2,3,5,6-Tetranitroso nitrobenzene (dry)* | Forbidden | | | | | | | | | | | |
| | Tetrapropylorthotitanate | 3 | UN2413 | III | FLAMMABLE LIQUID | B1, T8 | 150 | 202 | 242 | 5 L | 60 L | A | |
| | Tetrazene, *see* Guanyl nitrosaminoguanyltetrazene | | | | | | | | | | | | |
| | *Tetrazine (dry)* | Forbidden | | | | | | | | | | | |
| | Tetrazol-1-acetic acid | 1.4C | UN0407 | II | EXPLOSIVE 1.4C | | None | 62 | None | Forbidden | 75 kg | A | 1E, 5E |
| | *Tetrazolyl azide (dry)* | Forbidden | | | | | | | | | | | |
| | Tetryl, *see* Trinitrophenylmethylnitramine | | | | | | | | | | | | |
| | Thallium chlorate | 5.1 | UN2573 | II | OXIDIZER, POISON | | None | 212 | 242 | 5 kg | 25 kg | A | 56, 58, 106 |
| | Thallium compounds, n.o.s | 6.1 | UN1707 | II | POISON | | None | 212 | 242 | 25 kg | 100 kg | A | |
| | Thallium nitrate | 6.1 | UN2727 | II | POISON, OXIDIZER | | None | 212 | 242 | 5 kg | 25 kg | A | 89 |
| D | Thallium sulfate, solid | 6.1 | NA1707 | II | POISON | | None | 211 | 242 | 5 kg | 50 kg | A | |
| | Thia-4-pentanal | 6.1 | UN2785 | III | KEEP AWAY FROM FOOD | T8 | 153 | 203 | 241 | 60 L | 220 L | D | 25, 49 |
| | Thioacetic acid | 3 | UN2436 | II | FLAMMABLE LIQUID | T8 | 150 | 202 | 242 | 5 L | 60 L | B | |
| | *Thiocarbonylchloride*, *see* Thiophosgene | | | | | | | | | | | | |
| | Thioglycol | 6.1 | UN2966 | II | POISON | T8 | None | 202 | 243 | 5 L | 60 L | A | |
| | Thioglycolic acid | 8 | UN1940 | II | CORROSIVE | A7, B2, N34, T8 | 154 | 202 | 242 | 1 L | 30 L | A | |
| | Thiolactic acid | 6.1 | UN2936 | II | POISON | T8 | None | 212 | 242 | 25 kg | 100 kg | A | |
| + | Thionyl chloride | 8 | UN1836 | I | CORROSIVE, POISON | 2, A3, A10, B6, B9, B14, B32, B74, B77, N34, T38, T43, T45 | None | 227 | 244 | Forbidden | 2.5 L | C | 8, 40 |
| | Thiophene | 3 | UN2414 | II | FLAMMABLE LIQUID | T2 | 150 | 202 | 242 | 5 L | 60 L | B | 40 |
| + | Thiophosgene | 6.1 | UN2474 | II | POISON | 2, A7, B9, B14, B32, B74, N33, N34, T38, T43, T45 | None | 227 | 244 | Forbidden | 60 L | B | 26, 40 |
| | Thiophosphoryl chloride | 8 | UN1837 | II | CORROSIVE | A3, A7, B2, B8, B25, N34, T12 | None | 202 | 242 | Forbidden | 30 L | C | 8, 40 |
| | Thorium metal, pyrophoric | 7 | UN2975 | | RADIOACTIVE, SPONTANEOUSLY COMBUSTIBLE | | None | 418 | None | Forbidden | Forbidden | D | |
| | Thorium nitrate, solid | 7 | UN2976 | | RADIOACTIVE, OXIDIZER | | None | 419 | None | Forbidden | 15 kg | A | |
| | *Tin chloride, fuming*, *see* Stannic chloride, anhydrous | | | | | | | | | | | | |
| | *Tin perchloride or Tin tetrachloride*, *see* Stannic chloride, anhydrous | | | | | | | | | | | | |

## §172.101 HAZARDOUS MATERIALS TABLE, Continued

| Symbols | Hazardous materials descriptions and proper shipping names | Hazard class or Division | Identification Numbers | Packing group | Label(s) required (if not excepted) | Special provisions | (8) Packaging authorizations (§173.***) | | | (9) Quantity limitations | | (10) Vessel stowage requirements | |
|---|---|---|---|---|---|---|---|---|---|---|---|---|---|
| | | | | | | | Exceptions | Non-bulk packaging | Bulk packaging | Passenger aircraft or railcar | Cargo aircraft only | Vessel stowages | Other stowage provisions |
| (1) | (2) | (3) | (4) | (5) | (6) | (7) | (8A) | (8B) | (8C) | (9A) | (9B) | (10A) | (10B) |
| | Tinctures, medicinal | 3 | UN1293 | II | FLAMMABLE LIQUID | T8, T31 | 150 | 202 | 242 | 5 L | 60 L | B | |
| | | | | III | FLAMMABLE LIQUID | B1, T7, T30 | 150 | 203 | 241 | 60 L | 220 L | A | |
| | *Tinning flux see* Zinc chloride | | | | | | | | | | | | |
| | Titanium hydride | 4.1 | UN1871 | II | FLAMMABLE SOLID | A19, A20, N34 | None | 212 | 241 | 15 kg | 50 kg | E | |
| | Titanium powder, dry | 4.2 | UN2546 | I | SPONTANEOUSLY COMBUSTIBLE | | None | 211 | 242 | Forbidden | Forbidden | D | |
| | | | | II | SPONTANEOUSLY COMBUSTIBLE | A19, A20, N5, N34 | None | 212 | 241 | 15 kg | 50 kg | D | |
| | | | | III | SPONTANEOUSLY COMBUSTIBLE | | None | 213 | 241 | 25 kg | 100 kg | D | |
| | Titanium powder, wetted *with not less than 25 per cent water (a visible excess of water must be present) (a) mechanically produced, particle size less than 53 microns; (b) chemically produced, particle size less than 840 microns* | 4.1 | UN1352 | II | FLAMMABLE SOLID | A19, A20, N34 | None | 212 | 240 | 15 kg | 50 kg | E | |
| | Titanium sponge granules *or* Titanium sponge powders | 4.1 | UN2878 | III | FLAMMABLE SOLID | A1 | None | 213 | 240 | 25 kg | 100 kg | D | |
| D | Titanium sulfate solution | 8 | NA1760 | II | CORROSIVE | B2, B15 | None | 213 | 242 | 1 L | 30 L | B | 40 |
| + | Titanium tetrachloride | 8 | UN1838 | II | CORROSIVE, POISON | 2, A3, A6, B7, B9, B14, B32, B41, B74, B77, N41, T38, T43, T45 | None | 227 | 244 | Forbidden | 30 L | C | 8, 40 |
| | Titanium trichloride mixtures | 8 | UN2869 | II | CORROSIVE | A7, N34 | 154 | 212 | 240 | 15 kg | 50 kg | A | 40 |
| | Titanium trichloride, pyrophoric *or* Titanium trichloride mixtures, pyrophoric | 4.2 | UN2441 | I | SPONTANEOUSLY COMBUSTIBLE, CORROSIVE | A7, A8, A19, A20, N34 | None | 181 | 244 | Forbidden | Forbidden | D | 40 |
| | *TNT mixed with aluminum, see Tritonal* | | | | | | | | | | | | |
| | TNT *see* Trinitrotoluene, *etc* | | | | | | | | | | | | |
| | Toe puffs, *nitrocellulose base* | 4.1 | UN1353 | III | FLAMMABLE SOLID | A1 | None | 213 | 240 | 25 kg | 100 kg | D | |
| | Toluene | 3 | UN1294 | II | FLAMMABLE LIQUID | T1 | 150 | 202 | 242 | 5 L | 60 L | B | |
| | Toluene diisocyanate | 6.1 | UN2078 | II | POISON | T14 | None | 202 | 243 | 5 L | 60 L | B | 25, 40 |
| | Toluene sulfonic acid, *see* Alkyl, Aryl or Toluene sulfonic acid *etc* | | | | | | | | | | | | |
| | Toluidines *liquid* | 6.1 | UN1708 | II | POISON | T14 | None | 202 | 243 | 5 L | 60 L | A | |
| | Toluidines *solid* | 6.1 | UN1708 | II | POISON | | None | 212 | 241 | 25 kg | 100 kg | A | |
| | 2,4-Toluylenediamine *or* 2,4-Toluenediamine | 6.1 | UN1709 | III | KEEP AWAY FROM FOOD | T7 | 153 | 213 | 240 | 100 kg | 200 kg | A | |
| | Torpedoes, liquid fuelled, *with inert head* | 1.3J | UN0450 | II | EXPLOSIVE 1.3J | | 63(b) | 62 | None | Forbidden | Forbidden | E | 7E, 16E, 23E |
| | Torpedoes, liquid fuelled, *with or without bursting charge* | 1.1J | UN0449 | II | EXPLOSIVE 1.1J | | 63(b) | 62 | None | Forbidden | Forbidden | E | 7E, 16E, 23E |
| | Torpedoes *with bursting charge* | 1.1E | UN0329 | II | EXPLOSIVE 1.1E | | 63(b) | 62 | None | Forbidden | Forbidden | B | |
| | Torpedoes *with bursting charge* | 1.1F | UN0330 | II | EXPLOSIVE 1.1F | | 63(b) | 62 | None | Forbidden | Forbidden | B | |
| | Torpedoes *with bursting charge* | 1.1D | UN0451 | II | EXPLOSIVE 1.1D | | 63(b) | 62 | None | Forbidden | Forbidden | B | |
| | Tracers for ammunition | 1.3G | UN0212 | II | EXPLOSIVE 1.3G | | None | 62 | None | Forbidden | Forbidden | B | |
| | Tracers for ammunition | 1.4G | UN0306 | II | EXPLOSIVE 1.4G | | None | 62 | None | Forbidden | 75 kg | A | 24E |
| | *Tractors see* Vehicles, self propelled | | | | | | | | | | | | |
| | *Tri-(b-nitroxyethyl) ammonium nitrate* | Forbidden | | | | | | | | | | | |
| | Triallyl borate | 6.1 | UN2609 | III | KEEP AWAY FROM FOOD | | 153 | 203 | 241 | 60 L | 220 L | A | 13 |
| | Triallylamine | 3 | UN2610 | III | FLAMMABLE LIQUID | B1, T1 | 150 | 203 | 242 | 60 L | 220 L | A | 40 |
| | Triazine pesticides, liquid, flammable, toxic, n.o.s., *flash point less than 23 deg C* | 3 | UN2764 | I | FLAMMABLE LIQUID, POISON | | None | 201 | 243 | Forbidden | 30 L | B | |
| | | | | II | FLAMMABLE LIQUID, POISON | | None | 202 | 243 | 1 L | 60 L | B | |
| | Triazine pesticides, liquid, toxic, flammable, n.o.s. *flashpoint not less than 23 degrees C* | 6.1 | UN2997 | I | POISON, FLAMMABLE LIQUID | T42 | None | 201 | 243 | 1 L | 30 L | B | 40 |
| | | | | II | POISON, FLAMMABLE LIQUID | T14 | None | 202 | 243 | 5 L | 60 L | B | 40 |
| | | | | III | KEEP AWAY FROM FOOD, FLAMMABLE LIQUID | T14 | 153 | 203 | 242 | 60 L | 220 L | A | 40 |
| | Triazine pesticides, liquid, toxic, n.o.s. | 6.1 | UN2998 | I | POISON | T42 | None | 201 | 243 | 1 L | 30 L | B | 40 |
| | | | | II | POISON | T14 | None | 202 | 243 | 5 L | 60 L | B | 40 |
| | | | | III | KEEP AWAY FROM FOOD | T14 | 153 | 203 | 241 | 60 L | 220 L | A | 40 |
| | Triazine pesticides, solid, toxic, n.o.s. | 6.1 | UN2763 | I | POISON | | None | 211 | 242 | 5 kg | 50 kg | A | 40 |
| | | | | II | POISON | | None | 212 | 242 | 25 kg | 100 kg | A | 40 |
| | | | | III | KEEP AWAY FROM FOOD | | 153 | 213 | 240 | 100 kg | 200 kg | A | 40 |

## §172.101 HAZARDOUS MATERIALS TABLE, Continued

| Symbols | Hazardous materials descriptions and proper shipping names | Hazard class or Division | Identification Numbers | Packing group | Label(s) required (if not excepted) | Special provisions | (8) Packaging authorizations (§173.***) | | | (9) Quantity limitations | | (10) Vessel stowage requirements | |
|---|---|---|---|---|---|---|---|---|---|---|---|---|---|
| | | | | | | | Exceptions | Non-bulk packaging | Bulk packaging | Passenger aircraft or railcar | Cargo aircraft only | Vessel stowages | Other stowage provisions |
| (1) | (2) | (3) | (4) | (5) | (6) | (7) | (8A) | (8B) | (8C) | (9A) | (9B) | (10A) | (10B) |
| | Tributylamine | 8 | UN2542 | III | CORROSIVE | T2 | None | 203 | 241 | 5 L | 60 L | A | |
| D | mono-(Trichloro) tetra-(monopotassium dichloro)-penta-s-triazinetrione, dry *(containing over 39% available chlorine)* | 5.1 | NA2468 | II | OXIDIZER | | 152 | 212 | 240 | 5 kg | 25 kg | A | 13 |
| | *Trichloro-s-triazinetrione dry, containing over 39% available chlorine, see* Trichloroisocyanuric acid, dry | | | | | | | | | | | | |
| | Trichloroacetic acid | 8 | UN1839 | II | CORROSIVE | A7, N34 | 154 | 212 | 240 | 15 kg | 50 kg | A | |
| | Trichloroacetic acid, solution | 8 | UN2564 | II | CORROSIVE | A3, A6, A7, B2, N34, T8 | 154 | 202 | 242 | 1 L | 30 L | B | 8 |
| + | Trichloroacetyl chloride | 8 | UN2442 | II | CORROSIVE, POISON | 1, A3, A7, B9, B14, B30, B72, N34, T38, T43, T44 | None | 226 | 244 | Forbidden | Forbidden | D | 40 |
| | Trichlorobenzenes, liquid | 6.1 | UN2321 | III | KEEP AWAY FROM FOOD | T7 | 153 | 203 | 241 | 60 L | 220 L | A | |
| | Trichlorobutene | 6.1 | UN2322 | II | POISON | T8 | None | 202 | 243 | 5 L | 60 L | A | 25, 40, 48 |
| | 1,1,1-Trichloroethane | 6.1 | UN2831 | III | KEEP AWAY FROM FOOD | N36, T7 | 153 | 203 | 241 | 60 L | 220 L | A | 40 |
| | Trichloroethylene | 6.1 | UN1710 | III | KEEP AWAY FROM FOOD | N36, T1 | 153 | 203 | 241 | 60 L | 220 L | A | 40 |
| | Trichloroisocyanuric acid, dry | 5.1 | UN2468 | II | OXIDIZER | | 152 | 212 | 240 | 5 kg | 25 kg | A | 13 |
| | *Trichloromethyl perchlorate* | Forbidden | | | | | | | | | | | |
| | Trichlorosilane | 4.3 | UN1295 | I | DANGEROUS WHEN WET, FLAMMABLE LIQUID, CORROSIVE | A7, N34, T24, T26 | None | 201 | 244 | Forbidden | Forbidden | D | 21, 28, 40, 49, 100 |
| | Tricresyl phosphate *with more than 3 per cent ortho isomer* | 6.1 | UN2574 | II | POISON | A3, N33, N34, T8 | None | 202 | 243 | 5 L | 60 L | A | |
| | Triethyl phosphite | 3 | UN2323 | III | FLAMMABLE LIQUID | B1, T1 | 150 | 203 | 242 | 60 L | 220 L | A | |
| | Triethylamine | 3 | UN1296 | II | FLAMMABLE LIQUID | T8 | 150 | 202 | 242 | 5 L | 60 L | B | 40 |
| | Triethylenetetramine | 8 | UN2259 | II | CORROSIVE | B2, T8 | 154 | 202 | 242 | 1 L | 30 L | B | 40 |
| | Trifluoroacetic acid | 8 | UN2699 | I | CORROSIVE | A3, A6, A7, B4, N3, N34, T18, T27 | None | 201 | 242 | 0.5 L | 2.5 L | B | 12, 40, 48 |
| | Trifluoroacetylchloride | 2.2 | UN3057 | | NONFLAMMABLE GAS, CORROSIVE | B44 | None | 304 | 314, 315 | Forbidden | 25kg | D | 40 |
| | Trifluorochloroethylene, inhibited, *R1113* | 2.1 | UN1082 | | FLAMMABLE GAS | | 306 | 304 | 314, 315 | Forbidden | 150 kg | B | 40 |
| | Trifluoroethane, compressed, *R143* | 2.1 | UN2035 | | FLAMMABLE GAS | | 306 | 304 | 314, 315 | Forbidden | 150 kg | B | 40 |
| | Trifluoromethane | 2.2 | UN1984 | | NONFLAMMABLE GAS | | 306 | 304 | 314, 315 | 75 kg | 150 kg | A | |
| D | Trifluoromethane and chlorotrifluoromethane mixture *(constant boiling mixture) (R-503).* See Refrigerant gases, n.o.s. | | | | | | | | | | | | |
| | Trifluoromethane, refrigerated, liquid | 2.2 | UN3136 | | NONFLAMMABLE GAS | | 306 | None | 314, 315 | 50 kg | 500 kg | A | |
| | 3-Trifluoromethylaniline | 6.1 | UN2948 | II | POISON | T14 | None | 202 | 243 | 5 L | 60 L | A | 40 |
| | 2-Trifluoromethylaniline | 6.1 | UN2942 | III | KEEP AWAY FROM FOOD | | 153 | 203 | 241 | 60 L | 220 L | A | |
| | *Triformoxime trinitrate* | Forbidden | | | | | | | | | | | |
| | Triisobutylene | 3 | UN2324 | III | FLAMMABLE LIQUID | B1, T7, T30 | 150 | 203 | 242 | 60 L | 220 L | A | |
| | Triisocyanatoisocyanurate of isophoronediisocyanate, solution, *70 per cent, by mass* | 3 | UN2906 | III | FLAMMABLE LIQUID | B1, T1 | 150 | 203 | 242 | 60 L | 220 L | A | |
| | Triisopropyl borate | 3 | UN2616 | III | FLAMMABLE LIQUID | B1, T8, T31 | 150 | 202 | 242 | 60 L | 220 L | A | |
| D | Trimethoxysilane | 6.1 | NA9269 | I | POISON, FLAMMABLE LIQUID | 2, B9, B14, B32, B74, T38, T43, T45 | None | 227 | 244 | Forbidden | Forbidden | E | 40 |
| | Trimethyl borate | 3 | UN2416 | II | FLAMMABLE LIQUID | T14 | 150 | 202 | 242 | 5 L | 60 L | B | |
| | Trimethyl phosphite | 3 | UN2329 | III | FLAMMABLE LIQUID | T1 | 150 | 203 | 242 | 60 L | 220 L | A | |
| | *1,3,5-Trimethyl-2,4,6-trinitrobenzene* | Forbidden | | | | | | | | | | | |
| + | Trimethylacetyl chloride | 8 | UN2438 | II | CORROSIVE, FLAMMABLE LIQUID, POISON | 2, A3, A6, A7, B3, B9, B14, B32, B74, N34, T38, T43, T45 | None | 227 | 244 | 1 L | 30 L | D | 12, 21, 25, 40, 48 |
| | Trimethylamine, anhydrous | 2.1 | UN1083 | | FLAMMABLE GAS | | 306 | 304 | 314, 315 | Forbidden | 150 kg | B | 40 |
| | Trimethylamine, aqueous solutions *not more than 50 per cent trimethylamine by mass* | 3 | UN1297 | I | FLAMMABLE LIQUID | T42 | 150 | 202 | 242 | 5 L | 60 L | D | 40, 41 |
| | | | | II | FLAMMABLE LIQUID | B1, T14 | 150 | 203 | 242 | 60 L | 220 L | B | 40, 41 |
| | 1,3,5-Trimethylbenzene | 3 | UN2325 | III | FLAMMABLE LIQUID | B1, T1 | None | 203 | 242 | 60 L | 220 L | A | |

## §172.101 HAZARDOUS MATERIALS TABLE, Continued

| Symbols | Hazardous materials descriptions and proper shipping names | Hazard class or Division | Identification Numbers | Packing group | Label(s) required (if not excepted) | Special provisions | (8) Packaging authorizations (§173.***) | | | (9) Quantity limitations | | (10) Vessel stowage requirements | |
|---|---|---|---|---|---|---|---|---|---|---|---|---|---|
| | | | | | | | Exceptions | Non-bulk packaging | Bulk packaging | Passenger aircraft or railcar | Cargo aircraft only | Vessel stowages | Other stowage provisions |
| (1) | (2) | (3) | (4) | (5) | (6) | (7) | (8A) | (8B) | (8C) | (9A) | (9B) | (10A) | (10B) |
| | Trimethylchlorosilane | 3 | UN1298 | I | FLAMMABLE LIQUID, CORROSIVE | A3, A7, B77, N34, T14, T26 | None | 201 | 243 | Forbidden | 2.5 L | E | 40 |
| | Trimethylcyclohexylamine | 8 | UN2326 | III | CORROSIVE | T2 | 154 | 203 | 241 | 5 L | 60 L | A | 8 |
| | *Trimethylene glycol diperchlorate* | Forbidden | | | | | | | | | | | |
| | Trimethylhexamethylene diisocyanate | 6.1 | UN2328 | III | KEEP AWAY FROM FOOD | T8 | 153 | 203 | 241 | 60 L | 220 L | B | |
| | Trimethylhexamethylenediamines | 8 | UN2327 | III | CORROSIVE | T7 | 154 | 203 | 241 | 5 L | 60 L | A | 8 |
| | *Trimethylol nitromethane trinitrate* | Forbidden | | | | | | | | | | | |
| | Trinitro-meta-cresol | 1.1D | UN0216 | II | EXPLOSIVE 1.1D | | None | 62 | None | Forbidden | Forbidden | B | 1E, 5E |
| | *2,4,6-Trinitro-1,3-diazobenzene* | Forbidden | | | | | | | | | | | |
| | *2,4,6-Trinitro-1,3,5-triazido benzene (dry)* | Forbidden | | | | | | | | | | | |
| | *Trinitroacetic acid* | Forbidden | | | | | | | | | | | |
| | *Trinitroacetonitrile* | Forbidden | | | | | | | | | | | |
| | *Trinitroamine cobalt* | Forbidden | | | | | | | | | | | |
| | Trinitroaniline or Picramide | 1.1D | UN0153 | II | EXPLOSIVE 1.1D | | None | 62 | None | Forbidden | Forbidden | B | 1E, 5E, 19E |
| | Trinitroanisole | 1.1D | UN0213 | II | EXPLOSIVE 1.1D | | None | 62 | None | Forbidden | Forbidden | B | 1E, 5E, 19E |
| | Trinitrobenzene, *dry or wetted with less than 30 per cent water, by mass* | 1.1D | UN0214 | II | EXPLOSIVE 1.1D | | None | 62 | None | Forbidden | Forbidden | B | 1E, 5E, 19E |
| | Trinitrobenzene, *wetted with not less than 30 per cent water, by mass* | 4.1 | UN1354 | I | FLAMMABLE SOLID | A2, A8, A19, N41 | None | 211 | None | 0.5 kg | 0.5 kg | E | 28, 36 |
| | Trinitrobenzenesulfonic acid | 1.1D | UN0386 | II | EXPLOSIVE 1.1D | | None | 62 | None | Forbidden | Forbidden | E | 1E, 5E, 19E |
| | Trinitrobenzoic acid, *dry or wetted with less than 30 per cent water, by mass* | 1.1D | UN0215 | II | EXPLOSIVE 1.1D | | None | 62 | None | Forbidden | Forbidden | B | 1E, 5E |
| | Trinitrobenzoic acid, *wetted with not less than 30 per cent water, by mass* | 4.1 | UN1355 | I | FLAMMABLE SOLID | A2, A8, A19, N41 | None | 211 | None | 0.5 kg | 0.5 kg | E | 28, 36 |
| | Trinitrochlorobenzene or Picryl chloride | 1.1D | UN0155 | II | EXPLOSIVE 1.1D | | None | 62 | None | Forbidden | Forbidden | B | 1E, 5E |
| | *Trinitroethanol* | Forbidden | | | | | | | | | | | |
| | *Trinitroethylnitrate* | Forbidden | | | | | | | | | | | |
| | Trinitrofluorenone | 1.1D | UN0387 | II | EXPLOSIVE 1.1D | | None | 62 | None | Forbidden | Forbidden | B | 1E, 5E, 19E |
| | *Trinitromethane* | Forbidden | | | | | | | | | | | |
| | *1,3,5-Trinitronaphthalene* | Forbidden | | | | | | | | | | | |
| | Trinitronaphthalene | 1.1D | UN0217 | II | EXPLOSIVE 1.1D | | None | 62 | None | Forbidden | Forbidden | B | 1E, 5E, 19E |
| | Trinitrophenetole | 1.1D | UN0218 | II | EXPLOSIVE 1.1D | | None | 62 | None | Forbidden | Forbidden | B | 1E, 5E, 19E |
| | Trinitrophenol or Picric acid, *dry or wetted with less than 30 per cent water, by mass* | 1.1D | UN0154 | II | EXPLOSIVE 1.1D | | None | 62 | None | Forbidden | Forbidden | B | 1E, 5E |
| | Trinitrophenol, *wetted with not less than 30 per cent water, by mass* | 4.1 | UN1344 | I | FLAMMABLE SOLID | A8, A19, N41 | None | 211 | None | 1 kg | 15 kg | E | 28, 36, 48 |
| | *2,4,6-Trinitrophenyl guanidine (dry)* | Forbidden | | | | | | | | | | | |
| | *2,4,6-Trinitrophenyl nitramine* | Forbidden | | | | | | | | | | | |
| | *2,4,6-Trinitrophenyl trimethylol methyl nitramine trinitrate (dry)* | Forbidden | | | | | | | | | | | |
| | Trinitrophenylmethylnitramine or Tetryl | 1.1D | UN0208 | II | EXPLOSIVE 1.1D | | None | 62 | None | Forbidden | Forbidden | B | 1E, 5E |
| | Trinitroresorcinol or Styphnic acid, *dry or wetted with less than 20 per cent water, or mixture of alcohol and water, by mass* | 1.1D | UN0219 | II | EXPLOSIVE 1.1D | | None | 62 | None | Forbidden | Forbidden | B | 1E, 5E |
| | Trinitroresorcinol, wetted or Styphnic acid, *wetted with not less than 20 per cent water, or mixture of alcohol and water by mass* | 1.1D | UN0394 | II | EXPLOSIVE 1.1D | | None | 62 | None | Forbidden | Forbidden | B | 1E, 5E |
| | *2,4,6-Trinitroso-3-methyl nitraminoanisole* | Forbidden | | | | | | | | | | | |
| | *Trinitrotetramine cobalt nitrate* | Forbidden | | | | | | | | | | | |
| | Trinitrotoluene and Trinitrobenzene mixtures or Trinitrotoluene or TNT and trinitrobenzene mixtures or TNT and hexanitrostilbene mixturesand Hexanitrostilbene mixtures | 1.1D | UN0388 | II | EXPLOSIVE 1.1D | | None | 62 | None | Forbidden | Forbidden | B | 1E, 5E, 19E |

## §172.101 HAZARDOUS MATERIALS TABLE, Continued

| Symbols | Hazardous materials descriptions and proper shipping names | Hazard class or Division | Identification Numbers | Packing group | Label(s) required (if not excepted) | Special provisions | (8) Packaging authorizations (§173.***) | | | (9) Quantity limitations | | (10) Vessel stowage requirements | |
|---|---|---|---|---|---|---|---|---|---|---|---|---|---|
| | | | | | | | Exceptions | Non-bulk packaging | Bulk packaging | Passenger aircraft or railcar | Cargo aircraft only | Vessel stowages | Other stowage provisions |
| (1) | (2) | (3) | (4) | (5) | (6) | (7) | (8A) | (8B) | (8C) | (9A) | (9B) | (10A) | (10B) |
| | Trinitrotoluene mixtures containing Trinitrobenzene and Hexanitrostilbene *or* TNT mixtures containing trinitrobenzene and hexanitrostilbene | 1.1D | UN0389 | II | EXPLOSIVE 1.1D | | None | 62 | None | Forbidden | Forbidden | B | 1E, 5E |
| | Trinitrotoluene or TNT, *dry or wetted with less than 30 per cent water, by mass* | 1.1D | UN0209 | II | EXPLOSIVE 1.1D | | None | 62 | None | Forbidden | Forbidden | B | 1E, 5E |
| | Trinitrotoluene, wetted *with not less than 30 per cent water, by mass* | 4.1 | UN1356 | I | FLAMMABLE SOLID | A2, A8, A19, N41 | None | 211 | None | 0.5 kg | 0.5 kg | E | 28, 36 |
| | Tripropylamine | 3 | UN2260 | III | FLAMMABLE LIQUID, CORROSIVE | B1, T8 | None | 202 | 243 | 1 L | 5 L | A | 40 |
| | Tripropylene | 3 | UN2057 | II | FLAMMABLE LIQUID | T1 | 150 | 202 | 242 | 5 L | 60 L | B | |
| | | | | III | FLAMMABLE LIQUID | B1, T1 | 150 | 203 | 242 | 60 L | 220 L | A | |
| | Tris-(1-aziridinyl)phosphine oxide, solution | 6.1 | UN2501 | II | POISON | T8 | None | 202 | 243 | 5 L | 60 L | A | |
| | *Tris, bis-bifluoroamino diethoxy propane (TVOPA)* | Forbidden | | | | | | | | | | | |
| | Tritonal | 1.1D | UN0390 | II | EXPLOSIVE 1.1D | | None | 62 | None | Forbidden | Forbidden | B | 1E, 5E |
| | Tungsten hexafluoride | 2.3 | UN2196 | | POISON GAS | 3 | None | 338 | None | Forbidden | Forbidden | D | 40 |
| | Turpentine | 3 | UN1299 | III | FLAMMABLE LIQUID | B1, T1 | 150 | 203 | 242 | 60 L | 220 L | A | |
| I | Turpentine substitute | 3 | UN1300 | I | FLAMMABLE LIQUID | T1 | None | 201 | 243 | 1 L | 30 L | B | |
| | | | | II | FLAMMABLE LIQUID | T1 | 150 | 202 | 242 | 5 L | 60 L | B | |
| | | | | III | FLAMMABLE LIQUID | B1, T1 | 150 | 203 | 242 | 60 L | 220 L | A | |
| | Undecane | 3 | UN2330 | III | FLAMMABLE LIQUID | B1, T1 | 150 | 203 | 242 | 60 L | 220 L | A | |
| | Uranium hexafluoride, fissile (*containing more than 1% U-235)* | 7 | UN2977 | | RADIOACTIVE, CORROSIVE | | 453 | 417 | 417 | | | A | |
| | Uranium hexafluoride, *fissile excepted or non-fissile* | 7 | UN2978 | | RADIOACTIVE, CORROSIVE | | 421 | 425 | 425 | | | | |
| | Uranium metal, pyrophoric | 7 | UN2979 | | RADIOACTIVE, SPONTANEOUSLY COMBUSTIBLE | | None | 418 | None | | | D | |
| | Uranyl nitrate hexahydrate solution | 7 | UN2980 | | RADIOACTIVE, CORROSIVE | | 421, 425 | 415, 416, 417 | 415, 416, 417 | | | D | |
| | Uranyl nitrate, solid | 7 | UN2981 | | RADIOACTIVE, OXIDIZER | | None | 419 | None | Forbidden | 15 kg | A | |
| | Urea hydrogen peroxide | 5.1 | UN1511 | III | OXIDIZER | A1, A7, A29 | 152 | 213 | 240 | 25 kg | 100 kg | A | 13 |
| | Urea nitrate, *dry or wetted with less than 20 per cent water, by mass* | 1.1D | UN0220 | II | EXPLOSIVE 1.1D | | None | 62 | None | Forbidden | Forbidden | B | 1E, 5E |
| | Urea nitrate, wetted *with not less than 20 per cent water, by mass* | 4.1 | UN1357 | I | FLAMMABLE SOLID | A8, A19, N41 | None | 211 | None | 1 kg | 15 kg | A | 28 |
| | *Urea peroxide, see* Urea hydrogen peroxide | | | | | | | | | | | | |
| | Valeraldehyde | 3 | UN2058 | II | FLAMMABLE LIQUID | T1 | 150 | 202 | 242 | 5 L | 60 L | B | |
| | *Valeric acid, see* Corrosive liquids, n.o.s. | | | | | | | | | | | | |
| | Valeryl chloride | 8 | UN2502 | II | CORROSIVE | A3, A6, A7, B2, N34, T8 | 154 | 202 | 242 | 1 L | 30 L | C | 23, 40 |
| | Vanadium oxytrichloride | 8 | UN2443 | II | CORROSIVE | A3, A6, A7, B2, B16, N34, T8, T26 | 154 | 202 | 242 | Forbidden | 30 L | C | 40 |
| | Vanadium pentoxide, *nonfused form* | 6.1 | UN2862 | II | POISON | | None | 212 | 242 | 25 kg | 100 kg | A | |
| | Vanadium tetrachloride | 8 | UN2444 | I | CORROSIVE | A3, A6, A7, B4, N34, T8, T26 | None | 201 | 242 | Forbidden | 2.5 L | C | 8, 40 |
| | Vanadium trichloride | 8 | UN2475 | III | CORROSIVE | | 154 | 213 | 240 | 25 kg | 100 kg | A | 40 |
| | Vanadium trioxide, *nonfused form* | 6.1 | UN2860 | II | POISON | | None | 212 | 242 | 25 kg | 100 kg | A | |
| | Vanadyl sulfate | 6.1 | UN2931 | II | POISON | | None | 212 | 242 | 25 kg | 100 kg | A | |
| D | Vehicles, self-propelled *including internal combustion engines or other apparatus containing an internal combustion engine or electric storage battery(see also* Wheel chair, electric) | 9 | None | III | CLASS 9 | | 220 | 220 | None | No limit | No limit | A | |
| | *Very signal cartridge, see* Cartridges, signal | | | | | | | | | | | | |
| | Vinyl acetate, inhibited | 3 | UN1301 | II | FLAMMABLE LIQUID | T8 | 150 | 202 | 242 | 5 L | 60 L | B | |
| | Vinyl bromide, inhibited | 2.1 | UN1085 | | FLAMMABLE GAS | | 306 | 304 | None | Forbidden | 150 kg | B | 40 |
| | Vinyl butyrate, inhibited | 3 | UN2838 | II | FLAMMABLE LIQUID | T7 | 150 | 202 | 242 | 5 L | 60 L | B | |
| D | Vinyl chloride | 2.1 | NA1086 | | FLAMMABLE GAS | 21, B44 | 306 | 304 | 314, 315 | Forbidden | 150 kg | B | |
| | Vinyl chloride, inhibited | 2.1 | UN1086 | | FLAMMABLE GAS | B44 | 306 | 304 | 314, 315 | Forbidden | 150 kg | B | 40 |
| | Vinyl chloroacetate | 6.1 | UN2589 | II | POISON, FLAMMABLE LIQUID | T14 | None | 202 | 243 | 5 L | 60 L | A | |
| | Vinyl ethyl ether, inhibited | 3 | UN1302 | II | FLAMMABLE LIQUID | A3, T14 | None | 201 | 243 | 1 L | 30 L | E | |

## §172.101 HAZARDOUS MATERIALS TABLE, Continued

| Symbols | Hazardous materials descriptions and proper shipping names | Hazard class or Division | Identification Numbers | Packing group | Label(s) required (if not excepted) | Special provisions | (8) Packaging authorizations (§173.***) | | | (9) Quantity limitations | | (10) Vessel stowage requirements | |
|---|---|---|---|---|---|---|---|---|---|---|---|---|---|
| | | | | | | | Exceptions | Non-bulk packaging | Bulk packaging | Passenger aircraft or railcar | Cargo aircraft only | Vessel stowages | Other stowage provisions |
| (1) | (2) | (3) | (4) | (5) | (6) | (7) | (8A) | (8B) | (8C) | (9A) | (9B) | (10A) | (10B) |
| | Vinyl fluoride, inhibited | 2.1 | UN1860 | | FLAMMABLE GAS | B43 | 306 | 304 | 314, 315 | Forbidden | 150 kg | E | 40 |
| | Vinyl isobutyl ether, inhibited | 3 | UN1304 | II | FLAMMABLE LIQUID | T8 | 150 | 202 | 242 | 5 L | 60 L | B | |
| | Vinyl methyl ether, inhibited | 2 1 | UN1087 | | FLAMMABLE GAS | B44 | 306 | 304 | 314, 315 | Forbidden | 150 kg | B | 40 |
| | *Vinyl nitrate polymer* | Forbidden | | | | | | | | | | | |
| | Vinyl toluene, inhibited *mixed isomers* | 3 | UN2618 | III | FLAMMABLE LIQUID | B1, T1 | 150 | 203 | 242 | 60 L | 220 L | A | |
| | Vinylidene chloride, inhibited | 3 | UN1303 | I | FLAMMABLE LIQUID | T23, T29 | 150 | 201 | 243 | 1 L | 30 L | E | 40 |
| | Vinylpyridenes, inhibited | 6 1 | UN3073 | II | POISON, FLAMMABLE LIQUID | T8 | None | 212 | 243 | 5 L | 60 L | B | 40 |
| | Vinyltrichlorosilane | 3 | UN1305 | I | FLAMMABLE LIQUID, CORROSIVE | A3, A7, B6, N34, T14, T26 | None | 201 | 243 | Forbidden | 2 5 L | B | 40 |
| | Warheads, rocket *with burster or expelling charge* | 1.4D | UN0370 | II | EXPLOSIVE 1.4D | | None | 62 | None | Forbidden | 75 kg | A | 3E, 7E, 24E |
| | Warheads, rocket *with burster or expelling charge* | 1.4F | UN0371 | II | EXPLOSIVE 1.4F | | None | 62 | None | Forbidden | Forbidden | E | |
| | Warheads, rocket *with bursting charge* | 1.1D | UN0286 | II | EXPLOSIVE 1.1D | | None | 62 | None | Forbidden | Forbidden | B | 3E, 7E |
| | Warheads, rocket *with bursting charge* | 1.2D | UN0287 | II | EXPLOSIVE 1.2D | | None | 62 | None | Forbidden | Forbidden | B | 3E, 7E |
| | Warheads, rocket *with bursting charge* | 1.1F | UN0369 | II | EXPLOSIVE 1.1F | | None | 62 | None | Forbidden | Forbidden | E | |
| | Warheads, torpedo *with bursting charge* | 1.1D | UN0221 | II | EXPLOSIVE 1.1D | | None | 62 | None | Forbidden | Forbidden | B | 3E, 7E |
| | *Water reactive substances, n.o.s., see* Substances which in contact with water, *etc.* | | | | | | | | | | | | |
| AD | Wheel chair, electric *(spillable or non-spillable type batteries)* | 9 | None | III | CLASS 9 | | 222 | 222 | None | No limit | No limit | A | |
| | *White acid, see* Hydrofluoric acid mixtures | | | | | | | | | | | | |
| | White asbestos *(chrysotile, actinolite, anthophyllite, tremolite)* | 9 | UN2590 | III | CLASS 9 | | 155 | 216 | 240 | 200 kg | 200 kg | A | 34, 40 |
| | Wood preservatives, liquid | 3 | UN1306 | II | FLAMMABLE LIQUID | T7, T30 | 150 | 202 | 242 | 5 L | 60 L | B | 40 |
| | | | | III | FLAMMABLE LIQUID | B1, T7, T30 | 150 | 203 | 242 | 60 L | 220 L | A | 40 |
| | Xenon | 2.2 | UN2036 | | NONFLAMMABLE GAS | | 306 | 302 | None | 75 kg | 150 kg | A | |
| | Xenon, refrigerated liquid *(cryogenic liquids)* | 2.2 | UN2591 | | NONFLAMMABLE GAS | | 320 | None | None | 50 kg | 500 kg | B | |
| | Xylenes | 3 | UN1307 | II | FLAMMABLE LIQUID | T1 | 150 | 202 | 242 | 5 L | 60 L | B | |
| | | | | III | FLAMMABLE LIQUID | B1, T1 | 150 | 203 | 242 | 60 L | 220 L | A | |
| | Xylenols | 6.1 | UN2261 | II | POISON | T8 | None | 212 | 243 | 25 kg | 100 kg | A | |
| | Xylidines, solid *or* solution | 6.1 | UN1711 | II | POISON | T14 | None | 202 | 243 | 5 L | 60 L | A | |
| | Xylyl bromide | 6.1 | UN1701 | II | POISON | A3, A6, A7, N33 | None | 340 | None | Forbidden | 60 L | D | 40 |
| | *p-Xylyl diazide* | Forbidden | | | | | | | | | | | |
| | Zinc ammonium nitrite | 5.1 | UN1512 | II | OXIDIZER | | None | 212 | 242 | 5 kg | 25 kg | E | |
| | Zinc arsenate *or* Zinc arsenite *or* Zinc arsenate and Zinc arsenite mixtures. | 6.1 | UN1712 | II | POISON | | None | 212 | 242 | 25 kg | 100 kg | A | |
| | Zinc ashes | 4 3 | UN1435 | III | DANGEROUS WHEN WET | A1, A19 | None | 213 | 241 | 25 kg | 100 kg | A | |
| | Zinc bisulfite solution *see* Bisulfites, inorganic aqueous solutions, n.o.s. | | | | | | | | | | | | |
| | Zinc bromate | 5.1 | UN2469 | III | OXIDIZER | A1, A29 | 152 | 213 | 240 | 25 kg | 100 kg | A | 56, 58, 106 |
| | Zinc chlorate | 5.1 | UN1513 | II | OXIDIZER | A9, N34 | 152 | 212 | 242 | 5 kg | 25 kg | A | 56, 58, 106 |
| | Zinc chloride, anhydrous | 8 | UN2331 | III | CORROSIVE | | None | 213 | 240 | 25 kg | 100 kg | A | |
| | Zinc chloride, solution | 8 | UN1840 | III | CORROSIVE | T7 | 154 | 203 | 241 | 5 L | 60 L | A | |
| | Zinc cyanide | 6.1 | UN1713 | I | POISON | | None | 211 | 242 | 5 kg | 50 kg | A | 26 |
| | Zinc dithionite *or* Zinc hydrosulfite | 9 | UN1931 | III | None | | 155 | 204 | 240 | 100 kg | 200 kg | A | 13, 49, 102 |
| | *Zinc ethyl, see* Diethylzinc | | | | | | | | | | | | |
| | Zinc fluorosilicate | 6.1 | UN2855 | III | KEEP AWAY FROM FOOD | | 153 | 213 | 240 | 100 kg | 200 kg | A | 26 |
| | Zinc hydrosulfite, *see* Zinc dithionite | | | | | | | | | | | | |
| | *Zinc muriate solution, see* Zinc chloride, solution | | | | | | | | | | | | |

| Symbols | Hazardous materials descriptions and proper shipping names | Hazard class or Division | Identification Numbers | Packing group | Label(s) required (if not excepted) | Special provisions | (8) Packaging authorizations (§173.***) | | | (9) Quantity limitations | | (10) Vessel stowage requirements | |
|---|---|---|---|---|---|---|---|---|---|---|---|---|---|
| | | | | | | | Exceptions | Non-bulk packaging | Bulk packaging | Passenger aircraft or railcar | Cargo aircraft only | Vessel stowages | Other stowage provisions |
| (1) | (2) | (3) | (4) | (5) | (6) | (7) | (8A) | (8B) | (8C) | (9A) | (9B) | (10A) | (10B) |
| | Zinc nitrate | 5.1 | UN1514 | II | OXIDIZER | | 152 | 212 | 240 | 5 kg | 25 kg | A | |
| | Zinc permanganate | 5.1 | UN1515 | II | OXIDIZER | | 152 | 212 | 242 | 5 kg | 25 kg | D | 56, 58, 69, 106, 107 |
| | Zinc peroxide | 5.1 | UN1516 | II | OXIDIZER | | 152 | 212 | 242 | 5 kg | 25 kg | A | 13, 75, 106 |
| | Zinc phosphide | 4.3 | UN1714 | I | DANGEROUS WHEN WET, POISON | A19, N40 | None | 211 | None | Forbidden | 15 kg | E | 40, 85 |
| | Zinc powder *or* Zinc dust | 4.3 | UN1436 | I | DANGEROUS WHEN WET, SPONTANEOUSLY COMBUSTIBLE | A19, N40 | None | 211 | 242 | Forbidden | 15kg | A | |
| | | | | II | DANGEROUS WHEN WET, SPONTANEOUSLY COMBUSTIBLE | A19 | None | 212 | 242 | 15 kg | 50 kg | A | |
| | | | | III | DANGEROUS WHEN WET, SPONTANEOUSLY COMBUSTIBLE | | None | 213 | 242 | 25kg | 100kg | A | |
| | Zinc resinate | 4.1 | UN2714 | III | FLAMMABLE SOLID | A1 | 151 | 213 | 240 | 25 kg | 100 kg | A | |
| | Zinc selenate, *see* Selenates *or* Selenites | | | | | | | | | | | | |
| | Zinc selenite, *see* Selenates *or* Selenites | | | | | | | | | | | | |
| | *Zinc silicofluoride, see* Zinc fluorosilicate | | | | | | | | | | | | |
| | Zirconium, dry, *coiled wire, finished metal sheets, strip (thinner than 254 microns but not thinner than 18 microns)* | 4.1 | UN2858 | III | FLAMMABLE SOLID | A1 | 151 | 213 | 240 | 25 kg | 100 kg | A | |
| | Zirconium, dry, *finished sheets, strip or coiled wire* | 4.2 | UN2009 | III | SPONTANEOUSLY COMBUSTIBLE | A1, A19 | None | 213 | 240 | 25 kg | 100 kg | D | |
| | Zirconium hydride | 4.1 | UN1437 | II | FLAMMABLE SOLID | A19, A20, N34 | None | 212 | 240 | 15 kg | 50 kg | E | |
| | Zirconium nitrate | 5.1 | UN2728 | III | OXIDIZER | A1, A29 | 152 | 213 | 240 | 25 kg | 100 kg | A | |
| | Zirconium picramate, *dry or wetted with less than 20 per cent water, by mass* | 1.3C | UN0236 | II | EXPLOSIVE 1.3C | | None | 62 | None | Forbidden | Forbidden | B | 1E, 5E |
| | Zirconium picramate, wetted *with not less than 20 per cent water, by mass* | 4.1 | UN1517 | I | FLAMMABLE SOLID | N41 | None | 211 | None | 1 kg | 15 kg | D | 28, 36 |
| | Zirconium powder, dry | 4.2 | UN2008 | I | SPONTANEOUSLY COMBUSTIBLE | | None | 211 | 242 | Forbidden | Forbidden | D | |
| | | | | II | SPONTANEOUSLY COMBUSTIBLE | A19, A20, N5, N34 | None | 212 | 241 | 15 kg | 50 kg | D | |
| | | | | III | SPONTANEOUSLY COMBUSTIBLE | | None | 213 | 241 | 25kg | 100kg | D | |
| | Zirconium powder, wetted *with not less than 25 per cent water (a visible excess of water must be present) (a) mechanically produced, particle size less than 53 microns; (b) chemically produced, particle size less than 840 microns* | 4.1 | UN1358 | II | FLAMMABLE SOLID | A19, A20, N34 | None | 212 | 241 | 15 kg | 50 kg | E | |
| | Zirconium scrap | 4.2 | UN1932 | III | SPONTANEOUSLY COMBUSTIBLE | N34 | None | 213 | 240 | Forbidden | Forbidden | D | |
| D | Zirconium sulfate | 8 | NA9163 | III | CORROSIVE | N34 | None | 213 | 240 | 50 kg | No limit | A | |
| | Zirconium suspended in a liquid | 3 | UN1308 | II | FLAMMABLE LIQUID | | None | 202 | 242 | Forbidden | 60 L | B | |
| | Zirconium tetrachloride | 8 | UN2503 | III | CORROSIVE | | 154 | 213 | 240 | 25 kg | 100 kg | A | |

## Appendix A to §172.101 - List of Hazardous Substances and Reportable Quantities

1. This appendix lists materials and their corresponding reportable quantities (RQs) which are listed or designated as "hazardous substances" under section 101(14) of the Comprehensive Environmental Response, Compensation, and Liability Act (CERCLA; Pub. L. 96-510). This appendix is divided into 2 tables which are entitled "Table 1—Hazardous Substances Other Than Radionuclides" and "Table 2—Radionuclides". A material in this appendix is regulated as a hazardous material and a hazardous substance under this subchapter if it meets the definition of a hazardous substance in §171.8 of this subchapter.

2. The procedure for selecting a proper shipping name for a hazardous substance is set forth in §172.101(c)(8).

3. Column 1 of Table 1, entitled *"Hazardous substance"*, contains the names of those elements and compounds which are hazardous substances. Following the listing of elements and compounds is a listing of waste streams. These waste streams appear on the list in numerical sequence and are referenced by the appropriate "F" or "K" numbers. Column 2 of Table 1, entitled *"Synonyms"*, contains the names of synonyms for certain elements and compounds listed in Column 1. No synonyms are listed for waste streams. Synonyms are useful in identifying hazardous substances and in identifying proper shipping names. Column 3 of Table 1, entitled *"Reportable quantity (RQ)"*, contains the reportable quantity (RQ), in pounds and kilograms, for each hazardous substance listed in Column 1 of Table 1.

4. A series of notes are used throughout Table 1 and Table 2 to provide additional information concerning certain hazardous substances. These notes are explained at the end of each Table.

5. Table 2 lists radionuclides which are hazardous substances and their corresponding RQs. The RQs in Table 2 for radionuclides are expressed in units of curies and terabecquerels, whereas those in Table 1 are expressed in units of pounds. If a material is listed in both Table 1 and Table 2, the lowest RQ shall apply. Radionuclides are listed in alphabetical order. The RQs for radionuclides are given in the radiological unit of measure of curie, abbreviated "Ci", followed, in parentheses, by an equivalent unit measured in terabecquerels, abbreviated "TBq".

6. For mixtures of radionuclides, the following determinations shall be used in determining if a package contains an RQ of a hazardous substance: (i) If the identity and quantity (in curies or terabecquerels) of each radionuclide in a mixture or solution is known, the ratio between the quantity per package (in curies or terabecquerels) and the RQ for the radionuclide must be determined for each radionuclide. A package contains an RQ of a hazardous substance when the sum of the ratios for the radionuclides in the mixture or solution is equal to or greater than one; (ii) if the identity of each radionuclide in a mixture or solution is known but the quantity per package (in curies or terabecquerels) of one or more of the radionuclides is unknown, an RQ of a hazardous substance is present in a package when the total quantity (in curies or terabecquerels) of the mixture or solution is equal to or greater than the lowest RQ of any individual radionuclide in the mixture or solution; and (iii) if the identity of one or more radionuclides in a mixture or solution is unknown (or if the identity of a radionuclide by itself is unknown), an RQ of a hazardous substance is present when the total quantity (in curies or terabecquerels) in a package is equal to or greater than either one curie or the lowest RQ of any known individual radionuclide in the mixture or solution, whichever is lower.

**Table 1—Hazardous Substances Other Than Radionuclides**

| Hazardous Substances | Synonyms | Reportable Quantity (RQ) Pounds (Kilograms) |
|---|---|---|
| Acenaphthene | | 100 (45.4) |
| Acenaphthylene | | 5000 (2270) |
| Acetaldehyde* | Ethanal | 1000 (454) |
| Acetaldehyde, chloro- | Chloroacetaldehyde | 1000 (454) |
| Acetaldehyde, trichloro- | Chloral | 5000 (2270) |
| Acetamide, N-(aminothioxomethyl)- | 1-Acetyl-2-thiourea | 1000 (454) |
| Acetamide, N-(4-ethoxyphenyl)- | Phenacetin | 100 (45.4) |
| Acetamide, N-fluoren-2-yl- | 2-Acetylaminofluorene | 1 (0 454) |
| Acetamide, 2-fluoro- | Fluoroacetamide | 100 (45.4) |
| Acetic acid* | | 5000 (2270) |
| Acetic acid (2,4-dichlorophenoxy)- | 2,4-D, salts and esters<br>2,4-D acid | 100 (45.4) |
| Acetic acid, ethyl ester | Ethyl acetate* | 5000 (2270) |
| Acetic acid, fluoro-, sodium salt | Fluoroacetic acid, sodium salt | 10 (4.54) |
| Acetic acid, lead (2+)salt | Lead acetate | 5000 (2270) |
| Acetic acid, thallium(I+) salt | Thallium(I) acetate | 100 (45.4) |
| Acetic acid, (2,4,5-trichlorophenoxy) | 2,4,5-T*<br>2,4,5-T acid | 1000 (454) |
| Acetic anhydride* | | 5000 (2270) |
| Acetone* | 2-Propanone | 5000 (2270) |
| Acetone cyanohydrin* | Propanenitrile, 2-hydroxy-2-methyl-<br>2-Methyllactonitrile | 10 (4.54) |
| Acetonitrile* | Ethanenitrile | 5000 (2270) |
| Acetophenone | Ethanone, 1-phenyl- | 5000 (2270) |
| 2-Acetylaminofluorene | Acetamide, N-fluoren-2-yl- | 1 (0.454) |
| Acetyl bromide* | | 5000 (2270) |
| Acetyl chloride* | Ethanoyl chloride | 5000 (2270) |

## Table 1—Hazardous Substances Other Than Radionuclides, Continued

| Hazardous Substances | Synonyms | Reportable Quantity (RQ) Pounds (Kilograms) |
|---|---|---|
| 1-Acetyl-2-thiourea | Acetamide, N-(aminothioxomethyl)- | 1000 (454) |
| Acrolein* | 2-Propenal | 1 (0.454) |
| Acrylamide | 2-Propenamide | 5000 (2270) |
| Acrylic acid* | 2-Propenoic acid | 5000 (2270) |
| Acrylonitrile* | 2-Propenenitrile | 100 (45.4) |
| Adipic acid | | 5000 (2270) |
| Aldicarb | Propanal, 2-methyl-2-(methylthio)- | 1 (0.454) |
| | O-((methyl-amino) carbonyl)oxime | |
| Aldrin* | 1,2,3,4,10-10-Hexachloro-1,4,4a,5,8,8a- hexahydro- | 1 (0.454) |
| | 1,4,5,8-endo, exodimethanonaphthalene | |
| | 1,4,5,8-Dimethanonaphthalene, 1,2,3,4,10,10-hexachloro- | |
| | 1,4,4a,5,8,8a- hexahydro-,(1alpha,4alpha,4abeta,5alpha,8a.. | |
| Allyl alcohol* | 2-Propen-1-ol | 100 (45.4) |
| Allyl chloride* | | 1000 (454) |
| Aluminum phosphide* | | 100 (45.4) |
| Aluminum sulfate* | | 5000 (2270) |
| 5-(Aminomethyl)-3-isoxazolol | 3(2H)-isoxazolone, 5-(aminomethyl)- | 1000 (454) |
| | Muscimol | |
| 4-Aminopyridine | 4-Pyridinamine | 1000 (454) |
| Amitrole | 1H-1,2,4-Triazol-3-amine | 10 (4.54) |
| Ammonia* | | 100 (45.4) |
| Ammonium acetate | | 5000 (2270) |
| Ammonium benzoate | | 5000 (2270) |
| Ammonium bicarbonate | | 5000 (2270) |
| Ammonium bichromate | Ammonium dichromate @ | 10 (4.54) |
| Ammonium bifluoride* | | 100 (45.4) |
| Ammonium bisulfite* | | 5000 (2270) |
| Ammonium carbamate* | | 5000 (2270) |
| Ammonium carbonate* | | 5000 (2270) |
| Ammonium chloride | | 5000 (2270) |
| Ammonium chromate | | 10 (4.54) |
| Ammonium citrate, dibasic | | 5000 (2270) |
| Ammonium dichromate @ | Ammonium bichromate | 10 (4.54) |
| Ammonium fluoborate* | | 5000 (2270) |
| Ammonium fluoride* | | 100 (45.4) |
| Ammonium hydroxide* | | 1000 (454) |
| Ammonium oxalate* | | 5000 (2270) |
| Ammonium picrate* | Phenol, 2,4,6-trinitro-, ammonium salt | 10 (4.54) |
| Ammonium silicofluoride* | | 1000 (454) |
| Ammonium sulfamate | | 5000 (2270) |
| Ammonium sulfide* | | 100 (45.4) |
| Ammonium sulfite | | 5000 (2270) |
| Ammonium tartrate | | 5000 (2270) |
| Ammonium thiocyanate | | 5000 (2270) |
| Ammonium vanadate | Vanadic acid, ammonium salt | 1000 (454) |
| Amyl acetate* | | 5000 (2270) |
| iso-Amyl acetate | | |
| sec-Amyl acetate | | |
| tert-Amyl acetate | | |
| Aniline* | Benzenamine | 5000 (2270) |
| Anthracene | | 5000 (2270) |
| Antimony[c] | | 5000 (2270) |
| Antimony pentachloride* | | 1000 (454) |
| Antimony potassium tartrate* | | 100 (45.4) |
| Antimony tribromide* | | 1000 (454) |
| Antimony trichloride* | | 1000 (454) |
| Antimony trifluoride* | | 1000 (454) |
| Antimony trioxide | | 1000 (454) |
| Argentate(1-), bis(cyano-C)-, potassium | Potassium silver cyanide | 1 (0.454) |
| Aroclor 1016 | POLYCHLORINATED BIPHENYLS (PCBs) | 1 (0.454) |
| Aroclor 1221 | POLYCHLORINATED BIPHENYLS (PCBs) | 1 (0.454) |
| Aroclor 1232 | POLYCHLORINATED BIPHENYLS (PCBs) | 1 (0.454) |
| Aroclor 1242 | POLYCHLORINATED BIPHENYLS (PCBs) | 1 (0.454) |
| Aroclor 1248 | POLYCHLORINATED BIPHENYLS (PCBs) | 1 (0.454) |
| Aroclor 1254 | POLYCHLORINATED BIPHENYLS (PCBs) | 1 (0.454) |
| Aroclor 1260 | POLYCHLORINATED BIPHENYLS (PCBs) | 1 (0.454) |
| Arsenic*[c] | | 1 (0.454) |
| Arsenic acid* | Arsenic acid H3AsO4 | 1 (0.454) |
| Arsenic acid H3AsO4 | Arsenic acid* | 1 (0.454) |
| Arsenic disulfide* | | 1 (0.454) |

## Table 1—Hazardous Substances Other Than Radionuclides, Continued

| Hazardous Substances | Synonyms | Reportable Quantity (RQ) Pounds (Kilograms) |
|---|---|---|
| Arsenic oxide As203 | Arsenic trioxide* | 1 (0.454) |
| Arsenic oxide As205 | Arsenic pentoxide* | 1 (0.454) |
| Arsenic pentoxide* | Arsenic oxide As205 | 1 (0.454) |
| Arsenic trichloride* | | 1 (0.454) |
| Arsenic trioxide* | Arsenic oxide As203 | 1 (0.454) |
| Arsenic trisulfide* | | 1 (0.454) |
| Arsine, diethyl- | Diethylarsine | 1 (0.454) |
| Arsinic acid, dimethyl- | Cacodylic acid | 1 (0.454) |
| Arsonous dichloride, phenyl- | Dichlorophenylarsine | 1 (0.454) |
| | Phenyl dichloroarsine @ | |
| Asbestos* CC | | 1 (0.454) |
| Auramine | Benzenamine, 4,4'-carbonimidoylbis(N,N- dimethyl- | 100 (45.4) |
| Azaserine | L-Serine, diazoacetate (ester) | 1 (0.454) |
| Azinphos methyl @ | Guthion* | 1 (0.454) |
| Aziridine | Ethylenimine- | 1 (0.454) |
| | Ethylene imine@ | |
| Aziridine, 2-methyl- | 1,2-Propylenimine | 1 (0.454) |
| Azirino(2',3',3',4)pyrrolo(1,2-a)indole-4,7-dione, 6-amino-8-[((aminocarbonyl)oxy)methyl]-1,1a,2,8,8a,8b-hexahydro-8a-methoxy-5-methyl-, [1aS-(aalpha,8beta,8aalpha,8balpha)]-, | Mitomycin C | 10 (4.54) |
| Barium cyanide* | | 10 (4.54) |
| Benz[j]aceanthrylene, 1,2-dihydro-3-methyl- | 3-Methylcholanthrene | 10 (4.54) |
| Benz[c]acridine | 3,4-Benzacridine | 100 (45.4) |
| 3,4-Benzacridine | Benz[c]acridine | 100 (45.4) |
| Benzal chloride | Benzene, dichloromethyl- | 5000 (2270) |
| Benzamide,3,5-dichloro-N-(1,1-dimethyl-2-propynyl) | Pronamide | 5000 (2270) |
| Benz[a]anthracene | 1,2-Benzanthracene | 10 (4.54) |
| | Benzo[a]anthracene | |
| 1,2-Benzanthracene | Benz[a]anthracene | 10 (4.54) |
| | Benzo[a]anthracene | |
| Benz[a]anthracene, 7,12-dimethyl- | 7,12-Dimethylbenz[a]anthracene | 1 (0.454) |
| Benzenamine | Aniline* | 5000 (2270) |
| Benzenamine, 4,4'-carbonimidoylbis(N,N-dimethyl- | Auramine | 100 (45.4) |
| Benzenamine, 4-chloro- | p-Chloroaniline | 1000 (454) |
| Benzenamine, 4-chloro-2-methyl-,hydrochloride | 4-Chloro-o-toluidine, hydrochloride | 100 (45.4) |
| Benzenamine, N,N-dimethyl-4-(phenylazo)- | p-Dimethylaminoazobenzene | 10 (4.54) |
| Benzenamine, 2-methyl- | o-Toluidine | 100 (45.4) |
| Benzenamine, 4-methyl- | p-Toluidine | 100 (45.4) |
| Benzenamine, 4,4'-methylenebis(2-chloro- | 4,4'-Methylenebis(2-chloroaniline) | 10 (4.54) |
| Benzenamine,2-methyl-, hydrochloride | o-Toluidine hydrochloride | 100 (45.4) |
| Benzenamine,2-methyl-5-nitro- | 5-Nitro-o-toluidine | 100 (45.4) |
| Benzenamine,4-nitro- | p-Nitroaniline | 5000 (2270) |
| Benzene* | | 10 (4.54) |
| Benzene, 1-bromo-4-phenoxy- | 4-Bromophenyl phenyl ether | 100 (45.4) |
| Benzene, chloro- | Chlorobenzene* | 100 (45.4) |
| Benzene, chloromethyl- | Benzyl chloride* | 100 (45.4) |
| Benzene,1,2-dichloro- | o-Dichlorobenzene* | 100 (45.4) |
| | 1,2-Dichlorobenzene | |
| Benzene, 1,3-dichloro- | m-Dichlorobenzene | 100 (45.4) |
| | 1,3-Dichlorobenzene | |
| Benzene, 1,4-dichloro- | p-Dichlorobenzene* | 100 (45.4) |
| | 1,4-Dichlorobenzene | |
| Benzene, 1,1'-(2,2-dichloroethylidene)bis[4-chloro | DDD | 1 (0.454) |
| | TDE* | |
| | 4,4'-DDD | |
| Benzene, dichloromethyl- | Benzal chloride | 5000 (2270) |
| Benzene, 1,3-diisocyanatomethyl | Toluene diisocyanate* | 100 (45.4) |
| Benzene, dimethyl | Xylene* (mixed) | 1000 (454) |
| m-Benzene,dimethyl | m-Xylene | |
| o-Benzene,dimethyl | o-Xylene | |
| p-Benzene,dimethyl | p-Xylene | |
| Benzene, hexachloro- | Hexachlorobenzene | 10 (4.54) |
| Benzene, hexahydro- | Cyclohexane* | 1000 (454) |
| Benzene, hydroxy- | Phenol* | 1000 (454) |
| Benzene, methyl- | Toluene* | 1000 (454) |
| Benzene, 1-methyl-2,4-dinitro- | 2,4-Dinitrotoluene | 10 (4.54) |
| Benzene, 2-methyl-1,3-dinitro- | 2,6-Dinitrotoluene | 100 (45.4) |
| Benzene, 1-methylethyl- | Cumene | 5000 (2270) |
| Benzene, nitro- | Nitrobenzene* | 1000 (454) |
| Benzene, pentachloro- | Pentachlorobenzene | 10 (4.54) |
| Benzene, pentachloronitro- | Pentachloronitrobenzene (PCNB) | 100 (45.4) |
| Benzene, 1,2,4,5-tetrachloro- | 1,2,4,5-Tetrachlorobenzene | 5000 (2270) |
| Benzene, 1,1'-(2,2,2-trichloroethylidene)bis [4-chloro- | DDT* | 1 (0.454) |
| | 4,4'-DDT | |
| Benzene, 1,1'-(2,2,2-trichloroethylidene)bis [4-methoxy)- | Methoxychlor | 1 (0.454) |

## Table 1—Hazardous Substances Other Than Radionuclides, Continued

| Hazardous Substances | Synonyms | Reportable Quantity (RQ) Pounds (Kilograms) |
|---|---|---|
| Benzene, (trichloromethyl) | Benzotrichloride | 10 (4.54) |
| Benzene, 1,3,5-trinitro- | 1,3,5-Trinitrobenzene | 10 (4.54) |
| Benzeneacetic acid, 4-chloro-alpha-(4-chlorophenyl)-alpha-hydroxy-, ethyl ester | Chlorobenzilate | 10 (4.54) |
| Benzenebutanoic acid, 4-[bis(2-chloroethyl)amino]- | Chlorambucil | 10 (4.54) |
| Benzenediamine, ar-methyl- | Toluenediamine* | 10 (4.54) |
| 1,2-Benzenedicarboxylic acid,[bis(2-ethylhexyl)] ester | Bis(2-ethylhexyl)phthalate- | 100 (45.4) |
| | Diethylhexyl phthalate | |
| 1,2-Benzenedicarboxylic acid,dibutyl ester | n-Butyl phthalate | 10 (4.54) |
| | Dibutyl phthalate | |
| | Di-n-butyl phthalate | |
| 1,2-Benzenedicarboxylic acid, diethyl ester | Diethyl phthalate | 1000 (454) |
| 1,2-Benzenedicarboxylic acid, dimethyl ester | Dimethyl phthalate | 5000 (2270) |
| 1,2-Benzenedicarboxylic acid, dioctyl ester | Di-n-octyl phthalate | 5000 (2270) |
| 1,3-Benzenediol | Resorcinol | 5000 (2270) |
| 1,2-Benzenediol,4-[1- hydroxy-2-(methylamino) ethyl]- | Epinephrine | 1000 (454) |
| Benzeneethanamine, alpha,alpha-dimethyl- | alpha,alpha-Dimethylphenethylamine | 5000 (2270) |
| Benzeneethanamine, alpha,alpha-dimethyl- | alpha,alpha-Dimethylphenethylamine | 5000 (2270) |
| Benzenesulfonic acid chloride | Benzenesulfonyl chloride | 100 (45.4) |
| Benzenesulfonyl chloride | Benzenesulfonic acid chloride | 100 (45.4) |
| Benzenethiol | Phenyl mercaptan @ | 100 (45.4) |
| | Thiophenol | |
| Benzidine* | (1,1'-Biphenyl)-4,4' diamine | 1 (0.454) |
| 1,2-Benzisothiazol-3(2H)-one,1,1-dioxide | Saccharin and salts | 100 (45.4) |
| Benzo[a]anthracene | Benz[a]anthracene | 10 (4.54) |
| | 1,2-Benzanthracene | |
| 1,3-Benzodioxole, 5-(2-propenyl)- | Safrole | 100 (45.4) |
| 1,3-Benzodioxole, 5-(1-propenyl)- | Isosafrole | 100 (45.4) |
| 1,3-Benzodioxole, 5-propyl- | Dihydrosafrole | 10 (4.54) |
| Benzo[b]fluoranthene | | 1 (0.454) |
| Benzo[k]fluoranthene | | 5000 (2270) |
| Benzo[j,k]fluorene | Fluoranthene | 100 (45.4) |
| Benzoic acid | | 5000 (2270) |
| Benzonitrile* | | 5000 (2270) |
| Benzo[g,h,i]perylene | | 5000 (2270) |
| 2H-1-Benzopyran-2-one, 4-hydroxy-3-(3-oxo-1-phenyl-butyl) -, & salts, when present at concentrations greater than 0.3% | Warfarin, & salts, when present at concentrations greater than 0.3% | 100 (45.4) |
| Benzo[a]pyrene | 3,4-Benzopyrene | 1 (0.454) |
| 3,4-Benzopyrene | Benzo[a]pyrene | 1 (0.454) |
| p-Benzoquinone | 2,5-Cyclohexadiene-1,4-dione | 10 (4.54) |
| Benzo[rst]pentaphene | Dibenz[a,i]pyrene | 10 (4.54) |
| Benzotrichloride | Benzene, (trichloromethyl) | 10 (4.54) |
| Benzoyl chloride* | | 1000 (454) |
| 1,2-Benzphenanthrene | Chrysene | 100 (45.4) |
| Benzyl chloride* | Benzene, chloromethyl- | 100 (45.4) |
| Beryllium ¢ | Beryllium dust ¢ | 10 (4.54) |
| Beryllium chloride* | | 1 (0.454) |
| Beryllium dust ¢ | Beryllium ¢ | 10 (4.54) |
| Beryllium fluoride* | | 1 (0.454) |
| Beryllium nitrate* | | 1 (0.454) |
| alpha - BHC | | 10 (4.54) |
| beta - BHC | | 1 (0.454) |
| delta - BHC | | 1 (0.454) |
| gamma - BHC | Hexachlorocyclohexane (gamma isomer) | 1 (0.454) |
| | Lindane* | |
| | Cyclohexane,1,2,3,4,5,6-hexachloro-, (1alpha,2alpha,3beta,4alpha,5alpha,6beta)- | |
| 2,2'-Bioxirane | 1,2:3,4-Diepoxybutane | 10 (4.54) |
| (1,1'-Biphenyl)-4,4' diamine | Benzidine* | 1 (0.454) |
| (1,1'-Biphenyl)-4,4' diamine 3,3'dichloro- | 3,3'-Dichlorobenzidine | 1 (0.454) |
| (1,1'-Biphenyl)-4,4' diamine 3,3'-dimethoxy- | 3,3'-Dimethoxybenzidine | 10 (4.54) |
| (1,1'-Biphenyl)-4,4' diamine 3,3'-dimethyl- | 3,3'-Dimethylbenzidine | 10 (4.54) |
| Bis(2-chloroethoxy) methane | Ethane, 1,1'-[methylenebis (oxy)]bis(2-chloro- | 1000 (454) |
| | Dichloromethoxy ethane | |
| Bis(2-chloroethyl) ether | Dichloroethyl ether | 10 (4.54) |
| | Ethane, 1,1'-oxybis[2-chloro- | |
| Bis(2-ethylhexyl)phthalate | 1,2-Benzenedicarboxylic acid, [bis(2-ethylhexyl)] ester | 100 (45.4) |
| | Diethylhexyl phthalate | |
| Bromoacetone* | 2-Propanone, 1-bromo- | 1000 (454) |
| Bromoform | Methane, tribromo- | 100 (45.4) |

## Table 1—Hazardous Substances Other Than Radionuclides, Continued

| Hazardous Substances | Synonyms | Reportable Quantity (RQ) Pounds (Kilograms) |
|---|---|---|
| 4-Bromophenyl phenyl ether | Benzene, 1-bromo-4- phenoxy- | 100 (45.4) |
| Brucine | Strychnidin-10-one, 2,3-dimethoxy- | 100 (45.4) |
| 1,3-Butadiene, 1,1,2,3,4,4-hexachloro- | Hexachlorobutadiene* | 1 (0.454) |
| 1-Butanamine, N-butyl-N-nitroso- | N-Nitrosodi-n-butylamine | 10 (4.54) |
| 1-Butanol | n-Butyl alcohol* | 5000 (2270) |
| 2-Butanone | Ethyl methyl ketone @ | 5000 (2270) |
| | Methyl ethyl ketone(MEK)* | |
| 2-Butanone, 3,3-dimethyl-1-(methylthio)-,O-((methylamino) carbonyl)oxime | Thiofanox | 100 (45.4) |
| 2-Butanone peroxide | Methyl ethyl ketone peroxide* | 10 (4.54) |
| 2-Butenal | Crotonaldehyde* | 100 (45.4) |
| 2-Butene, 1,4-dichloro- | 1,4-Dichloro-2-butene | 1 (0.454) |
| 2-Butenoic acid, 2-methyl-,7((2,3-dihydroxy-2-(1-methoxyethyl)-3-methyl-1-oxobutoxy)methyl)-2,3,5,7a-tetrahydro-1H-pyrrolizin-1-ylester,(1S-(1alpha (Z),7(2S*,3R*),7aalpha))-, | Lasiocarpine | 10 (4.54) |
| Butyl acetate* | | 5000 (2270) |
| iso- Butyl acetate | | |
| sec- Butyl acetate | | |
| tert- Butyl acetate | | |
| n-Butyl alcohol* | 1-Butanol | 5000 (2270) |
| Butylamine* | | 1000 (454) |
| iso- Butylamine | | |
| sec- Butylamine | | |
| tert- Butylamine | | |
| Butyl benzyl phthalate | | 100 (45.4) |
| n-Butyl phthalate | Dibutyl phthalate | 10 (4.54) |
| | Di-n-butyl phthalate | |
| | 1,2-Benzenedicarboxylic acid,dibutyl ether | |
| Butyric acid* | | 5000 (2270) |
| iso-Butyric acid | | |
| Cacodylic acid | Arsenic acid, dimethyl- | 1 (0.454) |
| Cadmium ¢ | | 10 (4.54) |
| Cadmium acetate | | 10 (4.54) |
| Cadmium bromide | | 10 (4.54) |
| Cadmium chloride | | 10 (4.54) |
| Calcium arsenate* | | 1 (0.454) |
| Calcium arsenite* | | 1 (0.454) |
| Calcium carbide* | | 10 (4.54) |
| Calcium chromate | Chromic acid H2CrO4, calcium salt | 10 (4.54) |
| Calcium cyanide* | Calcium Cyanide Ca(CN)2 | 10 (4.54) |
| Calcium cyanide Ca(CN)2 | Calcium cyanide | 10 (4.54) |
| Calcium dodecylbenzene sulfonate | | 1000 (454) |
| Calcium hypochlorite* | | 10 (4.54) |
| Camphene, octachloro- | Toxaphene* | 1 (0.454) |
| Captan | | 10 (4.54) |
| Carbamic acid, ethyl ester | Ethyl carbamate (Urethan) | 100 (45.4) |
| Carbamic acid, methylnitroso-, ethyl ester | N-Nitroso-N-methylurethane | 1 (0.454) |
| Carbamic chloride, dimethyl- | Dimethylcarbamoyl chloride | 1 (0.454) |
| Carbamide, thio- | Thiourea | 10 (4.54) |
| Carbamimidoselenoic acid | Selenourea | 1000 (454) |
| Carbamothioic acid, bis (1-methylethyl)-, S-(2,3-dichloro-2-propenyl) ester | Diallate | 100 (45.4) |
| Carbaryl* | | 100 (45.4) |
| Carbofuran* | | 10 (4.54) |
| Carbon bisulfide* | Carbon disulfide* | 100 (45.4) |
| Carbon disulfide* | Carbon bisulfide* | 100 (45.4) |
| Carbonic acid, dithallium (I+) | Thallium(I) carbonate | 100 (45.4) |
| Carbonic dichloride | Phosgene* | 10 (4.54) |
| Carbonic difluoride | Carbon oxyfluoride | 1000 (454) |
| Carbonochloridic acid, methyl ester | Methyl chlorocarbonate* | 1000 (454) |
| | Methyl chloroformate * | |
| Carbon oxyfluoride | Carbonic difluoride | 1000 (454) |
| Carbon tetrachloride* | Methane, tetrachloro- | 10 (4.54) |
| Chloral | Acetaldehyde, trichloro- | 5000 (2270) |
| Chlorambucil | Benzenebutanoic acid, 4-(bis(2-chloroethyl)amino)- | 10 (4.54) |
| Chlordane* | Chlordane, technical* | 1 (0.454) |
| | 4,7-Methano-1H-indene, 1,2,4,5,6,7,8,8-octachloro-2,3,3a,4,7,7a-hexahydro- | |
| | Chlordane, alpha & gamma isomers | |

**Table 1—Hazardous Substances Other Than Radionuclides,** Continued

| Hazardous Substances | Synonyms | Reportable Quantity (RQ) Pounds (Kilograms) |
|---|---|---|
| Chlordane, alpha & gamma isomers | Chlordane, technical | 1 (0.454) |
| | Chlordane* | |
| | 4,7-Methano-1H-indene, 1,2,4,5,6,7,8,8-octachloro-2,3,3a,4,7,7a-hexahydro- | |
| Chlordane, technical* | Chlordane* | 1 (0.454) |
| | 4,7-Methano-1H-indene, 1,2,4,5,6,7,8,8-octachloro-2,3,3a,4,7,7a-hexahydro- | |
| | Chlordane, alpha & gamma isomers | |
| Chlorine* | | 10 (4.54) |
| Chlornaphazine | Naphthylamine, N,N'-bis (2-chloroethyl)- | 100 (45.4) |
| Chloroacetaldehyde | Acetaldehyde, chloro- | 1000 (454) |
| p-Chloroaniline | Benzenamine, 4-chloro- | 1000 (454) |
| Chlorobenzene* | Benzene, chloro- | 100 (45.4) |
| Chlorobenzilate | Benzeneacetic acid, 4-chloro-alpha-(4-chlorophenyl)-alpha-hydroxy-ethyl ester | 10 (4.54) |
| 4-Chloro-m-cresol | p-Chloro-m-cresol | 5000 (2270) |
| | Phenol, 4-chloro-3-methyl- | |
| p-Chloro-m-cresol | Phenol,4-chloro-3-methyl- | 5000 (2270) |
| | 4-Chloro-m-cresol | |
| Chlorodibromomethane | | 100 (45.4) |
| Chloroethane | Ethyl chloride @ | 100 (45.4) |
| 2-Chloroethyl vinyl ether | Ethene, 2-chloroethoxy- | 1000 (454) |
| Chloroform* | Methane, trichloro- | 10 (4.54) |
| Chloromethane | Methane, chloro | 100 (45.4) |
| | Methyl chloride* | |
| Chloromethyl methyl ether | Methane, chloromethoxy- | 1 (0.454) |
| | Methylchloromethyl ether @ | |
| beta-Chloronaphthalene | Naphthalene, 2-chloro- | 5000 (2270) |
| | 2-Chloronaphthalene | |
| 2-Chloronaphthalene | beta-Chloronaphthalene | 5000 (2270) |
| | Naphthalene, 2-chloro- | |
| 2-Chlorophenol | o-Chlorophenol | 100 (45.4) |
| | Phenol, 2-chloro- | |
| o-Chlorophenol | Phenol, 2-chloro- | 100 (45.4) |
| | 2-Chlorophenol | |
| 4-Chlorophenyl phenyl ether | | 5000 (2270) |
| 1-(o-Chlorophenyl)thiourea | Thiourea, (2-chlorophenyl)- | 100 (45.4) |
| 3-Chloropropionitrile | Propanenitrile, 3-chloro- | 1000 (454) |
| Chlorosulfonic acid* | | 1000 (454) |
| 4-Chloro-o-toluidine, hydrochloride | Benzenamine, 4-chloro-2-methyl hydrochloride | 100 (45.4) |
| Chlorpyrifos* | | 1 (0.454) |
| Chromic acetate | | 1000 (454) |
| Chromic acid* | | 10 (4.54) |
| Chromic acid H2CrO4, calcium salt | Calcium chromate | 10 (4.54) |
| Chromic sulfate | | 1000 (454) |
| Chromium ¢ | | 5000 (2270) |
| Chromous chloride | | 1000 (454) |
| Chrysene | 1,2-Benzphenanthrene | 100 (45.4) |
| Cobaltous bromide | | 1000 (454) |
| Cobaltous formate | | 1000 (454) |
| Cobaltous sulfamate | | 1000 (454) |
| Coke Oven Emissions | | 1 (0.454) |
| Copper ¢ | | 5000 (2270) |
| Copper chloride @ | Cupric chloride | 10 (4.54) |
| Copper cyanide* | Copper cyanide CuCN | 10 (4.54) |
| Copper cyanide CuCN | Copper cyanide* | 10 (4.54) |
| Coumaphos* | | 10 (4.54) |
| Creosote | | 1 (0.454) |
| Cresol(s)* | Cresylic acid | 1000 (454) |
| | Phenol,methyl- | |
| m-Cresol | m-Cresylic acid | |
| o-Cresol | o-Cresylic acid | |
| p-Cresol | p-Cresylic acid | |
| Cresylic acid | Cresols* | 1000 (454) |
| | Phenol,methyl- | |
| m-Cresol | m-Cresylic acid | |
| o-Cresol | o-Cresylic acid | |
| p-Cresol | p-Cresylic acid | |
| Crotonaldehyde* | 2-Butenal | 100 (45.4) |
| Cumene | Benzene, 1-methylethyl- | 5000 (2270) |

## Table 1—Hazardous Substances Other Than Radionuclides, Continued

| Hazardous Substances | Synonyms | Reportable Quantity (RQ) Pounds (Kilograms) |
|---|---|---|
| Cumene hydroperoxide @ | alpha,alpha-Dimethylbenzylhydroperoxide | 10 (4.54) |
| | Hydroperoxide, 1-methyl-1-phenylethyl- | |
| Cupric acetate | | 100 (45.4) |
| Cupric acetoarsenite* | | 1 (0.454) |
| Cupric chloride | Copper chloride@ | 10 (4.54) |
| Cupric nitrate* | | 100 (45.4) |
| Cupric oxalate | | 100 (45.4) |
| Cupric sulfate | | 10 (4.54) |
| Cupric sulfate ammoniated | | 100 (45.4) |
| Cupric tartrate | | 100 (45.4) |
| Cyanides (soluble cyanide salts), not elsewhere specified* | | 10 (4.54) |
| Cyanogen* | Ethanedinitrile | 100 (45.4) |
| Cyanogen bromide* | Cyanogen bromide (CN)Br | 1000 (454) |
| Cyanogen bromide (CN)Br | Cyanogen bromide* | 1000 (454) |
| Cyanogen chloride* | Cyanogen Chloride (CN)Cl | 10 (4.54) |
| Cyanogen chloride (CN)Cl | Cyanogen chloride* | 10 (4.54) |
| 2,5-Cyclohexadiene-1,4-dione | p-Benzoquinone | 10 (4.54) |
| Cyclohexane* | Benzene, hexahydro- | 1000 (454) |
| Cyclohexane, 1,2,3,4,5,6-hexachloro-, (1alpha,2alpha,3beta,4alpha,5alpha,6beta)- | gamma-BHC | 1 (0.454) |
| | Hexachlorocyclohexane(gamma isomer) | |
| | Lindane* | |
| Cyclohexanone | | 5000 (2270) |
| 2-Cyclohexyl-4,6-dinitrophenol | Phenol,2-cyclohexyl-4,6-dinitro- | 100 (45.4) |
| 1,3-Cyclopentadiene, 1,2,3,4,5,5-hexachloro- | Hexachlorocyclopentadiene* | 10 (4.54) |
| Cyclophosphamide | 2H-1,3,2-Oxazaphosphorin, 2-amine,N,N-bis((2-chloroethyl) tetrahydro-)2-oxide | 10 (4.54) |
| 2,4-D Acid | 2,4-D*, salts and esters | 100 (45.4) |
| | Acetic acid (2,4-dichlorophenoxy)- | |
| 2,4-D Ester | | 100 (45.4) |
| 2,4-D*, salts and esters | 2,4-D Acid | 100 (45.4) |
| | Acetic acid (2,4-dichloro-phenoxy)- | |
| Daunomycin | 5,12-Naphthacenedione, 8-acetyl-10-[3-amino,2,3,6-trideoxy-alpha-L- | 10 (4.54) |
| | lyxo-hexopyranosyl(oxy]-7,8,9,10-tetrahydro-6,8,11-trihydroxy-1-methoxy-(8S-cis)- | |
| DDD | Benzene,1,1'-(2,2-dichloroethylidene)bis[4-chloro | 1 (0.454) |
| | TDE* | |
| | 4,4'-DDD | |
| 4,4'-DDD | DDD | 1 (0.454) |
| | Dichlorodiphenyl dichloroethane | |
| | TDE* | |
| DDE | 4,4' DDE | 1 (0.454) |
| 4,4' DDE | DDE | 1 (0.454) |
| DDT* | Benzene, 1,1'-(2,2,2-trichloroethylidene)bis[4-chloro- | 1 (0.454) |
| | 4,4'-DDT | |
| 4,4' DDT | DDT* | 1 (0.454) |
| | Benzene,1,1'-(2,2,2-trichloroethylidene)bis[4-chloro- | |
| Diallate | Carbamothioic acid, bis (1-methylethyl)-, S-(2,3-dichloro-2-propenyl) ester | 100 (45.4) |
| Diamine | Hydrazine* | 1 (0.454) |
| Diazinon* | | 1 (0.454) |
| Dibenz[a,h]anthracene | Dibenzo[a,h]anthracene | 1 (0.454) |
| | 1,2:5,6-Dibenzanthracene | |
| 1,2:5,6-Dibenzanthracene | Dibenz[a,h]anthracene | 1 (0.454) |
| | Dibenzo[a,h]anthracene | |
| Dibenzo[a,h]anthracene | Dibenz[a,h]anthracene | 1 (0.454) |
| | 1,2:5,6-Dibenzanthracene | |
| Dibenz[a,i]pyrene | Bezo [rst]pentaphene | 10 (4.54) |
| 1,2-Dibromo-3-chloropropane | Propane, 1,2-dibromo-3-chloro- | 1 (0.454) |
| Dibutyl phthalate | Di-n-butyl phthalate | 10 (4.54) |
| | n-Butyl phthalate* | |
| | 1,2-Benzenedicarboxylic acid, dibutyl ester | |
| Di-n-butylphthalate | Dibutyl phthalate | 10 (4.54) |
| | n-butyl phthalate* | |
| | 1,2-Benzenedicarboxylic acid, dibutyl ester | |
| Dicamba | | 1000 (454) |
| Dichlobenil | | 100 (45.4) |
| Dichlone | | 1 (0.454) |
| Dichlorobenzene | | 100 (45.4) |

## Table 1—Hazardous Substances Other Than Radionuclides, Continued

| Hazardous Substances | Synonyms | Reportable Quantity (RQ) Pounds (Kilograms) |
|---|---|---|
| 1,2-Dichlorobenzene | Benzene, 1,2-dichloro-<br>o-Dichlorobenzene* | 100 (45.4) |
| 1,3-Dichlorobenzene | Benzene, 1,3-dichloro-<br>m-Dichlorobenzene | 100 (45.4) |
| 1,4-Dichlorobenzene | Benzene, 1,4-dichloro-<br>p-Dichlorobenzene* | 100 (45.4) |
| m-Dichlorobenzene | Benzene, 1,3-dichloro-<br>1,3-Dichlorobenzene | 100 (45.4) |
| o-Dichlorobenzene* | Benzene, 1,2-dichloro-<br>1,2-Dichlorobenzene | 100 (45.4) |
| p-Dichlorobenzene* | Benzene, 1,4-dichloro-<br>1,4-Dichlorobenzene | 100 (45.4) |
| 3,3'-Dichlorobenzidine | (1,1'-Biphenyl)-4,4'diamine,3,3' dichloro- | 1 (0.454) |
| Dichlorobromomethane | | 5000 (2270) |
| 1,4-Dichloro-2-butene | 2-Butene, 1,4-dichloro- | 1 (0.454) |
| Dichlorodifluoromethane* | Methane, dichlorodifluoro- | 5000 (2270) |
| 1,1-Dichloroethane | Ethane, 1,1-dichloro-<br>Ethylidene dichloride | 1000 (454) |
| 1,2-Dichloroethane | Ethane, 1,2-dichloro-<br>Ethylene dichloride* | 100 (45.4) |
| 1,1-Dichloroethylene | Ethane, 1,1-dichloro-<br>Vinylidene chloride* | 100 (45.4) |
| 1,2-Dichloroethylene | Ethene, 1,2-dichloro-(E) | 1000 (454) |
| 1,3-Dichloropropene | 1-Propene, 1,3-dichloro- | 100 (45.4) |
| Dichloroethyl ether | Bis (2-chloroethyl) ether<br>Ethane, 1,1'-oxybis(2-chloro- | 10 (4.54) |
| Dichloroisopropyl-ether | Propane,2,2'-oxybis [2-chloro- | 1000 (454) |
| Dichloromethane@ | Methane, dichloro-<br>Methylene chloride* | 1000 (454) |
| Dichloromethoxy ethane | Bis(2-chloroethoxy)methane<br>Ethane, 1,1'-[methylenebis(oxy)]bis(2-chloro- | 1000 (454) |
| Dichloromethyl ether | Methane, oxybis(chloro- | 1 (0.454) |
| 2,4-Dichlorophenol | Phenol, 2,4-dichloro- | 100 (45.4) |
| 2,6-Dichlorophenol | Phenol, 2,6-dichloro- | 100 (45.4) |
| Dichlorophenylarsine | Phenyl dichloroarsine@<br>Arsonous dichloride, phenyl- | 1 (0.454) |
| Dichloropropane* | | 1000 (454) |
| 1,1-Dichloropropane | | |
| 1,3-Dichloropropane | | |
| 1,2-Dichloropropane | Propylene dichloride*<br>Propane, 1,2-dichloro- | 1000 (454) |
| Dichloropropane - Dichloropropene (mixture) | | 100 (45.4) |
| Dichloropropene* | | 100 (45.4) |
| 2,3-Dichloropropene | | |
| 2,2-Dichloropropionic acid* | | 5000 (2270) |
| Dichlorvos* | | 10 (4.54) |
| Dicofol | | 10 (4.54) |
| Dieldrin* | 2,7:3,6-Dimethanonaphth[2,3-b]oxirene,3,4,5,6,9,9-hexachloro-1a,2,2a,3,6,6a,7,7a-octahydro-, (1alpha,2beta,2aalpha,3beta,6beta,6aalpha,7beta,7aalpha)- | 1 (0.454) |
| 1,2:3,4-Diepoxybutane | 2,2'-Bioxirane | 10 (4.54) |
| Diethylamine* | | 1000 (454) |
| Diethylarsine | Arsine, diethyl- | 1 (0.454) |
| 1,4-Diethylenedioxide | 1,4-Dioxane | 100 (45.4) |
| Diethylhexyl phthalate | 1,2-Benzenedicarbolic acid, [bis(2-ethylhexyl)] ester<br>Bis(2-ethylhexyl)phthalate | 100 (45.4) |
| N,N'-Diethylhydrazine | Hydrazine, 1,2-diethyl- | 10 (4.54) |
| O,O-Diethyl S-methyl dithiophosphate | Phosphorodithioic acid, O'O>-diethylS-methyl ester | 5000 (2270) |
| Diethyl-p-nitrophenyl phosphate | Phosphoric acid, diethyl p-nitrophenyl ester | 100 (45.4) |
| Diethyl phthalate | 1,2-Benzenedicarboxylic acid, diethyl ester | 1000 (454) |
| O,O-Diethyl O-pyrazinyl phosphorothioate | Phosphorothioic acid, O,O-diethyl O-pyrazinyl ester | 100 (45.4) |
| Diethylstilbestrol | Phenol, 4,4'-(1,2-diethyl-1,2-ethenediyl)bis-,(E) | 1 (0.454) |
| Dihydrosafrole | Benzene, 1,2- methylenedioxy-4-propyl- | 10 (4.54) |
| Diisopropyl fluorophosphate | Phosphorofluoridic acid, bis(1-methylethyl) ester | 100 (45.4) |
| 1,4,5,8-Dimethanonaphthalene,1,2,3,4,10,10-hexachloro-1,4,4a,5,8,8a-hexahydro,(1alpha,4alpha,4abeta,5abeta,8beta,8abeta)- | Isodrin | 1 (0.454) |

### Table 1—Hazardous Substances Other Than Radionuclides, Continued

| Hazardous Substances | Synonyms | Reportable Quantity (RQ) Pounds (Kilograms) |
|---|---|---|
| 1,4,5,8-Dimethanonaphthalene,1,2,3,4,10,10-10-hexachloro-1,4,4a,5,8,8a-hexahydro-, (1alpha,4alpha,4abeta,5alpha,8alpha,8abeta)- | Aldrin* | 1 (0.454) |
| | 1,2,3,4,10-10-Hexachloro-1,4,4a,5,8,8a-hexahydro-1,4:5,8-endo,exo-dimethanonapthalene | |
| 2,7:3,6-Dimethanonaphth[2,3-b]oxirene,3,4,5,6,9,9-hexachloro-1a,2,2a,3,6,6a,7,7a-octahydro-, (1aalpha,2beta,2abeta,3alpha,6alpha,6abeta,7beta,7aalpha)- | Endrin* | 1 (0.454) |
| | Endrin, and metabolites | |
| 2,7:3,6-Dimethanonaphth[2,3-b]oxirene, 3,4,5,6,9,9-hexachloro-1a,2,2a,3,6,6a,7,7a-octahydro-,(1aalpha,2beta,2aalpha,3beta,6beta,6aalpha,7beta,7aalpha)- | Dieldrin* | 1 (0.454) |
| Dimethoate | Phosphorodithioic acid, O,O- dimethyl S-[2(methyl amino)-2-oxoethyl]ester | 10 (4.54) |
| 3,3'-Dimethoxybenzidine | (1,1'-Biphenyl)-4,4' diamine,3,3' dimethoxy- | 10 (4.54) |
| Dimethylamine* | Methanamine, N-methyl- | 1000 (454) |
| p-Dimethylaminoazobenzene | Benzenamine, N,N-dimethyl-4-(phenylazo)- | 10 (4.54) |
| 7,12-Dimethylbenz[a] anthracene | Benz[a]anthracene,7,12-dimethyl- | 1 (0.454) |
| 3,3'-Dimethylbenzidine | (1,1' Biphenyl)-4,4'-diamine,3,3'-dimethyl- | 10 (4.54) |
| alpha,alpha-Dimethylbenzyl-hydroperoxide | Hydroperoxide,1-methyl-1-phenylethyl- | 10 (4.54) |
| | Cumene hydroperoxide@ | |
| Dimethylcarbamoyl chloride | Carbamic chloride, dimethyl- | 1 (0.454) |
| 1,1-Dimethylhydrazine | Dimethylhydrazine, unsymmetrical @ | 10 (4.54) |
| | Hydrazine, 1,1-dimethyl- | |
| 1,2-Dimethylhydrazine | Hydrazine, 1,2-dimethyl- | 1 (0.454) |
| Dimethylhydrazine,unsymmetrical@ | 1,1-Dimethylhydrazine | 10 (4.54) |
| | Hydrazine,1,1-dimethyl- | |
| alpha,alpha-Dimethylphenethylamine | Benezeneethanamine, alpha,alpha-dimethyl- | 5000 (2270) |
| 2,4-Dimethylphenol | Phenol, 2,4-dimethyl- | 100 (45.4) |
| Dimethyl phthalate | 1,2-Benzenedicarboxylic acid, dimethyl ester | 5000 (2270) |
| Dimethyl sulfate* | Sulfuric acid, dimethyl ester | 100 (45.4) |
| Dinitrobenzene* (mixed) | | 100 (45.4) |
| m- Dinitrobenzene | | |
| o- Dinitrobenzene | | |
| p- Dinitrobenzene | | |
| 4,6-Dinitro-o-cresol and salts | Phenol,2-methyl-4,6-dinitro- | 10 (4.54) |
| Dinitrophenol | | 10 (4.54) |
| 2,5- Dinitrophenol | | |
| 2,6- Dinitrophenol | | |
| 2,4-Dinitrophenol | Phenol, 2,4-dinitro- | 10 (4.54) |
| Dinitrotoluene | | 10 (4.54) |
| 3,4-Dinitrotoluene | | |
| 2,4-Dinitrotoluene | Benzene, 1-methyl-2,4-dinitro- | 10 (4.54) |
| 2,6-Dinitrotoluene | Benzene, 2-methyl-1,3-dinitro- | 100 (45.4) |
| Dinoseb | Phenol,2-(1-methylpropyl)-4,6-dinitro | 1000 (454) |
| Di-n-octyl phthalate | 1,2-Benzenedicarboxylic acid, dioctyl ester | 5000 (2270) |
| 1,4-Dioxane | 1,4-Diethylene dioxide | 100 (45.4)† |
| 1,2-Diphenylhydrazine | Hydrazine, 1,2-diphenyl- | 10 (4.54) |
| Diphosphoramide, octamethyl- | Octamethylpyrophosphoramide | 100 (45.4) |
| Diphosphoric acid,tetraethyl ester | Tetraethyl pyrophosphate* | 10 (4.54) |
| Dipropylamine | 1-Propanamine, N-propyl- | 5000 (2270) |
| Di-n-propylnitrosamine | 1-Propanamine, N-nitroso-N-propyl | 10 (4.54) |
| Diquat | | 1000 (454) |
| Disulfoton* | Phosphorodithioic acid, O,O-diethyl S-[2-(ethylthio) ethyl]ester | 1 (0.454) |
| Dithiobiuret | Thioimidodicarbonic diamide$[(H_2N)C(S)]_2NH$ | 100 (45.4) |
| Diuron | | 100 (45.4) |
| Dodecylbenzenesulfonic acid* | | 1000 (454) |
| Endosulfan* | 6-9-Methano-2,4,3-benzodioxathiepin, 6,7,8,9,10,10-hexachloro-1,5,5a,6,9,9a-hexahydro-, 3-oxide. | 1 (0.454) |
| alpha - Endosulfan | | 1 (0.454) |
| beta - Endosulfan | | 1 (0.454) |
| Endosulfan sulfate | | 1 (0.454) |
| Endothall | 7-Oxabicyclo[2.2.1]heptane-2,3-dicarboxylic acid | 1000 (454) |
| Endrin* | 2,7:3,6,Dimethanonaphth[2,3-b]oxirene, 3,4,5,6,9,9-hexachloro-1,a,2,2a,3,6,6a,7,7a-octa-hydro-,(1aalpha,2beta,2abeta,3alpha,6alpha,6abeta,7beta,7aalpha)- | 1 (0.454) |
| | Endrin, & metabolites | |
| Endrin, & metabolites | Endrin | 1 (0.454) |
| | 2,7:3,6-Dimethanonaphth[2,3-b]oxirene, 3,4,5,6,9,9-hexachloro-1a,2,2a,3,6,6a,7,7a oct-hydro-, (1aalpha,2beta,2abeta,3alpha,6alpha,6abeta,7beta,7aalpha)- | |

**Table 1—Hazardous Substances Other Than Radionuclides, Continued**

| Hazardous Substances | Synonyms | Reportable Quantity (RQ) Pounds (Kilograms) |
|---|---|---|
| Endrin aldehyde | | 1 (0.454) |
| Epichlorohydrin* | Oxirane, (chloromethyl)- | 100 (45.4) |
| Epinephrine | 1,2-Benzenediol, 4- [1-hydroxy-2-(methylamino)ethyl]- | 1000 (454) |
| Ethanal | Acetaldehyde* | 1000 (454) |
| Ethanamine, N-ethyl-N-nitroso- | N-Nitrosodiethylamine | 1 (0.454) |
| Ethane, 1,2-dibromo- | Ethylene dibromide* | 1 (0.454) |
| Ethane, 1,1-dichloro- | Ethylidene dichloride<br>1,1-Dichloroethane | 1000 (454) |
| Ethane, 1,2-dichloro- | Ethylene dichloride*<br>1,2-Dichloroethane | 100 (45.4) |
| Ethane, hexachloro- | Hexachloroethane* | 100 (45.4) |
| Ethane, 1,1'-[methylenebis(oxy)]bis(2-chloro- | Bis(2-chloroethoxy)methane<br>Dichloromethoxy ethane | 1000 (454) |
| Ethane, 1,1'-oxybis- | Ethyl ether* | 100 (45.4) |
| Ethane, 1,1'-oxybis(2-chloro- | Bis (2-chloroethyl) ether<br>Dichloroethyl ether | 10 (4.54) |
| Ethane, pentachloro- | Pentachloroethane | 10 (4.54) |
| Ethane, 1,1,1,2-tetrachloro- | 1,1,1,2-Tetrachloroethane-<br>Tetrachloroethane@ | 100 (45.4) |
| Ethane, 1,1,2,2-tetrachloro- | 1,1,2,2-Tetrachloroethane-<br>Tetrachlorethane@ | 100 (45.4) |
| Ethane, 1,1,2-trichloro- | 1,1,2-Trichloroethane | 100 (45.4) |
| Ethane, 1,1,1-trichloro- | Methyl chloroform<br>1,1,1-Trichloroethane* | 1000 (454) |
| 1,2-Ethanediamine, N,N-dimethyl-N'-2-pyridinyl-N'-(2-thienylmethyl)- | Methapyrilene | 5000 (2270) |
| Ethanedinitrile | Cyanogen* | 100 (45.4) |
| Ethanenitrile | Acetonitrile* | 5000 (2270) |
| Ethanethiobamide | Thioacetamide | 10 (4.54) |
| Ethanimidothioic acid, N-[[(methylamino)carbonyl] oxy]-, methyl ester | Methomyl | 100 (45.4) |
| Ethanol, 2-ethoxy- | Ethylene glycol monoethyl ether* | 1000 (454) |
| Ethanol, 2,2'-(nitrosoimino)bis- | N-Nitrosodiethanolamine | 1 (0.454) |
| Ethanone, 1-phenyl- | Acetophenone | 5000 (2270) |
| Ethanoyl chloride | Acetyl chloride* | 5000 (2270) |
| Ethene, chloro- | Vinyl chloride* | 1 (0.454) |
| Ethene,2-chloroethoxy- | 2-Chloroethyl vinyl ether | 1000 (454) |
| Ethene, 1,1-dichloro- | Vinylidene chloride*<br>1,1-Dichloroethylene | 100 (45.4) |
| Ethene, 1,2-dichloro-(E) | 1,2-Dichloroethylene | 1000 (454) |
| Ethene, tetrachloro- | Perchloroethylene<br>Tetrachloroethene<br>Tetrachloroethylene | 100 (45.4) |
| Ethene, trichloro- | Trichloroethene<br>Trichloroethylene | 100 (45.4) |
| Ethion* | | 10 (4.54) |
| Ethyl acetate* | Acetic acid, ethyl ester | 5000 (2270) |
| Ethyl acrylate* | 2-Propenoicacid, ethyl ester | 1000 (454) |
| Ethylbenzene* | | 1000 (454) |
| Ethyl carbamate (Urethan) | Carbamic acid, ethyl ester | 100 (45.4) |
| Ethyl chloride @ | Chloroethane | 100 (45.4) |
| Ethyl cyanide | Propanenitrile | 10 (4.54) |
| Ethylene dibromide* | Ethane, 1,2-dibromo- | 1 (0.454) |
| Ethylene dichloride* | 1,2-Dichloroethane<br>Ethane, 1,2-dichloro- | 100 (45.4) |
| Ethylene glycol monoethyl ether* | Ethanol, 2-ethoxy | 1000 (454) |
| Ethylene imine@ | Aziridine<br>Ethylenimine | 1 (0.454) |
| Ethylene oxide* | Oxirane | 10 (4.54) |
| Ethylenebisdithiocarbamic acid | Ethylenebisdithiocarbamic acid, salts and esters | 5000 (2270) |
| Ethylenebisdithiocarbamic acid, salts and esters | Ethylenebisdithiocarbamic acid | 5000 (2270) |
| Ethylenediamine* | | 5000 (2270) |
| Ethylenediamine tetraacetic acid (EDTA) | | 5000 (2270) |
| Ethylenethiourea | 2-Imidazolidinethione | 10 (4.54) |
| Ethylenimine | Aziridine-<br>Ethylene imine@ | 1 (0.454) |
| Ethyl ether* | Ethane, 1,1'-oxybis- | 100 (45.4) |
| Ethylidene dichloride | Ethane, 1,1-dichloro-<br>1,1-Dichloroethane | 1000 (454) |
| Ethyl methacrylate | 2-Propenoic acid, 2-methyl-, ethyl ester | 1000 (454) |

**Table 1—Hazardous Substances Other Than Radionuclides,** Continued

| Hazardous Substances | Synonyms | Reportable Quantity (RQ) Pounds (Kilograms) |
|---|---|---|
| Ethyl methanesulfonate | Methanesulfonic acid, ethyl ester | 1 (0.454) |
| Ethyl methyl ketone @ | 2-Butanone | 5000 (2270) |
| | Methyl ethyl ketone(MEK)* | |
| Famphur | Phosphorothioic acid, O,[4-[(dimethylamino)-sulfonyl]phenyl] O,O-dimethylester | 1000 (454) |
| Ferric ammonium citrate | | 1000 (454) |
| Ferric ammonium oxalate | | 1000 (454) |
| Ferric chloride | | 1000 (454) |
| Ferric fluoride | | 100 (45.4) |
| Ferric nitrate* | | 1000 (454) |
| Ferric sulfate | | 1000 (454) |
| Ferrous ammonium sulfate | | 1000 (454) |
| Ferrous chloride* | | 100 (45.4) |
| Ferrous sulfate | | 1000 (454) |
| Fluoranthene | Benzo[j,k]fluorene | 100 (45.4) |
| Fluorene | | 5000 (2270) |
| Fluorine* | | 10 (4.54) |
| Fluoroacetamide | Acetamide, 2-fluoro- | 100 (45.4) |
| Fluoroacetic acid, sodium salt | Acetic acid, fluoro-, sodium salt | 10 (4.54) |
| Formaldehyde* | Methylene oxide | 100 (45.4) |
| Formic acid* | Methanoic acid | 5000 (2270) |
| Fulminic acid, mercury(2+)salt | Mercury fulminate | 10 (4.54) |
| Fumaric acid | | 5000 (2270) |
| Furan* | Furfuran | 100 (45.4) |
| Furan, tetrahydro- | Tetrahydrofuran* | 1000 (454) |
| 2-Furancarboxaldehyde | Furfural* | 5000 (2270) |
| 2,5-Furandione | Maleic anhydride* | 5000 (2270) |
| Furfural* | 2-Furancarboxaldehyde | 5000 (2270) |
| Furfuran | Furan* | 100 (45.4) |
| Glucopyranose, 2-deoxy-2-(3-methyl-3-nitrosoureido)- | Steptozotocin | 1 (0.454) |
| | D-Glucose, 2-deoxy-2[[(methylnitrosoamino)-carbonyl]amino]- | |
| D-Glucose, 2-deoxy-2-[[(methylnitrosoamino)-carbonyl]amino]- | Streptozotocin | 1 (0.454) |
| | Glucopyranose, 2-deoxy-2-(3-methyl-3-nitrosoureido)- | |
| Glycidylaldehyde | Oxiranecarboxyaldehyde | 10 (4.54) |
| Guanidine, N-methyl-N'-nitro-N-nitroso- | MNNG | 10 (4.54) |
| Guthion* | Azinphos methyl @ | 1 (0.454) |
| Heptachlor | 4,7-Methano-1H-indene,1,4,5,6,7,8,8-heptachloro-3a,4,7,7a-tetrahydro- | 1 (0.454) |
| Heptachlor epoxide | | 1 (0.454) |
| Hexachlorobenzene | Benzene, hexachloro- | 10 (4.54) |
| Hexachlorobutadiene | 1,3-Butadiene, 1,1,2,3,4,4- hexachloro- | 1 (0.454) |
| Hexachlorocyclohexane (gamma isomer) | gamma - BHC | 1 (0.454) |
| | Lindane* | |
| | Cyclohexane,1,2,3,4,5,6-hexachloro-, (1alpha,2alpha,3beta,4alpha,5alpha,6beta)- | |
| Hexachlorocyclopentadiene* | 1,3-Cyclopentadiene, 1,2,3,4,5,5-hexachloro- | 10 (4.54) |
| Hexachloroethane* | Ethane, hexachloro- | 100 (45.4) |
| 1,2,3,4,10-10-Hexachloro-1,4,4a,5,8,8a-hexahydro-1,4:5,8-endo,exo-dimethanonaphthalene | Aldrin* | 1 (0.454) |
| | 1,4,5,8-Dimethanonaphthalene,1,2,3,4,10,10-hexachloro-1,4,4a,5,8,8a-hexahydro-, (1alpha,4alpha,4abeta,5alpha,8alpha,8abeta)- | |
| Hexachlorophene | Phenol,2,2'-methylenebis[3,4,6-trichloro | 100 (45.4) |
| Hexachloropropene | 1-Propene, 1,1,2,3,3,3-hexachloro- | 1000 (454) |
| Hexaethyl tetraphosphate* | Tetraphosphoric acid, hexaethyl ester | 100 (45.4) |
| 2,4-(1H,3H)-Pyrimidinedione, 5-[bis(2-chloroethyl) amino]- | Uracil mustard | 10 (4.54) |
| Hydrazine* | Diamine | 1 (0.454) |
| Hydrazine, 1,2-diethyl- | N,N'-Diethylhydrazine | 10 (4.54) |
| Hydrazine, 1,1-dimethyl- | 1,1-Dimethylhydrazine | 10 (4.54) |
| | Dimethylhydrazine, unsymmetrical @ | |
| Hydrazine, 1,2-dimethyl- | 1,2-Dimethylhydrazine | 1 (0.454) |
| Hydrazine, 1,2-diphenyl- | 1,2-Diphenylhydrazine | 10 (4.54) |
| Hydrazine, methyl- | Methyl hydrazine* | 10 (4.54) |
| Hydrazinecarbothioamide | Thiosemicarbazide | 100 (45.4) |
| Hydrochloric acid* | Hydrogen chloride* | 5000 (2270) |
| Hydrocyanic acid* | Hydrogen cyanide | 10 (4.54) |
| Hydrofluoric acid* | Hydrogen fluoride* | 100 (45.4) |
| Hydrogen chloride* | Hydrochloric acid* | 5000 (2270) |
| Hydrogen cyanide* | Hydrocyanic acid* | 10 (4.54) |
| Hydrogen fluoride* | Hydrofluoric acid* | 100 (45.4) |
| Hydrogen phosphide | Phosphine* | 100 (45.4) |

**Table 1—Hazardous Substances Other Than Radionuclides, Continued**

| Hazardous Substances | Synonyms | Reportable Quantity (RQ) Pounds (Kilograms) |
|---|---|---|
| Hydrogen sulfide* | Hydrogen sulfide H2S | 100 (45.4) |
| Hydrogen sulfide H2S | Hydrogen sulfide | 100 (45.4) |
| Hydroperoxide, 1-methyl-1-phenylethyl- | alpha,alpha-Dimethylbenzylhydroperoxide<br>Cumene hydroperoxide @ | 10 (4.54) |
| 2-Imidazolidinethione | Ethylenethiourea | 10 (4.54) |
| Indeno(1,2,3-cd)pyrene | 1,10-(1,2-Phenylene)pyrene | 100 (45.4) |
| 1,3-Isobenzofurandione | Phthalic anhydride | 5000 (2270) |
| Isobutyl alcohol | 1-Propanol, 2-methyl- | 5000 (2270) |
| Isodrin | 1,4,5,8-Dimethanonaphthalene, 1,2,3,4,10,10-hexachloro-1,4,4a,5,8,8a-hexahydro,(1alpha,4alpha,4abeta,5beta,8beta,8abeta)- | 1 (0.454) |
| Isophorone | | 5000 (2270) |
| Isoprene* | | 100 (45.4) |
| Isopropanolamine dodecylbenzene sulfonate | | 1000 (454) |
| Isosafrole | 1,3-Benzodioxole,5-(-1propenyl)- | 100 (45.4) |
| 3(2H)-Isoxazolone, 5-(aminomethyl)- | 5-(Aminomethyl)-3-isoxazolol<br>Muscimol | 1000 (454) |
| Kepone | 1,3,4-Metheno-2H-cyclobuta[cd]-pentalen-2-one,1,1a,3,3a,4,5,5,5a,5b,6-decachloroc-tahydro- | 1 (0.454) |
| Lasiocarpine | 2-Butenoic acid, 2-methyl-,7[[2,3-dihydroxy-2-(1-methoxyethyl)-3-methyl-1-oxobutoxy]methyl]-2,3,5,7a-tetrahydro-1H-pyrrolizin-1-yl ester,[1S-[1alpha(Z),7(2S*,3R*),7aalpha]]- | 10 (4.54) |
| Lead C | | 1 (0.454) |
| Lead acetate | Acetic acid, lead salt | 5000 (2270) |
| Lead arsenate* | | 1 (0.454) |
| Lead,bis(acetato-O)tetrahydroxytri | Lead subacetate | 100 (45.4) |
| Lead chloride* | | 100 (45.4) |
| Lead fluoborate* | | 100 (45.4) |
| Lead fluoride* | | 100 (45.4) |
| Lead iodide | | 100 (45.4) |
| Lead nitrate* | | 100 (45.4) |
| Lead phosphate | Phosphoric acid, lead(2+)salt(2:3) | 1 (0.454) |
| Lead stearate | | 5000 (2270) |
| Lead subacetate | Lead,bis(acetato-O)tetrahydroxytri | 100 (45.4) |
| Lead sulfate* | | 100 (45.4) |
| Lead sulfide | | 5000 (2270) |
| Lead thiocyanate | | 100 (45.4) |
| Lindane* | gamma - BHC<br>Hexachlorocyclohexane (gamma isomer)<br>Cyclohexene, 1,2,3,4,5,6-hexachloro-, (1alpha,2alpha,3beta,4alpha,5alpha,6beta)- | 1 (0.454) |
| Lithium chromate | | 10 (4.54) |
| Malathion* | | 100 (45.4) |
| Maleic acid* | | 5000 (2270) |
| Maleic anhydride* | 2,5-Furandione | 5000 (2270) |
| Maleic hydrazide | 3,6-Pyridazinedione, 1,2-dihydro- | 5000 (2270) |
| Malononitrile | Propanedinitrile | 1000 (454) |
| Melphalan | L-Phenylalenine, 4-[bis(2-chloroethyl)amino] | 1 (0.454) |
| Mercaptodimethur | | 10 (4.54) |
| Mercuric cyanide* | | 1 (0.454) |
| Mercuric nitrate* | | 10 (4.54) |
| Mercuric sulfate* | | 10 (4.54) |
| Mercuric thiocyanate | | 10 (4.54) |
| Mercurous nitrate* | | 10 (4.54) |
| Mercury* | | 1 (0.454) |
| Mercury, (acetato-O)phenyl- | Phenylmercuric acetate | 100 (45.4) |
| Mercury fulminate | Fulminic acid, mercury(2+)salt | 10 (4.54) |
| Methacrylonitrile | 2-Propenenitrile, 2-methyl- | 1000 (454) |
| Methanamine, N-methyl- | Dimethylamine* | 1000 (454) |
| Methanamine, N-methyl-N-nitroso | N-Nitrosodimethylamine | 10 (4.54) |
| Methane, bromo- | Methyl bromide* | 1000 (454) |
| Methane, chloro- | Chloromethane<br>Methyl chloride* | 100 (45.4) |
| Methane, chloromethoxy- | Chloromethyl methyl ether<br>Methylchloromethyl ether @ | 1 (0.454) |
| Methane, dibromo- | Methylene bromide | 1000 (454) |
| Methane, dichloro- | Methylene chloride*<br>Dichloromethane@ | 1000 (454) |
| Methane, dichlorodifluoro- | Dichlorodifluoromethane* | 5000 (2270) |
| Methane, iodo- | Methyl iodide | 100 (45.4) |
| Methane, isocyanato- | Methyl isocyanate* | 1 (0.454) |

## Table 1—Hazardous Substances Other Than Radionuclides, Continued

| Hazardous Substances | Synonyms | Reportable Quantity (RQ) Pounds (Kilograms) |
|---|---|---|
| Methane, oxybis(chloro- | Dichloromethyl ether | 1 (0.454) |
| Methane, tetrachloro- | Carbon tetrachloride* | 10 (4.54) |
| Methane, tetranitro- | Tetranitromethane* | 10 (4.54) |
| Methane, tribromo- | Bromoform | 100 (45.4) |
| Methane, trichloro- | Chloroform* | 10 (4.54) |
| Methane, trichlorofluoro- | Trichloromonofluoromethane | 5000 (2270) |
| Methenesulfenyl chloride, trichloro | Perchloromethyl mercaptan @ | 100 (45.4) |
| | Trichloromethane sulfenyl chloride | |
| Methanesulfonic acid, ethyl ester | Ethyl methanesulfonate | 1 (0.454) |
| Methanethiol | Methyl mercaptan* | 100 (45.4) |
| | Thiomethanol | |
| 6,9-Methano-2,4,3-benzodioxathiepin, 6,7,8,9,10,10-hexachloro-1,5,5a,6,9,9a-hexahydro-, 3-oxide | Endosulfan* | 1 (0.454) |
| Methanoic acid | Formic acid* | 5000 (2270) |
| 4,7-Methano-1H-indene, 1,4,5,6,7,8,8-heptachloro-3a,4,7,7a-tetrahydro- | Heptachlor | 1 (0.454) |
| 4,7-Methano-1H-indene,1,4,5,6,7,8,8-octachloro-2,3,3a,4,7,7a-hexahydro- | Chlordane* | 1 (0.454) |
| | Chlordane, technical* | |
| | Chlordane,alpha & gamma isomers | |
| Methanol* | Methyl alcohol* | 5000 (2270) |
| Methapyrilene | 1,2-Ethanediamine,N-N-dimethyl-N'-2-pyridinyl-N'-(2-thienylmethyl)- | 5000 (2270) |
| 1,3,4-Metheno-2H-cyclobuta[cd]-pentalen-2-one, 1,1a,3,3a,4,5,5,5a,5b,6-decachloroctahydro- | Kepone* | 1 (0.454) |
| Methomyl | Ethanimidothioic acid,N-[[(methylamino)carbonyl]oxy]-, methyl ester | 100 (45.4) |
| Methoxychlor | Benzene, 1,1'-(2,2,2-trichloroethylidene) bis[4-methoxy- | 1 (0.454) |
| Methyl alcohol* | Methanol* | 5000 (2270) |
| Methylamine @ | Monomethylamine | 100 (45.4) |
| Methyl bromide* | Methane, bromo- | 1000 (454) |
| 1-Methylbutadiene | 1,3-Pentadiene | 100 (45.4) |
| Methyl chloride* | Chloromethane | 100 (45.4) |
| | Methane, chloro- | |
| Methyl chlorocarbonate* | Carbonochloridic acid, methyl ester | 1000 (414) |
| | Methyl chloroformate * | |
| Methyl chloroform* | 1,1,1-Trichloroethane* | 1000 (454) |
| | Ethane, 1,1,1-trichloro- | |
| Methyl chloroformate * | Carbonochloridic acid, methyl ester | 1000 (454) |
| | Methyl chlorocarbonate* | |
| Methylchloromethyl ether @ | Chloromethyl methyl ether | 1 (0.454) |
| | Methene, chloromethoxy- | |
| 3-Methylcholanthrene | Benz[j]aceanthrylene, 1,2-dihydro-3-methyl- | 10 (4.54) |
| 4,4'-Methylenebis(2-chloroaniline) | Benzenamine, 4,4'-methylenebis(2-chloro- | 10 (4.54) |
| Methylene bromide | Methane, dibromo- | 1000 (454) |
| Methylene chloride* | Methane, dichloro- | 1000 (454) |
| | Dichloromethane@ | |
| Methylene oxide | Formaldehyde* | 100 (45.4) |
| Methyl ethyl ketone(MEK)* | 2-Butanone | 5000 (2270) |
| | Ethyl methyl ketone @ | |
| Methyl ethyl ketone peroxide* | 2-Butanone peroxide | 10 (4.54) |
| Methyl hydrazine* | Hydrazine, methyl- | 10 (4.54) |
| Methyl iodide | Methane, iodo- | 100 (45.4) |
| Methyl isobutyl ketone | 4-Methyl-2-pentanone | 5000 (2270) |
| Methyl isocyanate* | Methane, isocynato- | 1 (0.454) |
| 2-Methyllactonitrile | Acetone cyanohydrin* | 10 (4.54) |
| | Propanenitrile, 2-hydroxy-2- methyl- | |
| Methyl mercaptan* | Methanethiol | 100 (45.4) |
| | Thiomethanol | |
| Methyl methacrylate* | 2-Propenoic acid, 2-methyl-, methylester | 1000 (454) |
| Methyl parathion* | Phosphorothioic acid, O,O-dimethyl O-(4-nitrophenyl) ester | 100 (45.4) |
| 4-Methyl-2-pentanone | Methyl isobutyl ketone | 5000 (2270) |
| Methylthiouracil | 4(1H)-Pyrimidinone, 2,3-dihydro-6-methyl-2-thioxo- | 10 (4.54) |
| Mevinphos* | | 10 (4.54) |
| Mexacarbate* | | 1000 (454) |
| Mitomycin C | Azirino[2',3':3,4]pyrrolo [1,2-a]indole-4,7-dione,6-amino-8-[[(amino-carbonyl)oxy]methyl]-1,1a,2,8,8a,8b-hexahydro-8a-methoxy-5-methyl-, [1aS-(1aalpha,8beta,8aalpha,8balpha]- | 10 (4.54) |
| MNNG | Guanidine,N-methyl-N'-nitro-N-nitroso- | 10 (4.54) |
| Monoethylamine* | | 100 (45.4) |
| Monomethylamine | Methylamine @ | 100 (45.4) |

**Table 1—Hazardous Substances Other Than Radionuclides, Continued**

| Hazardous Substances | Synonyms | Reportable Quantity (RQ) Pounds (Kilograms) |
|---|---|---|
| Muscimol | 5-(Aminomethyl)-3-isoxazolol<br>3(2H)-Isoxazolone, 5-(aminomethyl)- | 1000 (454) |
| Naled | | 10 (4.54) |
| 5,12-Naphthacenedione, 8-acetyl-10-[3-amino-2,3,6-trideoxy-alpha-L-lyxo-hexopyranosyl)oxy]-7,8,9,10-tetrahydro-6,8,11-trihydroxy-1-methoxy-,(8S-cis)- | Daunomycin | 10 (4.54) |
| Naphthalenamine,N,N-bis(2-chloroethyl)- | Chlornaphazine | 100 (45.4) |
| Naphthalene* | | 100 (45.4) |
| Naphthalene, 2-chloro- | beta-Chloronaphthalene<br>2-Chloronaphthalene | 5000 (2270) |
| 1,4-Naphthalenedione | 1,4-Naphthoquinone | 5000 (2270) |
| 2,7-Naphthalenedisulfonic acid,3,3'-[(3,3'-dimethyl- (1,1'-biphenyl)-4,4'-diyl)-bis(azo)]bis(5-amino-4-hydroxy)-tetrasodium salt. | Trypan blue | 10 (4.54) |
| Naphthenic acid | | 100 (45.4) |
| 1,4-Naphthoquinone | 1,4-Naphthalenedione | 5000 (2270) |
| alpha-Naphthylamine | 1-Naphthylamine | 100 (45.4) |
| beta-Naphthylamine | 2-Naphthylamine | 1 (0.454) |
| 1-Naphthylamine | alpha-Naphthylamine | 100 (45.4) |
| 2-Naphthylamine | beta-Naphthylamine | 1 (0.454) |
| alpha-Naphthylthiourea | Thiourea, 1-naphthalenyl- | 100 (45.4) |
| Nickel ¢ | | 100 (45.4) |
| Nickel ammonium sulfate | | 100 (45.4) |
| Nickel carbonyl* | Nickel carbonyl Ni(CO)4,(T-4)- | 10 (4.54) |
| Nickel carbonyl Ni(CO)4,(T-4)- | Nickel carbonyl* | 10 (4.54) |
| Nickel chloride | | 100 (45.4) |
| Nickel cyanide* | Nickel cyanide Ni(CN)2 | 10 (4.54) |
| Nickel cyanide Ni(CN)2 | Nickel cyanide* | 10 (4.54) |
| Nickel hydroxide | | 10 (4.54) |
| Nickel nitrate* | | 100 (45.4) |
| Nickel sulfate | | 100 (45.4) |
| Nicotine* and salts* | Pyridine, 3-(1-methyl-2-pyrrolidinyl)-,(S)- | 100 (45.4) |
| Nitric acid* | | 1000 (454) |
| Nitric acid, thallium(1+)salt | Thallium(I) nitrate | 100 (45.4) |
| Nitric oxide* | Nitrogen oxide NO | 10 (4.54) |
| p-Nitroaniline* | Benzenamine, 4-nitro- | 5000 (2270) |
| Nitrobenzene* | Benzene, nitro- | 1000 (454) |
| Nitrogen dioxide* | Nitrogen oxide NO2<br>Nitrogen peroxide @<br>Nitrogen tetroxide @ | 10 (4.54) |
| Nitrogen oxide NO | Nitric oxide* | 10 (4.54) |
| Nitrogen oxide NO2 | Nitrogen dioxide*<br>Nitrogen peroxide @<br>Nitrogen tetroxide @ | 10 (4.54) |
| Nitrogen peroxide @ | Nitrogen dioxide*<br>Nitrogen oxide NO2<br>Nitrogen tetroxide @ | 10 (4.54) |
| Nitrogen tetroxide @ | Nitrogen dioxide*<br>Nitrogen oxide NO2<br>Nitrogen peroxide | 10 (4.54) |
| Nitroglycerine* | 1,2,3-Propanetriol, trinitrate- | 10 (4.54) |
| Nitrophenol (mixed) | | 100 (45.4) |
| m- | | |
| o- | 2-Nitrophenol | |
| p- | 4-Nitrophenol<br>Phenol, 4-nitro- | |
| o-Nitrophenol | 2-Nitrophenol | 100 (45.4) |
| p-Nitrophenol | Phenol, 4-nitro-<br>4-Nitrophenol | 100 (45.4) |
| 2-Nitrophenol | o-Nitrophenol | 100 (45.4) |
| 4-Nitrophenol | p-Nitrophenol<br>Phenol, 4-nitro- | 100 (45.4) |
| 2-Nitropropane | Propane, 2-nitro- | 10 (4.54) |
| N-Nitrosodi-n-butylamine | 1-Butanamine, N-butyl-N-nitroso- | 10 (4.54) |
| N-Nitrosodiethanolamine | Ethanol, 2,2'- (nitrosimino)bis- | 1 (0.454) |
| N-Nitrosodiethylamine | Ethanamine, N-ethyl-N- nitroso- | 1 (0.454) |
| N-Nitrosodimethylamine | Methanamine, N-methyl-N-nitroso- | 10 (4.54) |
| N-Nitrosodiphenylamine | | 100 (45.4) |
| N-Nitroso-N-ethylurea | Urea, N-ethyl-N-nitroso- | 1 (0.454) |
| N-Nitroso-N-methylurea | Urea, N-methyl-N-nitroso- | 1 (0.454) |
| N-Nitroso-N-methylurethane | Carbamic acid, methylnitroso-,ethyl ester | 1 (0.454) |

Table 1—Hazardous Substances Other Than Radionuclides, Continued

| Hazardous Substances | Synonyms | Reportable Quantity (RQ) Pounds (Kilograms) |
|---|---|---|
| 1-Propene, 1,1,2,3,3,3-hexachloro- | Hexachloropropene | 1000 (454) |
| 2-Propenenitrile | Acrylonitrile* | 100 (45.4) |
| 2-Propenenitrile, 2-methyl- | Methacrylonitrile | 1000 (454) |
| 2-Propenoic acid | Acrylic acid* | 5000 (2270) |
| 2-Propenoic acid, ethyl ester | Ethyl acrylate* | 1000 (454) |
| 2-Propenoic acid, 2-methyl-, ethyl ester | Ethyl methacrylate | 1000 (454) |
| 2-Propenoic acid, 2-methyl-, methyl ester | Methyl methacrylate* | 1000 (454) |
| 2-Propen-1-ol | Allyl alcohol* | 100 (45.4) |
| Propionic acid* | | 5000 (2270) |
| Propionic acid, 2-(2,4,5-trichlorophenoxy)- | Silvex (2,4,5-TP) | 100 (45.4) |
| | 2,4,5-TP @ | |
| | 2,4,5-TP acid | |
| Propionic anhydride | | 5000 (2270) |
| n-Propylamine* | 1-Propanamine | 5000 (2270) |
| Propylene dichloride* | 1,2-Dichloropropane | 1000 (454) |
| | Propane, 1,2-dichloro- | |
| Propylene oxide* | | 100 (45.4) |
| 1,2-Propylenimine* | Aziridine, 2-methyl- | 1 (0.454) |
| 2-Propyn-1-ol | Propargyl alcohol* | 1000 (454) |
| Pyrene | | 5000 (2270) |
| Pyrethrins | | 1 (0.454) |
| 3,6-Pyridazinedione, 1,2-dihydro- | Maleic hydrazide | 5000 (2270) |
| 4-Pyridinamine | 4-Aminopyridine | 1000 (454) |
| Pyridine* | | 1000 (454) |
| Pyridine, 2-methyl- | 2-Picoline | 5000 (2270) |
| Pyridine, 3-(1-methyl-2-pyrrolidinyl)-, (S) | Nicotine* and salts* | 100 (45.4) |
| 4(1H)-Pyrimidinone, 2,3-dihydro-6-methyl-2-thioxo- | Methylthiouracil | 10 (4.54) |
| Pyrrolidine, 1-nitroso- | N-Nitrosopyrrolidine | 1 (0.454) |
| Quinoline | | 5000 (2270) |
| RADIONUCLIDES | | See Table 2 |
| Reserpine | Yohimban-16-carboxylic acid,11,17-dimethoxy-18- [(3,4,5-trimethoxy-benzoyl)oxy-, methyl ester-, (3beta,16beta,17alpha,18beta,20alpha)- | 5000 (2270) |
| Resorcinol | 1,3-Benzenediol | 5000 (2270) |
| Saccharin and salts | 1,2-Benzisothiazol-3(2H)-one,1,1-dioxide | 100 (45.4) |
| Safrole | 1,3-Benzodioxole,5-(2-propenyl)- | 100 (45.4) |
| Selenious acid | | 10 (4.54) |
| Selenious acid, dithallium(1+)salt | Thallium selenite | 1000 (454) |
| Selenium ¢ | | 100 (45.4) |
| Selenium dioxide | Selenium oxide* | 10 (4.54) |
| Selenium oxide* | Selenium dioxide | 10 (4.54) |
| Selenium sulfide | Selenium sulfide SeS2 | 10 (4.54) |
| Selenium sulfide SeS2 | Selenium sulfide | 10 (4.54) |
| Selenourea | Carbamimidoselenoic acid | 1000 (454) |
| L-Serine, diazoacetate (ester) | Azaserine | 1 (0.454) |
| Silver ¢ | | 1000 (454) |
| Silver cyanide* | Silver cyanide Ag(CN) | 1 (0.454) |
| Silver cyanide Ag(CN) | Silver cyanide | 1 (0.454) |
| Silver nitrate* | | 1 (0.454) |
| Silvex(2,4,5-TP) | Propionic acid, 2-(2,4,5-trichlorophenoxy)- | 100 (45.4) |
| | 2,4,5-TP @ | |
| | 2,4,5-TP acid | |
| Sodium* | | 10 (4.54) |
| Sodium arsenate* | | 1 (0.454) |
| Sodium arsenite* | | 1 (0.454) |
| Sodium azide* | | 1000 (454) |
| Sodium bichromate | Sodium dichromate @ | 10 (4.54) |
| Sodium bifluoride* | | 100 (45.4) |
| Sodium bisulfite* | | 5000 (2270) |
| Sodium chromate | | 10 (4.54) |
| Sodium cyanide* | | 10 (4.54) |
| Sodium cyanide Na(CN) | Sodium cyanide | 10 (4.54) |
| Sodium dichromate @ | Sodium bichromate | 10 (4.54) |
| Sodium dodecylbenzene sulfonate | | 1000 (454) |
| Sodium fluoride* | | 1000 (454) |
| Sodium hydrosulfide* | | 5000 (2270) |
| Sodium hydroxide* | | 1000 (454) |
| Sodium hypochlorite* | | 100(45.4) |
| Sodium methylate* | | 1000 (454) |
| Sodium nitrite* | | 100 (45.4) |

## Table 1—Hazardous Substances Other Than Radionuclides, Continued

| Hazardous Substances | Synonyms | Reportable Quantity (RQ) Pounds (Kilograms) |
|---|---|---|
| Sodium phosphate, dibasic | | 5000 (2270) |
| Sodium phosphate, tribasic | | 5000 (2270) |
| Sodium selenite* | | 100 (45.4) |
| Streptozotocin | Glucopyranose, 2-deoxy-2-(3-methyl-3-nitrosoureido)- | 1 (0.454) |
| | D-Glucose, 2-deoxy-2-[[(methylnitrosoamino)-carbonyl]amino]- | |
| Strontium chromate | | 10 (4.54) |
| Strychnidin-10-one | Strychnine* and salts* | 10 (4.54) |
| Strychnidin-10-one, 2,3-dimethoxy- | Brucine | 100 (45.4) |
| Strychnine* and salts* | Strychnidin-10-one | 10 (4.54) |
| Styrene | | 1000 (454) |
| Sulfur chloride@ | Sulfur monochloride | 1000 (454) |
| Sulfur monochloride | Sulfur chloride@ | 1000 (454) |
| Sulfur phosphide | Phosphorus pentasulfide* | 100 (45.4) |
| | Phosphorus sulfide | |
| Sulfuric acid* | | 1000 (454) |
| Sulfuric acid, dimethyl ester | Dimethyl sulfate* | 100 (45.4) |
| Sulfuric acid, dithallium(I+)salt | Thallium(I) sulfate* | 100 (45.4) |
| 2,4,5-T* | 2,4,5-T acid | 1000 (454) |
| | Acetic acid,(2,4,5-trichlorophenoxy) | |
| 2,4,5-T acid | 2,4,5-T* | 1000 (454) |
| | Acetic acid, (2,4,5-trichlorophenoxy) | |
| 2,4,5-T amines | | 5000 (2270) |
| 2,4,5-T esters | | 1000 (454) |
| 2,4,5-T salts | | 1000 (454) |
| TDE* | DDD | 1 (0.454) |
| | Benzene, 1,1'-(2,2-dichloroethylidene)bis[4-chloro- | |
| | 4,4'-DDD | |
| 1,2,4,5-Tetrachlorobenzene | Benzene, 1,2,4,5-tetrachloro- | 5000 (2270) |
| 2,3,7,8-Tetrachlorodibenzo-p-dioxin (TCDD) | | 1 (0.454) |
| 1,1,1,2-Tetrachloroethane | Ethane,1,1,1,2-tetrachloro- | 100 (45.4) |
| | Tetrachloroethane@ | |
| 1,1,2,2-Tetrachloroethane | Ethane,1,1,2,2-tetrachloro- | 100 (45.4 |
| | Tetrachloroethane@ | |
| Tetrachloroethane@ | Ethane, 1,1,1,2-tetrachloro- | 100 (45.4) |
| | Ethane, 1,1,2,2-tetrachloro- | |
| | 1,1,1,2-Tetrachloroethane | |
| | 1,1,2,-Tetrachloroethane | |
| Tetrachloroethene | Ethene, tetrachloro- | 100 (45.4) |
| | Perchloroethylene* | |
| | Tetrachloroethylene* | |
| Tetrachloroethylene* | Ethene, tetrachloro- | 100 (45.4) |
| | Perchloroethylene* | |
| | Tetrachloroethene | |
| 2,3,4,6-Tetrachlorophenol | Phenol, 2,3,4,6-tetrachloro- | 10 (4.54) |
| Tetraethyl lead* | Plumbane, tetraethyl- | 10 (4.54) |
| Tetraethyl pyrophosphate* | Diphosphoric acid, tetraethyl ester | 10 (4.54) |
| Tetraethyldithiopyrophosphate | Thiodiphosphoric acid, tetraethyl ester | 100 (45.4) |
| Tetrahydrofuran* | Furan, tetrahydro- | 1000 (454) |
| Tetranitromethane* | Methane, tetranitro- | 10 (4.54) |
| Tetraphosphoricacid, hexaethyl ester | Hexaethyl tetraphosphate* | 100 (45.4) |
| Thallic oxide | Thallium oxide T1203 | 100 (45.4) |
| Thallium C | | 1000 (454) |
| Thallium(I) acetate | Acetic acid, thallium(I+)salt | 100 (45.4) |
| Thallium(I) carbonate | Carbonic acid, dithallium (I+) | 100 (45.4) |
| Thallium(I) chloride | Thallium chloride T1C1 | 100 (45.4) |
| Thallium chloride TlCl | Thallium(I) chloride | 100 (45.4) |
| Thallium(I) nitrate | Nitric acid, thallium(1+) salt | 100 (45.4) |
| Thallium oxide T1203 | Thallic oxide | 100 (45.4) |
| Thallium selenite | Selenious acid, dithallium(1+)salt | 1000 (454) |
| Thallium(I) sulfate* | Sulfuric acid, dithallium(I+) salt | 100 (45.4) |
| Thioacetamide | Ethanethioamide | 10 (4.54) |
| Thiodiphosphoric acid, tetraethyl ester | Tetraethyldithiopyrophosphate | 100 (45.4) |
| Thiofanox | 2-butanone, 3,3-Dimethyl-1-(methylthio)-, O-[(methylamino)carbonyl] oxime | 100 (45.4) |
| Thioimidodicarbonic diamide [(H2N)C(S)]2NH | Dithiobiuret | 100 (45.4) |
| Thiomethanol | Methanethiol | 100 (45.4) |
| | Methyl mercaptan* | |
| Thioperoxydicarbonic diamide [(H2N)C(S)]2S2, tetramethyl- | Thiram | 10 (4.54) |
| Thiophenol* | Benzenethiol | 100 (45.4) |
| | Phenyl mercaptan @ | |

**Table 1—Hazardous Substances Other Than Radionuclides, Continued**

| Hazardous Substances | Synonyms | Reportable Quantity (RQ) Pounds (Kilograms) |
|---|---|---|
| D040 Trichloroethylene | | 100 (45.4) |
| D041 2,4,5-Trichloroethylene | | 10 (4.54) |
| D042 2,4,6-Trichlorophenol | | 10 (4.54) |
| D043 Vinyl Chloride | | 1 (0.454) |
| F001 | | 10 (4.54) |
| The following spent halogenated solvents used in degreasing: all spent, solvent mixtures/blends used in degreasing containing, before use, a total of ten percent or more (by volume) of one or more of the below listed halogenated solvents or those solvents listed in F002, F004, and F005; and still bottoms from the recovery of these spent solvents and spent solvent mixtures. | | |
| (a) Tetrachloroethylene | | 100 (45.4) |
| (b) Trichloroethylene | | 100 (45.4) |
| (c) Methylene chloride | | 1000 (454) |
| (d) 1,1,1-Trichloroethane | | 1000 (454) |
| (e) Carbon tetrachloride | | 10 (4.54) |
| (f) Chlorinated fluorocarbons | | 5000 (2270) |
| F002 | | 10 (4.54) |
| The following spent halogenated solvents; all spent solvent mixtures/blends containing, before use, a total of ten percent or more (by volume) of one or more of the below listed halogenated solvents or those listed in F001, F004, F005; and still bottoms from the recovery of these spent solvents and spent solvent mixtures. | | |
| (a) Tetrachloroethylene | | 100 (45.4) |
| (b) Methylene chloride | | 1000 (454) |
| (c) Trichloroethylene | | 100 (45.4) |
| (d) 1,1,1-Trichloroethane | | 1000 (454) |
| (e) Chlorobenzene | | 100 (45.4) |
| (f) 1,1,2-Trichloro-1,2,2- trifluoroethane | | 5000 (2270) |
| (g) o-Dichlorobenzene | | 100 (45.4) |
| (h) Trichlorofluoromethane | | 5000 (2270) |
| (i) 1,1,2 Trichloroethane | | 100 (45.4) |
| F003 | | 100 (45.4) |
| The following spent non-halogenated solvents and solvents: | | |
| (a) Xylene | | 1000 (454) |
| (b) Acetone | | 5000 (2270) |
| (c) Ethyl acetate | | 5000 (2270) |
| (d) Ethylbenzene | | 1000 (454) |
| (e) Ethyl ether | | 100 (45.4) |
| (f) Methyl isobutyl ketone | | 5000 (2270) |
| (g) n-Butyl alcohol | | 5000 (2270) |
| (h) Cyclohexanone | | 5000 (2270) |
| (i) Methanol | | 5000 (2270) |
| F004 | | 1000 (454) |
| The following spent non-halogenated solvents and the still bottoms from the recovery of these solvents: | | |
| (a) Cresols/Cresylic acid | | 1000 (454) |
| (b) Nitrobenzene | | 1000 (454) |
| F005 | | 100 (45.4) |
| The following spent non-halogenated solvents and the still bottoms from the recovery of these solvents: | | |
| (a) Toluene | | 1000 (454) |
| (b) Methyl ethyl ketone | | 5000 (2270) |
| (c) Carbon disulfide | | 100 (45.4) |
| (d) Isobutanol | | 5000 (2270) |
| (e) Pyridine | | 1000 (454) |
| F006 | | 10 (4.54) |
| Wastewater treatment sludges from electroplating operations except from the following processes: (1) sulfuric acid anodizing of aluminum; (2) tin plating on carbon steel; (3) zinc plating (segregated basis) on carbon steel; (4) aluminum or zinc-aluminum plating on carbon steel; (5) cleaning/stripping associated with tin, zinc and aluminum plating on carbon steel; and (6) chemical etching and milling of aluminum. | | |
| F007 | | 10 (4.54) |
| Spent cyanide plating bath solutions from electroplating operations. | | |
| F008 | | 10 (4.54) |
| Plating bath residues from the bottom of plating baths from electroplating operations where cyanides are used in the process. | | |

**Table 1—Hazardous Substances Other Than Radionuclides, Continued**

| Hazardous Substances | Synonyms | Reportable Quantity (RQ) Pounds (Kilograms) |
|---|---|---|
| F009 Spent stripping and cleaning bath solutions from electroplating operations where cyanides are used in the process. | | 10 (4.54) |
| F010 Quenching bath residues from oil baths from metal heat treating operations where cyanides are used in the process. | | 10 (4.54) |
| F011 Spent cyanide solutions from salt bath pot cleaning from metal heat treating operations (except for precious metals heat treating spent cyanide solutions from salt bath pot cleaning). | | 10 (4.54) |
| F012 Quenching wastewater treatment sludges from metal heat treating operations where cyanides are used in the process. | | 10 (4.54) |
| F019 Wastewater treatment sludges from the chemical conversion coating of aluminum-except from zirconium phosphating in aluminum can washing when such phosphating is an exclusive conversion coating process. | | 10 (4.54) |
| F020 Wastes (except wastewater and spent carbon from hydrogen chloride purification from the production or manufacturing use (as a reactant, chemical intermediate, or component in a formulating process) of tri- or tetrachlorophenol, or of intermediates used to produce their pesticide derivatives. (This listing does not include wastes from the production of hexachlorophene from highly purified 2,4,5-trichlorophenol.) | | 1 (0.454) |
| F021 Wastes (except wastewater and spent carbon from hydrogen chloride purification) from the production or manufacturing use (as a reactant, chemical intermediate, or component in a formulating process) of pentachlorophenol, or of intermediates used to produce its derivatives. | | 1 (0.454) |
| F022 Wastes (except wastewater and spent carbon from hydrogen chloride purification) from the manufacturing use (as a reactant, chemical intermediate, or component in a formulating process) of tetra-, penta-, or hexachlorobenzenes under alkaline conditions. | | 1 (0.454) |
| F023 Wastes (except wastewater and spent carbon from hydrogen chloride purification) from the production of materials on equipment previously used for the production or manufacturing use (as a reactant, chemical intermediate, or component in a formulating process) of tri- and tetrachlorophenols. (This listing does not include wastes from equipment used only for the production or use of hexachlorophene from highly purified 2,4, 5-trichlorophenol.) | | 1 (0.454) |
| F024 Wastes, including but not limited to distillation residues, heavy ends, tars, and reactor cleanout wastes, from the production of chlorinated aliphatic hydrocarbons, having carbon content from one to five, utilizing free radical catalyzed processes. (This listing does not include light ends, spent filters and filter aids, spent dessicants(sic), wastewater, wastewater treatment sludges, spent catalysts, and wastes listed in Section 261.32.) | | 1 (0.454) |
| F025 Condensed light ends, spent filters and filter aids, and spent desiccant wastes from the production of certain chlorinated aliphatic hydrocarbons, by free radical catalyzed processes. These chlorinated aliphatic hydrocarbons are those having carbon chain lengths ranging from one to and including five, with varying amounts and positions of chlorine substitution. | | 1 (0.454) |
| F026 Wastes (except wastewater and spent carbon from hydrogen chloride purification) from the production of materials on equipment previously used for the manufacturing use (as a reactant, chemical intermediate, or component in a formulating process) of tetra-, penta-, or hexachlorobenzene under alkaline conditions. | | 1 (0.454) |

**Table 1—Hazardous Substances Other Than Radionuclides, Continued**

| Hazardous Substances | Synonyms | Reportable Quantity (RQ) Pounds (Kilograms) |
|---|---|---|
| K044. Wastewater treatment sludges from the manufacturing and processing of explosives. | | 10 (4.54) |
| K045. Spent carbon from the treatment of wastewater containing explosives. | | 10 (4.54) |
| K046. Wastewater treatment sludges from the manufacturing, formulation and loading of lead-based initiating compounds. | | 100 (45.4) |
| K047. Pink/red water from TNT operations. | | 10 (4.54) |
| K048. Dissolved air flotation (DAF) float from the petroleum refining industry. | | 1 (0.454) |
| K049. Slop oil emulsion solids from the petroleum refining industry. | | 1 (0.454) |
| K050. Heat exchanger bundle cleaning sludge from the petroleum refining industry. | | 10 (4.54) |
| K051. API separator sludge from the petroleum refining industry. | | 1 (0.454) |
| K052. Tank bottoms (leaded) from the petroleum refining industry. | | 10 (4.54) |
| K060. Ammonia still lime sludge from coking operations. | | 1 (0.454) |
| K061. Emission control dust/sludge from the primary production of steel in electric furnaces. | | 1 (0.454) |
| K062. Spent pickle liquor generated by steel finishing operations of facilities whithin the iron and steel industry. | | 1 (0.454) |
| K064. Acid plant blowdown slurry/sludge resulting from thickening of blowdown slurry from primary copper production. | | 1 (0.454) |
| K065. Surface impoundment solids contained in and dredged from surface impoundments at primary lead smelting facilities. | | 1 (0.454) |
| K066. Sludge from treatment of process wastewater and/or acid plant blowdown from primary zinc production. | | 1 (0.454) |
| K069. Emission control dust/sludge from secondary lead smelting. | | 1 (0.454) |
| K071. Brine purification muds from the mercury cell process in chlorine production, where separately prepurified brine is not used. | | 1 (0.454) |
| K073. Chlorinated hydrocarbon waste from the purification step of the diaphragm cell process using graphite anodes in chlorine production. | | 10 (4.54) |
| K083. Distillation bottoms from aniline extraction. | | 100 (45.4) |
| K084. Wastewater treatment sludges generated during the production of veterinary pharmaceuticals from arsenic or organo-arsenic compounds. | | 1 (0.454) |
| K085. Distillation or fractionation column bottoms from the production of chlorobenzenes. | | 10 (4.54) |
| K086. Solvent washes and sludges, caustic washes and sludges, or water washes and sludges from cleaning tubs and equipment used in the formulation of ink from pigments, driers, soaps, and stabilizers containing chromium and lead. | | 1 (0.454) |
| K087. Decanter tank tar sludge from coking operations. | | 100 (45.4) |

**Table 1—Hazardous Substances Other Than Radionuclides,** Continued.

| Hazardous Substances | Synonyms | Reportable Quantity (RQ) Pounds (Kilograms) |
|---|---|---|
| K088. Spent potliners from primary aluminum reduction. | | 1 (0.454) |
| K090. Emission control dust or sludge from ferrochromiumsilicon production. | | 1 (0.454) |
| K091. Emission control dust or sludge from ferrochromium production. | | 1 (0.454) |
| K093. Distillation light ends from the production of phthalic anhydride from ortho-xylene. | | 5000 (2270) |
| K094. Distillation bottoms from the production of phthalic anhydride from ortho-xylene. | | 5000 (2270) |
| K095. Distillation bottoms from the production of 1,1,1-trichloroethane. | | 100 (45.4) |
| K096. Heavy ends from the heavy ends column from the production of 1,1,1-trichloroethane. | | 100 (45.4) |
| K097. Vacuum stripper discharge from the chlordane chlorinator in the production of chlordane. | | 1 (0.454) |
| K098. Untreated process wastewater from the production of toxaphene. | | 1 (0.454) |
| K099. Untreated wastewater from the production of 2,4-D. | | 10 (4.54) |
| K100. Waste leaching solution from acid leaching of emission control dust/sludge from secondary lead smelting. | | 1 (0.454) |
| K101. Distillation tar residues from the distillation of aniline-based compounds in the production of veterinary phamaceuticals from arsenic or organo-arsenic compounds. | | 1 (0.454) |
| K102. Residue from the use of activated carbon for decolorization in the production of veterinary pharmaceuticals from arsenic or organo-arsenic compounds. | | 1 (0.454) |
| K103. Process residues from aniline extraction from the production of aniline. | | 100 (45.4) |
| K104. Combined wastewater streams generated from nitrobenzene/aniline chlorobenzenes. | | 10 (4.54) |
| K105. Separated aqueous stream from the reactor product washing step in the production of chlorobenzenes. | | 10 (4.54) |
| K106. Wastewater treatment sludge from the mercury cell process in chlorine production. | | 1 (0.454) |
| K107. Column bottoms from product seperation from the production of 1,1-dimethylhydrazine (UDMH) from carboxylic acid hydrazines. | | 10 (4.54) |
| K108. Condensed column overheads from product seperation and condensed reactor vent gases from the production of 1,1-dimethylhydrazine (UDMH) from carboxylic acid hydrazides. | | 10 (4.54) |
| K109. Spent filter cartridges from product purification from the production of 1,1-dimethylhydrazine (UDMH), from carboxylic acid hydrazides. | | 10 (4.54) |
| K110. Condensed column overheads from intermediate separation from the production of 1,1-dimethylhydrazines (UDMH) from carboxylic acid hydrazides. | | 10 (4.54) |
| K111. Product washwaters from the production of dinitrotoluene via nitration of toluene. | | 10 (4.54) |

Table 2—Radionuclides

| (1)—Radionuclides | (2)—Atomic number | (3)—Reportable quantity (RQ) Ci (TBq) |
|---|---|---|
| Actinium-224 | 89 | 100 (3.7) |
| Actinium-225 | 89 | 1 (.037) |
| Actinium-226 | 89 | 10 (.37) |
| Actinium-227 | 89 | 0.001 (.000037) |
| Actinium-228 | 89 | 10 (.37) |
| Aluminum-26 | 13 | 10 (.37) |
| Americium-237 | 95 | 1000 (37) |
| Americium-238 | 95 | 100 (3.7) |
| Americium-239 | 95 | 100 (3.7) |
| Americium-240 | 95 | 10 (.37) |
| Americium-241 | 95 | 0.01 (.00037) |
| Americium-242 | 95 | 100 (3.7) |
| Americium-242m | 95 | 0.01 (.00037) |
| Americium-243 | 95 | 0.01 (.00037) |
| Americium-244 | 95 | 10 (.37) |
| Americium-244m | 95 | 1000 (37) |
| Americium-245 | 95 | 1000 (37) |
| Americium-246 | 95 | 1000 (37) |
| Americium-246m | 95 | 1000 (37) |
| Antimony-115 | 51 | 1000 (37) |
| Antimony-116 | 51 | 1000 (37) |
| Antimony-116m | 51 | 100 (3.7) |
| Antimony-117 | 51 | 1000 (37) |
| Antimony-118m | 51 | 10 (.37) |
| Antimony-119 | 51 | 1000 (37) |
| Antimony-120 (16 min) | 51 | 1000 (37) |
| Antimony-120 (5.76 day) | 51 | 10 (.37) |
| Antimony-122 | 51 | 10 (.37) |
| Antimony-124 | 51 | 10 (.37) |
| Antimony-124m | 51 | 1000 (37) |
| Antimony-125 | 51 | 10 (.37) |
| Antimony-126 | 51 | 10 (.37) |
| Antimony-126m | 51 | 1000 (37) |
| Antimony-127 | 51 | 10 (.37) |
| Antimony-128 (10.4 min) | 51 | 1000 (37) |
| Antimony-128 (9.01 hr) | 51 | 10 (.37) |
| Antimony-129 | 51 | 100 (3.7) |
| Antimony-130 | 51 | 100 (3.7) |
| Antimony-131 | 51 | 1000 (37) |
| Argon-39 | 18 | 1000 (37) |
| Argon-41 | 18 | 10 (.37) |
| Arsenic-69 | 33 | 1000 (37) |
| Arsenic-70 | 33 | 100 (3.7) |
| Arsenic-71 | 33 | 100 (3.7) |
| Arsenic-72 | 33 | 10 (.37) |
| Arsenic-73 | 33 | 100 (3.7) |
| Arsenic-74 | 33 | 10 (.37) |
| Arsenic-76 | 33 | 100 (3.7) |
| Arsenic-77 | 33 | 1000 (37) |
| Arsenic-78 | 33 | 100 (3.7) |
| Astatine-207 | 85 | 100 (3.7) |
| Astatine-211 | 85 | 100 (3.7) |
| Barium-126 | 56 | 1000 (37) |
| Barium-128 | 56 | 10 (.37) |
| Barium-131 | 56 | 10 (.37) |
| Barium-131m | 56 | 1000 (37) |
| Barium-133 | 56 | 10 (.37) |
| Barium-133m | 56 | 100 (3.7) |
| Barium-135m | 56 | 1000 (37) |
| Barium-139 | 56 | 1000 (37) |
| Barium-140 | 56 | 10 (.37) |
| Barium-141 | 56 | 1000 (37) |
| Barium-142 | 56 | 1000 (37) |
| Berkelium-245 | 97 | 100 (3.7) |
| Berkelium-246 | 97 | 10 (.37) |
| Berkelium-247 | 97 | 0.01 (.00037) |
| Berkelium-249 | 97 | 1 (.037) |
| Berkelium-250 | 97 | 100 (3.7) |
| Beryllium-7 | 4 | 100 (3.7) |
| Beryllium-10 | 4 | 1 (.037) |

Table 2—Radionuclides, Continued

| (1)—Radionuclides | (2)—Atomic number | (3)—Reportable quantity (RQ) Ci (TBq) |
|---|---|---|
| Bismuth-200 | 83 | 100 (3.7) |
| Bismuth-201 | 83 | 100 (3.7) |
| Bismuth-202 | 83 | 1000 (37) |
| Bismuth-203 | 83 | 10 (.37) |
| Bismuth-205 | 83 | 10 (.37) |
| Bismuth-206 | 83 | 10 (.37) |
| Bismuth-207 | 83 | 10 (.37) |
| Bismuth-210 | 83 | 10 (.37) |
| Bismuth-210m | 83 | 0.1 (.0037) |
| Bismuth-212 | 83 | 100 (3.7) |
| Bismuth-213 | 83 | 100 (3.7) |
| Bismuth-214 | 83 | 100 (3.7) |
| Bromine-74 | 35 | 100 (3.7) |
| Bromine-74m | 35 | 100 (3.7) |
| Bromine-75 | 35 | 100 (3.7) |
| Bromine-76 | 35 | 10 (.37) |
| Bromine-77 | 35 | 100 (3.7) |
| Bromine-80m | 35 | 1000 (37) |
| Bromine-80 | 35 | 1000 (37) |
| Bromine-82 | 35 | 10 (.37) |
| Bromine-83 | 35 | 1000 (37) |
| Bromine-84 | 35 | 100 (3.7) |
| Cadmium-104 | 48 | 1000 (37) |
| Cadmium-107 | 48 | 1000 (37) |
| Cadmium-109 | 48 | 1 (.037) |
| Cadmium-113 | 48 | 0.1 (.0037) |
| Cadmium-113m | 48 | 0.1 (.0037) |
| Cadmium-115 | 48 | 100 (3.7) |
| Cadmium-115m | 48 | 10 (.37) |
| Cadmium-117 | 48 | 100 (3.7) |
| Cadmium-117m | 48 | 10 (.37) |
| Calcium-41 | 20 | 10 (.37) |
| Calcium-45 | 20 | 10 (.37) |
| Calcium-47 | 20 | 10 (.37) |
| Californium-244 | 98 | 1000 (37) |
| Californium-246 | 98 | 10 (.37) |
| Californium-248 | 98 | 0.1 (.0037) |
| Californium-249 | 98 | 0.01 (.00037) |
| Californium-250 | 98 | 0.01 (.00037) |
| Californium-251 | 98 | 0.01 (.00037) |
| Californium-252 | 98 | 0.1 (.0037) |
| Californium-253 | 98 | 10 (.37) |
| Californium-254 | 98 | 0.1 (.0037) |
| Carbon-11 | 6 | 1000 (37) |
| Carbon-14 | 6 | 10 (.37) |
| Cerium-134 | 58 | 10 (.37) |
| Cerium-135 | 58 | 10 (.37) |
| Cerium-137 | 58 | 1000 (37) |
| Cerium-137m | 58 | 100 (3.7) |
| Cerium-139 | 58 | 100 (3.7) |
| Cerium-141 | 58 | 10 (.37) |
| Cerium-143 | 58 | 100 (3.7) |
| Cerium-144 | 58 | 1 (.037) |
| Cesium-125 | 55 | 1000 (37) |
| Cesium-127 | 55 | 100 (3.7) |
| Cesium-129 | 55 | 100 (3.7) |
| Cesium-130 | 55 | 1000 (37) |
| Cesium-131 | 55 | 1000 (37) |
| Cesium-132 | 55 | 10 (.37) |
| Cesium-134 | 55 | 1 (.037) |
| Cesium-134m | 55 | 1000 (37) |
| Cesium-135 | 55 | 10 (.37) |
| Cesium-135m | 55 | 100 (3.7) |
| Cesium-136 | 55 | 10 (.37) |
| Cesium-137 | 55 | 1 (.037) |
| Cesium-138 | 55 | 100 (3.7) |
| Chlorine-36 | 17 | 10 (.37) |
| Chlorine-38 | 17 | 100 (3.7) |
| Chlorine-39 | 17 | 100 (3.7) |
| Chromium-48 | 24 | 100 (3.7) |

Table 2—Radionuclides, Continued

| (1)—Radionuclides | (2)—Atomic number | (3)—Reportable quantity (RQ) Ci (TBq) |
|---|---|---|
| Chromium-49 | 24 | 1000 (37) |
| Chromium-51 | 24 | 1000 (37) |
| Cobalt-55 | 27 | 10 (.37) |
| Cobalt-56 | 27 | 10 (.37) |
| Cobalt-57 | 27 | 100 (3.7) |
| Cobalt-58 | 27 | 10 (.37) |
| Cobalt-58m | 27 | 1000 (37) |
| Cobalt-60 | 27 | 10 (.37) |
| Cobalt-60m | 27 | 1000 (37) |
| Cobalt-61 | 27 | 1000 (37) |
| Cobalt-62m | 27 | 1000 (37) |
| Copper-60 | 29 | 100 (3.7) |
| Copper-61 | 29 | 100 (3.7) |
| Copper-64 | 29 | 1000 (37) |
| Copper-67 | 29 | 100 (3.7) |
| Curium-238 | 96 | 1000 (37) |
| Curium-240 | 96 | 1 (.037) |
| Curium-241 | 96 | 10 (.37) |
| Curium-242 | 96 | 1 (.037) |
| Curium-243 | 96 | 0.01 (.00037) |
| Curium-244 | 96 | 0.01 (.00037) |
| Curium-245 | 96 | 0.01 (.00037) |
| Curium-246 | 96 | 0.01 (.00037) |
| Curium-247 | 96 | 0.01 (.00037) |
| Curium-248 | 96 | 0.001 (.000037) |
| Curium-249 | 96 | 1000 (37) |
| Dysprosium-155 | 66 | 100 (3.7) |
| Dysprosium-157 | 66 | 100 (3.7) |
| Dysprosium-159 | 66 | 100 (3.7) |
| Dysprosium-165 | 66 | 1000 (37) |
| Dysprosium-166 | 66 | 10 (.37) |
| Einsteinium-250 | 99 | 10 (.37) |
| Einsteinium-251 | 99 | 1000 (37) |
| Einsteinium-253 | 99 | 10 (.37) |
| Einsteinium-254 | 99 | 0.1 (.0037) |
| Einsteinium-254m | 99 | 1 (.037) |
| Erbium-161 | 68 | 100 (3.7) |
| Erbium-165 | 68 | 1000 (37) |
| Erbium-169 | 68 | 100 (3.7) |
| Erbium-171 | 68 | 100 (3.7) |
| Erbium-172 | 68 | 10 (.37) |
| Europium-145 | 63 | 10 (.37) |
| Europium-146 | 63 | 10 (.37) |
| Europium-147 | 63 | 10 (.37) |
| Europium-148 | 63 | 10 (.37) |
| Europium-149 | 63 | 100 (3.7) |
| Europium-150 (12.6 hr) | 63 | 1000 (37) |
| Europium-150 (34.2 yr) | 63 | 10 (.37) |
| Europium-152 | 63 | 10 (.37) |
| Europium-152m | 63 | 100 (3.7) |
| Europium-154 | 63 | 10 (.37) |
| Europium-155 | 63 | 10 (.37) |
| Europium-156 | 63 | 10 (.37) |
| Europium-157 | 63 | 10 (.37) |
| Europium-158 | 63 | 1000 (37) |
| Fermium-252 | 100 | 10 (.37) |
| Fermium-253 | 100 | 10 (.37) |
| Fermium-254 | 100 | 100 (3.7) |
| Fermium-255 | 100 | 100 (3.7) |
| Fermium-257 | 100 | 1 (.037) |
| Fluorine-18 | 9 | 1000 (37) |
| Francium-222 | 87 | 100 (3.7) |
| Francium-223 | 87 | 100 (3.7) |
| Gadolinium-145 | 64 | 100 (3.7) |
| Gadolinium-146 | 64 | 10 (.37) |
| Gadolinium-147 | 64 | 10 (.37) |
| Gadolinium-148 | 64 | 0.001 (.000037) |
| Gadolinium-149 | 64 | 100 (3.7) |
| Gadolinium-151 | 64 | 100 (3.7) |
| Gadolinium-152 | 64 | 0.001 (.000037) |
| Gadolinium-153 | 64 | 10 (.37) |

Table 2—Radionuclides, Continued

| (1)—Radionuclides | (2)—Atomic number | (3)—Reportable quantity (RQ) Ci (TBq) |
|---|---|---|
| Gadolinium-159 | 64 | 1000 (37) |
| Gallium-65 | 31 | 1000 (37) |
| Gallium-66 | 31 | 10 (.37) |
| Gallium-67 | 31 | 100 (3.7) |
| Gallium-68 | 31 | 1000 (37) |
| Gallium-70 | 31 | 1000 (37) |
| Gallium-72 | 31 | 10 (.37) |
| Gallium-73 | 31 | 100 (3.7) |
| Germanium-66 | 32 | 100 (3.7) |
| Germanium-67 | 32 | 1000 (37) |
| Germanium-68 | 32 | 10 (.37) |
| Germanium-69 | 32 | 10 (.37) |
| Germanium-71 | 32 | 1000 (37) |
| Germanium-75 | 32 | 1000 (37) |
| Germanium-77 | 32 | 10 (.37) |
| Germanium-78 | 32 | 1000 (37) |
| Gold-193 | 79 | 100 (3.7) |
| Gold-194 | 79 | 10 (.37) |
| Gold-195 | 79 | 100 (3.7) |
| Gold-198 | 79 | 100 (3.7) |
| Gold-198m | 79 | 10 (.37) |
| Gold-199 | 79 | 100 (3.7) |
| Gold-200 | 79 | 1000 (37) |
| Gold-200m | 79 | 10 (.37) |
| Gold-201 | 79 | 1000 (37) |
| Hafnium-170 | 72 | 100 (3.7) |
| Hafnium-172 | 72 | 1 (.037) |
| Hafnium-173 | 72 | 100 (3.7) |
| Hafnium-175 | 72 | 100 (3.7) |
| Hafnium-177m | 72 | 1000 (37) |
| Hafnium-178m | 72 | 0.1 (.0037) |
| Hafnium-179m | 72 | 100 (3.7) |
| Hafnium-180m | 72 | 100 (3.7) |
| Hafnium-181 | 72 | 10 (.37) |
| Hafnium-182 | 72 | 0.1 (.0037) |
| Hafnium-182m | 72 | 100 (3.7) |
| Hafnium-183 | 72 | 100 (3.7) |
| Hafnium-184 | 72 | 100 (3.7) |
| Holmium-155 | 67 | 1000 (37) |
| Holmium-157 | 67 | 1000 (37) |
| Holmium-159 | 67 | 1000 (37) |
| Holmium-161 | 67 | 1000 (37) |
| Holmium-162 | 67 | 1000 (37) |
| Holmium-162m | 67 | 1000 (37) |
| Holmium-164 | 67 | 1000 (37) |
| Holmium-164m | 67 | 1000 (37) |
| Holmium-166 | 67 | 100 (3.7) |
| Holmium-166m | 67 | 1 (.037) |
| Holmium-167 | 67 | 100 (3.7) |
| Hydrogen-3 | 1 | 100 (3.7) |
| Indium-109 | 49 | 100 (3.7) |
| Indium-110 (69.1 min) | 49 | 100 (3.7) |
| Indium-110 (4.9 hr) | 49 | 10 (.37) |
| Indium-111 | 49 | 100 (3.7) |
| Indium-112 | 49 | 1000 (37) |
| Indium-113m | 49 | 1000 (37) |
| Indium-114m | 49 | 10 (.37) |
| Indium-115 | 49 | 0.1 (.0037) |
| Indium-115m | 49 | 100 (3.7) |
| Indium-116m | 49 | 100 (3.7) |
| Indium-117 | 49 | 1000 (37) |
| Indium-117m | 49 | 100 (3.7) |
| Indium-119m | 49 | 1000 (37) |
| Iodine-120 | 53 | 10 (.37) |
| Iodine-120m | 53 | 100 (3.7) |
| Iodine-121 | 53 | 100 (3.7) |
| Iodine-123 | 53 | 10 (.37) |
| Iodine-124 | 53 | 0.1 (.0037) |
| Iodine-125 | 53 | 0.01 (.00037) |
| Iodine-126 | 53 | 0.01 (.00037) |
| Iodine-128 | 53 | 1000 (37) |

**Table 2—Radionuclides, Continued**

| (1)—Radionuclides | (2)—Atomic number | (3)—Reportable quantity (RQ) Ci (TBq) |
|---|---|---|
| Iodine-129 | 53 | 0.001 (.000037) |
| Iodine-130 | 53 | 1 (.037) |
| Iodine-131 | 53 | 0.01 (.00037) |
| Iodine-132 | 53 | 10 (.37) |
| Iodine-132m | 53 | 10 (.37) |
| Iodine-133 | 53 | 0.1 (.0037) |
| Iodine-134 | 53 | 100 (3.7) |
| Iodine-135 | 53 | 10 (.37) |
| Iridium-182 | 77 | 1000 (37) |
| Iridium-184 | 77 | 100 (3.7) |
| Iridium-185 | 77 | 100 (3.7) |
| Iridium-186 | 77 | 10 (.37) |
| Iridium-187 | 77 | 100 (3.7) |
| Iridium-188 | 77 | 10 (.37) |
| Iridium-189 | 77 | 100 (3.7) |
| Iridium-190 | 77 | 10 (.37) |
| Iridium-190m | 77 | 1000 (37) |
| Iridium-192 | 77 | 10 (.37) |
| Iridium-192m | 77 | 100 (3.7) |
| Iridium-194 | 77 | 100 (3.7) |
| Iridium-194m | 77 | 10 (.37) |
| Iridium-195 | 77 | 1000 (37) |
| Iridium-195m | 77 | 100 (3.7) |
| Iron-52 | 26 | 100 (3.7) |
| Iron-55 | 26 | 100 (3.7) |
| Iron-59 | 26 | 10 (.37) |
| Iron-60 | 26 | 0.1 (.0037) |
| Krypton-74 | 36 | 10 (.37) |
| Krypton-76 | 36 | 10 (.37) |
| Krypton-77 | 36 | 10 (.37) |
| Krypton-79 | 36 | 100 (3.7) |
| Krypton-81 | 36 | 1000 (37) |
| Krypton-83m | 36 | 1000 (37) |
| Krypton-85 | 36 | 1000 (37) |
| Krypton-85m | 36 | 100 (3.7) |
| Krypton-87 | 36 | 10 (.37) |
| Krypton-88 | 36 | 10 (.37) |
| Lanthanum-131 | 57 | 1000 (37) |
| Lanthanum-132 | 57 | 100 (3.7) |
| Lanthanum-135 | 57 | 1000 (37) |
| Lanthanum-137 | 57 | 10 (.37) |
| Lanthanum-138 | 57 | 1 (.037) |
| Lanthanum-140 | 57 | 10 (.37) |
| Lanthanum-141 | 57 | 1000 (37) |
| Lanthanum-142 | 57 | 100 (3.7) |
| Lanthanum-143 | 57 | 1000 (37) |
| Lead-195m | 82 | 1000 (37) |
| Lead-198 | 82 | 100 (3.7) |
| Lead-199 | 82 | 100 (3.7) |
| Lead-200 | 82 | 100 (3.7) |
| Lead-201 | 82 | 100 (3.7) |
| Lead-202 | 82 | 1 (.037) |
| Lead-202m | 82 | 10 (.37) |
| Lead-203 | 82 | 100 (3.7) |
| Lead-205 | 82 | 100 (3.7) |
| Lead-209 | 82 | 1000 (37) |
| Lead-210 | 82 | 0.01 (.00037) |
| Lead-211 | 82 | 100 (3.7) |
| Lead-212 | 82 | 10 (.37) |
| Lead-214 | 82 | 100 (3.7) |
| Lutetium-169 | 71 | 10 (.37) |
| Lutetium-170 | 71 | 10 (.37) |
| Lutetium-171 | 71 | 10 (.37) |
| Lutetium-172 | 71 | 10 (.37) |
| Lutetium-173 | 71 | 100 (3.7) |
| Lutetium-174 | 71 | 10 (.37) |
| Lutetium-174m | 71 | 10 (.37) |
| Lutetium-176 | 71 | 1 (.037) |
| Lutetium-176m | 71 | 1000 (37) |
| Lutetium-177 | 71 | 100 (3.7) |
| Lutetium-177m | 71 | 10 (.37) |
| Lutetium-178 | 71 | 1000 (37) |
| Lutetium-178m | 71 | 1000 (37) |
| Lutetium-179 | 71 | 1000 (37) |
| Magnesium-28 | 12 | 10 (.37) |
| Manganese-51 | 25 | 1000 (37) |
| Manganese-52 | 25 | 10 (.37) |
| Manganese-52m | 25 | 1000 (37) |
| Manganese-53 | 25 | 1000 (37) |
| Manganese-54 | 25 | 10 (.37) |
| Manganese-56 | 25 | 100 (3.7) |
| Mendelevium-257 | 101 | 100 (3.7) |
| Mendelevium-258 | 101 | 1 (.037) |
| Mercury-193 | 80 | 100 (3.7) |
| Mercury-193m | 80 | 10 (.37) |
| Mercury-194 | 80 | 0.1 (.0037) |
| Mercury-195 | 80 | 100 (3.7) |
| Mercury-195m | 80 | 100 (3.7) |
| Mercury-197 | 80 | 1000 (37) |
| Mercury-197m | 80 | 1000 (37) |
| Mercury-199m | 80 | 1000 (37) |
| Mercury-203 | 80 | 10 (.37) |
| Molybdenum-90 | 42 | 100 (3.7) |
| Molybdenum-93 | 42 | 100 (3.7) |
| Molybdenum-93m | 42 | 10 (.37) |
| Molybdenum-99 | 42 | 100 (3.7) |
| Molybdenum-101 | 42 | 1000 (37) |
| Neodymium-136 | 60 | 1000 (37) |
| Neodymium-138 | 60 | 1000 (37) |
| Neodymium-139 | 60 | 1000 (37) |
| Neodymium-139m | 60 | 100 (3.7) |
| Neodymium-141 | 60 | 1000 (37) |
| Neodymium-147 | 60 | 10 (.37) |
| Neodymium-149 | 60 | 100 (3.7) |
| Neodymium-151 | 60 | 1000 (37) |
| Neptunium-232 | 93 | 1000 (37) |
| Neptunium-233 | 93 | 1000 (37) |
| Neptunium-234 | 93 | 10 (.37) |
| Neptunium-235 | 93 | 1000 (37) |
| Neptunium-236 (1.2 E 5 yr) | 93 | 0.1 (.0037) |
| Neptunium-236 (22.5 hr) | 93 | 100 (3.7) |
| Neptunium-237 | 93 | 0.01 (.00037) |
| Neptunium-238 | 93 | 10 (.37) |
| Neptunium-239 | 93 | 100 (3.7) |
| Neptunium-240 | 93 | 100 (3.7) |
| Nickel-56 | 28 | 10 (.37) |
| Nickel-57 | 28 | 10 (.37) |
| Nickel-59 | 28 | 100 (3.7) |
| Nickel-63 | 28 | 100 (3.7) |
| Nickel-65 | 28 | 100 (3.7) |
| Nickel-66 | 28 | 10 (.37) |
| Niobium-88 | 41 | 100 (3.7) |
| Niobium-89 (66 min) | 41 | 100 (3.7) |
| Niobium-89 (122 min) | 41 | 100 (3.7) |
| Niobium-90 | 41 | 10 (.37) |
| Niobium-93m | 41 | 100 (3.7) |
| Niobium-94 | 41 | 10 (.37) |
| Niobium-95 | 41 | 10 (.37) |
| Niobium-95m | 41 | 100 (3.7) |
| Niobium-96 | 41 | 10 (.37) |
| Niobium-97 | 41 | 100 (3.7) |
| Niobium-98 | 41 | 1000 (37) |
| Osmium-180 | 76 | 1000 (37) |
| Osmium-181 | 76 | 100 (3.7) |
| Osmium-182 | 76 | 100 (3.7) |
| Osmium-185 | 76 | 10 (.37) |
| Osmium-189m | 76 | 1000 (37) |
| Osmium-191 | 76 | 100 (3.7) |
| Osmium-191m | 76 | 1000 (37) |
| Osmium-193 | 76 | 100 (3.7) |
| Osmium-194 | 76 | 1 (.037) |
| Palladium-100 | 46 | 100 (3.7) |

Table 2—Radionuclides, Continued

| (1)—Radionuclides | (2)—Atomic number | (3)—Reportable quantity (RQ) Ci (TBq) |
|---|---|---|
| Palladium-101 | 46 | 100 (3.7) |
| Palladium-103 | 46 | 100 (3.7) |
| Palladium-107 | 46 | 100 (3.7) |
| Palladium-109 | 46 | 1000 (37) |
| Phosphorus-32 | 15 | 0.1 (.0037) |
| Phosphorus-33 | 15 | 1 (.037) |
| Platinum-186 | 78 | 100 (3.7) |
| Platinum-188 | 78 | 100 (3.7) |
| Platinum-189 | 78 | 100 (3.7) |
| Platinum-191 | 78 | 100 (3.7) |
| Platinum-193 | 78 | 1000 (37) |
| Platinum-193m | 78 | 100 (3.7) |
| Platinum-195m | 78 | 100 (3.7) |
| Platinum-197 | 78 | 1000 (37) |
| Platinum-197m | 78 | 1000 (37) |
| Platinum-199 | 78 | 1000 (37) |
| Platinum-200 | 78 | 100 (3.7) |
| Plutonium-234 | 94 | 1000 (37) |
| Plutonium-235 | 94 | 1000 (37) |
| Plutonium-236 | 94 | 0.1 (.0037) |
| Plutonium-237 | 94 | 1000 (37) |
| Plutonium-238 | 94 | 0.01 (.00037) |
| Plutonium-239 | 94 | 0.01 (.00037) |
| Plutonium-240 | 94 | 0.01 (.00037) |
| Plutonium-241 | 94 | 1 (.037) |
| Plutonium-242 | 94 | 0.01 (.00037) |
| Plutonium-243 | 94 | 1000 (37) |
| Plutonium-244 | 94 | 0.01 (.00037) |
| Plutonium-245 | 94 | 100 (3.7) |
| Polonium-203 | 84 | 100 (3.7) |
| Polonium-205 | 84 | 100 (3.7) |
| Polonium-207 | 84 | 10 (.37) |
| Polonium-210 | 84 | 0.01 (.00037) |
| Potassium-40 | 19 | 1 (.037) |
| Potassium-42 | 19 | 100 (3.7) |
| Potassium-43 | 19 | 10 (.37) |
| Potassium-44 | 19 | 100 (3.7) |
| Potassium-45 | 19 | 1000 (37) |
| Praseodymium-136 | 59 | 1000 (37) |
| Praseodymium-137 | 59 | 1000 (37) |
| Praseodymium-138m | 59 | 100 (3.7) |
| Praseodymium-139 | 59 | 1000 (37) |
| Praseodymium-142 | 59 | 100 (3.7) |
| Praseodymium-142m | 59 | 1000 (37) |
| Praseodymium-143 | 59 | 10 (.37) |
| Praseodymium-144 | 59 | 1000 (37) |
| Praseodymium-145 | 59 | 1000 (37) |
| Praseodymium-147 | 59 | 1000 (37) |
| Promethium-141 | 61 | 1000 (37) |
| Promethium-143 | 61 | 100 (3.7) |
| Promethium-144 | 61 | 10 (.37) |
| Promethium-145 | 61 | 100 (3.7) |
| Promethium-146 | 61 | 10 (.37) |
| Promethium-147 | 61 | 10 (.37) |
| Promethium-148 | 61 | 10 (.37) |
| Promethium-148m | 61 | 10 (.37) |
| Promethium-149 | 61 | 100 (3.7) |
| Promethium-150 | 61 | 100 (3.7) |
| Promethium-151 | 61 | 100 (3.7) |
| Protactinium-227 | 91 | 100 (3.7) |
| Protactinium-228 | 91 | 10 (.37) |
| Protactinium-230 | 91 | 10 (.37) |
| Protactinium-231 | 91 | 0.01 (.00037) |
| Protactinium-232 | 91 | 10 (.37) |
| Protactinium-233 | 91 | 100 (3.7) |
| Protactinium-234 | 91 | 10 (.37) |
| RADIONUCLIDES§† | | 1 (.037) |
| Radium-223 | 88 | 1 (.037) |
| Radium-224 | 88 | 10 (.37) |
| Radium-225 | 88 | 1 (.037) |
| Radium-226** | 88 | 0.1 (.0037) |

Table 2—Radionuclides, Continued

| (1)—Radionuclides | (2)—Atomic number | (3)—Reportable quantity (RQ) Ci (TBq) |
|---|---|---|
| Radium-227 | 88 | 1000 (37) |
| Radium-228 | 88 | 0.1 (.0037) |
| Radon-220 | 86 | 0.1 (.0037) |
| Radon-222 | 86 | 0.1 (.0037) |
| Rhenium-177 | 75 | 1000 (37) |
| Rhenium-178 | 75 | 1000 (37) |
| Rhenium-181 | 75 | 100 (3.7) |
| Rhenium-182 (12.7 hr) | 75 | 10 (.37) |
| Rhenium-182 (64.0 hr) | 75 | 10 (.37) |
| Rhenium-184 | 75 | 10 (.37) |
| Rhenium-184m | 75 | 10 (.37) |
| Rhenium-186 | 75 | 100 (3.7) |
| Rhenium-186m | 75 | 10 (.37) |
| Rhenium-187 | 75 | 1000 (37) |
| Rhenium-188 | 75 | 1000 (37) |
| Rhenium-188m | 75 | 1000 (37) |
| Rhenium-189 | 75 | 1000 (37) |
| Rhodium-99 | 45 | 10 (.37) |
| Rhodium-99m | 45 | 100 (3.7) |
| Rhodium-100 | 45 | 10 (.37) |
| Rhodium-101 | 45 | 10 (.37) |
| Rhodium-101m | 45 | 100 (3.7) |
| Rhodium-102 | 45 | 10 (.37) |
| Rhodium-102m | 45 | 10 (.37) |
| Rhodium-103m | 45 | 1000 (37) |
| Rhodium-105 | 45 | 100 (3.7) |
| Rhodium-106m | 45 | 10 (.37) |
| Rhodium-107 | 45 | 1000 (37) |
| Rubidium-79 | 37 | 1000 (37) |
| Rubidium-81 | 37 | 100 (3.7) |
| Rubidium-81m | 37 | 1000 (37) |
| Rubidium-82m | 37 | 10 (.37) |
| Rubidium-83 | 37 | 10 (.37) |
| Rubidium-84 | 37 | 10 (.37) |
| Rubidium-86 | 37 | 10 (.37) |
| Rubidium-88 | 37 | 1000 (37) |
| Rubidium-89 | 37 | 1000 (37) |
| Rubidium-87 | 37 | 10 (.37) |
| Ruthenium-94 | 44 | 1000 (37) |
| Ruthenium-97 | 44 | 100 (3.7) |
| Ruthenium-103 | 44 | 10 (.37) |
| Ruthenium-105 | 44 | 100 (3.7) |
| Ruthenium-106 | 44 | 1 (.037) |
| Samarium-141 | 62 | 1000 (37) |
| Samarium-141m | 62 | 1000 (37) |
| Samarium-142 | 62 | 1000 (37) |
| Samarium-145 | 62 | 100 (3.7) |
| Samarium-146 | 62 | 0.01 (.00037) |
| Samarium-147 | 62 | 0.01 (.00037) |
| Samarium-151 | 62 | 10 (.37) |
| Samarium-153 | 62 | 100 (3.7) |
| Samarium-155 | 62 | 1000 (37) |
| Samarium-156 | 62 | 100 (3.7) |
| Scandium-43 | 21 | 1000 (37) |
| Scandium-44 | 21 | 100 (3.7) |
| Scandium-44m | 21 | 10 (.37) |
| Scandium-46 | 21 | 10 (.37) |
| Scandium-47 | 21 | 100 (3.7) |
| Scandium-48 | 21 | 10 (.37) |
| Scandium-49 | 21 | 1000 (37) |
| Selenium-70 | 34 | 1000 (37) |
| Selenium-73 | 34 | 10 (.37) |
| Selenium-73m | 34 | 100 (3.7) |
| Selenium-75 | 34 | 10 (.37) |
| Selenium-79 | 34 | 10 (.37) |
| Selenium-81 | 34 | 1000 (37) |
| Selenium-81m | 34 | 1000 (37) |
| Selenium-83 | 34 | 1000 (37) |
| Silicon-31 | 14 | 1000 (37) |
| Silicon-32 | 14 | 1 (.037) |
| Silver-102 | 47 | 100 (3.7) |

Table 2—Radionuclides, Continued

| (1)—Radionuclides | (2)—Atomic number | (3)—Reportable quantity (RQ) Ci (TBq) |
|---|---|---|
| Silver-103 | 47 | 1000 (37) |
| Silver-104 | 47 | 1000 (37) |
| Silver-104m | 47 | 1000 (37) |
| Silver-105 | 47 | 10 (.37) |
| Silver-106 | 47 | 1000 (37) |
| Silver-106m | 47 | 10 (.37) |
| Silver-108m | 47 | 10 (.37) |
| Silver-110m | 47 | 10 (.37) |
| Silver-111 | 47 | 10 (.37) |
| Silver-112 | 47 | 100 (3.7) |
| Silver-115 | 47 | 1000 (37) |
| Sodium-22 | 11 | 10 (.37) |
| Sodium-24 | 11 | 10 (.37) |
| Strontium-80 | 38 | 100 (3.7) |
| Strontium-81 | 38 | 1000 (37) |
| Strontium-83 | 38 | 100 (3.7) |
| Strontium-85 | 38 | 10 (.37) |
| Strontium-85m | 38 | 1000 (37) |
| Strontium-87m | 38 | 100 (3.7) |
| Strontium-89 | 38 | 10 (.37) |
| Strontium-90 | 38 | 0.1 (.0037) |
| Strontium-91 | 38 | 10 (.37) |
| Strontium-92 | 38 | 100 (3.7) |
| Sulfur-35 | 16 | 1 (.037) |
| Tantalum-172 | 73 | 100 (3.7) |
| Tantalum-173 | 73 | 100 (3.7) |
| Tantalum-174 | 73 | 100 (3.7) |
| Tantalum-175 | 73 | 100 (3.7) |
| Tantalum-176 | 73 | 10 (.37) |
| Tantalum-177 | 73 | 1000 (37) |
| Tantalum-178 | 73 | 1000 (37) |
| Tantalum-179 | 73 | 1000 (37) |
| Tantalum-180 | 73 | 100 (3.7) |
| Tantalum-180m | 73 | 1000 (37) |
| Tantalum-182 | 73 | 10 (.37) |
| Tantalum-182m | 73 | 1000 (37) |
| Tantalum-183 | 73 | 100 (3.7) |
| Tantalum-184 | 73 | 10 (.37) |
| Tantalum-185 | 73 | 1000 (37) |
| Tantalum-186 | 73 | 1000 (37) |
| Technetium-93 | 43 | 100 (3.7) |
| Technetium-93m | 43 | 1000 (37) |
| Technetium-94 | 43 | 10 (.37) |
| Technetium-94m | 43 | 100 (3.7) |
| Technetium-96 | 43 | 10 (.37) |
| Technetium-96m | 43 | 1000 (37) |
| Technetium-97 | 43 | 100 (3.7) |
| Technetium-97m | 43 | 100 (3.7) |
| Technetium-98 | 43 | 10 (.37) |
| Technetium-99 | 43 | 10 (.37) |
| Technetium-99m | 43 | 100 (3.7) |
| Technetium-101 | 43 | 1000 (37) |
| Technetium-104 | 43 | 1000 (37) |
| Tellurium-116 | 52 | 1000 (37) |
| Tellurium-121 | 52 | 10 (.37) |
| Tellurium-121m | 52 | 10 (.37) |
| Tellurium-123 | 52 | 10 (.37) |
| Tellurium-123m | 52 | 10 (.37) |
| Tellurium-125m | 52 | 10 (.37) |
| Tellurium-127 | 52 | 1000 (37) |
| Tellurium-127m | 52 | 10 (.37) |
| Tellurium-129 | 52 | 1000 (37) |
| Tellurium-129m | 52 | 10 (.37) |
| Tellurium-131 | 52 | 1000 (37) |
| Tellurium-131m | 52 | 10 (.37) |
| Tellurium-132 | 52 | 10 (.37) |
| Tellurium-133 | 52 | 1000 (37) |
| Tellurium-133m | 52 | 1000 (37) |
| Tellurium-134 | 52 | 1000 (37) |
| Terbium-147 | 65 | 100 (3.7) |
| Terbium-149 | 65 | 100 (3.7) |

Table 2—Radionuclides, Continued

| (1)—Radionuclides | (2)—Atomic number | (3)—Reportable quantity (RQ) Ci (TBq) |
|---|---|---|
| Terbium-150 | 65 | 100 (3.7) |
| Terbium-151 | 65 | 10 (.37) |
| Terbium-153 | 65 | 100 (3.7) |
| Terbium-154 | 65 | 10 (.37) |
| Terbium-155 | 65 | 100 (3.7) |
| Terbium-156m (5.0 hr) | 65 | 1000 (37) |
| Terbium-156m (24.4 hr) | 65 | 1000 (37) |
| Terbium-156 | 65 | 10 (.37) |
| Terbium-157 | 65 | 100 (3.7) |
| Terbium-158 | 65 | 10 (.37) |
| Terbium-160 | 65 | 10 (.37) |
| Terbium-161 | 65 | 100 (3.7) |
| Thallium-194 | 81 | 1000 (37) |
| Thallium-194m | 81 | 100 (3.7) |
| Thallium-195 | 81 | 100 (3.7) |
| Thallium-197 | 81 | 100 (3.7) |
| Thallium-198 | 81 | 10 (.37) |
| Thallium-198m | 81 | 100 (3.7) |
| Thallium-199 | 81 | 100 (3.7) |
| Thallium-200 | 81 | 10 (.37) |
| Thallium-201 | 81 | 1000 (37) |
| Thallium-202 | 81 | 10 (.37) |
| Thallium-204 | 81 | 10 (.37) |
| Thorium (Irradiated) | 90 | *** |
| Thorium (Natural) | 90 | ** |
| Thorium-226 | 90 | 100 (3.7) |
| Thorium-227 | 90 | 1 (.037) |
| Thorium-228 | 90 | 0.01 (.00037) |
| Thorium-229 | 90 | 0.001 (.000037) |
| Thorium-230 | 90 | 0.01 (.00037) |
| Thorium-231 | 90 | 100 (3.7) |
| Thorium-232** | 90 | 0.001 (.000037) |
| Thorium-234 | 90 | 100 (3.7) |
| Thulium-162 | 69 | 1000 (37) |
| Thulium-166 | 69 | 10 (.37) |
| Thulium-167 | 69 | 100 (3.7) |
| Thulium-170 | 69 | 10 (.37) |
| Thulium-171 | 69 | 100 (3.7) |
| Thulium-172 | 69 | 100 (3.7) |
| Thulium-173 | 69 | 100 (3.7) |
| Thulium-175 | 69 | 1000 (37) |
| Tin-110 | 50 | 100 (3.7) |
| Tin-111 | 50 | 1000 (37) |
| Tin-113 | 50 | 10 (.37) |
| Tin-117m | 50 | 100 (3.7) |
| Tin-119m | 50 | 10 (.37) |
| Tin-121 | 50 | 1000 (37) |
| Tin-121m | 50 | 10 (.37) |
| Tin-123 | 50 | 10 (.37) |
| Tin-123m | 50 | 1000 (37) |
| Tin-125 | 50 | 10 (.37) |
| Tin-126 | 50 | 1 (.037) |
| Tin-127 | 50 | 100 (3.7) |
| Tin-128 | 50 | 1000 (37) |
| Titanium-44 | 22 | 1 (.037) |
| Titanium-45 | 22 | 1000 (37) |
| Tungsten-176 | 74 | 1000 (37) |
| Tungsten-177 | 74 | 100 (3.7) |
| Tungsten-178 | 74 | 100 (3.7) |
| Tungsten-179 | 74 | 1000 (37) |
| Tungsten-181 | 74 | 100 (3.7) |
| Tungsten-185 | 74 | 10 (.37) |
| Tungsten-187 | 74 | 100 (3.7) |
| Tungsten-188 | 74 | 10 (.37) |
| Uranium (Depleted) | 92 | *** |
| Uranium (Irradiated) | 92 | *** |
| Uranium (Natural) | 92 | ** |
| Uranium Enriched 20% or greater | 92 | *** |
| Uranium Enriched less than 20% | 92 | *** |
| Uranium-230 | 92 | 1 (.037) |
| Uranium-231 | 92 | 1000 (37) |

Table 2—Radionuclides, Continued

| (1)—Radionuclides | (2)—Atomic number | (3)—Reportable quantity (RQ) Ci (TBq) |
|---|---|---|
| Uranium-232 | 92 | 0.01 (.00037) |
| Uranium-233 | 92 | 0.1 (.0037) |
| Uranium-234** | 92 | 0.1 (.0037) |
| Uranium-235** | 92 | 0.1 (.0037) |
| Uranium-236 | 92 | 0.1 (.0037) |
| Uranium-237 | 92 | 100 (3.7) |
| Uranium-238** | 92 | 0.1 (.0037) |
| Uranium-239 | 92 | 1000 (37) |
| Uranium-240 | 92 | 1000 (37) |
| Vanadium-47 | 23 | 1000 (37) |
| Vanadium-48 | 23 | 10 (.37) |
| Vanadium-49 | 23 | 1000 (37) |
| Xenon-120 | 54 | 100 (3.7) |
| Xenon-121 | 54 | 10 (.37) |
| Xenon-122 | 54 | 100 (3.7) |
| Xenon-123 | 54 | 10 (.37) |
| Xenon-125 | 54 | 100 (3.7) |
| Xenon-127 | 54 | 100 (3.7) |
| Xenon-129m | 54 | 1000 (37) |
| Xenon-131m | 54 | 1000 (37) |
| Xenon-133 | 54 | 1000 (37) |
| Xenon-133m | 54 | 1000 (37) |
| Xenon-135 | 54 | 100 (3.7) |
| Xenon-135m | 54 | 10 (.37) |
| Xenon-138 | 54 | 10 (.37) |
| Ytterbium-162 | 70 | 1000 (37) |
| Ytterbium-166 | 70 | 10 (.37) |
| Ytterbium-167 | 70 | 1000 (37) |
| Ytterbium-169 | 70 | 10 (.37) |
| Ytterbium-175 | 70 | 100 (3.7) |
| Ytterbium-177 | 70 | 1000 (37) |
| Ytterbium-178 | 70 | 1000 (37) |
| Yttrium-86 | 39 | 10 (.37) |
| Yttrium-86m | 39 | 1000 (37) |
| Yttrium-87 | 39 | 10 (.37) |
| Yttrium-88 | 39 | 10 (.37) |
| Yttrium-90 | 39 | 10 (.37) |
| Yttrium-90m | 39 | 100 (3.7) |
| Yttrium-91 | 39 | 10 (.37) |
| Yttrium-91m | 39 | 1000 (37) |
| Yttrium-92 | 39 | 100 (3.7) |
| Yttrium-93 | 39 | 100 (3.7) |
| Yttrium-94 | 39 | 1000 (37) |
| Yttrium-95 | 39 | 1000 (37) |
| Zinc-62 | 30 | 100 (3.7) |
| Zinc-63 | 30 | 1000 (37) |
| Zinc-65 | 30 | 10 (.37) |
| Zinc-69 | 30 | 1000 (37) |
| Zinc-69m | 30 | 100 (3.7) |
| Zinc-71m | 30 | 100 (3.7) |
| Zinc-72 | 30 | 100 (3.7) |
| Zirconium-86 | 40 | 100 (3.7) |
| Zirconium-88 | 40 | 10 (.37) |
| Zirconium-89 | 40 | 100 (3.7) |
| Zirconium-93 | 40 | 1 (.037) |
| Zirconium-95 | 40 | 10 (.37) |
| Zirconium-97 | 40 | 10 (.37) |

§ The RQs for all radionuclides apply to chemical compounds containing the radionuclides and elemental forms regardless of the diameter of pieces of solid material.

† The RQ of one curie applies to all radionuclides not otherwise listed. Whenever the RQs in TABLE 1—HAZARDOUS SUBSTANCES OTHER THAN RADIONUCLIDES and this table conflict, the lowest RQ shall apply. For example, uranyl acetate and uranyl nitrate have RQs shown in TABLE 1 of 100 pounds, equivalent to about one-tenth the RQ level for uranium-238 in this table.

** The method to determine the RQs for mixtures or solutions of radionuclides can be found in paragraph 6 of the note preceding TABLE 1 of this Appendix. RQs for the following four common radionuclide mixtures are provided: radium-226 in secular equilibrium with its daughters (0.053 curie); natural uranium (0.1 curie); natural uranium in secular equilibrium with its daughters (0.052 curie); and natural thorium in secular equilibrium with its daughters (0.011 curie).

*** Indicates that the name was added by RSPA because it appears in the list of radionuclides in 49 CFR 173.435. The reportable quantity (RQ), if not specifically listed elsewhere in this Appendix, shall be determined in accordance with the procedures in Paragraph 6 of this Appendix.

## Appendix B to §172.101 - List of Marine Pollutants

1. This appendix lists potential marine pollutants as defined in §171.8 of this subchapter.

2. If a marine pollutant meets the definition of any hazard class or division as defined in this subchapter, other than Class 9, the class of the material must be determined in accordance with §173.2a of this subchapter.

3. This appendix contains two columns. The first column, entitled "S.M.P." (for severe marine pollutants), identifies whether a material is a severe marine pollutant. If the letters "PP" appear in this column for a material, the material is a severe marine pollutant, otherwise it is not. The second column, entitled "Marine Pollutant", lists the marine pollutants.

| S.M.P. | Marine Pollutant |
|---|---|
| (1) | (2) |
| | Acetone cyanohydrin, stabilized |
| | Acetylene dibromide |
| | Acetylene tetrabromide |
| | Acetylene tetrachloride |
| | Acraldehyde, inhibited |
| | Acrolein, inhibited |
| | Acrylic aldehyde, inhibited |
| | Alcohol C-12 - C-15 poly(1-3) ethoxylate |
| | Alcohol C-13 - C-15 poly (1-6) ethoxylate |
| | Alcohol C-6 C-17 (secondary)poly(3-6) ethoxylate |
| | Aldicarb |
| PP | Aldrin |
| | Alkylphenols, liquid, n.o.s. *(including C2-C8 homologues)* |
| | Alkylphenols, solid, n.o.s. *(including C2-C8 homologues)* |
| | Allyl bromide |
| | ortho-Aminoanisole |
| | Aminocarb |
| | Ammonium arsenate |
| | Ammonium dinitro-o-cresolate |
| | Amyl mercaptans |
| | ortho-Anisidines |
| | Arsenates, liquid, n.o.s. |
| | Arsenates, solid, n.o.s. |
| | Arsenic bromide |
| | Arsenic chloride |
| | Arsenical pesticides, liquid, toxic, flammable, n.o.s. |
| PP | Azenphos-methyl |
| PP | Azinphos-ethyl |
| | Barium compounds, soluble, n.o.s. |
| | Barium cyanide |
| | Bendiocarb |
| | Benquinox |
| | Benzyl chlorocarbonate |
| | Benzyl chloroformate |
| PP | Binapacryl |
| | Biphenyl phenyl ether and diphenyl oxide, mixtures |
| PP | Brodifacoum |
| | Bromine cyanide |
| | Bromoallylene |
| | ortho-Bromobenzyl cyanide |
| | Bromocyane |
| | Bromoform |
| PP | Bromophos-ethyl |
| | 3-Bromopropene |
| | Bromoxynil |
| | 2-Butenal, inhibited |
| | Butyl benzenes |
| | Butyl benzyl phthalate |
| | Butylphenols, liquid |
| | Butylphenols, solid |

| S.M.P. | Marine Pollutant |
|---|---|
| (1) | (2) |
| | para-tertiary-butyltoluene |
| PP | Cadmium compounds |
| | Cadmium sulphide |
| | Calcium arsenate |
| | Calcium arsenate and calcium arsenite, mixtures, solid |
| | Calcium cyanide |
| | Calcium naphthenate |
| PP | Camphechlor |
| | Carbaryl |
| | Carbofuran |
| | Carbon bisulphide |
| | Carbon tetrabromide |
| | Carbon tetrachloride |
| PP | Carbophenothion |
| | Cartap hydrochloride |
| PP | Chlordane |
| | Chlorfenvinphos |
| | Chlorinated paraffins (C-10 — C-13) |
| | Chlorine |
| | Chlorine cyanide, inhibited |
| | Chlomephos |
| | 4-Chloro-2-nitrotoluene |
| | Chloro-ortho-nitrotoluene |
| | 2-Chloro-5-trifluoromethylnitrobenzene |
| | para-Chlorobenzyl chloride, liquid or solid |
| | Chlorobenzylchlorides |
| | Chlorodinitrobenzenes |
| | 1-Chloroheptane |
| | 1-Chlorohexane |
| | Chloronitroanilines |
| | Chloronitrotoluenes, *liquid* |
| | Chloronitrotoluenes, *solid* |
| | 1-Chlorooctane |
| PP | Chlorophenates, liquid |
| PP | Chlorophenates, solid |
| | Chlorophenols, liquid |
| | Chlorophenols, solid |
| | Chlorophenyltrichlorosilane |
| | alpha-Chloropropylene |
| | 1-Chloropropylene |
| | 2-Chloropropylene |
| | Chlorotoluenes |
| PP | Chlorpyriphos |
| PP | Chlorthiophos |
| | Chromyl chloride |
| | Coal tar |
| | Coal tar naphtha |
| | Cocculus |
| | Copper acetoarsenite |
| | Copper arsenite |
| | Copper chloride |
| PP | Copper cyanide |
| | Coumachlor |
| PP | Coumaphos |
| | Creosote (coal tar) |
| | Creosote (wood tar) |
| | Cresols (*o-; m-; p-*) |
| PP | Cresyl diphenyl phosphate |
| | Cresylic acid |
| | Cresylic acid sodium salt |
| | Crotonaldehyde, inhibited |
| | Crotonic aldehyde |
| | Crotoxyphos |
| | Cumene |
| | Cupric arsenite |
| | Cupric chloride |
| PP | Cupric cyanide |
| | Cupriethylenediamine solution |
| | Cuprous chloride |
| | Cyanide mixtures |
| | Cyanide solutions |
| | Cyanides, inorganic, n.o.s. |
| | Cyanogen bromide |
| | Cyanogen chloride, inhibited |
| | Cyanophos |
| PP | Cyhexatin |
| PP | Cypermethrin |

| S.M.P. | Marine Pollutant |
|---|---|
| (1) | (2) |
| | 2,4-D |
| PP | DDT |
| | Decyl acrylate |
| | DEF |
| | Di-allate |
| | Di-n-Butyl phthalate |
| PP | Dialifos |
| PP | Diazinon |
| | 1,2-Dibromethene |
| | 1,2-Dibromoethane |
| PP | Dichlofenthion |
| | Dichloroanilines |
| | o-Dichlorobenzene |
| | p-Dichlorobenzene |
| | 1,3-Dichlorobenzene |
| | 1,2-Dichlorobenzene |
| | 1,4-Dichlorobenzene |
| | Dichlorobenzene (meta; ortho; para) |
| | 2,2-Dichlorodiethyl ether |
| | Dichloroether |
| | Dichloroethyl ether |
| | Dichloroethyl oxide |
| | 1,1-Dichloroethylene, inhibited |
| | 1,6-Dichlorohexane |
| | Dichlorophenols, liquid |
| | Dichlorophenols, solid |
| | 2,4-Dichlorophenoxyacetic acid (see also 2,4D) |
| | 2,4-Dichlorophenoxyacetic acid diethanolamine salt |
| | 2,4-Dichlorophenoxyacetic acid dimethylamine salt |
| | 2,4-Dichlorophenoxyacetic acid triisopropylamine salt |
| | Dichlorophenyltrichlorosilane |
| PP | Dichlorvos |
| | Dicrotophos |
| PP | Dieldrin |
| | Diisopropylbenzenes |
| PP | Dimethoate |
| | Dimethylarsinic acid |
| | Dimethylphenols, liquid or solid |
| | Dinitro-o-cresol, *solid* |
| | Dinitro-o-cresol, *solution* |
| | Dinitrochlorobenzenes, liquid or solid |
| | Dinitrophenol, *dry or wetted with less than 15 per cent water, by mass* |
| | Dinitrophenol solutions |
| | Dinitrophenol, wetted *with not less than 15 per cent water, by mass* |
| | Dinitrophenolates *alkali metals, dry or wetted with less than 15 per cent water, by mass* |
| | Dinitrophenolates, wetted *with not less than 15 per cent water, by mass* |
| | Dinobuton |
| | Dinoseb |
| | Dioxacarb |
| | Dioxathion |
| | Diphacinone |
| | Diphenyl |
| | Diphenyl ether |
| | Diphenyl oxide |
| | Diphenyl oxide and biphenyl phenyl ether mixtures |
| PP | Diphenylamine chloroarsine |
| PP | Diphenylchloroarsine, solid *or* liquid |
| | Disulfoton |
| | DNOC |
| | DNOC (pesticide) |
| PP | Dodecylphenol |
| | Drazoxolon |
| | Edifenphos |
| PP | Endosulfan |
| PP | Endrin |
| PP | EPN |
| PP | Ethion |
| | Ethoprophos |

| S.M.P. | Marine Pollutant |
|---|---|
| (1) | (2) |
| | Ethyl acrylate, inhibited |
| | Ethyl chlorothioformate |
| | Ethyl fluid |
| | 5-Ethyl-2-picoline |
| | Ethyl propenoate, inhibited |
| | 2-Ethyl-3-propylacrolein |
| | Ethyl tetraphosphate |
| | Ethyldichloroarsine |
| | Ethylene chloride |
| | Ethylene dibromide and methyl bromide mixtures, liquid |
| | Ethylene dichloride |
| | 2-Ethylhexenal |
| | Ethylidene dichloride |
| | Fenaminphos |
| PP | Fenitrothion |
| PP | Fenpropathrin |
| | Fensulfothion |
| PP | Fenthion |
| PP | Fentin acetate |
| PP | Fentin hydroxide |
| | Ferric arsenate |
| | Ferric arsenite |
| | Ferrous arsenate |
| PP | Fonofos |
| | Formetanate |
| PP | gamma-BHC |
| | Gasoline, leaded |
| PP | Heptachlor |
| | Heptenophos |
| | normal-Heptyl chloride |
| PP | Hexachlorobutadiene |
| | 1,3-Hexachlorobutadiene |
| | Hexaethyl tetraphosphate, *liquid* |
| | Hexaethyl tetraphosphate, *solid* |
| | normal-Hexyl chloride |
| | Hydrocyanic acid, anhydrous, stabilized |
| | Hydrocyanic acid, anhydrous, stabilized, absorbed in a porous inert material |
| | Hydrocyanic acid, aqueous solutions *not more than 20% hydrocyanic acid* |
| | Hydrogen cyanide, anhydrous, stabilized |
| | Hydrogen cyanide, anhydrous, stabilized, *absorbed in a porous inert material* |
| | Hydroxydimethylbenzenes, liquid or solid |
| | Ioxynil |
| | Isoamyl mercaptan |
| | Isodecyl acrylate |
| | Isodecyl diphenyl phosphate |
| | Isofenphos |
| | Isooctyl nitrate |
| | Isoprocarb |
| | Isopropenyl chloride |
| | Isopropenylbenzene |
| | Isopropyl chloride |
| | Isopropylbenzene |
| PP | Isoxathion |
| | Lead acetate |
| | Lead arsenates |
| | Lead arsenites |
| | Lead compounds, soluble, n.o.s. |
| | Lead cyanide |
| | Lead nitrate |
| | Lead perchlorate, solid or solution |
| | Lead tetraethyl |
| | Lead tetramethyl |
| PP | Lindane |
| | London Purple |
| | Magnesium arsenate |
| | Malathion |
| | Maneb or Maneb preparations with not less than 60% maneb |
| | Maneb *or* Maneb preparations *with not less than 60 per cent maneb* |
| | Maneb stabilized *or* Maneb preparations, stabilized *against self-heating* |
| | Manganese ethylene-1,2-bisdithiocarbamate |

| S.M.P. | Marine Pollutant |
|---|---|
| (1) | (2) |
| | Manganese ethylene-1,2-bisdithiocarbamate, stabilized against self-heating |
| | Mephosfolan |
| | Mercaptodimethur |
| | Mercarbam |
| PP | Mercuric acetate |
| PP | Mercuric ammonium chloride |
| PP | Mercuric arsenate |
| PP | Mercuric benzoate |
| PP | Mercuric bisulphate |
| PP | Mercuric bromide |
| PP | Mercuric chloride |
| PP | Mercuric cyanide |
| PP | Mercuric gluconate |
| | Mercuric iodide |
| PP | Mercuric nitrate |
| PP | Mercuric oleate |
| PP | Mercuric oxide |
| PP | Mercuric oxycyanide, desensitized |
| PP | Mercuric potassium cyanide |
| PP | Mercuric Sulphate |
| | Mercuric sulphide |
| PP | Mercuric thiocyanate |
| PP | Mercurol |
| PP | Mercurous acetate |
| PP | Mercurous bisuphate |
| PP | Mercurous bromide |
| PP | Mercurous chloride |
| PP | Mercurous nitrate |
| PP | Mercurous salicylate |
| PP | Mercurous sulphate |
| PP | Mercury acetates |
| PP | Mercury ammonium chloride |
| PP | Mercury based pesticides, liquid, flammable, toxic, n.o.s. |
| PP | Mercury based pesticides, liquid, toxic, flammable, n.o.s. |
| PP | Mercury based pesticides, liquid, toxic, n.o.s. |
| PP | Mercury based pesticides, solid, toxic, n.o.s. |
| PP | Mercury benzoate |
| PP | Mercury bichloride |
| PP | Mercury bisulphates |
| PP | Mercury bromides |
| PP | Mercury compounds, liquid, n.o.s. |
| PP | Mercury compounds, solid, n.o.s. |
| PP | Mercury compounds, solid, n.o.s. |
| PP | Mercury cyanide |
| PP | Mercury gluconate |
| PP | Mercury (I) (mercurous) compounds (pesticides) |
| PP | Mercury (II) (mercuric) compounds (pesticides) |
| | Mercury iodide |
| | Mercury iodide, *solution* |
| PP | Mercury nucleate |
| PP | Mercury oleate |
| PP | Mercury oxide |
| PP | Mercury oxycyanide, desensitized |
| PP | Mercury potassium cyanide |
| PP | Mercury potassium iodide |
| PP | Mercury salicylate |
| PP | Mercury sulfates |
| PP | Mercury thiocyanate |
| | Metaarsenic acid |
| | Metam-sodium |
| | Methamidophos |
| | Methanethiol |
| | Methidathion |
| | Methomyl |
| | ortho-Methoxyaniline |
| | Methyl bromide and ethylene dibromide mixtures, liquid |
| | 1-Methyl-4-ethylbenzene |
| | 2-Methyl-5-ethylpyridine |
| | Methyl mercaptan |
| | 2-Methyl-2-phenylpropane |
| | 3-Methyl pyridine |
| | Methyl salicylate |
| | 3-Methylacroleine, inhibited |
| | Methylchlorobenzenes |
| | Methylchloroform |
| | Methylene bromide |
| | Methylene dibromide |
| | Methylnaphthalenes, liquid |
| | Methylnaphthalenes, solid |
| | Methylnitrophenols |
| | alpha-Methylstyrene |
| | Methylstyrenes, inhibited |
| | Methyltrithion |
| | Methylvinylbenzenes, inhibited |
| PP | Mevinphos |
| | Mexacarbate |
| | Mirex |
| | Monocratophos |
| | Motor fuel anti-knock mixtures |
| | Motor fuel anti-knock mixtures or compunds |
| | Nabam |
| | Naled |
| | Naptha, coal tar |
| | Naphthalene, crude *or* refined |
| | Naphtanlene, molten |
| | Naphthenic acids, liquid |
| | Naphthenic acids, solid |
| PP | Nickel carbonyl |
| PP | Nickel cyanide |
| PP | Nickel tetracarbonyl |
| | Nitrates, inorganic, n.o.s. |
| | Nitrates, inorganic, n.o.s. |
| | 3-Nitro-4-chlorobenzotrifluoride |
| | Nitrobenzotrifluorides |
| | Nitrocresols |
| | Nitroxylenes, (o-; m-; p-) |
| | Nonylphenol |
| PP | Organotin compounds, liquid, n.o.s. |
| PP | Organotin compounds, (pesticides) |
| PP | Organotin compounds, solid, n.o.s. |
| PP | Organotin pesticides, liquid, flammable, toxic, n.o.s. *flash point less than 23 deg. C* |
| PP | Organotin pesticides, liquid, toxic, flammable, n.o.s. |
| PP | Organotin pesticides, liquid, toxic, n.o.s. |
| PP | Organotin pesticides, solid, toxic, n.o.s. |
| | Orthoarsenic acid |
| | Osmium tetroxide |
| | Oxamyl |
| | Oxydisulfoton |
| | Paraoxon |
| PP | Parathion |
| PP | Parathion-methyl |
| PP | PCBs*** |
| | Pentachloroethane |
| PP | Pentachlorophenol |
| | Pentalin |
| | Pentanethiols |
| | Perchloroethylene |
| | Perchloromethylmercaptan |
| | Petrol, leaded |
| PP | Phenarsazine chloride |
| PP | Phenthoate |
| | 1-Phenylbutane |
| | 2-Phenylbutane |
| | Phenylethylene, inhibited |
| PP | Phenylmercuric acetate |
| PP | Phenylmercuric compounds, n.o.s. |
| PP | Phenylmercuric hydroxide |
| PP | Phenylmercuric nitrate |
| | 2-Phenylpropene |
| PP | Phorate |
| PP | Phosalone |
| | Phosmet |
| PP | Phosphamidon |
| PP | Phosphorus, white *or* yellow dry *or* under water *or* in solution |
| PP | Phosphorus white, or yellow, molten |

| S.M.P. | Marine Pollutant |
|---|---|
| (1) | (2) |
| | Pindone (and salts of) |
| | alpha-Pinene |
| | Pirimicarb |
| PP | Pirimiphos-ethyl |
| PP | Polychlorinated biphenyls |
| PP | Polyhalogenated biphenyls, liquid *or* Terphenyls liquid |
| PP | Polyhalogenated biphenyls, solid *or* Terphenyls solid |
| PP | Potassium cuprocyanide |
| | Potassium cyanide |
| PP | Potassium cyanocuprate I |
| PP | Potasisum cyanomercurate |
| | Potassium dihydrogen arsenate |
| PP | Potassium mercuric iodide |
| | Promecarb |
| | Propaphos |
| | Propenal, inhibited |
| | Propenyl chloride (cis-; trans-) |
| | Propoxur |
| | Propylene dichloride |
| | Propylidene dichloride |
| | Prothoate |
| | Prussic acid, anhydrous, stabilized |
| | Prussic acid, anhydrous, stabilized, absorbed in a porous inert material |
| PP | Pyrazophos |
| | Quinalphos |
| | Rotenone |
| | Salithion |
| | Silver arsenite |
| | Silver cyanide |
| PP | Silver orthoarsenite |
| PP | Sodium copper cyanide, solid |
| PP | Sodium copper cyanide solution |
| PP | Sodium cuprocyanide, solid |
| PP | Sodium cuprocyanide, solution |
| | Sodium cyanide |
| | Sodium dinitro-o-cresolate, *dry or wetted with less than 15 per cent water, by mass* |
| | Sodium dinitro-ortho-cresolate, wetted *with not less than 15 per cent water, by mass* |
| | Sodium metaarsenite |
| | Sodium orthoarsenate |
| PP | Sodium pentachlorophenate |
| | Strontium orthoarsenite |
| | Strychnine *or* Strychnine salts |
| | Styrene monomer, inhibited |
| | Sulfotep |
| PP | Sulprophos |
| | Sym-Dichloroethyl ether |
| | Temephos |
| | TEPP |
| PP | Terbufos |
| | 1,1,2,2-Tetrabromoethane |
| | Tetrabromomethane |
| | Tetrachloroethane |
| | 1,1,2,2-Tetrachloroethylene |
| | Tetrachloromethane |
| | Tetrachlorophenol |
| | Tetraethyl dithiopyrophosphate |
| | Tetraethyl lead, liquid |
| | Tetramethyllead |
| | Thallium chlorate |
| | Thallium compounds, n.o.s. |
| | Thallium compounds (pesticides) |
| | Thallium nitrate |
| | Thallium sulfate |
| | Thallous chlorate |
| | Thiocarbonyl tetrachloride |
| | Triaryl phosphates, isopropylated |
| PP | Triaryl phosphates, n.o.s. |
| | Triazophos |
| | Tribromomethane |
| PP | Tributyltin compounds |
| | Trichlorfon |
| | Trichlorobenzenes, liquid |
| | Trichlorobutene |
| | Trichlorobutylene |
| | Trichloromethane sulphuryl chloride |
| | Trichloromethyl sulphochloride |
| | Trichloronat |
| | Tricresyl phosphate (less than 1% ortho-isomer) |
| PP | Tricresyl phosphate (not less than 1% ortho-isomer) |
| | Tricresyl phosphate *with more than 3 per cent ortho isomer* |
| | Triethylbenzene |
| | Trimethylene dichloride |
| PP | Triphenyltin compounds |
| | Tritolyl phosphate (less than 1% ortho-isomer) |
| PP | Tritolyl phosphate (not less than 1% ortho-isomer) |
| | Trixylenyl phosphate |
| | Turpentine |
| | Turpentine substitute |
| | Vinylbenzene, inhibited |
| | Vinylidene chloride, inhibited |
| | Vinyltoluenes, inhibited *mixed isomers* |
| | Warfarin (and salts of) |
| | White arsenic |
| PP | White phosphorus, dry |
| PP | White phosphorus, molten |
| PP | White phosphorus, wet |
| | White spirit, low (15-20%) aromatic |
| | Xylenols |
| PP | Yellow phosphorus, dry |
| PP | Yellow phosphorus, molten |
| PP | Yellow phosphorus, wet |
| | Zinc bromide |
| | Zinc cyanide |

## §172.102 Special Provisions

(a) *General.* When Column 7 of the §172.101 Table refers to a special provision for a hazardous material, the meaning and requirements of that provision are as set forth in this section. When a special provision specifies packaging or packaging requirements—

(1) The special provision is in addition to the standard requirements for all packagings prescribed in §173.24 of this subchapter and any other applicable packaging requirements in subparts A and B of part 173 of this subchapter; and

(2) To the extent a special provision imposes limitations or additional requirements on the packaging provisions set forth in Column 8 of the §172.101 Table, packagings must conform to the requirements of the special provision.

(b) *Description of codes for special provisions.* Special provisions contain packaging provisions, prohibitions, exceptions from requirements for particular quantities or forms of materials and requirements or prohibitions applicable to specific modes of transportation, as follows:

(1) A code consisting only of numbers (for example, "11") is multi-modal in application and may apply to bulk and non-bulk packagings.

(2) A code containing the letter "A" refers to a special provision which applies only to transportation by aircraft.

(3) A code containing the letter "B" refers to a special provision which applies only to bulk packagings requirements. Unless otherwise provided in this subchapter, these special provisions do not apply to IM portable tanks.

(4) A code containing the letter "H" refers to a special provision which applies only to transportation by highway.

(5) A code containing the letter "N" refers to a special provision which applies only to non-bulk packaging requirements.

(6) A code containing the letter "R" refers to a special provision which applies only to transportation by rail.

(7) A code containing the letter "T" refers to a special provision which applies only to transportation in IM portable tanks.

(8) A code containing the letter "W" refers to a special provision which applies only to transportation by water.

(c) *Tables of special provisions.* The following table list, and set forth the requirements of, the special provisions referred to in Column 7 of the §172.101 Table.

(1) *Numeric provisions.* These provisions are multi-modal and apply to bulk and non-bulk packagings:

*Code/Special Provisions*

1 This material is poisonous by inhalation (see §171.8 of this subchapter) in Hazard Zone A (see §173.116(a) or §173.133(a) of this subchapter), and must be described as an inhalation hazard under the provisions of this subchapter.

2 This material is poisonous by inhalation (see §171.8 of this subchapter) in Hazard Zone B (see §173.116(a) or §173.133(a) of this subchapter), and must be described as an inhalation hazard under the provisions of this subchapter.

3 This material is poisonous by inhalation (see §171.8 of this subchapter) in Hazard Zone C (see §173.116(a) of this subchapter), and must be described as an inhalation hazard under the provisions of this subchapter.

4 This material is poisonous by inhalation (see §171.8 of this subchapter) in Hazard Zone D (see §173.116(a) of this subchapter), and must be described as an inhalation hazard under the provisions of this subchapter.

5 If this material meets the definition for a material poisonous by inhalation (see §171.8 of this subchapter), a shipping name must be selected which identifies the inhalation hazard, in Division 2.3 or Division 6.1, as appropriate.

6 This material is poisonous-by-inhalation and must be described as an inhalation hazard under the provisions of this subchapter.

7 An ammonium nitrate fertilizer is a fertilizer formulation, containing 90% or more ammonium nitrate and no more than 0.2% organic combustible material (calculated as carbon), which does not meet the definition and criteria of a Class 1 (explosive) material (See §173.50 of this subchapter).

8 A hazardous substance that is not a hazardous waste may be shipped under the shipping description "Other regulated substances, liquid *or* solid, n.o.s.", as appropriate. In addition, for solid materials, special provision B54 applies.

9 Packaging for certain PCBs for disposal and storage is prescribed by EPA in 40 CFR 761.60 and 761.65.

10 An ammonium nitrate mixed fertilizer is a fertilizer formulation, containing less than 90% ammonium nitrate and other ingredients, which does not meet the definition and criteria of a Class 1 (explosive) material (See §173.50 of this subchapter).

11 The hazardous material must be packaged as either a liquid or a solid, as appropriate, depending on its physical form at 55°C (131° F) at atmospheric pressure.

12 In concentrations greater than 40 percent, this material has strong oxidizing properties and is capable of starting fires in contact with combustible materials. If appropriate, a package containing this material must conform with the additional labeling requirements of §172.402 of this subchapter.

13 The words "Inhalation Hazard" shall be entered on each shipping paper in association with the shipping description, shall be marked on each non-bulk package in association with the proper shipping name and identification number, and shall be marked on two opposing sides of each bulk package. Size of marking on bulk package must conform to §172.302(b) of this subchapter. The requirements

of §§172.203(m) and 172.505 of this subchapter do not apply.

14 Motor fuel antiknock mixtures is (mixture of one or more organic lead mixtures, such as tetraethyl lead, triethylmethyl lead, diethyldimethyl lead, ethyltrimethyl lead, and tetramethyl lead, with one or more halogen compounds such as ethylene dibromide and ethylene dichloride, hydrocarbon solvents or other equally efficient stabilizers) or tetraethyl lead.

15 Chemical kits include boxes, cases, etc., containing small amounts of various compatible dangerous goods which are used for medical, analytical or testing purposes and for which exceptions are provided in this subchapter. Inner packagings must not exceed 250 mL for liquids or 250 g for solids and must be protected from other materials in the kit. For transportation by aircraft, any dangerous goods forbidden in passenger aircraft may not be included in these kits. Kits must be packed in wooden boxes (4C1, 4C2), plywood boxes (4D), reconstituted wood boxes (4F), fiberboard boxes (4G) or plastic boxes (4H1, 4H2) which must be marked 'Chemical kits.' Each package shall be classified for the material contained therein; if two or more different classes of material are present the class of the package shall be determined in accordance with §173.2a of this subchapter. Each package must be labeled according to each substance contained in the package; this includes the primary hazard label and any subsidiary risk labels applicable to each individual substance within the kit. The total quantity of dangerous goods in any one kit must not exceed either 1 L or 1 kg. The total quantity of dangerous goods in any one package must not exceed either 10 L or 10 kg. The packing group assigned to the kit as a whole must be the most stringent packing group assigned to any individual substance contained in the kit.

16 This description applies only to materials that have been classed as Division 1.3.

17 Aqueous solutions of hydrogen peroxide containing less than 8 percent hydrogen peroxide are not subject to the requirements of this subchapter.

18 This article and its packaging must be approved by the Associate Administrator for Hazardous Materials Safety.

19 For domestic transportation only, the identification number "UN1075" may be used in place of the identification number specified in Column (4) of the §172.101 Table. The identification number used must be consistent on package markings, shipping papers and emergency response information.

20 The transport of this substance, when in concentrations of greater than 10% nitroglycerin, is prohibited. Concentrations of below 5% nitroglycerin may be transported as a Class 3 material; see UN 1204 and UN 3064.

21 This material must be stabilized by appropriate means (e.g., addition of chemical inhibitor, purging to remove oxygen) to prevent dangerous polymerization (see §173.21(f) of this subchapter).

22 If the hazardous material is in dispersion in organic liquid, the organic liquid must have a flash point above 50° C (122° F).

27 Sodium carbonate peroxhydrate is considered non-hazardous.

28 The dihydrated sodium salt of dichloroisocyanuric acid is not subject to the requirements of this subchapter.

31 Materials which have undergone sufficient heat treatment to render them non-hazardous are not subject to the requirements of this subchapter.

33 Ammonium nitrates and mixtures of an inorganic nitrite with an ammonium salt are prohibited.

41 When Packaging Method F1 or F5a (see §173.224(c) of this subchapter) are used, an EXPLOSIVE label is not required.

42 Fish meal or fish scrap may not be offered for transportation if the temperature of the material exceeds 49° C (120° F)

53 Packages of these materials must bear the subsidiary risk label, "EXPLOSIVE", unless otherwise provided in this subchapter or through an approval issued by the Associate Administrator for Hazardous Materials Safety, or the competent authority of the country of origin. A copy of the approval shall accompany the shipping papers.

101 The name of the particular substance or article must be specified.

102 The articles may be transported as in Division 1.4 Compatability Group D (1.4D) if all of the conditions specified in §173.63(a) of this subchapter are met.

103 Detonators which will not mass detonate and undergo only limited preparation in the shipping package may be assigned to 1.4B classification code. Mass detonate means that more than 90 percent of the devices tested in a package explode practically simultaneously. Limited propagation means that if one detonator near the center of a shipping package is exploded, the aggregate weight of explosives, excluding ignition and delay charges, in this and all additional detonators in the outside packaging that explode may not exceed 25 grams.

104 Detonators which meet the following conditions may be assigned to 1.4S classification code: Each detonator may contain no more than 1 g of explosive, excluding ignition and delay charges, and if one detonator near the center of a package detonates it will not cause functioning of any other device in the same or adjacent packages.

105 The word 'Agents' may be used instead of 'Explosives' when approved by the Associate Administrator for Hazardous Materials Safety.

106 The recognized name of the particular explosive may be specified in addition to the type.

107 The classification of the substance is expected to vary especially with the particle size and packaging but the border lines have not been experimentally determined; appropriate classifications should be verified following the test procedures in §§ 173.57 and 173.58 of this subchapter.

108 Fireworks must be so constructed and packaged that loose pyrotechnic composition will not be present in packages during transportation.

109 Rocket motors must be nonpropulsive in transportation. A rocket motor to be considered 'nonpropulsive' must be capable of unrestrained burning and must not appreciably move in any direction when ignited by any means.

110 Cartridges containing 3.2 grams or less of deflagrating (propellant) explosives installed in a fire extinguisher are not subject to the requirements of this subchapter.

111 Explosive substances of Division 1.1 Compatibility Group A (1.1A) are forbidden for transportation if dry or not desensitized, unless incorporated in a device.

112 "Cartridges, small arms" and "Cartridges, power devices (which are used to project fastening devices)" which have been classed in Division 1.4 Compatibility Group S (1.4S) may be reclassed and offered for domestic transportation as ORM-D material if they are offered for transportation and transported in accordance with the limitations and packaging requirements of §173.230 of this subchapter.

113 The sample must be given a tentative approval by an agency or laboratory in accordance with §173.56 of this subchapter.

114 Jet perforating guns, charged, oil well, without detonator may be reclassed to Division 1.4 Compatibility Group D (1.4D) if the following conditions are met:

a. The total weight of the explosive contents of the shaped charges assembled in the guns does not exceed 90.5 kg (200 pounds) per vehicle; and

b. The guns are packaged in accordance with Packing Method US006 as specified in §173.62 of this subchapter.

115 Boosters with detonator (detonating primers) in which the total explosive charge per unit does not exceed 25 g, and which will not mass detonate and undergo only limited propagation in the shipping package may be assigned to 1.4B classification code. Mass detonate means more than 90 percent of the devices tested in a package explode practically simultaneously. Limited propagation means that if one booster near the center of the package is exploded, the aggregate weight of explosives, excluding ignition and delay charges, in this and all additional boosters in the outside packaging that explode may not exceed 25 g.

116 Fuzes, detonating may be classed in Division 1.4 if the fuzes do not contain ore than 25 g of explosive per fuze and are made and packaged so that they will not cause functioning of other fuzes, explosives or other explosive devices if one of the fuzes detonates in a shipping packaging or in adjacent packages.

117 If shipment of the explosive substance is to take place at a time that freezing weather is anticipated, the water contained in the explosive substance must be mixed with denatured alcohol so that freezing will not occur.

(2) *"A" codes.* These provisions apply only to transportation by aircraft:

*Code/Special Provisions*

A1 Single packagings are not permitted on passenger aircraft.

A2 Single packagings are not permitted on aircraft.

A3 For combination packagings, if glass inner packagings (including ampoules) are used, they must be packed with absorbent material in tightly closed metal receptacles before packing in outer packagings.

A4 Liquids having an inhalation toxicity of Packing Group I are not permitted on aircraft.

A6 For combination packagings, if plastic inner packagings are used, they must be packed in tightly closed metal receptacles before packing in outer packagings.

A7 Steel packagings must be corrosion-resistant or have protection against corrosion.

A8 For combination packagings, if glass inner packagings (including ampoules) are used, they must be packed with cushioning material in tightly closed metal receptacles before packing in outer packagings.

A9 For combination packagings, if plastic bags are used, they must be packed in tightly closed metal receptacles before packing in outer packagings.

A10 When aluminum or aluminum alloy construction materials are used, they must be resistant to corrosion.

A11 For combination packagings, when metal inner packagings are permitted, only specification cylinders constructed of metals which are compatible with the hazardous material may be used.

A12 The quantity of lithium contained in any piece of equipment must not exceed 12 grams per cell and 500 grams per battery for cargo aircraft. No piece of equipment may contain more than 5 kilograms of lithium batteries. Lithium batteries are forbidden for transportation by passenger-carrying aircraft unless approved by the Associate Administrator for Hazardous Materials Safety.

A19 Combination packagings consisting of outer fiber drums or plywood drums, with inner plastic packagings, are not authorized for transportation by aircraft

A20 Plastic bags as inner receptacles of combination packagings are not authorized for transportation by aircraft.

A29 Combination packagings consisting of outer expanded plastic boxes with inner plastic bags are not authorized for transportation by aircraft.

A30 Ammonium permanganate is not authorized for transportation on aircraft.

A33 Ammmonium nitrites and mixtures of an inorganic nitrite with an ammonium salt are prohibited.

A34 Aerosols containing a corrosive liquid in Packing Group II charged with a gas are not permitted for transportation by aircraft.

(3) "B" codes. These provisions apply only to bulk packagings:

*Code/Special Provisions*

B1 If the material has a flash point at or above 38°C (100°F) and below 93°C (200°F), then the bulk packaging requirements of §173.241 of this subchapter are applicable. If the material has a flash point of less than 38°C (100°F), then the bulk packaging requirements of §173.242 of this subchapter are applicable.

B2 MC 300, MC 301, MC 302, MC 303, MC 305, and MC 306, and DOT 406 cargo tanks are not authorized.

B3 MC 300, MC 301, MC 302, MC 303, MC 305, and MC 306, and DOT 406 cargo tanks and DOT 57 portable tanks are not authorized.

B4 AAR 206 tank car tanks and MC 300, MC 301, MC 302, MC 303, MC 305, and MC 306, and DOT 406 cargo tanks are not authorized.

B5 Only ammonium nitrate solutions with 35 percent or less water that will remain completely in solution at a maximum lading temperature of 116°C (240°F) are authorized for the transport in the following bulk packagings: DOT 103 ALW, 111A60 ALW tank car tanks and MC 307, MC 312, DOT 407 and DOT 412 cargo tanks with at least 172 kPa (25 psig) design pressure. The packaging shall be designed for a working temperature of at least 121°C (250°F). Only Specifications MC 304, MC 307 or DOT 407 cargo tank motor vehicles are authorized for transportation by vessel.

B6 Packagings shall be made of steel.

B7 Safety relief devices are not authorized on multi-unit tank car tanks. Openings for safety relief devices shall be plugged or blank flanged.

B8 Packagings shall be made of nickel, stainless steel, or steel with nickel, stainless steel, lead orother suitable corrosion resistant metallic lining.

B9 Bottom outlets are not authorized.

B10 AAR 206 tank car tanks, MC 300, MC 301, MC 302, MC 303, MC 305 and MC 306 and DOT 406 cargo tanks, and DOT 57 portable tanks are not authorized.

B11 Tank car tanks must have a test pressure of at least 2,068.5 kPa (300 psi). Cargo and portable tanks must have a design pressure of at least 1,207 kPa (175 psig).

B12 Tank car tanks must be marked with the name of the lading in accordance with the requirements of §172.330 of this subchapter.

B13 A nonspecification cargo tank motor vehicle authorized in §173.247 of this subchapter must be at least equivalent in design and in construction to a DOT 406 cargo tank or MC 306 cargo tank (if constructed before September 1, 1993), except as follows:

a. Packagings equivalent to MC 306 cargo tanks are excepted from §§178.340-10, certification; 178.341-4, vents; and 178.341-5, emergency flow control.

b. Packagings equivalent to DOT 406 cargo tanks are excepted from §§178.345-7(d)(5), circumferential reinforcements, 178.345-14, marking; 178.345-15, certification; 178.346-10, pressure relief; and 178.346-11, outlets.

c. Packagings are excepted from the design stress limits at elevated temperatures, as described in the ASME Code. However, the design stress limits may not exceed 25 per cent of the stress, as specified in §178.65-5(b) of this subchapter, for 0 temper at the maximum design temperature of the cargo tank.

B14 Each tank, except a multi-unit tank car tank, must be insulated with at least 100 mm (3.9 inches) of cork or other suitable insulation material of sufficient thickness that the overall thermal conductance at 15.5°C (60°F) is not more than 1.533 kilojoules per hour per square meter per degree Celsius (0.075 Btu per hour per square foot per degree Fahrenheit) temperature differential. Insulation systems must not promote corrosion to steel when wet. Tank and jacket protective coatings are required. Additionally, all tank car tanks constructed after October 1, 1988 and tanks repaired after October 1, 1993, where the entire jacket is removed during the repair, must have tank and jacket protective coatings. The jacket must be flashed around all openings so as to be weather tight.

B15 Packagings must be protected with non-metallic linings impervious to the lading or have a suitable corrosion allowance.

B16 The lading must be completely covered with nitrogen, inert gas or other inert materials.

B17 Packagings must be made of aluminum.

B18 Open steel hoppers or bins are authorized.

B19 The hazardous material may not exceed 45 percent concentration in a non-volatile solvent.

B20 The hazardous material may not exceed 50 percent concentration in a non-volatile solvent.

B21 The hazardous material may not exceed 60 percent concentration in a non-volatile solvent.

B22 The hazardous material may not exceed 90 percent concentration in a non-volatile solvent.

B23 Tanks must be made of steel that is rubber lined or unlined. Unlined tanks must be passivated before being placed in service. If unlined tanks are washed out with water, they must be repassivated prior to return to service. Lading in unlined tanks must be inhibited so that the corrosive effect on steel is not greater than that of hydrofluoric acid of 65 percent concentration.

B24 Packagings must be made of stainless steel--in which the molybdenum content does not exceed 1.0 percent--or of aluminum.

B25 Packagings must be made from monel or nickel or monel-lined or nickel-lined steel.

B26 Tanks must be insulated. Insulation must be at least 100 mm (3.9 inches) except that the insulation thickness may be reduced to 51 mm (2 inches) over the exterior heater coils. Interior heating coils are not authorized. The packaging may not be loaded with a material outside of the packaging's design temperature range. In addition, the material also must be covered with an inert gas or the container must be filled with water to the tank's capacity. After unloading, the residual material also must be covered with an inert gas or the container must be filled with water to the tank's capacity .

B27 Tanks must have a service pressure of 1,034 kPa (150 psig). Tank car tanks must have a test pressure rating of 1,379 kPa (200 psi). Lading must be blanketed at all times with a dry inert gas at a pressure not to exceed 103 kPa (15 psig).

B28 Packagings must be made of stainless steel.

B29 When the lading is transported in a molten state, tanks may be equipped with heating coils except that interior heating coils are prohibited. Electric standpipe heaters for tank cars are permitted.

B30 MC 312, MC 330, MC 331 and DOT 412 cargo tanks and DOT 51 portable tanks must be made of stainless steel, except that steel other than stainless steel may be used in accordance with the provisions of §173.24b(b) of this subchapter. Thickness of stainless steel for tank shell and heads for cargo tanks and portable tanks must be the greater of 7.62 mm (0.300 inch) or the thickness required for a tank with a design pressure at least equal to 1.5 times the vapor pressure of the lading at 46°C (115° F). In addition, MC 312 and DOT 412 cargo tank motor vehicles must:

a. Be ASME Code (U) stamped for 100% radiography of all pressure-retaining welds;

b. Have accident damage protection which conforms with §178.345-8 of this subchapter;

c. Have a MAWP or design pressure of at least 87 psig: and

d. Have a bolted ASA manway cover.

B31 Bromine tank cars built prior to December 31, 1990 to DOT105A500W which have been stencilled and valved as DOT105A300W may continue in service.

B32 MC 312, MC 330, MC 331 DOT 412 cargo tanks and DOT 51 portable tanks must be made of stainless steel, except that steel other than stainless steel may be used in accordance with the provisions of §173.24b(b) of this subchapter. Thickness of stainless steel for tank shell and heads for cargo tanks and portable tanks must be the greater of 6.35 mm (0.250 inch) or the thickness required for a tank with a design pressure at least equal to 1.3 times the vapor pressure of the lading at 46°C (115° F). In addition, MC 312 and DOT 412 cargo tank motor vehicles must:

a. Be ASME Code (U) stamped for 100% radiography of all pressure-retaining welds;

b. Have accident damage protection which conforms with §178.345-8 of this subchapter;

c. Have a MAWP or design pressure of at least 87 psig; and

d. Have a bolted ASA manway cover.

B33 MC 300, MC 301, MC 302, MC 303, MC 305, MC 306, and DOT 406 cargo tanks equipped with a 1 psig normal vent used to transport gasoline are subject to the following requirements. Based on the volatility class determined by using ASTM D439 and the Reid vapor pressure (RVP) of the particular gasoline, the maximum lading pressure and maximum ambient temperature permitted during the loading of gasoline may not exceed that listed in Table I.

Table I—Maximum Ambient Temperature—Gasoline

| ASTM D439 volatility class | Maximum lading and ambient temperature (see note 1) |
|---|---|
| A (RVP<=9.0 psia) | 131° F |
| B (RVP<=10.0 psia) | 124° F |
| C (RVP<=11.5 psia) | 116° F |
| D (RVP<=13.5 psia) | 107° F |
| E (RVP<=15.0 psia) | 100° F |

Note 1: Based on maximum lading pressure of 1 psig at top of cargo tank.

B35 Tank cars containing hydrogen cyanide may be alternatively marked "Hydrocyanic acid, liquefied" if otherwise conforming to marking requirements in subpart D of this part.

B36 DOT 105J500W tank car tanks or Class DOT 106 or 110 tank car tanks are authorized.

B37 The amount of nitric oxide charged into any tank car tank may not exceed 1,379 kPa (200 psig) at 21°C (70°F).

B39 Mixtures with flash points less than 23°C (73°F) must bear FLAMMABLE placards as prescribed in Subpart F of part 172.

B41 Notwithstanding the periodic retest intervals specified in Retest Table 1 of 173.31 of this subchapter, the retest interval for safety relief valves on each single-unit tank car tank is 2 years and the retest interval on the tank and interior heater systems, if any, is as follows:

a. For a tank 10 years old or newer, 5 years;

b. For a tank older than 10 years but not older than 22 years, 3 years; and

c. For a tank older than 22 years, 1 year.

B42 Each DOT 105A, 105S and 105J tank car tank must be stenciled DOT 105A200W, DOT 105S200W, or DOT 105J200W, respectively. Each tank car tank must be equipped with a safety relief valve with a start-to-discharge pressure of 1,034 kPa (150 psig).

B43 For single unit tank car tanks built after June 30, 1982, tank anchor to tank shell fillet welds must be examined by a suitable non-destructive testing method to ensure that welds are free from cracks or other detrimental defects.

B44 All parts of valves and safety relief devices in contact with lading must be of a material which will not cause formation of acetylides.

B45 Safety relief valves must be equipped with stainless steel or platinum frangible discs approved by the AAR Committee on Tank Cars.

B46 The detachable protective housing for the loading and unloading valves of multi-unit tank car tanks must withstand tank test pressure and must be approved by the Associate Administrator for Hazardous Materials Safety.

B47 A safety relief device with a start-to-discharge pressure setting of 310 kPa (45 psig) is permitted.

B48 Portable tanks in sodium metal service may be visually inspected at least once every 5 years instead of being retested hydrostatically. Date of the visual inspection must be stenciled on the tank near the other required markings.

B49 Tanks equipped with interior heater coils are not authorized. Single unit tank car tanks must have a safety relief valve set at no more than 1551 kPa (225 psig).

B50 Each valve outlet of a multi-unit tank car tank must be sealed by a threaded solid plug or a threaded cap with inert luting or gasket material. Valves must be of stainless steel and the caps, plugs, and valve seats must be of a material that will not deteriorate as a result of contact with the lading.

B52 Notwithstanding the provisions of §173.24b of this subchapter, non-reclosing pressure relief devices are authorized on DOT 57 portable tanks.

B53 Packagings must be made of either aluminum or stainless steel.

B54 Open-top, sift-proof rail cars are also authorized.

B55 Water-tight, sift-proof, closed-top, metal-covered hopper cars, equipped with a venting arrangement (including flame arrestors) approved by the Associate Administrator for Hazardous Materials Safety are also authorized.

B56 Water-tight, sift-proof, closed-top, metal-covered hopper cars also authorized if the particle size of the hazardous material is not less than 149 microns.

B57 Class DOT 115A tank car tanks must be equiped with a safety vent of a diameter not less than 305mm (12 inches) complying with §179.221-1 of this subchapter on the outer shell must be stenciled "CHLOROPRENE" on both sides in letters not less than 102 mm (4 inches) high.

B58 Notwithstanding the provisions of §173.244(a) of this subchapter, only DOT 105J300W tank car tanks are authorized. Class 106 and 110 multi-unit tank car tanks are also authorized.

B59 AAR Specification 207A40W, 207A40W6, 207A48W, 207A60W, 207A80W tank car tanks are also authorized provided that the lading is covered in a nitrogen blanket.

B60 DOT Specification 106A500X multi-unit tank car tanks that are not equipped with a safety relief device of any type are authorized. For the transportation of phosgene, the outage must be sufficient to prevent tanks from becoming liquid full at 55°C (130°F).

B61 Written procedures covering details of tank car appurtenances, dome fittings, safety devices, and marking, loading, handling, inspection, and testing practices must be approved by the Associate Administrator for Hazardous Materials Safety before any single unit tank car tank is offered for transportation.

B62 Single unit tank car tanks must be equipped with a venting arrangement that is approved by the Associate Administrator for Hazardous Materials Safety.

B63 Notwithstanding the provisions of §173.314 of this subchapter, DOT Specification 105A100W, 111A100W4, 112A200W, and 114A340W tank car tanks, built before January 1, 1991, are also authorized. Specification 114A340W tank car tanks may not be equipped with any bottom outlet, but bottom washouts are permitted.

B64 Each single unit tank car tank built after December 31, 1990 must be equipped with a tank head puncture resistance system that conforms to §179.105-5 of this subchapter.

B65 DOT Specification 105A500W tank cars. Tank must be restenciled 105A300W and be equipped with safety valves of the type and size used on specification 105A300W tank cars. Tank car tank must be equipped with approved fittings and safety devices. The maximum filling density is 63 percent of the water capacity of the tank.

B66 Each tank must be equipped with gas tight valve protection caps. Outage must be sufficient to prevent tanks from becoming liquid full at 55;dgC (130;dgF). Specification 110A500W tanks must be stainless steel.

B67 Tanks must be lagged with not less than a 100 mm (3.9 inch) thickness of cork. All valves and fittings must be protected by a securely attached cover made of metal not subject to deterioration by the lading, and all valve openings, except safety valve, must be fitted with screw plugs or caps to prevent leakage in the event of valve failure.

B68 Sodium must be in a molten condition when loaded and allowed to solidify before shipment. Outage must be at least 5 percent at 98°C (208°F). Bulk packagings must have exterior heating coils fusion welded to the tank shell which have been properly stress relieved. The only tank car tanks authorized are Class DOT 105 tank cars having a test pressure of 2,069 kPa (300 psig) or greater.

B69 Dry sodium cyanide or potassium cyanide may be shipped in sift-proof weather-resistant metal covered hopper cars, covered motor vehicles, portable tanks or non-specification bins. Bins must be approved by the Associate Administrator for Hazardous Materials Safety. Flexible intermediate bulk containers (FIBCs) may also be used under conditions approved by the Associate Administrator for Hazardous Materials Safety.

B70 If DOT 103ANW tank car tank is used: All cast metal in contact with the lading must have 96.7 percent nickel content; and the lading must be anhydrous and free from any impurities.

B71 The only tank cars authorized are Class DOT 105, 112, and 114 tank car tanks with a test pressure of 2069 kPa (300 psig) or greater.

B72 Notwithstanding the provisions of §173.244(a) of this subchapter, only the following tank car tanks are authorized: DOT 105J500W tank car tanks and Class DOT 106 and 110 multi-unit tank car tanks.

B74 Notwithstanding the provisions of §§173.243(a) and 173.244(a) of this subchapter, only the following tank car tanks are authorized: DOT 105J300W, 112J340W, 112T340W, 114J340W, 114T340W and 105J300ALW tank car tanks; and Class DOT 106 and 110 multi-unit tank car tanks.

B76 DOT 105S500W tank car tanks restenciled to DOT 105S200W tank car tanks. Each tank car tank must must be equipped with a safety relief valve with a start-to-discharge pressure of 1,034 kPa (150 psig).

B77 Other packaging are authorized when approved by the Associate Administrator for Hazardous Materials Safety.

B78 Notwithstanding §173.240 of this subchapter, the only bulk packagings authorized for transportation by rail are Class DOT 103, 104, 105, 109, 111, 112, and 114 tank car tanks. Heater pipes must be of welded construction designed for a test pressure of 500 pounds per square inch. A 25 mm (1 inch) woven lining of asbestos or other approved material must be placed between the bolster slabbing and the bottom of the tank. If a tank car tank is equipped with a safety vent of the frangible disc type, the frangible disc must be perforated with a 3.2 mm (0.13 inch) diameter hole. If a tank car tank is equipped with a safety relief valve, the tank car tank must also be equipped with a vacuum relief valve.

B79 Tank car tanks shall have head puncture resistance and thermal protection in accordance with §§179.105-4 and 179.105-5 of this subchapter for tanks built before April 1, 1989.

B80 Each cargo tank must have a minimum design pressure of 276 kPa (40 psig).

B81 Venting and pressure relief devices for tank car tanks and cargo tanks must be approved by the Associate Administrator for Hazardous Materials Safety.

B82 Cargo tanks and portable tanks are not authorized.

B83 Bottom outlets are prohibited on tank car tanks transporting sulfuric acid in concentrations over 65.25 percent.

B84 Packagings must be protected with non-metallic linings impervious to the lading or have a suitable corrosion allowance for sulfuric acid or spent sulfuric acid in concentration up to 65.25 percent.

B85 Cargo tanks must be marked with the name of the lading accordance with the requirements of §172.302(b).

B86 Only DOT 105S600W tank car tanks are authorized.

B90 Steel tank conforming or equivalent to ASME specifications which contain solid or semisolid residual motor fuel antiknock mixture (including rust, scale, or other contaminants) may be shipped by rail freight or highway. The tank must have been designed and constructed to be capable of withstanding full vacuum. All openings must be closed with gasketed blank flanges or vapor tight threaded closures.

(4) *"H" codes.* These provisions apply only to transportation by highway. [Reserved]

(5) *"N" codes.* These provisions apply only to non-bulk packagings:

*Code/Special Provisions*

N3 Glass inner packagings are permitted in combination or composite packagings only if the hazardous material is free from hydrofluoric acid.

N4 For combination or composite packagings, glass inner packagings, other than ampoules, are not permitted.

N5 Glass materials of construction are not authorized for any part of a packaging which is normally in contact with the hazardous material.

N6 Battery fluid packaged with electric storage batteries, wet or dry, must conform to the packaging provisions of §173.159(g) or (h) of this subchapter.

N7 The hazard class or division number of the material must be marked on the package in accordance with §172.302 of this subchapter. However, the hazard label corresponding to the hazard class or division may be substituted for the marking.

N8 Nitroglycerin solution in alcohol may be transported under this entry only when the solution is packed in metal cans of not more than 1L capacity each, overpacked in a wooden box containing not more than 5L. Metal cans must be completely surrounded with absorbent cushioning material. Wooden boxes must be completely lined with a suitable material impervious to water and nitroglycerin.

N9 If the substance is impregnated with less than 5% oil, it is excepted from the labeling requirements of subpart D of this part and the packaging tests of subpart M of part 178 of this subchapter.

N10 Lighters and their inner packagings, which have been approved by the Associate Administrator for Hazardous Materials Safety (see §173.21(l) of this subchapter), must be packaged in one of the following outer packagings at the Packing Group II level: 4C1 or 4C2 wooden boxes; 4D plywood boxes; 4F reconstituted wood boxes; 4G fiberboard boxes; or 4H1 or 4H2 plastic boxes.

N11 This material is excepted for the specification packaging requirements of this subchapter if the material is packaged in strong, tight non-bulk packaging meeting the requirements of subparts A and B of Part 173 of this subchapter.

N12 Plastic packagings are not authorized.

N20 A 5M1 multi-wall paper bag is authorized if transported in a closed transport vehicle.

N25 Steel single packagings are not authorized.

N32 Aluminum materials of construction are not authorized for single packagings.

N33 Aluminum drums are not authorized.

N34 Aluminum construction materials are not authorized for any part of a packaging which is normally in contact with the hazardous material.

N36 Aluminum or aluminum alloy construction materials are permitted only for halogenated hydrocarbons that will not react with aluminum.

N37 This material may be shipped in a integrally-lined fiber drum (1G) which meets the general packaging requirements of subpart B of part 173 of this subchapter, the requirements of part 178 of this subchapter at the packing group assigned for the material and to any other special provisions of column 7 of the §172.101 table.

N40 This material is not authorized in the following packagings:

a. A combination packaging consisting of a 4G fiberboard box with inner receptacles of glass or earthenware;

b. A single packaging of a 4C2 sift-proof, natural wood box; or

c. A composite packaging 6PG2 (glass, porcelain or stoneware receptacles within a fiberboard box).

N41 Metal construction materials are not authorized for any part of a packaging which is normally in contact with the hazardous material.

N43 Metal drums are permitted as single packagings only if constructed of nickel or monel.

N45 Copper cartridges are authorized as inner packagings if the hazardous material is not in dispersion.

N50 A Class 9 material that meets the definition of a marine pollutant, but does not meet the definition of a hazardous substance or a hazardous waste or the definition in §173.140(a) of this subchapter, is excepted from the labeling requirements of this part.

N65 Outage must be sufficient to prevent cylinders or spheres from becoming liquid full at 55°C (130°F). The vacant space (outage) may be charged with a nonflammable nonliquefied compressed gas if the pressure in the cylinder or sphere at 55°C (130°F) does not exceed 125 percent of the marked service pressure.

N71 Combination packagings consisting of inner glass packagings of not over 1.0L (0.3 gallon) capacity each or inner metal packagings of not over 5.0L (1 gallon) capacity each, placed in strong outer packagings, are authorized. Packagings are not subject to the requirements of part 178 of this subchapter.

N72 Packagings must be examined by the Bureau of Explosives and approved by the Associate Administrator for Hazardous Materials Safety.

N73 Packagings consisting of outer wooden or fiberboard boxes with inner glass, metal or other strong containers; metal or fiber drums; kegs or barrels; or strong metal cans are authorized and need not conform to the requirements of part 178 of this subchapter.

N74 Packages consisting of tightly closed inner containers of glass, earthenware, metal or polyethylene, capacity not over 0.5kg (1.1 pounds) securely cushioned and packed in outer wooden barrels or wooden or fiberboard boxes, not over 15kg (33 pounds) net weight, are authorized and need not conform to the requirements of part 178 of this subchapter.

N75 Packages consisting of tightly closed inner packagings of glass, earthenware or metal, securely cushioned and packed in outer wooden barrels or wooden or fiberboard boxes, capacity not over 2.5kg (5.5 pounds) net weight, are authorized and need not conform to the requirements of part 178 of this subchapter.

N76 For materials of not more than 25 percent active ingredient by weight, packages consisting of inner metal packagings not greater than 250 ml (8 ounces) capacity each, packed in strong outer packagings together with sufficient absorbent material to completely absorb the liquid contents are authorized and need not conform to the requirements of part 178 of this subchapter.

N77 For materials of not more than two percent active ingredients by weight, packagings need not conform to the requirements of part 178 of this subchapter, if liquid contents are absorbed in an inert material.

N78 Packages consisting of inner glass, earthenware, or polyethylene or other nonfragile plastic bottles or jars not over 0.5 kg (1.1 pounds) capacity each, or metal cans not over five pounds capacity each, packed in outer wooden boxes, barrels or kegs, or fiberboard boxes are authorized and need not conform to the requirements of part 178 of this subchapter. Net weight of contents in fiberboard boxes may not exceed 29 kg (64 pounds). Net weight of contents in wooden boxes, barrels or kegs may not exceed 45 kg (99 pounds).

N79 Packages consisting of tightly closed metal inner packagings not over 0.5 kg (1.1 pounds) capacity each, packed in outer wooden or fiberboard boxes, or wooden barrels, are authorized and need not conform to the requirements of part 178 of this subchapter. Net weight of contents may not exceed 15 kg (33 pounds).

N80 Packages consisting of one inner metal can, not over 2.5 kg (5.5 pounds) capacity, packed in an outer wooden or fiberboard box, or a wooden barrel, are authorized and need not conform to the requirements of part 178 of this subchapter.

N81 Polychlorinated biphenyl items, as defined in 40 CFR 761.3, for which specification packagings are impractical, may be packaged in non-specification packagings meeting the general packaging requirements of subparts A and B of part 173 of this subchapter. Alternatively, the item itself may be used as a packaging if it meets the general packaging requirements of subparts A and B of part 173 of this subchapter.

N82 See §173.306 of this subchapter for classification criteria for flammable aerosols.

(6) *"R" codes*. These provisions apply only to transportation by rail. [Reserved]

(7) *"T" codes*. These provisions apply only to transportation in IM portable tanks. They are divided into two groupings, one of which apears as the IM Tank Configuration in paragraph (c)(7)(i) of this section, and the second of which imposes specific requirements and apears in paragraph (c)(7)(ii) of this section.

(i) *IM Tank Configurations*. Column 1 lists the code for the special provisions as specified in column 7 of the §172.101 table. Column 2 specifies the IM tank type, either IM 101 (§§178.270 and 178.271 of this subchapter) or IM 102 (§§178.270 and 178.272 of this subchapter). Column 3 specifies the minimum test pressure, in bars (1 bar = 14.5 psig), at which the periodic hydrostatic testing required by §178.32b of this subchapter must be conducted. Column 4 specifies either the section referenced for requirements for bottom openings or "Prohibited", which means bottom openings are prohibited. Column 5 specifies the section reference for requirements applicable to pressure relief devices.

IM TANK CONFIGURATIONS

| Code<br>(1) | IM tank type<br>(2) | Minimum test Pressure (bars)<br>(3) | Bottom Outlets<br>(4) | Pressure relief devices<br>(5) |
|---|---|---|---|---|
| T1 | 102 | 1.5 | §173.32c(g)(1) | §178.270-11(a)(1),(2) |
| T2 | 102 | 1.5 | §173.32c(g)(2) | §178.270-11(a)(1),(2) |
| T7 | 101 | 2.65 | §173.32c(g)(1) | §178.270-11(a)(1),(2) |
| T8 | 101 | 2.65 | §173.32c(g)(2) | §178.270-11(a)(1),(2) |
| T9 | 101 | 2.65 | Prohibited | §178.270-11(a)(1),(2) |
| T11 | 101 | 2.65 | §173.32c(g)(2) | §178.270-11(a)(3) |
| T12 | 101 | 2.65 | Prohibited | §178.270-11(a)(3) |
| T13 | 101 | 4 | §173.32c(g)(1) | §178.270-11(a)(1),(2) |
| T14 | 101 | 4 | §173.32c(g)(2) | §178.270-11(a)(1),(2) |
| T15 | 101 | 4 | Prohibited | §178.270-11(a)(1),(2) |
| T16 | 101 | 4 | §173.32c(g)(1) | §178.270-11(a)(3) |
| T17 | 101 | 4 | §173.32c(g)(2) | §178.270-11(a)(3) |
| T18 | 101 | 4 | Prohibited | §178.270-11(a)(3) |
| T20 | 101 | 6 | §173.32c(g)(2) | §178.270-11(a)(1),(2) |
| T21 | 101 | 6 | Prohibited | §178.270-11(a)(1),(2) |
| T22 | 101 | 6 | §173.32c(g)(1) | §178.270-11(a)(1)(2) |
| T23 | 101 | 6 | §173.32c(g)(2) | §178.270-11(a)(3) |
| T24 | 101 | 6 | Prohibited | §178.270-11(a)(3) |
| T28 | 101 | 10 | Prohibited | §178.270-11(a)(1),(2) |
| T39 | 101 | 10 | Prohibited | §178.270-11(a)(3) |
| T43 | 101 | 9 | Prohibited | §178.270-11(a)(3) |

(ii) *IM Tank special provisions.* These provisions apply only to transportation in IM portable tanks:

*Code/Special Provisions*

T25 This hazardous material is not permitted for transport in IM portable tanks.

T26 Each tank must have a minimum shell thickness of 6.35 mm (0.250 inch) mild steel.

T27 Each tank must have a minimum shell thichness of 8.0 mm (0.315 inch) mild steel.

T28 See entry for T28 in the IM Tank Configuration Table in paragraph (c)(7)(i) of this section.

T29 The lading must be completely covered with nitrogen, inert gas or other inert materials.

T30 IM 102 portable tanks without bottom openings or with bottom openings conforming to §173.32c(g)(1) of this subchapter are authorized for a hazardous material with a flash point of O°C (32°F) or greater and a vapor pressure not greater than 65.5 kPa (9.5 psia) at 65.6°C (150°F).

T31 IM 102 portable tanks without bottom openings or with bottom openings conforming to §173.32c(g)(2) of this subchapter are authorized for a hazardous material with a flash point of 0°C (32°F) or greater and a vapor pressure not greater than 65 kPa (9.4 psia) at 65°C (150°F).

T32 Each tank must have a minimum shell thickness of 10.0 mm (0.394 inch) mild steel with at least 5.0 mm (0.197 inch) lead lining.

T33 Dry phosphorus is not permitted. For transport in a molten state, the tank must be insulated in accordance with Note T38. Air must be eliminated from the interior of the tank. The tank may be heated, however, interior heating coils are prohibited.

T34 The IM Tank authorization is limited to aqueous solutions containing not more than 40% dimethylamine.

T35 Each tank must be equipped with reclosing (spring loaded) pressure relief valves set to discharge at pressures determined according to the pressure characteristics of the organic peroxide lading.

T36 Each tank must be equipped with pressure relief devices with sufficient venting capacity to prevent the tank from bursting.

T37 IM portable tanks are only authorized for the shipment of hydrogen peroxide solutions in water containing 72 percent or less hydrogen peroxide by weight. Pressure relief devices shall be designed to prevent the entry of foreign matter, the leakage of liquid and the development of any dangerous excess pressure. In addition, the tank shall be designed so that internal surfaces may be effectively cleaned and passivated. Each tank must be equipped with pressure relief devices conforming to the following requirements:

| Concentration of hydrogen peroxide solution | Total venting capacity in standard cubic feet per hour (S.C.F.H.) per pound of hydrogen peroxide solution |
|---|---|
| 52 percent or less | 11 |
| Over 52 percent but not greater than 60 percent | 22 |
| Over 60 percent but not greater than 72 percent | 32 |

T38 Each tank, except a multi-unit tank car tank, must be insulated with at least 100 mm (3.9 inches) of cork or other suitable insulation material of sufficient thickness that the overall thermal conductance at 15.5°C (60°F) is not more than 1.533 kilojoules per hour per square meter per degree celsius (0.075 Btu per hour per square foot per degree Fahrenheit) temperature differential. The exterior surface of a carbon steel tank and the interior surface of a carbon steel jacket must be given a protective coating. The jacket must be flashed around all openings so as to be weather-tight.

T39 See entry for T39 in the IM Tank Configuration Table in paragraph (c)(7)(i) of this section.

T40 Each tank must have a minimum shell thickness of 10.0 mm (0.39 inch) mild steel.

T41 Each tank must have a minimum shell thickness of 12.0 mm (0.47 inch) mild steel.

T42 Transport in IM portable tanks permitted only under conditions approved by the Associate Administrator for Hazardous Materials Safety.

T43 See entry for T43 in the IM Tank Configuration Table in paragraph (c)(7)(i) of this section.

T44 DOT Specification IM 101 portable tanks shall be made of stainless steel except that steel other than stainless steel may be used in accordance with the provisions of §173.24b(b) of this subchapter. Thickness of stainless steel for tank shell and heads must be the greater of 7.62 mm (0.300 inch) or the thickness required for a tank with a design pressure at least equal to 1.5 times the vapor pressure of the lading at 46°C (115°F).

T45 DOT Specification IM 101 portable tanks shall be made of stainless steel except that steel other than stainless steel may be used in accordance with the provisions of §713.24b(b) of this subchapter. Thickness of stainless steel for tank shell and heads must be the greater of 6.35 mm (0.250 inch) or the thickness required for a tank with a design pressure at least equal to 1.3 times the vapor pressure of the lading at 46°C (115°F).

T46 IM portable tanks in sodium metal service are not required to be hydrostatically retested.

(8) *"W" codes.* These provisions apply only to transportation by water:

*Code/Special Provisions*

W41 When offered for transportation by water, this material must be packaged in bales and be securely and tightly bound with rope, wire or similar means.

## Subpart C — Shipping Papers

### §172.200 Applicabilty.

(a) *Description of hazardous materials required.* Except as otherwise provided in this subpart, each person who offers a hazardous material for transportation shall describe the hazardous material on the shipping paper in the manner required by this subpart.

(b) This subpart does not apply to any material, other than a hazardous substance, hazardous waste or marine pollutant, that is —

(1) Identified by the letter "A" in Column 1 of the §172.101 Table, except when the material is offered or intended for transportation by air; or

(2) Identified by the letter "W" in Column 1 of the §172.101 Table, except when the material is offered or intended for transportation by water, or

(3) An ORM-D, except when the material is offered or intended for transportation by air.

### §172.201 General entries.

(a) *Contents.* When a description of hazardous material is required to be included on a shipping paper, that description must conform to the following requirements:

(1) When a hazardous material and a material not subject to the requirements of this subchapter are described on the same shipping paper, the hazardous material description entries required by §172.202 and those additional entries that may be required by §172.203.

(i) Must be entered first, or

(ii) Must be entered in a color that clearly contrasts with any description on the shipping paper of a material not subject to the requirements of this subchapter, except that a description on a reproduction of a shipping paper may be highlighted rather than printed, in a contrasting color (the provisions of this paragraph apply only to the basic description required by §§172.202(a)(1), (2), and (3)), or

(iii) Must be identified by the entry of an "X" placed before the proper shipping name in a column captioned "HM." (The "X" may be replaced by "RQ," if appropriate.)

(2) The required shipping description on a shipping paper and all copies thereof used for transportation purposes, must be legible and printed (manually or mechanically) in English.

(3) Unless it is specifically authorized or required in this subchapter, the required shipping description may not contain any code or abbreviation.

(4) A shipping paper may contain additional information concerning the material provided the information is not inconsistent with the required description. Unless otherwise permitted or required by this subpart, additional information must be placed after the basic description required by §172.202(a).

(b) *Name of shipper.* A shipping paper for a shipment by water must contain the name of the shipper.

(c) *Continuation page.* A shipping paper may consist of more than one page, if each page is consecutively numbered and the first page bears a notation specifying the total number of pages included in the shipping paper. For example, "Page 1 of 4 pages."

(d) *Emergency response telephone number.* A shipping paper must contain an emergency response telephone number, as prescribed in subpart G of part 172 of this subchapter. (Effective December 31, 1990)

### §172.202 Description of hazardous material on shipping papers.

(a) The shipping description of a hazardous material on a shipping paper must include —

(1) The proper shipping name prescribed for the material in §§172.101 or 172.102 (when authorized);

(2) The hazard class or division prescribed for the material as shown in Column 3 of the §172.101 Table (class names or subsidiary hazard class number may be entered following the numerical hazard class, or following the basic description). The hazard class need not be included for the entry "Combustible liquid, n.o.s.";

(3) The identification number prescribed for the material as shown in Column 4 of the §172.101 Table;

(4) The packing group, in Roman numerals, prescribed for the material in Column 5 of the §172.101 Table, if any. The packing group may be preceded by the letters "PG" (e.g., "PG II"); and

(5) Except for empty packagings (see §173.29 of this subchapter), cylinders for Class 2 (compressed gases) materials, and bulk packagings, the total quantity (by net or gross mass, capacity, or as otherwise appropriate), including the unit of measurement, of the hazardous material covered by the description (e.g., "800 lbs", "55 gal.", "3629 kg", or "208 L"). For cylinders for Class 2 (compressed gases) materials and bulk packagings, some indication of total quantity must be shown (e.g., "10 cylinders" or "1 cargo tank").

(b) Except as provided in this subpart, the basic description specified in paragraphs (a)(1), (2), (3) and (4) of this section must be shown in sequence with no additional information interspersed. For example: "Gasoline, 3, UN 1203, PG II".

(c) The total quantity of the material covered by one description must appear before or after, or both before and after, the description required and authorized by this subpart. The type of packaging and destination marks may be entered in any appropriate manner before or after the basic description. Abbreviations may be used to express units of measurement and types of packagings.

(d) Technical and chemical group names may be entered in parenthese between the proper shipping name and hazard class or following the basic description. An appropriate modifier, such as "contains" or "containing," and/or the percentage of the technical constituent may also be used. For example: "Flammable liquids, n.o.s. (contains Xylene and Benzene), 3, UN 1993, PG II".

(e) Except for those materials in the UN Recommendations, the ICAO Technical Instructions, or the IMDG Code, a material that is not a hazardous material according to this subchapter may not be offered for transportation or transported when its description on a shipping paper includes a hazard class or an identification number specified in §172.101.

### §172.203 Additional description requirements.

(a) *Exemptions.* Each shipping paper issued in connection with a shipment made under an exemption must bear the notation "DOT-E" followed by the exemption number assigned and so located that the notation is clearly associated with the description to which the exemption applies.

(b) *Limited quantites.* The description for a material offered for transportation as "limited quantity," as authorized by this subchapter, must include the words "Limited Quantity" or "Ltd Qty" following the basic description.

(c) *Hazardous substances.* (1) Except for radioactive materials described in accordance with paragraph (d) of this section, if the proper shipping name for a material that is a hazardous substance does not identify the hazardous substance by name, one of the following descriptions shall be entered, in parentheses, in association with the basic description:

(i) The name of the hazardous substance as shown in Appendix A to §172.101; or

(ii) For waste streams, the waste stream number; or

(iii) For wastes which exhibit an EPA characteristic of ignitability, corrosivity, reactivity, or EP toxicity, the letters "EPA" followed by the word "ignitability", or "corrosivity", or "reactivity", or "EP toxicity", as appropriate *or* the corresponding "D" number, as appropriate.

(2) The letters "RQ" shall be entered on the shipping paper either before or after, the basic description required by §172.202 for each hazardous substance (see definition in 171.8 of this subchapter). For example: "RQ, Allyl alcohol, 6.1, UN 1098, I"; or "Environmentally hazardous substance, solid, n.o.s., 9, UN 3077, III, RQ (Adipic acid)".

(d) *Radioactive material.* The description for a shipment of a Class 7 (radioactive) material must include the following additional entries as appropriate:

(1) The words "RADIOACTIVE MATERIAL" unless these words are contained in the proper shipping name.

(2) The name of each radionuclide in the radioactive material that is listed in §173.435 of this subchapter. Abbreviations, e.g., "$^{99}$Mo" are authorized.

(3) A description of the physical and chemical form of the material, if the material is not in special form (generic chemical description is acceptable for chemical form).

(4) The activity contained in each package of the shipment in terms of curies, millicuries, or microcuries. Abbreviations are authorized. For the shipment of a package containing a highway route controlled quantity of radioactive materials (see §173.403(l) of this subchapter), the words "Highway route controlled quantity" must be entered in association with the basic description.

(5) The category of label applied to each package in the shipment. For example: "RADIOACTIVE WHITE-I."

(6) The transport index assigned to each package in the shipment bearing RADIOACTIVE YELLOW-II or RADIOACTIVE YELLOW-III labels.

(7) For a shipment of fissile radioactive materials —

(i) The words "Fissile Exempt," if the package is exempt pursuant to §173.453 of this subchapter, or

(ii) If not exempt, the fissile class of each package in the shipment, pursuant to §173.455 of this subchapter; and

(iii) For a Fissile Class III shipment, the additional notation: "Warning-Fissile Class III Shipment. Do not Load More Than ***Packages per Vehicle." (Asterisks to be replaced by appropriate number.) "In loading and Storage Areas, Keep at Least 20 Feet (6 Meters) from Other Packages Bearing Radioactive Labels."

(iv) If a Fissile Class III shipment is to be transported by water, the supplementary notation must also include the following statement: "For shipment by water, only one Fissile Class III shipment is permitted in each hold."

(8) For a package approved by the U.S. Department of Energy (DOE) or U.S. Nuclear Regulatory Commission (USNRC), a notation of the package identification markings as prescribed in the applicable DOE or USNRC approval (See §173.471 of the subchapter.)

(9) For an export shipment or a shipment in a foreign made package, a notation of the package identification marking as prescribed in the applicable International Atomic Energy Agency (IAEA) Certificate of Competent Authority which has been issued for the package. (See §173.473 of the subchapter.)

(e) *Empty packagings.*

(1) The description on the shipping paper for a packaging containing the residue of a hazardous material may include the words "RESIDUE: Last Contained ***" in association with the basic description of the hazardous material last contained in the packaging.

(2) For a tank car containing the residue (as defined in 171.8) of a hazardous material, the requirements of §174.25(c) and paragraph (e)(3) of this section apply.

(3) If a packaging, including a tank car, contains a residue that is a hazardous substance, the

description on the shipping papers must be prefaced with the phrase "RESIDUE: Last Contained ***" and the letters "RQ" must be entered on the shipping paper either before or after the basic description.

(f) *Transportation by air.* When a package containing a hazardous material is offered for transportation by air and this subchapter prohibits its transportation aboard passenger-carrying aircraft, the words "Cargo aircraft only" must be entered after the basic description.

(g) *Transportation by rail.*

(1) The shipping paper for a rail car containing a hazardous material must contain the notation "Placarded" following by the name of the placard required for the rail car.

(2) The shipping paper for each Class DOT-113 tank car containing a flammable gas must contain an appropriate notation, such as "DOT-113A," and the statement "Do Not Hump or Cut Off Car While in Motion."

(3) When shipments of elevated temperature materials are transported under the exception permitted in §173.247(h)(3) of this subchapter, the shipping paper must contain an appropriate notation, such as "Maximum Operating Speed 15 mph."

(h) *Transportation by highway.* Following the basic description for a hazardous material in a specification MC 330 or MC 331 cargo tank, there must be entered for—

(1) *Anhydrous ammonia.*

(i) The words "0.2 PERCENT WATER" to indicate the suitability for shipping anhydrous ammonia in a cargo tank made of quenched and tempered steel as authorized by §173.315(a), Note 14 of this subchapter, or

(ii) The words "NOT FOR Q AND T TANKS" when the anhydrous ammonia does not contain 0.2 percent or more water by weight.

(2) *Liquefied petroleum gas.* (i) The word "NON-CORROSIVE" or "NONCOR" to indicate the suitability for shipping "NON-CORROSIVE" liquefied petroleum gas in a cargo tank made of quenched and tempered steel as authorized by §173.315(a), Note 15 to this subchapter, or

(ii) The words "NOT FOR Q AND T TANKS" for grades of liquefied petroleum gas other than "Noncorrosive".

(i) *Transportation by water.* Each shipment by water must have the following additional shipping paper entries:

(1) Identification of the type of packagings such as barrels, drums, cylinders, and boxes.

(2) The number of each type of package including those in a freight container or on a pallet.

(3) The gross mass of each type of package or the individual gross mass of each package.

(j) *Dangerous when wet material.* The words "Dangerous when wet" shall be entered on the shipping paper in association with the basic description for a material which meets the definition of a dangerous when wet material in §173.124(c) of this subchapter.

(k) *Technical names for "n.o.s." and other generic descriptions.* Unless otherwise excepted, if a material is described on a shipping paper by one of the proper shipping names listed in paragraph (k)(3) of this section, the technical name of the hazardous material must be entered in parentheses in association with the basic description. For example "Corrosive liquid, n.o.s. (Caprylyl chloride), 8, UN 1760; II", or "Corrosive liquid, n.o.s., 8, UN 1760, II (contains caprylyl chloride)". The word contains may be used in association with the technical name, if appropriate. For organic peroxides which may qualify for more than one generic listing depending on concentration, the technical name must include the actual concentration being shipped or the concentration range for the appropriate generic listing. For example, 'Organic peroxide type B, solid, 5.2, UN 3102 (dibenzoyl peroxide, 52-100%)', or 'Organic peroxide type E, solid, 5.2, UN 3108, (dibenzoyl peroxide, paste, <52%)'.

(1) In addition to the n.o.s. descriptions listed herein, the requirements of this section apply to all shipping descriptions for poisonous materials which are subject to the requirements of paragraph (m) of this section, and for which the proper shipping name does not specifically identify the poisonous constituent by technical name. For example, "Motor fuel antiknock mixtures (Tetraethyl lead), 6.1, UN 1649, I", or "Motor fuel antiknock mixtures, 6.1, UN 1649, I (Tetraethyl lead)".

(2) If a hazardous material is a mixture or solution of two or more hazardous materials, the technical names of at least two components most predominantly contributing to the hazards of the mixture or solution must be entered on the shipping paper as required by paragraph (k) of this section. For example, "Flammable liquid, corrosive, n.o.s., 3, UN 2924, II (contains Methanol, Potassium hydroxide)".

(3) Proper shipping names for which the provisions of this paragraph apply are as follows:

Alcohols, toxic, n.o.s.
Articles, explosive, n.o.s.
Caustic alkali liquids, n.o.s.
Combustible liquid, n.o.s.
Compounds, cleaning liquid, *corrosive*
Compounds, cleaning liquid, *flammable*
Compounds, tree *or* weed killing, liquid, *corrosive*
Compounds, tree *or* weed killing, liquid, *flammable*
Compounds, tree *or* weed killing, liquid, *poisonous*
Components, explosive train, n.o.s.
Compressed *or* Liquefied gases, n.o.s.
Compressed *or* Liquefied gases, flammable, n.o.s.
Compressed *or* Liquefied gases, flammable, toxic, n.o.s.
Compressed or Liquefied gases, toxic, n.o.s.
Contrivances, water-activated

Corrosive liquids, flammable, n.o.s.
Corrosive liquids, n.o.s.
Corrosive liquids, oxidizing, n.o.s.
Corrosive liquids, poisonous, n.o.s.
Corrosive liquids, which in contact with water emit flammable gases, n.o.s.
Corrosive solids, flammable, n.o.s.
Corrosive solids, n.o.s.
Corrosive solids, oxidizing, n.o.s.
Corrosive solids, poisonous, n.o.s.
Corrosive solids, self heating, n.o.s.
Corrosive solids, which in contact with water emit flammable gases, n.o.s.
Disinfectants, corrosive, liquid, n.o.s.
Disinfectants, liquid, n.o.s.
Disinfectants, solids, n.o.s.
Dispersant gas, n.o.s.
Dyes, liquid, n.o.s. *or* Dye intermediates, liquid, n.o.s.
Dyes, solid, n.o.s. *or* Dye intermediates, solid, n.o.s.
Environmentally hazardous substances, liquid, n.o.s.
Environmentally hazardous substances, solid, n.o.s.
Flammable liquids, corrosive, n.o.s.
Flammable liquids, n.o.s.
Flammable liquids, poisonous, n.o.s.
Flammable solids, corrosive, n.o.s.
Flammable solids, n.o.s.
Flammable solids, poisonous, n.o.s.
Halogenated irritating liquids, n.o.s.
Hazardous waste, liquid, n.o.s.
Hazardous waste, solid, n.o.s.
Infectious substances, affecting animals
Infectious substances, affecting humans
Insecticide gases, *flammable*, n.o.s.
Insecticide gases, n.o.s.
Insecticide gases, toxic, n.o.s.
Medicines, n.o.s.
Metal powders, flammable, n.o.s.
Organic peroxide type B, liquid
Organic peroxide type B, liquid, temperature controlled
Organic peroxide type B, solid
Organic peroxide type B, solid, temperature controlled
Organic peroxide type C, liquid
Organic peroxide type C, liquid, temperature controlled
Organic peroxide type C, solid
Organic peroxide type C, solid, temperature controlled
Organic peroxide type D, liquid
Organic peroxide type D, liquid, temperature controlled
Organic peroxide type D, solid
Organic peroxide type D, solid, temperature controlled
Organic peroxide type E, liquid
Organic peroxide type E, liquid, temperature controlled
Organic peroxide type E, solid
Organic peroxide type E, solid, temperature controlled
Organic peroxide type F, liquid
Organic peroxide type F, liquid, temperature controlled
Organic peroxide type F, solid
Organic peroxide type F, solid, temperature controlled
Other regulated substances, liquid, n.o.s.
Other regulated substances, solid, n.o.s.
Oxidizing substances, liquid, corrosive, n.o.s.
Oxidizing substances, liquid, n.o.s.
Oxidizing substances, liquid, poisonous, n.o.s.
Oxidizing substances, solid, corrosive, n.o.s.
Oxidizing substances, solid, n.o.s.
Oxidizing substances, solid, poisonous, n.o.s.
Pesticides, liquid, flammable, toxic, n.o.s.
Pesticides, liquid, toxic, flammable, n.o.s.
Pesticides, liquid, toxic, n.o.s.
Pesticides, solid, toxic, n.o.s.
Poisonous liquids, corrosive, n.o.s.
Poisonous liquids, flammable, n.o.s.
Poisonous liquids, n.o.s.
Poisonous liquids, oxidizing, n.o.s.
Poisonous liquids, which in contact with water emit flammable gases, n.o.s.
Poisonous solids, corrosive, n.o.s.
Poisonous solids, flammable, n.o.s.
Poisonous solids, n.o.s.
Poisonous solids, oxidizing, n.o.s.
Poisonous solids, self-heating, n.o.s.
Poisonous solids, which in contact with water emit flammable gases, n.o.s.
Pyrophoric liquids, n.o.s.
Pyrophoric solids, n.o.s.
Pyrophoric metals, n.o.s. *or* Pyrophoric alloys, n.o.s.
Refrigerant gas, n.o.s.
Refrigerant gases, n.o.s.
Rodenticides, n.o.s.
Samples, explosive (*other than initiating explosives*)
Self-heating substances, solid, corrosive, n.o.s.
Self-heating substances, solid, n.o.s.
Self-heating substances, solid, poisonous, n.o.s.
Self-reactive substances, Sample, n.o.s.
Self-reactive substances, Trial quantities, n.o.s.
Substances, explosive, n.o.s.
Substances, explosive, very insensitive (substances, EVI) n.o.s.".
Substances, which in contact with water emit flammable gases, liquid, corrosive, n.o.s.
Substances, which in contact with water emit flammable gases, liquid, n.o.s.
Substances, which in contact with water emit flammable gases, liquid, poisonous, n.o.s.
Substances, which in contact with water emit flammable gases, solid, corrosive, n.o.s.
Substances, which in contact with water emit flammable gases, solid, n.o.s.
Substances, which in contact with water emit flammable gases, solid, oxidizing, n.o.s.
Substances, which in contact with water emit flammable gases, solid, poisonous, n.o.s.
Substances, which in contact with water emit flammable gases, solid, self-heating, n.o.s.
Substances, which in contact with water emit flammable gases, solid, flammable n.o.s.
Tear gas substances, n.o.s. liquid *or* solid

(4) The provisions of this paragraph do not apply--

(i) To a material that is a hazardous waste and described using the proper shipping name "Hazardous waste, liquid *or* solid, n.o.s.", classed as a miscellaneous Class 9, provided the EPA hazardous waste number is included on the shipping paper in association with the basic description, or provided the material is described in accordance with the provisions of §172.203(c) of this part.

(ii) To a material for which the hazard class is to be determined by testing under the criteria in §172.101(c)(11).

(iii) If the n.o.s. description for the material (other than a mixture of hazardous materials of different classes meeting the definitions of more than one hazard class) contains the name of the chemical element or group which is primarily responsible for the material being included in the hazard class indicated. For example "Mercury compounds, solid, n.o.s., 6.1, UN 2025, II".

(iv) If the n.o.s. description for the material (which is a mixture of hazardous materials of different classes meeting the definition of more than one hazard class) contains the name of the chemical element or group responsible for the material meeting the definition of one of these classes. In such cases, only the technical name of the component that is not appropriately identified in the n.o.s. description shall be entered in parenthese. For example: 'Carbamate pesticides, liquid, flammable, toxic, n.o.s., *flash point less than 23°C* (contains Xylene) 3, 6.1, UN 2758, II'.

(l) *Marine pollutants.* (1) If the proper shipping name for a material which is a marine pollutant does not identify by name the component which makes the material a marine pollutant, the name of that component must appear in parentheses in association with the basic description. Where two or more components which make a material a marine pollutant are present, the names of at least two of the components most predominantly contributing to the marine pollutant designation must appear in parentheses in association with the basic description.

(2) The words "Marine Pollutant" shall be entered in association with the basic description for a material which is a marine pollutant.

(m) *Poisonous materials.* Notwithstanding the hazard class to which a material is assigned—(Effective December 31, 1990)

(1) If a liquid or solid material in a package meets the definition of a Division 6.1, Packing Group I or II, according to this subchapter, and the fact that it is a poison is not disclosed in the shipping name or class entry, the word "Poison" shall be entered on the shipping paper in association with the shipping description.

(2) If the technical name of the compound or principal constituent that causes a material to meet the definition of Division 6.1, Packing Group I or II (as defined in §173.132(a) of this subchapter), or Division 2.3 (as defined in §173.115(c) of this subchapter), is not included in the proper shipping name for the material, the technical name shall be entered on the shipping paper in the manner prescribed in paragraph (k) of this section.

(3) For materials which are poisonous by inhalation (see §171.8 of this subchapter), the words "Poison-Inhalation Hazard" and the words "Zone A", "Zone B", "Zone C", or "Zone D", for gases or "Zone A" or "Zone B" for liquids, as appropriate, shall be entered on the shipping paper immediately following the shipping description. The word "Poison" need not be repeated if it otherwise appears in the shipping description.

(n) *Elevated temperature materials.* Except for molten sulfur or molten aluminum, if a liquid material in a package meets the definition of an elevated temperature material in §171.8 of this subchapter, and the fact that it is an elevated temperature material is not disclosed in the shipping name, the word "HOT" must immediately precede the proper shipping name of the material on the shipping paper.

(o) *Oil.* Except for shipments in non-bulk packagings and petroleum distillates specifically identified in the §172.101 Table (e.g., gasoline, kerosene), if the fact that a hazardous material that meets the definition of oil is not communicated through the proper shipping name, the word "oil" must be entered, in parentheses, in association with the basic description.

### §172.204 Shipper's certification.

(a) *General.* Except as provided in paragraphs (b) and (c) of this section, each person who offers a hazardous material for transportation shall certify that the material is offered for transportation in accordance with this subchapter by printing (manually or mechanically) on the shipping paper containing the required shipping description the certification contained in paragraph (a)(1) of this section or the certification (declaration) containing the language contained in paragraph (a)(2) of this section.

(1) "This is to certify that the above-named materials are properly classified, described, packaged, marked and labeled, and are in proper condition for transportation according to the applicable regulations of the Department of Transportation."

Note.—In line one of the certification the words "herein-named" may be substituted for the words "above-named".

(2) "I hereby declare that the contents of this consignment are fully and accurately described above by proper shipping name and are classified, packed, marked and labeled, and are in all respects in proper condition for transport by [*] according to applicable international and national governmental regulations."

*Additional language indicating the modes of transportation to be used may be inserted at this point in the certification. All modes of transportation may be indicated provided that any mode not applicable to a specific shipment is deleted (lined out).

(b) *Exceptions.*

(1) Except for a hazardous waste, no certification is required for a hazardous material offered for transportation by motor vehicle and transported —

(i) In a cargo tank supplied by the carrier, or

(ii) By the shipper as a private carrier except for a hazardous material that is to be reshipped or transferred from one carrier to another.

(2) No certification is required for the return of an empty tank car which previously contained a hazardous material and which has not been cleaned or purged.

(c) *Transportation by air*—

(1) *General.* Certification containing the following language may be used in place of the certification required by paragraph (a) of this section:

I hereby certify that the contents of this consignment are fully and accurately described above by proper shipping name and are classified, packed, marked and labeled and in proper condition for carriage by air according to applicable national governmental regulations.

(2) *Certificate in duplicate.* Each person who offers a hazardous material to an aircraft operator for transportation by air shall provide two copies of the certification required in this section. (See §175.30 of this subchapter.)

(3) *Passenger and cargo aircraft.* Each person who offers for transportation by air a hazardous material authorized for air transportation shall add to the certification required in this section the following statement:

This shipment is within the limitations prescribed for passenger aircraft/cargo aircraft only (delete nonapplicable).

(4) *Radioactive material.* Each person who offers any radioactive material for transportation aboard a passenger-carrying aircraft shall sign (mechanically or manually) a printed certificate stating that the shipment contains radioactive material intended for use in, or incident to, reseach, or medical diagnosis or treatment.

(d) *Signature.* The certifications required by paragraph (a) or (c) of this section —

(1) Must be legibly signed by a principal, officer, partner, or employee of the shipper or his agent; and

(2) May be legibly signed manually, by typewriter, or by other mechanical means.

### §172.205 Hazardous waste manifest.

(a) No person may offer, transport, transfer, or deliver a hazardous waste (waste) unless an EPA Forms 8700-22 and 8700-22A (when necessary) hazardous waste manifest (manifest) is prepared in accordance with 40 CFR 262.20 and is signed, carried, and given as required of that person by this section.

(b) The shipper (generator) shall prepare the manifest in accordance with 40 CFR Part 262.

(c) The original copy of the manifest must be dated by, and bear the handwritten signature of, the person representing —

(1) The shipper (generator) of the waste at the time it is offered for transportation, and

(2) The initial carrier accepting the waste for transportation.

(d) A copy of the manifest must be dated by, and bear the handwritten signature of the person representing —

(1) Each subsequent carrier accepting the waste for transportation, at the time of acceptance, and

(2) The designated facility receiving the waste, upon receipt.

(e) A copy of the manifest bearing all required dates and signatures must be —

(1) Given to a person representing each carrier accepting the waste for transportation,

(2) Carried during transportation in the same manner as required by this subchapter for shipping papers,

(3) Given to a person representing the designated facility receiving the waste,

(4) Returned to the shipper (generator) by the carrier that transported the waste from the United States to a foreign destination with a notation of the date of departure from the United States, and

(5) Retained by the shipper (generator) and by the initial and each subsequent carrier for three years from the date the waste was accepted by the initial carrier. Each retained copy must bear all required signatures and dates up to and including those entered by the next person who received the waste.

(f) The requirements of paragraphs (d) and (e) of this section do not apply to a rail carrier when waste is delivered to a designated facility by railroad if —

(1) All of the information required to be entered on the manifest (except generator and carrier identification numbers and the generator's certification) is entered on the shipping paper carried in accordance with §174.26(c) of this subchapter;

(2) The delivering rail carrier obtains and retains a receipt for the waste that is dated by and bears the handwritten signature of the person representing the designated facility; and

(3) A copy of the shipping paper is retained for three years by each railroad transporting the waste.

(g) The person delivering a hazardous waste to an initial rail carrier shall send a copy of the manifest, dated and signed by a representative of the rail carrier, to the person representing the designated facility.

(h) A hazardous waste manifest required by 40 CFR Part 262, containing all of the information required by this subpart, may be used as the shipping paper required by this subpart.

## Subpart D — Marking

### §172.300 Applicability.

(a) Each person who offers a hazardous material for transportation shall mark each package, freight container, and transport vehicle containing the hazardous material in the manner required by this subpart.

(b) When assigned the function by this subpart, each carrier that transports a hazardous material shall mark each package, freight container, and transport vehicle containing the hazardous material in the manner required by this subpart.

### §172.301 General marking requirements for non-bulk packagings.

(a) *Proper shipping name and identification number.* (1) Except as otherwise provided by this subchapter, each person who offers for transportation a hazardous material in a non-bulk packaging shall mark the package with the proper shipping name and identification number (preceded by "UN" or "NA", as appropriate for the material as shown in the §172.101 table.

(2) The proper shipping name for a hazardous waste (as defined in §171.8 of this subchapter) is not required to include the word "waste" if the package bears the EPA marking prescribed by 40 CFR 262.32.

(b) *Technical names.* In addition to the marking required by paragraph (a) of this section, each non-bulk packaging containing hazardous materials subject to the provisions of §172.203(k) of this part shall be marked with the technical name in parentheses in association with the proper shipping name in accordance with the requirements and exceptions specified for display of technical descriptions on shipping papers in §172.203(k) of this part.

(c) *Exemption packagings.* The outside of each package authorized by an exemption shall be plainly and durably marked "DOT-E" followed by the exemption number assigned.

(d) *Consignee's or consignor's name and address.* Each person who offers for transportation a hazardous material in a non-bulk package shall mark that package with the name and address of the consignor or consignee except when the package is--

(1) Transported by highway only and will not be transferred from one motor carrier to another; or

(2) Part of a carload lot, truckload lot or freight container load, and the entire contents of the rail car, truck or freight container are shipped from one consignor to one consignee.

(e) *Previously marked packagings.* A package which has been previously marked as required for the material it contains and on which the marking remains legible, need not be remarked. (For empty packagings, see §173.29 of this subchapter.)

(f) *Marking exceptions.* (1) Identification numbers are not required on packages which contain only limited quantities, as defined in 171.8 of this subchapter, or ORM-D materials.

(2) The marking of technical names on non-bulk packagings filled for shipment prior to December 31, 1990 is not required until December 31, 1991.

**§172.302 General marking requirements for bulk packagings.**

(a) Identification numbers. Except as otherwise provided in this subpart, no person may offer for transportation or transport a hazardous material in a bulk packaging unless the packaging is marked as required by §172.332 with the identification number specified for the material in the §172.101 Table--

(1) On each side and each end, if the packaging has a capacity of 3,785L (1,000 gallons) or more;

(2) On two opposing sides, if the packaging has a capacity of less than 3,785L (1,000 gallons); or

(3) For cylinders permanently installed on a tube trailer motor vehicle, on each side and each end of the motor vehicle.

(b) Size of markings. Except as otherwise provided, markings required by this subpart on bulk packagings must have a width of at least 6.0 mm (0.24 inch) and a height of at least--

(1) 100mm (3.9 inches) for rail cars;

(2) 25 mm (one inch) for portable tanks with capacities of less than 3,785 L (1,000 gallons); and

(3) 50 mm (2.0 inches) for cargo tanks and other bulk packages.

(c) Exempton packagings. The outside of each bulk package used under the terms of an exemption shall be plainly and durably marked "DOT-E" followed by the exemption number assigned.

(d) Each bulk packaging marked with a proper shipping name, common name or identification number as required by this subpart must remain marked when it is emptied unless it is--

(1) Sufficiently cleaned of residue and purged of vapors to remove any potential hazard; or

(2) Refilled, with a material requiring different markings or no markings, to such an extent that any residue remaining in the packaging is no longer hazardous.

(e) Additional requirements for marking portable tanks, cargo tanks, tank cars, multi-unit tank car tanks, and other bulk packagings are prescribed in §§172.326, 172.328, 172.330, and 172.331, respectively, of this subpart.

(f) A bulk packaging marked prior to October 1, 1991, in conformance to the regulations of this subchapter in effect on September 30, 1991, need not be remarked if the key words of the proper shipping name are identical to those currently specified in the §172.101 Table. For example, a tank car marked "ANHYDROUS AMMONIA" need not be remarked "ANHYDROUS AMMONIA, LIQUEFIED".

**§172.303 Prohibited markings.**

(a) No person may offer for transportation or transport a packaging which is marked with the proper shipping name or identification number of a hazardous material unless the package contains the identified hazardous material or its residue.

(b) This section does not apply to—

(1) Transportation of a package in a transport vehicle or freight container if the package is not visible during transportation and is loaded by the shipper and unloaded by the shipper or consignee.

(2) Markings on a package which are securely covered in transportation.

(3) The marking of a shipping name on a package when the name describes a material not regulated under this subchapter.

**§172.304 Marking requirements.**

(a) The marking required in this subpart —

(1) Must be durable, in English and printed on or affixed to the surface of a package or on a label, tag, or sign;

(2) Must be displayed on a background of sharply contrasting color;

(3) Must be unobscured by labels or attachments; and

(4) Must be located away from any other marking (such as advertising) that could substantially reduce its effectiveness.

**§172.306 [Reserved]**

## §172.308 Authorized abbreviations.

(a) Abbreviations may not be used in a proper shipping name marking except as authorized in this section.

(b) For marking descriptions of ammunition, such as Ammunition for cannon without projectile, etc., the words "with" or "without" may be abbreviated as "W" or "W/O". For example: "Ammunition for cannon W/O projectile."

(c) The abbreviation "ORM" may be used in place of the words "Other Regulated Material."

(d) Abbreviations which appear as authorized descriptions in Column 2 of the §172.101 Table (e.g., "TNT" and "2,4-D") are authorized.

## §172.310 Radioactive materials.

(a) In addition to any other markings required by this subpart, each package containing radioactive materials must be marked as follows:

(1) Each package of radioactive materials in excess of 110 pounds (50 kilograms) must have its gross weight plainly and durably marked on the outside of the package.

(2) Each package of radioactive materials which conforms to the requirements for Type A or Type B packaging (§173.403 of this subchapter) must be plainly and durably marked on the outside of the package in letters at least ½-inch (13 mm) high, with the words "TYPE A" or "TYPE B" as appropriate. A packaging which is not in compliance with these requirements may not be so marked.

(3) Each package of radioactive material destined for export shipment must also be marked "USA" in conjunction with the specification marking, or other package certificate identification. (See §§173.471, 173.472, and 173.473 of this subchapter.)

## §172.312 Liquid hazardous materials in non-bulk packagings.

(a) Except as provided in this section, each non-bulk combination package having inner packagings containing liquid hazardous materials must be:

(1) Packed with closures upward, and

(2) Legibly marked, with package orientation markings that conform pictorially to ISO Standard 780-1985, on two opposite vertical sides of the package with the arrows pointing in the correct upright direction. Depicting a rectangular border around the arrows is optional.

**Package orientation**

(b) Arrows for purposes other than indicating proper package orientation may not be displayed on a package containing a liquid hazardous material.

(c) The requirements of paragraph (a) of this section do not apply to--

(1) A non-bulk package with inner packagings which are cylinders.

(2) Except when offered or intended for transportation by aircraft, packages containing flammable liquids in inner packagings of one liter or less prepared in accordance with §173.150(b) or (c) of this subchapter.

(3) When offered or intended for transportation by aircraft, packages containing flammable liquids in inner packagings of 120 ml (4 fluid oz.) or less prepared in accordance with §173.150 (b) or (c) of this subchapter when packed with sufficient absorption material between the inner and outer packagings to completely absorb the liquid contents.

(4) Liquids contained in manufactured articles (e.g., alcohol or mercury in thermometers) which are leak-tight in all orientations.

(5) A non-bulk package with hermetically-sealed inner packagings.

## §172.313 Poisonous hazardous materials.

(a) For materials poisonous by inhalation (see §171.8 of this subchapter), the package shall be marked "Inhalation Hazard" in association with the required labels or placards, as appropriate, or shipping name, when required. (See §172.302(b) of this subpart for size of markings on bulk packages.) Bulk packaging must be marked on two opposing sides.

(b) Each non-bulk plastic outer packaging used as a single or composite packaging for materials meeting the definition of Division 6.1 (in §173.132 of this subchapter) shall be permanently marked, by embossment or other durable means, with the word "POISON" in letters at least 6.3mm (0.25 inch) in height. Additional text or symbols related to hazard warning may be included in the marking. The marking shall be located within 150mm (6 inches) of the closure of the packaging.

### §172.316 Packagings containing materials classed as ORM-D.

(a) Each non-bulk packaging containing a material classed as ORM-D must be marked on at least one side or end with the ORM-D designation immediately following or below the proper shipping name of the material. The ORM designation must be placed within a rectangle that is approximately 6.3mm (0.25 inches) larger on each side than the designation. The designation for ORM-D must be:

(1) ORM-D-AIR for an ORM-D that is prepared for air shipment and packaged in accordance with the provisions of §173.27 of this subchapter.

(2) ORM-D for an ORM-D other than as described in paragraph (a)(1) of this section.

(b) When the ORM-D marking including the proper shipping name can not be affixed on the package surface, it may be on an attached tag.

(c) The marking ORM-D is the certification by the person offering the packaging for transportation that the material is properly described, classed, packaged, marked and labeled (when apropriate) and in proper condition for transportation according to the applicable regulations of this subchapter. This form of certification does not preclude the requirement for a certificate on a shipping paper when required by subpart C of this part.

### §172.320 Explosive hazardous materials.

(a) Except as otherwise provided in paragraphs (b), (c), (d) and (e) of this section, each package containing a Class 1 material must be marked with the EX-number for each substance, article or device contained therein.

(b) Except for fireworks approved in accordance with §173.56(j) of this subchapter, a package of Class 1 materials may be marked, in lieu of the EX-number required by paragraph (a) of this section, with a national stock number issued by the Department of Defense or identifying information required by regulations for commercial explosives specified in 27 CFR part 55, if the national stock number or identifying information can be specifically associated with the EX-number assigned.

(c) When more than five different Class 1 materials are packed in the same package, the package may be marked with only five of the EX-numbers, national stock numbers, product codes, or combination thereof.

(d) The requirements of this section do not apply if the EX-number, product code or national stock number of each explosive item described under a proper shipping description is shown in association with the shipping description required by §172.202(a) of this part. Product codes and national stock numbers must be traceable to the specific EX-number assigned by the Associate Administrator for Hazardous Materials Safety.

(e) The requirements of this section do not apply to the following Class 1 materials:

(1) Those being shipped to a testing agency in accordance with §173.56(d) of this subchapter;

(2) Those being shipped in accordance with §173.56(e) of this subchapter, for the purposes of developmental testing;

(3) Those which meet the requirements of §173.56(h) of this subchapter and therefore are not subject to the approval process of §173.56 of this subchapter;

(4) Until October 1, 1993, those which are shipped under §171.19 of this subchapter; and

(5) Those that are transported in accordance with §173.56(c)(2) of this subchapter and, therefore, are covered by a national security classification currently in effect.

### §172.322 Marine pollutants.

(a) For vessel transportation of each non-bulk packaging that contains a marine pollutant—

(1) If the proper shipping name for a material which is a marine pollutant does not identify by name the component which makes the material a marine pollutant, the name of that component must be marked on the package in parentheses in association with the marked proper shipping name. Where two or more components which make a material a marine pollutant are present, the names of at least two of the components most predominantly contributing to the marine pollutant designation must appear in parentheses in association with the marked proper shipping name; and

(2) The MARINE POLLUTANT mark shall be placed in association with the hazard warning labels required by Subpart E of this Part or, in the absence of any labels, in association with the marked proper shipping name.

(b) A bulk packaging that contains a marine pollutant must be marked on each end and each side with the MARINE POLLUTANT mark and must be visible from the direction it faces. This mark may be displayed in black lettering on a white square-on-point configuration having the same outside dimensions as a placard.

(c) A transport vehicle or freight container that contains a package subject to the marking requirements of paragraph (a) or (b) of this section must be marked with the MARINE POLLUTANT mark. The mark must appear on each side and each end of the transport vehicle or freight container, and must be visible from the direction it faces. This requirement may be met by the marking displayed on a freight container or portable tank loaded on a motor vehicle or rail car. This mark may be displayed in black lettering on a white square-on-point configuration having the same outside dimensions as a placard.

(d) The MARINE POLLUTANT mark is not required—

(1) On a combination package containing a severe marine pollutant (see appendix B to §172.101), in inner packagings each of which contains:

(i) 0.5 liters (17 ounces) or less net capacity for liquids; or

(ii) 500 grams (17.6 ounces) or less net capacity for solids.

(2) On a combination packaging containing a marine pollutant, other than a severe marine pollutant, in inner packagings each of which contains:

(i) 5 liters (1 gallon) or less net capacity for liquids; or

(ii) 5 kilograms (11 pounds) or less net capacity for solids.

(3) Except for transportation by vessel, on a bulk packaging, freight container or transport vehicle that bears a label or placard specified in Subparts E or F of this part.

(e) *MARINE POLLUTANT mark*. The MARINE POLLUTANT mark must conform to the following:

(1) Except for size, the MARINE POLLUTANT mark must appear as follows:

(2) The symbol, letters and border must be black and the background white, or the symbol, letters, border and background must be of contrasting color to the surface to which the mark is being affixed. For non-bulk packagings of marine pollutants, each side of the mark must be at least 100 mm (3.9 inches), except in the case of packagings which, because of their size, can only bear smaller marks. For bulk packagings, each side of the mark must be at least 250 mm (9.8 inches).

### §172.324 Hazardous substances in non-bulk packagings.

For each non-bulk package that contains a hazardous substance—

(a) Except for radioactive material in packages labeled in accordance with §172.403 of this subchapter, if the proper shipping name does not identify the hazardous substance by name, one of the following descriptions shall be marked on the package, in parentheses, in association with the proper shipping name:

(1) The name of the hazardous substance as shown in Appendix A to §172.101; or

(2) For waste streams, the waste stream number; or

(3) For wastes which exhibit an EPA characteristic of ignitability, corrosivity, reactivity, or EP toxicity, the letters "EPA" followed by the word "ignitability", or "corrosivity", or "reactivity", or "EP toxicity", as appropriate *or* the corresponding "D" number, as appropriate.

(b) The letters "RQ" shall be marked on the package in association with the proper shipping name.

### §172.325 Elevated temperature materials.

(a) Except as provided in paragraph (b) of this section, a bulk packaging containing an elevated temperature material must be marked on two opposing sides with the word "HOT" in black or white Gothic lettering on a contrasting background. The marking must be displayed on the packaging itself or in black lettering on a plain white square-on-point configuration having the same outside dimensions as a placard. (See §172.302(b) for size of markings on bulk packagings.)

(b) Bulk packagings containing molten aluminum or molten sulfur must be marked "MOLTEN ALUMINUM" or "MOLTEN SULFUR", respectively, in the same manner as prescribed in paragraph (a) of this section.

(c) If the identification number is displayed on a white-square-on-point display configuration, as prescribed in §172.336(b), the word "HOT" may be displayed in the upper corner of the same white-square-on-point display configuration. The word "HOT" must be in black letters having a height of at least 50 mm (2.0 inches). Except for size, these markings shall be as illustrated for an Elevated temperature material, liquid, n.o.s.:

### §172.326 Portable tanks.

(a) *Shipping name.* No person may offer for transportation or transport a portable tank containing a hazardous material unless it is legibly marked on two opposing sides with the proper shipping name specified for the material in the §172.101 Table.

(b) *Owner's name.* The name of the owner or of the lessee, if applicable, must be displayed on a portable tank that contains a hazardous material.

(c) *Identification numbers.* (1) If the identification number markings required by §172.302(a) are not visible, a transport vehicle or freight container used to transport a portable tank containing a hazardous material must be marked on each side and each end as required by §172.332 with the identification number specified for the material in the §172.101 Table.

(2) Each person who offers a portable tank containing a hazardous material to a motor carrier, for transportation in a transport vehicle or freight container, shall provide the motor carrier with the required identification numbers on placards, orange panels, or the white square-on-point configuration, as appropriate, for each side and each end of the transport vehicle or freight container from which identification numbers on the portable tank are not visible.

### §172.328 Cargo tanks.

(a) *Providing and affixing identification numbers.* Unless a cargo tank is already marked with the identification numbers required by this subpart, the identification numbers must be provided or affixed as follows:

(1) A person who offers a hazardous material to a motor carrier for transportation in a cargo tank shall provide the motor carrier the identification numbers on placards or shall affix orange panels containing the required identification numbers, prior to or at the time the material is offered for transportation.

(2) A person who offers a cargo tank containing a hazardous material for transportation shall affix the required identification numbers on panels or placards prior to or at the time the cargo tank is offered for transportation.

(b) *Required markings: Gases.* Except for certain nurse tanks which must be marked as specified in §173.315(m) of this subchapter, each cargo tank transporting a Class 2 material subject to this subchapter must be marked, in lettering no less than 50mm (2.0 inches), on each side and each end with--

(1) The proper shipping name specified for the gas in the §172.101 Table; or

(2) An appropriate common name for the material (e.g., "Refrigerant Gas").

(c) *QT/NQT markings.* Each MC 330 and MC 331 cargo tank must be marked near the specification plate, in letters no less than 50mm (2.0 inches) in height, with--

(1) "QT", if the cargo tank is constructed of quenched and tempered steel; or

(2) "NQT", if the cargo tank is constructed of other than quenched and tempered steel.

### §172.330 Tank cars and multi-unit tank car tanks.

(a) *Shipping name and identification number.* No person may offer for transportation or transport a hazardous material—

(1) In a tank car unless the tank car is:

(i) Marked on each side, when required by a special provision to the §172.101 Table or Part 173 of this subchapter, with the proper shipping name specified for the material in the §172.101 Table, or with a common name authorized for the material in this subchapter (e.g., "Refrigerant Gas"); and

(ii) Marked on each side and each end, as required by §172.302 of this subpart, with the identification number specified for the material in the §172.101 Table.

(2) In a multi-unit tank car tank, unless the tank is marked on two opposing sides, in letters and numerals no less than 50 mm (2.0 inches) high—

(i) With the proper shipping name specified for the material in the §172.101 Table or with a common name authorized for the material in this subchapter (e.g., "Refrigerant Gas"); and

(ii) With the identification number specified for the material in the §172.101 Table, unless marked in accordance with §§172.302(a) and 172.332 of this subpart.

(b) A motor vehicle or rail car used to transport a multi-unit tank car tank containing a hazardous material must be marked on each side and each end, as required by §172.332, with the identification number specified for the material in the §172.101 Table.

### §172.331 Bulk packagings other than portable tanks, cargo tanks, tank cars and multi-unit tank car tanks.

(a) Each person who offers a hazardous material to a motor carrier for transportation in a bulk packaging shall provide the motor carrier with the required identification numbers on placards or plain white square-on-point display configurations, as authorized, or shall affix orange panels containing the required identification numbers to the packaging prior to or at the time the material is offered for transportation, unless the packaging is already marked with the identification number as required by this subchapter.

(b) Each person who offers a bulk packaging containing a hazardous material for transportation shall affix to the packaging the required identification numbers on orange panels, square-on-point configurations or placards, as appropriate, prior to, or at the time the packaging is offered for transportation unless it is already marked with identification numbers as required by this subchapter.

### §172.332 Identification number markings.

(a) *General.* When required by §§172.302, 172.326, 172.328, 172.330, or 172.331 of this subpart, identification numbers shall be displayed on orange panels or placards as specified in this section or, when appropriate, on white square-on-point configurations as prescribed in §172.336(b).

(b) *Orange panels.* Display of an identification number on an orange panel shall be in conformance with the following:

(1) The orange panel must be 160mm (6.3 inches) high by 400mm (15.7 inches) wide with a 15 mm (0.6 inches) black outer border. The identification number shall be displayed in 100 mm (3.9 inches) black Helvetica Medium numerals on the orange panel. Measurements may vary from those specified plus or minus 5 mm (0.2 inches).

(2) The orange panel may be made of any durable material prescribed for placards in §172.519, and shall be of the orange color specified for labels or placards in Appendix A to this Part.

(3) The name and hazard class of a material may be shown in the upper left border of the orange panel in letters not more than 18 points high.

(4) Except for size and color, the orange panel and identification numbers shall be as illustrated for Liquefied petroleum gas:

(c) *Placards*: Display of an identification number on a hazard warning placard shall be in conformance with the following:

(1) The identification number shall be displayed across the center area of the placard in 88 mm (3.5 inches) black Alpine Gothic or Alternate Gothic No. 3 numerals on a white background 100 mm (3.9 inches) high and approximately 215 mm (8.5 inches) wide and may be outlined with a solid or dotted line border.

(2) The top of the 100 mm (3.9 inches) high white background shall be approximately 40 mm (1.6 inches) above the placard horizontal center line.

(3) An identification number may be displayed only on a placard corresponding to the primary hazard class of the hazardous material.

(4) For a COMBUSTIBLE placard used to display an identification number, the entire background below the white background for the identification number must be white during transportation by rail and may be white during transportation by highway.

(5) The name of the hazardous material and the hazard class may be shown in letters not more than 18 points high immediately within the upper border of the space on the placard bearing the identification number of the material.

(6) If an identification number is placed over the word(s) on a placard, the word(s) should be substantially covered to maximize the effectiveness of the identification number.

(d) Except for size and color, the display of an identification number on a placard shall be as illustrated for Acetone:

### §172.334 Identification numbers; prohibited display.

(a) No person may display an identification number on a RADIOACTIVE, EXPLOSIVES 1.1, 1.2, 1.3, 1.4, 1.5 or 1.6, DANGEROUS, or subsidiary hazard placard.

(b) No person may display an identification number on a placard, orange panel or white square-on-point display configuration unless—

(1) The identification number is specified for the material in §172.101;

(2) The identification number is displayed on the placard, orange panel or white square-on-point configuration authorized by §172.332 or §172.336(b), as appropriate, and any placard used for display of the identification number corresponds to the hazard class of the material specified in §172.504;

(3) Except as provided under §172.336(c)(4) or (c)(5) the package, freight container, or transport vehicle on which the number is displayed contains the hazardous material associated with that identification number in §172.101.

(c) Except as required by §172.332(c)(4) for a combustible liquid, the identification number of a material may be displayed only on the placards required by the tables in §172.504.

(d) Except as provided in §172.336, a placard bearing an identification number may not be used to meet the requirements of Subpart F of the Part unless it is the correct identification number for all hazardous materials of the same class in the transport vehicle or freight container on which it is displayed.

(e) Except as specified in §172.338, an identification number may not be displayed on an orange panel on a cargo tank unless affixed to the cargo tank by the person offering the hazardous material for transportation in the cargo tank.

(f) If a placard is required by §172.504, an identification number may not be displayed on an orange panel unless it is displayed in proximity to the placard.

(g) No person shall add any color, number, letter, symbol, or word other than as specified in this subchapter, to any identification number marking display which is required or authorized by this subchapter.

### §172.336 Identification numbers; special provisions.

(a) When not required or prohibited by this subpart, identification numbers may be displayed on a transport vehicle or a freight container in the manner prescribed by this subpart.

(b) For hazardous materials in hazard classes for which hazard warning placards are not specified, identification numbers, when required, must be displayed on either orange panels (see §172.332(b)) or on a plain white square-on-point display configuration having the same outside dimensions as a placard. In addition, for materials in hazard classes for which placards are specified and identification number displays are required, but for which identification numbers may not be displayed on the placards authorized for the material (see §172.334(a)), identification numbers must be displayed on orange panels or on the plain square-on-point display configuration in association with the required placards. An identification number displayed on a white square-on-point display configuration is not considered to be a placard.

(1) The 100 mm (3.9 inch) by 215 mm (8.5 inches) area containing the identification number shall be located as prescribed by §172.332(c)(1) and (c)(2) and may be outlined with a solid or dotted line border.

(c) Identification numbers are not required-

(1) On the ends of a portable tank, cargo tank or tank car having more than one compartment if hazardous materials having different identification numbers are being transported therein. In such a circumstance, the identification numbers on the sides of the tank shall be displayed in the same sequence as the compartments containing the materials they identify.

(2) On a cargo tank containing only gasoline, if the cargo tank is marked "Gasoline" on each side and rear in letters no less than 50 mm (2 inches) high, or is placarded in accordance with §172.542(c).

(3) On a cargo tank containing only fuel oil, if the cargo tank is marked "Fuel Oil" on each side and rear in letters no less the 50 mm (2 inches) high, or is placarded in accordance with §172.544(c).

(4) For each of the different liquid petroleum distillate fuels, including gasoline and gasohol in a compartmented cargo tank or tank car, if the identification number is displayed for the distillate fuel having the lowest flash point.

(5) For each of the different liquid petroleum distillate fuels, including gasoline and gasohol transported in a cargo tank, if the identification number is displayed for the liquid petroleum distillate fuel having the lowest flash point.

(6) On nurse tanks meeting the provisions of §173.315(m) of this subchapter.

### §172.338 Replacement of identification numbers.

If more than one of the identification number markings on placards, orange panels, or white square-on-point display configurations that are required to be displayed are lost, damaged or destroyed during transportation, the carrier shall replace all the missing or damaged identification numbers as soon as practicable. However, in such a case, the numbers may be entered by hand on the appropriate placard, orange panel or white square-on-point display configuration providing the correct identification numbers are entered legibly using an indelible marking material. When entered by hand, the identification numbers must be located in the white display area specified in §172.332. This section does not preclude required compliance with the placarding requirements of Subpart F of the subchapter.

## Subpart E — Labeling

### §172.400 General labeling requirements.

(a) Except as specified in §172.400a, each person who offers for transportation or transports a hazardous material in any of the following packages or containment devices, shall label the package or containment device with labels specified for the material in the §172.101 Table and in this subpart:

(1) A non-bulk package;

(2) A bulk packaging, other than a cargo tank, portable tank, or tank car, with a volumetric capacity of less than 18 $m^3$ (640 cubic feet), unless placarded in accordance with subpart F of this part;

(3) A portable tank of less than 3785 L (1000 gallons) capacity, unless placarded in accordance with subpart F of this part;

(4) A DOT Specification 106 or 110 multi-unit tank car tank, unless placarded in accordance with subpart F of this part; and

(5) An overpack, freight container or unit load device, of less than 18 $m^3$ (640 cubic feet), which contains a package for which labels are required, unless placarded or marked in accordance with §172.512 of this part.

(b) Labeling is required for a hazardous material which meets one or more hazard class definitions, in accordance with Column 6 of the §172.101 Table and the following table:

| Hazard class or division | Label name | Label design or section reference (§) |
|---|---|---|
| 1.1 | EXPLOSIVE 1.1 | 172.411 |
| 1.2 | EXPLOSIVE 1.2 | 172.411 |
| 1.3 | EXPLOSIVE 1.3 | 172.411 |
| 1.4 | EXPLOSIVE 1.4 | 172.411 |
| 1.5 | EXPLOSIVE 1.5 | 172.411 |
| 1.6 | EXPLOSIVE 1.6 | 172.411 |
| 2.1 | FLAMMABLE GAS | 172.417 |
| 2.2 | NON-FLAMMABLE GAS | 172.415 |
| 2.3 | POISON GAS | 172.416 |
| 3 (flammable liquid) | FLAMMABLE LIQUID | 172.419 |
| Combustible liquid | (none) | |
| 4.1 | FLAMMABLE SOLID | 172.420 |
| 4.2 | SPONTANEOUSLY COMBUSTIBLE | 172.422 |
| 4.3 | DANGEROUS WHEN WET | 172.423 |
| 5.1 | OXIDIZER | 172.426 |
| 5.2 | ORGANIC PEROXIDE | 172.427 |
| 6.1 (Packing Groups I and II) | POISON | 172.430 |
| 6.1 (Packing Group III) | KEEP AWAY FROM FOOD | 172.431 |
| 6.2 | INFECTIOUS SUBSTANCE[1] | 172.432 |
| 7 (see §172.403) | RADIOACTIVE WHITE-I | 172.436 |
| 7 | RADIOACTIVE YELLOW-II | 172.438 |
| 7 | RADIOACTIVE YELLOW-III | 172.440 |
| 7 (empty packages, see §173.427) | EMPTY | 172.450 |
| 8 | CORROSIVE | 172.442 |
| 9 | CLASS 9 | 172.446 |

[1]The ETIOLOGIC AGENT label specified in regulations of the Department of Health and Human Services at 42 CFR 72.3 may apply to packages of infectious substances.

### §172.400a Exceptions from labeling.

(a) Notwithstanding the provisions of §172.400, a label is not required on--

(1) A cylinder containing a Division 2.1 or Division 2.2 gas that is

(i) Not poisonous;

(ii) Carried by a private or contract motor carrier;

(iii) Not overpacked; and

(iv) Durably and legibly marked in accordance with CGA Pamphlet C-7, appendix A.

(2) A package or unit of military explosives (including ammunition) shipped by or on behalf of the DOD when in--

(i) Freight containerload, carload or truckload shipments, if loaded and unloaded by the shipper or DOD; or

(ii) Unitized or palletized break-bulk shipments by cargo vessel under charter to DOD if at least one required label is displayed on each unitized or palletized load.

(3) A package containing a hazardous material other than ammunition that is--

(i) Loaded and unloaded under the supervision of DOD personnel, and

(ii) Escorted by DOD personnel in a separate vehicle.

(4) A compressed gas cylinder permanently mounted in or on a transport vehicle.

(5) A freight container, aircraft unit load device or portable tank, which--

(i) Is placarded in accordance with Subpart F of this part, or

(ii) Conforms to paragraph (a)(3) or (b)(3) of §172.512.

(6) An overpack or unit load device in or on which labels representative of each hazardous material in the overpack or unit load device is visible.

(7) A package of low specific activity radioactive material, when transported under §173.425(b) of this subchapter.

(b) Certain exceptions to labeling requirements are provided for small quantities and limited quantities in applicable sections in part 173 of this subchapter.

**§172.401 Prohibited labeling.**

(a) Except as provided in paragraph (c) of this section, no person may offer for transportation or no carrier may transport any package bearing a label specified in this subpart unless —

(1) The package contains a material that is a hazardous material, and

(2) The label represents a hazard of the hazardous material in the package.

(b) No person may offer for transportation and no carrier may transport a package bearing any marking or label which by its color, design, or shape could be confused with or conflict with a label prescribed by this part.

(c) The restrictions in paragraphs (a) and (b) of this section, do not apply to packages labeled in conformance with —

(1) Any United Nations recommendation, including the class number (see §172.407), in the document entitled *"Transport of Dangerous Goods."*;

(2) The International Maritime Organization (IMO) requirements, including the class number (see §172.407), in the document entitled "International Maritime Dangerous Goods Code";

(3) The ICAO Technical Instructions, or

(4) The TDG Regulations.

**§172.402 Additional Labeling requirements.**

(a) *Subsidiary hazard labels.* Each package containing a hazardous material—

(1) Shall be labeled with primary and subsidiary hazard labels as specified in Column 6 of the §172.101 Table; and

(2) For other than Class 2 or Class 1 materials (for subsidiary labeling requirements for Class 1 materials see paragraph (e) of this section), if not already labeled under paragraph (a)(1) of this section, shall be labeled with subsidiary hazard labels in accordance with the following table:

SUBSIDIARY HAZARD LABELS

| Subsidiary hazard level (packing group) | Subsidiary Hazard (Class or Division) | | | | | | |
|---|---|---|---|---|---|---|---|
| | 3 | 4.1 | 4.2 | 4.3 | 5.1 | 6.1 | 8 |
| I | X | *** | *** | X | X | X | X |
| II | X | X | X | X | X | X | X |
| III | * | N | X | X | N | N | ** |

X—Required for all modes.
*—Required for transport by vessel only.
**—Required for transport by aircraft and vessel only.
***—Impossible as subsidiary hazard.
N—None required.

(b) *Display of hazard class on labels.* The appropriate hazard class or, for Division 5.1 or 5.2 the division number, shall be displayed in the lower corner of a primary hazard label and may not be displayed on a subsidiary label.

(c) *Cargo Aircraft Only label.* Each person who offers for transportation or transports by aircraft a package containing a hazardous material which is authorized on cargo aircraft only shall label the package with a CARGO AIRCRAFT ONLY label specified in §172.448 of this subpart.

(d) *Radioactive Materials.* Each package containing a radioactive material that also meets the definition of one or more additional hazards, except Class 9, shall be labeled as a radioactive material as required by §172.403 of this subpart and for each additional hazard.

(e) *Class 1 (explosive) Materials.* In addition to the label specified in Column 6 of the §172.101 Table, each package of Class 1 material that also meets the definition for:

(1) Division 6.1, Packing Groups I or II, shall be labeled POISON; or

(2) Class 7, shall be labeled in accordance with §172.403 of this subpart.

**§172.403 Radioactive material.**

(a) Unless excepted from labeling by §§173.421 or 173.425 of this subchapter, each package of radioactive material must be labeled as provided in this section.

(b) The proper label to affix to a package of radioactive material is based on the radiation level at the surface of the package, the transport index (§173.403 of this subchapter) and, if appropriate, the fissile characteristics of the package. The proper category of label shall be determined in accordance with paragraph (c) of this section. The label to be applied shall be the highest category required for any of the three determining conditions for the package. Radioactive White-I is the lowest category and Radioactive Yellow-III is the highest. For example: a package with a transport index of 0.8 and a maximum surface radiation level of 60 millirem per hour which contains no fissile material must bear a Radioactive Yellow-III label.

(c) Category of Label to be Applied to Radioactive Materials Packages:

| Transport index (T.I) | Radiation level at package surface (RL) | Fissile criteria | Label category.[1] |
|---|---|---|---|
| N/A ............ | RL < 0.5 millirem per hour (mrem/h). | Fissile class I only, no fissile class II or III. | White-I. |
| T.I. ≤ 1.0 ..... | 0.5 mrem/h < RL ≤ 50 mrem/h | Fissile class I, fissile class II, with T.I. ≤ 1.0, no fissile class III. | Yellow-II. |
| 1.0 < T.I. ..... | 50 mrem/h < RL. | Fissile class II with 1.0 < T.I., fissile class III. | Yellow-III. |

[1]Any package containing a "highway route controlled quantity" (§173.403 of this subchapter) must be labeled as Radioactive Yellow-III.

(d) EMPTY label. See §173.427(d) of this subchapter for EMPTY labeling requirements.

(e) [Reserved].

(f) Each package required by this section to be labeled with a RADIOACTIVE label must have two of these labels, affixed to opposite sides of the package. (See §172.406(e)(3) for freight container label requirements).

(g) The following applicable items of information must be entered in the blank spaces on the RADIOACTIVE label by legible printing (manual or mechanical), using durable weather resistant means of marking:

(1) "Contents." The name of the radionuclides as taken from the listing of radionuclides in §173.435 (symbols which conform to established radiation protection terminology are authorized, i.e., Mo, Co, etc.). For mixtures of radionuclides, the most restrictive radionuclides on the basis of radiotoxicity must be listed as space on the label allows.

(2) "Activity." Units shall be expressed in appropriate curie units, ie., curies (Ci), millicuries (mCi) or microcurries (μCi) (abbreviations are authorized). For a fissile material, the weight in grams or kilograms of the fissile radiosotope also may be inserted.

(3) "Transportation index." (See §173.403.)

### §172.404 Labels for mixed and consolidated packaging.

(a) *Mixed packaging.* When hazardous materials having different hazard classes are packed within the same packaging, or within the same outside container or overpack as described in §173.25 and authorized by §173.21 of this subchapter, the packaging, outside container or overpack must be labeled as required for each class of hazardous material contained therein.

(b) *Consolidated packaging.* When two or more packages containing compatible hazardous material (see §173.21) are placed within the same outside container or overpack, the outside container or overpack must be labeled as required for each class of hazardous material contained therein.

### §172.405 Authorized label modifications.

(a) For Classes 1, 2, 3, 4, 5, 6, and 8, text indicating a hazard (for example FLAMMABLE LIQUID) is not required on a primary or subsidiary label when--

(1) The label otherwise conforms to the provisions of this subpart, and

(2) The hazard class or, for Division 5.1 or 5.2 the division number, is displayed in the lower corner of the label, if the label corresponds to the primary hazard class of the hazardous material.

(b) For a package containing Oxygen, compressed, or Oxygen, refrigerated liquid, the OXIDIZER label specified in §172.426 of this subpart, modified to display the word "OXYGEN" instead of "OXIDIZER", and the class number "2" instead of "5.1", may be used in place of the NON-FLAMMABLE GAS and OXIDIZER labels. Notwithstanding the provisions of paragraph (a) of this section, the word "OXYGEN" must appear on the label.

### §172.406 Placement of labels.

(a) *General.* (1) Except as provided in paragraphs (b) and (e) of this section, each label required by this subpart must--

(i) Be printed on or affixed to a surface (other than the bottom) of the package or containment device containing the hazardous material; and

(ii) Be located on the same surface of the package and near the proper shipping name marking, if the package dimensions are adequate.

(2) Except as provided in paragraph (e) of this section, duplicate labeling is not required on a package or containment device (such as to satisfy redundant labeling requirements).

(b) *Exceptions.* A label may be printed on or placed on a securely affixed tag, or may be affixed by other suitable means to:

(1) A package that contains no radioactive material and which has dimensions less than those of the required label;

(2) A cylinder; and

(3) A package which has such an irregular surface that a label cannot be satisfactorily affixed.

(c) *Placement of multiple labels.* When primary and subsidiary hazard labels are required, they must be displayed next to each other. Placement conforms to this requirement if labels are within 150 mm (6 inches) of one another.

(d) *Contrast with background.* Each label must be printed on or affixed to a background of contrasting color, or must have a dotted or solid line outer border.

(e) *Duplicate labeling.* Generally, only one of each different required label must be displayed on a package. However, duplicate labels must be displayed on at least two sides or two ends (other than the bottom) of—

(1) Each non-bulk package or overpack having a volume of 1.8 $m^3$ (64 cubic feet) or more;

(2) Each non-bulk package containing a radioactive material;

(3) Each DOT 106 or 110 multi-unit tank car tank. Labels must be displayed on each end;

(4) Each portable tank of less than 3,785 L (1000 gallons) capacity; and

(5) Each freight container or aircraft unit load device having a volume of 1.8 $m^3$ (64 cubic feet) or more, but less than 18 $m^3$ (640 cubic feet). One of each required label must be displayed on or near the closure.

(f) *Visibility.* A label must be clearly visible and may not be obscured by markings or attachments.

## §172.407 Label specifications.

(a) *Durability.* Each label, whether printed on or affixed to a package, must be durable and weather resistant. A label on a package must be able to withstand, without deterioration or a substantial change in color, a 30-day exposure to conditions incident to transportation that reasonably could be expected to be encountered by the labeled package.

(b) *Design.* (1) Except for size and color, the printing, inner border, and symbol on each label must be as shown in §§172.411 through 172.448 of this subpart, as appropriate.

(2) The dotted line border shown on each label is not part of the label specification, except when used as an alternative for the solid line outer border to meet the requirements of §172.406(d) of this subpart.

(c) *Size.* (1) Each diamond (square-on-point) label prescribed in this subpart must be at least 100 mm (3.9 inches) on each side with each side having a solid line inner border 5.0 to 6.3 mm (0.2 to 0.25 inches) from the edge.

(2) The CARGO AIRCRAFT ONLY label must be a rectangle measuring at least 110 mm (4.3 inches) in height by 120 mm (4.7 inches) in width. The word "DANGER" must be shown in letters measuring at least 12.7 mm (0.5 inches) in height.

(3) Except as otherwise provided in this subpart, the hazard class number, or division number, as appropriate, must be at least 6.3 mm (0.25 inches) and not greater than 12.7 mm (0.5 inches).

(4) When text indicating a hazard is displayed on a label, the label name must be shown in letters measuring at least 7.6 mm (0.3 inches) in height except that--

(i) For a SPONTANEOUSLY COMBUSTIBLE or DANGEROUS WHEN WET label, respectively, the words "Spontaneously" and "When Wet" must be shown in letters measuring at least 5.1 mm (0.2 inches) in height.

(ii) For a KEEP AWAY FROM FOOD label, the word "HARMFUL" must be shown in letters measuring at least 7.6 mm (0.3 inches) in height.

(5) The symbol on each label must be proportionate in size to that shown in the appropriate section of this subpart.

(d) *Color.* (1) The background color on each label must be as prescribed in §§172.411 through 172.448 of this subpart, as appropriate.

(2) The symbol, text, numbers, and border must be shown in black on a label except that--

(i) White may be used on a label with a one color background of green, red or blue; and

(ii) White must be used for the text and class number for the CORROSIVE label.

(3) Black and any color on a label must be able to withstand, without substantial change, a 72-hour fadeometer test (for a description of equipment designed for this purpose, see ASTM G 23-69 (1975) or ASTM G 26-70).

(4)(i) A color on a label, upon visual examination, must fall within the color tolerances—

(A) Displayed on color charts conforming to the technical specifications for charts set forth in Table 1 or 2 in appendix A to this part; or

(B) For labels printed on packaging surfaces, specified in Table 3 in appendix A to this part.

(ii) Color charts conforming to appendix A to this part are on display in Room 8421, Nassif Building, 400 Seventh Street, SW., Washington DC 20590-0001.

(5) The specified label color must extend to the edge of the label in the area designated on each label except the CORROSIVE, RADIOACTIVE YELLOW-II AND RADIOACTIVE YELLOW-III labels on which the color must extend only to the inner border.

(e) *Form identification.* A label may contain form identification information, including the name of its maker, provided that information is printed outside the solid line inner border in no larger than 10-point type.

(f) *Exceptions.* A label conforming to specifications in the UN Recommendations may be used in place of a corresponding label which conforms to the requirements of this subpart.

### §172.411 EXPLOSIVE 1.1, 1.2, 1.3, 1.4, 1.5 and 1.6 labels, and EXPLOSIVE Subsidiary label.

(a) Except for size and color, the EXPLOSIVE 1.1, EXPLOSIVE 1.2 and EXPLOSIVE 1.3 labels must be as follows:

(b) In addition to complying with §172.407, the background color on the EXPLOSIVE 1.1, EXPLOSIVE 1.2 and EXPLOSIVE 1.3 labels must be orange. The "**" shall be replaced with the appropriate division number and compatibility group. The compatibility group letter must be the same size as the division number and must be shown as a capitalized Roman letter.

(c) Except for size and color, the EXPLOSIVE 1.4, EXPLOSIVE 1.5, EXPLOSIVE 1.6 labels, and EXPLOSIVE Subsidiary labels must be as follows:

EXPLOSIVE SUBSIDIARY LABEL:

(d) In addition to complying with §172.407, the background color on the EXPLOSIVE 1.4, EXPLOSIVE 1.5, EXPLOSIVE 1.6, and EXPLOSIVE subsidiary label must be orange. Except for the EXPLOSIVE subsidiary label, the "*" shall be replaced with the appropriate compatibility group. The compatibility group letter must be shown as a capitalized Roman letter measuring at least 12.7 mm (0.5 inches) in height. Except for the EXPLOSIVE subsidiary label, division numerals must measure at least 30 mm (1.2 inches) in height and at least 5 mm (0.2 inches) in width.

**§172.415 NON-FLAMMABLE GAS label.**

(a) Except for size and color, the NON-FLAMMABLE GAS label must be as follows:

(b) In addition to complying with §172.407, the background color on the NON-FLAMMABLE GAS label must be green.

**§172.416 POISON GAS Label.**

(a) Except for size and color, the POISON GAS label must be as follows:

(b) In addition to complying with §172.407, the background on the POISON GAS label must be white.

**§172.417 FLAMMABLE GAS Label.**

(a) Except for size and color, the FLAMMABLE GAS label must be as follows:

(b) In addition to complying with §172.407, the background color on the FLAMMABLE GAS label must be red.

**§172.419 FLAMMABLE LIQUID label.**

(a) Except for size and color the FLAMMABLE LIQUID label must be as follows:

(b) In addition to complying with §172.407, the background color on the FLAMMABLE LIQUID label must be red.

## §172.420 FLAMMABLE SOLID label.

(a) Except for size and color, the FLAMMABLE SOLID label must be as follows:

(b) In addition to complying with §172.407, the background on the FLAMMABLE SOLID label must be white with vertical red stripes equally spaced on each side of a red stripe placed in the center of the label. The red vertical stripes must be spaced so that, visually, they appear equal in width to the white spaces between them. The symbol (flame) and text (when used) must be overprinted. The text "FLAMMABLE SOLID" may be placed in a white rectangle.

## §172.422 SPONTANEOUSLY COMBUSTIBLE label.

(a) Except for size and color, the SPONTANEOUSLY COMBUSTIBLE label must be as follows:

(b) In addition to complying with §172.407, the background color on the lower half of the SPONTANEOUSLY COMBUSTIBLE label must be red and the upper half must be white.

## §172.423 DANGEROUS WHEN WET label.

(a) Except for size and color, the DANGEROUS WHEN WET label must be as follows:

(b) In addition to complying with §172.407, the background color on the DANGEROUS WHEN WET label must be blue.

## §172.426 OXIDIZER label.

(a) Except for size and color, the OXIDIZER label must be as follows:

(b) In addition to complying with §172.407, the background color on the OXIDIZER label must be yellow.

## §172.427 ORGANIC PEROXIDE label.

(a) Except for size and color, the ORGANIC PEROXIDE Label must be as follows:

(b) In addition to complying with §172.407, the background color on the ORGANIC PEROXIDE label must be yellow.

**§172.430 POISON label.**

(a) Except for size and color, the POISON label must be as follows:

(b) In addition to complying with §172.407, the background on the POISON label must be white.

**§172.431 KEEP AWAY FROM FOOD label.**

(a) Except for size and color, the KEEP AWAY FROM FOOD label must be as follows:

(b) In addition to complying with §172.407, the background on the KEEP AWAY FROM FOOD label must be white.

**§172.432 INFECTIOUS SUBSTANCE label.**

(a) Except for size and color, the INFECTIOUS SUBSTANCE label must be as follows:

(b) In addition to complying with §172.407, the background on the INFECTIOUS SUBSTANCE label must be white.

**§172.436 RADIOACTIVE WHITE-I label.**

(a) Except for size and color, the RADIOACTIVE WHITE-I label must be as follows:

(b) In addition to complying with §172.407, the background on the RADIOACTIVE WHITE-I label must be white. The printing and symbol must be black, except for the "I" which must be red.

**§172.438 RADIOACTIVE YELLOW-II label.**

(a) Except for size and color, the RADIOACTIVE YELLOW-II must be as follows:

(b) In addition to complying with §172.407, the background color on the RADIOACTIVE YELLOW-II label must be yellow in the top half and white in the lower half. The printing and symbol must be black, except for the "II" which must be red.

**§172.440 RADIOACTIVE YELLOW-III label.**

(a) Except for size and color, the RADIOACTIVE YELLOW-III label must be as follows:

(b) In addition to complying with §172.407, the background color on the RADIOACTIVE YELLOW-III label must be yellow in the top half and white in the lower half. The printing and symbol must be black, except for the "III" which must be red.

**§172.442 CORROSIVE label.**

(a) Except for size and color, the CORROSIVE label must be as follows:

(b) In addition to complying with §172.407, the background on the CORROSIVE label must be white in the top half and black in the lower half.

**§172.446 CLASS 9 label.**

(a) Except for size and color, the "CLASS 9" (miscellaneous hazardous materials) label must be as follows:

(b) In addition to complying with §172.407, the background on the CLASS 9 label must be white with seven black vertical stripes on the top half. The black vertical stripes must be spaced, so that, visually, they appear equal in width to the six white spaces between them. The lower half of the label must be white with the class number "9" underlined and centered at the bottom.

**§172.448 CARGO AIRCRAFT ONLY Label.**

(a) Except for size and color, the CARGO AIRCRAFT ONLY label must be as follows:

(b) The CARGO AIRCRAFT ONLY label must be black on an orange background.

**§172.450 EMPTY label.**

(a) Each EMPTY label, except for size, must be as follows:

(1) Each side must be at least 6 inches (152 mm) with each letter at least 1 inch (25.4 mm) in height.

(2) The label must be white with black printing.
[See Sec. 173.427]

**Subpart F — Placarding**

**§172.500 Applicability of placarding requirements.**

(a) Each person who offers for transportation or transports any hazardous material subject to this subchapter shall comply with the applicable placarding requirements of this subpart.

(b) This subpart does not apply to--

(1) Infectious substances;

(2) Hazardous materials classed as ORM-D;

(3) Hazardous materials authorized by this subchapter to be offered for transportation as Limited Quantities when identified as such on shipping papers in accordance with §172.203(b);

(4) Hazardous materials which are packaged as small quantities under the provisions of §173.4 of this subchapter; and

(5) Combustible liquids in non-bulk packagings.

**§172.502 Prohibited and permissive placarding.**

(a) *Prohibited placarding.* Except as provided in paragraph (b) of this section, no person may affix or display on a packaging, freight container, unit load device, motor vehicle or rail car--

(1) Any placard described in this subpart unless—

(i) The material being offered or transported is a hazardous material;

(ii) The placard represents a hazard of the hazardous material being offered or transported; and

(iii) Any placarding conforms to the requirements of this subpart.

(2) Any sign or other device that, by its color, design, shape or content, could be confused with any placard prescribed in this subpart.

(b) *Exceptions.* (1) The restrictions in paragraph (a) of this section do not apply to a bulk packaging, freight container, unit load device, transport vehicle or rail car which is placarded in conformance with the TDG Regulations, the IMDG Code or the UN Recommendations.

(2) The restrictions of paragraph (a) of this section do not apply to the display of an identification number on a white square-on-point configuration in accordance with §172.336(b) of this part.

(c) *Permissive placarding.* Placards may be displayed for a hazardous material, even when not required, if the placarding otherwise conforms to the requirements of this subpart.

### §172.503 Identification number display on placards.

For procedures and limitations pertaining to the display of identification numbers on placards, see §172.334.

### §172.504 General placarding requirements.

(a) *General.* Except as otherwise provided in this subchapter, each bulk packaging, freight container, unit load device, transport vehicle or rail car containing any quantity of a hazardous material must be placarded on each side and each end with the type of placards specified in Tables 1 and 2 of this section and in accordance with other placarding requirements of this subpart, including the specifications for the placards named in the tables and described in detail in §§172.519 through 172.558.

(b) *DANGEROUS placard.* A freight container, unit load device, transport vehicle or rail car which contains non-bulk packagings with two or more categories of hazardous materials that require different placards specified in Table 2 may be placarded with DANGEROUS placards instead of the separate placarding specified for each of the materials in Table 2. However, when 2,268 kg (5,000 pounds) or more of one category of material is loaded therein at one loading facility, the placard specified in Table 2 of paragraph (e) of this section for that category must be applied.

(c) *Exception for less than 454 kg (1,001 pounds).* Except for bulk packagings and hazardous materials subject to §172.505, when hazardous materials covered by Table 2 of this section are transported by highway or rail, placards are not required on —

(1) A transport vehicle or freight container which contains less than 454 kg (1001 pounds) aggregate gross weight of hazardous materials covered by Table 2 of paragraph (e) of this section; or

(2) A rail car loaded with transport vehicles or freight containers, none of which is required to be placarded.

The exceptions provided in paragraph (c) of this section do not prohibit the display of placards in the manner prescribed in this subpart, if not otherwise prohibited (see §172.502), on transport vehicles or freight containers which are not required to be placarded.

(d) *Exception for empty non-bulk packages.* A non-bulk packaging that contains only the residue of a hazardous material covered by Table 2 of paragraph (e) of this section need not be included in determining placarding requirements.

(e) *Placarding tables.* Placards are specified for hazardous materials in accordance with the following tables:

Table 1

| Category of material (Hazard class or division number and additional description, as appropriate) | Placard name | Placard design section reference (§) |
|---|---|---|
| 1.1 | EXPLOSIVES 1.1 | 172.522 |
| 1.2 | EXPLOSIVES 1.2 | 172.522 |
| 1.3 | EXPLOSIVES 1.3 | 172.522 |
| 2.3 | POISON GAS | 172.540 |
| 4.3 | DANGEROUS WHEN WET | 172.548 |
| 6.1 (PG I, inhalation hazard only) | POISON | 172.554 |
| 7 (Radioactive Yellow III label only) | RADIOACTIVE[1] | 172.556 |

[1] RADIOACTIVE placard also required for exclusive use shipments of low specific activity material in accordance with §173.425(b) or (c) of this subchapter.

Table 2

| Category of material (Hazard class or division number and additional description, as appropriate) | Placard name | Placard design section reference (§) |
|---|---|---|
| 1.4 | EXPLOSIVES 1.4 | 172.523 |
| 1.5 | EXPLOSIVES 1.5 | 172.524 |
| 1.6 | EXPLOSIVES 1.6 | 172.525 |
| 2.1 | FLAMMABLE GAS | 172.532 |
| 2.2 | NON-FLAMMABLE GAS | 172.528 |
| 3 | FLAMMABLE | 172.542 |
| Combustible liquid | COMBUSTIBLE | 172.544 |
| 4.1 | FLAMMABLE SOLID | 172.546 |
| 4.2 | SPONTANEOUSLY COMBUSTIBLE | 172.547 |
| 5.1 | OXIDIZER | 172.550 |
| 5.2 | ORGANIC PEROXIDE | 172.552 |
| 6.1 (PG I or II, other than PG I inhalation hazard) | POISON | 172.554 |
| 6.1 (PG III) | KEEP AWAY FROM FOOD | 172.553 |
| 6.2 | (None) | |
| 8 | CORROSIVE | 172.558 |
| 9 | CLASS 9 | 172.560 |
| ORM-D | (None) | |

(f) *Additional placarding exceptions.* (1) When more than one division placard is required for Class 1 materials on a transport vehicle, rail car, freight container or unit load device, only the placard representing the lowest division number must be displayed.

(2) A FLAMMABLE placard may be used in place of a COMBUSTIBLE placard on--

(i) A cargo tank or portable tank.

(ii) A compartmented tank car which contains both flammable and combustible liquids.

(3) A NON-FLAMMABLE GAS placard is not required on a transport vehicle which contains non-flammable gas if the transport vehicle also contains flammable gas or oxygen and it is placarded with FLAMMABLE GAS or OXYGEN placards, as required.

(4) OXIDIZER placards are not required for Division 5.1 materials on freight containers, unit load devices, transport vehicles or rail cars which also contain Division 1.1 or 1.2 materials and which are placarded with EXPLOSIVES 1.1 or 1.2 placards, as required.

(5) For transportation by transport vehicle or rail car only, an OXIDIZER placard is not required for Division 5.1 materials on a transport vehicle, rail car or freight container which also contains Division 1.5 explosives and is placarded with EXPLOSIVES 1.5 placards, as required.

(6) The EXPLOSIVE 1.4 placard is not required for those Division 1.4 Compatibility Group S (1.4S) materials that are not required to be labeled 1.4S.

(7) For domestic transportation of oxygen, compressed or oxygen, refrigerated liquid, the OXYGEN placard in §172.530 of this subpart may be used in place of a NON-FLAMMABLE GAS placard.

(8) Except for a material classed as a combustible liquid that also meets the definition of a Class 9 material, a COMBUSTIBLE placard is not required for a material classed as a combustible liquid when transported in a non-bulk packaging. For a material in a non-bulk packaging classed as a combustible liquid that also meets the definition of a Class 9 material, the CLASS 9 placard may be substituted for the COMBUSTIBLE placard.

(9) For domestic transportation, a Class 9 placard is not required. A bulk packaging containing a Class 9 material must be marked with the appropriate identification number displayed on a Class 9 placard, an orange panel or a white-square-on-point display configuration as required by subpart D of this part.

(10) For domestic transportation of Division 6.1, PG III materials, a POISON placard may be used in place of a KEEP AWAY FROM FOOD placard.

(g) For shipments of Class 1 (explosive) materials by aircraft or vessel, the applicable compatibility group letter must be displayed on the placards required by this section.

**§172.505 Placarding for subsidiary hazards.**

(a) Each transport vehicle, freight container, portable tank and unit load device that contains a poisonous material subject to the "Poison-Inhalation Hazard" shipping description of §172.203(m)(3) must be placarded with a POISON or POISON GAS placard, as appropriate, on each side and each end, in addition to any other placard required for that material in §172.504. Duplication of the POISON or POISON GAS placard is not required.

(b) In addition to the RADIOACTIVE placard which may be required by §172.504(e) of this subpart, each transport vehicle, portable tank or freight container that contains 454 kg (1001 pounds) or more gross weight of fissile or low specific activity uranium hexafluoride shall be placarded with a CORROSIVE placard on each side and each end.

(c) Each transport vehicle, portable tank, freight container or unit load device that contains a material which has a subsidiary hazard of being dangerous when wet, as defined in §173.124 of this subchapter, shall be placarded with DANGEROUS WHEN WET placards, on each side and each end, in addition to the placards required by §172.504.

(d) Hazardous materials that possess secondary hazards may exhibit subsidiary placards that correspond to the placards described in this part, even when not required by this part (see also §172.519(b)(4) of this subpart).

### §172.506 Providing and affixing placards: Highway.

(a) Each person offering a motor carrier a hazardous material for transportation by highway shall provide to the motor carrier the required placards for the material being offered prior to or at the same time the material is offered for transportation, unless the carrier's motor vehicle is already placarded for the material as required by this subpart.

(1) No motor carrier may transport a hazardous material in a motor vehicle unless the placards required for the hazardous material are affixed thereto as required by this subpart.

### §172.507 Special placarding provisions: Highway.

(a) Each motor vehicle used to transport a package of highway route controlled quantity radioactive materials (see §173.403(1) of this subchapter) must have the required RADIOACTIVE warning placard placed on a square background as described in §172.527.

(b) A nurse tank, meeting the provisions of §173.315(m) of this subchapter, is not required to be placarded on an end containing valves, fittings, regulators or gauges when those appurtenances prevent the markings and placard from being properly placed and visible.

### §172.508 Providing and affixing placards: rail.

(a) Each person offering a hazardous material for transportation by rail shall affix to the rail car containing the material, the placards specified by this subpart. Placards displayed on motor vehicles transport containers, or portable tanks may be used to satisfy this requirement, if the placards otherwise conform to the provisions of this subpart.

(b) No rail carrier may accept a rail car containing a hazardous material for transportation unless the placards for the hazardous material are affixed thereto as required by this subpart.

### §172.510 Special placarding provisions: rail.

(a) *Square background required.* (1) A material classed in Division 1.1 or 1.2 transported by rail, and which require EXPLOSIVES 1.1 or EXPLOSIVES 1.2 placards affixed to a rail car, must have the placard placed on a square background as described in §172.527.

(2) A material classed in Division 2.3, Hazard Zone A or 6.1, Packing Group I, Hazard Zone A which is transported by rail and which requires POISON GAS or POISON placards affixed to a rail car, must have the placards placed on a square background as described in §172.527.

(3) A tank car which contains a residue of a Division 2.3, Hazard Zone A or 6.1, Packing Group I, Hazard Zone A material, and which require POISON GAS-RESIDUE or POISON-RESIDUE placards affixed to a railcar, must have the RESIDUE placard placed on a square background as described in §172.527.

(b) [Reserved]

(c) *RESIDUE and subsidiary placards.* Each tank car containing the residue of a hazardous material must have each primary placard changed to the corresponding RESIDUE placard. See §172.505 of this subpart for display requirements for subsidiary hazards.

(d) *FUMIGATION placard.* Each transport vehicle and freight container containing lading that has been fumigated or treated with poisonous liquid, solid, or gas, and that is offered for transportation by rail must have the placard specified in §173.9 of this subchapter affixed on or near each door.

(e) *Chemical ammunition.* Each rail car containing Division 1.1 or 1.2 (explosive) ammunition which also meets the definition of a material poisonous by inhalation (see §171.8 of this subchapter) must be placarded EXPLOSIVES 1.1 or EXPLOSIVES 1.2 and POISON GAS or POISON.

### §172.512 Freight containers and aircraft unit load devices.

(a) *Capacity of 18 $m^3$ (640 cubic feet) or more.* Each person who offers for transportation, and each person who loads and transports, a hazardous material in a freight container or aircraft unit load device having a capacity of 640 cubic feet or more shall affix to the freight container or aircraft unit load device the placards specified for the material in accordance with §172.504. However, —

(1) The placarding exception provided in § 172.504(c) applies to motor vehicles transporting freight containers and aircraft unit load devices,

(2) The placarding exception provided in §172.504(c), applies to each freight container and aircraft unit load device being transported for delivery to a consignee immediately following an air or water shipment, and,

(3) Placarding is not required on a freight container or aircraft unit load device if it is only transported by air and is identified as containing

a hazardous material in the manner provided in Part 5, Chapter 2, Section 2.7, of the ICAO Technical Instructions.

(b) *Capacity less than 640 cubic feet.* Each person who offers for transportation by air, and each person who loads and transports by air, a hazardous material in a freight container or aircraft unit load device having a capacity of less than 640 cubic feet shall affix one placard of the type specified by paragraph (a) of this section unless the freight container or aircraft unit load device —

(1) Is labeled in accordance with subpart E of this part, including;

(2) Contains radioactive materials requiring the Radioactive Yellow III label and is placarded with one Radioactive placard and is labeled in accordance with §172.406(e); or,

(3) Is identified as containing a hazardous material in the manner provided in Part 5, Chapter 2, Section 2.7, of the ICAO Technical Instructions. When hazardous materials are offered for transportation, not involving air transportation, in a freight container having a capacity of less than 640 cubic feet the freight container need not be placarded. However, if not placarded it must be labeled subject to the inspection requirements of §175.30 of this chapter.

**§172.514 Bulk packagings other than tank cars.**

(a) Except as provided in paragraph (c) of this section, each person who offers for transportation a bulk packaging, other than a tank car, which contains a hazardous material, shall affix the placards specified for the material in §§172.504 and 172.505.

(b) Each bulk packaging, other than a tank car, that is required to be placarded when it contains a hazardous material, must remain placarded when it is emptied, unless it is-

(1) Sufficiently cleaned of residue and purged of vapors to remove any potential hazard; or

(2) Refilled, with a material requiring different placards or no placards, to such an extent that any residue remaining in the packaging is no longer hazardous.

(c) Exceptions. The following packagings may be placarded on only two opposite sides or, alternatively, may be labeled instead of placarded in accordance with subpart E of this part:

(1) A portable tank having a capacity of less than 3,785 L (1000 gallons);

(2) A DOT 106 or 110 multi-unit tank car tank;

(3) A bulk packaging other than a portable tank, cargo tank, or tank car (e.g., a bulk bag or box) with a volumetric capacity of less than 18 $m^3$ (640 cubic feet).

**§172.516 Visibility and display of placards.**

(a) Each placard on a motor vehicle and each placard on a rail car must be readily visible from the direction it faces except from the direction of another motor vehicle or rail car to which the motor vehicle or rail car is coupled. This requirement may be met by the placards displayed on the freight containers or portable tanks loaded on a motor vehicle or rail car.

(b) The required placarding of the front of a motor vehicle may be on the front of a truck-tractor instead of or in addition to the placarding on the front of the cargo body to which a truck-tractor is attached.

(c) Each placard on a transport vehicle, bulk packaging, freight container or aircraft unit load device must—

(1) Be securely attached or affixed thereto or placed in a holder thereon. (See Appendix C to this part.)

(2) Be located clear of appurtenances and devices such as ladders, pipes, doors, and tarpaulins;

(3) So far as practicable, be located so that dirt or water is not directed to it from the wheels of the transport vehicle;

(4) Be located away from any marking (such as advertising) that could substantially reduce its effectiveness, and in any case at least 3 inches (76.0 mm) away from such marking.

(5) Have the words or identification number (when authorized) printed on it displayed horizontally, reading from left to right;

(6) Be maintained by the carrier in a condition so that the format, legibility, and color, and visibility of the placard will not be substantially reduced due to damage, deterioration, or obscurement by dirt or other matter.

(7) Be affixed to a background of contrasting color, or must have a dotted or solid line outer border which contrasts with the background color.

(d) Recommended specifications for a placard holder are set forth in Appendix C of this part. Except for a placard holder similar to that contained in Appendix C to this part, the means used to attach a placard may not obscure any part of its surface other than the borders.

(e) A placard or placard holder may be hinged provided the required format, color, and legibility of the placard are maintained.

**§172.519 General specifications for placards.**

(a) *Strength and durability.* Placards must conform to the following:

(1) A placard may be made of any plastic, metal or other material capable of withstanding, without deterioration or a substantial reduction in effectiveness, a 30-day exposure to open weather conditions.

(2) A placard made of tagboard must be at least equal to that designated commercially as white tagboard. Tagboard must have a weight of at least 80 kg (176 pounds) per ream of 610 by 910 mm (24 by 36-inch) sheets, waterproofing materials included. In addition, each placard made of tagboard must be able to pass a 414 kPa (60 p.s.i.) Mullen test.

(3) Reflective or retroreflective materials may be used on a placard if the prescribed colors, strength and durability are maintained.

(b) *Design.* (1) Except as provided in §172.332 of this part, each placard must be as described in this subpart, and except for size and color, the printing, inner border and symbol must be as shown in §§172.521 through 172.558 of this subpart, as appropriate.

(2) The dotted line border shown on each placard is not part of the placard specification. However, a dotted or solid line outer border may be used when needed to indicate the full size of a placard that is part of a larger format or is on a background of a non-contrasting color.

(3) For other than Class 7 or the OXYGEN placard, text indicating a hazard (for example, "FLAMMABLE") is not required.

(4) For a placard corresponding to the primary hazard class of a material, the hazard class or division number must be displayed in the lower corner of the placard. However, no hazard class or division number may be displayed on a placard corresponding to a subsidiary hazard of the material.

(c) *Size.* (1) Each placard prescribed in this subpart must measure at least 273 mm (10.8 inches) on each side and must have a solid line inner border approximately 12.7 mm (0.5 inches) from each edge.

(2) Except as otherwise provided in this subpart, the hazard class or division number, as appropriate, must be shown in numerals measuring at least 41 mm (1.6 inches) in height.

(3) Except as otherwise provided in this subpart, when text indicating a hazard is displayed on a placard, the printing must be in letters measuring at least 41 mm (1.6 inches) in height.

(d) *Color.* (1) The background color, symbol, text, numerals and inner border on a placard must be as specified in §§172.521 through 172.558 of this subpart, as appropriate.

(2) Black and any color on a placard must be able to withstand, without substantial change-

(i) A 72-hour fadeometer test (for a description of equipment designed for this purpose, see ASTM G 23-69 or ASTM G 26-70); and

(ii) A 30-day exposure to open weather.

(3) Upon visual examination, a color on a placard must fall within the color tolerances displayed on the approprite Hazardous Materials Label and Placard Color Tolerance Chart (see §172.407(d)(4)).

(4) The placard color must extend to the inner border and may extend to the edge of the placard in the area designated on each placard except the color on the CORROSIVE and RADIOACTIVE placards (black and yellow, respectively) must extend only to the inner border.

(e) *Form identification.* A placard may contain form identification information, including the name of its maker, provided that information is printed outside of the solid line inner border in no larger than 10-point type.

(f) *Exceptions.* For a shipment under the provisions of §§171.11, 171.12 or §171.12a of this subchapter, a placard conforming to specifications in the ICAO Technical Instructions, the IMDG Code, or the TDG Regulations, respectively, may be used in place of a corresponding placard which conforms to the requirements of this subpart.

### §172.521 DANGEROUS placard.

(a) Except for size and color, the DANGEROUS placard must be as follows:

(b) In addition to meeting the requirements of §172.519, and Appendix B to this part, the DANGEROUS placard must have a red upper and lower triangle. The placard center area and ½-inch (12.7 mm) border must be white. The inscription must be black with the 1/8-inch (3.2 mm) border marker in the white area at each end of the inscription red.

### §172.522 EXPLOSIVES 1.1, EXPLOSIVES 1.2 and EXPLOSIVES 1.3 placards.

(a) Except for size and color, the EXPLOSIVES 1.1, EXPLOSIVES 1.2 and EXPLOSIVES 1.3 placards must be as follows:

(b) In addition to complying with §172.519 of this subpart, the background color on the EXPLOSIVES 1.1, EXPLOSIVES 1.2, and EXPLOSIVES 1.3 placards must be orange. The "*" shall be replaced with the appropriate division number and, when required, appropriate compatibility group letter. The symbol, text, numerals and inner border must be black.

**§172.523 EXPLOSIVES 1.4 placard.**

(a) Except for size and color, the EXPLOSIVES 1.4 placard must be as follows:

(b) In addition to complying with §172.519 of this subpart, the background color on the EXPLOSIVES 1.4 placard must be orange. The "*" shall be replaced, when required, with the appropriate compatibility group letter. The division numeral, 1.4, must measure at least 64 mm (2.5 inches) in height. The text, numerals and inner border must be black.

**§172.524 EXPLOSIVES 1.5 placard.**

(a) Except for size and color, the EXPLOSIVES 1.5 placard must be as follows:

(b) In addition to complying with §172.519 of this subpart, the background color on the EXPLOSIVES 1.5 placard must be orange. The "*" shall be replaced, when required, with the appropriate compatibility group letter. The division numeral, 1.5, must measure at least 64 mm (2.5 inches) in height. The text, numerals and inner border must be black.

**§172.525 EXPLOSIVES 1.6 placard.**

(a) Except for size and color the EXPLOSIVES 1.6 placard must be as follows:

(b) In addition to complying with §172.519 of this subpart, the background color on the EXPLOSIVES 1.6 placard must be orange. The "*" shall be replaced, when required, with the appropriate compatibility group letter. The division numerals, 1.6, must measure at least 64 mm (2.5 inches) in height. The text, numerals and inner border must be black.

## §172.526 Standard requirements for the RESIDUE placard.

(a) Each RESIDUE placard must be as follows:

(1) Except as provided in paragraph (a)(3) of this section, the lower triangle of the RESIDUE placard must be black and the word "RESIDUE" must be in white letters approximately 25 mm (1 inch) high, made with approximately 6.3 mm (0.25 inch) stroke.

(2) Except for the RADIOACTIVE, EXPLOSIVES 1.1, 1.2, 1.3, 1.4, 1.5 or 1.6, DANGEROUS, or subsidiary placard required by §172.505 of this subpart, the RESIDUE placard may be used to display the appropriate identification number in accordance with the provisions of Subpart D of this part.

(3) For a combustible liquid residue, the lower triangle of the RESIDUE placard must be white and the word "RESIDUE" must be in black letters.

(4) Otherwise, the RESIDUE placard must be as specified in §§172.519, 172.528, 172.530, 172.532, 172.540, 172.542, 172.544, 172.546, 172.547, 172.548, 172.550, 172.552, 172.553, 172.554, 172.558 and 172.560 as appropriate for the residue of the hazardous material being transported and required by this subchapter to be placarded. No other placard may be used as a RESIDUE placard.

(b) The lower part of each placard must be specified in Appendix B to this Part and as illustrated on the FLAMMABLE-RESIDUE placard which, except for size and color, must be as follows:

(c) The RESIDUE placard must be as shown in paragraph (b) of this section and may be —

(1) A separate placard,

(2) On the reverse side of a placard, or

(3) A composite made according to the specifications in this section. The lower triangle of the appropriate placard should have a black triangle bearing the word RESIDUE in white letters with the appropriate hazard class number in white.

## §172.527 Background requirements for certain placards.

(a) Except for size and color, the square background required by §172.510(a) for certain placards on rail cars, and §172.507 for placards on motor vehicles containing a package of highway route controlled quantity radioactive materials, must be as follows:

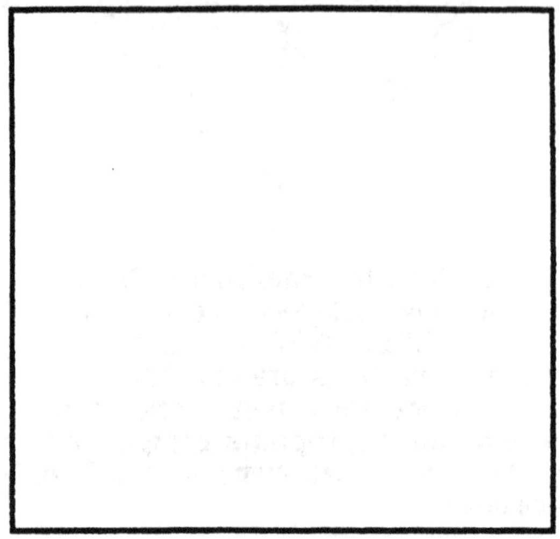

(b) In addition to meeting the requirements of §172.519 for minimum durability and strength, the square background must consist of a white square measuring 14¼ inches (362.0 mm) on each side surrounded by a black border extending to 15¼ inches (387.0 mm) on each side.

## §172.528 NON-FLAMMABLE GAS placard.

(a) Except for size and color, the NON-FLAMMABLE GAS placard must be as follows:

(b) In addition to complying with §172.519, the background color on the NON-FLAMMABLE GAS placard must be green. The letters in both words must be at least 38 mm (1.5 inches) high. The symbol, text, class number and inner border must be white.

**§172.530 OXYGEN placard.**

(a) Except for size and color, the OXYGEN placard must be as follows:

(b) In addition to complying with §172.519, the background color on the OXYGEN placard must be yellow. The symbol, text, class number and inner border must be black.

**§172.532 FLAMMABLE GAS placard.**

(a) Except for size and color, the FLAMMABLE GAS placard must be as follows:

(b) In addition to complying with §712.519, the background color on the FLAMMABLE GAS placard must be red. The symbol, text, class number and inner border must be white.

**§172.540 POISON GAS placard.**

(a) Except for size and color, the POISON GAS placard must be as follows:

(b) In addition to complying with §172.519, the background color on the POISON GAS placard must be white. The symbol, text, class number and inner border must be black.

**§172.542 FLAMMABLE placard.**

(a) Except for size and color, the FLAMMABLE placard must be as follows:

(b) In addition to complying with §172.519, the background color on the FLAMMABLE placard must be red. The symbol, text, class number and inner border must be white.

(c) The word "GASOLINE" may be used in place of the word "FLAMMABLE" on a placard that is displayed on a cargo tank or a portable tank being used to transport gasoline by highway. The word "GASOLINE" must be shown in white.

**§172.544 COMBUSTIBLE placard.**

(a) Except for size and color, the COMBUSTIBLE placard must be as follows:

(b) In addition to complying with §172.519, the background color on the COMBUSTIBLE placard must be red. The symbol, text, class number and inner border must be white. On a COMBUSTIBLE placard with a white bottom as prescribed by §172.332(c)(4), the class number must be red or black.

(c) The words "FUEL OIL" may be used in place of the word "COMBUSTIBLE" on a placard that is displayed on a cargo tank or portable tank being used to transport by highway fuel oil that is not classed as a flammable liquid. The words "FUEL OIL" must be shown in white.

**§172.546 FLAMMABLE SOLID placard.**

(a) Except for size and color, the FLAMMABLE SOLID placard must be as follows:

(b) In addition to complying with §172.519, the background on the FLAMMABLE SOLID placard must be white with seven vertical red stripes. The stripes must be equally spaced, with one red stripe placed in the center of the label. Each red stripe and each white space between two red stripes must be 25 mm (0.98 inches) wide. The letters in the word "SOLID" must be at least 38.1 mm (1.5 inches) high. The symbol, text, class number and inner border must be black.

**§172.547 SPONTANEOUSLY COMBUSTIBLE placard.**

(a) Except for size and color, the SPONTANEOUSLY COMBUSTIBLE placard must be as follows:

(b) In addition to complying with §172.519, the background color on the SPONTANEOUSLY COMBUSTIBLE placard must be red in the lower half and white in upper half. The letters in the word "SPONTANEOUSLY" must be at least 25 mm (0.98 inches) high. The symbol, text, class number and inner border must be black.

**§172.548 DANGEROUS WHEN WET placard.**

(a) Except for size and color, the DANGEROUS WHEN WET placard must be as follows:

(b) In addition to complying with §172.519, the background color on the DANGEROUS WHEN WET placard must be blue. The letters in the words "WHEN WET" must be at least 25 mm (0.98 inches) high. The symbol, text, class number and inner border must be white.

**§172.550 OXIDIZER placard.**

(a) Except for size and color, the OXIDIZER placard must be as follows:

(b) In addition to complying with §172.519, the background color on the OXIDIZER placard must be yellow. The symbol, text, division number and inner border must be black.

**§172.552 ORGANIC PEROXIDE placard.**

(a) Except for size and color, the ORGANIC PEROXIDE placard must be as follows:

(b) In addition to complying with §172.519, the background color on the ORGANIC PEROXIDE placard must be yellow. The symbol, text, division number and inner border must be black.

**§172.553 KEEP AWAY FROM FOOD placard.**

(a) Except for size and color, the KEEP AWAY FROM FOOD placard must be as follows:

(b) In addition to complying with §172.519, the background on the KEEP AWAY FROM FOOD placard must be white. The size of the lettering below the word "HARMFUL" must be proportional to that shown. The symbol, text, class number and inner border must be black.

**§172.554 POISON placard.**

(a) Except for size and color, the POISON placard must be as follows:

(b) In addition to complying with §172.519, the background on the POISON placard must be white. The symbol, text, class number and inner border must black.

**§172.556 RADIOACTIVE placard.**

(a) Except for size and color, the RADIOACTIVE placard must be as follows:

(b) In addition to complying with §172.519, the background color on the RADIOACTIVE placard must be white in the lower portion with a yellow triangle in the upper portion. The base of the yellow triangle must be 29 mm ± 5 mm (1.1 inches ± 0.2 inches) above the placard horizontal center line. The symbol, text, class number and inner border must be black.

**§172.558 CORROSIVE placard.**

(a) Except for size and color, the CORROSIVE placard must be as follows:

(b) In addition to complying with §172.519, the background color on the CORROSIVE placard must be black in the lower portion with a white triangle in the upper portion. The base of the white triangle must be 38 mm ± 5 mm (1.5 inches ± 0.2 inches) above the placard horizontal center line. The text and class number must be white. The symbol and inner border must be black.

**§172.560 CLASS 9 placard.**

(a) Except for size and color the CLASS 9 (miscellaneous hazardous materials) placard must be as follows:

(b) In addition to conformance with §172.519, the background on the CLASS 9 placard must be white with seven black vertical stripes on the top half extending from the top of the placard to one inch above the horizontal centerline. The black vertical stripes must be spaced so that, visually, they appear equal in width to the six white spaces between them. The space below the vertical lines must be white with the class number 9 underlined and centered at the bottom.

## Subpart G—Emergency Response Information

**§172.600 Applicability and general requirements.**

(a) *Scope.* Except as provided in paragraph (d) of this section, this subpart prescribes requirements for providing and maintaining emergency response information during transportation and at facilities where hazardous materials are loaded for transportation, stored incidental to transportation or otherwise handled during any phase of transportation.

(b) *Applicability.* This subpart applies to persons who offer for transportation, accept for transportation, transfer or otherwise handle hazardous materials during transportation.

(c) *General requirements.* No person to whom this subpart applies may offer for transportation, accept for transportation, transfer, store or otherwise handle during transportation a hazardous material unless:

(1) Emergency response information conforming to this subpart is immediately available for use at all times the hazardous material is present; and

(2) Emergency response information, including the emergency response telephone number, required by this subpart is immediately available to any person who, as a representative of a Federal, state or local government agency,

responds to an incident involving a hazardous material, or is conducting an investigation which involves a hazardous material.

(d) *Exception.* The requirements of this subpart do not apply to hazardous materials which are excepted from the shipping paper requirements of this subchapter.

**§172.602 Emergency response information.**

(a) *Information required.* For purposes of this subpart, the term "emergency response information" means information that can be used in the mitigation of an incident involving hazardous materials and, as a minimum, must contain the following information:

(1) The basic description and technical name of the hazardous material as required by §§172.202 and 172.203(k), the ICAO Technical Instructions, the IMDG Code, or the TDG Regulations, as appropriate;

(2) Immediate hazards to health;

(3) Risks of fire or explosion;

(4) Immediate precautions to be taken in the event of an accident or incident;

(5) Immediate methods for handling fires;

(6) Initial methods for handling spills or leaks in the absence of fire; and

(7) Preliminary first aid measures.

(b) *Form of information.* The information required for a hazardous material by paragraph (a) of this section must be:

(1) Printed legibly in English;

(2) Available for use away from the package containing the hazardous material; and

(3) Presented—

(i) On a shipping paper;

(ii) In a document, other than a shipping paper, that includes both the basic description and technical name of the hazardous material as required by §§172.202 and 172.203(k), the ICAO Technical Instructions, the IMDG Code or the TDG Regulations, as appropriate, and the emergency response information required (e.g., a material safety data sheet); or

(iii) Related to the information on a shipping paper, a written notification to pilot-in-command, or a dangerous cargo manifest, in a separate document (e.g, an emergency response guidance document), in a manner that cross-references the description of the hazardous material on the shipping paper with the emergency response information contained in the document. Aboard aircraft, the ICAO "Emergency Response Guidance for Aircraft Incidents Involving Dangerous Goods" and aboard vessels, the IMO "Emergency Procedures for Ships Carrying Dangerous Goods", or equivalent documents, may be used to satisfy the requirements of this section for a separate document.

(c) *Maintenance of information.* Emergency response information shall be maintained as follows:

(1) *Carriers.* Each carrier who transports a hazardous material shall maintain the information specified in paragraph (a) of this section in the same manner as prescribed for shipping papers, except that the information must be maintained in the same manner aboard aircraft as the notification to pilot-in-command, and aboard vessels in the same manner as the dangerous cargo manifest. This information must be immediately accessible to train crew personnel, drivers of motor vehicles, flight crew members, and bridge personnel on vessels for use in the event of incidents involving hazardous materials.

(2) *Facility operators.* Each operator of a facility where a hazardous material is received, stored or handled during transportation, shall maintain the information required by paragraph (a) of this section whenever the hazardous material is present. This information must be in a location that is immediately accessible to facility personnel in the event of an incident involving the hazardous material.

**§172.604 Emergency response telephone number.**

(a) A person who offers a hazardous material for transportation must provide a 24-hour emergency response telephone number (including the area code or international access code) for use in the event of an emergency involving the hazardous material. The telephone number must be—

(1) Monitored at all times the hazardous material is in transportation, including storage incidental to transportation;

(2) The number of a person who is either knowledgeable of the hazards and characteristics of the hazardous material being shipped and has comprehensive emergency response and incident mitigation information for that material, or has immediate access to a person who possesses such knowledge and information; and

(3) Entered on a shipping paper, as follows:

(i) Immediately following the description of the hazardous material required by subpart C of this subchapter; or

(ii) Entered once on the shipping paper in a clearly visible location. This provision may be used only if the telephone number applies to each hazardous material entered on the shipping paper, and if it is indicated that the telephone number is for emergency response information (for example: "EMERGENCY CONTACT: * * *).

(b) The telephone number required by paragraph (a) of this section must be the number of the person offering the hazardous material for transportation or the number of an agency or organization capable of, and accepting responsibility for, providing the detailed information concerning the hazardous material. A person offering a hazardous material for transportation who lists the telephone number of an agency or organization shall insure that agency or organization has received current information on the material, as required by paragraph (a)(2), before it is offered for transportation.

## Subpart H—Training

### §172.700 Purpose and scope.

(a) *Purpose.* This subpart prescribes requirements for training hazmat employees.

(b) *Scope.* Training as used in this subpart means a systematic program that ensures a hazmat employee has familiarity with the general provisions of this subchapter, is able to recognize and identify hazardous materials, has knowledge of specific requirements of this subchapter applicable to functions performed by the employee, and has knowledge of emergency response information, self-protection measures and accident prevention methods and procedures (see §172.704).

(c) *Modal-specific training requirements.* Additional training requirements for the individual modes of transportation are prescribed in parts 174, 175, 176, and 177 of this subchapter.

### §172.701 Federal/State relationship.

This subpart and the parts referenced in §172.700(c) prescribe minimum training requirements for the transportation of hazardous materials. For motor vehicle drivers, however, a State may impose more stringent training requirements only if those requirements—

(a) Do not conflict with the training requirements in this subpart and in part 177 of this subchapter; and

(b) Apply only to drivers domiciled in that State.

### §172.702 Applicability and responsibility for training and testing.

(a) A hazmat employer shall ensure that each of its hazmat employees is trained in accordance with the requirements prescribed in this subpart.

(b) A hazmat employee who performs any function subject to the requirements of this subchapter may not perform that function unless trained in accordance with the requirements prescribed in this subpart. It is the duty of each hazmat employer to comply with the applicable requirements of this subchapter and to thoroughly instruct each hazmat employee in relation thereto.

(c) Training may be provided by the hazmat employer or other public or private sources.

(d) A hazmat employer shall ensure that each of its hazmat employees is tested by appropriate means on the training subjects covered in §172.704.

### §172.704 Training requirements.

(a) Hazmat employee training shall include the following:

(1) *General awareness/familiarization training.* Each hazmat employee shall be provided general awareness/familiarization training designed to provide familiarity with the requirements of this subchapter, and to enable the employee to recognize and identify hazardous materials consistent with the hazard communication standards of this subchapter.

(2) *Function-specific training.* (i) Each hazmat employee shall be provided function-specific training concerning requirements of this subchapter, or exemptions issued under subchapter B of this chapter, which are specifically applicable to the functions the employee performs.

(ii) As an alternative to function-specific training on the requirements of this subchapter, training relating to the requirements of the ICAO Technical Instructions and the IMDG Code may be provided to the extent such training addresses functions authorized by §§171.11 and 171.12 of this subchapter.

(3) *Safety training.* Each hazmat employee shall receive safety training concerning—

(i) Emergency response information required by subpart G of part 172;

(ii) Measures to protect the employee from the hazards associated with hazardous materials to which they may be exposed in the work place, including specific measures the hazmat employer has implemented to protect employees from exposure; and

(iii) Methods and procedures for avoiding accidents, such as the proper procedures for handling packages containing hazardous materials.

(b) *OSHA or EPA Training.* Training conducted by employers to comply with the hazard communication programs required by the Occupational Safety and Health Administration (OSHA) of the Department of Labor (29 CFR 1910.120) or the Environmental Protection Agency (EPA)(40 CFR 311.1), to the extent that training addresses the training specified in paragraph (a) of this section, may be used to satisfy the training requirements in paragraph (a) of this section, in order to avoid unnecessary duplication of training.

(c) *Initial and recurrent training.*—(1)*Initial training.* Each hazmat employer shall train each hazmat employee as follows:

(i) Training for a hazmat employee employed on or before July 2, 1993, shall be completed prior to October 1, 1993.

(ii) Training for a hazmat employee employed after July 2, 1993, shall be completed within 90 days after employment.

(iii) A hazmat employee who changes hazardous materials job functions shall complete training in the new job function(s) within 90 days after the change.

(iv) A hazmat employee described in paragraph (c)(1)(ii) or (iii) of this section, may perform new hazardous materials job functions prior to the completion of training provided the employee performs those functions under the supervision of a properly trained and knowledgeable hazmat employee.

(2) *Recurrent Training.* A hazmat employee shall receive the training required by this subpart at least once every two years.

(3) *Relevant Training.* Relevant training received from a previous employer or other source may be used to satisfy the requirements of this subpart provided a current record of training is obtained from hazmat employees' previous employer.

(4) *Compliance.* Each hazmat employer is responsible for compliance with the requirements of this subchapter regardless of whether the training required by this subpart has been completed.

(d) *Recordkeeping.* A record of current training, inclusive of the preceding two years, in accordance with this subpart shall be created and retained by each hazmat employer for each hazmat employee for as long as that employee is employed by that employer as a hazmat employee and for 90 days thereafter. The record shall include:

(1) The hazmat employee's name;

(2) The most recent training completion date of the hazmat employee's training;

(3) A description, copy, or the location of the training materials used to meet the requirements in paragraph (a) of this section;

(4) The name and address of the person providing the training; and

(5) Certification that the hazmat employee has been trained and tested, as required by this subpart.

(e) *Limitation.* A hazmat employee who repairs, modifies, reconditions, or tests packagings as qualified for use in the transportation of hazardous materials, and who does not perform any other function subject to the requirements of this subchapter, is not subject to the safety training requirement of paragraph (a)(3) of this section.

## APPENDIX A — OFFICE OF HAZARDOUS MATERIALS REGULATION COLOR TOLERANCE CHARTS AND TABLES

The following are Munsell notations and Commission Internationale de L'Eclariage (CIE) coordinates which describe the Office of Hazardous Materials Transportation Label and Placard Color Tolerance Charts in Tables 1 and 2, and the CIE coordinates for the color tolerances specified in Table 3. Central colors and tolerances described in Table 2 approximate those described in Table 2 while allowing for differences in production methods and materials used to manufacture labels and placards surfaced with printing inks. Primarily, the color charts based on Table 1 are for label or placard colors applied as opaque coatings such as paint, enamel or plastic, whereas color charts based on Table 2 are intended for use with labels and placards surfaced only with inks.

For labels printed directly on packaging surfaces, Table 3 may be used, although compliance with either Table 1 or Table 2 is sufficient. However, if visual reference indicates that the colors of labels printed directly on package surfaces are outside the Table 1 or 2 tolerances, a spectrophotometer or other instrumentation may be required to insure compliance with Table 3.

**Table 1**
**Specifications for Color Tolerance Charts for Use With Labels and Placards Surfaced With Paint, Lacquer, Enamel, Plastic, Other Opague Coatings, or Ink[1]**

| Color | Munsell notations | CIE data for source C | | |
|---|---|---|---|---|
| | | Y | x | y |
| **Red:** | | | | |
| Central color | 7.5R 4.0/14 | 12.00 | .5959 | .3269 |
| Orange | 8.5 4.0/14 | 12.00 | .6037 | .3389 |
| Purple and vivid | 6.5 4.0/14 | 12.00 | .5869 | .3269 |
| Grayish | 7.5R 4.0/12 | 12.00 | .5603 | .3321 |
| Vivid | 7.5R 4.0/16 | 12.00 | .6260 | .3192 |
| Light | 7.5R 4.5/14 | 15.57 | .5775 | .3320 |
| Dark | 7.5R 3.5/14 | 09.00 | .6226 | .3141 |
| **Orange:** | | | | |
| Central color | 5.0YR 6.0/15 | 30.05 | .5510 | .4214 |
| Yellow and Grayish | 6.25YR 6.0/15 | 30.05 | .5452 | .4329 |
| Red and vivid | 3.75YR 6.0/15 | 30.05 | .5552 | .4091 |
| Grayish | 5.0YR 6.0/13 | 30.05 | .5311 | .4154 |
| Vivid | 5.0YR 6.0/16 | 30.05 | .5597 | .4239 |
| Light | 5.0YR 6.5/15 | 36.20 | .5427 | .4206 |
| Dark | 5.0YR 5.5/15 | 24.58 | .5606 | .4218 |
| **Yellow:** | | | | |
| Central color | 5.0Y 8.0/12 | 59.10 | .4562 | .4788 |
| Green | 6.5Y 8.0/12 | 59.10 | .4498 | .4865 |
| Orange and vivid | 3.5Y 8.0/12 | 59.10 | .4632 | .4669 |
| Grayish | 5.0Y 8.0/10 | 59.10 | .4376 | .4601 |
| Vivid | 5.0Y 8.0/14 | 59.10 | .4699 | .4920 |
| Light | 5.0Y 8.5/12 | 68.40 | .4508 | .4754 |
| Dark | 5.0Y 7.5/12 | 50.68 | .4620 | .4823 |
| **Green:** | | | | |
| Central color | 7.5G 4.0/9 | 12.00 | .2111 | .4121 |
| Bluish | 0.5BG 4.0/9 | 12.00 | .1974 | .3809 |
| Green-yellow | 5.0G 4.0/9 | 12.00 | .2237 | .4399 |
| Grayish A | 7.5G 4.0/7 | 12.00 | .2350 | .3922 |
| Grayish B[2] | 7.5G 4.0/6 | 12.00 | .2467 | .3822 |
| Vivid | 7.5G 4.0/11 | 12.00 | .1848 | .4319 |
| Light | 7.5G 4.5/9 | 15.57 | .2204 | .4060 |
| Dark | 7.5G 3.5/9 | 09.00 | .2027 | .4163 |
| **Blue:** | | | | |
| Central color | 2.5PB 3.5/10 | 09.00 | .1691 | .1744 |
| Purple | 4.5PB 3.5/10 | 09.00 | .1796 | .1711 |
| Green and vivid | 10.0B 3.5/10 | 09.00 | .1557 | .1815 |
| Grayish | 2.5PB 3.5/8 | 09.00 | .1888 | .1964 |
| Vivid | 2.5PB 3.5/12 | 09.00 | .1516 | .1547 |
| Light | 2.5PB 4.0/10 | 12.00 | .1805 | .1888 |
| Dark | 2.5PB 3.0/10 | 06.55 | .1576 | .1600 |

**Table 1, Continued**

| Color | Munsell notations | CIE data for source C | | |
|---|---|---|---|---|
| | | Y | x | y |
| **Purple:** | | | | |
| Central color | 10.0P 4.5/10 | 15.57 | .3307 | .2245 |
| Reddish purple | 2.5RP 4.5/10 | 15.57 | .3584 | .2377 |
| Blue purple | 7.5P 4.5/10 | 15.57 | .3068 | .2145 |
| Reddish gray | 10.0P 4.5/8 | 15.57 | .3280 | .2391 |
| Gray[2] | 10.0P 4.5/6.5 | 15.57 | .3254 | .2519 |
| Vivid | 10.0P 4.5/12 | 15.57 | .3333 | .2101 |
| Light | 10.0P 5.0/10 | 19.77 | .3308 | .2328 |
| Dark | 10.0P 4.0/10 | 12.00 | .3306 | .2162 |

[1]Maximum chroma is not limited.
[2]For the colors green and purple, the minimum saturation (chroma) limits for porcelain enamel on metal are lower than for most other surface coatings. Therefore, the minimum chroma limits of these two colors as displayed on the Charts for comparison to porcelain enamel on metal is low, as shown for green (grayish B) and purple (gray).
Note. — CIE = Commission Internationale di L'Eclairage.

**Table 2**
**Specifications for Color Tolerance Charts for Use With Labels and Placards Surfaced With Ink[1]**

| Color/Series | Munsell notations | CIE data for source C | | |
|---|---|---|---|---|
| | | Y | x | y |
| **Red:** | | | | |
| **Central series:** | | | | |
| Central color | 6.8R 4.47/12.8 | 15.34 | .5510 | .3286 |
| Grayish | 7.2R 4.72/12.2 | 17.37 | .5368 | .3348 |
| Purple | 6.4R 4.49/12.7 | 15.52 | .5442 | .3258 |
| Purple and vivid | 6.1R 4.33/13.1 | 14.25 | .5529 | .3209 |
| Vivid | 6.7R 4.29/13.2 | 13.99 | .5717 | .3253 |
| Orange | 7.3R 4.47/12.8 | 15.34 | .5572 | .3331 |
| Orange and grayish | 7.65R 4.70/12/4 | 17.20 | .5438 | .3382 |
| **Light Series:** | | | | |
| Light | 7.0R 4.72/13.2 | 17.32 | .5511 | .3322 |
| Light and orange | 7.4R 4.96/12.6 | 19.38 | .5365 | .3382 |
| Light and purple | 6.6R 4.79/12.9 | 17.94 | .5397 | .3289 |
| **Dark Series:** | | | | |
| Dark A | 6.7R 4.19/19.5 | 13.30 | .5566 | .3265 |
| Dark B | 7.0R 4.25/12/35 | 13.72 | .5522 | .3294 |
| Dark and purple | 7.5R 4.23/12.4 | 13.58 | .5577 | .3329 |
| **Orange:** | | | | |
| **Central series:** | | | | |
| Central color | 5.0YR 6.10/12.15 | 31.27 | .5193 | .4117 |
| Yellow and grayish A | 5.8YR 6.22/11.7 | 32.69 | .5114 | .4155 |
| Yellow and grayish B | 6.1YR 6.26/11.85 | 33.20 | .5109 | .4190 |
| Vivid | 5.1YR 6.07/12.3 | 30.86 | .5226 | .4134 |
| Red and vivid A | 3.9YR 5.87/12.75 | 28.53 | .5318 | .4038 |
| Red and vivid B | 3.6YR 5.91/12.6 | 29.05 | .5291 | .4021 |
| Grayish | 4.9YR 6.10/11.9 | 31.22 | .5170 | .4089 |
| **Light series:** | | | | |
| Light and vivid A | 5.8YR 6.78/12.7 | 39.94 | .5120 | .4177 |
| Light and yellow | 6.0YR 6.80/12.8 | 40.20 | .5135 | .4198 |
| Light and vivid B | 4.9YR 6.60/12.9 | 37.47 | .5216 | .4126 |
| **Dark series:** | | | | |
| Dark and yellow | 5.8YR 5.98/11.0 | 29.87 | .5052 | .4132 |
| Dark A | 5.1YR 5.80/11.1 | 27.80 | .5127 | .4094 |
| Dark B | 5.0YR 5.80/11.0 | 27.67 | .5109 | .4068 |

# Appendix B
# HM-181 49 CFR Part 173

# Appendix B

## 49 CFR PART 173--SHIPPERS--GENERAL REQUIREMENTS FOR SHIPMENTS AND PACKAGINGS

### Subpart A--General

| Sec. | |
|---|---|
| 173.1 | Purpose and scope. |
| 173.2 | Hazardous materials classes and index to hazard class definitions. |
| 173.2a | Classification of a material having more than one hazard. |
| 173.3 | Packaging and exceptions. |
| 173.4 | Exceptions for small quantities. |
| 173.5 | Agricultural operations: |
| 173.5a | Oilfield service vehicles. |
| 173.6 | [Removed] |
| 173.7 | US. Government material. |
| 173.9 | Cars, truck bodies, freight containers, or trailers containing lading which has been fumigated or treated with Class 3, Division 2.1, 2.3, or 6.1 materials. |
| 173.10 | Tank car shipments. |
| 173.11 | Shipper's registration statement; flammable cryogenic liquids. |
| 173.12 | Exceptions for shipment of waste materials. |

### Subpart B--Preparation of Hazardous Materials for Transportation

| Sec. | |
|---|---|
| 173.21 | Forbidden materials and packages. |
| 173.22 | Shipper's responsibility. |
| 173.22a | Use of packagings authorized under exemptions. |
| 173.23 | Previously authorized packaging. |
| 173.24 | General requirements for packagings and packages. |
| 173.24a | Additional general requirements for non-bulk packagings and packages. |
| 173.24b | Additional general requirements for bulk packagings. |
| 173.25 | Authorized packages and overpacks. |
| 173.26 | Quantity limitations. |
| 173.27 | General requirements for transportation by aircraft. |
| 173.28 | Reuse reconditioning and remanufacture of packagings. |
| 173.29 | Empty packagings. |
| 173.30 | Loading and unloading of transport vehicles. |
| 173.31 | Qualification, maintenance, and use of tank cars. |

**Subpart C--Definitions, Classification and Packaging for Class 1**



**Subpart D--Definitions, Classification, Packing Group Assignments and Exceptions for Hazardous Material Other Than Class 1 and Class 7**

**Subpart E--Non-bulk Packaging for Hazardous Materials Other Than Class 1 and Class 7**

| | |
|---|---|
| | methyl bromide or methyl chloride mixtures, etc. |
| 173.194 | Gas identification sets. |
| 173.195 | Hydrogen cyanide, anhydrous, stabilized (hydrocyanic acid, aqueous solution). |
| 173.196 | Infectious substances (etiologic agents). |
| 173.197 | Regulated medical waste. |
| 173.198 | Nickel carbonyl. |
| 173.201 | Non-bulk packagings for liquid hazardous materials in Packing Group I. |
| 173.202 | Non-bulk packagings for liquid hazardous materials in Packing Group II. |
| 173.203 | Non-bulk packagings for liquid hazardous materials in Packing Group III. |
| 173.204 | Non-bulk, non-specifications packagings for certain hazardous materials. |
| 173.205 | Specification cylinders for liquid hazardous materials. |
| 173.211 | Non-bulk packagings for solid hazardous materials in Packing Group I. |
| 173.212 | Non-bulk packagings for solid hazardous materials in Packing Group II. |
| 173.213 | Non-bulk packagings for solid hazardous materials in Packing Group III. |
| 173.214 | Packagings which require approval by the Associate Administrator for Hazardous Materials Safety. |
| 173.216 | Asbestos, blue, brown, or white. |
| 173.217 | Carbon dioxide, solid (dry ice). |
| 173.218 | Fish meal or fish scrap. |
| 173.219 | Life-saving appliances. |
| 173.220 | Internal combustion engines, self-propelled vehicles, and mechanical equipment containing internal combustion engines or wet batteries. |
| 173.221 | Polystyrene beads, expandable. |
| 173.222 | Wheelchairs equipped with wet electric storage batteries. |
| 173.224 | Packaging and control and emergency temperatures for self-reactive materials. |
| 173.225 | Packaging requirements and other provisions for organic peroxides. |
| 173.226 | Materials poisonous by inhalation, Division 6.1, Packing Group I, Hazard Zone A. |
| 173.227 | Materials poisonous by inhalation, Division 6.1, Packing Group I, Hazard Zone B. |
| 173.228 | Bromine pentaflouride or bromine trifluoride. |

173.229 Chloric acid solution or chlorine dioxide hydrate, frozen.

173.230 Non-bulk packagings for ORM-D materials.

## Subpart F--Bulk Packaging for Hazardous Materials Other Than Class 1 and Class 7

Sec.

173.240 Bulk packaging for certain low hazard solid materials.

173.241 Bulk packaging for certain low hazard liquid and solid materials.

173.242 Bulk packagings for certain medium hazard liquids and solids, including solids with dual hazards.

173.243 Bulk packaging for certain high hazard liquids and dual hazard liquids which pose a moderate hazard.

173.244 Bulk packaging for certain pyrophoric liquids (Division 4.2), dangerous when wet (Division 4.3) materials, and poisonous liquids with inhalation hazards (Division 6.1).

173.245 Bulk packaging for extremely hazardous materials such as poisonous gases (Division 2.3).

173.247 Bulk packaging for certain elevated temperature materials (Class 9) and certain flammable elevated temperature materials (Class 3).

173.249 Bromine.

## Subpart G--Gases; Preparation and Packaging

Sec.

173.300 [Removed and Reserved]

173.300a Approval of independent inspection agency.

173.300b Approval of non-domestic chemical analyses and tests.

173.300c Termination of approval.

173.301 General requirements for shipment of compressed gases in cylinders.

173.302 Charging of cylinders with non-liquefied compressed gases.

173.303 Charging of cylinders with compressed gas in solution (acetylene).

173.304 Charging of cylinders with liquefied compressed gas.

173.305 Charging of cylinders with a mixture of compressed gas and other material.

173.306 Limited quantities of compressed gases.

173.307 Exceptions for compressed gases.

173.308 Cigarette lighter or other similar device charged with fuel.

**Subpart H--[Removed and Reserved]**

**Subpart I--Radioactive Materials**

173.465 Type A packaging tests.

Plastic Packaging
and Receptacles

APPENDIX C 173.9990C--
Procedure for Base-level Vibration Testing

APPENDIX D 173.9990D--Test Methods for Dynamite (Explosive, Blasting, Type A)

APPENDIX E 173.9990E--
Guidelines for the Classification and Packing Group Assignment of Class 4 Materials

APPENDIX F 173.9990F--
Guidelines for the Classification and Packing Group Assignment of Division 5.1 Materials

APPENDIX G 173.9990G--
Definition of Regulated Medical Waste.

**AUTHORITY:** 49 U.S.C. App. 1803, 1804, 1805, 1806, 1807, 1808, 1817; 49 CFR Part 1.

## §173.1 Purpose and scope.

(a) This part includes:

(1) Definitions of hazardous materials for transportation purposes;

(2) Requirements to be observed in preparing hazardous materials for shipment by air, highway, rail, or water, or any combination thereof; and

(3) Inspection, testing, and retesting responsibilities for persons who retest, recondition, maintain, repair and rebuild containers used or intended for use in the transportation of hazardous materials.

(b) A shipment of hazardous materials that is not prepared in accordance with this subchapter may not be offered for transportation by air, highway, rail, or water. It is the responsibility of each hazmat employer subject to the requirements of this subchapter to ensure that each hazmat employee is trained in accordance with the requirements prescribed in this subchapter. It is the duty of each person who offers hazardous materials for transportation to instruct each of his officers, agents, and employees having any responsibility for preparing hazardous materials for shipment as to applicable regulations in this subchapter.

(c) When a person other than the person preparing a hazardous material for shipment performs a function required by this part, that person shall perform the function in accordance with this part.

(d) In general, the Hazardous Materials Regulations (HMR) contained in this subchapter are based on the Recommendations of the United Nations Committee of Experts on the Transport of Dangerous Goods and are consistent with international regulations issued by the International Civil Aviation Organization (ICAO Technical Instructions) and the International Maritime Organization (IMDG Code). However, the HMR are not consistent in all respects with the UN Recommendations, the ICAO Technical Instructions or the IMDG Code, and compliance with the HMR will not guarantee acceptance by regulatory bodies outside of the United States.

## §173.2 Hazardous materials classes and index to hazard class definitions.

The hazard class of a hazardous material is indicated either by its class (or division) number, its class name, or by the letters "ORM-D". The following table lists class numbers, division numbers, class or division names and those sections of this subchapter which contain definitions for classifying hazardous materials, including forbidden materials.

| Class No. | Division No. (if any) | Name of class or division | 49 CFR reference for definitions |
|---|---|---|---|
| None | | Forbidden materials | 173.21 |
| None | | Forbidden explosives | 173.53 |
| 1 | 1.1 | Explosives (with a mass explosion hazard) | 173.50 |
| 1 | 1.2 | Explosives (with a projection hazard) | 173.50 |
| 1 | 1.3 | Explosives (with predominately a fire hazard) | 173.50 |
| 1 | 1.4 | Explosives (with no significant blast hazard) | 173.50 |
| 1 | 1.5 | Very insensitive explosives; blasting agents | 173.50 |
| 1 | 1.6 | Extremely insensitive detonating substances | 173.50 |
| 2 | 2.1 | Flammable gas | 173.115 |
| 2 | 2.2 | Non-flammable compressed gas | 173.115 |
| 2 | 2.3 | Poisonous gas | 173.115 |
| 3 | | Flammable and combustible liquid | 173.120 |
| 4 | 4.1 | Flammable solid | 173.124 |
| 4 | 4.2 | Spontaneously combustible material | 173.124 |
| 4 | 4.3 | Dangerous when wet material | 173.124 |
| 5 | 5.1 | Oxidizer | 173.127 |
| 5 | 5.2 | Organic peroxide | 173.128 |
| 6 | 6.1 | Poisonous materials | 173.132 |
| 6 | 6.2 | Infectious substance (Etiologic agent) | 173.134 |
| 7 | | Radioactive material | 173.403 |
| 8 | | Corrosive material | 173.136 |
| 9 | | Miscellaneous hazardous material | 173.140 |
| None | | Other regulated material: ORM-D | 173.144 |

## §173.2a Classification of a material having more than one hazard.

(a) Classification of a material having more than one hazard. Except as provided in paragraph (c) of this section, a material not specifically listed in the §172.101 Table that meets the definition of more than one hazard class or division as defined in this part, shall be classed according to the highest applicable hazard class of the following hazard classes, which are listed in descending order of hazard:

(1) Class 7 (radioactive materials, other than limited quantities).

(2) Division 2.3 (poisonous gases).

(3) Division 2.1 (flammable gases).

(4) Division 2.2 (nonflammable gases).

(5) Division 6.1 (poisonous liquids), Packing Group I, poisonous-by-inhalation only.

(6) A material that meets the definition of a pyrophoric material in 173.124(b)(1) of this subchapter (Division 4.2).

(7) A material that meets the definition of a self-reactive material in 173.124(a)(2). of this subchapter (Division 4.1).

(8) Class 3 (flammable liquids), Class 8 (corrosive materials), Division 4.1 (flammable solids), Division 4.2 (spontaneously combustible materials), Division 4.3 (dangerous when wet materials), Division 5.1 (oxidizers) or Division 6.1 (poisonous liquids or solids other than Packing Group I, poisonous-by-inhalation). The hazard class and packing group for a material meeting more than one of these hazards shall be determined using the precedence table in paragraph (b) of this section.

(9) Combustible liquids.

(10) Class 9 (miscellaneous hazardous materials).

(b) Precedence of hazard table for Classes 3 and 8 and Divisions 4.1, 4.2, 4.3, 5.1 and 6.1. The following table ranks those materials that meet the definition of Classes 3 and 8 and Divisions 4.1, 4.2, 4.3, 5.1 and 6.1:

**PRECEDENCE OF HAZARD TABLE**
(Hazard Class & Packing Group)

| | 4.2 | 4.3 | 5.1 I[1] | 5.1 II[1] | 5.1 III[1] | 6.1, I dermal | 6.1, I oral | 6.1, II | 6.1 III | 8, I liquid | 8, I solid | 8, II liquid | 8, II solid | 8, III liquid | 8, III solid |
|---|---|---|---|---|---|---|---|---|---|---|---|---|---|---|---|
| 3 I | | | | | | 3 | 3 | 3 | 3 | 3 | ([3]) | 3 | ([3]) | 3 | ([3]) |
| 3 II | | | | | | 3 | 3 | 3 | 3 | 8 | ([3]) | 3 | ([3]) | 3 | ([3]) |
| 3 III | | | | | | 6.1 | 6.1 | 6.1 | 3[4] | 8 | ([3]) | 8 | ([3]) | 3 | ([3]) |
| 4.1 II[2] | 4.2 | 4.3 | 5.1 | 4.1 | 4.1 | 6.1 | 6.1 | 4.1 | 4.1 | ([3]) | 8 | ([3]) | 4.1 | ([3]) | 4.1 |
| 4.1 III[2] | 4.2 | 4.3 | 5.1 | 4.1 | 4.1 | 6.1 | 6.1 | 6.1 | 4.1 | ([3]) | 8 | ([3]) | 8 | ([3]) | 4.1 |
| 4.2 II | | 4.3 | 5.1 | 4.2 | 4.2 | 6.1 | 6.1 | 4.2 | 4.2 | ([3]) | 8 | ([3]) | 4.2 | ([3]) | 4.2 |
| 4.2 III | | 4.3 | 5.1 | 5.1 | 4.2 | 6.1 | 6.1 | 6.1 | 4.2 | ([3]) | 8 | ([3]) | 8 | ([3]) | 4.2 |
| 4.3 I | | | 5.1 | 4.3 | 4.3 | 6.1 | 4.3 | 4.3 | 4.3 | 4.3 | 4.3 | 4.3 | 4.3 | 4.3 | 4.3 |
| 4.3 II | | | 5.1 | 4.3 | 4.3 | 6.1 | 4.3 | 4.3 | 4.3 | 8 | 8 | 8 | 4.3 | 4.3 | 4.3 |
| 4.3 III | | | 5.1 | 5.1 | 4.3 | 6.1 | 6.1 | 6.1 | 4.3 | 8 | 8 | 8 | 8 | 4.3 | 4.3 |
| 5.1 I[1] | | | | | | 5.1 | 5.1 | 5.1 | 5.1 | 5.1 | 5.1 | 5.1 | 5.1 | 5.1 | 5.1 |
| 5.1 II[1] | | | | | | 6.1 | 5.1 | 5.1 | 5.1 | 8 | 8 | 8 | 5.1 | 5.1 | 5.1 |
| 5.1 III[1] | | | | | | 6.1 | 6.1 | 6.1 | 5.1 | 8 | 8 | 8 | 8 | 5.1 | 5.1 |
| 6.1, I Dermal | | | | | | | | | | 8 | 6.1 | 6.1 | 6.1 | 6.1 | 6.1 |
| 6.1 I Oral | | | | | | | | | | 8 | 6.1 | 6.1 | 6.1 | 6.1 | 6.1 |
| 6.1 II Inhalation | | | | | | | | | | 8 | 6.1 | 6.1 | 6.1 | 6.1 | 6.1 |
| 6.1 II Dermal | | | | | | | | | | 8 | 6.1 | 8 | 6.1 | 6.1 | 6.1 |
| 6.1 II Oral | | | | | | | | | | 8 | 8 | 8 | 6.1 | 6.1 | 6.1 |
| 6.1 III | | | | | | | | | | 8 | 8 | 8 | 8 | 8 | 8 |

[1]There are at present no established criteria for determining Packing Groups for liquids in Division 5.1. For the time being, the degree of hazard is to be assessed by analogy with listed substances, allocating the substances to Packing Group I, great; II, medium; or III, minor danger.

[2]Substances of Division 4.1 other than self-reactive substances.

[3]Denotes an impossible combination.

[4]For pesticides only, where a material has the hazards of Class 3, Packing Group III, and Division 6.1, Packing Group III, the primary hazard is Division 6.1, Packing Group III.

(c) The following materials are not subject to the provisions of paragraph (a) of this section because of their unique properties:

(1) A Class 1 (explosive) material that meets any other hazard class or division as defined in this part shall be assigned a division in Class 1. Class 1 materials shall be classed and approved in accordance with §173.56 of this part;

(2) A Division 5.2 (organic peroxide) material that meets the definition of any other hazard class or division as defined in this part, shall be classed as Division 5.2;

(3) A Division 6.2 (infectious substance) material that meets the definition of any other hazard class or division as defined in this part shall be classed as Division 6.2;

(4) A material that meets the definition of a wetted explosive in §173.124(a)(1) of this subchapter (Division 4.1). Wetted explosives are either specifically listed in the §172.101 Table or are approved by the Associate Administrator for Hazardous Materials Safety (see §173.124(a)(1) of this subchapter); and

(5) A limited quantity of a Class 7 (radioactive) material that meets

the definition for more than one hazard class or division shall be classed in accordance with §173.421-2.

## 173.3 Packaging and exceptions.

(a) The packaging of hazardous materials for transportation by air, highway, rail, or water must be as specified in this part. Methods of manufacture, packing, and storage of hazardous materials, that affect safety in transportation, must be open to inspection by a duly authorized representative of the initial carrier or of the Department. Methods of manufacture and related functions necessary for completion of a DOT specification or U.N. standard packaging must be open to inspection by a representative of the Department.

(b) The regulations setting forth packaging requirements for a specific material apply to all modes of transportation unless otherwise stated, or unless exceptions from packaging requirements are authorized.

(c) Salvage drums. Packages of hazardous materials that are damaged, defective, or found leaking and hazardous materials that have spilled or leaked may be placed in a metal or plastic removable head salvage drum that is compatible with the lading and shipped for repackaging or disposal under the following conditions:

(1) The drum must be a UN 1A2, 1B2, 1N2 or 1H2 tested and marked for Packing Group III or higher performance standards for liquids or solids and a leakproofness test of 20 kPa (3 psi). Alternatively, a drum manufactured and marked prior to October 1, 1993 as a salvage drum, in accordance with the provisions of this section in effect on September 30, 1991, is authorized. Capacity of the drum may not exceed 450 L (119 gallons).

(2) Each drum shall be provided when necessary with sufficient cushioning and absorption material to prevent excessive movement of the damaged package and to eliminate the presence of any free liquid at the time the salvage drum is closed. All cushioning and absorbent material used in the drum must be compatible with the hazardous material

(3) Each drum shall be marked with the proper shipping name of the material inside the packaging and the name and address of the consignee. In addition, the drum shall be marked "Salvage Drum."

(4) Each drum shall be labeled as prescribed for the respective material.

(5) The shipper shall prepare shipping papers in accordance with subpart C of Part 172 of this subchapter.

(6) The overpack requirements of §173.25 do not apply to drums used in accordance with this paragraph.

## §173.4 Exceptions for small quantities.

(a) Small quantities of Class 3, Division 4.1, Division 5.1, Division 5.2, Class 8, Division 6.1 materials, and Class 7 materials, and Class 9 materials that also meet the definition of one or more of these hazard classes, are not subject to any other requirements of this subchapter if:

(1) The maximum quantity of material per inner receptacle is limited to:

(i) Thirty (30) ml (1 ounce) for authorized liquids, other than Division 6.1, Packing Group I, materials;

(ii) Thirty (30) g (1 ounce) for authorized solids, other than Division 6.1, Packing Group I, materials;

(iii) One (1) g (0.04 ounce) for authorized materials classed as Division 6.1, Packing Group I; and

(iv) An activity level not exceeding that specified in §§173.421, 173.422 or 173.424, as appropriate, for a package containing a Class 7 material.

(2) With the exception of temperature sensing devices, each inner receptacle:

(i) Is not liquid-full at 55 °C (131 °F), and

(ii) Is constructed of plastic having a minimum thickness of no less than 0.2 mm (0.008 inch), or earthenware, glass, or metal ;

(3) Each inner receptacle with a removable closure has its closure held securely in place with wire, tape, or other positive means;

(4) Unless equivalent cushioning and absorbent material surrounds the inside packaging, each inner receptacle is securely packed in an inside packaging with cushioning and absorbent material that:

(i) Will not react chemically with the material, and

(ii) Is capable of absorbing the entire contents (if a liquid) of the receptacle;

(5) The inside packaging is securely packed in a strong outside packaging;

(6) The completed package, as demonstrated by prototype testing, is capable of sustaining--

(i) Each of the following free drops made from a height of 1.8 m (5.9 feet) directly onto a solid unyielding surface without breakage or leakage from any inner receptacle and without a substantial reduction in the effectiveness of the package:

(A) One drop flat on bottom;

(B) One drop flat on top;

(C) One drop flat on the long side;

(D) One drop flat on the short side; and

(E) One drop on a corner at the junction of three intersecting edges; and

(ii) A compressive load as specified in §178.606(c) of this subchapter.

Note: Each of the tests in paragraph (6) of this section may be performed on a different, but identical, package; i.e., all tests need not be performed on the same package.

(7) Placement of the material in the package or packing different materials in the package does not result in a violation of §173. 21;

(8) The gross mass of the completed package does not exceed 29 kg (64 pounds);

(9) The shipper certifies conformance with this section by marking the outside of the package with the statement: "This package conforms to conditions and limitations specified in 49 CFR 173.4";

(10) The package is not opened or otherwise altered until it is no longer in commerce; and

(11) The package, unless specifically approved by the Associate Administrator for

Hazardous Materials Safety does not contain a material assigned any of the following identification numbers associated with the hazardous materials description in the §172.101 Table:

1092 1131 1259 1380 1397 1419 1422 1432 1433 1491 1504 1749 1798 1831 1873 2031 2032 2495 2626 2813 2845 2924 2925 9191 9193

(b) A package containing a Class 7 material also must conform to the requirements of §173.421(a) through (e), or §173.422(a) through (f), as appropriate.

**173.7 U.S. Government material.**

(a) Hazardous materials offered for transportation by, for, or to the Department of Defense (DOD) of the U.S. Government, including commercial shipments pursuant to a DOD contract, must be packaged in accordance with the regulations in this subchapter or in packagings of equal or greater strength and efficiency as certified by DOD pursuant to the "Policies and Procedures for Hazardous Materials Packaging Certification, AFLCR 800-29/AFSCR 800-29/DARCOM-R 700-103/NAVMATINST 4030.11/DLAR 4145.37." Hazardous materials offered for transportation by DOD under this provision may be reshipped by any shipper to any consignee provided the original packaging has not been damaged or altered in any manner.

(1) Hazardous materials sold by the DOD in packagings that are not marked in accordance with the requirements of this subchapter may be shipped from DOD installations if the DOD certifies in writing that the packagings are equal to or greater in strength and efficiency than the packaging prescribed in this subchapter. The shipper shall obtain such a certification in duplicate for each shipment. He shall give one copy to the originating carrier and retain the other for no less than 1 year.

(b) Shipments of Class 7 materials, made by or under the direction or supervision of the U.S. Department of Energy or the Department of Defense, and which are escorted by personnel specifically designated by or under the authority of those agencies, for the purpose of national security, are not subject to the regulations in Parts 170-189 of this subchapter.

(c) Shipments of explosive samples, not exceeding one gram net weight, offered by and consigned to the Bureau of Alcohol, Tobacco and Firearms (ATF) of the Department of the Treasury are not otherwise subject to the regulations in Parts 110-189 of this subchapter when placed in a specifically designed multi-unit assembly packed in a strong outside packaging. The packaging must be of a type accepted by ATF as capable of precluding a propagation of any explosion outside the packaging. The second component from the outside of the packaging must be marked or tagged to indicate the presence of an explosive.

(d) Notwithstanding the requirements of §§173.416 and 173.417 of this subchapter, packagings made by or under the direction of the U.S. Department of

Energy may be used for the transportation of Class 7 materials when evaluated, approved, and certified by the Department of Energy against packaging standards equivalent to those specified in 10 CFR Part 71. Packages shipped in accordance with this paragraph shall be marked and otherwise prepared for shipment in a manner equivalent to that required by this subchapter for packagings approved by the Nuclear Regulatory Commission.

### 173.12 Exceptions for shipment of waste materials.

(a) Open head drums. If a hazardous material that is a hazardous waste is required by this subchapter to be shipped in a closed head drum (i.e., a drum with a 7.0 cm (3 inches) or less bung opening) and the hazardous waste contains solids or semisolids that make its placement in a closed head drum impracticable, an equivalent (except for closure) open head drum may be used for the hazardous waste.

(b) Lab packs. (1) Waste materials classed as Class or Division 3, 4.1, 4.2, 4.3, 5.1, 6.1, 8, or 9 are excepted from the specification packaging requirements of this subchapter for combination packagings if packaged in accordance with this paragraph and transported for disposal or recovery by highway only. In addition, a generic description from the §172.101 Table may be used in place of specific chemical names, when two or more chemically compatible waste materials in the same hazard class are packaged in the same outside packaging.

(2) Additional packaging requirements are as follows:

(i) The outer packaging must be a UN 1A2 or UN 1B2 metal drum, a UN 1D plywood drum, a UN 1G fiber drum or a UN 1H2 plastic drum tested and marked at least for the Packing Group III performance level for liquids or solids;

(ii) The inner packagings must be either glass, not exceeding 4 L (1 gallon) rated capacity, or metal or plastic, not exceeding 20 L (5.3 gallons) rated capacity;

(iii) Each outer packaging may contain only one class of hazardous material;

(iv) Inner packagings containing liquid must be surrounded by a chemically compatible absorbent material in sufficient quantity to absorb the total liquid contents; and

(v) Gross weight of the complete package may not exceed 205 kg (452 lbs).

(3) Prohibited materials. Materials meeting the definition of Division 6.1, Packing Group I, or Division 4.2, Packing Group I, and bromine pentafluoride; bromine trifluoride; chloric acid; and oleum (fuming sulfuric acid) may not be packaged or described under the provisions of this paragraph.

(c) Reuse of packagings. A previously used packaging may be reused for the shipment of hazardous waste to designated facilities, not subject to the reconditioning and reuse provisions contained in §173.28 and Part 178 of this subchapter, under the following conditions:

(1) Except as authorized by this paragraph, the waste must be packaged in accordance with this part and offered for transportation in accordance with the requirements of this subchapter.

(2) Transportation is performed by highway only.

(3) A package is not offered for transportation less than 24 hours after it is finally closed for transportation, and each package is inspected for leakage and is found to be free from leaks immediately prior to being offered for transportation.

(4) Each package is loaded by the shipper and unloaded by the consignee, unless the motor carrier is a private or contract carrier.

(5) The packaging may be used only once under this paragraph and may not be used again for shipment of hazardous materials except in accordance with §173.28.

(d) Technical names for n.o.s. descriptions. The requirements for the inclusion of technical names for n.o.s. descriptions on shipping papers and package markings, §§172.203 and 172.301 of this subchapter, respectively, do not apply to packagings prepared in accordance with paragraph (b) of this section, except as follows:

(1) Packages containing materials meeting the definition of a hazardous substance must be described as required in §172.203(c) of this subchapter and marked as required in §172.324 of this subchapter;

(2) Packages containing hazardous materials subject to the provisions of §172.203(m) of this subchapter must be described in accordance with §172.203(m) of this subchapter; and

(3) Packagings containing marine pollutants must be described as required in §172.203(1) of this subchapter and marked as required in §172.322 of this subchapter.

**173.24 General requirements for packagings and packages.**

(a) Applicability. Except as otherwise provided in this subchapter, the provisions of this section apply to--

(1) Bulk and non-bulk packagings;

(2) New packagings and packagings which are reused; and

(3) Specification and non-specification packagings.

(b) Each package used for the shipment of hazardous materials under this subchapter shall be designed, constructed, maintained, filled, its contents so limited, and closed, so that under conditions normally incident to transportation--

(1) Except as otherwise provided in this subchapter, there will be no identifiable (without the use of instruments) release of hazardous materials to the environment;

(2) The effectiveness of the package will not be substantially reduced; for example, impact resistance, strength, packaging compatibility, etc. must be maintained for the minimum and maximum temperatures encountered during transportation;

(3) There will be no mixture of gases or vapors in the package which could, through any credible spontaneous increase of heat or pressure, significantly reduce the effectiveness of the packaging.

(c) Authorized packagings. A packaging is authorized for a hazardous material only if--

(1) The packaging is prescribed or permitted for the hazardous material in a packaging section specified for that material in Column 8 of the §172.101 Table and conforms to applicable requirements in the special provisions of Column 7 of the §172.101 Table and, for specification packagings (including U.N. standard packagings), the specification requirements in Parts 178 and 179 of this subchapter; or

(2) The packaging is permitted under, and conforms to, provisions contained in §§171.11, 171.12, 171.12a, 173.3, 173.4, 173.5, 173. 7, 173.27, or 176.11 of this subchapter.

(d) DOT specification and U.N. standard packagings. For DOT specification packagings (including U.N. standard packagings), conformance to the applicable specifications in Parts 178 and 179 of this subchapter is required in all details. For performance-oriented packagings covered by subpart L of Part 178 of this subchapter, each packaging must be capable of meeting the performance test requirements specified in subpart M of Part 178 of this subchapter for the applicable packing group shown in Column 5 of the §172.101 Table.

(e) Compatibility. (1) Even though certain packagings are specified in this part, it is, nevertheless, the responsibility of the person offering a hazardous material for transportation to ensure that such packagings are compatible with their lading. This particularly applies to corrosivity, permeability, softening, premature aging and embrittlement.

(2) Packaging materials and contents must be such that there will be no significant chemical or galvanic reaction between the materials and contents of the package.

(3) Plastic packagings and receptacles. (i) Plastic used in packagings and receptacles must be of a type compatible with the lading and may not be permeable to an extent that a hazardous condition is likely to occur during transportation, handling or refilling .

(ii) Each plastic packaging or receptacle which is used for liquid hazardous materials must be capable of withstanding without failure the procedure specified in Appendix B of this part ("Procedure for Testing Chemical Compatibility and Rate of Permeation in Plastic Packagings and Receptacles"). The procedure specified in appendix B of this part must be performed on each plastic packaging or receptacle used for Packing Group I materials. The maximum rate of permeation of hazardous lading through or into the plastic packaging or receptacles may not exceed 0.5 percent for materials meeting the definition of a Division 6.1 material according to §173 .132 and 2.0 percent for other hazardous materials, when subjected to a temperature no lower than--

(A) 18 °C (64 °F) for 180 days in accordance with Test Method 1 in appendix B of this part;

(B) 50 °C (122 °F) for 28 days in accordance with Test Method 2 in appendix B of this part; or

(C) 60 °C (140 °F) for 14 days in accordance with Test Method 3 in appendix B of this part.

(iii) Alternative procedures or rates of permeation are permitted if they yield a level of safety equivalent to or greater than that provided by paragraph (e)(3)(ii) of this section and are specifically approved by the Associate Administrator for Hazardous Materials Safety.

(4) Mixed contents. Hazardous materials may not be packed or mixed together in the same outer packaging with other hazardous or nonhazardous materials if such materials are capable of reacting dangerously with each other and causing--

(i) Combustion or dangerous evolution of heat;

(ii) Evolution of flammable or poisonous gases; or

(iii) Formation of unstable or corrosive materials.

(5) Packagings used for solids, which may become liquid at temperatures likely to be encountered during transportation, must be capable of containing the hazardous material in the liquid state.

(f) Closures. (1) Closures on packagings shall be so designed and closed that under conditions (including the effects of temperature and vibration) normally incident to transportation--

(i) Except as provided in paragraph (g) of this section, there is no identifiable release of hazardous materials to the environment from the opening to which the closure is applied; and

(ii) The closure is secure and leakproof.

(2) Except as otherwise provided in this subchapter, a closure (including gaskets or other closure components, if any) used on a specification packaging must conform to all applicable requirements of the specification.

(g) Venting. Venting of packagings, to reduce internal pressure which may develop by the evolution of gas from the contents, is permitted only when--

(1) Transportation by aircraft is not involved;

(2) Except as otherwise provided in this subchapter, the evolved gases are not poisonous, likely to create a flammable mixture with air or be an asphyxiant under normal conditions of transportation;

(3) The packaging is designed so as to preclude an unintentional release of hazardous materials from the receptacle; and

(4) For shipments in bulk packagings, venting is authorized for the specific hazardous material by a special provision in the §172.101 Table or by the applicable bulk packaging specification in Part 178 of this subchapter.

(h) Outage and filling limits--(1) General. When filling packagings and receptacles for liquids, sufficient ullage (outage) must be left to ensure that neither leakage nor permanent distortion of the packaging or receptacle will occur as a result of an expansion of the liquid caused by temperatures likely to be encountered during transportation. Requirements for outage and filling limits for non-bulk and bulk packagings are specified in §§173.24a(d) and 173.24b(a), respectively.

(2) Compressed gases and cryogenic liquids. Filling limits for compressed gases and cryogenic liquids are specified in §173.301 through §173.306 for cylinders and §173.314 through §173.319 for bulk packaging.

(i) Air transportation. Packages offered or intended for transportation by aircraft must conform to the general requirements for transportation by aircraft in §173.27, except as provided in §171.11 of this subchapter.

**173.24a Additional general requirements for non-bulk packagings and packages.**

(a) Packaging design. Except as provided in §172.312 of this subchapter:

(1) Inner packaging closures. A combination packaging containing liquid hazardous materials must be packed so that closures on inner packagings are upright.

(2) Friction. The nature and thickness of the outer packaging must be such that friction during transportation is not likely to generate an amount of heat sufficient to alter dangerously the chemical stability of the contents.

(3) Securing and cushioning. Inner packagings of combination packagings must be so packed, secured and cushioned to prevent their breakage or leakage and to control their movement within the outer packaging under conditions normally incident to transportation. Cushioning material must not be capable of reacting dangerously with the contents of the inner packagings.

(4) Metallic devices. Nails, staples, and other metallic devices shall not protrude into the interior of the outer packaging in such a manner as to be likely to damage inner packagings or receptacles.

(5) Vibration. Each non-bulk package must be capable of withstanding, without rupture or leakage, the vibration test procedure specified in §178.608 of this subchapter.

(b) Non-bulk packaging filling limits.

(1) A single or composite non-bulk packaging may be filled with a liquid hazardous material only when the specific gravity of the material does not exceed that marked on the packaging, or a specific gravity of 1.2 if not marked, except as follows:

(i) A Packing Group I packaging may be used for a Packing Group II material with a specific gravity not exceeding the greater of 1.8 or 1.5 times the specific gravity marked on the packaging, provided all the performance criteria can still be met with the higher specific gravity material;

(ii) A Packing Group I packaging may be used for a Packing Group III material with a specific gravity not exceeding the greater of 2.7 or 2.25 times the specific gravity marked on the packaging, provided all the performance criteria can still be met with the higher specific gravity material; and

(iii) A Packing Group II packaging may be used for a Packing Group III material with a specific gravity not exceeding the greater of 1.8, or 1.5 times the specific gravity marked on the packaging, provided all the performance criteria can still be met with the higher specific gravity material.

(2) Except as otherwise provided in this section, a single or composite non-bulk packaging may not be filled with a solid hazardous material to a gross mass greater than the maximum gross mass marked on the packaging.

(3) A single or composite non-bulk packaging which is tested and marked for liquid hazardous materials may be filled with a solid hazardous material to a gross mass, in kilograms, not exceeding the rated capacity of the packaging in liters, multiplied by the specific gravity marked on the packaging, or 1.2 if not marked. In addition:

(i) A single or composite non-bulk packaging which is tested and marked for Packing Group I liquid hazardous materials may be filled with a solid Packing Group II hazardous material to a gross mass, in kilograms, not exceeding the rated capacity of the packaging in liters, multiplied by 1.5, multiplied by the specific gravity marked on the packaging, or 1.2 if not marked.

(ii) A single or composite non-bulk packaging which is tested and marked for Packing Group I liquid hazardous materials may be filled with a solid Packing Group III hazardous material to a gross mass, in kilograms, not exceeding the rated capacity of the packaging in liters, multiplied by 2.25, multiplied by the specific gravity marked on the packaging, or 1.2 if not marked.

(iii) A single or composite non-bulk packaging which is tested and marked for Packing Group II liquid hazardous materials may be filled with a solid Packing Group III hazardous material to a gross mass, in kilograms, not exceeding the rated capacity of the packaging in liters, multiplied by 1.5, multiplied by the specific gravity marked on the packaging, or 1.2 if not marked.

(4) Packagings tested as prescribed in §178.605 of this subchapter and marked with the hydrostatic test pressure as prescribed in §178.503(a)(5) of this subchapter may be used for liquids only when the vapor pressure of the liquid conforms to one of the following:

(i) The vapor pressure must be such that the total pressure in the packaging (i.e., the vapor pressure of the liquid plus the partial pressure of air or other inert gases, less 100 kPa (15 psi)) at 55 °C (131 °F), determined on the basis of a maximum degree of filling in accordance with paragraph (b)(1) of this section and a filling temperature of 15 °C (59 °F)), will not exceed two-thirds of the marked test pressure;

(ii) The vapor pressure at 50 °C (122 °F) must be less than four-sevenths of the sum of the marked test pressure plus 100 kPa (15 psi); or

(iii) The vapor pressure at 55 °C (131 °F) must be less than two-thirds of the sum of the marked test pressure plus 100 kPa (15 psi).

(5) No hazardous material may remain on the outside of a package after filling.

(c) Mixed contents. (1) An outer non-bulk packaging may contain more than one hazardous material only when--

(i) The inner and outer packagings used for each

hazardous material conform to the relevant packaging sections of this part applicable to that hazardous material;

(ii) The package as prepared for shipment meets the performance tests prescribed in Part 178 of this subchapter for the packing group indicating the highest order of hazard for the hazardous materials contained in the package;

(iii) Corrosive materials (except ORM-D) in bottles are further packed in securely closed inner receptacles before packing in outer packagings; and

(iv) For transportation by aircraft, the total net quantity does not exceed the lowest permitted maximum net quantity per package as shown in Column 9a or 9b, as appropriate, of the §172.101 Table. The permitted maximum net quantity must be calculated in kilograms if a package contains both a liquid and a solid.

(2) A packaging containing inner packagings of Division 6.2 materials may not contain other hazardous materials, except dry ice.

(d) Liquids must not completely fill a receptacle at a temperature of 55 °C (131 °F) or less.

## 173.24b Additional general requirements for bulk packagings.

(a) Outage and filling limits. (1) Liquids and liquefied gases must be so loaded that the outage is at least one percent of the total capacity of a cargo or portable tank, or compartment thereof, or at least one percent of the total capacity of the tank and dome for tank car and multi-unit tank car tanks at the reference temperature of 46 °C (115 °F) for uninsulated tanks and 41 °C (105 °F) f or insulated tanks.

(2) Hazardous materials may not be loaded into the dome of a tank car. If the dome of the tank car does not provide sufficient outage, vacant space must be left in the shell to provide the required outage.

(3) Bulk packagings for materials poisonous by inhalation. For a material which meets the definition of poisonous by inhalation (see §171.8 of this subchapter), the outage in a bulk packaging must be at least five percent of the total capacity of the tank or compartment at the reference temperature of 46 °C (115 °F) for uninsulated tanks and 41 °C (105 °F) for insulated tanks.

(b) Equivalent steel. For the purposes of this section, stainless steel is steel with a guaranteed minimum tensile strength of 51.7 deka newtons per square millimeter (75,000 psi) and a guaranteed elongation of 40 percent or greater. Where the regulations permit steel other than stainless steel to be used in place of a specified stainless steel (for example, as in §172.102 of this subchapter, special provision B30), the minimum thickness for the steel must be obtained from one of the following formulas, as appropriate:

Formula for metric units:

$e_1 = (12.74e_0)/(Rm_1\ A_1)\ 1/3$

Formula for non-metric units:

$e_1 = (144.2e_0)/(Rm_1\ A_1)\ 1/3$

where:

$e_0$ = Required thickness of the reference stainless steel in millimeters or inches respectively;

$e_1$ = Equivalent thickness of the steel used in millimeters or inches respectively;

$Rm_1$ = Specified minimum tensile strength of the steel used in deka-newtons per square millimeter or pounds per square inch respectively; and

$A_1$ = Specified minimum percentage elongation of the steel used multiplied by 100 (for example, 20 percent times 100 equals 20). Elongation values used must be determined from a 50 mm or 2 inch test specimen.

(c) Air pressure in excess of ambient atmospheric pressure may not be used to load or unload any lading which may create an air-enriched mixture within the flammability range of the lading in the vapor space of the tank.

(d) A bulk packaging may not be loaded with a hazardous material that:

(1) Is at a temperature outside of the packaging's design temperature range; or

(2) Exceeds the maximum weight of lading marked on the specification plate.

### 173.28 Reuse reconditioning and remanufacture of packagings.

(a) *General.* Packagings and receptacles used more than once must be in such condition, including closure devices and cushioning materials, that they conform in all respects to the prescribed requirements of this subchapter. Before reuse, each packaging must be inspected and may not be reused unless free from incompatible residue, rupture, or other damage which reduces its structural integrity.

(b) *Reuse of non-bulk packaging.* A non-bulk packaging used more than once must conform to the following provisions and limitations:

(1) A non-bulk packaging which, upon inspection, shows evidence of a reduction in integrity may not be reused unless it is reconditioned in accordance with paragraph (c) of this section.

(2) Before reuse, packagings subject to the leakproofness test with air prescribed in §178.604 of this subchapter shall be--

(i) Retested without failure in accordance with §178.604 of this subchapter using an internal air pressure (gauge) of at least 48 kPa (7.0 psi) for Packing Group I and 20 kPa (3.0 psi) for Packing Group II and Packing Group III; and

(ii) Marked with the letter "L", with the name and address or symbol of the person conducting the test, and the month and last two digits of the year the test was conducted. Symbols, if used, must be registered with the Associate Administrator for Hazardous Materials Safety.

(3) Packagings made of paper, plastic film, or textile are not authorized for reuse;

(4) Metal and plastic drums and jerricans used as single packagings or the outer packagings of composite packagings are authorized for reuse only when they are marked in a permanent manner (e.g., embossed) in millimeters with the minimum thickness of the packaging material and conform to the following minimum construction criteria:

| Maximum capacity not over | Minimum thickness of packaging material | |
|---|---|---|
| | Metal drum or jerrican | Plastic drum or jerrican |
| 20 L | 0.6 mm (0.024 inch). | 1.1 mm (0.043 inch). |
| 30 L | 0.7 mm (0.028 inch). | 1.1 mm (0.043 inch). |
| 40 L | 0.7 mm (0.028 inch). | 1.8 mm (0.071 inch). |
| 60 L | 0.9 mm (0.035 inch). | 1.8 mm (0.071 inch). |
| 120 L | 0.9 mm (0.035 inch). | 2.2 mm (0.087 inch). |
| 220 L | 0.96 mm (0.038 inch)[1]. | 2.2 mm (0.087 inch). |
| 450 L | 1.8 mm (0.071 inch). | 5.0 mm (0.197 inch). |

[1]Metal drums or jerricans constructed with a minimum thickness of 0.8 mm (0.03 inch) body and 1.1 mm (0.043 inch) heads are authorized.

(5) Plastic inner receptacles of composite packagings must have a minimum thickness of 1.0 mm (0.039 inch).

(6) A previously used non-bulk packaging may be reused for the shipment of hazardous waste, not subject to the reconditioning and reuse provisions of this section, in accordance with §173.12(c).

(c) *Reconditioning of non-bulk packaging.* (1) For the purpose of this subchapter, reconditioning of metal drums is:

(i) Cleaning to base material of construction, with all former contents, internal and external corrosion, and any coatings and labels removed;

(ii) Restoring to original shape and contour, with chimes (if any) straightened and sealed, and all non-integral gaskets replaced: and

(iii) Inspecting after cleaning but before painting, Packagings that have visible pitting, significant reduction in material thickness, metal fatigue, damaged threads or closures, or other significant defects, must be rejected.

(2) For the purpose of this subchapter, reconditioning of a non-bulk packaging other than a metal drum is restoring the packaging by repair, or replacement of non-integral packaging components (such as removable gaskets, closure devices, cushioning material, etc.) to a condition such that it conforms in all respects with the prescribed requirements of this subchapter. Packagings which have significant defects which cannot be repaired may not be reused.

(3) A person who reconditions a packaging manufactured and marked under the provisions of subpart L of Part 178 of this subchapter, shall mark that packaging as required by §178.503(c) of this subchapter. The marking is the certification of the reconditioner that the packaging conforms to the standard for which it is marked and that all functions performed by the reconditioner which are prescribed by this subchapter have been performed in compliance with this subchapter.

(d) *Remanufacture of non-bulk packagings.* For the purpose of this subchapter, remanufacture is the conversion of a non-specification, non-bulk packaging to a DOT specification or U.N. standard, the conversion of a packaging meeting one specification or standard to another specification or standard (for example, conversion of 1A1 non-removable head drums to 1A2 removable head drums) or the replacement of integral structural packaging components (such as non-removable heads on drums). A person who remanufactures a non-bulk packaging to conform to a specification or standard in Part 178 of this subchapter is subject to the requirements of Part 178 of this subchapter as a manufacturer.

(e) *Non-reusable containers.* A packaging marked as NRC according to the DOT specification or UN standard requirements of part 178 of this subchapter may be reused for the shipment of any material not required by this subchapter to be shipped in a DOT specification or UN standard packaging.

**173.29 Empty packagings.**

(a) General. Except as otherwise provided in this section, an empty packaging containing only the residue of a hazardous material shall be offered for transportation and transported in the same manner as when it previously contained a greater quantity of that hazardous material.

(b) Notwithstanding the requirements of paragraph (a) of this section, an empty packaging is not subject to any other requirements of this subchapter if it conforms to the following provisions:

(1) Any hazardous material shipping name and identification number markings, and any hazard warning labels or placards are removed, obliterated, or securely covered in transportation. This provision does not apply to transportation in a transport vehicle or a freight container if the packaging is not visible during transportation and the packaging is loaded by the shipper and unloaded by the shipper or consignee;

(2) The packaging--

(i) Is unused;

(ii) Is sufficiently cleaned of residue and purged of vapors to remove any potential hazard;

(iii) Is refilled with a material which is not hazardous to such an extent that any residue remaining in the packaging no longer poses a hazard; or

(iv) Contains only the residue of--

(A) An ORM-D material; or

(B) A non-flammable gas with no subsidiary hazard at an absolute pressure less than 276 kPa (40 psia); at 21 °C (70 °F); and

(3) Any material contained in the packaging does not meet the definitions in §171.8 of this subchapter for a hazardous substance, a hazardous waste, or a marine pollutant.

(c) A non-bulk packaging containing only the residue of a hazardous material covered by Table 2 of §172.504 of this subchapter--

(1) Does not have to be included in determining the applicability of the placarding requirements of subpart F of Part 172 of this subchapter; and

(2) Is not subject to the shipping paper requirements of this subchapter when collected and transported by a contract or private carrier for reconditioning, remanufacture or reuse.

(d) Notwithstanding the stowage requirements in Column 10a of the §172.101 Table for transportation by vessel, an empty drum or cylinder may be stowed on deck or under deck.

(e) Specific provisions for describing an empty packaging on a shipping paper appear in §172.203(e) of this subchapter.

(f) An empty tank car must conform to the placarding requirements specified in §172.510(c) of this subchapter.

(g) A package which contains a residue of an elevated temperature material may remain marked in the same manner as when it contained a greater quantity of the material even though it no longer meets the definition in §171.8 of this subchapter for an elevated temperature material.

**173.115 Class 2, Divisions 2.1, 2.2, and 2.3--Definitions.**

(a) Division 2.1 (Flammable gas). For the purpose of this subchapter, a "flammable gas" (Division 2.1) means any material which is a gas at 20°C (68°F) or less and 101.3 kPa (14.7 psi) of pressure (a material which has a boiling point of 20°C (68°F) or less at 101 .3 kPa (14.7 psi)) which--

(1) Is ignitable at 101.3 kPa (14.7 psi) when in a mixture of 13 percent or less by volume with air; or

(2) Has a flammable range at 101.3 kPa (14.7 psi) with air of at least 12 percent regardless of the lower limit. Except for aerosols, the limits specified in paragraphs (a)(1) and (a)(2) of this section shall be determined at 101.3 kPa (14.7 psi ) of pressure and a temperature of 20°C (68°F) in accordance with ASTM E681-85, Standard Test Method for Concentration Limits of Flammability of Chemicals. The flammability of aerosols is determined by the tests specified in §173.306(i)(2), (3), and (4) of this part.

(b) Division 2.2 (non-flammable, non-poisonous compressed gas--including compressed gas, liquefied gas, pressurized cryogenic gas and compressed gas in solution). For the purpose of this subchapter, a "non-flammable, non-poisonous compressed gas" (Division 2.2) means any material (or mixture) which--

(1) Exerts in the packaging an absolute pressure of 280 kPa (41 psia) or greater at 20 °C (68 °F), and

(2) Does not meet the definition of Division 2.1 or 2.3.

(c) Division 2.3 (Gas poisonous by inhalation). For the purpose of this subchapter, a "gas poisonous by inhalation" (Division 2.3) means a material which is a gas at 20°C (68°F) or less and a pressure of 101.3 kPa (14.7 psi) (a material which has a boiling point o f 20°C (68°F) or less at 101.3 kPa (14.7 psi)) and which--

(1) Is known to be so toxic to humans as to pose a hazard to health during transportation, or

(2) In the absence of adequate data on human toxicity, is presumed to be toxic to humans because when tested on laboratory animals it has an $LC_{50}$ value of not more than 5000 ml/m3 (see §173.116(a) of this subpart for assignment of Hazard Zones A, B, C or D). $LC_{50}$ values for mixtures may be determined using the formula in §173.133(b)(1)(i) of this subpart.

(d) Non-liquefied compressed gas. A "non-liquefied compressed gas" means a gas, other than in solution, which in a packaging under the charged pressure is entirely gaseous at a temperature of 20 °C (68 °F).

(e) Liquefied compressed gas. A "liquefied compressed gas" means a gas which in a packaging under the charged pressure, is partially liquid at a temperature of 20 °C (68 °F).

(f) Compressed gas in solution. A "compressed gas in solution" is a non-liquefied compressed gas which is dissolved in a solvent.

(g) Cryogenic liquid. A "cryogenic liquid" means a refrigerated liquefied gas having a boiling point colder than -90 °C (-130 °F) at 101.3 kPa (14.7 psi) absolute. A material meeting this definition is subject to requirements of this subchapter without regard to whether it meets the definition of a non-flammable, non-poisonous compressed gas in paragraph (b) of this section.

(h) Flammable range. The term "flammable range" means the difference between the minimum and maximum volume percentages of the material in air that forms a flammable mixture.

(i) Service pressure. The term "service pressure" means the authorized pressure marking on the packaging. For example, for a cylinder marked "DOT 3A1800", the service pressure is 12410 kPa (1800 psi).

(j) Refrigerant gas or Dispersant gas. The terms "Refrigerant gas" or "Dispersant gas" apply to all non-poisonous refrigerant gases, dispersant gases (fluorocarbons) listed in §§172.101, 173.304(a)(2), 173.314(c), 173.315(a)(1) and 173.315(h), and mixtures thereof, or any other compressed gas having a vapor pressure not exceeding 1792 kPa (260 psi) at 54 °C (130 °F), and restricted for use as a refrigerant, dispersant or blowing agent.

**173.116 Class 2--Assignment of hazard zone.**

(a) The hazard zone of a Class 2, Division 2.3 material is assigned in Column 7 of the §172.101 Table. There are no hazard zones for

Divisions 2.1 and 2.2. When the §172.101 Table provides more than one hazard zone for a Division 2.3 material, or indicates that the hazard zone be determined on the basis of the grouping criteria for Division 2.3, the hazard zone shall be determined by applying the following criteria:

| Hazard zone | Inhalation toxicity |
|---|---|
| A | $LC_{50}$ less than or equal to 200 ppm. |
| B | $LC_{50}$ greater than 200 ppm and less than or equal to 1000 ppm. |
| C | $LC_{50}$ greater than 1000 ppm and less than or equal to 3000 ppm. |
| D | $LC_{50}$ greater than 3000 ppm or less than or equal to 5000 ppm. |

(b) The criteria specified in paragraph (a) of this section are represented graphically in §173.133, Figure 1.

**173.117--173.119 [Reserved]**

**173.120 Class 3--Definitions.**

(a) Flammable liquid. (1) For the purpose of this subchapter, a "flammable liquid" (Class 3) means a liquid having a flash point of not more than 60.5 °C (141 °F), or any material in a liquid phase with a flash point at or above 37.8 °C (100 °F) that is intentionally heated and offered for transportation or transported at or above its flash point in a bulk packaging, with the following exceptions:

(i) Any liquid meeting one of the definitions specified in 173.115 of this part;

(ii) Any mixture having one or more components with a flash point of 60.5 °C (141 °F) or higher, that makes up at least 99 per cent of the total volume of the mixture, if the mixture is not offered for transportation or transported at or above its flash point .

(2) For the purposes of this subchapter, a distilled spirit of 140 proof or lower is considered to have a flash point of no lower than 23 °C (73 °F).

(b) Combustible liquid. (1) For the purpose of this subchapter, a "combustible liquid" means any liquid that does not meet the definition of any other hazard class specified in this subchapter and has a flash point above 60.5 °C (141 °F) and below 93 °C (200 °F).

(2) A flammable liquid with a flash point at or above 38°C (100°F) that does not meet the definition of any other hazard class ma y be reclassed as a combustible liquid. This provision does not apply to transportation by vessel or aircraft, except where other means of transportation is impracticable.

(c) Flash point. (1) "Flash point" means the minimum temperature at which a liquid gives off vapor within a test vessel in sufficient concentration to form an ignitable mixture with air near the surface of the liquid. It shall be determined as follows:

(i) For a homogeneous, single-phase, liquid having a viscosity less than 45 S.U.S. at 38 °C (100 °F) that does not form a surface film while under test, one of the following test procedures shall be used:

(A) Standard Method of Test for Flash Point by Tag Closed Tester, (ASTM D56-79); or

(B) Standard Methods of Test for Flash Point of Liquids by Setaflash Closed Tester, (ASTM D3278-78).

(ii) For a liquid other than one meeting all of the criteria of paragraph (c)(1)(i) of this section, one of the following test procedures shall be used:

(A) Standard Method of Test for Flash Point by Pensky--Martens Closed Tester, (ASTM D93-80). For cutback asphalt, use Method B of ASTM D93-80 or alternate tests authorized in this standard; or

(B) Standard Methods of Test for Flash Point of Liquids by Setaflash Closed Tester (ASTM D3278-78).

(2) For a liquid that is a mixture of compounds that have different volatility and flash points, its flash point shall be determined as specified in paragraph (c)(1) of this section, on the material in the form in which it is to be shipped. If it is determined by this test that the flash point is higher than -7 °C (20 °F) a second test shall be made as follows: a portion of the mixture shall be placed in an open beaker (or similar container) of such dimensions that the height of the liquid can be adjusted so that the ratio of the volume of the liquid to

the exposed surface area is 6 to one. The liquid shall be allowed to evaporate under ambient pressure and temperature (20 to 25 °C (68 to 77 °F)) for a period of 4 hours or until 10 percent by volume has evaporated, whichever comes first. A flash point is then run on a portion of the liquid remaining in the evaporation container and the lower of the two flash points shall be the flash point of the material.

(3) For flash point determinations by Setaflash closed tester, the glass syringe specified need not be used as the method of measurement of the test sample if a minimum quantity of 2 ml (0.1 ounce) is assured in the test cup.

(d) If experience or other data indicate that the hazard of a material is greater or less than indicated by the criteria specified in paragraphs (a) and (b) of this section, the Associate Administrator for Hazardous Materials Safety may revise the classification or make the material subject or not subject to the requirements of Parts 170-189 of this subchapter.

**173.121 Class 3--Assignment of packing group.**

(a) The packing group of a Class 3 material is assigned in Column 5 of the §172.101 Table. When the §172.101 Table provides more than one packing group for a hazardous material, or indicates that the packing group is to be determined on the basis of the grouping criteria for Class 3, the packing group shall be determined by applying the following criteria:

| Packing group | Flash point (closed-cup) | Initial boiling point |
|---|---|---|
| I | | ≤ 35°C (95°F) |
| II | < 23°C (73°F) | > 35°C (95°F) |
| III | ≥ 23°C, ≤ 60.5°C (141°F) | > 35°C (95°F) |

(b) Criteria for inclusion of viscous Class 3 materials in Packing Group III.

(1) Viscous Class 3 materials in Packing Group II with a flash point of less than 23 °C (73 °F) may be grouped in Packing Group III provided that--

(i) Less than 3 percent of the clear solvent layer separates in the solvent separation test;

(ii) The mixture contains not more than 5 percent of substances in Packing Group I or II of Division 6.1 or Class 8, or not more than 5 percent of substances in Packing Group I of Class 3 requiring a POISON or CORROSIVE subsidiary label;

(iii) The capacity of the packaging is not more than 30 L (7.9 gallons); and

(iv) The viscosity and flash point are in accordance with the following table:

| Flowtime in seconds | | Flash point in degrees C |
|---|---|---|
| 4 mm cup | 8 mm cup | |
| Over 20 | | Over 17. |
| Over 60 | | Over 10. |
| Over 100 | | Over +5. |
| Over 160 | | Over -1. |
| Over 220 | Over 17 | Over -5. |
| | Over 40 | No lower limit. |

(2) The methods by which the tests referred to in paragraph (b)(1) of this section shall be performed are as follows:

(i) Viscosity Test. The flowtime in seconds is determined at 23 °C (73 °F) using the ISO Standard cup with a 4.0 mm (0.2 inch) jet (ISO-2431-72). Where the flowtime exceeds 200 seconds, a second test is carried out using the ISO standard cup but modified to take a jet of 8 mm (0.3 inch) diameter.

(ii) Solvent Separation Test. This test is carried out at 23 °C (73 °F) using a 100.0 ml (3 ounces) measuring cylinder of the stoppered type of approximately 25.0 cm (9.8 inches) total height and of a uniform internal diameter of approximately 30 mm (1.2 inches) over the calibrated section. The sample should be stirred to obtain a uniform consistency, and poured in up to the 100 ml (3 ounces) mark. The stopper should be inserted and the cylinder left standing undisturbed for 24 hours. After 24 hours, the height of the upper separated layer should be measured and the percentage of this layer as compared with the total height of the sample calculated.

## 173.124 Class 4 Divisions 4.1, 4.2 and 4.3--Definitions.

(a) Division 4.1 (Flammable Solid). For the purposes of this subchapter, "flammable solid" (Division 4.1) means any of the following three types of materials:

(1) Wetted explosives that--

(i) When dry are Explosives of Class 1 other than those of compatibility group A, which are wetted with sufficient water, alcohol, or plasticizer to suppress explosive properties; and

(ii) Are specifically authorized by name either in the §172.101 Table or have been assigned a shipping name and hazard class by the Associate Administrator for Hazardous Materials Safety under the provisions of--

(A) An exemption issued under subchapter A of this chapter; or

(B) An approval issued under §173.56(i) of this part.

(2) Self-reactive materials are materials that are liable to undergo, at normal or elevated temperatures, a strongly exothermal decomposition caused by excessively high transport temperatures or by contamination; and

(3) Readily combustible solids are materials that--

(i) Are solids which may cause a fire through friction, such as matches;

(ii) Show a burning rate faster than 2.2 mm (0.087 inches) per second when tested in accordance with paragraph 2.c.(2) of appendix E to this part; or

(iii) Any metal powders that can be ignited and react over the whole length of a sample in 10 minutes or less, when tested in accordance with paragraph 2.c.(2) of appendix E to this part.

(b) Division 4.2 (Spontaneously Combustible Material). For the purposes of this subchapter, "spontaneously combustible material" (Division 4.2) means--

(1) A pyrophoric material. A pyrophoric material is a liquid or solid that, even in small quantities and without an external ignition source, can ignite within five (5) minutes after coming in contact with air when tested according to

paragraph 3.a.(1) or 3.a. (2), as appropriate, of appendix E to this part.

(2) A self-heating material. A self-heating material is a material that, when in contact with air and without an energy supply, is liable to self-heat. A material of this type which exhibits spontaneous ignition or if the temperature of the sample exceeds 200 °C (392 °F) during the 24=hour test period when tested in accordance with paragraph 3.b.(1) of appendix E to this part, is classed as a Division 4.2 material.

(c) Division 4.3 (Dangerous when wet material). For the purposes of this chapter, "dangerous when wet material" (Division 4.3) means a material that, by contact with water, is liable to become spontaneously flammable or to give off flammable or toxic gas at a rate greater than 1 liter per kilogram of the material, per hour, when tested in accordance with paragraph 4 of appendix E to this part.

**173.125 Class 4--Assignment of packing group.**

(a) The packing group of a Class 4 material is as assigned in Column 5 of the §172.101 Table. When the §172.101 Table indicates that the packing group of a hazardous material is to be determined on the basis of test results following test methods given in appendix E of this part, the packing group shall be determined by applying the appropriate criteria given in this section.

(b) Packing group criteria for readily combustible materials of Division 4.1 is as follows:

(1) For materials other than metal powders, a material is assigned to--

(i) Packing Group II, if the burning rate is greater than 2.2 mm/s and the flame passes the wetted zone; or

(ii) Packing Group III, if the burning rate is greater than 2.2 mm/s and the wetted zone stops the flame.

(2) For metal powders, a material is assigned to--

(i) Packing Group II, if the zone of reaction spreads over the whole length of the sample in 5 minutes or less; or

(ii) Packing Group III, if the zone of reaction spreads over the whole length of the sample in more than 5 but not more than 10 minutes.

(3) Solids which may cause a fire through friction are assigned to packing groups by analogy with existing entries in the §172.101 Table.

(c) Packing group criteria for Division 4.2 materials is as follows:

(1) Pyrophoric liquids and solids of Division 4.2 are assigned to Packing Group I.

(2) A self-heating material is assigned to--

(i) Packing Group II, if the material gives a positive test result when tested with the 2.5-cm cube size sample; or

(ii) Packing Group III, if the material gives a positive test result when tested with the 10-cm cube size sample but a negative test result with the 2.5-cm cube size sample.

(d) A Division 4.3 dangerous when wet material is assigned to--

(1) Packing Group I, if spontaneous ignition occurs, or

demonstrates a tendency of spontaneous ignition, or the rate of evolution of flammable gases is equal or greater than 10 liters per kilogram of material over any one minute; or

(2) Packing Group II, if the rate of evolution of flammable gases is equal to or greater than 20 liters per kilogram of material per hour, and which does not meet the criteria for Packing Group I; or

(3) Packing Group III, if the rate of evolution of flammable gases is greater than 1 liter per kilogram of material per hour, and which does not meet the criteria for Packing Group I or II.

**173.127 Class 5, Division 5.1--Definition and assignment of packing groups.**

(a) Definition. For the purpose of this subchapter, "oxidizer" (Division 5.1) means a material that may, generally by yielding oxygen, cause or enhance the combustion of other materials. A solid material is classed as a Division 5.1 material if, when tested in accordance with Appendix F to this part, in either concentration tested, the mean burning time of the test mixture, is equal to or less than that of the average of the three tests with ammonium persulfate mixture. A liquid is classed as a Division 5.1 material by analogy to existing entries in the §172.101 Table.

(b) Assignment of packing groups. (1) The packing group of a Division 5.1 material shall be as assigned in Column 5 of the §172.101 Table.

(2) When the §172.101 Table indicates that the packing group of a solid oxidizer is to be determined on the basis of the test results following test method given in Appendix F to this part, the packing group shall be assigned by the following criteria:

(i) Packing Group I, for a material which, in either concentration tested, exhibits a burning time equal to or less than that of potassium bromate;

(ii) Packing Group II, for a material which, in either concentration tested, exhibits a burning time between that of potassium bromate and that of potassium perchlorate: or

(iii) Packing Group III, for a material which, in either concentration tested, exhibits a burning time between that of potassium perchlorate and that of ammonium persulphate.

(3) Liquid oxidizers are assigned to packing groups by analogy with existing entries in the §172.101 Table.

**173.128 Class 5, Division 5.2--Definitions and types.**

(a) Definitions. For the purposes of this subchapter, "organic peroxide" (Division 5.2) means any organic compound containing oxygen (O) in the bivalent -O-O- structure and which may be considered a derivative of hydrogen peroxide, where one or more of the hydrogen atoms have been replaced by organic radicals, unless any of the following paragraphs apply:

(1) The material meets the definition of an explosive as prescribed in subpart C of this part, in which case it must be classed as an explosive;

(2) The material is forbidden from being offered for

transportation according to §172.101 of this subchapter or §173.21;

(3) The Associate Administrator for Hazardous Materials Safety has determined that the material does not present a hazard which is associated with a Division 5.2 material; or

(4) The material meets one of the following conditions:

(i) For materials containing no more than 1.0 percent hydrogen peroxide, the available oxygen, as calculated using the equation in paragraph (a)(4)(ii) of this section, is not more than 1.0 percent, or

(ii) For materials containing more than 1.0 percent but not more than 7.0 percent hydrogen peroxide, the available oxygen, content (Oa) is not more than 0.5 percent, when determined using the equation:

$$O_a = 16 \times \sum_{i=1}^{k} \frac{n_i c_i}{m_i}$$

where, for a material containing k species of organic peroxides:

$n_i$=number of -O-O- groups per molecule of the *i*th species
$c_i$=concentration (mass percent) of the *i*th species
$m_i$=molecular mass of the *i*th species

(b) Generic types. Division 5.2 organic peroxides are assigned to a generic system which consists of seven types. An organic peroxide identified by technical name in the Organic Peroxides Table in §173.225 is assigned to a generic type in accordance with that Table. Organic peroxides not identified in the Organic Peroxides Table are assigned to generic types under the procedures of paragraph (c) of this section.

(1) Type A. Organic peroxide type A is an organic peroxide which can detonate or deflagrate rapidly as packaged for transport. Transportation of type A organic peroxides is forbidden.

(2) Type B. Organic peroxide type B is an organic peroxide which, as packaged for transport, neither detonates nor deflagrates rapidly, but can undergo a thermal explosion.

(3) Type C. Organic peroxide type C is an organic peroxide which, as packaged for transport, neither detonates nor deflagrates rapidly and cannot undergo a thermal explosion.

(4) Type D. Organic peroxide type D is an organic peroxide which--

(i) Detonates only partially, but does not deflagrate rapidly and is not affected by heat when confined;

(ii) Does not detonate, deflagrates slowly, and shows no violent effect if heated when confined; or

(iii) Does not detonate or deflagrate, and shows a medium effect when heated under confinement.

(5) Type E. Organic peroxide type E is an organic peroxide which neither detonates nor deflagrates and shows low, or no, effect when heated under confinement.

(6) Type F. Organic peroxide type F is an organic peroxide which will not detonate in a cavitated state, does not deflagrate, shows only a low, or no, effect if heated when confined, and has low, or no, explosive power.

(7) Type G. Organic peroxide type G is an organic peroxide

which will not detonate in a cavitated state, will not deflagrate, shows no effect when heated under confinement, has no explosive power, is thermally stable (self-accelerating decomposition temperature above 60°C (140°F)), and, for desensitized liquid formulations, is desensitized with a compatible organic liquid which boils above 150 °C (300 °F) (diluent type A, see §173.225(b)).

(c) Procedure for assigning an organic peroxide to a generic type. An organic peroxide shall be assigned to a generic type based on- -

(1) Its physical state (i.e., liquid or solid), in accordance with the definitions for liquid and solid in §171.8 of this subchapter;

(2) A determination as to its control temperature and emergency temperature, if any, under the provisions of §173.21(f);

(3) Performance of the organic peroxide under the test procedures specified in the United Nations Recommendations on the Transport of Dangerous Goods, Tests and Criteria, part III, and the provisions of paragraph (d) of this section; and

(4) Except for an organic peroxide which is identified by technical name in the Organic Peroxides Table in §173.225(b) or an organic peroxide which may be shipped as a sample under the provisions of §173.225(c), the organic peroxide is approved, in writing, by the Associate Administrator for Hazardous Materials Safety, including assignment of a generic type and shipping description. The person requesting approval shall submit: (i) all relevant data concerning physical state, temperature controls, and test results or (ii) an approval issued for the organic peroxide by the competent authority of a foreign government, to the Associate Administrator for Hazardous Materials Safety.

(d) Tests. The generic type for an organic peroxide shall be determined using the testing protocol from Figure 11.1 (Classification and Flow Chart Scheme for Organic Peroxides) from the UN Recommendations, Tests and Criteria, part III.

**173.129 Class 5, Division 5.2-- Assignment of packing group.**

All Division 5.2 materials are assigned to Packing Group II in Column 5 of the §172.101 Table.

**173.132 Class 6, Division 6.1-- Definitions.**

(a) For the purpose of this subchapter, "poisonous material" (Division 6.1) means a material, other than a gas, which is known to be so toxic to humans as to afford a hazard to health during transportation, or which, in the absence of adequate data on human toxicity:

(1) Is presumed to be toxic to humans because it falls within any one of the following categories when tested on laboratory animals whenever possible, animal test data that has been reported in the chemical literature should be used:

(i) Oral Toxicity. A liquid with an $LD_{50}$ for acute oral toxicity of not more than 500 mg/kg or a solid with an $LD_{50}$ for acute oral toxicity of not more than 200 mg/kg.

(ii) Dermal Toxicity. A material with an $LD_{50}$ for acute dermal toxicity of not more than 1000 mg/kg.

(iii) Inhalation Toxicity. (A) A dust or mist with an $LC_{50}$ for acute toxicity on inhalation of not more than 10 mg/L; or

(B) A material with a saturated vapor concentration in air at 20 °C (68 °F) of more than one-fifth of the $LC_{50}$ for acute toxicity on inhalation of vapors and with an $LC_{50}$ for acute toxicity on inhalation of vapors of not more than 5000 ml/m3; or

(2) Is an irritating material, with properties similar to tear gas, which causes extreme irritation, especially in confined spaces.

(b) For the purposes of this subchapter--

(1) $LD_{50}$ for acute oral toxicity means that dose of the material administered which to both male and female young adult albino rats which causes death within 14 days in half the animals tested. The number of animals tested must be sufficient to give statistically valid results and be in conformity with good pharmacological practices. The result is expressed in mg/kg body mass.

(2) $LD_{50}$ for acute dermal toxicity means that dose of the material which, administered by continuous contact for 24 hours with the shaved intact skin (avoiding abrading) of an albino rabbit, causes death within 14 days in half of the animals tested. The number of animals tested must be sufficient to give statistically valid results and be in conformity with good pharmacological practices. The result is expressed in mg/kg body mass.

(3) $LC_{50}$ for acute toxicity on inhalation means that concentration of vapor, mist, or dust which, administered by continuous inhalation for one hour to both male and female young adult albino rats, causes death within 14 days in half of the animals tested. If the material is administered to the animals as a dust or mist, more than 90 percent of the particles available for inhalation in the test must have a diameter of 10 microns or less if it is reasonably foreseeable that such concentrations could be encountered by a human during transport. The result is expressed in mg/L of air for dusts and mists or in mL/m3 of air (parts per million) for vapors . See § 173.133(b) for $LC_{50}$ determination for mixtures and for limit tests.

(i) When provisions of this subchapter require the use of the $LC_{50}$ for acute toxicity on inhalation of dusts and mists based on a one-hour exposure and such data is not available, the $LC_{50}$ for acute toxicity on inhalation based on a four-hour exposure may be multiplied by four and the product substituted for the one-hour $LC_{50}$ for acute toxicity on inhalation.

(ii) When the provisions of this subchapter require the use of the $LC_{50}$ for acute toxicity on inhalation of vapors based on a one-hour exposure and such data is not available, the $LC_{50}$ for acute toxicity on inhalation based on a four-hour exposure may be multiplied by two and the product substituted for the one-hour $LC_{50}$ for acute toxicity on inhalation.

(c) The foregoing categories shall not apply if the Associate Administrator for Hazardous Materials Safety has determined that the physical characteristics of the material or its probable hazards to humans as shown by documented experience indicate that the material will not cause serious sickness or death.

**173.133 Assignment of packing group and hazard zones for Division 6.1 materials.**

(a) The packing group of Division 6.1 materials shall be as assigned in Column 5 of the § 172.101 Table. When the § 172.101 Table provides more than one packing group and hazard zone for a hazardous material, the packing group and hazard zone shall be determined by applying the following criteria:

(1) The packing group assignment for routes of administration other than inhalation of vapors shall be in accordance with the following table:

| Packing Group | Oral toxicity $LD_{50}$ (mg/kg) | Dermal toxicity $LD_{50}$ (mg/kg) | Inhalation toxicity by dusts and mists $LC_{50}$ (mg/L) |
|---|---|---|---|
| I | ≤ 5 | ≤ 40 | ≤ 0.5 |
| II | > 5, ≤ 50 | > 40, ≤ 200 | > 0.5 ≤ 2 |
| III | solids: > 50, ≤ 200; liquids: > 50, ≤ 500 | > 200, ≤ 1000 | > 2 ≤ 10 |

**(2)(i) The packing group and hazard zone assignments for liquids (see §173.115(c) of this subpart for gases) based on inhalation of vapors shall be in accordance with the following Table:**

| Packing Group | Vapor concentration and toxicity |
|---|---|
| I (Hazard Zone A). | $V \geq 500\ LC_{50}$ and $LC_{50} \leq 200$ mL/M³. |
| I (Hazard Zone B). | $V \geq 10\ LC_{50}$; $LC_{50} \leq 1000$ mL/m³; and the criteria for Packing Group I, Hazard Zone A are not met. |
| II | $V \geq LC_{50}$; $LC_{50} \leq 3000$ mL/m³; and the criteria for Packing Group I, are not met. |
| III | $V \geq .2\ LC_{50}$; $LC_{50} \leq 5000$ mL/m³; and the criteria for Packing Groups I and II, are not met. |

Note 1: V is the saturated vapor concentration in air of the material in mL/m³ at 20°C and standard atmospheric pressure.

Note 2: A liquid in Division 6.1 meeting criteria for Packing Group I, Hazard Zones A or B stated in paragraph (a)(2) of this section is a material poisonous by inhalation subject to the additional hazard communication requirements in §§172.203(m)(3), 172.313 and Table 1 of §172.504(e) of this subchapter.

**(ii) These criteria are represented graphically in Figure 1:**

**Figure 1**

**Inhalation Toxicity: Packing Group and Hazard Zone Borderlines**

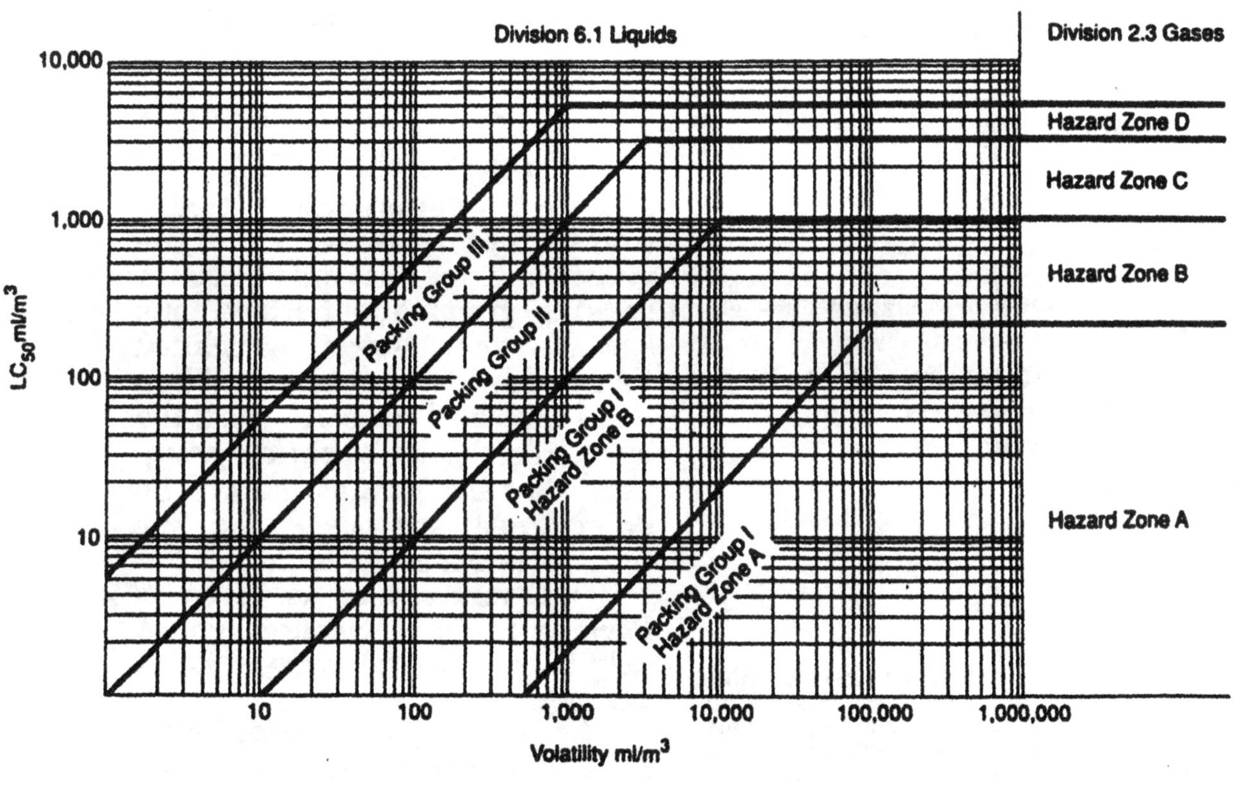

(3) When the packing group determined by applying these criteria is different for two or more (oral, dermal or inhalation) routes of administration, the packing group assigned to the material shall be that indicated for the highest degree of toxicity for any of the routes of administration.

(4) Notwithstanding the provisions of this paragraph, the packing group and hazard zone of a tear gas substance is as assigned in Column 5 of the §172.101 Table.

(b) The packing group and hazard zone for Division 6.1 mixtures that are poisonous (toxic) by inhalation may be determined by one of the following methods:

(1) Where $LC_{50}$ data is available on each of the poisonous (toxic) substances comprising the mixture--

(i) The $LC_{50}$ of the mixture is estimated using the formula:

$$LC_{50}\text{ (mixture)} = \frac{1}{\sum_{i=1}^{n} \frac{f_i}{LC_{50i}}}$$

where fi=mole fraction of the ith component substance of the liquid.

LC50i=mean lethal concentration of the ith component substance in ml/m3

(ii) The volatility of each component substance is estimated using the formula:

$$V_i = P_i \times \frac{10^6}{101.3} \text{ ml/m}^3$$

where:

Pi=partial pressure of the ith component substance in kPa at 20°C and one atmospheric pressure. Pi may be calculated according to Raoult's Law using appropriate activity coefficients. Where activity coefficients are not available, the coefficient may be assumed to be 1.0.

(iii) The ratio of the volatility to the $LC_{50}$ is calculated using the formula:

$$R = \sum_{i=1}^{n} \frac{V_i}{LC_{50i}}$$

(iv) Using the calculated values $LC_{50}$ (mixture) and R, the packing group for the mixture is determined as follows:

| Packing Group (Hazard Zone) | Ratio of volatility and $LC_{50}$ |
|---|---|
| I (Hazard Zone A). | $R \geq 500$ and $LC_{50}$ (mixture) $\leq 200$ ml/m³. |
| I (Hazard Zone B). | $R \geq 10$ and $LC_{50}$ (mixture) $\leq 1000$ ml/m³; and the criteria for Packing Group I, Hazard Zone A, are not met. |
| II (Hazard Zone C). | $R \geq 1$ and $LC_{50}$ (mixture) $\leq 3000$ ml/m³; and the criteria for Packing Group I, Hazard Zones A and B are not met. |
| III (Hazard Zone D). | $R \geq 1/5$ and $LC_{50}$ (mixture) $\leq 5000$ ml/m³; and the criteria for Packing Groups I and II, Hazard Zones A, B and C are not met. |

(2) In the absence of $LC_{50}$ data on the poisonous (toxic) constituent substances, the mixture may be assigned a packing group and hazard zone based on the following simplified threshold toxicity tests. When these threshold tests are used, the most restrictive packing group and hazard zone must be determined and used for the transportation of the mixture.

(i) A mixture is assigned to Packing Group I, Hazard Zone A only if both the following criteria are met:

(A) A sample of the liquid mixture is vaporized and diluted with air to create a test atmosphere of 200 ml/m3 vaporized mixture in air. Ten albino rats (five male and five female) are exposed to the test atmosphere as determined by an analytical method appropriate for the material being classified for one hour and observed for fourteen days. If five or more of the animals die within the fourteen-day observation period, the mixture is presumed to have an $LC_{50}$ equal to or less than 200 ml/m3.

(B) A sample of the vapor in equilibrium with the liquid mixture is diluted with 499 equal volumes of air to form a test atmosphere. Ten albino rats (five male and five female) are exposed to the test atmosphere for one hour and observed for fourteen days. If five or more of the animals die within the fourteen-day observation period, the mixture is presumed to have a volatility equal to or greater than 500 times the mixture $LC_{50}$.

(ii) A mixture is assigned to Packing Group I, Hazard Zone B only if both the following criteria are met, and the mixture does not meet the criteria for Packing Group I, Hazard Zone A:

(A) A sample of the liquid mixture is vaporized and diluted with air to create a test atmosphere of 1000 ml/m3 vaporized mixture in air. Ten albino rats (five male and five female) are exposed to the test atmosphere for one hour and observed for fourteen days. If five or more of the animals die within the fourteen-day observation period, the mixture is presumed to have an $LC_{50}$ equal to or less than 1000 ml/m3.

(B) A sample of the vapor in equilibrium with the liquid mixture is diluted with 9 equal volumes of air to form a test atmosphere. Ten albino rats (five male and five female) are exposed to the test atmosphere for one hour and observed for fourteen days . If five or more of the animals die within the fourteen-day observation period, the mixture is presumed to have a volatility equal to or greater than 10 times the mixture $LC_{50}$.

(iii) A mixture is assigned to Packing Group II only if both the following criteria are met, and the mixture does not meet the criteria for Packing Group I (Hazard Zones A or B):

(A) A sample of the liquid mixture is vaporized and diluted with air to create a test atmosphere of 3000 ml/m3 vaporized mixture in air. Ten albino rats (five male and five female) are exposed to the test atmosphere for one hour and observed for fourteen days. If five or more of the animals die within the fourteen-day observation period, the mixture is presumed to have an $LC_{50}$ equal to or less than 3000 ml/m3.

(B) A sample of the vapor in equilibrium with the liquid mixture is used to form a test atmosphere. Ten albino rats (five male and five female) are exposed to the test atmosphere for one hour and observed for fourteen days. If five or more of the animals die within the fourteen-day observation period, the mixture is presumed to have a volatility equal to or greater than the mixture $LC_{50}$.

(iv) A mixture is assigned to Packing Group III only if both the following criteria are met, and the mixture does not meet the criteria for Packing Groups I (Hazard Zones A or B) or Packing Group II (Hazard Zone C):

(A) A sample of the liquid mixture is vaporized and diluted with air to create a test atmosphere of 5000 ml/m3 vaporized mixture in air. Ten albino rats (five male and five female) are exposed to the test atmosphere for one hour and observed for fourteen days. If five or more of the animals die within the fourteen-day observation period, the mixture is presumed to have an $LC_{50}$ equal to or less than 5000 ml/m3.

(B) The vapor pressure of the liquid mixture is measured and if the vapor concentration is equal to or greater than 1000 ml/m3, the mixture is presumed to have a volatility equal to or greater than 1/5 the mixture $LC_{50}$.

### 173.134 Class 6, Division 6.2--Definitions.

(a) For the purpose of this subchapter--

(1) An "infectious substance" (Division 6.2) means a viable microorganism, or its toxin, which causes or may cause disease in humans or animals, and includes those agents listed in 42 CFR 72.3 of the regulations of the Department of Health and Human Services or any other agent that causes or may cause severe, disabling or fatal disease. The terms "infectious substances" and "etiologic agent" are synonymous.

(2) A "diagnostic specimen" means any human or animal material including, but not limited to, excreta, secreta, blood, and its components, tissue, and tissue fluids, being shipped for purposes of diagnosis.

(3) A "biological product" means a material prepared and manufactured in accordance with the provisions of 9 CFR part 102 (Licenses for biological products), 9 CFR part 103 (Biological products for experimental treatment of animals), 9 CFR part 104 (Permits for biological products), 21 CFR part 312 (Investigational new drug application), or 21 CFR parts 600 to 680 (Biologics), and which, in accordance with these provisions, may be shipped in interstate commerce.

(4) A "regulated medical waste" as defined in Appendix G of this part.

(b) The requirements of this subpart supplement the requirements of the Department of Health and Human Services contained in 42 CFR part 72.

(c) Packing groups are not assigned to Division 6.2 materials.

### 173.136 Class 8--Definitions.

(a) For the purpose of this subchapter, "corrosive material" (Class 8) means a liquid or solid that causes visible destruction or irreversible alterations in human skin tissue at the site of contact, or a liquid that has a severe corrosion rate on steel or aluminum, in accordance with the following criteria:

(1) A material is considered to be destructive or to cause irreversible alteration in human skin tissue if, when tested on the intact skin of an albino rabbit by the technique described in Appendix A to this part, the structure of the tissue at the site of

contact is destroyed or changed irreversibly after an exposure period of 4 hours or less.

(2) A liquid is considered to have a severe corrosion rate if its corrosion rate exceeds 6.25 mm (0.246 inches) a year on steel (SAE 1020) or aluminum (nonclad 7075-T6) at a test temperature of 55 °C (131 °F). An acceptable test is described in NACE Standard TM-01-69.

(b) If human experience or other data indicate that the hazard of a material is greater or less than indicated by the results of the tests specified in paragraph (a) of this section, RSPA may revise its classification or make the determination that the material is not subject to the requirements of this subchapter.

**173.137 Class 8--Assignment of packing group.**

The packing group of Class 8 material is indicated in Column 5 of the §172.101 Table. When the §172.101 Table provides more than one packing group for a hazardous material, the packing group shall be determined by applying the following criteria:

(a) Packing Group 1. Substances that cause visible destruction or irreversible alterations of the skin tissue at the site of contact when tested on the intact skin of an animal for a period of not more than 3 minutes.

(b) Packing Group II. Substances, other than those meeting Packing Group I criteria, that cause visible destruction or irreversible alterations of the skin tissue at the site of contact when tested on the intact skin of an animal for a period of not more than 60 minutes.

(c) Packing Group III. Substances, other than those meeting Packing Group I or II criteria--

(1) That cause visible destruction or irreversible alterations of the skin tissue at the site of contact when tested on the intact skin of an animal for a period of not more than 4 hours; or

(2) Which have a corrosion rate on steel or aluminum surfaces exceeding 6.25 mm (0.246 inch) a year at a test temperature of 55 ° C (131 °F).

**173.140 Class 9--Definitions.**

For the purposes of this subchapter, "miscellaneous hazardous material" (Class 9) means a material which presents a hazard during transportation but which does not meet the definition of any other hazard class. This class includes:

(a) Any material which has an anesthetic, noxious or other similar property which could cause extreme annoyance or discomfort to a flight crew member so as to prevent the correct performance of assigned duties; or

(b) Any material that meets the definition in 171.8 of this subchapter for an elevated temperature material, a hazardous substance, a hazardous waste, or a marine pollutant.

**173.141 Class 9--Assignment of packing group.**

The packing group of a Class 9 material is as indicated in Column 5 of the §172.101 Table.

**173.144 Other Regulated Materials (ORM)--Definitions.**

For the purpose of this subchapter, "ORM-D material" means a material such as a consumer commodity, which, although otherwise subject to the regulations of this subchapter, presents a limited hazard during transportation due to its form, quantity and packaging. It must be a material for which exceptions are provided in the §172.101 Table. Each ORM-D material and category of ORM-D material is listed in the §172.101 Table.

**173.145 Other Regulated Materials--Assignment of packing group.**

Packing groups are not assigned to ORM-D materials.

**173.150 Exceptions for Class 3 (flammable and combustible liquids).**

(a) General. Exceptions for hazardous materials shipments in the following paragraphs are permitted only if this section is referenced for the specific hazardous material in the §172.101 Table of this subchapter and the material does not meet the definition of another hazard class.

(b) Limited quantities. Limited quantities of flammable liquids (Class 3) are excepted from labeling requirements, unless offered for transportation or transported by aircraft, and the specification packaging requirements of this subchapter when packaged in combination packagings according to this paragraph. In addition, shipments of limited quantities are not subject to subpart F (Placarding ) of Part 172 of this subchapter. Each package must conform to the packaging requirements of subpart B of this part and may not exceed 30 kg (66 pounds) gross weight. The following combination packagings are authorized:

(1) For flammable liquids in Packing Group I, inner packagings not over 0.5 L (0.1 gallon) net capacity each, packed in strong outer packagings;

(2) For flammable liquids in Packing Group II, inner packagings not over 1.0 L (0.3 gallon) net capacity each, packed in strong outer packaging; and

(3) For flammable liquids in Packing Group III, inner packagings not over 5.0 L (1 gallon) net capacity each, packed in strong outer packagings.

(c) Consumer commodities. A limited quantity which conforms to the provisions of paragraph (b) of this section and is a "consumer commodity" as defined in §171.8 of this subchapter, may be renamed "Consumer commodity" and reclassed as ORM-D material. In addition to the exceptions provided by paragraph (b) of this section, shipments of ORM-D materials are not subject to the shipping paper requirements of subpart C of Part 172 of this subchapter, unless the material meets the definition of a hazardous substance, a hazardous waste, or a marine pollutant or unless offered for transportation or transported by aircraft, and are eligible for the exceptions provided in §173.156.

(d) Alcoholic beverages. Alcoholic beverages (wine and

distilled spirits as defined in 27 CFR 4.10 and 5.11) in packagings of four liters or less are not subject to the requirements of this subchapter.

(e) Aqueous solutions of alcohol. An aqueous solution containing 24 percent or less alcohol by volume and no other hazardous material--

(1) May be reclassed as a combustible liquid; and

(2) Is not subject to the requirements of this subchapter if it contains no less than 50 percent water.

(f) Combustible liquids. (1) A flammable liquid with a flash point at or above 38 °C (100°F) that does not meet the definition of any other hazard class, may be reclassed as a combustible liquid. This provision does not apply to transportation by vessel or aircraft, except where other means of transportation is impracticable.

(2) The requirements in this subchapter do not apply to a material classed as a combustible liquid in a non-bulk packaging unless the combustible liquid is a hazardous substance, a hazardous waste, or a marine pollutant.

(3) A combustible liquid that is in a bulk packaging or a combustible liquid that is a hazardous substance, a hazardous waste, or a marine pollutant is not subject to the requirements of this subchapter except those pertaining to:

(i) Shipping papers, waybills, switching orders, and hazardous waste manifests;

(ii) Marking of packages;

(iii) Display of identification numbers on bulk packages;

(iv) Placarding requirements of Subpart F of Part 172 of this subchapter;

(v) Carriage aboard aircraft and vessels (for packaging requirements for transport by vessel, see §176.340 of this subchapter);

(vi) Reporting incidents as prescribed by §§171.15 and 171.16 of this subchapter;

(vii) Packaging requirements of Subpart B of this part and, in addition, non-bulk packagings must conform with requirements of §173.203; and

(viii) The requirements of §§173.1, 173.21, 173.24, 173.24a, 173.24b, 174.1, 177.804, 177.817, and 177.834 of this subchapter.

(4) A combustible liquid that is not a hazardous substance, a hazardous waste, or a marine pollutant is not subject to the requirements of this subchapter if it is a mixture of one or more components that--

(i) Has a flash point at or above 93 °C (200 °F),

(ii) Comprises at least 99 percent of the volume of the mixture, and

(iii) Is not offered for transportation or transported as a liquid at a temperature at or above its flash point.

### 173.151 Exceptions for Division 4.1 (flammable solids).

(a) General. Exceptions for hazardous materials shipments in the following paragraphs are permitted only if this section is referenced for the specific hazardous material in the §172.101 Table of this subchapter.

(b) Limited quantities of Division 4.1 flammable solids.

Limited quantities of flammable solids (Division 4.1) in Packing Groups II and III are excepted from labeling, unless offered for transportation or transported by aircraft, and the specification packaging requirements of this subchapter when packaged in combination packagings according to this paragraph. In addition, shipments of limited quantities are not subject to subpart F (Placarding) of Part 172 of this subchapter. Each package must conform to the packaging requirements of subpart B of this part and may not exceed 30 kg (66 pounds) gross weight. The following combination packagings are authorized:

(1) For flammable solids in Packing Group II, inner packagings not over 1.0 kg (2.2 pounds) net capacity each, packed in strong outer packagings; and

(2) For flammable solids in Packing Group III, inner packagings not over 5.0 kg (11 pounds) net capacity each, packed in strong outer packagings

(c) Consumer commodities. A limited quantity which conforms to the provisions of paragraph (b) of this section, and charcoal briquettes in packagings not exceeding 30 kg (66 pounds) gross weight, may be renamed "Consumer commodity" and reclassed as ORM-D material, if the material is a "consumer commodity" as defined in §171.8 of this subchapter. In addition to the exceptions provided by paragraph (b) of this section, shipments are not subject to the shipping paper requirements of subpart C of Part 172 of this subchapter, unless the material meets the definition of a hazardous substance, a hazardous waste, or a marine pollutant or unless offered for transportation or transported by aircraft, and are eligible for the exceptions provided in §173.156.

**173.152 Exceptions for Division 5.1 (oxidizers) and Division 5.2 (organic peroxides).**

(a) General. Exceptions for hazardous materials shipments in the following paragraphs are permitted only if this section is referenced for the specific hazardous material in the §172.101 Table of this subchapter.

(b) Limited quantities. Limited quantities of oxidizers (Division 5.1) in Packing Groups II and III and organic peroxides (Division 5.2) are excepted from labeling, unless offered or intended for transportation by aircraft, and the specification packaging requirements of this subchapter when packaged in combination packagings according to this paragraph. In addition, shipments of these limited quantities are not subject to subpart F of Part 172 (Placarding) of this subchapter. Each package must conform to the packaging requirements of subpart B of this part and may not exceed 30 kg (66 pounds) gross weight. The following combination packagings are authorized:

(1) For oxidizers in Packing Group II, inner packagings not over 1.0 L (0.3 gallon) net capacity each for liquids or not over 1.0 kg (2.2 pounds) net capacity each for solids, packed in strong outer packagings.

(2) For oxidizers in Packing Group III, inner packagings not over 4.0 L (1 gallon) net capacity

each for liquids or not over 5.0 kg (11 pounds) net capacity each for solids, packed in strong outer packagings.

(3) For organic peroxides, inner packagings not over 30 ml (1.0 ounce) net capacity for liquids or 30 g (1.1 ounces) net capacity for solids, packed in strong outer packagings.

(c) Consumer commodities. A limited quantity which conforms to the provisions of paragraph (b) of this section and is a "consumer commodity" as defined in §171.8 of this subchapter, may be renamed "Consumer commodity" and reclassed as ORM-D material. In addition to the exceptions provided by paragraph (b) of this section, shipments are not subject to the shipping paper requirements of subpart C of Part 172 of this subchapter, unless the material meets the definition of a hazardous substance, a hazardous waste, or a marine pollutant or unless offered for transportation or transported by aircraft, and are eligible for the exceptions provided in §173.156.

### 173.153 Exceptions for Division 6.1 (poisonous materials).

(a) General. Exceptions for hazardous materials shipments in the following paragraphs are permitted only if this section is referenced for the specific hazardous material in the §172.101 Table of this subchapter.

(b) Limited quantities of Division 6.1 materials. Limited quantities of poisonous materials (Division 6.1) in Packing Group III are excepted from the specification packaging requirements of this subchapter when packaged in combination packagings according to this paragraph. In addition, shipments of these limited quantities are not subject to subpart F of Part 172 (Placarding) of this subchapter. Each package must conform to the packaging requirements of subpart B of this part and may not exceed 30 kg (66 pounds) gross weight. The following combination packagings are authorized:

(1) For poisonous liquids, inner packagings not over 4.0 L (1 gallon) net capacity each, packed in strong outer packagings; and

(2) For poisonous solids, inner packagings not over 5.0 kg (11 pounds) net capacity each, packed in strong outer packagings.

(c) Consumer commodities. The following provisions apply to consumer commodities:

(1) A limited quantity which conforms to the provisions of paragraph (b) of this section and is a "consumer commodity" as defined in §171.8 of this subchapter, may be renamed "Consumer commodity" and reclassed as ORM-D material.

(2) A poisonous material which is a drug or medicine and is a "consumer commodity" as defined in §171.8 of this subchapter, may be renamed "Consumer commodity" and reclassed as ORM-D material if packaged in a combination packaging not exceeding 30 kg (66 pounds) with inner packagings not over 250 ml (8 ounces) net capacity for liquids or 250 g (8.8 ounces) net capacity for solids packed in strong outer packagings. Each package must conform to the packaging requirements of subpart B of this part.

(3) Packages of ORM-D material are excepted from the specification packaging requirements of this subchapter and from the labeling requirements of subpart E of Part 172 of this subchapter. Shipments of ORM-D material are eligible for the exceptions provided in § 173.156 and in paragraph (b) of this section and are not subject to the shipping paper requirements of subpart C of Part 172 of this subchapter, unless the material meets the definition of a hazardous substance, a hazardous waste, or a marine pollutant or unless offered for transportation or transported by aircraft.

### 173.154 Exceptions for Class 8 (corrosive materials).

(a) General. Exceptions for hazardous materials shipments in the following paragraphs are permitted only if this section is referenced for the specific hazardous material in the § 172.101 Table of this subchapter.

(b) Limited quantities. Limited quantities of corrosive materials (Class 8) in Packing Groups II and III are excepted from labeling, unless offered or intended for transportation by aircraft, and the specification packaging requirements of this subchapter when packaged in combination packagings according to this paragraph. In addition, shipments of these limited quantities are not subject to subpart F (Placarding) of Part 172 of this subchapter. Each package must conform to the packaging requirements of subpart B of this part and may not exceed 30 kg (66 pounds) gross weight. The following combination packagings are authorized:

(1) For corrosive materials in Packing Group II, in inner packagings not over 1.0 L (0.3 gallon) net capacity each for liquids or not over 1.0 kg (2.2 pounds) net capacity each for solids, packed in strong outer packagings.

(2) For corrosive materials in Packing Group III, in inner packagings not over 4.0 L (1 gallon) net capacity each for liquids or not over 5.0 kg (11 pounds) net capacity each for solids, packed in strong outer packagings.

(c) Consumer commodities. A limited quantity which conforms to the provisions of paragraph (b) of this section and is a "consumer commodity" as defined in § 171.8 of this subchapter may be renamed "Consumer commodity" and reclassed as ORM-D material. In addition to the exceptions provided by paragraph (b) of this section, shipments of ORM-D materials are not subject to the shipping paper requirements of subpart C of Part 172 of this subchapter, unless the material meets the definition of a hazardous substance, a hazardous waste, or a marine pollutant or unless offered or intended for transportation by aircraft, and are eligible for the exceptions provided in § 173.156.

(d) Materials corrosive to aluminum or steel only. Except for a hazardous substance, a hazardous waste, or a marine pollutant, a material classed as a Class 8, Packing Group III, material solely because of its corrosive effect--

(1) On aluminum is not subject to any other requirements of this subchapter when transported by motor vehicle or rail car in a packaging constructed of materials that will not react dangerously with or be degraded by the corrosive material; or

(2) On steel is not subject to any other requirements of this subchapter when transported by motor vehicle or rail car in a bulk packaging constructed of materials that will not react dangerously with or be degraded by the corrosive material.

**173.155 Exceptions for Class 9 (miscellaneous hazardous materials).**

(a) General. Exceptions for hazardous materials shipments in the following paragraphs are permitted only if this section is referenced for the specific hazardous material in the §172.101 Table of this subchapter.

(b) Limited quantities. Limited quantities of miscellaneous hazardous materials (Class 9) are excepted from labeling, unless offered or intended for transportation by aircraft, and the specification packaging requirements of this subchapter when packaged in combination packagings according to this paragraph. In addition, shipments of these limited quantities are not subject to subpart F (Placarding) of Part 172 of this subchapter. Each package must conform to the packaging requirements of subpart B of this part and may not exceed 30 kg (66 pounds) gross weight. The following combination packagings are authorized:

(1) For liquids, inner packagings not over 4.0 L (1 gallon) net capacity each. packed in strong outer packagings.

(2) For solids, inner packagings not over 5.0 kg (11 pounds) net capacity each, packed in strong outer packagings.

(c) Consumer commodities. A limited quantity which conforms to the provisions of paragraph (b) of this section and is a "consumer commodity" as defined in §171.8 of this subchapter, may be renamed "Consumer commodity" and reclassed as ORM-D material. In addition to the exceptions provided by paragraph (b) of this section, shipments of ORM-D materials are not subject to the shipping paper requirements of Subpart C of Part 172 of this subchapter, unless the material meets the definition of a hazardous substance, a hazardous waste, or a marine pollutant or unless offered for transportation or transported by aircraft, and are eligible for the exceptions provided in §173.156.

(d) [Removed]

**173.156 Exceptions for ORM materials.**

(a) Exceptions for hazardous materials shipments in the following paragraphs are permitted only if this section is referenced for the specific hazardous material in the §172.101 Table or in a packaging section in this part.

(b) ORM-D. Packagings for ORM-D materials are specified according to hazard class in §§173.150 through 173.155 and in §173.306. In addition to other exceptions specified for ORM-D materials in this part, strong outer packagings as specified in this part and the marking requirements specified in §172.316 of this subchapter are not required for materials classed as ORM-D when unitized in cages, carts, boxes or similar overpacks and when offered for transportation, or transported, by a private or contract motor carrier or a common carrier vehicle under exclusive use for such service, from a manufacturer to a distribution center, from a manufacturer or a distribution center to a retail outlet, or return. Packagings of ORM-D materials shipped in this manner are not subject to the 30 kg (66 pound) gross weight limitation.

## Subpart E--Non-bulk Packaging for Hazardous Materials Other Than Class 1 and Class 7

### 173.201 Non-bulk packagings for liquid hazardous materials in Packing Group I.

(a) When §172.101 of this subchapter specifies that a liquid hazardous material be packaged under this section, only non-bulk packagings prescribed in this section may be used for its transportation. Each packaging must conform to the general packaging requirements of subpart B of Part 173, to the requirements of Part 178 of this subchapter at the Packing Group I performance level, and to the requirements of the special provisions of Column 7 of the §172.101 Table.

(b) The following combination packagings are authorized:

Outer packagings:

- Steel drum: 1A1 or 1A2
- Aluminum drum: 1B1 or 1B2
- Metal drum other than steel or aluminum: 1N1 or 1N2
- Plywood drum: 1D
- Fiber drum: 1G
- Plastic drum: 1H1 or 1H2
- Steel jerrican: 3A1 or 3A2
- Plastic jerrican: 3H1 or 3H2
- Steel box: 4A1 or 4A2
- Aluminum box: 4B1 or 4B2
- Natural wood box: 4C1 or 4C2
- Plywood box: 4D
- Reconstituted wood box: 4F
- Fiberboard box: 4G
- Expanded plastic box: 4H1
- Solid plastic box: 4H2

Inner packagings:

- Glass or earthenware receptacles
- Plastic receptacles
- Metal receptacles
- Glass ampoules

(c) Except for transportation by passenger aircraft, the following single packagings are authorized:

- Steel drum: 1A1 or 1A2
- Aluminum drum: 1B1 or 1B2
- Metal drum other than steel, or aluminum: 1N1 or 1N2
- Plastic drum: 1H1 or 1H2
- Steel jerrican: 3A1 or 3A2
- Plastic jerrican: 3H1 or 3H2
- Plastic receptacle in steel, aluminum, fiber or plastic drum: 6HA1, 6HB1, 6HG1, 6HH
- Plastic receptacle in steel, aluminum, wooden, plywood or fiberboard box: 6HA2, 6HB2, 6HC, 6HD2 or 6HG2
- Glass, porcelain or stoneware in steel' aluminum or fiber

drum: 6PA1, 6PB1 or 6PG1
Glass, porcelain or stoneware in steel, aluminum, wooden or fiberboard box: 6PA2, 6PB2, 6PC or 6PG2
Glass, porcelain or stoneware in solid or expanded plastic packaging: 6PH1 or 6PH2
Cylinders, specification, as prescribed for any compressed gas, except for Specifications 8 and 3HT

**173.202 Non-bulk packagings for liquid hazardous materials in Packing Group II.**

(a) When §172.101 of this subchapter specifies that a liquid hazardous material be packaged under this section, only non-bulk packagings prescribed in this section may be used for its transportation. Each packaging must conform to the general packaging requirements of subpart B of Part 173, to the requirements of Part 178 of this subchapter at the Packing Group I or II performance level (unless otherwise excepted), and to the particular requirements of the special provisions of Column 7 of the §172.101 Table.

(b) The following combination packagings are authorized:

Outer packagings:
Steel drum: 1A1 or 1A2
Aluminum drum: 1B1 or 1B2
Metal drum other than steel or aluminum: 1N1 or 1N2
Plywood drum: 1D
Fiber drum: 1G
Plastic drum: 1H1 or 1H2
Wooden barrel: 2C2
Steel jerrican: 3A1 or 3A2
Plastic jerrican: 3H1 or 3H2
Steel box: 4A1 or 4A2
Aluminum box: 4B1 or 4B2
Natural wood box: 4C1 or 4C2
Plywood box: 4D
Reconstituted wood box: 4F
Fiberboard box: 4G
Expanded plastic box: 4H1
Solid plastic box: 4H2

Inner packagings:
Glass or earthenware receptacles
Plastic receptacles
Metal receptacles
Glass ampoules

(c) Except for transportation by passenger aircraft, the following single packagings are authorized:

Steel drum: 1A1 or 1A2
Aluminum drum: 1B1 or 1B2
Metal drum other than steel or aluminum: 1N1 or 1N2
Plastic drum: 1H1 or 1H2
Fiber drum: 1G (with liner)
Wooden barrel: 2C1
Steel jerrican: 3A1 or 3A2
Plastic jerrican: 3H1 or 3H2
Plastic receptacle in steel, aluminum, fiber or plastic drum: 6HA1, 6HB1, 6HG1 or 6HH
Plastic receptacle in steel, aluminum, wooden, plywood or fiberboard box: 6HA2, 6HB2, 6HC, 6HD2 or 6HG2
Glass, porcelain or stoneware in steel, aluminum or fiber drum: 6PA1, 6PB1 or 6PG1
Glass, porcelain or stoneware in steel, aluminum, wooden or fiberboard box: 6PA2, 6PB2, 6PC or 6PG2

Glass, porcelain or stoneware in solid or expanded plastic packaging: 6PH1 or 6PH2
Plastic receptacle in plywood drum: 6HD1
Glass, porcelain or stoneware in plywood drum or wickerwork hamper: 6PDI or 6PD2
Cylinders, specification, as prescribed for any compressed gas, except for Specifications 8 and 3HT

**173.203 Non-bulk packagings for liquid hazardous materials in Packing Group III.**

(a) When §172.101 of this subchapter specifies that a liquid hazardous material be packaged under this section, only non-bulk packagings prescribed in this section may be used for its transportation. Each packaging must conform to the general packaging requirements of subpart B of Part 173, to the requirements of Part 178 of this subchapter at the Packing Group I, II or III performance level, and to the requirements of the special provisions of Column 7 of the §172.101 Table.

(b) The following combination packagings are authorized:

Outer packagings:
Steel drum: 1A1 or 1A2
Aluminum drum: 1B1 or 1B2
Metal drum other than steel or aluminum: 1N1 or 1N2
Plywood drum: 1D
Fiber drum: 1G
Plastic drum: 1H1 or 1H2
Wooden barrel: 2C2
Steel jerrican: 3A1 or 3A2
Plastic jerrican: 3H1 or 3H2
Steel box: 4Al or 4A2
Aluminum box: 4B1 or 4B2
Natural wood box: 4C1 or 4C2
Plywood box: 4D
Reconstituted wood box: 4F
Fiberboard box: 4G
Expanded plastic box: 4H1
Solid plastic box: 4H2

Inner packagings:
Glass or earthenware receptacles
Plastic receptacles
Metal receptacles
Glass ampoules

(c) The following single packagings are authorized:

Steel drum: 1A1 or 1A2
Aluminum drum: 1B1 or 1B2
Metal drum other than steel or aluminum: 1N1
Plastic drum: 1H1 or 1H2
Fiber drum: 1G (with liner)
Wooden barrel: 2C1
Steel jerrican: 3A1 or 3A2
Plastic jerrican: 3H1 or 3H2
Plastic receptacle in steel, aluminum, fiber or plastic drum: 6HA1, 6HB1, 6HG1 or 6HH
Plastic receptacle in steel, aluminum, wooden, plywood or fiberboard box: 6HA2, 6HB2, 6HC, 6HD2 or 6HG2
Glass, porcelain or stoneware in steel, aluminum or fiber drum: 6PA1, 6PB1 or 6PG1
Glass, porcelain or stoneware in steel, aluminum, wooden or fiberboard box: 6PA2, 6PB2, 6PC or 6PG2
Glass, porcelain or stoneware in solid or expanded

plastic packaging: 6PH1 or 6PH2[NL] Plastic receptacle in plywood drum: 6HD1

Glass, porcelain or stoneware in plywood drum or wickerwork hamper: 6PD1 or 6PD2

Cylinders, as prescribed for any compressed gas, except for Specifications 8 and 3HT

**173.204 Non-bulk, non-specification packagings for certain hazardous materials.**

When §172.101 of this subchapter specifies that a liquid or solid hazardous material be packaged under this section, any appropriate non-bulk packaging which conforms to the general packaging requirements of subpart B of part 173 may be used for its transportation. Packagings need not conform to the requirements of Part 178 of this subchapter.

**173.205 Specification cylinders for liquid hazardous materials.**

When §172.101 of this subchapter specifies that a hazardous material be packaged under this section, any specification cylinder, except those specified for acetylene, is authorized. Cylinders used for poisonous materials (Division 6.1 or 2.3) must conform to the requirements of §173.40.

**173.211 Non-bulk packagings for solid hazardous materials in Packing Group I.**

(a) When §172.101 of this subchapter specifies that a solid hazardous material be packaged under this section, only non-bulk packagings prescribed in this section may be used for its transportation. Each package must conform to the general packaging requirements of subpart B of Part 173, to the requirements of Part 178 of this subchapter at the Packing Group I performance level, and to the requirements of the special provisions of Column 7 of the §172.101 Table.

(b) The following combination packagings are authorized:

Outer packagings:

Steel drum: 1A1 or 1A2
Aluminum drum: 1B1 or 1B2
Metal drum other than steel or aluminum: 1N1 or 1N2
Plywood drum: 1D
Fiber drum: 1G
Plastic drum: 1H1 or 1H2
Wooden barrel: 2C2
Steel jerrican: 3A1 or 3A2
Plastic jerrican: 3H1 or 3H2
Steel box: 4A1 or 4A2
Aluminum box: 4B1 or 4B2
Natural wood box: 4C1 or 4C2
Plywood box: 4D
Reconstituted wood box: 4F
Fiberboard box: 4G
Solid plastic box: 4H2

Inner packagings:

Glass or earthenware receptacles
Plastic receptacles
Metal receptacles
Glass ampoules

(c) Except for transportation by passenger aircraft, the following single packagings are authorized:

Steel drum: 1A1 or 1A2
Aluminum drum: 1B1 or 1B2
Metal drum other than steel or aluminum: 1N1 or 1N2
Plastic drum: 1H1 or 1H2
Fiber drum: 1G
Steel jerrican: 3A1 or 3A2
Plastic jerrican: 3H1 or 3H2
Steel box with liner: 4A2
Aluminum box with liner: 4B2
Natural wood box, sift proof: 4C2
Plastic receptacle in steel, aluminum, plywood, fiber or plastic drum: 6HA1, 6HB1, 6HD1, 6HG1 or 6HH
Plastic receptacle in steel, aluminum, wooden, plywood or fiberboard box: 6HA2, 6HB2, 6HC, 6HD2 or 6HG2
Glass, porcelain or stoneware in steel, aluminum, plywood or fiber drum: 6PA1, 6PB1, 6PD1 or 6PG1
Glass, porcelain or stoneware in steel, aluminum, wooden or fiberboard box: 6PA2, 6PB2, 6PC or 6PG2
Glass, porcelain or stoneware in expanded or solid plastic packaging: 6PH1 or 6PH2

**173.212 Non-bulk packagings for solid hazardous materials in Packing Group II.**

(a) When §172.101 of this subchapter specifies that a solid hazardous material be packaged under this section, only non-bulk packagings prescribed in this section may be used for its transportation. Each package must conform to the general packaging requirements of subpart B of Part 173, to the requirements of Part 178 of this subchapter at the Packing Group I or II performance level, and to t he requirements of the special provisions of Column 7 of the §172.101 Table.

(b) The following combination packagings are authorized:

Outer packagings:
Steel drum: 1A1 or 1A2
Aluminum drum: 1B1 or 1B2
Metal drum other than steel or aluminum: 1N1 or 1N2
Plywood drum: 1D
Fiber drum: 1G
Plastic drum: 1H1 or 1H2
Wooden barrel: 2C2
Steel jerrican: 3A1 or 3A2
Plastic jerrican: 3H1 or 3H2
Steel box: 4A1 or 4A2
Aluminum box: 4B1 or 4B2
Natural wood box: 4C1 or 4C2
Plywood box: 4D
Reconstituted wood box: 4F
Fiberboard box: 4G
Solid plastic box: 4H2

Inner packagings:
Glass or earthenware receptacles
Plastic receptacles
Metal receptacles
Glass ampoules

(c) Except for transportation by passenger aircraft, the following single packagings are authorized:

Steel drum: 1A1 or 1A2
Aluminum drum: 1B1 or 1B2
Plywood drum: 1D
Plastic drum: 1H1 or 1H2
Fiber drum: 1G
Metal drum other than steel or aluminum: 1N1 or 1N2
Wooden barrel: 2C1 or 2C2
Steel jerrican: 3A1 or 3A2
Plastic jerrican: 3H1 or 3H2
Steel box: 4A1
Steel box with liner: 4A2
Aluminum box: 4B1
Aluminum box with liner: 4B2
Natural wood box: 4C1
Natural wood box, sift proof: 4C2
Plywood box: 4D
Reconstituted wood box: 4F
Fiberboard box: 4G
Expanded plastic box: 4H1
Solid plastic box: 4H2
Bag, woven plastic: 5H1, 5H2 or 5H3
Bag, plastic film: 5H4
Bag, textile: 5L1, 5L2 or 5L3
Bag, paper, multiwall, water resistant: 5M2
Plastic receptacle in steel, aluminum, plywood fiber or plastic drum: 6HA1, 6HB1, 6HD1, 6HG1 or 6HH
Plastic receptacle in steel aluminum, wood, plywood or fiberboard box: 6HA2, 6HB2, 6HC, 6HD2 or 6HG2
Glass, porcelain or stoneware in steel, aluminum, plywood or fiber drum: 6PA1, 6PB1, 6PD1 or 6PG1
Glass, porcelain or stoneware in steel, aluminum, wooden or fiberboard box: 6PA2, 6PB2, 6PC or 6PG2
Glass, porcelain or stoneware in expanded or solid plastic packaging: 6PH1 or 6PH2

**173.213 Non-bulk packagings for solid hazardous materials in Packing Group III.**

(a) When §172.101 of this subchapter specifies that a solid hazardous material be packaged under this section, only non-bulk packagings prescribed in this section may be used for its transportation. Each package must conform to the general packaging requirements of subpart B of Part 173, to the requirements of Part 178 of this subchapter at the Packing Group I, II or III performance level, and to the requirements of the special provisions of Column 7 of the §172.101 Table.

(b) The following combination packagings are authorized:

Outer packagings:
Steel drum: 1A1 or 1A2
Aluminum drum: 1B1 or 1B2
Metal drum other than steel or aluminum: 1N1 or 1N2
Plywood drum: 1D
Fiber drum: 1G
Plastic drum: 1H1 or 1H2
Wooden barrel: 2C2
Steel jerrican: 3A1 or 3A2
Plastic jerrican: 3H1 or 3H2
Steel box: 4A1 or 4A2
Aluminum box: 4B1 or 4B2
Natural wood box: 4C1 or 4C2
Plywood box: 4D
Reconstituted wood box: 4F
Fiberboard box: 4G
Solid plastic box: 4H2

Inner packagings:
- Glass or earthenware receptacles
- Plastic receptacles
- Metal receptacles
- Glass ampoules

(c) The following single packagings are authorized:

- Steel drum: 1A1 or 1A2
- Aluminum drum: 1B1 or 1B2
- Plywood drum: 1D
- Fiber drum: 1G
- Plastic drum: 1H1 or 1H2
- Metal drum other than steel or aluminum: 1N1 or 1N2
- Wooden barrel: 2C1 or 2C2
- Steel jerrican: 3A1 or 3A2
- Plastic jerrican: 3H1 or 3H2
- Steel box with liner: 4A2
- Steel box: 4A1
- Aluminum box with liner: 4B2
- Natural wood box: 4C1
- Natural wood box, sift proof: 4C2
- Plywood box: 4D
- Reconstituted wood box: 4F
- Fiberboard box: 4G
- Expanded plastic box: 4H1
- Solid plastic box: 4H2
- Bag, woven plastic: 5H1, 5H2 or 5H3
- Bag, plastic film: 5H4
- Bag, textile: 5L1, 5L2 or 5L3
- Bag, paper, multiwall, water resistant: 5M2
- Plastic receptacle in steel, aluminum, plywood, fiber or plastic drum: 6HA1, 6HB1, 6HD1, 6HG1 or 6HH
- Plastic receptacle in steel, aluminum, wooden, plywood or fiberboard box: 6HA2, 6HB2, 6HC, 6HD2 or 6HG2
- Glass, porcelain or stoneware in steel, aluminum, plywood or fiber drum: 6PA1, 6PB1, 6PD1 or 6PG1
- Glass, porcelain or stoneware in steel, aluminum, wooden or fiberboard box: 6PA2, 6PB2, 6PC or 6PG2
- Glass, porcelain or stoneware in expanded or solid plastic packaging: 6PH1 or 6PH2

**173.240 Bulk packaging for certain low hazard solid materials.**

When §172.101 of this subchapter specifies that a hazardous material be packaged under this section, only the following bulk packagings are authorized, subject to the requirements of subparts A and B of part 173 of this subchapter and the special provisions specified in Column 7 of the §172.101 Table.

(a) Rail cars: Class DOT 103, 104, 105, 109, 111, 112, 114, or 115 tank car tanks; Class 106 or 110 multi-unit tank car tanks; and metal non-DOT specification, sift-proof tank car tanks and sift-proof closed cars.

(b) Motor vehicles: Specification MC 300, MC 301, MC 302, MC 303, MC 304, MC 305, MC 306, MC 307, MC 310, MC 311, MC 312, MC 330, MC 331, DOT 406, DOT 407, and DOT 412 cargo tank motor vehicles; non-DOT specification, sift-proof cargo tank motor vehicles; and sift -proof closed vehicles.

(c) Portable tanks and closed bulk bins: DOT 51, 52, 53, 56, 57 and 60 portable tanks; IMO type 1, 2 and 5, and IM 101 and IM 102 portable tanks; marine portable tanks conforming to 46 CFR part

64; and sift-proof non-DOT specification portable tanks and closed bulk bins.

### 173.242 Bulk packagings for certain medium hazard liquids and solids, including solids with dual hazards.

When §172.101 of this subchapter specifies that a hazardous material be packaged under this section, only the following bulk packagings are authorized, subject to the requirements of subparts A and B of Part 173 of this subchapter and the special provisions specified in Column 7 of the §172.101 Table.

(a) Rail cars: Class DOT 103, 104, 105, 109, 111, 112, 114, or 115 tank car tanks; Class 106 or 110 multi-unit tank car tanks and A AR Class 206W tank car tanks.

(b) Cargo tanks: Specification MC 300, MC 301, MC 302, MC 303, MC 304, MC 305, MC 306, MC 307, MC 310, MC 311, MC 312, MC 330, MC 3 31, DOT 406, DOT 407, and DOT 412 cargo tank motor vehicles. Cargo tanks used to transport Class 3, Packing Group I or II, or Packing Group III with a flash point of less than 38 °C (100 °F); Class 6, Packing Group I or II; and Class 8, Packing Group I or II materials must conform to the following special requirements:

(1) Pressure relief system: Except as provided by §173.33(d), each cargo tank must be equipped with a pressure relief system meeting the requirements of §178.346-10 or §178.347-10 of this subchapter. However, pressure relief devices on MC 310, MC 311 and MC 312 cargo tanks must meet the requirements for a Specification MC 307 cargo tank. Pressure relief devices on MC 330 and MC 331 cargo tanks must meet the requirement in §178.337-9 of this subchapter.

(2) Bottom outlets: DOT 406, DOT 407 and DOT 412 must be equipped with stop-valves meeting the requirements of §178.345-11 of this subchapter; MC 304, MC 307, MC 310, MC 311, and MC 312 cargo tanks must be equipped with stop-valves capable of being remotely closed within 30 seconds of actuation by manual or mechanic means and (except for Class 8, Packing Group I and II) by a closure activated at a temperature not over 121 °C (250 °F); MC 330 and MC 331 cargo tanks must be equipped with internal self-closing stop-valves meeting the requirements in §178.337-11 of this subchapter.

(c) Portable tanks: DOT 51, 52, 53, 56, 57 and 60 portable tanks; and marine portable tanks conforming to 46 CFR part 64. DOT 57 portable tanks used for the transportation by vessel of Class 3, Packing Group II, materials must conform to the following:

(1) Each tank must have a minimum design pressure of 62 kPa (9 psig) and be equipped in accordance with §178.253-4 of this subchapter, except that frangible devices are not authorized; and

(2) No pressure relief device may open at less than 34.4 kPa (5 psig).

### 173.243 Bulk packaging for certain high hazard liquids and dual hazard liquids which pose a moderate hazard.

When §172.101 of this subchapter specifies that a hazardous material be packaged under this section, only the following bulk packagings are authorized, subject to the requirements of subparts A and B of Part 173 of this subchapter and the special provisions specified in Column 7 of the §172.101 Table.

(a) Rail cars: Class DOT 103, 104, 105, 109, 111, 112, 114, or 115 fushion-welded tank car tanks; and Class 106 or 110 multi-unit tank car tanks.

(b) Cargo tanks. Specification MC 304, MC 307, MC 330, MC 331 cargo tank motor vehicles; and MC 310, MC 311, MC 312, DOT 407, and DOT 412 cargo tank motor vehicles with tank design pressure of at least 172.4 kPa (25 psig). Cargo tanks used to transport Class 3 or Division 6.1 materials, or Class 8, Packing Group I or II materials must conform to the following special requirements:

(1) Pressure relief system: Except as provided by §173.33(d), each cargo tank must be equipped with a pressure relief system meeting the requirements of §178.346-10 or §178.347-10 of this subchapter. However, pressure relief devices on MC 310, MC 311 and MC 312 cargo tanks must meet the requirements for a Specification MC 307 cargo tank. Pressure relief devices on MC 330 and MC 331 cargo tanks must meet the requirement in §178.337-9 of this subchapter.

(2) Bottom outlets: DOT 407 and DOT 412 must be equipped with stop-valves meeting the requirements of §178.345-11 of this subchapter; MC 304, MC 307, MC 310, MC 311, and MC 312 cargo tanks must be equipped with stop-valves capable of being remotely closed within 30 seconds of actuation by manual or mechanic means and (except for Class 8, Packing Group I and II) by a closure activated at a temperature not over 121 °C (250 °F); MC 330 and MC 331 cargo tanks must be equipped with internal self-closing stop-valves meeting the requirements in §178.337-11 of this subchapter.

(c) Portable tanks: DOT 51 and DOT 60 portable tanks; and marine portable tanks conforming to 46 CFR 64 with design pressure of at least 172.4 kPa (25 psig).

### 173.245 Bulk packaging for extremely hazardous materials such as poisonous gases (Division 2.3).

When §172.101 of this subchapter specifies that a hazardous material be packaged under this section, only the following bulk packagings are authorized, subject to the requirements of subparts A and B of Part 173 of this subchapter and the special provisions specified in Column 7 of the §172.101 Table.

(a) Tank car tanks and multi-unit tank car tanks, when approved by the Associate Administrator for Hazardous Materials Safety.

(b) Cargo tank motor vehicles and portable tanks, when approved by the Associate

Administrator for Hazardous Materials Safety.

### 173.247 Bulk packaging for certain elevated temperature materials (Class 9) and certain flammable elevated temperature materials (C lass 3).

When §172.101 of this subchapter specifies that a hazardous material be packaged under this section, only the following bulk packagings are authorized, subject to the requirements of subparts A and B of part 173 of this subchapter and the special provisions in Column 7 of the §172.101 Table. On or after October 1, 1993, authorized packagings must meet all requirements in paragraph (g) of this section, unless otherwise excepted.

(a) *Rail cars*: Class DOT 103, 104, 105, 109, 111, 112, 114, 115 tank car tanks; Class DOT 106, 110 multi-unit tank car tanks; AAR Class 203W, 206W, 211W tank car tanks; and non-DOT specification tank car tanks equivalent in structural design and accident damage resistance to specification packagings.

(b) *Cargo tanks*: Specification MC 300, MC 301, MC 302, MC 303, MC 304, MC 305, MC 306, MC 307, MC 310, MC 311, MC 312, MC 330, MC 331 cargo tank motor vehicles; DOT 406, DOT 407, DOT 412 cargo tank motor vehicles; and non-DOT specification cargo tank motor vehicles equivalent in structural design and accident damage resistance to specification packagings. A non-DOT specification cargo tank motor vehicle constructed of carbon steel which is in elevated temperature material service is excepted from §178.345-7(d)(5) of this subchapter.

(c) *Portable tanks*: Specification 51, 52, 53, 56, 57, 60 portable tanks; IM 101, 102 portable tanks; marine portable tanks conforming to 46 CFR Part 64; and non-specification portable tanks equivalent in structural design and accident damage resistance to specification packagings.

(d) *Crucibles*: Nonspecification crucibles designed and constructed such that the stress in the packaging does not exceed one fourth (0.25) of the ultimate strength of the packaging material at any temperature within the design temperature range. Stress is determined under a load equal to the sum of the static or working pressure in combination with the loads developed from accelerations and decelerations incident to normal transportation. For highway transportation, these forces are assumed to be "1.7g" vertical, "0.75g" longitudinal, and "0.4g" transverse, in reference to the axes of the transport vehicle. Each accelerative or decelerative load may be considered separately.

(e) *Kettles*: A kettle, for the purpose of this section, is a bulk packaging (portable tank or cargo tank) having a capacity not greater than 5678 L (1500 gallons) with an integral heating apparatus used for melting various bituminous products such as asphalt. Kettles used for the transport of asphalt or bitumen are subject to the following requirements:

(1) *Low stability kettles*. Kettles with a ratio of track-width to fully loaded center of gravity (CG)

height less than 2.5 must meet all requirements of paragraph (g) of this section (track-width is the distance measured between the outer edge of the kettle tires; CG height is measured perpendicular from the road surface).

(2) *High stability kettles*. (i) Kettles with a total capacity of less than 2650 L (700 gallons) and a ratio of track-width to fully loaded CG height of 2.5 or more are excepted from all requirements of paragraph (g)(2) of this section and the rollover protection requirements of paragraph (g)(6) of this section, if closures meet the requirements of paragraph (e)(2)(iii) of this section.

(ii) Kettles with a total capacity of 2650 L (700 gallons) or more and a ratio of track-width to fully loaded CG height of 2.5 or more are excepted from the "substantially leak tight" requirements of paragraph (g)(2) of this section and the rollover protection requirements of paragraph (g)(6) of this section if closures meet the requirements of paragraph (e)(2)(iii) of this section.

(iii) Closures must be securely closed during transportation. Closures also must be designed to prevent opening and the expulsion of lading in a rollover accident.

(f) *Other bulk packagings*: Bulk packagings, other than those specified in paragraphs (a) through (e) of this section, which are used for the transport of elevated temperature materials, must conform to all requirements of paragraph (g) of this section on or after October 1, 1993.

(g) *General requirements*. Bulk packagings authorized or used for transport of elevated temperature materials must conform to the following requirements:

(1) *Pressure and vacuum control equipment*. When pressure or vacuum control equipment is required on a packaging authorized in this section, such equipment must be of a self-reclosing design, must prevent package rupture or collapse due to pressure, must prevent significant release of lading due to packaging overturn or splashing or surging during normal transport conditions, and may be external to the packaging.

(i) Pressure control equipment is not required if pressure in the packaging would increase less than 10 percent as a result of heating the lading from the lowest design operating temperature to a temperature likely to be encountered if the packaging were engulfed in a fire. When pressure control equipment is required, it must prevent rupture of the packaging from heating, including fire engulfment.

(ii) Vacuum control equipment is not required if the packaging is designed to withstand an external pressure of 100 Kpa (14.5 psig) or if pressure in the packaging would decrease less than 10 percent as a result of the lading cooling from the highest design operating temperature to the lowest temperature incurred in transport. When vacuum control equipment is required, it must prevent collapse of the packaging from a cooling-induced pressure differential.

(iii) When the regulations require a reclosing pressure relief device, the lading must not render the devices inoperable (i.e. from clogging, freezing, or fouling). If the lading affects the proper operation of the device, the packaging must have:

(A) A safety relief device incorporating a frangible disc or a permanent opening, each having a maximum effective area of 22 $cm^2$ (3.4 $in.^2$), for transportation by highway;

(B) For transportation of asphalt by highway, a safety relief device incorporating a frangible disc or a permanent opening, each having a maximum effective area of 46 $cm^2$ (7.1 $in.^2$); or

(C) For transportation by rail, a safety relief device incorporating a frangible disc, meeting the requirements of § 179.20 0-18 of this subchapter.

(iv) Reclosing pressure relief devices, frangible discs or permanent openings must not allow the release of lading during normal transportation conditions (i.e., due to splashing or surging).

(2) *Closures*. All openings, except permanent vent openings authorized in paragraph (g)(1)(iii) of this section, must be securely closed during transportation. Packagings must be substantially leak-tight so as not to allow any more than dripping or trickling of a non-continuous flow when overturned. Closures must be designed and constructed to withstand, without exceeding the yield strength of the packaging, twice the static loading produced by the lading in any packaging orientation and at all operating temperatures.

(3) *Strength*. Each packaging must be designed and constructed to withstand, without exceeding the yield strength of the packaging, twice the static loading produced by the lading in any orientation and at all operating temperatures.

(4) *Compatibility*. The packaging and lading must be compatible over the entire operating temperature range.

(5) *Markings*. In addition to any other markings required by this subchapter, each packaging must be durably marked in a place readily accessible for inspection in characters at least 4.8 mm (3/16 inch) with the manufacturer's name, date of manufacture, design temperature range, and maximum product weight (or "load limit" for tank cars) or volumetric capacity.

(6) *Accident damage protection*. For transportation by highway, external loading and unloading valves and closures must be protected from impact damage resulting from collision or overturn. Spraying equipment and the road oil application portion of a packaging are excepted from this requirement.

(7) *New construction*. Specification packagings that are being manufactured for the transport of elevated temperature materials must be authorized for current construction.

(h) *Exceptions*.

(1) *General*. Packagings manufactured for elevated temperature materials service prior to October 1, 1993, which are not in full compliance with the requirements in paragraph (g) of this section, may continue in service if they meet the applicable requirements of subparts A and B of this part and meet the closure requirements in paragraph (g)(2) of this section by March 30, 1995.

(2) *Kettles*. Kettles in service prior to October 1, 1993, which are

used to transport asphalt or bitumen, are excepted from specific provisions of this section as follows:

(i) Kettles with a total capacity of less than 2650 L (700 gallons), which are not in full compliance with the requirements of paragraph (g) of this section, may continue in elevated temperature material service if they meet the applicable requirements of subparts A and B of this part and if, after March 30, 1995, closures are secured during transport to resist opening in an overturn.

(ii) Kettles with a total capacity of 2650 L (700 gallons) or more, which are not in full compliance with the requirements of paragraph (g) of this section, may continue in elevated temperature material service if they meet the applicable requirements of subparts A and B of this part and if, after March 30, 1995, closures are secured during transport to resist opening in an overturn and no opening exceeds 46 $cm^2$ (7.1 $in^2$).

(3) Molten metals and molten glass. This section does not apply to packagings used for transportation of molten metals and molten glass by rail when movement is restricted to operating speeds less than 15 miles per hour. (See §172.203(g)(3) of this subchapter for shipping paper requirements.)

(4) Solid elevated temperature materials. A material which meets the definition of a solid elevated temperature material is excepted from all requirements of this subchapter except §172.325 of this subchapter.

# Appendix C
# DOT Regulatory Glossary

## DOT REGULATORY GLOSSARY

In the DOT regulations for shippers, transporters, and manufacturers, the terms given in italics below have the following meanings:

*Approved*: approval issued or recognized by the Department unless otherwise specifically indicated in the DOT regulations for shippers, transporters, and manufacturers.

*Atmospheric gases*: air, nitrogen, oxygen, argon, krypton, neon and xenon.

*Authorized Inspector*: an Inspector who is currently commissioned by the National Board of Boiler and Pressure Vessel Inspectors and employed as an Inspector by an Authorized Inspection Agency.

*Authorized Inspection Agency*: (1) A jurisdiction which has adopted and administers one or more sections of the ASME Boiler and Pressure Vessel Code as a legal requirement and has a representative serving as a member of the ASME Conference Committee; or (2) an insurance company which has been licensed or registered by the appropriate authority of a State of the United States or a Province of Canada to underwrite boiler and pressure vessel insurance in such State or Province.

*Bag*: a flexible packaging made of paper, plastic film, textiles, woven material or other similar materials.

*Bar*: a metric-related unit of pressure, close to one atmosphere. 1 bar = 100 kPa (14.5 psi).

*Barge*: a non-selfpropelled vessel.

*Bottle*: an inner packaging having a neck of relatively smaller cross section than the body and an opening capable of holding a closure for retention of the contents.

*Bottom shell*: that portion of a tank car surface, excluding the head ends of the tank car tank, that lies within two feet, measured circumferentially, of the bottom longitudinal center line of the tank car tank.

*Box*: a packaging with complete rectangular or polygonal faces, made of metal, wood, plywood, reconstituted wood, fiberboard, plastic, or other suitable material.

*Break-bulk*: packages of hazardous materials that are handled individually, palletized, or unitized for purposes of transportation as opposed to bulk and containerized freight.

*Btu*: British thermal unit.

*Bulk packaging*: a packaging, other than a vessel or a barge, including a transport vehicle or freight container, in which hazardous materials are loaded with no intermediate form of containment and which has—

(1) A maximum capacity greater than 450 L (119 gallons) as a receptacle for a liquid;

(2) A maximum net mass greater than 400 kg (882 pounds) and a maximum capacity greater than 450 L (119 gallons) as a receptacle for a solid; or

(3) A water capacity greater than 454 kg (1,000 pounds) as a receptacle for a gas as defined in 49 CFR 173.115.

*Bureau of Explosives*: the Bureau of Explosives (B of E) of the Association of American Railroads.

*C*: Celsius or Centigrade.

*Captain of the Port* (COTP): the officer of the Coast Guard, under the command of a District Commander, so designated by the Commandant for the purpose of giving immediate direction to Coast Guard law enforcement activities within an assigned area. As used in the DOT regulations for shippers, transporters, and manufacturers, the term "Captain of the Port" includes an authorized representative of the Captain of the Port.

*Carfloat*: a vessel that operates on a short run on an highway over water, and does not operate in ocean, coastwise, or ferry service.

*Cargo aircraft only*: an aircraft that is used to transport cargo and is not engaged in carrying passengers. For purposes of the DOT regulations for shippers, transporters, and manufacturers, the terms "cargo aircraft only," "cargo-only aircraft," and "cargo aircraft" have the same meaning.

*Cargo tank*: a bulk packaging which:

(1) Is a tank intended primarily for the carriage of liquids or gases and includes appurtenances, reinforcements, fittings and closures (See "tank");

(2) Is permanently attached to or forms a part of a motor vehicle, or is not permanently attached to a motor vehicle but which, by reason of its size, construction or attachment to a motor vehicle is loaded or unloaded without being removed from the motor vehicle; and

(3) Is not fabricated under a specification for cylinders, portable tanks, tank cars or multi-unit tank car tanks.

*Cargo tank motor vehicle*: a motor vehicle with one or more cargo tanks permanently attached to or forming an integral part of the motor vehicle.

*Cargo vessel*:

(1) Any vessel other than a passenger vessel; and

(2) Any ferry being operated under authority of a change of character certificate issued by a Coast Guard Officer-in-Charge, Marine Inspection.

*Carrier*: a person engaged in the transportation of passengers or property by:

(1) Land or water, as a common, contract, or private carrier, or

(2) Civil aircraft.

*CC*: closed-cup.

*Character of vessel*: the type of service in which the vessel is engaged at the time of carriage of a hazardous material.

*Class*: hazard class. See "hazard class."

*Class 1* (explosive): any substance or article, including a device, which is designed to function by explosion (i.e., an extremely rapid release of gas and heat) or which, by chemical reaction within itself, is able to function in a similar manner even if not designed to function by explosion, unless the substance or article is otherwise classed under the provisions of DOT regulations.

*Class 2*: Div. 2.1 (flammable gas), Div. 2.2 (non-flammable, non-poisonous compressed gas--including compressed gas, liquefied gas, pressurized cryogenic gas and compressed gas in solution), or Div. 2.3 (gas poisonous by inhalation). (See 49 CFR 173.115 for detailed definitions of each division).

*Class 3* (flammable liquid): a liquid having a flash point of not more than 60.5 °C (141 °F), or any material in a liquid phase with a flash point at or above 37.8 °C (100 °F) that is intentionally heated and offered for transportation or transported at or above its flash point in a bulk packaging,with the exceptions detailed in 49 CFR 173.120.

*Class 4*: Div. 4.1 (flammable solid), Div. 4.2 (spontaneously combustible material), or Div. 4.3 (dangerous when wet material). See 49 CFR 173.124.

*Class 5*: Div. 5.1 (oxidizing material) or Div. 5.2 (organic peroxide). See 49 CFR 173.124.

*Class 6*: Div. 6.1 (poisonous material), or Div. 6.2 (infectious substance). See 49 CFR 173.132.

*Class 7* (radioactive material): any material having a specific activity greater than 0.002 microcuries per gram (μCi/g) (see definition of "specific activity" and other technical definitions at 49 CFR 173.403).

*Class 8* (corrosive material): a liquid or solid that causes visible destruction or irreversible alterations in human skin tissue at the site of contact, or a liquid that has a severe corrosion rate on steel or aluminum, in accordance with the criteria given at 49 CFR 173.136(a).

*Class 9* (miscellaneous hazardous material): a material which has an anesthetic, noxious or other similar property which could cause extreme annoyance or discomfort to a flight crew member so as to prevent the correct performance of assigned duties; and meets the definition in 49 CFR 171.8 for a hazardous substance or a hazardous waste (see definitions below).

*Closure*: a device which closes an opening in a receptacle.

*COFC*: container-on-flat-car.

*Combination packaging*: a combination of packaging, for transport purposes, consisting of one or more inner packagings secured in a non-bulk outer packaging. It does not include a composite packaging.

*Combustible liquid*: any liquid that does not meet the definition of any other hazard class specified in the DOT regulations, except Class 9, and has a flash point above 60.5 °C (141 °F) and below 93 °C (200 °F).

*Compatibility group letter*: a designated alphabetical letter used to categorize different types of explosive substances and articles for purposes of stowage and segregation. See 49 CFR 173.52.

*Competent Authority*: a national agency responsible under its national law for the control or regulation of a particular aspect of the transportation of hazardous materials (dangerous goods). The term "Appropriate Authority", as used in the ICAO Technical Instructions, has the same meaning as "Competent Authority". For purposes of the DOT regulations for shippers, transporters, and manufacturers, the Associate Administrator for Hazardous Materials Safety is the Competent Authority for the United States.

*Composite packaging*: a packaging consisting of an outer packaging and an inner receptacle, so constructed that the inner receptacle and outer packaging form an integral packaging. Once assembled it remains thereafter an integrated single unit; it is filled, stored, shipped and emptied as such.

*Compressed gas*: non-flammable, non-poisonous compressed gas—including compressed gas, liquefied gas, pressurized cryogenic gas and compressed gas in solution—that exerts in the packaging an absolute pressure of 280 kPa (41 psia) at 20 °C (68 °F), and does not meet the definition of a gas poisonous by inhalation or non-liquefied compressed gas. See 49 CFR 173.115 for detailed definitions.

*Consumer commodity*: a material that is packaged and distributed in a form intended or suitable for sale through retail sales agencies or instrumentalities for consumption by individuals for purposes of personal care or household use. This term also includes drugs and medicines.

*Containership*: a cargo vessel designed and constructed to transport, within specifically designed cells, portable tanks and freight containers which are lifted on and off with their contents intact.

*Corrosive material*: a liquid or solid that causes visible destruction or irreversible alterations in human skin tissue at the site of contact, or a liquid that has a severe corrosion rate on steel or aluminum, in accordance with the criteria given at 49 CFR 173.136(a).

*Crate*: an outer packaging with incomplete surfaces.

*Crewmember*: a person assigned to perform duty in an aircraft during flight time.

*Cryogenic liquid*: a refrigerated liquefied gas having a boiling point colder than −90 °C (−130 °F) at 101.3 kPa (14.7 psi) absolute. A material meeting this definition is subject to requirements of the DOT regulations for shippers, transporters, and manufacturers without regard to whether it meets the definition of a non-flammable, non-poisonous compressed gas at 49 CFR173.115(b).

*Cylinder*: a pressure vessel designed for pressures higher than 40 psia and having a circular cross section. It does not include a portable tank, multi-unit tank car tank, cargo tank, or tank car.

*Design Certifying Engineer*: a person registered with the Department in accordance with 49 CFR Part 107, Subpart F who has the knowledge and ability to perform stress analysis of pressure vessels and otherwise determine if a cargo tank design and construction meets the applicable DOT specification and has an engineering degree and one year of work experience in structural or mechanical design (or meets criteria at 49 CFR 107.502(f)). Persons registered as professional engineers by appropriate authority of a State of the United States, or a Province of Canada, who have the requisite experience may be registered under this program.

*Dangerous when wet material*: a material that, by contact with water, is liable to become spontaneously flammable or to give off flammable or toxic gas at a rate greater than 1 liter per kilogram of the material, per hour, when tested in accordance with paragraph 4 of Appendix E to 49 CFR Part 173.

*District Commander*: the District Commander of the Coast Guard, or his authorized representative, who has jurisdiction in the particular geographical area.

*DOD*: the U.S. Department of Defense.

*Domestic transportation*: transportation between places within the United States other than through a foreign country.

*Designated facility*: a hazardous waste treatment, storage, or disposal facility that has been designated on the manifest by the generator.

*Division*: a subdivision of a hazard class.

*Drum*: a flat-ended or convex-ended cylindrical packaging made of metal, fiberboard, plastic, plywood, or other suitable materials. This definition also includes packagings of other shapes made of metal or plastic (e.g., round taper-necked packagings or pail-shaped packagings) but does not include cylinders, jerricans, wooden barrels or bulk packagings.

*Elevated temperature material*: a material which, when offered for transportation or transported in a bulk packaging:

(1) Is in a liquid phase and at a temperature at or above 100 °C (212 °F);

(2) Is in a liquid phase with a flash point at or above 37.8 °C (100 °F) that is intentionally heated and offered for transportation or transported at or above its flash point; or

(3) Is in a solid phase and at a temperature at or above 240 °C (464 °F).

*Engine*: a locomotive propelled by any form of energy and used by a railroad.

*EPA*: U.S. Environmental Protection Agency.

*EX number*: a number, preceded by the prefix "Ex–", which is assigned by the Associate Administrator for Hazardous Materials Safety to identify an explosive which has been approved. See 49 CFR 173.56.

*°F*: degree Fahrenheit.

*Ferry vessel*: a vessel which is limited in its use to the carriage of deck passengers or vehicles or both, operates on a short run on a frequent schedule between two points over the most direct water route, other than in ocean or coastwise service, and is offered as a public service of a type normally attributed to a bridge or tunnel.

*Filling density* has the following meanings:

(1) For compressed gases in cylinders, the percent ratio of the weight of gas in a container to the weight of water that the container will hold at 60 °F (1 lb. of water = 27.737 cubic inches at 60 °F).

(2) For compressed gases in tank cars, the percent ratio of the weight of gas in the tank to the weight of water that the tank will hold. For determining the water capacity of the tank in pounds, the weight of one gallon of water at 15.55 °C (60 °F) in air is 8.32828 pounds.

(3) For compressed gases in cargo tanks and portable tanks, the percent ratio of the weight of gas in the tank to the weight of water that the tank will hold. For determining the water capacity of the tank in pounds, the weight of a gallon (231 cubic inches) of water at 60 °F. in air shall be 8.32828 pounds.

(4) For cryogenic liquids in cylinders, except hydrogen, the percent ratio of the weight of lading in the packaging to the weight of water that the packaging will hold at 60 °F (1 lb. of water = 27.737 cubic inches at 60 °F).

(5) For hydrogen, cryogenic liquid in cylinders, the percent ratio of the weight of lading in a packaging to the weight of water that the packaging will hold at minus 423 °F. The volume of the packaging at minus 423 °F is determined in cubic inches. The volume is converted to pounds of water (1 lb. of water = 27.737 cubic inches).

(6) For cryogenic liquids in cargo tanks, the percent ratio of the weight of lading in the tank to the weight of water that the tank will hold at the design service temperature (one pound of water = 27.737 cubic inches at 60 °F, or one gallon of water = 231 cubic inches at 60 °F and weighs 8.32828 pounds).

(7) For cryogenic liquids in tank cars, the percent ratio of the weight of lading in the tank to the weight of water that the tank will hold at the design service temperature (one pound of water = 27.737 cubic inches at 60 °F, or one gallon of water = 231 cubic inches at 60 °F and weighs 8.32828 pounds).

*Flammable gas*: any material which is a gas at 20 °C (68 °F) or less and 101.3 kPa (14.7 psi) of pressure, which is ignitable at 101.3 kPa (14.7 psi) when in a mixture of 13 percent or less by volume with air; or has a flammable range at 101.3 kPa (14.7 psi) with air of at least 12 percent regardless of the lower limit; subject to conditions at 49 CFR 173.115(a).

*Flammable liquid*: a liquid having a flash point of not more than 60.5 °C (141 °F), or any material in a liquid phase with a flash point at or above 37.8 °C (100 °F) that is intentionally heated and offered for transportation or transported at or above its flash point in a bulk packaging, with the exceptions listed at 49 CFR 173.120.

*Flammable solid*: wetted explosives, self-reactive materials, and readily combustible materials, as defined at 49 CFR 173.124.

*Flash point*: the minimum temperature at which a substance gives off flammable vapors which, in contact with sparks or flame, will ignite. (For criteria, see 49 CFR 173.121.)

*Freight container*: a reusable container having a volume of 64 cubic feet or more, designed and constructed to permit being lifted with its contents intact and intended primarily for containment of packages (in unit form) during transportation.

*Fuel tank*: a tank other than a cargo tank, used to transport flammable or combustible liquid, or compressed gas for the purpose of supplying fuel for propulsion of the transport vehicle to which it is attached, or for the operation of other equipment on the transport vehicle.

*Gross weight* or *Gross mass*: the weight of a packaging plus the weight of its contents.

*Hazard class*: the category of hazard assigned to a hazardous material under the definitional criteria of 49 CFR Part 173 and the provisions of the 49 CFR 172.101 Table. A material may meet the defining criteria for more than one hazard class but is assigned to only one hazard class.

*Hazard zone*: one of four levels of hazard (Hazard Zones A through D) assigned to gases, as specified in 49 CFR 173.116(a), and one of two levels of hazards (Hazard Zones A and B) assigned to liquids that are poisonous by inhalation, as specified in 49 CFR 173.133(a). A hazard zone is based on the $LC_{50}$ value for acute inhalation toxicity of gases and vapors, as specified in 49 CFR 173.133(a).

*Hazardous material*: a substance or material, which has been determined by the Secretary of Transportation to be capable of posing an unreasonable risk to health, safety, and property when transported in commerce, and which has been so designated. The term includes hazardous substances, hazardous wastes, marine pollutants, and elevated temperature materials as defined in this section, materials designated as hazardous under the provisions of 49 CFR §§172.101 and 172.102, and materials that meet the defining criteria for hazard classes and divisions in 49 CFR Part 173.

*Hazardous substance*: for the purposes of the DOT regulations for shippers, transporters, and manufacturers, a material, including its mixtures and solutions, that—

(1) Is listed in Appendix A to 49 CFR 172.101;

(2) Is in a quantity, in one package, which equals or exceeds the reportable quantity (RQ) listed in Appendix A to §172.101; and

(3) When in a mixture or solution—

(i) For radionuclides, conforms to paragraph 6 of Appendix A to §172.101.

(ii) For other than radionuclides, is in a concentration by weight which equals or exceeds the concentration corresponding to the RQ of the material, as shown in the following table:

| RQ pounds (kilograms) | Concentration by weight | |
|---|---|---|
| | Percent | PPM |
| 5,000 (2270) | 10 | 100,000 |
| 1,000 (454) | 2 | 20,000 |
| 100 (45.4) | 0.2 | 2,000 |
| 10 (4.54) | 0.02 | 200 |
| 1 (0.454) | 0.002 | 20 |

This definition does not apply to petroleum products that are lubricants or fuels (see 40 CFR 300.6)

*Hazardous waste*, for the purposes of regulations of RSPA: any material that is subject to the Hazardous Waste Manifest requirements of the U.S. Environmental Protection Agency specified in 40 CFR Part 262.

*Hazmat employee*: a person who is employed by a hazmat employer and who in the course of employment directly affects hazardous materials transportation safety. This term includes an owner-operator of a motor vehicle which transports hazardous materials in commerce. This term includes an individual, including a self-employed individual, employed by a hazmat employer who, during the course of employment:

(1) Loads, unloads, or handles hazardous materials;

(2) Tests, reconditions, repairs, modifies, marks, or otherwise represents containers, drums, or packagings as qualified for use in the transportation of hazardous materials;

(3) Prepares hazardous materials for transportation;

(4) Is responsible for safety of transporting hazardous materials; or

(5) Operates a vehicle used to transport hazardous materials.

*Hazmat employer*: a person who uses one or more of its employees in connection with: transporting hazardous materials in commerce; causing hazardous materials to be transported or shipped in commerce; or representing, marking, certifying, selling, offering, reconditioning, testing, repairing, or modifying containers, drums, or packagings as qualified for use in the transportation of hazardous materials. This term includes an owner-operator of a motor vehicle which transports hazardous materials in commerce. This term also includes any department, agency, or instrumentality of the United States, a State, a political subdivision of a State, or an Indian tribe engaged in an activity described in the first sentence of this definition.

*Hermetically sealed*: closed by fusion, gasketing, crimping, or equivalent: so that no gas or vapor can enter or escape.

*IAEA*: International Atomic Energy Agency.

*IATA*: International Air Transport Association.

*ICAO*: International Civil Aviation Organization.

*IM portable tank*: See *Intermodal portable tank.*

*IMDG Code*: International Maritime Dangerous Goods Code.

*IMO:* International Maritime Organization.

*Infectious substance* (etiologic agent): a viable microorganism, or its toxin, which causes or may cause disease in humans or animals, and includes those agents listed in 42 CFR 72.3 of the regulations of the Department of Health and Human Services; or any other agent that causes or may cause severe, disabling or fatal disease. The terms "infectious substances" and "etiologic agent" are synonymous. See 49 CFR 173.134.

*Inner packaging*: a packaging for which an outer packaging is required for transport. It does not include the inner receptacle of a composite packaging.

*Inner receptacle*: a receptacle which requires an outer packaging in order to perform its containment function. The inner receptacle may be an inner packaging of a combination packaging or the inner receptacle of a composite packaging.

*Intermodal container*: a freight container designed and constructed to permit it to be used interchangeably in two or more modes of transport.

*Intermodal portable tank* or *IM portable tank*: a specific class of portable tanks designed primarily for international intermodal use.

*International transportation*: transportation—

(1) Between any place in the United States and any place in a foreign country;

(2) Between places in the United States through a foreign country; or

(3) Between places in one or more foreign countries through the United States.

*Irritating material*: a material with properties similar to tear gas, which causes extreme irritation, especially in confined spaces.

*Jerrican*: a metal or plastic packaging of rectangular or polygonal cross-section.

*Limited quantity*, when specified as such in a section applicable to a particular material,: the maximum amount of a hazardous material for which there is a specific labeling or packaging exception.

*Liquid*: a material that has a vertical flow of over 2 inches (50 mm) within a three minute period, or a material having one gram (1 g) or more liquid separation, when determined in accordance with the procedures specified in ASTM D 4359-84, "Standard Test Method for Determining whether a Material is a Liquid or Solid," 1984 edition.

*Liquid phase*: a material that meets the definition of "liquid" when evaluated at the higher of the temperature at which it is offered for transportation or at which it is transported, not at the 37.8 °C (100 °F) temperature specified in ASTM D 4359-84.

*Magazine vessel*: a vessel used for the receiving, storing, or dispensing of explosives.

*Magnetic material*: for carriage by aircraft, any package which has a magnetic field of more than 0.00525 gauss measured at 4.5 m (15 feet) from any surface of the package.

*Marking*: the descriptive name, identification number, instructions, cautions, weight, specification, or UN marks or combinations thereof, required by the DOT regulations for shippers, transporters, and manufacturers, on outer packagings of hazardous materials.

*Marine pollutant*: a hazardous material that is listed in appendix B to 49 CFR 172.101 and, when in a solution or mixture of one or more marine pollutants, is packaged in a concentration equal to or exceeding one of the following:

(1) Ten percent by weight of the solution or mixture for materials listed in Appendix B

(2) One percent by weight of the solution or mixture for materials that are identified as severe marine pollutants ("PP" in the S.M.P. column) in Appendix B.

*Material poisonous by inhalation*:

(1) A gas meeting the defining criteria in 49 CFR 173.115(c) and assigned to Hazard Zone A, B, C, or D in accordance with 49 CFR 173.116(a);

(2) A liquid (other than as a mist) meeting the defining criteria in 49 CFR 173.132(a)(1)(iii) and assigned to Hazard Zone A or B in accordance with 49 CFR 173.133(a); or

(3) Any material identified as an inhalation hazard by a special provision in Column 7 of the 49 CFR 172.101 Table.

*Maximum capacity*: the maximum inner volume of receptacles or packagings.

*Maximum net mass*: the allowable maximum net mass of contents in a single packaging, or as used in 49 CFR Part 178, Subpart M, the maximum combined mass of inner packaging, and the contents thereof.

*Maximum Allowable Working Pressure* or *MAWP*: for DOT specification cargo tanks used to transport liquid hazardous materials, see specifications at 49 CFR 178.345-1(k).

*Mixture*: a material composed of more than one chemical compound or element.

*Mode*: any of the following transportation methods; rail, highway, air, or water.

*Motor vehicle* includes a vehicle, machine, tractor, trailer, or semitrailer, or any combination thereof, propelled or drawn by mechanical power and used upon the highways in the transportation of passengers or property. It does not include a vehicle, locomotive, or car operated exclusively on a rail or rails, or a trolley bus operated by electric power derived from a fixed overhead wire, furnishing local passenger transportation similar to street-railway service.

*Name of contents*: the proper shipping name as specified in 49 CFR 172.101.

*Navigable waters*: for the purposes of the DOT regulations for shippers, transporters, and manufacturers, waters of the United States, including the territorial seas.

*Non-bulk packaging*: a packaging that has—

(1) A maximum capacity of 450 L (119 gallons) or less as a receptacle for a liquid;

(2) A maximum net mass of 400 kg (882 pounds) or less and a maximum capacity of 450 L (119 gallons) or less as a receptacle for a solid; or

(3) A water capacity of 454 kg (1,000 pounds) or less as a receptacle for a gas as defined in 49 CFR 173.115.

*N.O.S.*: not otherwise specified.

*n.o.s. description:* a shipping description from the 49 CFR 172.101 Table which includes the abbreviation "n.o.s.", and as contained in the lists in 49 CFR 172.203(k)(3), regarding additional description requirements.

*NPT*: an American Standard taper pipe thread in compliance with the requirements of Federal Standard H28, Part II, Section VII (see 49 CFR 171.7(d)(12)).

*NRC* (non-reusable container): a packaging (container) whose reuse is restricted in accordance with the provisions of 49 CFR 173.28.

*Occupied caboose*: a rail car being used to transport non-passenger personnel.

*Officer in Charge, Marine Inspection*: a person from the civilian or military branch of the Coast Guard designated as such by the Commandant and who under the supervision and direction of the Coast Guard District Commander is in charge of a designated inspection zone for the performance of duties with respect to the enforcement and administration of Title 52, Revised Statutes, acts amendatory thereof or supplemental thereto, rules and regulations thereunder, and the inspection required thereby.

*Offshore supply vessel*: a cargo vessel of less than 500 gross tons that regularly transports goods, supplies or equipment in support of exploration or production of offshore mineral or energy resources.

*Oil*: oil of any kind or in any form, including, but not limited to, petroleum, vegetable oil, animal oil, fuel oil, sludge, oil refuse, and oil mixed with wastes other than dredged spoil.

*Operator*: a person who controls the use of an aircraft, vessel, or vehicle.

*Organic peroxide*: any organic compound containing oxygen (O) in the bivalent -O-O- structure and which may be considered a derivative of hydrogen peroxide, where one or more of the hydrogen atoms have been replaced by organic radicals; unless any of the exceptions listed at 49 CFR 173.128(a) apply.

*ORM*: other regulated material. See 49 CFR 173.144.

*Outage* or *ullage*: the amount by which a packaging falls short of being liquid full, usually expressed in percent by volume.

*Outer packaging*: the outermost enclosure of a composite or combination packaging together with any absorbent materials, cushioning and any other components necessary to contain and protect inner receptacles or inner packagings.

*Overpack* except when referenced to a packaging specified in 49 CFR Part 178: an enclosure that is used by a single consignor to provide protection or convenience in handling of a package or to consolidate two or more packages. "Overpack" does not include a freight container.

*Oxidizer*: a material that may, generally by yielding oxygen, cause or enhance the combustion of other materials.

*Package* or *Outside Package*: a packaging plus its contents. For radioactive materials, the packaging together with its radioactive contents as presented for transport.

*Packaging*: a receptacle and any other components or materials necessary for the receptacle to perform its containment function in conformance with the minimum packing requirements of the DOT regulations for shippers, transporters, and manufacturers. For radioactive materials packaging, the assembly of components necessary to ensure compliance with the packaging requirements of 49 CFR Part 173, Subpart I. It may consist of one or more receptacles, absorbent materials, spacing structures, thermal insulation, radiation shielding, and devices for cooling or absorbing mechanical shocks. The conveyance, tie-down system, and auxiliary equipment may sometimes be designated as part of the packaging.

*Packing group*: a grouping according to the degree of danger presented by hazardous materials. Packing Group I indicates great danger; Packing Group II, medium danger; Packing Group III, minor danger. In the 49 CFR 172.101 Table, Column 5 specifies one or more packing groups assigned to a material corresponding to the proper shipping name and hazard class for that material. Classes 2 and 7 materials and ORM-D materials do not have packing groups. If more than one packing group is indicated for an entry, the packing group for the hazardous material is determined using the criteria for assignment of packing groups specified in subpart D of Part 173. See 49 CFR 172.101(f).

*Passenger* (With respect to vessels and for the purposes of 49 CFR Part 176 only): a person being carried on a vessel other than:

(1) The owner or his representative;

(2) The operator;

(3) A bona fide member of the crew engaged in the business of the vessel who has contributed no consideration for his carriage and who is paid for his services; or

(4) A guest who has not contributed any consideration directly or indirectly for his carriage.

*Passenger-carrying aircraft*: an aircraft that carries any person other than a crewmember, company employee, an authorized representative of the United States, or a person accompanying the shipment.

*Passenger vessel*:

(1) A vessel subject to any of the requirements of the International Convention for the Safety of Life at Sea, 1974, which carries more than 12 passengers;

(2) A cargo vessel documented under the laws of the United States and not subject to the Convention, which carries more than 16 passengers;

(3) A cargo vessel of any foreign nation that extends reciprocal privileges and is not subject to that Convention and which carries more than 16 passengers; and

(4) A vessel engaged in a ferry operation and which carries passengers.

*Person*: an individual, firm, copartnership, corporation, company, association, joint-stock association, including any trustee, receiver, assignee, or similar representative thereof, or government, Indian tribe, or agency or instrumentality of any government or Indian tribe when it offers hazardous materials for transportation in commerce or transports hazardous materials in furtherance of a commercial enterprise, but such term does not include:

(1) The United States Postal Service; or

(2) For the purposes of sections 110 and 111 of the Hazardous Materials Transportation Act (49 App. U.S.C. 1809-1810), any agency or instrumentality of the Federal Government.

*Placarded car*: a rail car which is placarded in accordance with the requirements of 49 CFR Part 172 except those cars displaying only the FUMIGATION placards as required by 49 CFR 172.510.

*Poisonous materials*: material, other than gases, which are known to be so toxic to humans as to afford a hazard to health during transportation, or which, in the absence of adequate data on human toxicity, meet toxicity testing criteria detailed at 49 CFR 173.132(a).

*Portable tank*: a bulk packaging (except a cylinder having a water capacity of 1000 pounds or less) designed primarily to be loaded onto, or on, or temporarily attached to a transport vehicle or ship and equipped with skids, mountings, or accessories to facilitate handling of the tank by mechanical means. It does not include a cargo tank, tank car, multi-unit tank car tank, or trailer carrying 3AX, 3AAX, or 3T cylinders.

*Preferred route* or *Preferred highway* is a highway for shipment of "highway route controlled quantities" of radioactive materials so designated by a State routing agency, and any Interstate System highway for which an alternative highway has not been designated by such State agency as provided by 49 CFR 177.825(b).

*Primary hazard*: the hazard class of a material as assigned in the 49 CFR 172.101 Table.

*Private track* or *Private siding*: track located outside of a carrier's right-of-way, yard, or terminals where the carrier does not own the rails, ties, roadbed, or right-of-way and includes track or portion of track which is devoted to the purpose of its user either by lease or written agreement, in which case the lease or written agreement is considered equivalent to ownership.

*Proper shipping name*: the name of the hazardous material shown in Roman print (not italics) in 49 CFR 172.101.

*P.s.i.* or *psi*: pounds per square inch.

*P.s.i.a.* or *psia*: pounds per square inch absolute.

*P.s.i.g.* or *psig*: pounds per square inch gauge.

*Public vessel*: a vessel owned by and being used in the public service of the United States. It does not include a vessel owned by the United States and engaged in a trade or commercial service or a vessel under contract or charter to the United States.

*Pyrophoric liquid*: a liquid or solid that, even in small quantities and without an external ignition source, can ignite within five (5) minutes after coming in contact with air when tested according to paragraph 3.a.(1) or 3.a.(2), as appropriate, of appendix E to 49 CFR Part 173.

*Radioactive materials*: See 49 CFR 173.403 for detailed definitions relating to radioactive materials.

*Railroad*: a person engaged in transportation by rail.

*Rail freight car*: a car designed to carry freight or non-passenger personnel by rail, and includes a box car, flat car, gondola car, hopper car, tank car, and occupied caboose.

*Receptacle*: a containment vessel for receiving and holding materials, including any means of closing.

*Registered Inspector*: a person registered with the Department in accordance with 49 CFR Part 107, Subpart F who has the knowledge and ability to determine if a cargo tank conforms with the applicable DOT specification and has, at a minimum, any of the following combinations of education and work experience in cargo tank design, construction, inspection, or repair (or meets criteria at 49 CFR 107.502(f)):

(1) An engineering degree and one year of work experience,

(2) An associate degree in engineering and two years of work experience, or

(3) A high school diploma (or General Equivalency Diploma) and three years of work experience.

*Reportable quantity* (RQ) for the purposes of the DOT regulations for shippers, transporters, and manufacturers: the quantity specified in Column 3 of the Appendix to 49 CFR 172.101 for any material identified in Column 1 of the Appendix.

*Research*: investigation of experimentation aimed at the discovery of new theories or laws and the discovery and interpretation of facts or revision of accepted theories or laws in the light of new facts.

*Residue*: the hazardous material remaining in a packaging, including a tank car, after its contents have been unloaded to the maximum extent practicable and before the packaging is either refilled or cleaned of hazardous material and purged to remove any hazardous vapors.

*RSPA*: the Research and Special Programs Administration, U.S. Department of Transportation, Washington, DC 20590.

*SCF* (standard cubic foot): one cubic foot of gas measured at 60 °F. and 14.7 psia.

*Self-heating material*: a material that, when in contact with air and without an energy supply, is liable to self-heat. A material of this type which exhibits spontaneous ignition, or if the temperature of the sample exceeds 200 °C (392 °F) during the 24-hour test period when tested in accordance with paragraph 3.b.(1) of appendix E to 49 CFR Part 173, is classed as a Division 4.2 material.

*Sheathing*: a covering consisting of a smooth layer of wood placed over metal and secured to prevent any movement.

*Shipping paper*: a shipping order, bill of lading, manifest or other shipping document serving a similar purpose and containing the information required by §§ 172.202, 172.203 and 172.204.

*Single packaging*: a non-bulk packaging other than a combination packaging.

*Solid*: a material which has a vertical flow of two inches (50 mm) or less within a three-minute period, or a separation of one gram (1 g) or less of liquid when determined in accordance with the procedures specified in ASTM D 4359-84 "Standard Test Method for Determining Whether a Material is a Liquid or Solid," 1984 edition.

*Solution*: any homogeneous liquid mixture of two or more chemical compounds or elements that will not undergo any segregation under conditions normal to transportation.

*Specification packaging*: a packaging conforming to one of the specifications or standards for packagings in Part 178 or Part 179.

*Spontaneously combustible material*: A pyrophoric material or self-heating material.

*State*: a State of the United States, the District of Columbia, the Commonwealth of Puerto Rico, the Commonwealth of the Northern Mariana Islands, the Virgin Islands, American Samoa, Guam, or any other territory or possession of the United States designated by the Secretary.

*State-designated route*: a preferred route selected in accordance with U.S. DOT "Guidelines for Selecting Preferred Highway Routes for Highway Route Controlled Quantities of Radioactive Materials" or an equivalent routing analysis which adequately considers overall risk to the public.

*State routing agency*: an entity (including a common agency of more than one State such as one established by Interstate compact) which is authorized to use State legal process pursuant to 49 CFR 177.825 to impose routing requirements, enforceable by State agencies, on carriers of radioactive materials without regard to intrastate jurisdictional boundaries. This term also includes Indian tribal authorities which have police powers to regulate and enforce highway routing requirements within their lands.

*Stowage*: the act of placing hazardous materials on board a vessel.

*Strong outside container*: the outermost enclosure which provides protection against the unintentional release of its contents under conditions normally incident to transportation.

*Subsidiary hazard*: a hazard of a material other than the primary hazard. (See "primary hazard").

*Table in § 172.101* or *§ 172.101 Table:* the Hazardous Materials Table in 49 CFR 172.101.

*Tank*: container, consisting of a shell and heads, that forms a pressure tight vessel having openings designed to accept pressure tight fittings or closures, but excluding any appurtenances, reinforcements, fittings, or closures. For tank specification purposes, "tank" means inner vessel and "jacket" means either the outer shell or insulation cover.

*Technical name*: a recognized chemical name or microbiological name currently used in scientific and technical handbooks, journals, and texts. Generic descriptions are authorized for use as technical names provided they readily identify the general chemical group, or microbiological group. Examples of acceptable generic descriptions are organic phosphate compounds, petroleum aliphatic hydrocarbons, and tertiary amines. For proficiency testing only, generic microbiological descriptions such as bacteria, mycobacteria, fungus, and viral samples may be used. Except for names which appear in 49 CFR Part 172, Subpart B, trade names may not be used as technical names.

*TOFC*: trailer-on-flat-car.

*Top shell*: the tank car tank surface, excluding the head ends and bottom shell of the tank car tank.

*Trailership*: a vessel, other than a carfloat, specifically equipped to carry motor transport vehicles and fitted with installed securing devices to tie down each vehicle. The term "trailership" includes "Roll-on/Roll-off (RO/RO)" vessels.

*Train*: one or more engines coupled with one or more rail cars, except during switching operations or where the operation is that of classifying and assembling rail cars within a railroad yard for the purpose of making or breaking up trains.

*Trainship*: a vessel other than a rail car ferry or carfloat, specifically equipped to transport railroad vehicles, and fitted with installed securing devices to tie down each vehicle.

*Transport vehicle*: a cargo-carrying vehicle such as an automobile, van, tractor, truck, semitrailer, tank car or rail car used for the transportation of cargo by any mode. Each cargo-carrying body (trailer, rail car, etc.) is a separate transport vehicle.

*UFC*: Uniform Freight Classification.

*Unit load device*: any type of freight container, aircraft container, aircraft pallet with a net, or aircraft pallet with a net over an igloo.

*United States*: a State of the United States, the District of Columbia, the Commonwealth of Puerto Rico, the Commonwealth of the Northern Mariana Islands, the Virgin Islands, American Samoa, Guam, or any other territory or possession of the United States designated by the Secretary.

*UN*: United Nations.

*UN standard packaging*: a specification packaging conforming to the requirements in of 49 CFR Part 178, Subparts L and M.

*Vessel* includes every description of watercraft, used or capable of being used as a means of transportation on the water.

*Viscous liquid*: a liquid material which has a measured viscosity in excess of 2500 centistokes at 25° C (77° F.) when determined in accordance with the procedures specified in ASTM Method D 445-72 "Kinematic Viscosity of Transparent and Opaque Liquids (and the Calculation of Dynamic Viscosity)" or ASTM Method D 1200-70 "Viscosity of Paints, Varnishes, and Lacquers by Ford Viscosity Cup."

*Volatility* refers to the relative rate of evaporation of materials to assume the vapor state.

*Water reactive material*: a material that, by contact with water, is liable to become spontaneously flammable or to give off flammable or toxic gas at a rate greater than 1 liter per kilogram of the material, per hour, when tested in accordance with paragraph 4 of appendix E to 49 CFR Part 173.

*Water resistant*: having a degree of resistance to permeability by and damage caused by water in liquid form.

*Wooden barrel*: a packaging made of natural wood, of round cross-section, having convex walls, consisting of staves and heads and fitted with hoops.

*W.T.*: watertight.

## Conversion Factors Used in Hazardous Materials Measurements
### (to 7 significant digits)

| Dimension | SI to Customary | Customary to SI |
|---|---|---|
| Length | 1 cm = 0.3937008 in<br>1 m = 3.280840 ft | 1 in = 2.540000 cm<br>1 ft = 0.3048000 m |
| Thickness | 1 mm = 0.03937008 in | 1 in = 25.40000 mm |
| Mass (Weight) | 1 kg = 2.204622 lb<br>1 g = 0.03527397 oz | 1 lb = 0.4535924 kg<br>1 oz = 28.34952 g |
| Pressure[a] | 1 kPa = 0.1450377 psi | 1 psi = 6.894757 kPa |
| Volume (Liquid) | 1 L[b] = 0.2641720 gal<br>1 mL = 0.03381402 oz<br>1 $m^3$ = 35.31466 $ft^3$ | 1 gal = 3.785412 L<br>1 oz = 29.57353 mL<br>1 $ft^3$ = 0.02831685 $m^3$ |
| Density | 1kg/$m^3$ =0.06242797 lb/$ft^3$ | 1 lb/$ft^3$ =16.01846 kg/$m^3$ |

[a] 1 standard atmosphere is 101.3250 kPa (14.69594 psi).

[b] The international standard abbreviation for the liter is a lower-case *l*. Many international publications use a script *ℓ* to avoid confusion with the numeral *1*, but the U.S. standard is to use an upper-case *L*.

# Index

CPSIA information can be obtained
at www.ICGtesting.com
Printed in the USA
BVHW010210070821
613303BV00003B/11

9 780471 288442